Handbook of
Molecular and Cellular Methods in Biology and Medicine

SECOND EDITION

Handbook of Molecular and Cellular Methods in Biology and Medicine

SECOND EDITION

EDITED BY

**Leland J. Cseke
Peter B. Kaufman
Gopi K. Podila
Chung-Jui Tsai**

CRC PRESS

Boca Raton   London   New York   Washington, D.C.

Library of Congress Cataloging-in-Publication Data

Handbook of molecular and cellular methods in biology and medicine / edited by Leland J. Cseke
... [et al.].—2nd ed.
    p. cm.
Includes bibliographical references and index.
ISBN 0-8493-0815-1
  1. Molecular biology—Laboratory manuals. 2. Cytology—Laboratory manuals. I. Title:
Molecular and cellular methods in biology and medicine. II. Cseke, Leland J. III. Title.

QH506.H36 2003
572.8—dc21
                                          2003051459

This book contains information obtained from authentic and highly regarded sources. Reprinted material is quoted with permission, and sources are indicated. A wide variety of references are listed. Reasonable efforts have been made to publish reliable data and information, but the authors and the publisher cannot assume responsibility for the validity of all materials or for the consequences of their use.

Neither this book nor any part may be reproduced or transmitted in any form or by any means, electronic or mechanical, including photocopying, microfilming, and recording, or by any information storage or retrieval system, without prior permission in writing from the publisher.

All rights reserved. Authorization to photocopy items for internal or personal use, or the personal or internal use of specific clients, may be granted by CRC Press LLC, provided that $1.50 per page photocopied is paid directly to Copyright Clearance Center, 222 Rosewood Drive, Danvers, MA 01923 USA The fee code for users of the Transactional Reporting Service is ISBN 0-8493-0815-1/04/$0.00+$1.50. The fee is subject to change without notice. For organizations that have been granted a photocopy license by the CCC, a separate system of payment has been arranged.

The consent of CRC Press LLC does not extend to copying for general distribution, for promotion, for creating new works, or for resale. Specific permission must be obtained in writing from CRC Press LLC for such copying.

Direct all inquiries to CRC Press LLC, 2000 N.W. Corporate Blvd., Boca Raton, Florida 33431.

Trademark Notice: Product or corporate names may be trademarks or registered trademarks, and are used only for identification and explanation, without intent to infringe.

Visit the CRC Press Web site at www.crcpress.com

© 2004 by CRC Press LLC

No claim to original U.S. Government works
International Standard Book Number 0-8493-0815-1
Library of Congress Card Number 2003051459
Printed in the United States of America 1 2 3 4 5 6 7 8 9 0
Printed on acid-free paper

Preface

Since publication of the first edition of the *Handbook of Molecular and Cellular Methods in Biology and Medicine* in 1995, there have been several milestones in the field of biology. Genome sequencing of higher eukaryotes has progressed at an unprecedented speed. Starting with baker's yeast (*Saccharomyces cerevisiae*) in 1996, the organisms whose genetic code has been completely sequenced now include human, *Arabidopsis* (2000), and rice (*Oryza sativa*) (2002), with new completions being added on a regular basis. The invention of DNA microarray technology and advances in bioinformatics have generated vast amounts of genomic data, significantly boosting the throughput of biological research. While the molecular and cellular methods presented in the first edition remain as valuable tools, it is clear that today's researchers need an updated tool kit that incorporates conventional as well as modern approaches to tackle biological and medicinal research in the postgenomics era.

We have significantly revised this CRC Press handbook in the second edition in order to address these recent changes. All protocols have been evaluated, revised, and sometimes replaced with more efficient, reliable, or simpler ones. Preparation and hybridization of DNA, RNA, and protein blots are now presented in separate chapters (**Chapter 5, Chapter 6,** and **Chapter 8**). Additional methods for DNA footprinting and gel-retardation assay (**Chapter 13**), mRNA differential display (**Chapter 14**), functional genomics and DNA microarray analysis (**Chapter 15**), microscopy (**Chapter 21**), and bioseparation techniques and their applications (**Chapter 23**) have also been introduced. New chapters dealing with inhibition of gene expression (**Chapter 20**), localization of gene expression (**Chapter 22**), combinatorial techniques (**Chapter 24**), and computational data-mining methods applied to combinatorial chemistry libraries (**Chapter 25**) are presented.

As in the first edition of this book, we have included, within each chapter, various notes and cautionary considerations for potentially hazardous reagents. However, we also strongly recommend that the reader take the time to observe the precautions detailed for each chemical on individual material safety data sheets (MSDS) prior to performing each procedure. Observing correct safety procedures and developing good safety habits are perhaps the most important steps the reader can take toward avoiding long-term health concerns for the user as well as the people working around him or her.

We thank all those from CRC Press involved in this endeavor for their patience and help with the preparation of the second edition of this handbook. We specifically would like to thank Marsha Hecht, project editor at CRC Press, for her superb work helping to make this book accurate and easy for the reader to use.

Leland J. Cseke
Peter B. Kaufman
Gopi K. Podila
Chung-Jui Tsai

Editors

Leland J. Cseke, Ph.D., received a doctorate in plant cellular and molecular biology through the Department of Molecular, Cellular, and Developmental Biology at the University of Michigan, Ann Arbor. His dissertation research included the molecular biology, evolution, and biotechnological applications of terpenoid scent compound production in *Clarkia* and *Oenothera* species in the laboratory of Dr. Eran Pichersky. Currently, Dr. Cseke is a research assistant professor in the Department of Biological Sciences at the University of Alabama, Huntsville, where he works with Dr. Gopi K. Podila to investigate the activity of aspen (*Populus tremuloides*) MADS-box genes in wood development. Similarly, Dr. Cseke spent several years as a research assistant professor at Michigan Technological University, Houghton, working to discover the functionality of floral-specific MADS-box genes in aspen flower development. Dr. Cseke has also been a postdoctoral fellow in the Department of Plant Sciences at the University of Arizona, Tucson, in the laboratory of Dr. Rich Jorgensen. There, he worked to elucidate the factors involved in functional sense and antisense suppression of genes involved in anthocyanin biosynthesis. Dr. Cseke's interests include the biosynthesis of plant chemical products, their uses by humans, and the study of the global affects of transgenes on plant metabolism. This led to his coauthoring another book, *Natural Products from Plants* (CRC Press, 1999). In addition, Dr. Cseke has done some work in the study of possible methods for improving the separation and enhancing the biosynthesis of the cancer-fighting diterpene, taxol, in *Taxus* species in the laboratory of Dr. Peter Kaufman, and his knowledge of such subjects has been directed toward the teaching of classes emphasizing biotechnology and the chemical principles of biology.

Peter B. Kaufman, Ph.D., received his bachelor of science degree in plant science from Cornell University, Ithaca, NY, and his doctorate in plant biology from the University of California, Davis. He is currently professor emeritus of biology at the University of Michigan, Ann Arbor, and research scientist with the University of Michigan Complementary and Alternative Medicine Research Center (CAMRC). He is the author of nine books and over 225 research publications. His most recent books include *Natural Products from Plants* (CRC Press, 1999), published with Harry Brielmann, Leland Cseke, Sara Warber, James Hoyt, and James Duke; and *Creating a Sustainable Future: Living in Harmony with the Earth* (Researchco Book Centre, New Delhi, India, 2002), published with James Hoyt, Christopher Coon, Barbara Madsen, Sara Warber, J.N. Govil, and Casey R. Lu. He is a fellow of the American Association for the Advancement of Science (AAAS). He is also past president of the Michigan Botanical Club and past chairman of the Michigan Natural Areas Council. He served as secretary-treasurer of the American Society of Gravitational and Space Biology (ASGSB) and was the recipient of ASGSB's Orr Reynolds Distinguished Service Award. He is currently doing research on natural products of medicinal value in plants with support from CAMRC and the National Institutes of Health, and on the mechanism of signal transduction in graviresponding cereal grass and cut flower shoots with support from a USDA/BARD grant with colleagues at the Volcani Institute in Israel. He has performed research with colleagues at Lund

University, Lund, Sweden; University of Calgary, in Alberta, Canada; Nagoya and Kyushu Universities, Nagoya and Fukuoka, Japan; International Rice Research Institute, Los Baños, Philippines; Michigan State University, East Lansing; University of Colorado, Boulder; North Carolina State University, Raleigh; Purdue University, West Lafayette, Indiana; USDA Plant Hormone Lab, Beltsville, Maryland; Hawaiian Sugarcane Planters' Association, Aiea Heights, Honolulu; and Yerevan State University, Yerevan, Armenia.

Gopi K. Podila, Ph.D., is professor and chair of the Department of Biological Sciences at the University of Alabama, Huntsville. Until May 2002, Dr. Podila served as professor of biochemistry and molecular biology in the Department of Biological Sciences at Michigan Tech University, Houghton. Dr. Podila received a bachelor of science in biological sciences from Nagarjuna University in India, a master of science in plant pathology from Louisiana State University (1983), and a doctorate from Indiana State University (1987) in molecular biology. Dr. Podila's research deals with plant-fungus interactions, molecular biology of plant development, plant biotechnology, and functional genomics. Dr. Podila's research is supported through USDA, USFS, NSF, DOE, CPBR, and industry. Dr. Podila has published over 80 papers in peer-reviewed journals and books. He has also published a book on current advances in mycorrhizae research. Dr. Podila serves on the editorial boards of *Symbiosis*, *New Phytologist*, and *Physiology and Molecular Biology of Plants*. He serves as *ad hoc* reviewer for many journals, USDA, NSF, USDA-BARD, Italian Ministry of Education, and research grant proposals. He also serves on NIH-MBRS genetics panel. Dr. Podila had visiting professor appointments at INRA-Nancy, France; University of Torino and University of Urbino, Italy; and University of Helsinki, Finland. Dr. Podila has organized and chaired several international symposia on plant-fungus interactions and currently serves as councilor-at-large for the International Symbiosis Society.

Chung-Jui Tsai, Ph.D., is an associate professor of forest biotechnology in the School of Forest Resources and Environmental Science at Michigan Technological University, Houghton. Dr. Tsai received her bachelor and master of science degrees from the National Taiwan University in Taiwan, and her doctorate in forest science from Michigan Technological University in 1995. Dr. Tsai's research concerns plant secondary metabolism, lignin and flavonoid biosynthesis, vascular cambium differentiation, wood formation, functional genomics, and metabolic profiling. She has received research grants from the U.S. Department of Agriculture, the U.S. Department of Energy, the Consortium for Plant Biotechnology Research, and the Michigan Life Sciences Corridor. Dr. Tsai has published more than 16 papers in peer-reviewed journals and books as well as numerous reports in conference proceedings. She teaches an introductory course in plant biotechnology and tree biotechnology at Michigan Tech.

Contributors

Harry Brielmann, Ph.D., is a senior scientist level II at Neurogen Corp., Branford, CT, which he joined in 1997. He received a B.Sc. degree in chemistry in 1981 from the University of Connecticut, an M.Sc. degree in chemistry from the University of Massachusetts in 1989, and a Ph.D. degree in chemistry from Wesleyan University in 1994. He has held postdoctoral positions at the University of Hawaii (1994–1995) and at Wayne State University, Detroit, Michigan. Since joining Neurogen, his research has focused on medicinal and combinatorial chemistry.

Soo Chul Chang, Ph.D., is research professor at the Center for Cell Signaling Research, Ewha Women's University, Seoul, Korea. He obtained a B.Sc. degree in biology in 1984, a M.Sc. degree in biology in 1986, and a Ph.D. in biology in 1995, all at Yonsei University. He performed research on the gravitropic response mechanism in oat-shoot pulvini with Dr. Kaufman at the Department of Biology, University of Michigan, Ann Arbor, from 1995 to 1998. He is currently doing research on the signal-transduction pathway in plants in relation to the gravitropic response mechanism and on brassinosteroid plant hormones. He has taught general biology and plant physiology at Yonsei University and Hanyang University, respectively. His writing contributions in **Chapter 3, Chapter 8,** and **Chapter 22** were supported by Korea Research Foundation (KRF-2001-050-D00042).

Feng Chen, Ph.D., is a research fellow of molecular, cellular, and developmental biology at the University of Michigan, Ann Arbor. Dr. Chen received a B.Sc. in molecular biology from Nankai University, a M.Sc. in genetics from the Institute of Genetics, Chinese Academy of Sciences, China, and a Ph.D. in plant biology from the University of California, Davis, in 2000. Dr. Chen wrote his Ph.D. thesis on seed germination, in particular to understand the physiological roles of expansin and xyloglucan endotransglycosylase in tomato seed germination. His current research concerns the molecular genetics, biochemistry, and evolution of floral scent biosynthesis. Dr. Chen has published six peer-reviewed journal papers and two book chapters in *Seed Biology: Advances and Applications* (2000) and *Recent Advances in Phytochemistry* (2003).

Daotian Fu, Ph.D., is currently the scientific director of bioanalytical development at Genzyme Corp. He received a B.Sc. degree in microbiology from Shandong University in China and a Ph.D. in biochemistry from Iowa State University, Ames. He also had postdoctoral training at the Complex Carbohydrate Research Center at the University of Georgia. His research interests have been in the areas of protein chemistry and carbohydrate structure analysis, particularly in developing new analytical technologies to support the development of new protein- and carbohydrate-based therapeutics. Dr. Fu has held several positions in the life science industry, including manager of analytical chemistry and quality control at Neose Technologies, Inc., and director of protein chemistry at Charles River Laboratories. During this time, he has established several analytical and quality control laboratories, and he has developed multiple analytical paradigms under the FDA regulations to support over ten different therapeutic proteins and carbohydrates.

Myeon Haeng Cho, Ph.D., is associate professor of plant physiology, Department of Biology, Yonsei University, Seoul, Korea. He received a B.Sc. degree in biology in 1984, a M.Sc. degree in biology in 1986 at Yonsei University, and a Ph.D. in botany in 1994 at North Carolina State University, Raleigh. His research interests are in membrane transport and signal transduction, molecular biology, and stress physiology in plants.

Scott A. Harding, Ph.D., is a research assistant professor in the School of Forest Resources and Environmental Science at Michigan Technological University, Houghton. Dr. Harding received a Ph.D. in agronomy at Kansas State University, Manhattan, in 1990. His research interests include plant secondary metabolism, stress physiology, and calcium-mediated signaling.

Dr. Shekhar P. Joshi, Ph.D., is an assistant professor of plant molecular genetics at the Plant Biotechnology Research Center, School of Forest Resources and Environmental Science at Michigan Technological University. Dr. Joshi received his B.Sc. and M.Sc. degrees in botany and his Ph.D degree in biochemistry from The University of Poona in India. He held several positions in India, Germany, and the U.S. before joining Michigan Tech University. Dr. Joshi's research is supported through USDA, DOE, NSF and industry. It deals with the genomics, biotechnology, and bioinformatics of cellulose and lignin biosynthesis in trees. He also works to assess the impact of long-term stand management on genetic diversity in northern hardwoods through the molecular analysis and bioinformatics of tree growth and development. Dr. Joshi is a recipient of prestigious NSF Early Career Award. He has published more than 49 papers in peer-reviewed journals and books. Dr. Joshi serves as an ad hoc reviewer for several journals and funding agencies, and he has published highly-cited seminal papers on plant bioinformatics.

Bin Goo Kang, Ph.D., received a B.Sc. degree in biology from Yonsei University, Seoul, Korea, a M.Sc. degree in biology from Tufts University, Medford, Massachusetts, and a Ph.D. in botany from the University of Michigan, Ann Arbor. He is currently professor emeritus of biology at Yonsei University and CEO of BioNox Inc., Seoul, Korea. He has published over 80 research publications, mainly on hormone action in plant physiology. He is a fellow of the Korean Academy of Science and Technology and president of the Humboldt Club of Korea. He is also past president of the Korean Society for Molecular Biology and vice president of the Botanical Society of Korea. He served as editorial board member for the *Journal of Biochemistry and Molecular Biology*, *Journal of the Korean History of Science and Society*, and *Korean Journal of Botany*. In addition, he served on the research program committees of the Ministry of Science and Technology, the Ministry of Education, and the Korean Science and Engineering Foundation. He was the recipient of the Huan Prize for Biological Sciences, Yonsei Faculty Award for Research Excellence in Basic Science, and the Korean Federation of Science and Technology Societies Award for Research Excellence in Plant Biology. He has performed research at the MSU-DOE Plant Research Laboratory, East Lansing, Michigan; Fairchild Tropical Garden Research Center, Miami, Florida; University of Freiburg, Institute for Biology III, Freiburg, Germany; Smithsonian Institution, Environmental Research Center, Rockville, Maryland; and at the Research Institute for Biochemical Regulation, Nagoya University, Nagoya, Japan.

Donghern Kim, Ph.D., received B.Sc. and M.Sc. degrees in agricultural chemistry from Seoul National University, Korea, and a Ph.D. in plant cellular and molecular biology through the Department of Molecular, Cellular, and Developmental Biology at the University of Michigan, Ann Arbor. His dissertation research in the laboratory of Dr. Peter Kaufman included biochemical interpretation of changes in auxin sensitivity of *Avena sativa* (oat) shoot pulvinus tissue during the gravitropic response. He is currently a senior researcher at the National Institute of Agricultural Biotechnology in the Rural Development Administration, Korea. He is also currently doing research on metabolic engineering of rice to modify carbon metabolism, including photosynthesis, photosynthate translocation, and starch biosynthesis, and on the characterization of auxin receptors responsible for determining tissue-specific auxin in rice.

Ara Kirakosyan, Ph.D., is a member of the Faculty of Biology, Yerevan State University, Yerevan, Armenia. He received his Ph.D. degree in molecular biology at Yerevan State University in 1993. His research interests are in molecular biology, natural products of medicinal value in plants, and biophysics. He has conducted natural products research at Gifu Pharmaceutical University in Gifu, Japan; at Heinrich Heine University in Düsseldorf, Germany; at Cornell University, Ithaca, New York; at the Institute of Plant Genetics and Crop Plant Research at Gatersleben, Germany (where he was a DAAD fellow); and at the University of Michigan, Ann Arbor, where he was a Fulbright fellow and is currently a botanical fellow in the Complementary and Alternative Medicine Research Center and the Department of Molecular, Cellular, and Developmental Biology.

Eun Kyeong Lee, M.Sc., is a Ph.D. student in the Department of Biology, Yonsei University, Seoul, Korea. She received a B.Sc. degree in biology in 1994 at Yonsei University, Wonju, Korea, and a M.Sc. degree in biology in 1997 at Yonsei University, Seoul, Korea. She is studying the ABC transporters in *Arabidopsis thaliana*.

June Seung Lee, Ph.D., is a professor in the Department of Biological Science and the director of the Center for Cell Signaling Research, Ewha Women's University, Seoul, Korea. He received a B.Sc. degree in biology in 1975, a M.Sc. degree in biology in 1977, and a Ph.D. in biology in 1981, all at Yonsei University, Seoul, Korea. He performed research on gravitropic response in maize root in the Department of Biology, Ohio State University, Columbus. During that period, he found that calcium exerted important roles in root gravitropic responses. He performed research at the University of Bonn, Bonn, Germany, and Tohoku University, Sendai, Japan. He has published more than 40 papers based on his research. His current research interest pertains to the roles of reactive oxygen species in plant physiology.

Junho Lee, Ph.D., is an associate professor in the Department of Biology, Yonsei University, Seoul, Korea. He received a B.Sc. degree in biology in 1986, a M.Sc. degree in biology in 1989 at Seoul National University, Seoul, Korea, and a Ph.D. in biology in 1994 at the California Institute of Technology, Pasadena. He has received several awards, including a National Research Laboratory grant (2001) from the Ministry of Science and Technology, Korea, and an outstanding research paper award from the Science Foundation in Korea (2001). He also received an outstanding professor award from Yonsei University (2002). He has taught developmental biology, advanced developmental genetics, and developmental physiology at Yonsei University. He has presented papers at meetings of the *C. elegans* conferences and in the international journals including PNAS, MCB, and MBC. His research interests include clathrin-associated proteins, telomere-binding proteins, alcohol-sensitive genes, neuron-specific gene expression, and rapsyn function in the nematode, *C. elegans*.

Shirley Louise-May, Ph.D., is a senior scientist, level II, at Neurogen Corp., Branford, Connecticut, which she joined in February 1998. She received a B.Sc. degree in chemistry in 1987 from the University of Maine, Orono, a M.Sc. degree in chemical physics from Columbia University in 1988, and a Ph.D. degree in biophysical chemistry and chemical physics from Wesleyan University in 1994. She has held postdoctoral positions at the Institut de Biologie Moleculaire et Cellulaire du CNRS at the Université Louis Pasteur in Strasbourg, France, as a Chateaubrian fellow (1994–1996) and at Parke Davis/Warner Lambert Pharmaceutical Research, Inc. (now Pfizer Corp.), in Ann Arbor, Michigan, conducting computational modeling and simulation on ribonucleic acids (1996–1997). Since joining Neurogen, her research has focused on medicinal chemistry; computational modeling and simulation of small molecules; chemoinformatics and data mining applied to combinatorial libraries and high-throughput pharmacology data; predictive modeling of biophysical, ADME, and druglike properties; and pharmacophore model development.

Casey R. Lu, Ph.D., is associate professor of biology in the Department of Biological Sciences, Humboldt State University, Arcata, CA. He received a B.Sc. degree (with honors) in biology in 1980, a M.Sc. degree in biology in 1987, and a Ph.D. in education and biology in 1993, all at the University of Michigan, Ann Arbor. Dr. Lu has received several awards, including a University of Michigan biology block grant, a University of Michigan School of Education Merit Award and Dean's Fellowship Award, a Humboldt State University (HSU) scholarly and creative activities grant to develop computer-assisted quiz sets, and two Dean of CNRS Awards to develop multimedia CD study aids for introductory biology students. Recently (1999–2001), he received a National Science Foundation grant to assist with the purchase of a new transmission electron microscope and equipment for Humboldt State University, Arcata, California, and two grants from the University of California Office of the President for establishing and directing a California science project site at HSU, known as Redwood Science Project. Dr. Lu has taught introductory botany and practical botany at the University of Michigan; was a naturalist-intern at Stony Kill Farm Environmental Education Center in Wappingers Falls, New York, and at Big Cypress Nature Center in Naples, Florida; and was a high school science teacher at Ithaca High School (biology and physical science) in Ithaca, New York. At HSU, he teaches introductory cellular and molecular biology, plant physiology, secondary science methods, plant tissue culture, and transmission electron microscopy. He has presented papers at meetings of the California Science Teachers Association in 1998 at San Jose, California, and at the National Science Teachers Association in 1998 at Seattle, Washington. His research interests include plant cell ultrastructural changes in plants challenged with heavy metals and physiology of the gravitropic response in plants.

Jaegal Shim, Ph.D., is a research fellow in the Department of Biology, Yonsei University, Seoul, Korea. He received a B.Sc. degree in biology, a M.Sc. degree in biology, and a Ph.D. in biology, all at Yonsei University. His research interests include clathrin-associated protein functions, and heat-shock proteins and their response to hypoxia.

Darren H. Touchell, Ph.D., is a postdoctoral research scientist in the Plant Biotechnology Research Center in the School of Forest Resources and Environmental Science at Michigan Technological University, Houghton. Dr. Touchell received a B.Sc. degree in biology in 1991 from Curtin University of Technology, Australia, and a Ph.D. in botany in 1996 from the University of Western Australia. He has held postdoctoral positions at Kings Park and Botanic Garden, Australia (1996–1998), and at the USDA National Seed Storage Laboratory (1998–2000), conducting research on plant tissue culture and cryobiological methods for conserving plant germ plasm. His current research focus is on assessing differential gene expression in transgenic aspen.

Yuh-Shuh Wang, Ph.D., graduated in 2002 with a doctoral degree in forest molecular genetics and biotechnology from Michigan Technological University, Houghton. Dr. Wang received B.Sc. and M.Sc. degrees through the Department of Botany at the National Taiwan University, Taipei, Taiwan. Her research has focused on molecular mechanisms of plant development. Using cDNA-AFLP and differential display, she isolated and characterized suites of cDNAs that are differentially regulated during tuberization of sweet potato and vascular development of quaking aspen.

Sara Warber, M.D., is the co-director of the University of Michigan's Complementary and Alternative Medicine Center, is a lecturer in the University of Michigan's Department of Family Medicine, and is board certified in family medicine and holistic medicine. She received her medical degree from Michigan State University, East Lansing, and completed her internship and residency with the Department of Family Medicine at the University of Michigan. She has recently completed a fellowship in the Robert Wood Johnson Clinical Scholars Program at the University of Michigan. She studied herbalism and spiritual healing for 14 years with a Native American healer. Dr. Warber has been instrumental in the current process of designing an integrative medicine clinic at the University of Michigan Health System and in the development of a complementary and alternative medicine curriculum at the University of Michigan Medical School. Dr. Warber sits on the boards of Womenheart, the National Coalition for Women with Heart Disease, and the Nichols Arboretum Academic Advisory Committee. Her research interests focus on health promotion; the use of herbs, energy healing, and other complementary therapies. An accomplished public speaker in the area of complementary and alternative medicine, Dr. Warber is a member of the American Holistic Medical Association, the American Academy of Family Physicians, and the North American Primary Care Research Group.

William Wu, Ph.D., is currently assistant research professor, Department of Cell and Developmental Biology, University of Michigan, Ann Arbor. Before this, he was senior research scientist at Esperion Therapeutics, Ann Arbor. He is also professor of biology at the Hunan Normal University, Changsha, Hunan, the Peoples Republic of China. He received a M.Sc. degree in biology from Hunan Normal University, Changsha, Hunan, the Peoples Republic of China, in 1984. In 1992, he received a Ph.D. degree in molecular and cellular biology at Ohio University, Athens. He has completed three years of postdoctoral training in molecular biology at the University of Michigan, Ann Arbor. Dr. Wu is an internationally recognized expert in molecular biology. He has presented and published over 35 research papers at scientific meetings and in national and international journals. As a senior author, Dr. Wu has contributed 34 chapters of molecular and cellular biology methodologies in three books. He has broad knowledge of molecular biology, biotechnology, protein biochemistry, cellular biology, and molecular genetics. He is highly experienced and has extensive hands-on expertise in a variety of current molecular biology techniques. Dr. Wu is a member of the American Association for the Advancement of Science.

Kefei Zheng, Ph.D., is a senior scientist at Applied Biosystems Group, a business of Applera Corp. She received B.Sc. and M.Sc. degrees in physical chemistry from Fudan University in China and a Ph.D. in analytical chemistry from the University of Michigan, Ann Arbor, in 1998. After completing her postdoctoral training at the University of California at San Diego, she worked as a senior applications chemist at Waters Corp. for 2 years. She is currently doing research on analysis and characterization of proteins and peptides using analytical tools like mass spectrometry and liquid chromatography.

Contents

Chapter 1. Isolation and Purification of DNA 1
Leland J. Cseke, William Wu, and Chung-Jui Tsai

Chapter 2. Isolation and Purification of RNA 25
Chung-Jui Tsai, Leland J. Cseke, and Scott A. Harding

Chapter 3. Extraction and Purification of Proteins 45
Soo Chul Chang and Leland J. Cseke

Chapter 4. Preparation of Nucleic Acid Probes..67
William Wu and Leland J. Cseke

Chapter 5. Southern Blot Hybridization..85
Chung-Jui Tsai and Leland J. Cseke

Chapter 6. Northern Blot Hybridization..105
Myeon Haeng Cho, Eun Kyeong Lee, Bin Goo Kang, and Leland J. Cseke

Chapter 7. Preparation of Monoclonal and Polyclonal Antibodies
against Specific Protein(s) 113
Leland J. Cseke and William Wu

Chapter 8. Western Blot Hybridization 127
Leland J. Cseke, William Wu, June Seung Lee, and Soo Chul Chang

Chapter 9. cDNA Libraries..147
Leland J. Cseke, William Wu, and Peter B. Kaufman

Chapter 10. Genomic DNA Libraries 193
Leland J. Cseke, William Wu, and Peter B. Kaufman

Chapter 11. DNA Sequencing and Analysis 237
Leland J. Cseke, William Wu, and Peter B. Kaufman

Chapter 12. DNA Site-Directed and Deletion Mutagenesis 271
Leland J. Cseke and William Wu

Chapter 13. DNA Footprinting and Gel-Retardation Assay..........................291
William Wu and Leland J. Cseke

Chapter 14. Differential Display..305
Darren H. Touchell, Yuh-Shuh Wang, Scott A. Harding, and Chung-Jui Tsai

Chapter 15. Functional Genomics and DNA Microarray Technology.................319
Gopi K. Podila, Chandrashakhar P. Joshi, and Peter B. Kaufman

Chapter 16. *In Vitro* Translation of mRNA(s) and Analysis of Protein by Gel Electrophoresis 347
William Wu and Leland J. Cseke

Chapter 17. Plant Tissue and Cell Culture 357
Donghern Kim, Leland J. Cseke, Ara Kirakosyan, and Peter B. Kaufman

Chapter 18. Gene Transfer and Expression in Animals 377
William Wu, Peter B. Kaufman, and Leland J. Cseke

Chapter 19. Gene Transfer and Expression in Plants 409
William Wu, Leland J. Cseke, and Peter B. Kaufman

Chapter 20. Inhibition of Gene Expression 433
Jaegal Shim, Junho Lee, Leland J. Cseke, and William Wu

Chapter 21. Microscopy: Light, Scanning Electron, Environmental Scanning Electron, Transmission Electron, and Confocal 461
Casey R. Lu, Peter B. Kaufman, and Leland J. Cseke

Chapter 22. Localization of Gene Expression 487
Scott A. Harding, Chung-Jui Tsai, Leland J. Cseke, Peter B. Kaufman, Soo Chul Chang, and Feng Chen

Chapter 23. Bioseparation Techniques and Their Applications 509
Kefei Zheng, Daotian Fu, Leland J. Cseke, Ara Kirakosyan, Sara Warber, and Peter B. Kaufman

Chapter 24. Combinatorial Techniques 535
Harry Brielmann

Chapter 25. Computational Data-Mining Methods Applied to Combinatorial Chemistry Libraries 549
Shirley Louise-May

Index 565

Chapter 1

Isolation and Purification of DNA

Leland J. Cseke, William Wu, and Chung-Jui Tsai

Contents

I.	Isolation of Plasmid DNA	2
	A. Isolation of Plasmid DNA by Alkaline Lysis	2
	1. Mini-Preparation of Plasmid DNA	2
	2. Quick Mini-Preparation	3
	3. Large-Scale Preparation of Plasmid DNA	4
	B. Isolation of Plasmid DNA by CsCl Gradient Centrifugation	5
II.	Isolation of λDNA	7
	A. Phage Lysate Preparation by the Plate Method	7
	B. Phage Lysate Preparation by the Liquid Method	8
	C. Mini-Preparation of λDNA	9
	D. Large-Scale Purification of λDNA by the CsCl Gradient Centrifugation	9
III.	Isolation of Genomic DNA from Animals	10
	A. Extraction of Genomic DNA with Organic Solvents	10
	B. Extraction of Genomic DNA with Formamide	11
IV.	Isolation of Genomic or Organelle DNA from Plants	12
	A. Isolation of Plant Genomic DNA	12
	B. Isolation of Chloroplast DNA by Sucrose Gradient	14
V.	Purification of DNA Fragments from Agarose Gels	15
	A. Elution of DNA Fragments by the Freeze-and-Thaw Method	15
	B. Elution of DNA Fragments in Wells of Agarose Gel	16
VI.	Determination of DNA Quality and Quantity	17
	A. Spectrophotometric Measurement	17
	B. Fluorometric Measurement	18
VII.	Troubleshooting Guide	19
	Reagents Needed	19
References		23

Deoxyribonucleic acid (DNA) isolation and purification are essential techniques for molecular biology studies. The quality and integrity of the isolated DNA directly affect the results of all subsequent scientific research.[1-4] A number of methods have been developed for extraction and purification of DNA from different tissue types and organisms. We present several well-developed protocols here for the isolation of DNA from bacteria, λ phages, animals, and plants, and for the purification of DNA fragments from agarose gels.

I. Isolation of Plasmid DNA

Plasmids such as the pGEM series, pBR322, pBIN series, pBluescript series, and pUC series are commonly used vectors for DNA subcloning and sequencing. Therefore, extraction of plasmid DNA is a very useful technique in modern molecular biology studies. Plasmids are usually purified from liquid cultures by inoculating an appropriate volume of Luria-Bertaini (LB) medium with a single bacterial colony taken from an agar streak plate. Many of the commercial plasmids are high-copy-number vectors. They can replicate up to 300 copies per cell, so that they can be purified from cultures in high yield. However, some vectors, such as pBR322, are low-copy-number plasmids that need to be selectively amplified by chloramphenicol treatment for several hours. Chloramphenicol can inhibit host-protein synthesis and prevent replication of the bacterial chromosome, thereby increasing plasmid DNA copy numbers. Many methods have been developed for isolation of plasmid DNA. We describe in detail several protocols, largely based on the alkaline lysis method, that are very commonly used and successful in many laboratories. If one's budget allows, a number of plasmid DNA extraction kits (mini- to maxi-scale) are available commercially that can be conveniently and reliably used to obtain good quality plasmid DNA. The readers should refer to the manufacturers' instructions for the extraction protocols.

A. Isolation of Plasmid DNA by Alkaline Lysis

The principle of this procedure is to take advantage of the alkaline denaturation of plasmid and chromosomal DNA and of the selective renaturation of plasmid DNA following neutralization of the solution. The isolated plasmid DNA is suitable for use in restriction enzyme digestion, *in vitro* transcription, and DNA sequencing by either the manual procedures or automated sequence analysis.

1. Mini-Preparation of Plasmid DNA

1. Inoculate a single colony containing the plasmid of interest in 5 ml of LB medium and 50 µg/ml of appropriate antibiotics using an inoculation loop or needle. Culture the bacteria at 37°C for 7 h to overnight with shaking at 250 rpm. For convenience, when preparing multiple cultures, select isolated colonies with autoclaved toothpicks (cut into ~1-cm pieces) and inoculate cultures by dropping the toothpick directly into the culture tubes. Number each colony and tube correspondingly to ensure accurate tracking. If necessary, reincubate the plate at 37°C for 3 to 5 h and store the plate at 4°C.

Notes: *(1) Use sterile techniques. Handle the toothpicks with flamed forceps. Alternatively, sterile pipette tips can be used for inoculation. Wipe the shaft of the pipette clean with ethanol before use. (2) Prolonged culture will increase cell density, but at the same time, the number of aged or dead cells increases, and reduced plasmid yields can result. Rich media, such as terrific broth (TB), can be used to shorten the culture time required to reach the desired cell density.*

Isolation and Purification of DNA

2. Add 1.5 ml of the culture to a microcentrifuge tube and centrifuge at 12,000 g for 30 sec. Remove the liquid and invert the tube on a paper towel to dry the bacterial pellet for 4 min.
3. Resuspend the pellet by adding 0.1 ml of ice-cold plasmid lysis buffer and vortex for 2 min. Incubate the tube for 5 min at room temperature. This step lyses the bacteria by hyperlytic osmosis and releases the DNA and other contents.
4. Add 0.2 ml of freshly prepared alkaline solution and mix by inversion. Never vortex. Incubate the tube on ice for 5 min. The function of this step is to denature the plasmid and chromosomal DNAs and proteins.
5. Add 0.15 ml of ice-cold potassium acetate solution. Mix by inversion for 20 sec and incubate on ice for 5 min. The purpose of this step is to selectively renature the plasmid DNA. Some chromosomal DNA may be partially renatured and bound by proteins, which will be extracted by phenol/chloroform in later steps.
6. Centrifuge at 12,000 g for 5 min and carefully transfer the supernatant to a fresh tube.
7. Add *RNase* A to the supernatant to a final concentration of 20 $\mu g \cdot ml^{-1}$. Incubate the tube at 37°C for 20 min. *RNase* A degrades total ribonucleic acids (RNA) from the sample.
8. Add an equal volume of buffered phenol:chloroform:isoamyl alcohol (25:24:1) to each tube and mix by vortexing for 1 min.

Caution: *Phenol is very toxic and should be handled in a fume hood. After use, the waste phenol must be collected in a glass toxic-waste container with a clear label and disposed of properly in compliance with the local hazardous chemical disposal procedures.*

9. Centrifuge at 12,000 g for 2 min and transfer the upper, aqueous phase to a fresh tube.
10. Add an equal volume of chloroform:isoamyl alcohol (24:1) and mix by vortexing for 1 min.
11. Centrifuge at 12,000 g for 2 min and transfer the supernatant to a fresh tube.
12. Add two volumes of 100% ethanol and mix by inversion to precipitate plasmid DNA.
13. Centrifuge at 12,000 g for 10 min at room temperature. Decant the supernatant and rinse the pellet thoroughly with 1 ml of prechilled 70% ethanol. Centrifuge briefly, decant the solution, and dry the plasmid DNA pellet.
14. Dissolve the plasmid DNA in 25 μl of sterile distilled, deionized H_2O (ddH_2O) or TE buffer.
15. Determine the DNA quality and quantity as described in **Section VI** and store the sample at −20°C until use.

Notes: *(1) Ethanol precipitation of DNA is traditionally carried out at a cold temperature (−20 or −70°C). However, this is now known to be unnecessary.[2] DNA at concentrations as low as 20 $ng \cdot \mu l^{-1}$ can be effectively precipitated, in the absence of carrier, by incubation at 0°C and centrifugation in a microcentrifuge.[2] (2) A longer centrifugation time, or centrifugation at 4°C, may be necessary when handling smaller DNA fragments (<100 bp [basepairs]) or dilute DNA solutions.*

2. Quick Mini-Preparation

The following protocol was modified from the original miniprep method described above with several shortcuts designed to create as few waste materials as possible. The entire protocol can be completed within 30 min, and the DNA is suitable for restriction enzyme digestion. If desired (usually after initial screening), the DNA from selected clones can be further treated with *RNase* A, followed by phenol/chloroform and chloroform extraction. DNA purified this way is suitable for both manual and automated sequencing.

1. Prepare 1-ml liquid cultures in 1.5-ml eppendorf tubes using procedures similar to those described above. Culture the bacteria at 37°C for 12 to 16 h without shaking. This "standing culture" procedure, however, is not recommended for larger culture volumes.
2. **Alternatively**, grow bacteria as described above in 3 ml liquid media for 6 to 7 h with vigorous shaking (250 to 300 rpm).

3. Spin the tubes containing 1.5 ml of the bacterial cells at top speed for 30 sec. Pour off the liquid and, with the tube inverted, suck away as much excessive liquid as possible with a piece of Kimwipe®.
4. Add 0.1 ml of ice-cold plasmid lysis buffer and resuspend the pellet by vortexing.
5. **Immediately** add 0.2 ml of alkaline solution and mix the contents gently by inverting the tube five times. Multiple tubes can be mixed by placing a second rack on top of the first rack and inverting everything in unison.
6. **Immediately** add 0.15 ml of ice-cold potassium acetate solution. Mix by inverting the tube five times.
7. **Immediately** add 0.45 ml of buffered phenol:chloroform:isoamyl alcohol to the tube and mix the contents by inverting the tube gently several times.
8. Centrifuge at top speed for 5 min. The phenol:chloroform:isoamyl alcohol phase helps bind the cellular and protein debris at the interface. During this spin, label a new set of microcentrifuge tubes. As soon as the spin stops, carefully transfer 0.4 ml of the supernatant to a fresh tube using a Pasteur pipette. This pipette can be used repeatedly without the risk of contamination if it is rinsed with ddH$_2$O between each sample. However, this is **NOT** recommended if the plasmid samples will be used for *in vitro* transcription or **polymerase chain reaction (PCR)**.
9. Precipitate plasmid DNA in the aqueous phase with an equal volume of isopropanol. Mix well and centrifuge at top speed for 5 min. The isopropanol can simply be added using a squirt bottle.
10. Wash the pellet with 70% ethanol (again added using a squirt bottle) and repellet the plasmid if necessary by centrifugation at top speed for 30 sec. Decant the supernatant and use a piece of Kimwipe to remove as much liquid as possible. Dry briefly under vacuum.
11. Dissolve the plasmid DNA in 20 μl of TE buffer or sterile ddH$_2$O containing 20 μg·ml^{-1} *RNase* A. Check 1 to 2 μl on a minigel. Store the sample at $-20°C$. The DNA is suitable for restriction enzyme digestion.

Note: *It is necessary to degrade RNA before or during electrophoresis; otherwise, RNA molecules will scavenge most of the ethidium bromide (EtBr), leaving DNA molecules invisible on the stained gel. However, no incubation is required, as RNase A digestion occurs quickly.*

12. (**Optional**) For sequencing purposes, dilute the plasmid DNA of selected clones (after minigel or restriction digestion verification) to 100 μl. Extract with an equal volume of phenol:chloroform:isoamyl alcohol to remove *RNase* A, and twice with chloroform:isoamyl alcohol to remove residual phenol.
13. Carefully transfer the supernatant to a fresh tube and precipitate the DNA with 0.1 volume of 3 *M* sodium acetate (pH 5.2) and two volumes of 100% ethanol. Centrifuge at top speed for 10 min at room temperature. Wash the DNA with 70% ethanol and dry the pellet briefly under vacuum.
14. Dissolve the pellet in 20 μl of sterile ddH$_2$O or TE buffer. Check 1 μl in a minigel and measure the DNA concentration as described in **Section VI**. Store the sample at $-20°C$ until use.

3. Large-Scale Preparation of Plasmid DNA

The principle in this protocol is the same as for the minipreparation method.

1. Inoculate a single colony or dilute 1:100 from a small-scale liquid culture of *E. coli* containing the plasmid of interest in 100 ml of LB medium and 50 μg·ml^{-1} of appropriate antibiotics. Incubate the bacteria at 37°C for 8 to 12 h with shaking at 250 rpm.
2. Harvest the cells in two 50-ml centrifuge tubes and centrifuge at 12,000 g for 1 min at 4°C. Remove the liquid and invert the tubes on a paper towel to dry the bacterial pellet.
3. Resuspend the pellet by adding 2 ml of ice-cold plasmid lysis buffer and vortex for 2 min. Incubate the tubes for 5 min at room temperature (optional).
4. Add 4 ml of a freshly prepared alkaline solution and mix by inversion. Do not vortex. Incubate the tubes on ice for 5 min (optional).
5. Add 3 ml of ice-cold potassium acetate solution and mix by inversion five times. Incubate on ice for 5 min (optional).
6. Centrifuge at 12,000 g for 5 min at 4°C and gently transfer the supernatant to fresh tubes.

Isolation and Purification of DNA

7. Add *RNase A* to the supernatant to a final concentration of 20 $\mu g \cdot ml^{-1}$. Incubate the tubes at 37°C for 30 min.
8. Extract with one volume of buffered phenol:chloroform:isoamyl alcohol (25:24:1) and mix by vortexing for 30 sec.
9. Centrifuge at 12,000 g for 5 min at 4°C. Carefully transfer the upper, aqueous phase to fresh tubes.
10. Extract with one volume of chloroform:isoamyl alcohol (24:1), mix by vortexing for 30 sec, and centrifuge as in step 9. Transfer the supernatant to a fresh tube.
11. Add 2 to 2.5 volumes of 100% ethanol. Mix by inversion to precipitate the plasmid DNA.
12. Centrifuge at 12,000 g for 10 min at 4°C. Remove the supernatant, rinse the pellet thoroughly with 6 ml of prechilled 70% ethanol, and centrifuge again. Dry the plasmid DNA briefly under vacuum.
13. Dissolve the plasmid DNA pellet in 200 μl of TE buffer or ddH_2O.
14. Determine the DNA quality and quantity as described in **Section VI** and store the samples at −20°C until use.

B. Isolation of Plasmid DNA by CsCl Gradient Centrifugation

This method, as compared with the protocols described above, is a relatively complicated, time-consuming, and expensive. It requires a large-scale culture of bacteria, and it is not suitable for minipreparation of plasmid DNA. The researchers also need access to an ultracentrifuge. Its use is often associated with the need for large amounts of very high quality DNA that cannot be obtained by using simple procedures.

1. Inoculate a single colony or dilute 1:100 from a small-scale liquid culture of *E. coli* containing the plasmid of interest in 200 ml of LB medium containing 2 ml of 20% of maltose, 2 ml of 1 M $MgSO_4$, and 50 $\mu g \cdot ml^{-1}$ of appropriate antibiotics. Incubate the bacteria at 37°C overnight with shaking at 250 rpm.
2. Pellet the bacteria by centrifuging at 12,000 g for 1 min at 4°C. Remove the liquid and invert the tubes on a paper towel to dry the bacterial pellet for 4 min.
3. Proceed with plasmid DNA extraction, as described above, using 5 ml plasmid lysis buffer, 10 ml alkaline solution, and 7.5 ml potassium acetate.
4. Centrifuge at 12,000 g for 10 min at 4°C and carefully transfer the supernatant to fresh tubes.
5. Add 0.8 volume of isopropanol to the supernatant and mix well to precipitate the DNA.
6. Centrifuge at 12,000 g for 10 min at 4°C, decant the supernatant, and gently rinse the pellet with 6 ml of prechilled 70% ethanol. Dry the plasmid DNA briefly under vacuum.
7. Dissolve the plasmid DNA pellet in 4 ml of TE buffer or ddH_2O. Continue to step 8 for CsCl gradient purification or store at −20°C.
8. Accurately measure the volume of the DNA sample and add 1 g of solid CsCl per milliliter of DNA sample. The CsCl can be dissolved by pipetting the solution up and down with a Pasteur pipette. The solution can also be warmed to 30°C to help bring the CsCl into solution.
9. Add 0.35 ml of 10-$mg \cdot ml^{-1}$ EtBr for every 5 ml of the DNA-CsCl solution. Immediately and gently mix the solution. The final density of the mixture should be 1.55 to 1.6 $g \cdot ml^{-1}$.

Caution: *EtBr is a potentially carcinogenic compound. Gloves should always be worn when working with the solutions. The used EtBr solution must be collected in a special container to gradually inactivate the chemical before it is discarded.[5] Stock solutions of EtBr should be stored in light-tight bottles at 4°C.*

10. Centrifuge at 8000 rpm for 5 min at room temperature. Bacterial debris, if any, will be pelleted at the bottom of the tube, while bacterial proteins complexed with EtBr will appear as a red scum floating on the top.
11. Gently transfer the clear, red DNA-CsCl mixture under the surface scum into Quick-Seal® ultracentrifuge tubes (Beckman Coulter, Fullerton, CA) using a Pasteur pipette or a disposable syringe fitted with a large-gauge needle. A number of different tubes and rotors are available for ultracentrifugation,

and the protocols associated with their use may work equally well. Please see manufacturer's description.

Notes: Fill the tubes right up to the neck of each tube (about 5.2 ml per tube). If the tube is underfilled, the large air bubble present after sealing the tube may cause the tube to collapse. Carefully balance the tubes with CsCl solution ($1 \text{ g} \cdot \text{ml}^{-1}$ in TE buffer).

12. Turn on and warm the sealing machine for 5 min. Place the tubes on a metal sealing plate and cover each tube with an appropriate metal cap. Place each tube, one at a time, right under the sealing rack and push and hold the hot rack against the metal cap of the tube for 30 to 60 sec. The neck of the tube gradually melts until a sound is heard from the sealing machine. Quickly remove the tube from the melt rack and place the cap of the tube against a cold metal strip. Press and hold the strip against the cap for about 1 to 2 min to completely seal the tube and then remove the cap.
13. Attach a cap and a metal adapter to each sealed tube for ultracentrifugation and carefully balance, using Scotch™ tape if necessary. Symmetrically place the tubes into the Beckman vertical rotor and cover the tubes with their own caps and adapters.
14. Centrifuge the CsCl density gradients at 45,000 rpm for 16 h (VTi65), 45,000 rpm for 48 h (Ti50), 60,000 rpm for 24 h (Ti65), or 60,000 rpm for 24 h (Ti70.1) at 20°C.

Note: Never centrifuge at 4°C. CsCl precipitates out of solution at temperatures lower than 15°C, and this will reduce the density of CsCl in the gradient.

15. Stop centrifugation using either a very low brake setting or no brake at all. Slowly remove the tubes from the rotor and place them in a plastic rack. Use long-wavelength ultraviolet (UV) light (360 nm) to visualize and locate the relaxed and supercoiled plasmid DNA bands (Figure 1.1).

Notes: Short-wavelength UV light can cause nicking of the DNA. Two bands of DNA are visible in the center of the gradient (see Figure 1.1). The upper band usually contains linear bacterial chromosomal DNA and nicked circular plasmid DNA. The lower (and usually larger) band consists of closed circular plasmid DNA. The deep-red pellet on the bottom of the tube consists of EtBr-stained RNA. The materials on the top surface are usually proteins.

16. Carefully clamp the tube in a ring-stand rack. Insert a 16- to 20-gauge needle near the top of the tube to allow air to enter when DNA band is withdrawn. Wipe the outside of the tube with ethanol and place a piece of Scotch tape to one side of the tube covering the DNA band area. The tape is used as a seal to prevent leaks after needle puncture.[2] For removal of the plasmid DNA, slowly and carefully puncture the tube with a second sterile needle (16 to 20 gauge), beveled side up, through the tape 1 to 2 mm below the closed circular plasmid DNA band. Carefully angle the needle in order to withdraw the band of DNA and transfer it to a microcentrifuge tube (Figure 1.1). Be sure to adjust the needle angle to maintain contact with the DNA band during withdrawal. To avoid contamination, **do not** attempt to recover all traces of the DNA band. Discard the ultracentrifuge tube and the contents as EtBr waste according to local safety procedures.
17. Extract EtBr from the DNA-CsCl solution with one volume of TE-saturated isopropanol or n-butanol. Mix the aqueous and organic phases by inverting the tube several times and allow the phases to separate by gravity or by centrifugation at 1000 rpm for 3 min at room temperature. The EtBr partitions into the upper organic phase, which appears pink, and is removed with a pipette. Repeat the extraction four to five times until no pink color is seen in both the aqueous and organic phases.
18. Dilute the lower aqueous phase containing DNA-CsCl solution with two volumes of TE buffer and remove the CsCl from the DNA-CsCl solution by precipitation with six volumes of chilled 100% ethanol. Incubate at 4°C for 30 min to overnight.

Note: Do not incubate at −20 or −80°C; otherwise, the CsCl will coprecipitate.

Isolation and Purification of DNA

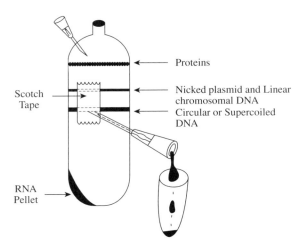

FIGURE 1.1
Illustration for recovery of closed circular or supercoiled plasmid DNA following purification by the CsCl gradient centrifugation method. (From Kaufman, P.B., Wu, W., Kim, D., and Cseke, L.J., *Handbook of Molecular and Cellular Methods in Biology and Medicine*, 1st ed., CRC Press, Boca Raton, FL, 1995.)

19. Centrifuge at 12,000 g for 15 min at room temperature. Wash the pellet with 5 ml of 70% ethanol. Dry the DNA pellet under vacuum briefly.
20. Resuspend the DNA in 0.5 ml of sterile ddH$_2$O or TE buffer.
21. Determine the DNA quality and quantity as described in **Section VI** and store the sample at −20°C until use.

II. Isolation of λDNA

λDNAs are commonly used vectors in cDNA and genomic DNA library construction because of their ability to host large DNA inserts stably. After library screening, the recombinant λDNA of positive clones that harbor the putative DNA of interest needs to be extracted from the phage for subcloning or further DNA analysis and verification. We describe here two phage amplification protocols, the plate method and the liquid method, for phage lysate preparation. With the former, it is easier to monitor the growth of phage, whereas the latter method is easier to scale-up. λDNA can be isolated and purified by either minipreparation or large-scale preparation procedures, or by CsCl gradient centrifugation. Many modifications of the traditional methods for λDNA preparation have emerged in recent years. Among these are a variety of kits produced by several different companies. The kits have the advantage of being quite simple and rapid; however, they are also more costly.

A. Phage Lysate Preparation by the Plate Method

1. Prepare LB plates, in a laminar-flow hood, by slowly pouring 25 ml of LB agar medium into each 95- or 100-mm sterile plastic petri plate. Remove any bubbles with a sterile pipette tip and keep the plates open for 10 to 15 min before covering them. Allow the medium to harden for about 1 to 2 h. The plates can be stored at room temperature for up to 10 days or at 4°C for up to 2 months. Before use, dry the plates in a laminar-flow hood for 15 to 30 min with the lids open or for 1 to 2 h with the covers on.

Notes: *(1) Do not use agar-plates that are too wet or too dry because bacteria and bacteriophage grow poorly under these conditions. (2) The top agarose separates easily from the supporting agar if applied to wet plates.*

2. Based on the titering experiment of the plaque-forming units (pfu) of the plaque eluate (see **Chapter 9**), place about 1×10^5 to 1×10^6 pfu of the eluate into a microcentrifuge tube and bring the volume to 50 to 100 μl with SM buffer for each 100-mm plate. Mix the bacteriophage sample with 100 μl of a fresh, overnight culture of appropriate plating bacterial strains, such as *E. coli* Y1089, Y1090, or LE 392. Incubate the tubes at 37°C for 20 to 30 min to allow the phage to absorb the bacteria.
3. Add 2.8 ml of melted top agarose into each glass tube and keep in a 50°C water bath until use. Add the absorbed bacteria (about 150 to 200 μl) prepared in step 2 to each of the glass tubes containing the top agarose solution. Mix and immediately pour onto the center of a LB plate. Quickly and evenly distribute the top agarose mixture over the surface of the supporting agar.

Notes: *This step should be performed carefully, but rapidly. If done too slowly, the top agarose mixture will harden before being poured onto the LB plate. Uneven distribution of the top agarose mixture will affect the growth of bacteria and bacteriophage.*

4. Allow the top agarose to harden for 15 min, invert the plates, and incubate at 37°C for 10 to 12 h or until the plaques become confluent.
5. Overlay each plate with 4 ml of SM buffer. Carefully scrape the top agarose with a clean spatula into a 35- or 50-ml centrifuge tube. Break up the agarose with the spatula and incubate at room temperature with slow shaking for 30 min.

Note: *Never scrape the bottom agar, which contains potential inhibitors of restriction enzymes.*

6. Centrifuge at 12,000 g for 10 min at 4°C to pellet the agarose and bacterial debris.
7. Carefully transfer the supernatant to a fresh tube. Add chloroform to 0.2% if long-term storage is desired. Store at 4°C until use for isolation of λDNA (see **Sections II.C** and **II.D**).

B. Phage Lysate Preparation by the Liquid Method

1. For each 100 ml of culture, place 20 to 25 μl of 1×10^5 pfu of plaque eluate into a microcentrifuge tube and add 480 to 475 μl of SM buffer. Mix well with 0.5 ml of fresh, overnight-cultured bacteria, and allow the bacteria and phage to adhere to each other by incubating at 37°C for 30 min.
2. Add the above mixture to a 250- or 500-ml sterile flask containing 100 ml of prewarmed (37°C) LB medium supplemented with 10 mM MgSO$_4$. Incubate at 37°C with shaking at 250 rpm until lysis occurs, which usually takes 8 to 11 h. The medium should be cloudy after several hours of culture, and then it becomes clear upon cell lysis. Cellular debris also becomes visible in the lysed culture.

Notes: *Using a proper bacteria-to-bacteriophage ratio is very important for successful lysate preparation. If the bacteria density is much over that of the bacteriophage, it will take longer for lysis to occur. In contrast, if the bacteriophage concentration is much higher than that of the bacteria, lysis will quickly become visible at the beginning of the incubation and stop prematurely. In addition, careful observations should be made after about 8 h of culture because the lysis may occur rapidly after that time. Cultures should be stopped after lysis occurs. Otherwise, the bacteria grow continuously and the cultures become cloudy again. If that happens, it will take a long time to see lysis again, or no lysis takes place at all.*

3. Immediately centrifuge the culture at 12,000 g for 10 min at 4°C. Transfer the supernatant to a fresh tube and add chloroform as described, if necessary. Store at 4°C until use.

Isolation and Purification of DNA

C. Mini-Preparation of λDNA

1. Add *RNase* A and *DNase* I to the λ lysate supernatant obtained from **Sections II.A** or **II.B**, each to a final concentration of 1 $\mu g \cdot ml^{-1}$. Place at 37°C for 30 min.

Notes: *RNase A functions to hydrolyze the RNA. DNase I will digest chromosomal DNA but not the packed phage DNA.*

2. Precipitate the phage particles by adding one volume of polyethylene glycol (PEG) precipitation buffer and incubate for at least 1 h on ice.
3. Centrifuge at 12,000 g for 15 min at 4°C and allow the pellet to dry at room temperature for 5 min.
4. Resuspend the phage particles with 1 ml of SM buffer per 10 ml of initial phage lysate and mix by vortexing briefly.
5. Centrifuge at 12,000 g for 4 min at 4°C to remove debris, and then carefully transfer the supernatant to a fresh tube.
6. Extract the phage particle proteins with an equal volume of buffered phenol:chloroform:isoamyl alcohol. Mix for 1 min and centrifuge at 12,000 g for 5 min.
7. Transfer the top, aqueous phase to a fresh tube and extract the supernatant one more time as in step 6.
8. Transfer the top, aqueous phase to a fresh tube and extract once with one volume of chloroform:isoamyl alcohol. Vortex and centrifuge as in step 6.
9. Carefully transfer the upper, aqueous phase to a fresh tube and add 0.1 volume of 3 M sodium acetate and 1 volume of isopropanol or 2 volumes of chilled 100% ethanol. Mix well and allow the DNA to precipitate at room temperature for 5 to 10 min.
10. Centrifuge at 12,000 g for 10 min at 4°C and decant the supernatant. Wash the pellet with 2 ml of 70% ethanol, centrifuge, and dry the pellet briefly under vacuum. Resuspend the DNA in 50 to 100 μl of TE buffer. Measure the DNA concentration and store at −20°C until use.

D. Large-Scale Purification of λDNA by the CsCl Gradient Centrifugation

1. To hydrolyze chromosomal DNA and RNA, add *RNase* A and pancreatic *DNase* I to the lysate, each to a final concentration of 1 $\mu g \cdot ml^{-1}$. Incubate for 30 min at 37°C.
2. Add NaCl powder to a final concentration of 1 M and dissolve. Place the sample on ice for 1 h. This helps the dissociation of bacteriophage particles from cellular debris and increases the precipitation of bacteriophage particles in the PEG solution used later.
3. Centrifuge at 12,000 g for 10 min at 4°C. Transfer the supernatant to a fresh tube and discard the bacterial debris.
4. Add solid PEG (mol wt 8000) to a final concentration of 10% (w/v) and dissolve by stirring on a magnetic stirrer at room temperature.
5. Place on ice for 2 h to allow the bacteriophage particles to precipitate.
6. Centrifuge at 12,000 g for 10 min at 4°C and decant the supernatant. Allow the pellet (bacteriophage particles) to dry at room temperature to remove the remaining fluid.
7. Resuspend the pellet in 0.2 ml of SM buffer per 10 ml of initial phage lysate.
8. Extract the PEG and cellular debris with one volume of chloroform and vortex for 1 min.
9. Centrifuge at 3000 g for 15 min at 4°C and carefully transfer the top, aqueous phase containing the bacteriophage particles to a clean tube.
10. Add exactly 0.5 $g \cdot ml^{-1}$ of solid CsCl to the aqueous phase and gently mix to dissolve. Carefully layer the mixture onto the top of the prepared CsCl step gradients (see **Reagents Needed**, for recipe) in Beckman SW41 or SW28 (or equivalent) clear polypropylene or polycarbonate centrifuge tubes.
11. Centrifuge at 22,000 rpm for 2 h at 4°C in a Beckman SW41 or SW28. Slowly remove the centrifuge tubes and locate the bands. A bluish band of the bacteriophage particles is usually visible at the interface between the 1.45- and 1.50-$g \cdot ml^{-1}$ layers. Place the tube against a black background and

shine a light from above to help examine the band. The cellular debris band should be at the interface between the 1.45 g·ml^{-1} layer and the sample layer.

12. Wipe the outside of the tube clean with ethanol and attach a piece of Scotch tape to the outside of the tube. Insert a needle to the top of the tube, as shown in Figure 1.1 and described in **Section I.B**. Remove the particle band by carefully puncturing the side of the tube through the tape using a 21-gauge needle and slowly collect the band as described. Alternatively, the band can be collected using a sterile pipette or a Pasteur pipette. Starting from the sample layer, carefully remove the sample layer, cellular debris, and the 1.45 g·ml^{-1} layer. The particle band can then be easily collected with a fresh pipette tip.
13. Place the suspension of the particle band in an ultracentrifuge tube and fill two-thirds of the tube with CsCl solution (1.5 g·ml^{-1} in SM). Centrifuge at 38,000 rpm for 24 h at 4°C (Beckman rotor Ti50) or at 35,000 rpm for 24 h at 4°C (Beckman rotor SW50.1 or equivalent).
14. Remove the band of bacteriophage particles as in step 12 and store at 4°C until use.
15. Remove CsCl from the bacteriophage particle suspension by dialysis for 2 h at room temperature against a 1000-fold volume of dialysis buffer A, with one buffer change.
16. Transfer the dialyzed sample to a fresh tube and add *Proteinase K* to a final concentration of 50 µg·ml^{-1}. Incubate at 56°C for 1 h.
17. Extract *Proteinase K* with one volume of buffered phenol:chloroform:isoamyl alcohol. Mix and centrifuge at 12,000 g for 10 min.
18. Transfer the upper, aqueous phase to a fresh tube and add an equal volume of chloroform:isoamyl alcohol. Mix by vortexing and centrifugation as in step 17.
19. Carefully transfer the upper, aqueous phase to a fresh tube and add 0.1 volume of 3 *M* sodium acetate and two volumes of chilled 100% ethanol to precipitate.
20. Centrifuge at 12,000 g for 10 min and carefully remove the supernatant. Dry the DNA pellet under vacuum and resuspend the DNA in 50 to 100 µl sterile ddH$_2$O or TE buffer. Determine the DNA quality and quantity as described in **Section VI** and store DNA at −20°C until use.

III. Isolation of Genomic DNA from Animals

Because animal cells lack cell walls and chloroplasts, as compared with plant cells, it is relatively easy for the cells to be lysed and DNA to be extracted without chloroplast DNA contamination. The following protocols describe in detail the isolation of animal DNA that is suitable for restriction enzyme digestion, genomic library construction in λ or cosmid DNA vectors, and Southern blot analysis. In order to obtain high-molecular-weight DNA, certain precautions must be taken. All glassware, plastic pipette tips, centrifuge tubes, cell scrapers or policemen, solutions, and buffers should be autoclaved or filter-sterilized in order to avoid *DNase* contamination. Molecular biology-grade or ultrapure chemicals and reagents are strongly recommended. Gloves should be worn during isolation procedures, and vigorous shaking should be avoided to prevent DNA from shearing.

A. Extraction of Genomic DNA with Organic Solvents

1. Harvest tissue or collect cells and add genomic lysis buffer to the samples.
   Extraction from fresh blood cells: Collect cells from fresh blood (20 ml per extraction) by centrifuging at 1000 rpm for 15 min at room temperature and carefully decanting the supernatant containing lysed red cells without nuclei. Add one volume (20 ml) of phosphate-buffered saline (PBS) to the remainder containing white blood cells and centrifuge at 1000 rpm for 15 min at room temperature. Carefully decant the supernatant and resuspend the cells in 15 ml of acid citrate dextrose (ACD) solution. Incubate at 37°C for 1 to 1.5 h prior to step 2.
   Extraction from frozen blood cells: Thaw frozen blood sample (20 ml per extraction) and transfer the sample to a fresh centrifuge tube. Add an equal volume of PBS and centrifuge at 1000 rpm for 15 min at room temperature. Decant the supernatant containing lysed red cells and resuspend

the white blood cells in 15 ml of ACD solution. Incubate the sample at 37°C for 1 to 1.5 h. Proceed to step 2.

Extraction from cells grown in monolayers: Harvest the confluent monolayers of cells with ice-cold PBS by adding 10 to 15 ml ice-cold PBS and carefully scrape the cells, using a sterile policeman or cell scraper, into a clean tube. Repeat once and combine the cells into the same tube. Collect the cells by centrifugation at 1000 rpm for 10 min at 4°C. Resuspend the cells in eight volumes of ice-cold PBS and centrifuge at 1000 rpm for 10 min at 4°C. Resuspend the cells at 4×10^7 cells·ml^{-1} in TE buffer (pH 8.0). Transfer the cell suspension to a clean tube or a flask and add 9 ml of genomic lysis buffer per milliliter of the cell suspension. Incubate the mixture at 37°C for 1 to 1.5 h. Continue to step 2.

Extraction from cells grown in suspension: Collect the cells by centrifugation at 1000 rpm for 10 min at 4°C and wash them twice with one volume of ice-cold PBS. Again collect the cells by centrifugation at 1000 rpm for 10 min at 4°C. Resuspend the cells at a concentration of 4×10^7 cells·ml^{-1} in TE buffer (pH 8.0). Add 10 volumes of genomic lysis buffer and incubate at 37°C for 1 to 1.5 h. Proceed to step 2.

Extraction from tissue: Harvest fresh and soft tissue using sterile scissors or a razor blade and immediately freeze in liquid nitrogen. Store at −80°C until use. Grind the tissue in liquid nitrogen using a prechilled mortar and pestle until a fine powder is obtained. Using a chilled spatula, transfer the powder, little by little, to a centrifuge tube containing eight to ten volumes of genomic lysis buffer per gram of tissue. Swirl the tube gently to mix after each transfer. Incubate the tube at 37°C for 1 to 1.5 h. Proceed to step 2.

2. Add *Proteinase K* (20 mg·ml^{-1}) to a final concentration of 100 mg·ml^{-1} of the lysed cell suspension from step 1 and gently mix.
3. Incubate the mixture in a water bath at 50°C for 2 to 4 h. Gently swirl the viscous mixture every 20 min.
4. Allow the mixture to cool to room temperature and transfer the mixture into a fresh centrifuge tube. Add one volume of buffered phenol and mix well by gently inverting the tube for 5 min.
5. Centrifuge at 6000 g at room temperature for 15 min and carefully transfer the top, viscous, aqueous phase to a fresh tube with a wide-bore pipette.
6. Repeat phenol extraction two to three times, as in steps 4 and 5, until no visible proteins and cellular debris occur at the interface.
7. To the supernatant, add 0.1 volume of 3 *M* sodium acetate and 2 volumes of chilled 100% ethanol to precipitate the DNA at room temperature for 15 to 30 min.
8. Carefully "fish" out the DNA into a fresh tube with a glass hook made from a flame-bent Pasteur pipette. This is the most desirable method without copurifying fragmented DNA. Alternatively, pellet the DNA by centrifugation at 5000 g for 5 min at 4°C. Gently rinse the DNA with 4 ml of 70% ethanol and air-dry the sample. **Do not** overdry the sample, especially if vacuum is applied. Otherwise, the DNA will be difficult to resuspend.
9. Resuspend the DNA in an appropriate amount of TE buffer (1 ml per 4×10^7 cells). To facilitate the resuspension, the DNA sample can be placed at 45 to 50°C for 15 to 30 min with gentle and occasional shaking until it is dissolved. Keep the tube open to evaporate remaining ethanol.
10. Determine the quality and quantity of the DNA as described in **Section VI**. Store the genomic DNA samples at 4°C.

B. Extraction of Genomic DNA with Formamide

This is a relatively simple and low-cost method as compared with the previous protocol. The DNA isolated by this method is generally of high molecular weight (>200 Kbp [kilobase-pairs]), and suitable for genomic library construction in cosmic vectors, restriction enzyme digestion, and Southern blot analysis. The procedures are described as follows.

1. Conduct steps 1 to 3 as described in the previous protocol.
2. Allow the suspension of lysed cells to cool to 15°C, add 0.7 volumes of the denaturation buffer to the suspension, and shake gently to mix. Incubate at 15°C overnight.

3. Pool and dialyze the viscous suspension in a collodion bag (Sartorius SM 13200E or equivalent) three times against 4 L of dialysis buffer A and five times against 4 L of dialysis buffer B.
4. Transfer the dialyzed solution to a fresh tube. If cell debris remains in the solution, centrifuge at 6000 g at room temperature for 15 min and transfer the supernatant to a fresh tube.
5. Determine the DNA quality and quantity as described in **Section VI** and store the DNA at 4°C.

IV. Isolation of Genomic or Organelle DNA from Plants

Plant cells have walls and chloroplasts. These features make the procedures to isolate and purify DNA from plants more difficult than from animals. Traditional methods, based on CsCl gradient centrifugation, are time-consuming and relatively expensive to perform, as described above. DNA can be extracted with high purity, but the yield is very low. CTAB (hexadecyltrimethylammonium bromide)-based extraction methods are now widely used for extraction of genomic DNA from a broad range of plant tissues. CTAB is a nonionic detergent and was originally used to differentially solubilize nucleic acids and polysaccharides under various salt (i.e., NaCl) concentrations.[3,6] The protocol presented here is a simplified CTAB extraction procedure[7] that has been successfully used in our laboratories for a wide variety of plant species. The yield of DNA depends on the source of tissue (i.e., leaf, flower, stem, or root) and plant species. Generally, young tissues with meristematic cells produce higher DNA yield, whereas certain highly differentiated tissues, such as xylem, which is programmed for early cell death, are a poor source for DNA extraction. For genomic library construction or Southern blot analysis, we strongly recommend that etiolated (dark-grown) plants be used to avoid chloroplast DNA contamination and to reduce polysaccharide content. All of the common molecular biology practices should be followed. Care should be taken during the extraction to avoid vigorous shaking and shearing of the genomic DNA.

A. Isolation of Plant Genomic DNA

1. Harvest the tissue (1 to 5 g per extraction) and quickly freeze in liquid N_2. The frozen tissue can be stored in aluminum packets at –80°C until use.
2. Set a water bath to 65°C. Dispense 7 ml of DNA extraction buffer per g of plant tissue in a polypropylene or Falcon centrifuge tube and prewarm the tube in the 65°C water bath. Prepare an appropriate number of tubes if multiple sample extractions are desired.
3. Prechill a mortar and a pestle with liquid N_2. Place frozen tissue in the mortar and add liquid N_2 to immerse tissue. While the liquid N_2 is evaporating, very gently crush the tissue clumps into small pieces with the pestle. Add a spatula of acid-washed sea sand to aid in grinding and more liquid N_2 if necessary. Addition of liquid N_2 from this point on should be done very carefully by slowly pouring liquid N_2 along the wall of the mortar to prevent frozen tissue from bouncing out. Do not pour liquid nitrogen directly onto the tissue. When the liquid N_2 has nearly evaporated, vigorously grind the tissue to a fine powder.
4. Quickly transfer frozen powder to a centrifuge tube containing prewarmed extraction buffer with a liquid-nitrogen-precooled spatula. Swirl the tube frequently to mix contents during the transfer. Add an appropriate volume of 2-mercaptoethanol in a fume hood, cap the tube, and vortex for 30 sec to homogenize.

Notes: *(1) It is very important to prechill the mortar with liquid N_2 to prevent tissue from thawing out. Plant tissues are rich in nuclease activities, so the frozen powder should not be thawed until disrupted in warm extraction buffer. (2) For tissues rich in phenolic compounds (such as older leaves, bark, or root tissues), a small amount of insoluble PVPP (0.1 g per g of tissue) can be added to the frozen tissue during grinding. PVPP binds polyphenols and prevents them from copurifying with DNA.[8] (3) The vortex/homogenization step is necessary*

Isolation and Purification of DNA

for effective tissue dispersal. However, the solution should be handled gently in all subsequent steps in order to avoid shearing of the DNA.

5. Incubate the tube at 65°C for 30 min with occasional gentle swirling. If multiple samples are to be extracted, rinse the mortar and pestle thoroughly with 70% ethanol several times, wipe dry and repeat step 2 above. Each tube should be marked with the time when placed into the 65°C water bath in order to keep track of the incubation time.
6. Bring the samples to room temperature and let stand for 5 min. Extract with one volume of chloroform:isoamyl alcohol in the fume hood. Mix gently but thoroughly by inverting the tubes several times. Centrifuge at 12,000 g for 10 min at room temperature.

Note: *CTAB will precipitate out of solution at temperatures lower than 15°C.*

7. Carefully transfer the top, aqueous phase to a fresh tube with a wide-bore pipette. Avoid contamination with the interface. If the solution remains cloudy, repeat the centrifugation step or repeat the extraction.
8. Add an equal volume of isopropanol to precipitate DNA at room temperature for 10 min. Pellet the DNA by centrifugation for 10 min at 12,000 g at room temperature. Wash with 70% ethanol and dry briefly by inverting the tubes over a paper towel. **Do not** overdry the DNA or it will be difficult to resuspend.
9. Dissolve the DNA pellet in 3 ml TE buffer. Treat the DNA with *RNase A* at a final concentration of 20 $\mu g \cdot ml^{-1}$ for 30 min at 37°C.
10. **(Optional)** For tissues rich in polysaccharides, add 0.1 volume of 3 M sodium acetate (pH 5.2) and incubate at 4°C for 30 min. Centrifuge at 12,000 g for 10 min at room temperature to pellet the gel-like polysaccharides. Transfer the DNA-containing supernatant to a fresh tube.
11. Reprecipitate the DNA with two volumes of cold 100% ethanol (and 0.1 volume of 3 M sodium acetate if step 10 was omitted). It is best to try to hook the precipitated DNA and spool it onto a flame-sealed and bent Pasteur pipette. This will leave behind fragmented DNA. However, if difficulties are encountered with the spooling procedure, the DNA can be pelleted by centrifugation at 12,000 g for 10 min at 4°C.
12. Wash the spooled DNA by soaking the entire hook in 70% ethanol for 10 min. Air-dry the spooled DNA attached to the hook for a short period of time to avoid excess ethanol. If the DNA was pelleted, wash as usual with 70% ethanol followed by centrifugation. Decant the supernatant and air-dry to remove excess ethanol.
13. Resuspend the DNA in 0.9 ml of TE buffer. Spooled DNA will fall off the hook soon after the hook is placed in TE. Incubate the tube at 65 to 70°C for 5 min if pellet is difficult to dissolve. Transfer the DNA to a 2-ml microcentrifuge tube.
14. Extract with an equal volume of buffered phenol:chloroform:isoamyl alcohol, mix by gently inverting the tubes several times, and centrifuge at 12,000 g for 5 min at 4°C. Repeat this step two to three times until the interface is clear of debris; then, further extract once with chloroform:isoamyl alcohol.
15. Carefully transfer the aqueous phase to a fresh tube. Add 0.1 volume of 3 M sodium acetate and 1 volume of isopropanol to precipitate the DNA as described above. Hooking the DNA a second time is not required unless contamination with fragmented DNA is intolerable.
16. Dissolve the DNA in TE buffer (100 μl per gram of tissue). Heat the DNA at 65 to 70°C briefly to aid resuspension, if necessary.
17. Check 1 to 2 μl of DNA on a 0.8% agarose minigel alongside undigested λDNA (0.2 to 0.5 μg) as a molecular-weight standard. Ideally, the plant genomic DNA will migrate at a similar rate as, or more slowly than, the undigested λDNA, with an apparent molecular size >50 kb. If predigested λDNA markers are used, the plant genomic DNA should migrate more slowly than the largest (21 to 23 kb) DNA fragments. Good quality genomic DNA migrates as a high-molecular-weight mass, with very little or no smearing.
18. Measure DNA concentration as described in **Section VI**. For genomic Southern analysis, a concentration of 0.4 to 1 $\mu g \cdot \mu l^{-1}$ is preferred. If the DNA concentration is too low, consider reprecipitation and resuspension in a smaller volume. Store the sample at 4°C.

B. Isolation of Chloroplast DNA by Sucrose Gradient

Chloroplasts are organelles unique to plant and algal cells. The chloroplast genome resembles both bacterial and eukaryotic nuclear DNAs in its organization. It is composed of a single, circular molecule of double-strand DNA (cpDNA) with a molecular weight of about 85 to 300 kb in green algae and 120 to 190 kb in most higher plants. While there are several different types of plastids known in plants, chloroplasts are abundant in green, photosynthetic tissues such as leaves, and consequently, these tissues are recommended for cpDNA extraction. The protocol described below represents a modified, easy, and successful procedure.[9,10]

1. Grow plants in the dark for 24 to 36 h before harvesting tissue. Dark incubation helps reduce starch accumulation in the chloroplasts.
2. Collect young green leaves from the plants (10 g per extraction) with a clean razor blade and place the leaves in ice-cold water. Rinse the tissue in cold, 5% commercial bleach, such as Clorox™, for 4 min followed by two washes with cold distilled water to remove the bleach.
3. **Isolation of Chloroplasts:** Briefly dry the tissue with a clean paper towel and slice the leaves into strips (1 to 2 mm) with a razor blade. Transfer the leaf strips to a blender containing six volumes of ice-cold chloroplast isolation buffer per gram of tissue (fresh weight). Grind the tissue at low speed using four 4-sec bursts in a cold room until a good suspension is obtained.
4. Immediately filter the homogenate through five layers of cheesecloth, followed by filtration using one layer of Miracloth. Collect the filtrate in a centrifuge tube and spin at 3000 g for 15 min at 4°C.
5. Discard the supernatant and gently resuspend the green pellet in 5 ml of chloroplast washing buffer. It is best to use a small brush to resuspend the pellets in order to avoid damaging intact plastids. Layer the resuspended chloroplasts over a 30/45/60% sucrose step gradient (see **Reagents Needed**, for preparation of the gradient). Cover the top of the gradient with 1 ml of washing buffer.
6. Centrifuge in an ultracentrifuge (e.g., Beckman, SW27) at 25,000 rpm for 1 h at 4°C. Remove the intact, dark green chloroplast band, usually at or near the 30/45% interface, using a Pasteur pipette or a wide-bore pipette.
7. Add three to four volumes of cold washing buffer to the chloroplast sample, mix gently, and centrifuge at 6000 g for 10 min at 4°C. Gently resuspend the pellet in 5 ml cold washing buffer and store at 4°C until use.
8. **Extraction of cpDNA:** Pipette 2.5 ml of chloroplast sample to a fresh centrifuge tube and add 0.25 ml of 10 mg·ml^{-1} *Proteinase K* solution. Gently mix by inversion and incubate at room temperature for 15 to 20 min.
9. Add 0.25 ml of 20% sarkosyl solution and gently mix by inversion to avoid shearing the cpDNA. Incubate the sample at room temperature for 50 min with occasional gentle swirling.
10. Add 9.1 ml of 0.672 g·ml^{-1} CsCl stock solution and gently mix by inversion to dissolve. Allow the mixture to stand for at least 60 min at room temperature.
11. Centrifuge at 12,000 g for 20 min. Gently transfer the supernatant to a fresh graduated tube and discard the pellet of membranes and starch particles.
12. Adjust the supernatant to 4.8 ml with TE buffer and add 6.7 ml of 1.24 g·ml^{-1} CsCl stock solution and 0.3 ml of 10 mg·ml^{-1} EtBr solution. Slowly and gently mix the sample by inversion.
13. Transfer the cpDNA-CsCl mixture to a Quick-Seal ultracentrifuge tube and proceed to ultracentrifugation as described in **Section I.B** (steps 12 to 16).
14. Remove EtBr by extraction with TE-saturated isopropanol three times as described. CsCl is then removed by ethanol precipitation, as described in **Section I.B**, above, or by dialysis as in the following step.
15. Transfer the colorless aqueous phase containing cpDNA-CsCl to a dialysis bag and dialyze against at least three changes of 2 L of dialysis buffer B for 1 to 2 days.
16. Transfer the dialyzed sample to a microcentrifuge tube and centrifuge at top speed for 5 min at room temperature to remove any particles. Determine the DNA quality and quantity and store at −20°C until use.

V. Purification of DNA Fragments from Agarose Gels

Isolation of specific DNA fragment(s) is necessary for preparation of DNA probe(s) in nucleic acid hybridization and for preparation of insert and vector fragments during subcloning or construction of expression constructs. Fragment isolation is commonly accomplished by first separating DNA fragments (derived from restriction enzyme digestion or PCR) by agarose-gel electrophoresis followed by identification and excision of the target band(s). DNA is then eluted from the gel slice. A large number of methods have been published, and many commercial kits are now available for DNA elution and purification from agarose gels. These kits have the advantages of being very quick and easy to use. However, they are also more expensive than traditional methods, and some are not suitable for isolation of large (>10 kb) or small (<100 bp) DNA fragments. Here, we describe two protocols that are simple and relatively inexpensive to perform for elution of DNA fragments from agarose gel slices. DNA purified by these methods has an adequate yield and purity for subsequent ligation, cloning, and labeling.

A. Elution of DNA Fragments by the Freeze-and-Thaw Method

1. Set up a standard restriction enzyme digestion in a microcentrifuge tube using DNA (usually plasmid DNA) containing the sequence of interest in the following order:
   - ddH$_2$O, to 50 µl
   - DNA, 0.2 to 1 µg
   - 10X restriction enzyme buffer, 5 µl
   - Restriction enzyme, one to three units
   - RNase A (if necessary), 20 µg·ml^{-1}
2. Mix well and spin briefly to collect the contents. Incubate the reaction at an appropriate temperature for 2 h to overnight.

Notes: *(1) One unit of restriction enzyme activity is defined as the amount required to completely digest 1 µg of DNA in a 50-µl reaction in 1 h at the appropriate temperature. Most laboratory-prepared DNA (e.g., miniprep plasmid DNA) contains impurities and often requires prolonged incubation (up to overnight) to achieve complete digestion. (2) Most commercial restriction enzymes are available at concentrations of 5 to 10 units per µl. It may be necessary to dilute the enzymes before use. Do not store diluted enzymes unless a proper dilution buffer (see manufacturer's instructions) is used. (3) If digestion with two or more restriction enzymes is desired, make sure that the reaction buffer selected is compatible with all enzymes.*

3. Carry out electrophoresis as described in **Chapter 5**, using 0.8 to 1% low melting-temperature agarose and 1X TAE buffer.
4. Transfer the gel to a UV transilluminator and visualize the DNA bands under long wavelength (305 to 327 nm). Excise the band(s) of interest with a clean razor blade or spatula, trim as much excess, unstained gel area as possible, and place the slices in a microcentrifuge tube.

Notes: *(1) TAE buffer is preferred to TBE buffer if eluted DNA fragments are to be used for ligation, as borate ions inhibit the ligation reaction.[2] (2) Short wavelength (e.g., <270 nm) UV source should not be used to visualize DNA, as it may damage the DNA and compromise subsequent manipulations (e.g., subcloning). Alternatively, add 1 mmol·l^{-1} guanosine to the electrophoresis buffer to protect the DNA from UV damage.[11]*

Caution: *UV light is harmful. Protective eyeglasses and gloves should be worn when using UV light.*

5. Add two gel volumes of TE buffer to the gel slice and completely melt the gel in a 60 to 70°C water bath.

Note: *The gel slices can be directly melted without adding any TE buffer. The DNA concentration is usually high, but the yield of the DNA fragments will be much lower than when TE buffer is added.*

6. Immediately chill the melted gel solution on dry ice or liquid N_2 and place the tube at −70°C for at least 10 min.
7. Thaw the gel mixture by tapping the tube vigorously. It takes about 5 to 10 min to thaw the gel into resuspension.
8. Centrifuge at 12,000 g for 5 min at room temperature.
9. Carefully transfer the liquid phase containing the eluted DNA fragment into a fresh tube. The DNA solution can be used directly for ligation or labeling. Proceed to the following if higher DNA quality or concentration is desired.
10. Extract the solution with one equal volume of buffered phenol, vortex, and centrifuge at 12,000 g for 5 min at room temperature.
11. Transfer the top, aqueous phase to a fresh tube, taking care not to touch the white substance (agarose) at the interface. Extract the aqueous phase once with buffered phenol:chloroform:isoamyl alcohol and once with chloroform:isoamyl alcohol.
12. Precipitate the aqueous phase containing DNA by adding 0.2 volume of 10 *M* ammonium acetate and 2 volumes of 100% ethanol at 4°C for 20 min. Centrifuge at 12,000 g for 10 min at 4°C.

Notes: *(1) Ammonium acetate is used to reduce unwanted coprecipitation of agarose with DNA. (2) Carriers can be used to improve the recovery of small quantities of DNA by ethanol precipitation. Commonly used carriers are yeast tRNA (transfer RNA) at 10 to 20 µg·ml^{-1} or glycogen at 50 µg·ml^{-1}. tRNA is not suitable for recovery of DNA that will be used in reactions involving terminal transferase or polynucleotide kinase, whereas glycogen should be avoided if the DNA is to be used for DNA-protein interaction studies.*[2]

13. Discard the supernatant and briefly rinse the pellet with 1 ml of 70% ethanol. Dry the DNA briefly and dissolve the DNA in an appropriate volume of ddH$_2$O or TE buffer. Store at −20°C until use.

B. Elution of DNA Fragments in Wells of Agarose Gel

1. Carry out DNA digestion and gel electrophoresis as described above and in **Chapter 5**, but add the running buffer up to the upper edges of the gel instead of covering the gel.
2. During electrophoresis, monitor the separation of DNA bands stained by EtBr on the gel using a handheld long-wavelength UV lamp. When it is clear that the band of interest is separated from other undesirable band(s), stop the electrophoresis, and use a spatula to make a well in front of the band of interest. Add 20 to 60 µl of running buffer into the well.
3. Continue electrophoresis while monitoring the band of interest until it migrates into the well.
4. Stop electrophoresis and transfer the solution containing the DNA of interest from the well to a fresh tube.
5. Extract the solution with an equal volume of buffered phenol, followed by one equal volume of chloroform:isoamyl alcohol as described above (optional). Precipitate the DNA with 0.2 volume of 10 *M* ammonium acetate and 2 volumes of 100% ethanol as described above. Dissolve the DNA in ddH$_2$O or TE and store at −20°C.

VI. Determination of DNA Quality and Quantity

Assessment of the quality and concentration of the isolated DNA is essential to all subsequent manipulations. DNA integrity following extraction, or the efficiency of restriction enzyme digestion, can only be monitored by agarose-gel electrophoresis, as described in **Chapter 5**. A rough estimate of the concentration of a DNA band of interest can be obtained on an EtBr-stained gel by loading molecular weight markers of known DNA concentration in an adjacent lane. This procedure works well for most recombinant DNA applications where an estimate of DNA concentration is sufficient. One should avoid overloading DNA, as the concentration of EtBr-saturated bands is very difficult to estimate by eye. DNA yield, concentration, and/or purity can be determined spectrophotometrically by measuring UV light absorbance (A) of samples at wavelengths of 260 and 280 nm. Ratios of the absorbances at these wavelengths (A_{260}/A_{280}) provide a gauge of protein contamination in a DNA sample, while the absorbance at 260 (A_{260}) is used to calculate the concentration of DNA in the sample. UV light absorbance, however, cannot be used to distinguish DNA from contaminating RNA, and it yields poor estimates of DNA concentration in preparations contaminated with protein, phenol, oligo- or polysaccharides. Applications that depend on accurate quantitation of very small amounts of DNA require fluorometry. Several DNA-specific fluorescent dyes are available, and measurement is more sensitive than with UV absorbance spectrophotometry.

A. Spectrophotometric Measurement

The protocol described below is for general reference. Readers should refer to the instruction manual for proper operation of their spectrophotometer.

1. Turn on the UV spectrophotometer and set the wavelengths at 260 and 280 nm according to the instructions. Some newer models of spectrophotometers contain a computerized monitoring unit that can automatically and simultaneously measure a DNA sample at wavelengths of 260 and 280 nm, and calculate and display the DNA concentration together with the ratio of A_{260}/A_{280}. If such functions are not available, program the spectrophotometer to measure and record the A_{260} and A_{280} readings for each sample manually.
2. Set up a reference using a blank solution. Depending on the DNA sample, add 0.5 ml of TE buffer (pH 8.0) or ddH$_2$O to a clean 1-ml cuvette or to each of two 1-ml cuvettes (one-half to three-quarters full), depending on whether one or two cells are available in the spectrophotometer. Insert the cuvette(s) into cuvette holder(s) in the sample compartment with the optical (clear) sides of the cuvette(s) facing the light path. Close the sample compartment cover and adjust the number to 0.000 (either manually or by pressing the "Auto Zero" button, according to the manufacturer's instructions).

Note: *Gloves should be worn when handling cuvettes. Cuvettes should be rinsed with 95% ethanol followed by ddH$_2$O, and then carefully dried with lens paper prior to reuse. Do not rub the cuvettes with anything other than lens paper or they may become scratched. Unclean or scratched cuvettes will interfere with sample readings and affect the accuracy.*

3. In a clean microcentrifuge tube, dilute the DNA sample in blank solution to a total volume of 0.5 ml for measurement. Briefly vortex the sample to mix and transfer the sample to a clean cuvette by simply pouring or using a disposable Pasteur pipette.
4. Insert the sample cuvette and read the absorbance of the sample cuvette against the blank cuvette according to the spectrophotometer instructions. If only one cell is available, remove the reference cuvette after adjusting the absorbance to 0.000 and insert the sample cuvette. Record the reading.
5. After measuring DNA sample(s) at 260 and 280 nm, calculate the ratio of A_{260}/A_{280} and concentration for each DNA sample. A clean DNA preparation should have an A_{260}/A_{280} ratio of 1.7 to 2.0. The concentration of a DNA sample is calculated as follows:

$$\text{DNA } (\mu g \cdot \mu l^{-1}) = A_{260} \text{ reading} \times 50\ \mu g \cdot ml^{-1} \times \text{dilution factor}$$

For example, if 2 µl of DNA is diluted to 500 µl for measuring, and its A_{260} reading is 0.4000, then:

$$\begin{aligned}
\text{DNA } (\mu g \cdot \mu l^{-1}) &= A_{260} \text{ reading} \times 50\ \mu g \cdot ml^{-1} \times \text{dilution factor} \\
&= 0.4000 \times 50\ \mu g \cdot ml^{-1} \times 500 \div 2 \\
&= 5000\ \mu g \cdot ml^{-1} \\
&= 5\ \mu g \cdot \mu l^{-1}
\end{aligned}$$

B. Fluorometric Measurement

The method introduced here is based on the use of a DNA-specific dye, the PicoGreen® dsDNA quantitation reagent (Molecular Probes, Eugene, OR), in conjunction with the TD-700 minifluorometer (Turner Designs, Sunnyvale, CA) with a minicell (100-µl assay volume). With the PicoGreen reagent, as little as 25 ng·ml⁻¹ can be detected in the presence of RNA and free nucleotides.[12] Below is a generic description of the protocol based on the high-range assay using a minicell (100 µl assay volume). Readers should refer to the product information sheet of PicoGreen when developing their own fluorometry assay.

1. Turn on the fluorometer (with ~480-nm excitation filter and ~520-nm emission filter) and allow the lamp to warm up for 10 min.
2. Prepare PicoGreen reagent by diluting an aliquot of the PicoGreen stock 1:200 in TE buffer in a microcentrifuge tube protected from light. Prepare enough working solution for the number of assays desired (50 µl needed per assay).

Note: *The PicoGreen reagent should be freshly diluted and used within a few hours of its preparation.*

3. Prepare DNA standards for calibration. Dilute DNA standards to a final concentration of 1000, 500, 100, and 10 ng·ml⁻¹ in 100 µl of TE and 100 µl of PicoGreen reagent (200 µl final). Mix well and protect from light.
4. Measure the sample fluorescence in a minicell according to fluorometer instructions. Subtract the blank value and generate a standard curve (fluorescence vs. DNA concentration). The standard curve is automatically generated when using the TD-700 minifluorometer. Follow the fluorometer's instructions.
5. Dilute a small aliquot of each DNA sample (usually 1 µl) in 50 µl TE buffer and mix with 50 µl of PicoGreen reagent. The upper detection limit of the high-range assay is 1 µg·ml⁻¹ DNA in the assay mixture (i.e., 100 ng in 100-µl assay). Therefore, it is necessary to dilute highly concentrated DNA samples (usually 10- to 100-fold) before preparing the assay mixture. Protect the mixture from light and incubate at room temperature for 2 to 5 min.
6. Measure the fluorescence of the sample mixture as quickly as possible to avoid photobleaching artifacts.
7. The concentration of the sample mixture is automatically calculated and displayed by the TD-700 minifluorometer. Use the dilution factor to calculate the DNA sample concentration. Alternatively, determine sample concentration according to the standard curve. DNA concentration determined in a 100-µl assay can be calculated as follows:

For example, if 1 µl of 10-fold diluted DNA is used in a 100-µl assay mixture for measurement, and an assay mixture concentration of 800 ng·ml⁻¹ is obtained (based on standard curve), then:

$$\begin{aligned}
\text{Total amount of DNA present in assay mixture} &= 800\ \text{ng} \cdot ml^{-1} \times 0.1\ ml \\
&= 80\ \text{ng}
\end{aligned}$$

Therefore, 80 ng DNA was present in 1 µl of 10-fold-diluted sample:

$$\text{DNA concentration of original undiluted sample} = 80 \text{ ng} \cdot \mu l^{-1} \times 10$$
$$= 800 \text{ ng} \cdot \mu l^{-1}$$

VII. Troubleshooting Guide

Symptom	Solutions
DNA degradation	Make certain that all labware and solutions are autoclaved or filter-sterilized.
	When processing frozen tissues, **do not** allow frozen tissue/powder to thaw until disrupted in extraction buffer. Use prechilled mortar and pestle. Transfer of frozen powder to extraction buffer should be done as quickly as possible.
	Avoid vortexing or vigorous shaking of genomic DNA-containing mixture following the homogenization step.
	RNase A contains *DNase*. Boil *RNase* A and aliquot before use (see **Reagents Needed**).
Low DNA yield	Insufficient grinding or tissue disruption. Use acid-washed sea sand to aid grinding and homogenize thoroughly.
	Avoid using mature tissues for DNA extraction.
	DNA pellet may be inadvertently lost after ethanol precipitation.
Resuspended DNA is viscous	Indication of polysaccharide contamination (for plant genomic DNA extraction). Polysaccharides can be precipitated in 0.3 M of sodium acetate (pH 5.2) at 4°C for 30 min and centrifuged at 12,000 rpm for 10 min at 4°C. The DNA is then precipitated with two volumes of ethanol.
	Etiolate the plants for 2 to 4 days prior to tissue harvesting.

Reagents Needed

ACD Solution

Citric acid (4.8 g)
Trisodium citrate (13.2 g)
Glucose (14.7 g)
Adjust to a final volume of 1 L with ddH$_2$O and filter-sterilize.

Alkaline Solution

0.2 N NaOH
1% SDS
Freshly diluted from 10 N NaOH and 10% SDS solutions.

10 M Ammonium Acetate

Dissolve 77 g ammonium acetate in 50 ml ddH$_2$O and adjust volume to 100 ml.

CTAB DNA Extraction Buffer

1.4 M NaCl
20 mM EDTA (ethylenediaminetetraacetic acid)
100 mM Tris-HCl, pH 8.0
3% (w/v) CTAB (hexadecyltrimethylammonium bromide)
1% (v/v) 2-Mercaptoethanol, **add before use**

CTAB dissolves quickly in NaCl solution; however, heating (but not to boiling) may also be required. Autoclave and store at room temperature.

Chloroform:Isoamyl Alcohol (24:1)

48 ml Chloroform
2 ml Isoamyl alcohol
Mix and store at room temperature.

Chloroplast Isolation Buffer

350 mM Sorbitol (233.3 ml of 1.5 M sorbitol solution)
50 mM Tris, pH 8.0 (50 ml of 1 M Tris)
5 mM EDTA (10 ml of 0.5 M EDTA)
0.1% (w/v) BSA (Fraction V)
0.1% (v/v) 2-Mercaptoethanol, **add before use**
0.1% (w/v) Sodium ascorbate, **add before use**
Prepare 1.5 M sorbitol stock by dissolving 273.3 g sorbitol in ddH$_2$O and store at −20°C. Dilute Sorbitol, Tris and EDTA stock solutions in ddH$_2$O and add 1 g BSA. Adjust the volume to 1 L and filter sterilize. Store at 4°C.

Chloroplast Washing Buffer

350 mM Sorbitol (23.3 ml of 1.5 M sorbitol solution)
50 mM Tris, pH 8.0 (5 ml of 1 M Tris)
20 mM EDTA (4 ml of 500 mM EDTA)
Dilute from stock solutions and adjust final volume to 100 ml with ddH$_2$O. Autoclave and store at 4°C.

0.672 g·ml^{-1} CsCl Solution

Dissolve 53.76 g CsCl in 40 ml TE and adjust the volume to 80 ml with TE.

1.24 g·ml^{-1} CsCl Solution

Dissolve 99.2 g CsCl in 40 ml TE and adjust the volume to 80 ml with TE.

CsCl Step Gradient

Density (g/ml)	CsCl (g)	SM (ml)	Refractive index (η)
1.45	60	85	1.3768
1.50	67	82	1.3815
1.70	95	75	1.3990

The step gradient can be made in Beckman SW41 or SW28 clear polypropylene or polycarbonate centrifuge tubes, either by carefully layering equal volumes of each of the above solutions of decreasing density on top of one another or by layering the solutions of increasing density under one another. Store the gradient at 4°C until use.

Denaturation Buffer

80% Deionized formamide (v/v)
0.8 M NaCl
20 mM Tris, pH 8.0

Isolation and Purification of DNA

Dialysis Buffer A
0.1 M NaCl
20 mM Tris, pH 8.0
10 mM EDTA, pH 8.0

Dialysis Buffer B
10 mM NaCl
10 mM Tris, pH 8.0
1 mM EDTA, pH 8.0

0.5 M EDTA
Dissolve 93.06 g Na$_2$EDTA·2H$_2$O in 300 ml ddH$_2$O. Adjust pH to 8.0 with 2 N NaOH solution. Adjust final volume to 500 ml with ddH$_2$O and store at room temperature.

10 mg·ml^{-1} Ethidium Bromide (EtBr)
Dissolve 200 mg EtBr in 20 ml ddH$_2$O. Store in a light-tight bottle at 4°C.

Genomic Lysis Buffer
10 mM Tris-HCl, pH 8.0
100 mM EDTA, pH 8.0
20 μg·ml^{-1} RNase A
0.5% SDS

LB (Luria-Bertaini) Medium
Tryptone (10 g)
Yeast extract (5 g)
NaCl (5 g)
Dissolve in ddH$_2$O and adjust the volume to 1 L. Adjust pH to 7.5 with 2 N NaOH. Autoclave.

LB Top Agarose
Tryptone (2 g)
Yeast extract (1 g)
NaCl (1 g)
Agarose (0.6 g)
Dissolve in ddH$_2$O and adjust the volume to 200 ml. Adjust pH to 7.5 with 2 N NaOH and autoclave. When the solution has cooled to below 60°C, add 2 ml of 1 M MgSO$_4$.

LB Plates
Add 15 g of agar to 1 L of LB medium and autoclave. When the medium cools to 50 to 60°C, add appropriate antibiotics, mix well and pour 20 to 25 ml of medium into 100-mm petri dishes. Allow the media to harden in a laminar-flow hood. Store at room temperature for 10 days or at 4°C for up to 2 months.

20% Maltose
Dissolve 20 g maltose in ddH$_2$O and adjust volume to 100 ml. Filter sterilize and store at 4°C.

1 M MgSO$_4$
Dissolve 24.6 g MgSO$_4$·7H$_2$O in ddH$_2$O and adjust volume to 100 ml. Filter sterilize and store at room temperature.

PEG Precipitation Solution

Dissolve 30% (w/v) PEG (mol. wt. 8000) in 2 M NaCl solution and store at 4°C.

Buffered Phenol

We strongly recommend purchase of saturated phenol (pH 8.0) from a commercial vendor. Most buffered phenol comes with a separate bottle of equilibration buffer to adjust the phenol phase to pH 8.0. To use, add the entire bottle of equilibration buffer to the phenol bottle and 0.1% (w/v) hydroxyquinoline (as an antioxidant). Stir the mixture for 15 min and allow the phases to separate (requires several hours) before use. Store at 4°C.

Caution: Phenol is highly corrosive and can be absorbed through skin. Wear gloves and work in a fume hood when handling phenol.

Phenol:Chloroform:Isoamyl Alcohol

Mix 1 part of buffered phenol with 1 part of chloroform:isoamyl alcohol (24:1). Top the organic phase with TE buffer (about 1 cm height) and allow the phases to separate. Store at 4°C in a light-tight bottle.

Phosphate-Buffered Saline (PBS)

10 mM Sodium phosphate, pH 7.4
137 mM NaCl
2.7 mM KCl

Prepare 1 M stock solutions of Na_2HPO_4 and NaH_2PO_4 using ddH_2O. Prepare 1 M sodium phosphate buffer, pH 7.4, by mixing 22.6 ml of 1 M NaH_2PO_4 and 77.4 ml of 1 M Na_2HPO_4 (autoclave and store at 4°C). Prepare the PBS buffer by dissolving 8.07 g NaCl and 0.2 g KCl in 800 ml ddH_2O. Add 10 ml of 1 M sodium phosphate (pH 7.4) buffer and adjust the volume to 1 L. Autoclave and store at room temperature.

Plasmid Lysis Buffer

25 mM Tris, pH 8.0
10 mM EDTA, pH 8.0
50 mM Glucose
Autoclave and store at 4°C.

Potassium Acetate Solution (pH 4.8)

Dilute 29.5 ml of glacial acetic acid in ~50 ml of ddH_2O and add KOH pellets to pH 4.8. Adjust the volume to 100 ml and store at 4°C.

10 mg·ml^{-1} Proteinase K Solution

Dissolve 100 mg *Proteinase K* in 10 ml of sterile ddH_2O, PBS, or 10 mM Tris buffer (pH 7.5). Store at −20°C.

DNase-Free RNase A

To make *DNase*-free *RNase A*, prepare a 10 mg·ml^{-1} solution of *RNase* A in sterile 10 mM Tris (pH 7.5) and 15 mM NaCl. Boil for 15 min and slowly cool to room temperature. Aliquot and store at −20°C.

20% (w/v) Sarkosyl

Dissolve 40 g sarkosyl (*N*-lauroylsarcosine) in 100 ml of ddH_2O. Adjust the volume to 200 ml and store at room temperature.

SM Buffer

50 mM Tris, pH 7.5
100 mM NaCl
8 mM MgSO$_4$
Autoclave.
Add 0.01% gelatin (from 2% stock, filter-sterilized). Store at 4°C.

3 M Sodium Acetate Solution (pH 5.2)

Dissolve sodium acetate in ddH$_2$O and adjust pH to 5.2 with 3 M glacial acetic acid. Autoclave and store at room temperature.

30/45/60% Sucrose Step Gradient

Make 30, 45, and 60% (w/v) sucrose solutions, respectively, by dissolving 30, 45, and 60 g sucrose in a final volume of 100 ml ddH$_2$O.

Add an appropriate volume of 60% sucrose solution into a centrifuge tube and freeze on dry ice or place at −80°C. Add the same volume of 45% sucrose solution on the top of frozen 60% sucrose and freeze. Overlay the same volume of 30% sucrose solution on the top of 45% sucrose.

Place the centrifuge tube at 4°C overnight prior to use. The sucrose solutions will gradually thaw and diffuse, forming a gradient.

50X TAE

Dissolve 242 g Tris-base in 800 ml sterile ddH$_2$O. Add 57.1 ml glacial acetic acid and 100 ml 0.5 M EDTA (pH 8.0). Adjust volume to 1 L and store at room temperature.

TB (Terrific Broth) Medium

Tryptone (12 g)
Yeast extract (24 g)
Dissolve in ddH$_2$O and adjust volume to 1 L. Adjust pH to 7.5 and autoclave.

5X TBE

225 mM Tris-base
225 mM Boric acid
5 mM EDTA, pH 8
Dissolve 54 g Tris-base and 27.5 g boric acid in ∼500 ml ddH$_2$O. Add 20 ml of 0.5 M EDTA stock and adjust the volume to 1 L.

TE Buffer

10 mM Tris, pH 8.0
1 mM EDTA, pH 8.0

1 M Tris (pH 7.5 or 8.0)

Dissolve 121.1 g Tris-HCl in 800 ml ddH$_2$O. Adjust pH with 2 N HCl to desired pH. Adjust final volume to 1 L, autoclave and store at room temperature.

References

1. **Taylor, B.H., Manhart, J.R., and Amasino, R.M.,** Isolation and characterization of plant DNAs, in *Methods in Plant Molecular Biology and Biotechnology*, Glick, B.R. and Thompson, J.E., Eds., CRC Press, Boca Raton, FL, 1993.

2. **Sambrook, J. and Russell, D.W.,** *Molecular Cloning: A Laboratory Manual,* 3rd ed., Cold Spring Harbor Laboratory Press, New York, 2001.
3. **Murray, M.G. and Thompson, W.F.,** Rapid isolation of high molecular weight DNA, *Nucleic Acids Res.,* 8, 4321, 1980.
4. **Wu, L.-L., Song, I., Kim, D., and Kaufman, P.B.,** Molecular basis of the increase in invertase activity elicited by gravistimulation of oat-shoot pulvini, *J. Plant Physiol.,* 142, 179, 1993.
5. **Lunn, G. and Sansone, E.B.,** Ethidium bromide: destruction and decontamination of solutions, *Anal. Biochem.,* 162, 453, 1987.
6. **Rogers, S.O. and Bendich, A.J.,** Extraction of DNA from milligram amounts of fresh, herbarium and mummified plant tissues, *Plant Mol. Biol.,* 5, 69, 1985.
7. **Aitchitt, M., Ainsworth, C.C., and Thangavelu, M.,** A rapid and efficient method for the extraction of total DNA from mature leaves of the date palm (*Phoenix dactylifera* L.), *Plant Mol. Biol. Rep.,* 11, 317, 1993.
8. **Lodhi, M.A., Ye, G.-N., Weeden, N.F., and Reisch, B.I.,** A simple and efficient method for DNA extraction from grapevine cultivars and *Vitis* species, *Plant Mol. Biol. Rep.,* 12, 6, 1994.
9. **Milligan, B.G.,** Purification of chloroplast DNA using hexadecyltrimethylammonium bromide, *Plant Mol. Biol. Rep.,* 7, 144, 1989.
10. **Palmer, J.D.,** Isolation and structural analysis of chloroplast DNA, *Methods Enzymol.,* 118, 11, 1986.
11. **Grundemann, D. and Schomig, E.,** Protection of DNA during preparative agarose gel electrophoresis against damage induced by ultraviolet light, *BioTechniques,* 21, 898, 1996.
12. **Singer, V.L., Jones, L.J., Yue, S.T., and Haugland, R.P.,** Characterization of PicoGreen reagent and development of a fluorescence-based solution assay for double-stranded DNA quantitation, *Anal. Biochem.,* 249, 228, 1997.

Chapter 2

Isolation and Purification of RNA

Chung-Jui Tsai, Leland J. Cseke, and Scott A. Harding

Contents

I. Extraction of Total RNA from Animals 26
   A. Acid Guanidinium Thiocyanate Method 26
   B. LiCl/Urea Method 27
   C. Guanidinium Thiocyanate–CsCl Gradient Method 28
   D. Phenol-Chloroform/LiCl Method...29
II. Extraction of Total RNA from Plants..29
   A. Alkaline Tris/Sarkosyl Method 29
   B. CTAB Method 31
   C. Commercial RNA Extraction Kits 31
III. Purification of Poly(A)+ RNA from Total RNA...32
   A. Purification of Poly(A)+ RNA by Oligo(dT)-Cellulose Column 32
   B. Mini-Purification of Poly(A)+ RNA Using Oligo(dT)-Cellulose Columns 33
   C. Purification of Poly(A)+ RNA Using Magnetic Oligo(dT)-Beads 33
IV. Fractionation of RNA Using Denaturing Sucrose Gradients..................................35
V. Determination of RNA Quality and Quantity 36
   A. Electrophoresis 36
   B. Spectrophotometric Measurements..37
   C. Fluorometric Measurements...38
VI. Troubleshooting Guide...39
   Reagents Needed 39
References 44

Ribonucleic acid (RNA) extraction and handling are technically more challenging than deoxyribonucleic acid (DNA) manipulation due to the ubiquity and stability of *RNases* and to the presence of a reactive 2′-OH on RNA ribose residues.[1,2] Consequently, significant measures are taken to preserve RNA integrity throughout purification. Rapid inactivation of *RNases* released during cell lysis is critical for the isolation of undegraded RNA. All RNA extraction buffers, therefore, contain powerful inactivating denaturants such as guanidinium salts or detergents. Since airborne bacteria and the skin (e.g., hands) are also sources of *RNases*, all solutions and labware used in RNA isolation should be treated to remove *RNases*, and gloves must be worn at all times. Successful RNA extraction relies on good laboratory practice and *RNase*-free technique. We recommend the following precautions for RNA work:

1. All glassware and reuseable plasticware used in solution preparation, storage, and RNA extraction should be treated to remove residual *RNase* activity and be autoclaved before use. Most protocols call for treatment of labware with 0.1% DEPC (diethylpyrocarbonate) in distilled, deionized H_2O (ddH_2O) (stir to mix) for several hours to overnight, followed by several rinses with autoclaved ddH_2O and autoclaving. However, *DEPC is a suspected carcinogen and should be handled carefully and used only in a fume hood.* Alternatively, we have found AbSolve™ (Cat. #NEF971, NEN Life Science Products, Boston, MA), a liquid-concentrate cleaner formulated to destroy *RNases*, to be equally effective and more user-friendly. To use, immerse all labware in 2% AbSolve in ddH_2O for several hours to overnight and rinse with autoclaved ddH_2O prior to autoclaving.
2. Maintain separate stock solutions, buffers, and ddH_2O for RNA work. Prepare all solutions using treated glassware as described above. See notes for reagents and solution preparation below.
3. Disposable plasticware such as centrifuge tubes and pipette tips are usually free of *RNase* contamination and can be used as is (from sterile packages) or simply autoclaved without any pretreatment. However, it is advised that opened bags of RNA plasticware be tightly closed and sequestered from general lab circulation if they are to be used for future RNA work.
4. Gloves must be worn at all times when working with RNA, and they should be changed frequently when touching freezer/refrigerator handles, doorknobs, or any surfaces that are not *RNase*-free.
5. Decontaminate workbench area and pipetter shafts and tip ejectors by wiping thoroughly with diluted AbSolve or any other commercially available *RNase* decontaminants before RNA extraction. Electrophoresis apparatus (tanks, trays, and combs) should also be treated with 2% AbSolve and rinsed with autoclaved ddH_2O before use. DEPC will destroy materials made from polycarbonate or polystyrene and should not be used to decontaminate such materials.

I. Extraction of Total RNA from Animals

A. Acid Guanidinium Thiocyanate Method

1. For RNA isolation from tissues, harvest tissue and immediately freeze in liquid nitrogen. Store at –70°C until use. Grind 1 g tissue in a clean blender using liquid nitrogen. Keep adding liquid nitrogen and grinding until a fine powder is obtained. Immediately transfer the powder with a sterile, chilled spatula into a clean 15- or 50-ml polypropylene tube. Add 10 ml of guanidinium thiocyanate RNA extraction buffer,[3] mix, and keep the tube on ice. Proceed to step 5.
2. Alternatively, homogenize the tissue in 10 ml of guanidinium thiocyanate RNA extraction buffer in a sterile polypropylene tube on ice by using a glass-Teflon homogenizer or equivalent polytron at top speed for two 30-sec bursts. Transfer the homogenate to a fresh polypropylene tube and proceed to step 5.
3. For RNA isolation from cultured cells grown in suspension or monolayer cultures, collect 1 to 2×10^8 cells in a sterile 50-ml polypropylene tube by centrifugation at 400 g for 5 min at 4°C. Wash the cell pellet with 25 ml of ice-cold, sterile 1X PBS and centrifuge at 400 g for 5 min at 4°C. Repeat washing and centrifugation to remove all traces of serum that contains *RNase*. Remove the supernatant and resuspend in 10 ml guanidinium thiocyanate RNA extraction buffer. Keep the tube on ice and homogenize the cell suspension using a polytron with a clean microprobe for two 0.5- to 1-min bursts at top speed and carefully transfer homogenate to a fresh tube. Proceed to step 5.

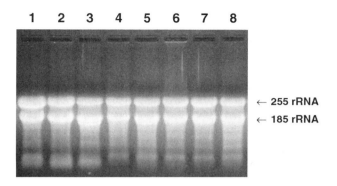

FIGURE 2.1
Total RNA was isolated from various tissues of aspen using the CTAB method and fractionated on a denaturing formaldehyde gel containing 1% agarose. Each lane contained 20 µg of total RNA from (1) apices, (2) young leaves, (3) mature leaves, (4) young shoots, (5) developing xylem, (6) developing phloem, (7) mature female flowers, or (8) root tips. RNA integrity is assessed by the appearance of two major rRNA bands with little tailing. In undegraded RNA, the upper rRNA band will stain at least as brightly as the lower one.

4. Alternatively, wash and resuspend the cells as described above. Keep the suspension of cells on ice. In order to shear the DNA and lower the viscosity, sonicate the suspension cells using a clean microtip at the maximum power for two 30-sec to 1-min bursts. A properly sonicated solution should be thin enough to pass freely through the opening of a Pasteur pipette. Proceed to step 5.

Caution: Guanidinium thiocyanate is a potent chaotropic agent and irritant.

5. Add the following to the sample in the following order: 0.1 volume of 3 M sodium acetate (pH 4.0), 1 volume of acid phenol, and 0.2 volume of chloroform:isoamyl alcohol (24:1). Cap the tube and mix thoroughly by vortexing for 20 sec.

Caution: Phenol is poisonous and can cause severe burns. Wear a lab coat, gloves, and goggles, and work in a fume hood when handling this reagent. If phenol contacts your skin, wash the area immediately with large volumes of water, but do not rinse with ethanol! Phenol waste must be collected in a glass container with a clear label and disposed of properly in compliance with the local procedures for disposal of hazardous chemicals.

6. Centrifuge at 12,000 g for 20 min at 4°C to separate the phases. Carefully transfer the top, aqueous phase to a fresh tube.
7. Add 10 ml of isopropanol to precipitate the RNA at −20°C for 1.5 to 2 h. Centrifuge at 12,000 g for 20 min at 4°C. Resuspend the RNA pellet in 3 ml of guanidinium thiocyanate RNA extraction buffer and add 3 ml of isopropanol. Mix and incubate at −20°C for 2 h to overnight.
8. Centrifuge at 12,000 g for 20 min at 4°C. Briefly rinse the RNA pellet with 4 ml of 70% ethanol. Dry the pellet briefly.
9. Dissolve the total RNA in 200 µl of DEPC-ddH$_2$O or 0.5% SDS (sodium dodecyl sulfate) (w/v in DEPC-ddH$_2$O). Determine the RNA quality and quantity as described in **Section V**. Store the rest in small aliquots at −80°C.
10. For long-term storage, RNA can be stored in an ethanol precipitation solution (0.1 volume of 3 M sodium acetate and 2.5 volumes of 100% ethanol) at −80°C.

B. LiCl/Urea Method

1. Homogenize the tissue or cultured cells as described in **Section I.A** above using the LiCl/urea RNA extraction buffer.

2. Chill the homogenate on ice for at least 4 h and centrifuge at 3000 g for 25 min at 4°C.
3. Remove the supernatant and add 5 ml of cold LiCl/urea RNA extraction buffer. Resuspend the pellet thoroughly by vortexing and centrifuge as in step 2.
4. Carefully decant the supernatant and dissolve the pellet in LiCl/urea buffer.
5. Extract proteins, carbohydrates, DNA, and cellular debris with one volume of phenol:chloroform:isoamyl alcohol (25:24:1). Vortex for 1 min.
6. Centrifuge at 12,000 g for 10 min at 4°C. Carefully transfer the top, aqueous phase to a fresh tube without disturbing the interface. Repeat steps 5 to 6 one more time.
7. Add 0.1 volume of 3 M sodium acetate and 2.5 volumes of chilled 100% ethanol. Mix well and allow the RNA to precipitate at –20°C for 2 h or at –80°C for 20 min.
8. Centrifuge at 12,000 g for 10 min at 4°C. Discard the supernatant and wash the RNA pellet with 4 ml of 70% ethanol. Dry the pellet under vacuum briefly and dissolve the total RNA in 50 µl of DEPC-ddH$_2$O or 0.5% SDS.
9. Determine the RNA quality and quantity as described in **Section V**. Store the RNA in small aliquots at –80°C.

C. Guanidinium Thiocyanate–CsCl Gradient Method

1. Collect liquid-nitrogen-powdered tissue or cultured cells in a centrifuge tube. Immediately add guanidinium thiocyanate extraction buffer to the tube (5 ml per g of tissue, or 1 ml per 2×10^7 cells) and mix thoroughly to achieve lysis of the tissue/cells.
2. Homogenize the mixture on ice with a Polytron homogenizer (*RNase*-free) at top speed for two 1-min bursts.
3. Centrifuge at 5000 g for 10 min at room temperature. Transfer the supernatant to a fresh tube.
4. Carefully layer the supernatant onto a cushion of CsCl/EDTA (ethylenediaminetetraacetic acid) solution in a sterile, clear ultracentrifuge tube without disrupting the CsCl/EDTA cushion. Mark the position of the cushion on the outside of the tube. If required, add more guanidinium thiocyanate buffer to fill the tube (to ~2 mm from the top) and to balance the rotor.
5. Proceed to ultracentrifugation at 20°C, at the speed and for the time given in Table 2.1, using a swinging-bucket rotor to pellet the RNA.
6. Carefully remove the tubes from the centrifuge. A DNA band may be visible in the lower third of the tube. Use a sterile Pasteur pipette to gently remove all DNA at the interface; then, remove the upper CsCl/EDTA layer close to the bottom of the tube. Gently invert the tube to drain away any remaining fluid with Kimwipes™. Mark the position of the RNA pellet.
7. Briefly wash the RNA pellet with 2 ml of 70% ethanol to remove residual CsCl and air-dry the RNA pellet at room temperature. Dissolve the RNA in 100 µl of DEPC-ddH$_2$O. Transfer the RNA to a microcentrifuge tube, wash the ultracentrifuge tube with another 100 µl water, and pool the RNA solutions together.
8. Precipitate the RNA with 0.1 volume of 3 M sodium acetate and 2.5 volumes of chilled 100% ethanol. Mix well and incubate at –20°C for at least 2 h.

TABLE 2.1
Centrifugation Speed and Time for Swinging-Bucket Rotors

	Rotor Type		
	SW60 (7/16 in. × 23 in.)	SW40 (9/16 in. × 32 in.)	SW28 (12 in. × 31 in.)
Volume of CsCl/EDTA (ml)	1.2	3.5	12.0
Volume of homogenate (ml)	3.1	9.7	26.5
Time (h)	12	24	26
Speed (rpm)	40,000	32,000	23,500

9. Centrifuge at 12,000 g for 10 min at 4°C. Discard the supernatant. Briefly rinse the pellet with 70% ethanol. Dry the pellet briefly.
10. Dissolve the RNA in 50 µl of DEPC-treated ddH$_2$O or TE buffer or in 0.5% SDS.
11. Determine the RNA quality and quantity as described in **Section V**. Store the RNA in small aliquots at –80°C.

D. Phenol-Chloroform/LiCl Method

1. This protocol is recommended for RNA isolation from oocytes or eggs. Collect fresh oocytes, fertilized eggs, or embryos from specific organisms of interest and rapidly wash twice with serum-free medium (e.g., 1X PBS buffer). Transfer the material to a clean centrifuge tube and decant the buffer after the cells or tissues settle.
2. Add eight volumes of homogenization buffer and homogenize the tissue on ice until a fine suspension is obtained. Incubate the homogenate for 40 to 60 min at 37°C with occasional mixing.
3. Add one volume of phenol:chloroform (1:1) and vortex for 30 sec.
4. Centrifuge at 12,000 g for 15 min at 4°C. Carefully transfer the top, aqueous phase to a fresh tube.
5. Repeat the phenol:chloroform extraction as in steps 3 to 4.
6. Precipitate RNA by adding 0.1 volume of 3 M sodium acetate and 2.5 volumes of chilled 100% ethanol. Place the tube at –20°C for 2 h.
7. Centrifuge at 12,000 g for 10 min at 4°C and discard the supernatant. Briefly rinse the pellet with 1 ml of 70% ethanol and dry the pellet briefly under vacuum.
8. Resuspend the RNA in 500 µl of DEPC-treated TE buffer. Add 500 µl of 8 M LiCl solution and place at 4°C for 3 to 5 h to precipitate the RNA. LiCl helps to remove glycoproteins and other contaminants from the sample.
9. Centrifuge at 12,000 g for 15 min at 4°C and carefully decant the supernatant. Briefly rinse the pellet with 2 ml of 70% ethanol and dry the pellet.
10. Dissolve the RNA in 100 µl of water. Add 0.1 volume of 3 M sodium acetate and 2.5 volumes of chilled ethanol. Place at –20°C for at least 2 h.
11. Repeat step 9. Dissolve the RNA in 50 µl of DEPC-treated TE buffer or ddH$_2$O. Determine the RNA quality and quantity as described in **Section V**. Store RNA in small aliquots at –80°C.

II. Extraction of Total RNA from Plants

Several protocols are available for total RNA extraction from plant tissues. Chaotropic agents like guanidinium hydrochloride or guanidinium isothiocyanate are commonly used for RNA extraction from a wide variety of tissue types and plant species. However, they are not suitable for RNA extraction from tissues of woody plants or tissues that are rich in polysaccharides and polyphenols.[4,5] The two RNA extraction protocols[4,6] presented here effectively extract high-quality RNA from tissues containing high levels of polysaccharides and phenolic compounds, and these should be suitable for most plant tissues. The protocols have been successfully used in our laboratories for total RNA extraction from the following tissues: leaves; apices; roots; stems; developing secondary xylem, phloem, and floral buds of *Populus*;[7] developing secondary xylem of conifers (pine, spruce, and Douglas-fir),[6] eucalypts, and acacia; and leaves and stems of tobacco. For tissues containing high levels of polyphenols, the CTAB (hexadecyltrimethylammonium bromide) protocol is recommended. Both protocols can be scaled to efficiently extract RNA from as little as 50 mg of tissue.

A. Alkaline Tris/Sarkosyl Method

1. Harvest and immediately freeze tissue in liquid nitrogen to avoid RNA degradation. The frozen tissue can be stored in aluminum packets at –80°C or liquid nitrogen until use.

2. Prechill a mortar and a pestle with liquid nitrogen. In a fume hood, mix 3 ml alkaline Tris/sarkosyl RNA extraction buffer, 3 ml buffered phenol, 0.6 ml chloroform:isoamyl alcohol (24:1), and 30 µl 2-mercaptoethanol per g of plant tissue in an *RNase*-free polypropylene centrifuge tube and leave on ice. Prepare as many tubes as needed (up to six) for extraction of multiple samples. For each sample, have two clean *RNase*-free polypropylene tubes on hand, one for sample lysis and another for isopropanol precipitation.
3. Place frozen tissue in the mortar and add liquid nitrogen to immerse tissue. While the liquid nitrogen is evaporating, gently crush the tissue clumps into small pieces with the pestle. Add a spatula of acid-washed sea sand to aid grinding and add additional liquid nitrogen if necessary. Addition of liquid nitrogen from this point on should be done very carefully to prevent dispersal of frozen tissue from the mortar. Do not pour liquid nitrogen directly onto the tissue. When the liquid nitrogen is nearly evaporated, vigorously grind the tissue to a fine powder.
4. Use a liquid nitrogen-prechilled spatula to transfer frozen powder to the centrifuge tube containing extraction buffer. Swirl the tube frequently to mix contents during the transfer.
5. Cap the tube and vortex at high speed for 1 min to homogenize. Add 200 µl 3 *M* sodium acetate and vortex for another 30 sec.

Note: *(1) It is very important to prechill the mortar with liquid nitrogen to prevent tissue from thawing out. Plant tissues are rich in nuclease activities, so the frozen powder should not be thawed until disrupted in extraction buffer. (2) The vortex step is necessary for effective tissue dispersal. Alternatively, a Polytron homogenizer precleaned with diluted AbSolve solution overnight and rinsed with autoclaved ddH$_2$O can be used. Rinse the Polytron thoroughly with autoclaved ddH$_2$O in between samples to avoid carryover contamination.*

6. Incubate on ice for 15 min to 1 h if multiple samples are to be extracted. Centrifuge at 12,000 g for 10 min at 4°C. While experienced researchers can easily handle four to six extractions at a time before centrifugation, extraction of two samples at a time is probably a good starting point for beginners.
7. Carefully transfer the aqueous phase to a new polypropylene centrifuge tube (*RNase*-free) and precipitate with an equal volume of isopropanol at −70°C for 20 min. Centrifuge at 12,000 g for 10 min at 4°C.
8. Wash the pellet with 70% ethanol; centrifuge at 12,000 g for 5 min at 4°C; and briefly dry the pellet.
9. Optional for tissues rich in polysaccharides (such as leaves and floral buds); otherwise, proceed to step 9. Resuspend pellet in 800 µl alkaline Tris/sarkosyl RNA extraction buffer and transfer to a 2-ml microcentrifuge tube. Add 800 µl buffered phenol and 160 µl chloroform:isoamyl alcohol. Repeat steps 5 to 8.
10. Resuspend pellet in 900 µl DEPC-ddH$_2$O. Extract once with an equal volume of acid phenol:chloroform:isoamyl alcohol (25:24:1) and once with chloroform:isoamyl alcohol (24:1).
11. Transfer the aqueous phase to a new microcentrifuge tube. Save 10 µl to check RNA quality in a minigel. Adjust the volume to 900 µl with DEPC-ddH$_2$O. Add 300 µl of 8 *M* LiCl to a final concentration of 2 *M* to precipitate the RNA at 4°C for 3 h to overnight.
12. Centrifuge at 12,000 g in a microcentrifuge for 10 min at 4°C; wash with 70% ethanol; and air-dry the pellet for 10 min.

Notes: *(1) Do not allow the pellet to dry completely, or the RNA will be difficult to resuspend. (2) The LiCl-precipitation step can be repeated for more effective removal of DNA, if desired.*

13. Resuspend the RNA in approximately 100 µl DEPC-ddH$_2$O or TE buffer per g of starting tissue. Centrifuge to remove any insoluble material. Determine the RNA quality and quantity as described in **Section V**. Store the rest in small aliquots at −80°C. For long-term storage, RNA can be stored in sodium acetate-ethanol precipitation solution at −80°C and harvested before use.

B. CTAB Method

1. Prechill a mortar and a pestle with liquid nitrogen. Dispense 5 ml CTAB extraction buffer with 50 mM ascorbic acid per 1 g of plant tissue into an *RNase*-free polypropylene or Falcon centrifuge tube and prewarm the tube in a 60 to 65°C water bath. For each sample to be extracted, have three clean centrifuge tubes on hand, one for lysis and two for phase separation and precipitation steps.
2. Grind the tissue to a fine powder with liquid nitrogen as described above. Quickly transfer frozen powder to the prewarmed extraction buffer, and swirl the centrifuge tube frequently to mix the contents during the transfer. Add 2% (v/v) 2-mercaptoethanol in a fume hood, cap the tube, and vortex at high speed for 1 min.
3. Extract with an equal volume of chloroform:isoamyl alcohol (24:1) by vortexing for 1 min. Centrifuge at room temperature for 10 min at 12,000 g to separate the phases. If Falcon tubes are used, a lower centrifugation speed (according to the manufacturer) should be used. Centrifuge longer if phases are not well separated.

Note: *When working with multiple samples, let the first sample stand at room temperature after adding chloroform:isoamyl alcohol and vortexing. Prepare up to six samples this way before centrifugation. While researchers experienced with RNA work may process up to six samples at a time, it is highly recommended that only one or two samples be handled at a time by beginners.*

4. Carefully transfer the aqueous phase containing RNA into a fresh centrifuge tube. Repeat the chloroform:isoamyl alcohol extraction once.
5. Add ⅓ volume of 8 M LiCl to the supernatant. Mix well and precipitate RNA for 3 h to overnight at 4°C. Centrifuge at 12,000 g for 20 min at 4°C.
6. Resuspend pellet in 500 μl SSTE and transfer the contents to a microcentrifuge tube. Extract twice with an equal volume of chloroform:isoamyl alcohol.
7. Precipitate the RNA with 2.5 volumes of ethanol at –80°C for 30 min or 2 h at –20°C.
8. Harvest RNA by centrifugation at top speed in a microcentrifuge for 10 min at 4°C; wash with 70% ethanol; and air-dry the pellet.
9. Resuspend the RNA in approximately 100 μl DEPC-H_2O or TE per g of starting tissue. Centrifuge to remove any insoluble material. Determine the RNA quality and quantity as described in **Section V**. Store the rest in small aliquots at –80°C.

C. Commercial RNA Extraction Kits

A number of RNA extraction kits are available commercially that are very convenient for RNA extraction from limited sample sizes, when budget allows. However, these kits need to be empirically evaluated regarding their RNA extraction efficiency, as they may vary from tissue to tissue, which may render them unsuitable for certain tissue types (e.g., polysaccharide-rich tissues). Examples of commercial kits include: the RNAgents® Total RNA Isolation System and the SV Total RNA Isolation System from Promega (Madison, WI), the NucleoBond RNA/DNA Kit from Clontech (Palo Alto, CA), and the various RNaesy kits from Qiagen (Valencia, CA). The readers should refer to the manufacturers' instructions for RNA-extraction protocols and to **Chapter 14** for an example of using commercial RNA extraction kits for gene-expression analysis by messenger RNA (mRNA) differential display.

III. Purification of Poly(A)⁺ RNA from Total RNA

The term poly(A)⁺ RNA refers to populations of eukaryotic mRNAs enriched in 3′ polyadenylated mRNA transcripts. These mRNAs serve as templates for protein translation, and as such, their isolation and quantitation has formed the basis of numerous strategies for the study of cellular activities. Applications ranging from Northern blot detection of rare mRNA transcripts to enhancement of rare message representation in cDNA (complementary DNA) libraries depend on the isolation of high-quality poly(A)⁺ RNA from total RNA. Poly(A)⁺ mRNAs make up 1 to 2% of total RNA and can be separated from ribosomal RNA (rRNA) and transfer RNA (tRNA) by selective hybridization of their poly(A) tails onto oligo(dT) sequences affixed to various support matrices. Improvements in the design of support matrices to reduce nonspecific RNA binding, and in particular, the development of spherical beads that can be magnetically or otherwise retrieved from complex mixtures after hybridization, have vastly simplified poly(A)⁺ RNA purification in recent years.

A. Purification of Poly(A)⁺ RNA by Oligo(dT)-Cellulose Column

1. Add 4 ml of binding buffer to 0.5 g of oligo(dT)-cellulose type 7 (Collaborative Research Inc., Bedford, MA) in a clean tube and mix well. Transfer the suspension to a 10-ml poly-prep chromatography column (Bio-Rad, Hercule, CA) or equivalent. Vertically clamp the column in a holder and place at 4°C for 2 h to equilibrate the cellulose resin.
2. Bring the column setup to room temperature and wash the column by adding 2 ml of 0.1 N NaOH solution through the top of the resin. The color of the resin changes from white to yellow. Allow the washing solution to drain away by gravity and repeat washing the column five times.
3. Neutralize the column by adding 4 ml of binding buffer to the top of the resin and allow the binding buffer to drain away by gravity. The color of the resin returns to white. Repeat five times and store at 4°C until use.
4. Add one volume of loading buffer to the total RNA sample prepared. Heat the mixture for 10 min at 65°C to denature the secondary structure of RNA, and then allow the sample to cool to room temperature.

Note: *The concentration of SDS in the loading buffer should never exceed 0.5%; otherwise, the SDS will precipitate and block the column. We found that 0.2% SDS in the binding buffer works well.*

5. Slowly load the total RNA sample to the top of the column capped at the bottom and allow the sample to run through the column by gravity. Gently loosen the resin in the column using a fine needle and let the column stand for 5 min. Remove the bottom cap to drain away the fluid from the column and collect the fluid containing some unbound poly(A)⁺ RNA. Reload the fluid onto the column to allow the unbound poly(A)⁺ RNA to bind to the oligo(dT)-cellulose column and collect the eluate. Repeat once.
6. Wash the column with 2 ml of washing buffer and drain away the buffer. Repeat the wash step three times.
7. Cap the bottom of the column and add 1 ml of elution buffer to the column to elute bound poly(A)⁺ RNA. Gently loosen the resin in the column using a clean needle and allow the column to stand for 4 min. Remove the bottom cap and collect the eluate in an *RNase*-free tube. Elute the column one more time and collect the eluate.
8. Pool the eluates together and add 0.1 volume of 3 M sodium acetate and 2.5 volumes of 100% chilled ethanol to precipitate the mRNA. Mix and place at −20°C overnight.
9. Centrifuge at 12,000 g for 15 min at 4°C. Decant the supernatant and briefly wash the mRNA pellet with 2 ml of 70% ethanol. Dry the pellet briefly to remove ethanol and dissolve the mRNA in 20 to 50 µl of DEPC-treated TE buffer.

Isolation and Purification of RNA

10. Take 2 or 4 µl of the sample to measure the concentration and purity of the mRNA as described in **Section V**. The yield of poly(A)$^+$ RNA may be expected to be approximately 6% of the total RNA. Store the sample at –70°C until use for Northern blot analysis, *in vitro* transcription, or cDNA synthesis (see **Chapters 6, 9,** and **14**).

B. Mini-Purification of Poly(A)$^+$ RNA Using Oligo(dT)-Cellulose Columns

1. Add 1 ml of binding buffer to 0.3 g of oligo(dT)-cellulose type 7 (Collaborative Research Inc., Bedford, MA) in an eppendorf tube. Mix and place at 4°C for 30 min.
2. Wash the resin with 1 ml of 0.1 N NaOH solution and mix gently for a few minutes. Centrifuge at 1500 g for 2 min and discard the supernatant. Repeat this step eight times.

Note: *Centrifugation greater than 1500 g may cause damage to the oligo(dT) beads.*

3. Neutralize the resin with 1 ml of binding buffer and gently loosen the beads with a sterile needle. Centrifuge at 1500 g for 2 min and carefully decant the supernatant. Repeat this step eight times. Resuspend the resin with 1 ml of binding buffer.
4. Add one volume of loading buffer to ~200 µg of the total RNA sample (about 2 µg/µl). Heat the mixture for 10 min to 65°C and allow to cool to room temperature.
5. Load the total RNA sample to the 1 ml slurry of oligo(dT)-cellulose beads prepared in step 3. Gently shake for 2 min and place at room temperature for 15 min to allow the RNA to bind to the beads.
6. Centrifuge at 1500 g for 5 min and carefully remove the supernatant.
7. Add 0.5 ml of wash buffer to the resin and gently shake for 10 sec. Centrifuge at 1500 g for 2 min. Carefully remove the supernatant. Repeat washing the resin three more times.
8. Add 0.2 ml of elution buffer to the resin and gently shake for 4 min. Centrifuge at 1500 g for 5 min. Carefully transfer the eulate to a fresh tube. Repeat this twice.
9. Pool the eluates and add 0.1 volume of 3 M sodium acetate and 2.5 volumes of 100% chilled ethanol to precipitate the mRNA. Mix and place at –20°C overnight.
10. Centrifuge at 12,000 g for 15 min and briefly wash the mRNA pellet with 1 ml of 70% ethanol. Dry the pellet under vacuum and dissolve the mRNA in 10 µl of DEPC-treated TE buffer. Take 2 µl of the sample in order to measure the concentration and quality of the mRNA as described below. Store at –70°C until use.

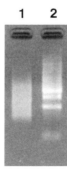

FIGURE 2.2
Poly(A)$^+$ RNA, purified from plant total RNA using the magnetic oligo(dT) beads, and unbound RNA were ethanol-precipitated and resuspended in water prior to electrophoresis on a 1% agarose/TAE gel. Lane 1, poly(A)$^+$ RNA (0.4 µg), appears as a 0.5- to 3-kb smear; lane 2, unbound RNA, is equivalent to 0.4 µg of total RNA.

C. Purification of Poly(A)$^+$ RNA Using Magnetic Oligo(dT)-Beads

Magnetic oligo(dT) beads provide a very simple and reliable alternative to other oligo(dT) matrices for the purification of poly(A)$^+$ RNA from samples of total RNA. Magnetic Dynabeads™ supplied by Dynal Inc. (Lake Success, NY) exhibit superior monodispersity and durability compared with other manufacturers' beads, and 1 ml of Dynabeads can be reconditioned several times to isolate up to 50 µg mRNA. To use these or similar beads, it is necessary to purchase a magnetic microcentrifuge tube holder for the various wash and elution steps. (Promega provides a cheap alternative to the holder sold by Dynal.) The procedure described is based on our experience purifying poly(A)$^+$ RNA from CTAB- or alkaline Tris/Sarkosyl-isolated total RNA (1 to 5 mg), and it can be scaled down to obtain poly(A)$^+$ RNA from 10 µg of total RNA. The following protocol is for purification of poly(A)$^+$ RNA from 1 to 2 mg total RNA stored in 200 to 400 µl DEPC-ddH$_2$O. If followed completely (steps 1 through 10) and successfully, the procedure normally yields 2 to 4 µg of uncontaminated, rRNA-free poly(A)$^+$ RNA. When analyzed on an EtBr stained denaturing gel, most of the poly(A)$^+$ RNA is between 0.7 and 2.5 kb (kilobase-pairs) in size. If the shortened protocol is chosen (steps 1 through 6), the yield of poly(A)$^+$ RNA, though less pure, will range from 20 to 40 µg.

1. Wash and equilibrate the magnetic beads. Completely resuspend unused or stored Dynabeads by gentle tapping. Use a wide-bore 200 µl pipette tip to transfer 200 µl of bead slurry to an *RNase*-free, 1.5-ml microcentrifuge tube. Place tube on a magnetic rack for 30 sec, or however long it takes for all of the beads to adhere to the magnet-facing wall of the tube. Without removing the tube from the rack, pipette away **all** storage buffer (faint brown discoloration of the buffer indicates bead degradation or insufficient time allowed for complete magnetic removal of beads).
2. Remove tube from the rack, add 500 µl equilibrating/binding buffer (EB), and gently roll or tap tube to resuspend the beads. Remove buffer as in step 1 and repeat this procedure three times. Keep beads in 200 µl EB buffer while preparing total RNA (step 3).
3. Prepare total RNA by adding one volume of 2X EB to aliquot of total RNA. Before adding EB, save a small aliquot of total RNA for follow-up troubleshooting. Up to 0.2% SDS can be included in this and subsequent steps to prevent RNA degradation as long as the sample is not chilled. Remove EB buffer from beads (step 2) and mix RNA with equilibrated beads.

Note: *Frothing due to careless handling of samples containing SDS will interfere with subsequent steps. As long as careful RNase-free technique is used, SDS is not required.*

4. Incubate RNA bead suspension for 2 min at 65°C and cool to room temperature for 5 min, tapping occasionally to suspend beads. Place suspension in magnetic rack for 30 sec and pipette unbound RNA back to its original sample tube.

Note: *Do not discard unbound RNA! It will be reused, and later, it will be available for troubleshooting.*

5. Wash beads to remove contaminating RNAs. Resuspend beads in 600 µl EB with gentle tapping and incubate 1 min. Remove and discard EB wash. Repeat three times. Repeat one more time using 600 µl 0.5X EB. Carefully remove all traces of the final wash before elution (step 6).
6. Elute bound RNA. Resuspend beads in 100 µl elution buffer. Incubate at 70°C for 2 min and immediately place tube on magnetic rack. Transfer the eluted RNA to a clean *RNase*-free microcentrifuge tube and freeze at –80°C, or ethanol precipitate overnight (see step 8).

Note: *At this stage the eluted RNA is highly poly(A)$^+$ RNA-enriched and suitable for many purposes. However, it is likely to contain substantial ribosomal RNA contamination. Ethanol precipitation and a second round of oligo(dT) purification (step 10) is recommended for maximum yield of highly purified poly(A)$^+$ RNA.*

7. To improve poly(A)$^+$ RNA yield, repeat steps 4 through 6 at least three times by reusing the aliquot of total RNA saved after step 4. Increasing the starting amount of beads and reducing the number of selections saves time but may increase surface losses and rRNA carryover.

8. Mix each 100 µl aliquot of eluted RNA with 10 µl 3 M sodium acetate (pH 5.2) and 300 µl ethanol. Precipitate 4 h to overnight at –20°C. Centrifuge at 20,000 g for 30 min at 4°C and carefully remove supernatant. Layer 500 µl of 70% ethanol over pellet and centrifuge at 20,000 g for 2 min at 4°C. Carefully remove ethanol.

Note: *After preparing all of the eluted RNA aliquots for precipitation, resuspend oligo(dT) beads in 100 µl EB and store at 4°C for step 10. Do not use new beads or new tubes at step 10, as RNA will be lost to surfaces.*

9. Air dry RNA pellets for 10 min. Resuspend each pellet in 25 µl DEPC-ddH_2O and pool resuspended RNA aliquots into one microcentrifuge tube.
10. Repurify pooled RNAs by repeating steps 3 through 6 and step 8. Resuspend in an appropriate volume of DEPC-ddH_2O and store at –80°C.
11. Used beads can be stored at 4°C in oligo(dT) bead storage buffer.

Note: *Commercial mRNA extraction kits, based on similar poly(A)$^+$ RNA purification principles, can also be used. Examples include the PolyATtract® mRNA Isolation System from Promega (Madison, WI) and the Oligotex mRNA kit and the Oligotex Direct mRNA kit from Qiagen (Valencia, CA).*

IV. Fractionation of RNA Using Denaturing Sucrose Gradients

This is an efficient way to obtain specific size fractions of RNA. RNA fractionated in this way can be used for cDNA synthesis and library construction. The frequency of positive cDNA clones can be increased significantly when screening libraries synthesized from fractionated mRNA. The RNA obtained from sucrose gradients can also be used for *in vitro* translation.

1. Prepare sucrose gradients (10 to 30% w/v) in $^9/_{16}$-in. × $3^1/_2$-in. ultracentrifuge tubes (Beckman® SW41 or equivalent). To prepare gradients using methylmercuric hydroxide as a denaturant, dissolve sucrose in DEPC-treated water, add 1 M Tris-HCl (pH 7.5) and 0.5 M EDTA to final concentrations of 10 mM Tris-HCl and 1 mM EDTA, using *RNase*-free techniques. Add methylmercuric hydroxide to a final concentration of 10 mM. Alternatively, a less toxic procedure is to prepare the sucrose density gradient in 4% formamide/95% DMSO/1% Buffer A.[8,9] The gradients can be poured by using a gradient maker or by adding the solutions, stepwise, in order of decreasing sucrose density. A convenient way to pour step gradients is to place the tube on dry ice and layer sucrose solutions stepwise after freezing each layer. Finally, thaw gradients in an upright position overnight at 4°C.

Caution: *Methylmercuric hydroxide is extremely toxic and volatile. Wear gloves when handling such solutions and perform these operations in a fume hood.*

2. Add methylmercuric hydroxide to a final concentration of 20 mM to 100 µl RNA sample (1 µg/µl) and carefully load the RNA sample onto the gradients. If using DMSO/formamide as the denaturant, heat RNA for 5 min at 65°C in DEPC-treated H_2O. Dilute heated sample 1:10 in 99% DMSO/1 mM EDTA and layer onto the gradient.
3. Centrifuge the gradients at 34,000 rpm for 15 to 18 h at 4°C in a Beckman SW41 rotor (or its equivalent).
4. Collect 0.2- to 0.3-ml fractions through a hypodermic needle inserted into the bottom of the centrifuge tube as described in **Chapter 1**. Dilute each of the fractions with one volume of sterile, DEPC-treated water containing 5 mM 2-mercaptoethanol. Add 60 µl of 3 M sodium acetate buffer and place the tubes at 0°C for 1 h. If using DMSO/formamide, dilute RNA fraction with two volumes of

DEPC-ddH$_2$O. Precipitate several hours at −20°C in 0.1 volume of 3 M sodium acetate and 2.5 volumes of 100% ethanol.
5. Centrifuge at 12,000 g for 15 min at 0°C and carefully decant the supernatants into a fresh tube. Wash the RNA pellet with 2 ml of 70% ethanol and allow to dry for 20 min at room temperature.
6. Resuspend the RNA in 20 µl of DEPC–ddH$_2$O and add 0.1 volume of 3 M sodium acetate buffer and 3 volumes of chilled 100% ethanol. Mix well and allow to precipitate at −70°C for 30 min.
7. Centrifuge at 12,000 g for 10 min at 4°C. Wash with 70% ethanol and dry the RNA briefly under vacuum. Dissolve the RNA in 20 µl of DEPC-treated TE buffer. Determine the RNA quality and quantity as described in **Section V**. Store the rest in small aliquots at −80°C.

V. Determination of RNA Quality and Quantity

The integrity of purified RNA, and of RNA from intermediate stages of purification, can only be confirmed by agarose gel electrophoresis on denaturing (formaldehyde, glyoxal or methyl mercuric) or, more conveniently, on nondenaturing TAE minigels. RNA yield and purity is spectrophotometrically determined by measuring ultraviolet (UV) light absorbance (A) of samples at wavelengths of 230, 260, and 280 nm. Ratios of the absorbances at these wavelengths provide a gauge of protein (A_{260}/A_{280}) or carbohydrate (A_{260}/A_{230}) contamination in an RNA sample. Very small amounts of RNA or RNA that is contaminated with DNA can sometimes be more accurately quantified fluorometrically in the presence of RNA-specific dyes.

A. Electrophoresis

A number of minigel formats are available for rapid visualization of RNA integrity or DNA contamination in as little as 1 or 2 µl of sample. We use an EmbiTech™ (San Diego, CA) mini-electrophoresis system requiring less than 10 ml agarose gel to analyze 6 to 17 samples in a 10-min run. Denaturing formaldehyde minigels must be set up and run in a hood, while nondenaturing Tris-acetate-EDTA (TAE) minigels can be run without taking this precaution. Prior to running RNA samples, the minigel trays, combs, and electrophoresis tank should be washed with warm (60°C) 2% AbSolve, rinsed with autoclaved ddH$_2$O, and Kimwiped dry. It may be helpful to reserve a dedicated piece of apparatus for RNA work.

1. Nondenaturing TAE gel preparation: In an *RNase*-free 125-ml flask, prepare a stock of 1% agarose (w/v) by adding 0.5 g agarose to 50 ml of 1X TAE buffer. Melt the agarose by intermittent microwaving (10 to 30 sec) and manual swirling. Care during this part of the procedure will prevent geysering of the gel solution from the flask. Cool gel solution to ~60°C and pour into the gel casting tray with an appropriate comb to a depth of ~3 mm. Thin gels can improve visualization of small amounts of RNA. Unused gel solution can be stored in a 50 to 60°C oven for a few days or at room temperature and carefully melted in a microwave before use.
2. Denaturing formaldehyde gel preparation: In an *RNase*-free 50-ml flask, prepare a volume of 1% agarose gel solution that can be poured as soon as it is made. Formaldehyde gel solutions should not be stored or remelted for use unless heating is done in a hood. To prepare a 20-ml gel solution, melt 0.2 g agarose in 16 ml autoclaved ddH$_2$O in a microwave oven. Allow solution to cool to ~60°C, add 2 ml of 10X MOPS buffer and 2 ml of 37% formaldehyde in a fume hood, and mix by swirling. A clean gel-casting unit should be set up in the hood for pouring. Fill a gel tray to a depth of 3 to 5 mm with gel solution.
3. Electrophoresis: Prepare sufficient electrophoresis buffer (1X TAE for nondenaturing gel, 1X MOPS for denaturing gel) to fill electrophoresis tank and submerge the gel. Cut out a small parafilm strip, and on its protected surface, distribute an array of spots, corresponding to the number of RNA samples to be analyzed, of 1- to 2-µl droplets of RNA sample buffer. To each droplet, dispense 1 to 2 µl of RNA sample. With minimal delay, load samples into wells of a submerged minigel. Delays of 10 min

Isolation and Purification of RNA

or more before loading will allow droplets to evaporate and will compromise the analysis. Run the gel at a constant voltage of 100 V. After the dye front (bromophenol blue) migrates at least 3 cm into the gel, examine the gel under UV. RNA integrity is assessed by the appearance of two major rRNA bands. In undegraded RNA, these bands exhibit little tailing, and the upper band will stain at least as brightly as the lower one. RNA from leaf tissues contains several smaller rRNAs in addition to the two major bands described.

Notes: *(1) The protocol described here is for rapid visualization of RNA quality only and cannot be applied to substitute for denaturing agarose gel electrophoresis during Northern blot analysis as described in* **Chapter 6**. *(2) It is highly recommended that samples from intermediate steps of RNA preparation be checked in a minigel to evaluate the RNA integrity and aid in troubleshooting.*

B. Spectrophotometric Measurements

1. Turn on a UV spectrophotometer and program it to measure sample absorbance at $\lambda = 230$, 260, and 280 nm. Allow 5 min for the deuterium UV lamp to equilibrate. Use matched quartz cuvettes. Cold reagents may cause fogging and condensation on cuvette surfaces, which will interfere with absorbance measurements.
2. Most dual-cell machines display the absorbance difference between a cuvette containing diluted sample and a reference cuvette containing the solvent (such as ddH_2O) used to dilute the sample. The displayed value represents the sample absorbance. Program the instrument to display zero absorbance (i.e., background correction) at the 230-, 260-, and 280-nm wavelengths when the reference and sample cuvettes contain only solvent. If sample RNA is to be diluted in 1 ml of ddH_2O for measurement, the reference cuvette should contain 1 ml ddH_2O. Additives like DEPC will affect the UV absorbance spectrum of a sample. So, if the RNA sample is stored in DEPC-ddH_2O, it should be diluted in DEPC-ddH_2O, and the reference cuvette should contain DEPC-ddH_2O.
3. Remove the reference solution from the sample cuvette and dispense 1 to 2 µl (depending upon the RNA concentration) of sample into one corner of the cuvette. Wash the sample into the cuvette by dispensing 1 ml solvent onto the same spot and mix by gentle pipetting. Record absorbance at the three wavelengths, remove, rinse, and drain cuvette on a Kimwipe and repeat the procedure for the next sample. Adjust measurement volumes for samples with highly concentrated or very dilute RNA.

Note: *Gloves should be worn when handling cuvettes. Cuvettes should be rinsed with 95% ethanol followed by ddH_2O, and wiped dry with Kimwipe prior to reuse. Unclean cuvettes may contain contaminated material that can affect readings between samples.*

4. After measuring RNA sample absorbance (A) at 230, 260 and 280 nm, calculate the A_{260}/A_{280} and A_{260}/A_{230} ratios and use A_{260} to calculate RNA concentration. Ideally, A_{260}/A_{280} and A_{260}/A_{230} should both be ~2.0. This has been empirically determined to indicate low protein and carbohydrate contamination.[1] A_{260}/A_{280} ratios less than 1.8 for RNA samples may be an indication that the RNA is contaminated with phenolic or proteinaceous material. This can interfere with stability and suitability of the RNA for downstream use in cDNA library construction or *in vitro* transcription. A_{260}/A_{280} ratios or A_{260}/A_{230} ratios less than 1.5 are an indication that A_{260} cannot be assumed to reflect RNA absorbance, but instead, that it reflects a contribution to absorbance by contaminating protein or carbohydrate. RNA concentration in a 1.0-ml sample can be calculated as follows:

$$\text{total RNA (µg)} = A_{260} \times 40 \text{ µg/ml} \times \frac{\text{total volume}}{\text{µl RNA used}}$$

The absorbance of 40 µg total RNA/ml = 1.0 when measured in a 1-cm path length.
For example, if 2 µl of total RNA is diluted to 1.0 ml for measurement and $A_{260} = 0.2$, then:

$$\text{total RNA (μg/ml)} = A_{260} \times 40 \text{ μg/ml}$$
$$= 0.2 \times 40 \text{ μg/ml}$$
$$= 8 \text{ μg/ml}$$

Sample RNA concentration = 8 μg RNA/ml ÷ 2-μl sample/ml = 4 μg/μl

C. Fluorometric Measurements

Applications that depend on accurate quantitation of very small amounts of RNA call for RNA-selective fluorometry. RNA-specific fluorescent dyes like RiboGreen® available from Molecular Probes (Eugene, OR), can be used for this purpose. Used in conjunction with the TD-700 minifluorometer (Turner Designs, Sunnyvale, CA), such dyes can be used to quantify as little as 1 ng RNA from a sample contaminated with small amounts of DNA. This permits quantitation of, for example, a 100-μl sample containing 50 ng poly(A)⁺ RNA. Whereas 1 μl of this sample would be more than sufficient for fluorimetric RNA determination, it would be necessary to place the entire sample into a 100-μl microcuvette to obtain an A_{260} spectrophotometric reading of 0.01 representing the sum of RNA plus contaminating DNA. The fluorometric method is also useful for quantitation of RNA probes used for *in situ* hybridization (see **Chapter 22**). Below is a generic description of the protocol based on the high-range assay using a minicell (100-μl assay volume). The reader should refer to the product information sheet of RiboGreen and the instruction manual of the specific fluorometer used to develop the exact protocol.

1. Turn on the fluorometer (with ~480-nm excitation filter and ~520-nm emission filter) and allow the lamp to warm up for 10 min.
2. Prepare RiboGreen reagent by diluting an aliquot of the RiboGreen stock 1:200 in TE buffer in a microcentrifuge tube protected from light. Prepare enough working solution for the number of assays desired (50 μl needed per assay).
3. Prepare RNA standards for calibration. Dilute RNA standards to final concentrations of 1000, 500, 100, and 10 ng/ml in 100 μl of TE and 100 μl of RiboGreen reagent (200 μl final). Mix well and protect from light.
4. Measure the sample fluorescence in a minicell according to fluorometer instructions. Subtract the blank value and generate a standard curve (fluorescence vs. RNA concentration).
5. Dilute a small aliquot of the RNA samples (usually 1 μl) in 50 μl of TE buffer and mix with 50 μl of RiboGreen reagent. The upper detection limit of the high-range assay is 100 ng/μl RNA. Therefore, it is necessary to dilute highly concentrated RNA samples (usually 10- to 100-fold) before preparing the assay mixture. Protect the mixture from light and incubate at room temperature for 2 to 5 min.
6. Measure the fluorescence of the sample mixture as soon, and as consistently, as possible among samples to avoid photobleaching effects.
7. Determine sample concentration according to the standard curve. RNA concentration determined in a 100-μl assay can be calculated as follows:

   For example, if 1 μl RNA is diluted in a 100-μl assay mixture for measurement, and if the assay mixture concentration = 500 ng/ml (based on standard curve), then

   Total amount of RNA present in assay mixture = 500 ng/ml and per 0.1 ml
   = 50 ng

   Sample RNA concentration = 50 ng ÷ 1-μl sample
   = 50 ng/μl

VI. Troubleshooting Guide

Symptom	Solutions

Total RNA extraction

RNA degradation	Make certain that all labware and solutions are *RNase*-free.
	Do not allow frozen tissue/powder to thaw until disrupted in extraction buffer. Use prechilled mortar and pestle. Transfer of frozen powder to extraction buffer should be done as quickly as possible.
Low RNA yield	Insufficient grinding or tissue disruption. Use acid-washed sea sand to aid grinding and homogenize thoroughly. Low RNA yield could also be due to the use of mature tissues.
Low A_{260}/A_{280} or A_{260}/A_{230} ratios	High absorbance at 230 and 280 nm indicates contamination by phenolic compounds and proteins. Try alternative protocol, perform multiple phenol/chloroform:isoamyl alcohol extractions, and perform multiple rounds of LiCl precipitation. Inclusion of antioxidant such as ascorbic acid in the extraction buffer may help improve the ratios.
Resuspended RNA is viscous	Indication of polysaccharide contamination.
	Polysaccharides can be precipitated in 0.3 *M* of DEPC-treated sodium acetate (pH 5.2) at 4°C for 30 min and centrifuged at 10,000 rpm for 10 min at 4°C. The supernatant containing RNA is then precipitated with two volumes of ethanol. Alternatively, try different protocols presented in **Section II**.

Poly(A)⁺ RNA purification

RNA degradation	The utmost care needs to be taken to avoid introducing *RNase* contamination.
Contamination with rRNA	Remember that due to its secondary structure and well-defined sizes, rRNA intercalates more EtBr than poly(A)⁺ RNA, and therefore it is more visible on a stained gel. Increase stringency of washing steps by lowering salt concentration, increasing wash volume, or increasing wash temperature to 37°C.

Reagents Needed

All RNA solutions should be made in DEPC-ddH$_2$O using *RNase*-free labware and filter-sterilized or autoclaved.

Caution: *DEPC is a carcinogen and should be handled with care. Gloves should be worn when working with this reagent.*

Alkaline Tris/Sarkosyl RNA Extraction Buffer

100 m*M* Tris, pH 9.0
200 m*M* NaCl
15 m*M* EDTA, pH 8.0
0.5% (w/v) Sarkosyl
1% (v/v) 2-Mercaptoethanol, add before use
Prepare extraction buffer by adding appropriate amounts of *RNase*-free stock solutions (see recipes below) and make up the volume with DEPC-ddH$_2$O. Autoclave and store at room temperature.

Binding Buffer

    10 mM Tris-HCl, pH 7.5
    0.5 M NaCl
    1 mM EDTA
    0.5% SDS
    Dilute from stock solutions with DEPC-ddH$_2$O, autoclave, and store at room temperature.

Buffer A (50X)

    0.5 M Tris, pH 7.5
    50 mM EDTA
    0.5 M LiCl

Buffered Phenol (pH 4 or pH 8)

    We strongly recommend purchase of saturated phenol from commercial vendors. Buffered phenol at pH 8.0 is suitable for DNA (or nucleic acid in general) extraction, whereas acid phenol (pH 4.0) is specifically used for RNA purification, as the DNA partitions into the organic phase at lower pH. Unless described as *acid* phenol, *buffered* phenol is used in this chapter to refer to phenol equilibrated to pH 8.0

    Most buffered phenol comes with a separate bottle of equilibration buffer to adjust the phenol phase to pH 8.0. To use, add the entire bottle of equilibration buffer to the phenol bottle and 0.1% (w/v) hydroxyquinoline (as an antioxidant). Stir the mixture for 15 min and allow the phases to separate (several hours) before use. Acid phenol can be used as is, with the addition of 0.1% hydroxyquinoline. Store phenol solutions at 4°C.

Caution: *Phenol is a dangerous reagent. Wear gloves and work in a fume hood when handling phenol.*

CTAB RNA Extraction Buffer

    2 M NaCl
    25 mM EDTA, pH 8.0
    0.1 M Tris, pH 9.0
    2% (w/v) PVP (K-30)
    2% (w/v) CTAB
    50 mM Ascorbic acid, add before use
    2% (v/v) 2-Mercaptoethanol, add before use
    Prepare 2 M ascorbic acid stock in DEPC-ddH$_2$O using *RNase*-free labware. Store solution in small aliquots at –20°C.
    Heat to dissolve CTAB in DEPC-ddH$_2$O with stirring and add PVP and other stock solutions. Adjust the volume with DEPC-ddH$_2$O, autoclave, and store at room temperature.

Chloroform:Isoamyl Alcohol (24:1)

    48 ml Chloroform
    2 ml Isoamyl alcohol
    Mix and store at room temperature.

CsCl/EDTA Solution

    Dissolve 96.0 g of CsCl in 0.01 M EDTA (pH 8.0) to a final volume of 100 ml (i.e., 5.7 M CsCl).
    Add DEPC to a final concentration of 0.1% and stir vigorously in a fume hood for 1 h. Autoclave.
    Allow the solution to cool to room temperature and adjust the volume to 100 ml with DEPC-ddH$_2$O to ensure a final CsCl concentration of 5.7 M.

DEPC-ddH$_2$O

Add 0.1% DEPC to ddH$_2$O in an *RNase*-free glass bottle. Stir vigorously in a fume hood overnight. Autoclave and store at room temperature.

Equilibrating/Binding (EB) Buffer

0.5 M NaCl
100 mM Tris, pH 7.5

0.5 M EDTA (pH 8.0)

Dissolve 18.61 g of EDTA-disodium salt in 70 ml ddH$_2$O. Adjust pH to 8.0 with 10 N NaOH and add ddH$_2$O to 100 ml.

Treat the solutions with 0.1% DEPC overnight with stirring in a fume hood. Autoclave and store at room temperature.

Elution Buffer

10 mM Tris-HCl, pH 7.5

Ethanol (100% or 70%)

100% Ethanol: use directly from the vendor.
70% Ethanol: mix 70 ml of 100% ethanol with 30 ml DEPC-ddH$_2$O.
Store both at −20°C. We strongly recommend that ethanol solutions used for RNA work be maintained separately from other routine molecular biology work.

10X Formaldehyde Loading Dye

50% (v/v) Glycerol
10 mM EDTA, pH 8.0
0.25% Bromophenol blue
0.25% Xylene cyanol FF
Prepare in DEPC-ddH$_2$O and store at −80°C.

Guanidinium Thiocyanate RNA Extraction Buffer

4 M Guanidinium thiocyanate
25 mM Sodium citrate, pH 7.0
0.5% Sarkosyl
100 mM 2-Mercaptoethanol, add before use
Prepare extraction buffer by adding appropriate amounts of stock solutions and make up the volume with DEPC-ddH$_2$O. Filter sterilize and store in a light-tight bottle at room temperature for up to 3 months.

Homogenization Buffer

50 mM Tris-HCl, pH 7.5
50 mM NaCl
5 mM EDTA, pH 8.0
0.5% SDS
200 μg/ml Proteinase K (from a stock solution of 20 mg/ml)

8 M LiCl Solution

Dissolve LiCl in DEPC-ddH$_2$O, autoclave, and store at room temperature.

LiCL/Urea RNA Extraction Solution

3 M LiCl
6 M Urea

Dissolve 63 g LiCl in ~300 ml DEPC-ddH$_2$O thoroughly prior to adding 180 g urea. Adjust the volume to 500 ml DEPC-ddH$_2$O. Filter sterilize.

Loading Buffer

20 mM Tris-HCl, pH 7.5
1 M NaCl
2 mM EDTA
0.2% SDS

10X MOPS

0.2 M MOPS
20 mM Sodium acetate
10 mM EDTA, pH 8.0

Dissolve 41.8 g of MOPS and 2.72 g of sodium acetate trihydrate in DEPC-ddH$_2$O using *RNase*-free labware. Add 20 ml of DEPC-treated 5 M EDTA (pH 8.0). Adjust pH to 7.0 with NaOH using an *RNase*-free electrode (e.g., treated with AbSolve). Adjust the volume to 1 L using DEPC-ddH$_2$O. Store at room temperature and protect from light.

5 M NaCl

Dissolve 146 g of NaCl in ddH$_2$O and adjust the volume to 500 ml. Treat the solutions with 0.1% DEPC overnight with stirring in a fume hood. Autoclave and store at room temperature.

10 N NaOH

Dissolve 40 g NaOH in 100 ml DEPC-ddH$_2$O using *RNase*-free labware. Store at room temperature. Dilute with DEPC-ddH$_2$O to prepare 0.1 N NaOH solution.

Oligo(dT) Bead Storage Buffer

250 mM Tris, pH 7.5
20 mM EDTA, pH 8.0
0.1% Tween-20
0.02% (w/v) NaN$_3$

Dilute Tris and EDTA stocks and Tween-20 in DEPC-ddH$_2$O in a sterile Falcon centrifuge tube, add NaN$_3$, and shake to dissolve.

Caution: *NaN$_3$ is highly poisonous and should be handled with care. Wear gloves and goggles.*

1X PBS Buffer

10 mM Sodium phosphate, pH 7.2
130 mM NaCl

Prepare 1 M stock solutions of Na$_2$HPO$_4$ and NaH$_2$PO$_4$ using DEPC-ddH$_2$O.
Prepare 1 M sodium phosphate buffer, pH 7.2 by mixing 68.4 ml Na$_2$HPO$_4$ and 31.6 ml NaH$_2$PO$_4$.
Dilute the stocks in DEPC-ddH$_2$O to make 1X PBS. Autoclave and store at room temperature.

RNA Sample Buffer

110 μl 10X MOPS
250 μl Formaldehyde
538 μl Formamide
100 μl 10X Formaldehyde gel loading dye (recipe above)
2 μl EtBr (10 mg/ml)
Store at –80°C.

10% Sarkosyl

Dissolve 10 g sarkosyl (*N*-lauroylsarcosine) in 100 ml of DEPC-ddH$_2$O using *RNase*-free labware and store at room temperature.

20% SDS

Dissolve 20 g sodium dodecyl sulfate in 100 ml of DEPC-ddH$_2$O using *RNase*-free labware and store at room temperature.

Caution: *SDS is an irritant and toxic. Handle carefully when weighing the powder. Do not breathe the dust.*

3 M Sodium Acetate Solution (pH 4.0 or pH 5.2)

Dissolve sodium acetate in ddH$_2$O and adjust pH to 4.0 or 5.2 with 3 *M* acetic acid. Treat the solutions with 0.1% DEPC overnight with stirring in a fume hood. Autoclave and store at room temperature.

SSTE

1.0 *M* NaCl
0.5% SDS
10 m*M* Tris, pH 8.0
1 m*M* EDTA, pH 8.0

1 M Tris Buffer (pH 7.5, pH 8.0, or pH 9.0)

Dissolve 121 g of Tris-base in 800 ml DEPC-ddH$_2$O using *RNase*-free labware. Adjust to desired pH with 0.1 *M* HCl and autoclave. Store at room temperature.

50X TAE Buffer

Dissolve 242 g Tris-base in 800 ml sterile ddH$_2$O. Add 57.1 ml glacial acetic acid and 100 ml 0.5 *M* EDTA (pH 8.0). Adjust volume to 1 L and store at room temperature.

TE Buffer

10 m*M* Tris-HCl, pH 8.0
1 m*M* EDTA, pH 8.0

Wash Buffer

10 m*M* Tris-HCl, pH 7.5
100 m*M* NaCl
1 m*M* EDTA

References

1. **Sambrook, J. and Russell, D.W.,** *Molecular Cloning, A Laboratory Manual,* 3rd ed., Cold Spring Harbor Press, Cold Spring Harbor, NY, 2001.
2. **Tesniere, C. and Vayda, M.E.,** Method for the isolation of high-quality RNA from grape berry tissues without contaminating tannins or carbohydrates, *Plant Mol. Biol. Rep.,* 9, 242, 1991.
3. **Chomczynski, P. and Sacchi, N.,** Single-step method of RNA isolation by acid guanidinium thiocyanate phenol-chloroform extraction, *Anal. Biochem.,* 162, 156, 1987.
4. **Chang, S., Puryear, J., and Cairney, J.,** A simple and efficient method for isolating RNA from pine trees, *Plant Mol. Biol. Rep.,* 11, 113, 1993.
5. **Dong, J.-Z. and Dunstan, D.I.,** A reliable method for extraction of RNA from various conifer tissues, *Plant Cell Rep.,* 15, 516, 1996.
6. **Bugos, R.C., Chiang, V.L., Zhang, X.-H., Campbell, E.R., Podila, G.K., and Campbell, W.H.,** RNA isolation from plant tissues recalcitrant to extraction in guanidine, *BioTtechniques,* 19, 734, 1991.
7. **Kao, Y.-Y., Harding, S.A., and Tsai, C.-J.,** Differential expression of two distinct phenylalanine ammonia-lyase genes in condensed tannin-accumulating and lignifying cells of quaking aspen, *Plant Physiol.,* 130, 796, 2002.
8. **Burr, B., Burr, F.A., Rubenstein, I., and Simon, M.N.,** Purification and translation of zein messenger RNA from maize endosperm protein bodies, *Proc. Natl. Acad. Sci. U.S.A.,* 75, 696, 1978.
9. **Beachy, R.N., Tompson, J.F., and Madison, J.T.,** Isolation and characterization of messenger RNAs that code for the subunits of soybean seed protein, In *The Plant Seed: Development, Preservation and Germination,* Rubenstein, I. et al., Eds., Academic Press, New York, 1979, p. 67.

Chapter 3

Extraction and Purification of Proteins

Soo Chul Chang and Leland J. Cseke

Contents

I.	General Considerations	46
II.	Protein Extraction Protocols	47
	A. Extraction of Proteins from Fresh Animal or Plant Tissues	47
	B. Extraction of Proteins from Frozen Animal or Plant Tissues	49
	C. Extraction of Proteins from Cultures of *E. coli*	49
	D. Extraction of Cytosolic Proteins from Eukaryotic Cells	50
	E. Isolation of Proteins from Cellular Membranes in Cell Suspensions as well as Animal or Plant Tissues	50
III.	Protein Concentration Determination	51
	A. Bicinchoninic Acid (BCA) Assay	51
	B. Bradford Assay	52
IV.	Purification of Protein(s) of Interest from Protein Mixtures	52
	A. Gel Filtration	52
	B. Ion-Exchange Chromatography	54
	C. Affinity Chromatography	56
	D. Immunoprecipitation of Proteins	56
	E. Purification of Proteins/Enzymes by Nondenaturing Gel Electrophoresis	56
	1. Preparation of the Separating Gel	57
	2. Preparation of the Stacking Gel	57
	3. Loading the Samples and Protein Standard Markers into the Gel	58
	4. Electrophoresis	58
	5. Staining and Destaining of the Gel Using Coomassie Blue (CB)	59
	6. Identifying the Band(s) in the Unstained Gels	59
	7. Electroelution of Protein/Enzyme of Interest from Gel Slices (Using the Model 422 Electro-Eluter from Bio-Rad, Inc.)	60
	Reagents Needed	60
References		65

Proteins and enzymes are the final products of gene expression. Although deoxyribonucleic acid (DNA) stores the genetic information, it is the proteins that determine the shape and structure of a cell, a tissue, an organ, and the intact organism. Enzymes, which are usually proteins in nature, control the expression of genes and the development of an organism. Overall, proteins constitute half of the dry weight of the cell. In recent years, proteins and enzymes have become one of the core parts of molecular biology studies. This "hot field" primarily comes from the concept of "reverse genetics." A number of molecular biology studies start from specific protein(s) that determine the particular phenotype of an organism, and go back to clone, identify, and characterize the specific gene(s) expressing the protein(s) of interest. In addition, proteins/enzymes are involved in broad molecular biology studies such as the interaction between the *cis*-element and *trans*-factor, DNA/protein interaction in gel retardation and footprinting, screening of cDNA libraries as probes, immunoblotting or precipitation, and *in vitro* translation.[1-7]

In the study of proteins and enzymes, extraction and purification of these macromolecules are fundamentally important.[1,8,9] The present chapter describes the procedures for extraction and purification in detail.

I. General Considerations

Proteins and enzymes are relatively unstable macromolecules as compared with DNA. Their structures or activities are sensitive to a variety of factors or parameters involved in isolation and purification procedures. The following factors should be considered when handling proteins and enzymes.

1. Buffer conditions: A buffer is a solution consisting of a conjugate base and a conjugate acid group that is able to resist pH change to varying degrees. It is strongly recommended that specific buffer conditions be maintained during the extraction and purification, since the structures or activities of proteins/enzymes are very sensitive to environmental pH changes. Some investigators use distilled, deionized water to replace the extraction buffer. This may work well for some proteins or enzymes, but other proteins/enzymes may be degraded or lose their enzymatic activities. Therefore, buffer conditions should be optimized for specific cells, tissues, and protein types. In the preparation of buffers, we recommend preparation of a buffer stock solution (100X or 10X), which can be diluted to the desired working buffer concentration prior to use. After being autoclaved or sterile-filtered, the stock buffer solution is quite stable and resists contamination at 4°C or at room temperature.
2. Water purity: Distilled, deionized water (ddH$_2$O) is highly recommended to be used as the primary ingredient in solutions or buffer systems. Unpurified water contains a lot of microbial organisms or proteases that cause protein degradation.
3. Temperature: Temperature is an important factor that causes protein degradation or loss of enzyme activity. Whenever possible, a cold environment, such as 0 to 4°C in an ice bucket with ice or in a cold room, should be maintained during the extraction and purification of proteins and enzymes in order to achieve the desired quality and integrity of proteins.
4. Protease inhibitors: Proteases hydrolyze proteins or enzymes. They may exist in solutions and buffers or be released from cells upon their physical disruption. To maximally inhibit those proteases, protease inhibitors are recommended to be added to specific solutions or buffers used for protein/enzyme isolation and purification. Commonly used protease inhibitors include the following:
    a. Phenylmethanesulfonyl fluoride (PMSF) serves to inhibit serine proteases such as chymotrypsin, trypsin, thrombin, or thiol protease (papain). PMSF is soluble in isopropanol at 10 mg/ml and is quite stable at room temperature as a stock solution.
    b. Leupeptin functions to inhibit serine and thiol proteases such as plasmin, cathepsin B, and papain. A stock solution (10 mg/ml in ddH$_2$O) is stable for 1 week at 4°C or 1 year at −20°C.
    c. Ethylenediaminetetraacetic acid (EDTA) is a metalloprotease inhibitor and can also inhibit the activities of other proteases. An EDTA stock solution (0.5 M, pH 8.0) is stable for 6 months at 4°C. Use NaOH solution to adjust the pH, or the EDTA will remain insoluble.

Extraction and Purification of Proteins

   d. Pepstatin A acts as an inhibitor of acid proteases such as pepsin, renin, cathepsin D, and chymosin. A stock solution (1 mg/ml in methanol) is stable for 1 year at –20°C.
   e. Others reagents such as aprotinin and benzamidine are also effective protease inhibitors.
5. Detergents: The extraction and purification of membrane proteins usually requires detergents. Detergents are amphipathic molecules consisting of a hydrophobic portion of a linear or branched hydrocarbon "tail" and a hydrophilic "head." They can form micelles with the hydrophilic head portions facing outward. Membrane proteins are solubilized by the detergent and form mixed micelles with the detergent. The common detergents used for protein extraction and purification are as follows:
   a. Ionic detergents: These detergents contain charged head groups (+/–) and serve to denature proteins in molecular-size separations. For example, sodium dodecyl sulfate (SDS) can denature proteins into their monomeric moieties and make proteins negatively charged. These properties can then be separated, based on their molecular weights (MW), by SDS-PAGE (polyacrylamide gel electrophoresis).
   b. Nonionic detergents: These compounds have uncharged hydrophilic head groups and are less likely to disrupt protein–protein interactions. They are less denaturing in action than ionic detergents and may cause proteins to aggregate. Common nonionic detergents include Triton X-100, Triton X-114, Nonidet P-40, octylglucoside, and Tween-20. These are used to block nonspecific protein interactions in solid-phase immunochemistry such as enzyme-linked immunosorbent assay (ELISA), radioimmunoassay (RIA), and immunoblotting.
6. Reducing agents: Many proteins or enzymes may lose activity when oxidized by O_2. This can be avoided by adding a reducing agent to the solution or buffer. 2-Mercaptoethanol can reduce the intramolecular disulfide bonds of proteins. This may result in protein inactivation. Dithiothreitol (DTT) oxidation may cause the formation of stable disulfide bonds in proteins.
7. Salts, metal ions, and ionic strength: These should be considered for the extraction and purification of particular proteins and enzymes.
8. Storage of proteins and enzymes: The half-life of a protein largely depends on the storage temperature. The best conditions are at 4, –20, or –80°C, or in liquid nitrogen (–196°C), depending on particular uses. Frequently, thawing and freezing may cause protein degradation. Addition of glycerol (20 to 30 or 50%) is often recommended for the storage of enzymes. Glycerol maintains the protein solution at very low temperature without freezing. Also, addition of 0.02% (3 mM) sodium azide is ideal for preventing contamination by bacteria or fungi during long-term storage of proteins or enzymes.

II. Protein Extraction Protocols

A. Extraction of Proteins from Fresh Animal or Plant Tissues

1. Prepare appropriate extraction buffer solutions and place in ice water for at least 1 h prior to use. The type of buffer varies with the particular protein(s) of interest. The commonly used homogenizing buffers are sodium phosphate, Tris, Tris-sucrose, and phosphate-buffered saline (PBS, see **Reagents Needed** at the end of this chapter).
2. Rinse the tissue (1 to 5 g per extraction for animal tissue, and 10 to 50 g per extraction for plant tissue) three times with an appropriate homogenizing buffer to wash away traces of blood or other extracellular materials.
3. Chop the washed tissue into small pieces (2 to 5 mm) with a clean razor blade or equivalent knife.
4. Transfer the chopped tissues into a homogenization jar or a glass mortar or blender. Add four to five volumes of ice-cold homogenizing buffer to it. Place on ice.
5. Homogenize the tissue on ice to a fine homogenate.

Notes: *(1) If using a power-driven Polter-Elvehjem glass-Teflon homogenizer, set the speed at 500 to 1500 rpm and pass the homogenizer through the sample four to six times at 5 to 10 sec per stroke. (2) If using a blender or a polytron, homogenize the tissue at top speed for 1.5 to 2 min, pausing for 4 sec between each pulse of 20 sec. (3) If using a hand homogenizer, pass the sample through 10 to 20 times until a fine homogenate is obtained.*

6. Quickly vacuum filter the homogenate through four layers of cheesecloth into a beaker on ice.
7. Save the filtrate and directly use it for total proteins (animal tissue) or for total soluble proteins (plant tissue) precipitation in step 8. Residues filtered from plant tissue can be used for cell-wall-bound (insoluble) protein extraction as follows:
    a. Resuspend the residue in two volumes of appropriate buffer containing 1 M NaCl.
    b. Keep at 4°C for 12 to 24 h with stirring.
    c. Carry out vacuum-filtration as in step 6 and save the filtrate for insoluble protein precipitation. Proceed to step 8.
8. Centrifuge the filtrate (soluble and insoluble) at 4000 to 5000 g for 5 min at 4°C to remove cellular debris. Carefully transfer the supernatant into a fresh beaker or flask and place it on ice.
9. Measure the volume of the supernatant and precipitate proteins as follows:
    a. Precipitation of total proteins:
        i. Add 80% (w/v) solid ammonium sulfate [$(NH_4)_2SO_4$] to the supernatant prepared in step 8.
        ii. Mix well and allow to precipitate at 4°C for 30 to 60 min.
        iii. Centrifuge at 12,000 to 15,000 g at 4°C for 20 min.
        iv. Discard the supernatant and resuspend the protein pellet in 2 ml (for 1 to 2 g of animal tissue) or in 10 ml (for 25 to 50 g plant tissue) of appropriate dialysis buffer such as 50 mM sodium acetate buffer (pH 4.5 to 5.5). Keep the sample at 4°C until use.
        v. Soak a dialysis bag in diluted dialysis buffer (1:5 dilution in ddH_2O) for 30 min. Seal the bottom end of the bag by making a tight knot or use clamps.

Note: *The exclusion size of the dialysis bag should be small enough to allow $(NH_4)_2SO_4$ or other ions to pass through. Total proteins should remain in the bag in order to prevent loss of some proteins.*

        vi. In a cold room, carefully transfer the protein sample in step iv into a presoaked dialysis bag and seal the other end of the bag. Place the bag and a magnetic stir bar in a large beaker containing 2 L of 1:5 diluted dialysis buffer. Put the beaker on a stir plate and turn on the stir bar. Allow dialysis to occur for 48 h with fresh buffer replacements at 12- or 24-h intervals.

Notes: *After dialysis, the volume of the sample increases due to water moving into the bag during dialysis. Therefore, leave some space (1 to 2 cm^3) when sealing the bag for dialysis.*

        vii. Carefully cut the top end of the dialysis bag with a pair of scissors and transfer the dialyzed sample into a fresh tube. Store the sample at 4 or –20°C until use for protein concentration measurement and purification of specific protein(s) or enzyme(s) of interest.
    b. Fractional precipitation of proteins:
        In order to take advantage of different proteins being precipitated at different levels of $(NH_4)_2SO_4$ saturation, total proteins can be divided into several fractions. This will make the purification of specific proteins/enzymes much easier later on. The number of fractions and $(NH_4)_2SO_4$ saturation range of fractions depend on one's particular interest. The procedure below is designed for three $(NH_4)_2SO_4$ precipitation fractions at 0 to 30, 30 to 50, and 50 to 80% saturation.
        i. 0 to 30% $(NH_4)_2SO_4$ precipitation:
            - Add 30% (w/v) solid $(NH_4)_2SO_4$ to the supernatant prepared in step 8.
            - Carry out steps 9.a.ii to 9.a.vii in the procedure for total protein precipitation; however, save the supernatant in step 9.a.iv.
        ii. 30 to 50% $(NH_4)_2SO_4$ precipitation:
            - Add 50% (w/v) solid $(NH_4)_2SO_4$ to the supernatant saved from the 0 to 30% $(NH_4)_2SO_4$ precipitation.
            - Carry out steps 9.a.ii to 9.a.vii in the procedure for total protein precipitation; however save the supernatant in step 9.a.iv.
        iii. 50 to 80% $(NH_4)_2SO_4$ precipitation:

Extraction and Purification of Proteins

- Add 80% (w/v) solid $(NH_4)_2SO_4$ to the supernatant saved from the 30 to 50% $(NH_4)_2SO_4$ precipitation.
- Carry out steps 9.a.ii to 9.a.vii in the procedure for total protein precipitation.

Note: *Different protein samples, such as soluble and insoluble proteins, should be labeled during the above procedures.*

B. Extraction of Proteins from Frozen Animal or Plant Tissues

1. Thaw the tissues on ice.
2. Carry out steps 1 to 9 described in **Section II.A**.

Notes: *An alternative method is as follows. (1) Grind the frozen tissues in liquid N_2 to a fine powder and briefly warm the powder for 3 min at room temperature. (2) Transfer the powder to a beaker on ice containing five volumes of an appropriate extraction buffer. Thoroughly suspend the powder into the buffer and allow it to set for 30 min on ice. (3) Carry out steps 6 to 9 in **Section II.A**.*

C. Extraction of Proteins from Cultures of *E. coli*

1. Harvest cells by centrifuging 500 to 1000 ml of *E. coli* liquid culture at 900 g for 15 min at 4°C.
2. Discard the supernatant and resuspend the pellet in lysis buffer I (3.5 ml/g wet cells).

Notes: *For extraction of secretory proteins, save the supernatant and discard the cell pellet. The proteins can be directly precipitated as described in step 9 in **Section II.A**.*

3. To the suspension, add 5 µl of 0.1 M PMSF as a protease inhibitor, and then add 0.1 ml of lysozyme (10 mg/ml) per 4 ml of cell suspension.

Caution: *PMSF is extremely toxic to the eyes, skin, and membranes, and it may be fatal if swallowed. Care must be taken when handling this chemical. Gloves should be worn.*

4. Incubate at 25 to 37°C for 20 min with shaking at 60 rpm.
5. Add 4 mg of deoxycholic acid per 4 ml of cell suspension.
6. Incubate at 37°C and stir with a glass rod until the lysate is viscous. Add 2 µl of *DNase I* (10 mg/ml) per 4 ml of cell suspension.
7. Place the lysate at room temperature until it is no longer viscous (approximately 30 min).
8. Centrifuge the cell lysate at 12,000 g for 15 min at 4°C.
9. Use the supernatant for the crude extraction of part of proteins from *E. coli* by carrying out step 9 in **Section II.A**. The pellet contains cytoplasmic granules that usually include a high level of expression of proteins in *E. coli*. Therefore, further extraction of the majority of the proteins from these granules or inclusion bodies must be taken.

Note: *In the crude extract, lysozymes and DNase I are included in the protein mixture, and they should be excluded when analyzing the protein in the crude extract.*

10. Resuspend the pellet from step 9 in five volumes of extraction buffer.
11. Incubate at room temperature for 10 min and centrifuge at 15,000 g for 20 min at 4°C.
12. Save the supernatant on ice. Resuspend the pellet in 0.2 ml of lysis buffer II containing 8 M urea and 0.1 mM PMSF.

13. Incubate at room temperature for 1 to 1.5 h.
14. Add nine volumes of potassium buffer.
15. Incubate at room temperature for 30 min and adjust the pH to 8.0 with 3 N HCl.
16. Continue to incubate at room temperature for 30 min.
17. Centrifuge at 15,000 g for 15 min at room temperature.
18. Pool the supernatant with the supernatant from step 12 and carry out protein precipitation as described in step 9 in **Section II.A**.

Notes: *The pellet can be resuspended in 2X loading buffer and analyzed by SDS-PAGE. If a lot of the protein of interest is still in the pellet, further lysis and extraction may be needed.*

D. Extraction of Cytosolic Proteins from Eukaryotic Cells

1. Centrifuge the cell culture suspension at 500 to 900 g at 4°C for 10 min.
2. Wash the cells by resuspending the cell pellet in four volumes of PBS or TBS (Tris-buffered saline) and centrifuge as in step 1 and repeat washing twice.
3. Carry out lysis of the cells by either of the following methods.
   a. Nitrogen cavitation:
      i. Resuspend the cells in ten volumes of cavitation buffer (pH 7.4).
      ii. Subject the cells to a stream of nitrogen gas at 375 psi and 4°C for 20 to 25 min in a cell-disruption chamber (Parr Instrument Co., Moline, IL).
      iii. Collect dropwise the cavitate in a tube and centrifuge at 1000 g for 5 min at 4°C to remove cellular debris.
      iv. Transfer the supernatant to a fresh tube on ice and prepare to carry out protein precipitation.
   b. Lysis with detergent:
      i. Resuspend the cell pellet in ten volumes of lysis buffer II.
      ii. Incubate at 0 to 4°C for 1 to 2 h with occasional shaking.
      iii. Centrifuge at 1000 g to remove cellular debris and save the supernatant for protein extraction.
4. Carry out protein precipitation as described in step 9 in **Section II.A**.

E. Isolation of Proteins from Cellular Membranes in Cell Suspensions as well as Animal or Plant Tissues

1. Harvest cells by centrifuging the cell suspension or culture at 900 g for 10 min at 4°C.
2. Resuspend the cell pellet in ten volumes of PBS or TBS buffer and centrifuge as in step 1. Repeat washing twice. For animal and plant tissues, wash the tissues three times with ten volumes of ddH$_2$O and slice the tissues into pieces.
3. Resuspend the washed cells in ten volumes of N-[2-Hydroxyethyl]piperazine-N'-[2-ethane sulfic acid] (HEPES)-KOH buffer. Place the animal tissue (2 to 5 g per extraction) or plant tissue (10 to 25 g per extraction) in five volumes of HEPES-KOH buffer.
4. Homogenize the tissues or cell suspension on ice to a fine homogenate using an appropriate cell/tissue homogenizer.
5. Centrifuge at 9000 g for 15 min at 4°C.
6. Transfer the supernatant into fresh ultracentrifuge tubes and discard the pellet.
7. Carefully balance the centrifuge tubes and carry out ultracentrifugation at 50,000 to 55,000 g at 4°C for 60 min.
8. Discard the supernatant, briefly air dry, and save the membrane pellet (enriched plasma membrane and intracellular membrane fractions).

Notes: *The membrane pellet can be directly resuspended in 1X loading buffer for protein analysis by SDS-PAGE (see Western blotting protocol in **Chapter 8**). If solubilization of membrane proteins is desired, proceed to the next step.*

Extraction and Purification of Proteins

9. Resuspend the membrane pellet in three volumes of membrane extraction buffer.
10. Incubate at room temperature for 1 to 2 h with occasional shaking.
11. Add seven volumes of potassium buffer (pH 10.7) to the lysis mixture and incubate at room temperature for 30 min.
12. Adjust the pH to 8.0 with 3 N HCl and continue to incubate at room temperature for 30 min.
13. Centrifuge at 12,000 g for 5 min at room temperature.
14. Transfer the supernatant to a fresh tube or beaker, and carry out protein precipitation as described in step 9 in **Section II.A**.

III. Protein Concentration Determination

Prior to protein analysis by SDS-PAGE or equivalent gels, or protein purification, the concentration of total or fractional proteins extracted should be determined by one of the following methods that are the most reliable methods used in our laboratory.

A. Bicinchoninic Acid (BCA) Assay

1. Prepare the solutions such as Reagent A and B, working solution, deoxycholate (DOC) solution, and trichloroacetic acid (TCA) solution (see **Reagents Needed** at the end of this chapter).
2. Precipitate the protein sample using the DOC-TCA method in order to avoid interfering substances.
   a. Dilute protein sample to 1 ml with ddH$_2$O and add 0.1 ml of 0.15% DOC solution.
   b. Mix well by vortexing and let stand at room temperature for 10 min.
   c. Add 0.1 ml of 72% TCA solution, vortex, set for 1 min at room temperature, and centrifuge at 12,000 g for 15 min using a microcentrifuge.
   d. Decant the supernatant and completely air dry at room temperature by inverting the tube on a 3MM Whatman® filter paper.
   e. Dissolve the protein pellet in 70 µl ddH$_2$O. Proceed to step 4.
3. Prepare the BSA (bovine serum albumin) standard curve:
   a. Prepare the BSA solution (1 mg/ml) in ddH$_2$O. Store at 4°C until use.
   b. Individually add 0, 10, 20, 30, 40, 50, and 60 µl of the BSA solution to microcentrifuge tubes containing 1, 0.99, 0.98, 0.97, 0.96, 0.95, and 0.94 ml ddH$_2$O, respectively.
   c. Add 0.1 ml of 0.15% DOC solution to each tube.
   d. Carry out steps 2.b to 2.e as described in the protein sample precipitation protocol with DOC-TCA.
4. Add 1.4 ml of working solution to each tube (1 volume of protein sample with 20 volumes of working solution).
5. Mix well and incubate at room temperature for at least 2 h or at 37°C for 30 min.
6. Let cool to room temperature in the case of 37°C incubation.
7. Transfer the liquid into spectrophotometer cuvettes and measure absorbance at 562 nm.
   a. Turn on the spectrophotometer and set the wavelength to 562 nm according to the instructions.
   b. Individually measure the absorbance of each of BSA standards, starting from zero and proceeding to higher concentrations of BSA.
      i. Transfer the liquid from the tube lacking BSA into a cuvette and insert the cuvette into the cell of the spectrophotometer. Use this tube as a blank or a reference and set the reading at 0.0000.
      ii. Replace the blank/reference cuvette with the one containing 10 µg BSA and record the absorbance.
      iii. In the same way, read and record the absorbance for 20, 30, 40, 50, and 60 µg BSA standards, respectively.
   c. Directly measure OD$_{562}$ for protein samples and record the absorbances.

Note: If $OD_{562} > 3.000$, dilute the sample and read again.

d. Draw a standard curve and calculate the concentrations of protein samples. If a UV-visible recording spectrophotometer (e.g., UV 160U, Shimadzu) is available, the instrument can automatically record, on the computer monitor screen, each OD_{562} and make a standard curve as well as calculate the concentration of each protein sample. If the spectrophotometer cannot simultaneously record OD_{562}, be sure to record each absorbance value and draw a standard curve on graph paper using BSA concentration as the ordinate and OD_{562} as abscissa. To determine the concentration of each protein sample, find its OD_{562} and draw a horizonal line crossing to the standard curve. Starting from the cross-point on the standard curve, draw a vertical line crossing to the ordinate. This point of crossing on the ordinate is the concentration of BSA equal to the concentration of protein sample.

B. Bradford Assay

This is a rapid and reliable method.

1. Prepare the solutions and reagents: Bradford stock solution and Bradford working solution.
2. Prepare the BSA solution (1 mg/ml) in ddH_2O. Add 0, 10, 20, 30, 40, 50, and 60 μl of BSA solution to each of seven microcentrifuge tubes, respectively. Add ddH_2O to a final volume of 0.1 ml for each tube.
3. Take 0.1 ml of each of the protein samples and transfer to another set of microcentrifuge tubes.
4. Add 1 ml of Bradford working buffer to each tube, mix by vortexing, and let stand for 2 min but less than 1 h.
5. Carry out steps 7.a to 7.d in **Section III.A**, but set the spectrophotometer's wavelength to 595 nm.

IV. Purification of Protein(s) of Interest from Protein Mixtures

Protein(s) of interest can be purified from an extracted protein mixture according to size, charge, and binding affinity.[1-4,6,8] Several techniques have been developed to do this. The commonly used methods described in the present chapter are gel filtration, ion-exchange chromatography, affinity chromatography, and gel electrophoresis (Figure 3.1). The high-pressure/performance liquid chromatography (HPLC) method is described in **Chapter 23**. However, the specific methods and the order of chromatography protocols used in protein purification depend on the particular protein or enzyme of interest. The detailed protocols described below have been used successfully in our laboratory for the purification of proteins as well as enzymes. One requirement for protein/enzyme purification is that a quick, easy, and inexpensive assay method must be known for enzyme purification so that a known molecular weight (MW), specific affinity, or immunoaffinity of non-enzymatic protein(s) of interest can be detected using the appropriate methods.

A. Gel Filtration

Gel filtration or molecular-exclusion chromatography separates proteins according to molecular sizes. A number of commercial gel matrixes are available, such as Sephadex (G-10, G-25, G-50, and G-75), Sepharose, Sephacryl, Sepharose CL, and Bio-Gel. Sephadex is composed of polysaccharide that is readily contaminated by bacteria. Bio-Gel is made of polyacrylamide and can resist bacterial contamination, but it is very toxic. Different gel resins have different protein size-exclusion ranges. By choosing an appropriate gel resin as a column matrix, suspending the appropriate buffer, and transferring to a chromatography column, proteins can be separated by running them through the column. Because there are tiny holes in billions and billions of gel beads, small-sized proteins will pass through those holes and take a longer time to run out of the column as compared with

Extraction and Purification of Proteins

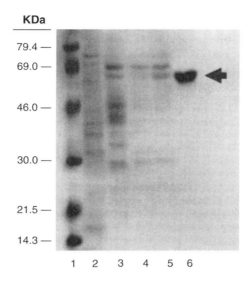

FIGURE 3.1
Purification of invertase from a protein mixture subjected to SDS-PAGE. Lane 1: standard protein molecular weight markers; Lane 2: crude extract; Lane 3: CM Sephadex eluate; Lane 4: Sephadex G-75 fraction; Lane 5: ConA-Sepharose affinity column eluate; Lane 6: nondenaturing PAGE eluate. (From Kaufman, P.B., Wu, W., Kim, D., and Cseke, L.J., *Handbook of Molecular and Cellular Methods in Biology and Medicine*, 1st ed., CRC Press, Boca Raton, FL, 1995.)

large-sized proteins that cannot get into those holes, but instead run directly out of the column through void space in the column.

Procedure

1. Suspend the appropriate amount of gel powder (e.g., Sephadex) in a clean beaker containing about 20 volumes of the appropriate buffer and allow to equilibrate for at least 1 to 2 h at 4°C.

Notes: *(1) All the procedures should be carried out at 4°C for enzyme purification. For other proteins, 4°C is also better than room temperature to prevent bacterial contamination. (2) The selection of the kind of gel-powder to be used as the column matrix depends on the particular protein(s) of interest. For example, if the protein of interest is 70 kDa, an appropriate Sephadex or Bio-Gel product that can exclude proteins larger than 65 kDa should be chosen. Therefore, these proteins will run through the void space in the column faster. (3) The equilibrating buffer used for the column varies with the specific protein/enzyme to be purified. Usually, ddH_2O (pH 7.0) is satisfactory for gel filtration.*

2. Transfer the matrix suspension into a plastic or glass column (2 × 60 cm to 2 × 100 cm), setting it vertically with the bottom valve closed. The column should be filled with the gel suspension up to a level about 5 cm from the top. Add more buffer or ddH_2O to the top and let stand for 30 min.
3. Drain the fluid by opening the bottom valve and close the valve afterward.
4. Add 0.5 to 1.0 ml of an appropriate dye solution, such as Blue Dextran, to the top of the column, open the bottom valve, and collect the eluate in a beaker. Continue to add the buffer to the top of the column until the dye runs out of the column. The volume of collected eluate is the estimated void volume of the column.

Note: *The MW of the dye should be in the exclusion-size range of the gel matrix in order to accurately estimate the void volume of the column.*

5. In a cold room, assemble an automatic fraction collector with about 100 collection tubes and place the collector under the column. Set the collection volume per tube at 1 to 2.5 ml. Above the column, connect a beaker or equivalent bottle containing 400 to 500 ml of the appropriate elution buffer. The bottle or beaker should have a plastic pipe connected to the top of the column. If an automatic protein-peak UV detector is available, connect the detector to the bottom of the column and then to the collector, according to the instructions.
6. Pretest the flow of the assembled components by running the buffer through the column at a flow rate of 0.5 to 1.0 ml/min.
7. Stop the addition of elution buffer to the column. Carefully add the extracted protein sample (1 to 10 ml) to the top of the column and start to collect the eluate. When the protein sample solution subsides into the bed matrix, reconnect the elution buffer to the top of the column and allow the buffer to fill the top headspace of the column. Then adjust the flow rate at the top of the column so that it is the same as the flow rate at the base of the column. Allow the filtration through the column to occur for 10 to 20 h at 4°C.
8. Stop running the column and transfer the tubes in appropriate order to an ice water bath until analysis can be performed.
9. Carry out the appropriate enzyme assay for each of the tubes and pool the active fractions for further purification. For nonenzymatic proteins, appropriate analysis methods should be used to identify the positive fraction(s) containing the protein(s) of interest. These methods include immunoassay, immunoblotting, and MW determination by SDS-PAGE or by elution profile of standard protein markers chromatographed on the same column under the same conditions. The commonly used protein markers are thyroglobulin, bovine (669 kDa); apoferritin, horse spleen (443 kDa); α-amylase, sweet potato (200 kDa); alcohol dehydrogenase, yeast (150 kDa); albumin, bovine serum (66 kDa); and carbonic anhydrase, bovine erythrocytes (29 kDa). After chromatography, make a standard curve using Ve/Vo (Ve: elution volume, Vo: void volume) as the ordinate and log MW as the abscissa. For the protein sample, calculate the Ve/Vo for the particular fraction and determine the MW of proteins in the fraction from the standard curve. Pool the fraction(s) containing the expected proteins for further purification.
10. If necessary, dialyze the pooled samples against diluted elution buffer (1:5 dilution) to reduce the ionic strength prior to the next purification procedure.
11. Concentrate the pooled or dialyzed sample using Amicon centrifuge tubes or equivalent tubes. Add 5 to 7 ml to each tube, assemble the tubes according to the instructions, and centrifuge at 2000 to 3000 g for 20 min at 4°C. Stop the centrifugation, decant the fluid from the collection tube, and repeat centrifugation twice. Transfer the concentrated protein sample from the inner tube into a fresh tube and proceed to the next step of purification.

B. Ion-Exchange Chromatography

Unlike gel filtration, which separates proteins according to different molecular sizes, ion-exchange chromatography separates proteins based on their charge. The basic principle is that at a given pH, most proteins have an overall negative or positive charge depending on their pI value. This makes it possible for them to interact with an oppositely charged chromatographic matrix. Different proteins have different amounts of charge, causing differential retardation in chromatography, which facilitates separation of proteins. There are two type of columns commonly used. One is diethylaminoethyl (DEAE) cellulose for binding to net negatively charged proteins. The other is carboxymethyl (CM) Sephadex for binding to net positively charged proteins.

Procedure 1: DEAE Cellulose Chromatography

1. Prepare 2 L of 10 to 20 mM phosphate buffer (pH 6.0) containing 1 mM EDTA, 1 mM benzamidine, and 0.1 mM PMSF. Store at 4°C.
2. Suspend an appropriate amount of DEAE cellulose in 20 volumes of phosphate buffer and allow it to equilibrate for 1 to 2 h at 4°C.

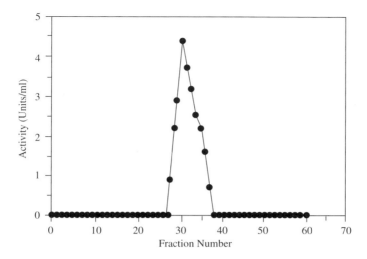

FIGURE 3.2
Purification of invertase from a protein mixture by diethylaminoethyl (DEAE) cellulose chromatography. A single peak is evident between fractions 27 and 38. (From Kaufman, P.B., Wu, W., Kim, D., and Cseke, L.J., *Handbook of Molecular and Cellular Methods in Biology and Medicine*, 1st ed., CRC Press, Boca Raton, FL, 1995.)

3. Carry out steps 2 to 5 in **Section IV.A** except for the following:
   a. The column size is 1.5 × 20 cm to 2 × 40 cm.
   b. The flow rate is set at 2 ml/min and 3 ml per fraction.
   c. The elution buffer is a linear gradient of 0 to 0.6 M NaCl in phosphate buffer or ddH_2O or appropriate buffer. This can be produced using a commercial gradient maker consisting of two chambers. First, close the channel between the two chambers. Add a volume (e.g., 250 ml) of phosphate buffer lacking NaCl to the inner chamber. Add an equal volume (e.g., 250 ml) of phosphate buffer containing 0.6 M NaCl to the outer chamber. Drop a stir bar in the inner chamber and place the entire gradient maker on a stir plate that is set above the DEAE cellulose column.
   d. Add the active-fraction pool purified by gel filtration onto the surface of the gel matrix and start to collect the eluate. Wash the column with two to three bed volumes of phosphate buffer (pH 6.0) and collect the eluate.
   e. Elute the bound proteins with a 0 to 0.6 M NaCl linear gradient by opening the valve of the gradient maker to the column and then opening the channel between the two chambers with the magnetic stir bar rotating.

Note: *The flow rate to the column is set at about 3 to 4 ml/min until the top space in the column is filled. Then it is reduced to the same flow rate as that from the bottom of the column. The channel between the two chambers should be open to obtain about the same flow rate as that from the gradient maker to the column. Record the fractions before elution and those fractions obtained during elution.*

4. Transfer the tubes in the proper order to an ice water bath.
5. Carry out an appropriate analysis to find the active or positive fractions containing the protein/enzyme of interest (Figure 3.2). Pool those fractions.
6. Carry out steps 10 (optional) and 11 in **Section IV.A**.

Procedure 2: CM Sephadex Chromatography

This chromatography method separates proteins based on their net charge as well as MW. The procedures are almost the same as those used in DEAE cellulose chromatography except for the following:

1. Use CM Sephadex as the bed matrix for the column.
2. The buffer pH should be 4.5 to 5.5 or an appropriate value depending on the particular protein of interest.

C. Affinity Chromatography

This is a powerful means of purifying proteins. The principle of this technique is based on the fact that some proteins have a very high affinity for specific chemical groups (ligands) covalently attached to a chromatographic bed material, the matrix. After loading and running the protein mixture through the column, only those protein(s) having a high affinity to the column can bind to the column matrix. Other proteins, in contrast, run directly through the column. The bound proteins(s) will then be eluted from the column by a solution containing a high concentration of the soluble form of the ligand or the specific residue that is recognized and bound by the proteins(s). A number of affinity columns are commercially available depending on the particular protein purification desired. One example is ConA-Sepharose affinity chromatography, which is specific for glycoprotein purification. The procedures are almost the same as in **Section IV.B** except for the following:

1. Use ConA-Sepharose as the bed matrix.
2. The buffer mixture should consist of 20 mM sodium phosphate or sodium acetate (pH 6.0) that contains 0.5 to 1 M NaCl, 1 mM MgCl$_2$, 1 mM MnCl$_2$, and 1 mM CaCl$_2$.
3. The linear-gradient elution solution is 0 to 250 mM α-methyl mannoside.
4. The flow rate is 1 ml/min and 2 ml per fraction.

D. Immunoprecipitation of Proteins

This technique is very powerful when a monoclonal antibody is available for a specific antigen of interest.

1. Mix the protein sample with an excess of monoclonal antibody solution in an appropriate binding buffer that depends on the specific conditions required for the antibody–antigen interaction.
2. Incubate for 3 to 6 h at 4°C.
3. Add the appropriate reagent or resin to help precipitate the antigen–antibody complex and then allow to incubate for another 1 to 2 h at 4°C.
4. Centrifuge at 10,000 g for 15 min at 4°C to pellet the precipitate. Unprecipitated proteins are still soublized in the buffer.
5. Decant the supernatant and dissolve the pellet in an appropriate volume of buffer. SDS-PAGE analysis should be carried out to check the antigen of interest using a positive antigen as a control.

E. Purification of Proteins/Enzymes by Nondenaturing Gel Electrophoresis

The partially purified protein sample obtained from the previous chromatographic separations can be further purified with nondenaturing polyacrylamide gel electrophoresis (PAGE), or native gel electrophoresis. This procedure separates proteins according to their size and charge properties. The acrylamide pore size serves as a molecular sieve to separate different sizes of proteins. Further, proteins that are more highly charged at the pH of the separating gel migrate faster than those with less-charged molecules. The major merit of PAGE is that it minimizes the denaturation of proteins in contrast to sodium dodecyl sulfate (SDS)-PAGE, which does denature proteins. Thus, many

Extraction and Purification of Proteins

enzymes still have biological activities after running PAGE. These include, for example, esterase, dehydrogenase alkaline phosphatase, α-amylase, transferases, hydrolases, lyases, β-fructofuranosidase (invertase), and isomerases. The enzyme activity can be assayed either directly within the gel or following protein elution from the gel. Procedures for carrying out discontinuous native gel electrophoresis can be exactly the same as for SDS-PAGE, except that there is no SDS component. The following procedure is a modified method that works well in our laboratory.

1. Preparation of the Separating Gel
   1. Thoroughly clean glass plates and spacers with detergent; then wash with tap water, rinse with distilled water several times, and air dry.
   2. Wear gloves and rinse the glass plates and spacer with 100% ethanol, and wipe dry with Kimwipe™ paper tissues. Assemble a vertical slab gel unit, such as the SE 600 Vertical Slab Gel Unit (Amersham Biosciences, San Francisco, CA), in the casting mode according to the instructions. Place one spacer (1.5 mm thick) at each of the two sides between the two glass plates and clamp, forming a sandwich. Repeat for the second sandwich. The size of the separating gel varies with individuals. A standard size is $120 \times 140 \times 1.5$ mm.

Note: *Apply a little stopcock grease to the spacer areas at both the top and bottom ends of the sandwiches to prevent potential leaking.*

   3. Check potential leaking by pipetting some distilled water into the sandwich and drain away the water afterward by inverting the unit. Set the unit at an even level using a water balancer or its equivalent.
   4. Prepare the separating gel solution in a clean 100-ml beaker or 125-ml flask in the order shown below (for two pieces of standard-sized gels):
      a. 12.5 ml Monomer solution for the separating gel
      b. 16.25 ml 4X Separating gel buffer
      c. 36.85 ml ddH$_2$O
      Mix well after each addition.
   5. Degas the solution to remove any air bubbles by applying a vacuum for 5 min (optional). Add 0.2 ml of freshly prepared 10% ammonium persulfate solution (APS) and 80 µl tetramethylethylenediamine (TEMED). Mix by gently swirling. Do not generate air bubbles.
   6. Immediately pipette the mixture with a 50-ml syringe into the assembled sandwich up to a level 4 cm from the top.
   7. Take up 0.8 ml of ddH$_2$O in a 1-ml syringe equipped with a 2-in. 22-gauge needle or its equivalent and carefully load 0.3 ml of distilled water, starting from one top corner next to the spacer, onto the surface of the acrylamide gel solution. Repeat on the other side of the slab next to the other spacer. The water will layer evenly across the surface of the gel mixture. The purpose of applying water is to make the surface of the gel very even. A very sharp gel–water interface will be visible after the gel has polymerized.
   8. After polymerization, drain away the water layer by gently tilting the casting unit, and rinse with 2 ml of overlay solution once. Add 1 ml of overlay solution on the top of the gel and allow the gel to sit for 30 min.

2. Preparation of the Stacking Gel
   1. Prepare the stacking gel solution in a clean 25-ml beaker or 25-ml flask in the order shown below (for two pieces of standard-sized gels):
      a. 12 ml Monomer solution for the stacking gel
      b. 6 ml 4X Stacking gel buffer
      c. Degas (optional)
   2. Drain the overlay solution on the separating gel by tilting the casting unit and rinse with 2 ml of the 4X stacking gel mixture. Drain away the stacking gel solution and insert the comb (1.5 mm thick) into the top of the glass sandwich.

3. Add 98 µl of 10% AP solution and 20 µl of TEMED to the gel mixture. Gently mix well and add the gel mixture by pipette or its equivalent into the sandwich from both upper corners next to the spacers, filling up to the top edge. Allow the gel to polymerize for 30 min at room temperature.

Notes: *The stacking gel solution may be added prior to inserting the comb into the sandwich. However, air bubbles are easily generated and trapped around the teeth of the comb. We recommend that the comb be inserted into the sandwich prior to filling with the stacking gel mixture.*

3. Loading the Samples and Protein Standard Markers into the Gel

1. While the stacking gel is polymerizing, add 0.2 to 0.5 volume of sample loading buffer to each of the partially purified protein samples and protein standard markers.

Notes: *The amount of proteins loaded into one well should be 10 to 35 µg in a total volume of 5 to 35 µl. The amount of protein standard marker for one well should be 2 to 10 µg.*

2. Slowly and vertically pull the comb straight up from the gel. Rinse each well with upper tank buffer using a syringe or pipette to add the buffer. Carefully invert the casting stand to drain the wells and repeat twice.
3. Place the casting stand upright and fill each well with 10 µl of upper tank buffer. Take up sample or markers with a syringe equipped with a 2-in. 22-gauge needle or pipette with an equivalent tip attached. Carefully insert the needle into the buffer in each well and underlay the sample or standard marker in the well. Repeat until all the samples are loaded.

Notes: *The volume loaded into each well should be less than 40 µl for standard-sized gels or less than 15 µl for minigels. Overloading may cause samples to float and cause contamination among wells. It is recommended that markers be loaded into the front left or right well, or both wells.*

4. Carefully overlay the wells with upper tank buffer, using a syringe or pipette, until the buffer reaches the top of the gel.

Notes: *The samples and markers containing blue dye should be visible at the bottom of each well. Mark the orientation of samples and record in lab notebook.*

4. Electrophoresis

1. Carefully place the upper buffer chamber in its proper position according to the instructions. Remove lower clamps and clamp the sandwiches to the bottom of the upper buffer chamber. Do not disturb the wells. Check any potential leaking by filling 10 to 20 ml of upper-tank buffer into the chamber.

Note: *Apply a little stopcock grease on the spacer areas at the top corners of the sandwiches to prevent potential leaking.*

2. Carefully transfer the assembled unit into the lower buffer chamber according to the instructions. Place the entire tank apparatus on a magnetic stirrer. Slowly fill the lower chamber with 3 L of lower-tank buffer and add 1 L of upper-tank buffer to the upper chamber, or use an appropriate volume depending on the sizes of the chambers.

Notes: *The bottom-chamber buffer should be of sufficient volume to cover two-thirds of the slabs; otherwise, heating will not occur evenly, thus causing distortion of the band patterns of the gel. For the upper chamber, gently fill the running buffer from one corner into the* chamber. *In order to avoid washing out and/or mixing the samples, do not pour the buffer into the well area.*

3. Before running the gel, put a stir bar in the lower chamber and stir the buffer for 1 min to remove any air bubbles trapped under the ends of the sandwiches.
4. Add a few drops of dye solution, such as phenol red, into the upper-chamber buffer if the amount of dye in the samples is not sufficient.
5. Place the lid on top of the unit and connect the power supply. The cathode (negative pole) should be connected to the upper buffer chamber and the anode (positive pole) should be connected to the lower buffer chamber.
6. Set the power supply at constant power or current and turn on the power supply. Adjust the current to 25 to 30 mA/1.5-mm-thick standard-sized gel. For two pieces of gel, set the current at 50 to 60 mA. The voltage will increase during the time the gel is running. Allow the gel to run for several hours at 0 to 4°C until the dye reaches the bottom of the gel.
7. Turn off the power supply and disconnect the power cables. Loosen and remove the clamps at both sides of the sandwiches. Place the gel sandwiches on the table or bench. Slowly remove the spacers. Use an extra spacer to carefully separate the two glass plates starting from one corner. Remove the top plate; the gel should be on the bottom glass plate. Using a razor blade, make a small cut at the upper left or right corner of the gel to record the orientation.

5. Staining and Destaining of the Gel Using Coomassie Blue (CB)

1. Cut a small strip about 1 cm wide from each of the gels and transfer into a glass or plastic tray containing 100 ml of CB solution. Stain for 4 to 8 h at room temperature with slow shaking at 60 rpm. Alternatively, tightly cover the tray and stain for 20 to 30 min in a 50°C water bath. Wrap the rest of the gels with SaranWrap™ and keep at 4°C.
2. Remove the staining solution and replace it with destaining solution I. Allow destaining for 1 to 1.5 h.
3. Gently transfer the stained gel and carefully place it between two sheets of matrix. Gently roll the sandwich and insert it into a cylinder holder and cover the two ends of the holder. An alternative way is to keep the gel in the tray and replace the destaining solution I with destaining solution II. Destain the gel with shaking at 60 rpm until a clear background is obtained.
4. Place the cylinder into the SE 530 Destainer (Amersham Biosciences) or its equivalent filled with destaining solution II. Put a stir bar into the tank and place the tank on a magnetic stirrer. Turn on the stir bar and allow the gel to destain overnight up to a few days until a clear background is achieved.

Notes: A fast destaining method is to destain the gel with 0.5 to 1% (v/v) Clorox® (commercial sodium hypochlorite bleach) in distilled water. Constantly monitor the destaining of the background. When relatively clear protein bands are visible, immediately rinse the gel with a large volume of distilled water several times until the blue color is stable. The disadvantage of this procedure is that if the gel is over-destained, some bands may disappear. Never use >1.5% Clorox.

6. Identifying the Band(s) in the Unstained Gels

1. After staining and destaining, carefully match the stained and unstained portions of the gel on a glass plate.
2. Cut individual gel strips from the unstained portions of the gel that are equivalent to the same areas as the stained bands.
3. Slice individually the cut gel strips into pieces using a clean razor blade. Use part of the slices (approximately one-third) to directly assay the enzyme activity of interest using an appropriate enzymatic method or other detecting method such as an immunoassay. Keep the rest of the slices at 4°C until protein/enzyme elution is performed.
4. Based on the assay data, identify the positive band(s) of interest and proceed to protein/enzyme elution.

7. Electroelution of Protein/Enzyme of Interest from Gel Slices (Using the Model 422 Electro-Eluter from Bio-Rad, Inc.)

1. Prepare the elution buffer as follows:
   - 25 mM Tris base
   - 192 mM Glycine
   - 0.1% SDS (omitted for enzyme elution)
2. Soak the membrane caps in the elution buffer at 60°C for 2 h prior to use.
3. Wear gloves and carefully attach the cap, filled with the fresh elution buffer, to the bottom of the glass tube and insert the tube into the rack of the eluter. There is a dialysis lid at the bottom of the tube to hold the gel slices in place. The protein/enzyme can pass through the lid by eletroelution down to the cap containing a membrane at the bottom that only allows the electrons to pass through.

Note: *Avoid any air bubbles in the membrane cap.*

4. Fill the tube with elution buffer and carefully transfer the gel slices containing the protein/enzyme of interest into the elution tube. Avoid any air bubbles inside the tube.
5. Assemble the eluter according to the instructions. Fill the upper and lower chambers with appropriate volumes of elution buffer. Place a stir bar in the lower chamber.
6. Connect to the power supply with the positive pole at the bottom tank and the negative pole at the upper tank. Carry out the elution of protein/enzyme at 9 mA per glass tube (constant current) for 1 to 3 h at 4°C or at room temperature.
7. Reverse the positive and negative electrodes and continue to elute for 1 to 5 min to loosen the protein/enzyme adhering to the membrane of the cap, allowing it to enter the elution buffer.
8. Transfer the rack containing the elution tubes onto the lab bench. Avoid shaking during this process. Gently remove some of the elution buffer from the tube to a level slightly above the bottom of the tube and carefully remove the cap from the bottom of the tube. Suspend the eluted solution in the cap by sucking it up and down with a pipette.
9. Transfer the solution (approximately 0.6 ml) to a fresh tube and add 50 to 100 µl fresh elution buffer to wash the cap and collect the solution. Pool the solutions and measure the concentration of the elution protein(s). Assay enzyme activity if necessary.

Reagents Needed

10% Ammonium Persulfate Solution (APS)
Ammonium persulfate (0.2 g)
Dissolve well in 2.0 ml ddH$_2$O.
Store at 4°C for up to 10 days.

Bradford Stock Solution
95% Ethanol (100 ml)
85% Phosphoric acid (200 ml)
Serva Blue G (350 mg)
Store at room temperature for up to 1 year.

Note: *Serva Blue G is used for protein measurement and Serva Blue R is used for protein gel staining.*

Bradford Working Buffer
ddH$_2$O (425 ml)
95% Ethanol (15 ml)
85% Phosphoric acid (30 ml)
Bradford stock solution (30 ml)

Filter through Whatman No. 1 paper and store in a brown bottle at room temperature for up to 4 weeks.

Cavitation Buffer

130 mM KCl
0.1 mM MgCl$_2$
1 mM EDTA
1 mM DTT
10 mM HEPES
0.1 mM PMSF
10 mg/ml Each of leupeptin, soybean trypsin inhibitor, aprotinin, and pepstatin

1% Coomassie Blue (CB) Solution

Coomassie Blue R-250 (2 g)
Dissolve well in 200 ml ddH$_2$O.
Filter.

0.15% (w/v) Deoxycholate (DOC) Solution

Deoxycholate (0.15 g)
Dissolve in 100 ml ddH$_2$O.
Store at 4°C.

Destaining Solution I

Methanol (500 ml)
Acetic acid (100 ml)
Add ddH$_2$O to a final volume of 1000 ml.

Destaining Solution II

Acetic acid (245 ml)
Methanol (175 ml)
Add ddH$_2$O to a final volume of 3.5 L.

Extraction Buffer

10 mM EDTA, pH 8.0
0.5% Triton X-100

HEPES-KOH Buffer

0.25 M Sucrose
30 mM HEPES
3 mM EDTA
0.1 mM MgCl$_2$
14 mM 2-Mercaptoethanol
0.1 mM PMSF
Adjust the pH to 7.5 with 1 N KOH.

Lower-Tank Buffer (pH 7.5)

63 mM Tris (22.7 g)
50 mM HCl (1 N) (150 ml)
Add ddH$_2$O to a final volume of 3 L.

Lysis Buffer I

    50 mM Tris-HCl, pH 8.0
    1 mM EDTA, pH 8.0
    100 mM NaCl

Lysis Buffer II

    50 mM Tris-HCl, pH 8.0
    150 mM NaCl
    1% Nonidet P-40 or 1% Triton X-100
    0.1 mM PMSF
    0.05% (w/v) SDS (should be omitted for enzyme isolation)

Membrane Extraction Buffer

    50 mM Tris-HCl, pH 8.0
    1 mM EDTA, pH 8.0
    100 mM NaCl
    0.1 mM PMSF
    8 M Urea

Monomer Solution for Separating Gel

    [38% (w/v) Acrylamide, 2% *bis*-acrylamide]
    Acrylamide (76 g)
    N,N-Methylene-*bis*-acrylamide (4 g)
    Dissolve well after each addition in 150 ml ddH$_2$O.
    Add ddH$_2$O to a final volume of 200 ml.
    Wrap the bottle with aluminum foil and store at 4°C in the dark.

Caution: *Acrylamide is neurotoxic. Gloves should be worn when handling this chemical.*

Monomer Solution for the Stacking Gel

    [5% (w/v) Acrylamide, 1.25% *bis*-acrylamide]
    Acrylamide (10 g)
    N,N-Methylene-*bis*-acrylamide (2.5 g)
    Dissolve well after each addition in 150 ml ddH$_2$O.
    Add ddH$_2$O to a final volume of 200 ml.
    Wrap the bottle with aluminum foil and store at 4°C in the dark.

Overlay Buffer

    Separating gel buffer (25 ml)
    Add ddH$_2$O to a final volume of 100 ml.
    Store at 4°C.

Phosphate-Buffered Saline (PBS), pH 7.4

    NaCl (8 g)
    KCl (0.2 g)
    KH$_2$PO$_4$ (0.24 g)
    Na$_2$HPO$_4$ (1.44 g)
    Dissolve well after each addition in ddH$_2$O. Adjust the pH to 7.4 with 3 N HCl. Add ddH$_2$O to 1 L. Autoclave.

Extraction and Purification of Proteins

Phenylmethane sulfonyl fluoride (PMSF) Stock Solution
10 mg/ml dissolved in 100% isopropanol
Store at 4°C or room temperature.

Potassium Buffer
50 mM KH_2PO_4, pH 10.7
1 mM EDTA, pH 8.0
50 mM NaCl
Adjust the pH to 10.7 with KOH.

Reagent A (1 L)
BCA (10 g)
$Na_2CO_3 \cdot H_2O$ (20 g)
$Na_2C_4H_4O_6(2H_2O)$ (1.6 g)
NaOH (4 g)
$NaHCO_3$ (9.5 g)
Dissolve well after each addition in ddH_2O.
Adjust the pH to 11.25 with NaOH or solid $NaHCO_3$ (optional).
Add ddH_2O to 1 L.
Store at room temperature for up to 1 year.

Reagent B (100 ml)
$CuSO_4 \cdot 5H_2O$ (4 g)
Add ddH_2O to 100 ml.
Store at room temperature for up to 1 year.

Sample Loading Solution
Bromophenol blue (0.05 g)
50% Glycerol (25 ml)
Dissolve well after each addition in 15 ml ddH_2O.
Add ddH_2O to a final volume of 50 ml and store at –20°C.

4X Separating Gel Buffer
[11.47% Tris (w/v), 28.92% 1 N HCl (w/v)]
Tris (23 g)
Dissolve well in 80 ml ddH_2O.
1 N HCl (57.84 ml)
Add ddH_2O to a final volume of 200 ml.
Store at 4°C.

0.1 M Sodium Phosphate Buffer, pH 6.1
85% (v/v) 0.2 M NaH_2PO_4 solution
15% (v/v) 0.2 M Na_2HPO_4 solution
1 mM EDTA
1 mM DTT
0.1 mM PMSF
1 mM Benzamidine
Dissolve or mix well after each addition in ddH_2O. Sterile filtering is optional.

4X Stacking Gel Buffer

    0.158 mM Tris (1.92 g)
    Dissolve well in 50 ml ddH$_2$O.
    1 N Phosphoric acid (H$_3$PO$_4$, 85%) (25.6 ml)
    Add ddH$_2$O to a final volume of 100 ml.
    Store at 4°C.

Staining Solution

    1% CB solution (125 ml)
    Methanol (500 ml)
    Acetic acid (100 ml)
    Add ddH$_2$O to a final volume of 1000 ml.

TBS Buffer

    10 mM Tris-HCl, pH 7.5
    150 mM NaCl

72% (w/v) Trichloroacetic Acid (TCA) Solution

    TCA (72 g)
    Dissolve in 100 ml ddH$_2$O.
    Store at 4°C.

1 M Tris Buffer, pH 7.6

    1 M Tris-HCl, pH 7.6
    4 mM EDTA
    0.2% (v/v) Mercaptoethanol (not for enzyme extraction)
    2% (w/v) Polyvinylpyrrolidone
    1 mM Benzamidine
    Dissolve well after each addition in ddH$_2$O. Sterile-filtering is optional.

Tris-Sucrose Buffer

    0.1 mM Tris-HCl, pH 7.0
    250 mM Sucrose
    1 mM MgCl$_2$
    50 mM KCl
    0.1 mM PMSF
    Dissolve well after each addition in ddH$_2$O. Sterile-filtering is optional.

Upper-Tank Buffer (pH 8.9)

    37.6 mM Tris (4.56 g)
    40 mM Glycine (3 g)
    Add ddH$_2$O to a final volume of 1 L.

Working Solution

    50 volumes of Reagent A
    1 volume of Reagent B
    Store at room temperature for up to 1 week.

References

1. **Deutcher, M.P., Ed.,** *Guide to Protein Purification. Methods in Enzymology,* Vol. 182, Academic Press, New York, 1990.
2. **Dean, P.D.G., Johnson, W.S., and Middle, F.A., Eds.,** *Affinity Chromatography: A Practical Approach,* IRL Press, Oxford, 1985.
3. **Anonymous,** *Affinity Chromatography: Principles and Methods Handbook,* Pharmacia Fine Chemicals, Ljungforetagen AB, Orebro, Sweden, 1983.
4. **Scopes, R.K.,** *Protein Purification: Principles and Practice,* 2nd ed., Springer-Verlag, New York, 1987, chaps. 5 and 7.
5. **Davis, L.G., Dibner, M.D., and Battey, J.F.,** *Basic Methods in Molecular Biology,* Elsevier, New York, 1986, chap. 19.
6. **Wu, L.-L., Mitchell, J.P., Cohn, N.S., and Kaufman, P.B.,** Gibberellin (GA_3) enhances cell wall invertase activity and mRNA levels in elongating dwarf pea *(Pisum sativum)* shoots, *Intl. J. Plant Sci.,* 154, 280, 1993.
7. **Smith, B.J.,** SDS polyacrylamide gel electrophoresis of proteins, in *Methods in Molecular Biology, Proteins,* Vol. 1, Walker, J.M, Ed., Humana Press, Clifton, NJ, 1984, chap. 6.
8. **Harrington, M.G.,** Elution of protein from gels, in *Guide to Protein Purification. Methods in Enzymology,* Vol. 182, Deutscher, M.P., Ed., Academic Press, San Diego, CA, 1990, chap. 37.
9. **Merril, C.R.,** Gel-staining techniques, in *Guide to Protein Purification. Methods in Enzymology,* Vol. 182, Deutscher, M.P., Ed., Academic Press, San Diego, CA, 1990, chap. 36.

Chapter 4

Preparation of Nucleic Acid Probes

William Wu and Leland J. Cseke

Contents

I. Radioactive Labeling Methods 68
   A. Nick-Translation Labeling of dsDNA..68
   B. Random-Primer Labeling of dsDNA..71
   C. Labeling of ssDNA 72
   D. 3′-End Labeling to Fill Recessed 3′ Ends of dsDNA 72
   E. 3′-End Labeling of ssDNA with Terminal Transferase................................74
   F. 5′-End Labeling Using Bacteriophage T4 Polynucleotide Kinase 74
   G. Labeling of RNA by *In Vitro* Transcription 75
   Reagents Needed for Radioactive Labeling 77
II. Nonradioactive Labeling Methods..79
   A. Random-Primer Labeling of dsDNA 80
   B. Nick-Translation Labeling of dsDNA with Digoxigenin-11-dUTP............81
   C. 3′-End Labeling with DIG-11-ddUTP..81
   D. 3′ Tailing of DNA with DIG-11-dUTP/dATP 81
   E. 5′-End Labeling with DIG-NHS-Ester 82
   F. RNA Labeling 82
   Reagents Need for Nonradioactive Labeling...82
References 84

Molecular biology studies usually involve molecular cloning, characterization, and analysis of gene expression. These procedures rely heavily on nucleic acid hybridization, such as Southern blots, Northern blots, and dot blots. One of the most important aspects of these protocols is the use of either radioactively or nonradioactively labeled nucleic acids (deoxyribonucleic acid [DNA] or ribonucleic acid [RNA]) as probes that specifically hybridize with their complementary DNA or RNA strands. The quality of the probe plays an essential role in detecting specific DNA or RNA sequences of interest. Therefore, the preparation of a probe with high specific activity is critical in nucleic acid hybridization.[1] The present chapter describes in detail reliable methods for the labeling of DNA and RNA.[2-5] These methods are well established and have been routinely used in our laboratory.

I. Radioactive Labeling Methods

A. Nick-Translation Labeling of dsDNA

The principle of nick translation is that one strand of a double-strand DNA (dsDNA) molecule is nicked with *DNase I*, generating free 3'-hydroxyl ends within the unlabeled DNA. *E. coli* DNA polymerase I, by virtue of its 5' to 3' exonucleolytic activity, removes nucleotides from the 5' side of the nick and simultaneously adds new nucleotides to the 3'-hydroxyl terminus of the nick. During the incorporation of new nucleotides, one of four deoxyribonucleotides is radioactively labeled (e.g., [α-^{32}P]dATP or [α-^{32}P]dCTP, commercially available) and is incorporated into the new strand by a base complementary to the template. In this way, a high specific activity (10^8 cpm/μg) of labeled DNA can be obtained using [α-^{32}P]dATP or [α-^{32}P]dCTP. The protocol given below works well in our hands.

1. Set up a reaction on ice as follows:
   - 10X Nick-translation buffer (5 μl)
   - DNA sample (0.4 to 1 μg in <2 μl)
   - Mixture of three unlabeled dNTPs (10 μl)
   - [α-^{32}P]dATP or [α-^{32}P]dCTP (>3000 Ci/mmol; 4 to 7 μl)
   - Diluted *DNase I* (10 ng/ml) (5 μl)
   - *E. coli* DNA polymerase I (2.5 to 5 units)
   - Add distilled, deionized water (ddH$_2$O) to a final volume of 50 μl.

Notes: *(1) Three unlabeled dNTPs can be made by mixing equal volumes of the three nucleotides (each is 1.5 mM stock solution from a commercial source) and minus [α-^{32}P]dATP or [α-^{32}P]dCTP selected as a label. (2) The use of [α-^{32}P]dATP is better for the analysis of sequences rich in A/T nucleotides, while the use of [α-^{32}P]dCTP is better for the analysis of sequences rich in G/C. (3) The amount of DNase I used in the reaction mixture should be optimized. Normally, a 10^5-fold dilution of a stock solution of pancreatic DNase I (1 mg/ml) in ice-cold 1X nick-translation buffer containing 50% of glycerol is ready to use. (4) The amount of sample DNA can be as low as 25 ng in a reaction of 5 to 20 μl.*

2. Incubate the reaction for 1 h at 15°C.

Notes: *(1) The temperature should not be higher than 18°C, which can generate "snapback" DNA by E. coli DNA polymerase I. "Snapback" DNA can lower the efficiency of hybridization. (2) Longer incubation periods (1 to 2 h) are acceptable. However, too-long an incubation may reduce the overall length of the labeled DNA.*

3. Stop the reaction by adding 2 μl of 0.5 *M* EDTA (pH 8.0) stop solution. Place the tube on ice.
4. Determine the percentage of [α-^{32}P]dCTP incorporated into the DNA.
   a. DE-81 filter-binding assay:
      i. Dilute 1 μl of the labeled mixture in 99 μl (1:100) of 0.5 *M* EDTA solution. Spot 3 μl of the diluted sample, in duplicate, on Whatman DE-81 circular filters (2.3 cm diameter). Dry the filters under a heat lamp.
      ii. Wash one of the two filters in 50 ml of 0.5 *M* sodium phosphate buffer (pH 6.8) for 5 min to remove unincorporated cpm. Repeat washing once. The other filter will be used directly for total cpm in the sample.
      iii. Add an appropriate volume of scintillation fluid (about 5 ml) to each tube containing one of the filters. Count the cpm in a scintillation counter according to the counter's instructions.

Preparation of Nucleic Acid Probes

b. Trichloroacetic acid (TCA) precipitation:
  i. Dilute 1 µl of the labeled reaction in 99 µl (1:100) of 0.5 M EDTA solution. Spot 3 µl of the diluted sample on a glass-fiber filter or a nitrocellulose filter for determinating the total cpm in the sample. Air dry the filter.
  ii. Add 3 µl of the same diluted sample into a tube containing 100 µl of 0.1 mg/ml carrier DNA or acetylated BSA (bovine serum albumin) and 20 mM EDTA. Mix well.
  iii. Add 1.3 ml of ice-cold 10% TCA and 1% sodium pyrophosphate to the mixture. Mix well and incubate the tube on ice for 20 to 25 min to precipitate the DNA.
  iv. Filter the precipitated DNA on a glass-fiber filter or a nitrocellulose filter under vacuum. Wash the filter with 5 ml of ice-cold 10% TCA four times under vacuum. Rinse the filter with 5 ml acetone (for glass-fiber filters only) or 5 ml of 95% ethanol. Air dry the filter.
  v. Transfer the filters to two cpm-counting tubes and add 5 to 10 ml of scintillation fluid to each tube. Count the total cpm and incorporated cpm in a scintillation counter according to the instructions.

5. Calculate the specific activity of the probe.
   a. Calculate theoretical yield as follows:

$$\text{Theoretical yield (ng)} = \frac{\text{dNTP added} \times 4 \times 330 \text{ ng / nmol}}{\text{Specific activity of the labeled dNTP } (\mu\text{Ci / nmol})}$$

   b. Calculate the percentage of incorporation:

$$\text{Percent incorporation} = \frac{\text{cpm incorporated}}{\text{total cpm}} \times 100$$

   c. Calculate the amount of DNA synthesized:

$$\text{DNA synthesized (ng)} = \text{percent incorporation} \times 0.01 \times \text{theoretical yield}$$

   d. Calculate the specific activity of the prepared probe:

$$\text{Specific activity (cpm / }\mu\text{g)} = \frac{\text{total cpm incorporated}}{(\text{DNA synthesized} + \text{input DNA}) \text{ (ng)} \times 0.001 \text{ }\mu\text{g / ng}}$$

Notes: The total cpm incorporated is equal to the cpm incorporated × 33.3 × 50. The factor 33.3 comes from using 3 µl of 1:100 dilution for the filter binding or TCA precipitation assay. The factor 50 is derived from using 1 µl of the 50 µl reaction for 1:100 dilution.

Example

Given that 25 ng DNA is to be labeled and that 50 µCi [α-^{32}P]dCTP (3000 Ci/mmol) is used in 50 µl of a standard reaction, and assuming that 4.92×10^4 cpm is precipitated by TCA and that 5.28×10^4 cpm is the total cpm in the sample, the calculations are made as follows:

$$\text{Theoretical yield} = \frac{50 \text{ }\mu\text{Ci} \times 4 \times 330 \text{ ng / nmol}}{3000 \text{ }\mu\text{Ci / nmol}}$$

$$= 22 \text{ ng}$$

$$\text{Percent incorporation} = \frac{4.92 \times 10^4}{5.28 \times 10^4} \times 100$$

$$= 93\%$$

$$\text{DNA synthesized} = 93 \times 0.01 \times 22$$

$$= 20.5 \text{ ng}$$

$$\text{Specific activity} = \frac{4.92 \times 10^4 \times 33.3 \times 50}{(20.5 \text{ ng} + 25 \text{ ng}) \times 0.001 \text{ µg / ng}}$$

$$= 1.8 \times 10^9 \text{ cpm / µg DNA}$$

Note: *The specific activity of 2×10^8 to 2×10^9 cpm/µg is considered to be a high specific activity that should be used in Southern, Northern, or equivalent hybridizations.*

6. Purify the probe by removing unincorporated isotope. This is an optional step, but we strongly recommend that it be carried out because the unpurified probe may produce a background (unexpected black spots) on the hybridization filter. Chromatography on Sephadex G-50 spin columns Sephadex G-50 or Bio-Gel P-60 spin column is very effective for separating labeled DNA from unincorporated radioactive precursor such as [α-^{32}P]dCTP or [α-^{32}P]dATP and oligomers that are retained in the column. This is very useful when an optimal signal-to-noise ratio probe, 150 to 1500 bases in length, is generated for optimal hybridization.
   a. Resuspend 2 to 4 g Sephadex G-50 (Figure 4.1) or Bio-Gel P-60 in 50 to 100 ml of TEN buffer (see "Reagents Needed for Radioactive Labeling" after **Section I.G**) and allow to equilibrate for at least 1 h. Store at 4°C until use.
   b. Insert a small amount of sterile glass wool in the bottom of a 1-ml disposable syringe using the barrel of the syringe to tamp the glass wool in place.
   c. Fill the syringe completely with the Sephadex G-50 or Bio-Gel P-60 suspension.
   d. Insert the syringe containing the suspension into a 15-ml disposable plastic tube and place the tube in a swinging-bucket rotor in a bench-top centrifuge. Centrifuge at 1600 g for 4 min at room temperature.

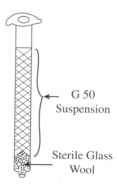

FIGURE 4.1
Preparation of a G-50 column for purification of labeled probe from unincorporated, radioactive nucleotides. (From Kaufman, P.B., Wu, W., Kim, D., and Cseke, L.J., *Handbook of Molecular and Cellular Methods in Biology and Medicine*, 1st ed., CRC Press, Boca Raton, FL, 1995.)

e. Repeat adding the suspended resin to the syringe and centrifuging at 1600 g for 4 min until the packaged volume reaches 0.9 ml in the syringe and remains unchanged after centrifugation (Figure 4.1).
f. Add 100 μl of 1X TEN buffer to the top of the column and centrifuge as above. Repeat this step two to three times.

Note: *This procedure can also be done without the centrifugation steps. However, a longer period of time is required to prepare the column using only gravity. Still, the use of gravity only may be somewhat less tedious. We leave this up to the user.*

g. Transfer the spin column to a fresh 15-ml disposable tube. Carefully add the labeled DNA sample onto the top of the resin dropwise using a pipette so as not to disturb the column bed.

Notes: *If the labeled DNA sample is less than 0.1 ml, dilute it to 0.1 ml in 1X TEN buffer.*

h. Centrifuge at 1600 g for 4 min at room temperature. Remove and discard the column containing unincorporated radioactive label in a radioactive waste container. Carefully transfer the effluent (about 0.1 ml) from the bottom of the 15-ml tube to a fresh microcentrifuge tube. Cap it and store at –20°C until use.

B. Random-Primer Labeling of dsDNA

Random hexanucleotides are a population of oligonucleotide hexamers synthesized by an automatic DNA synthesizer that contains all four bases in every position. Such a mixture of random primers is available from commercial companies such as Promega Corp., Pharmacia, and Boehringer Mannhein Chemicals. The primers are used to prime DNA synthesis *in vitro* from any denatured, closed circular, or linear dsDNA as template using the Klenow fragment of *E. coli* DNA polymerase I. Because this enzyme lacks 5′ to 3′ exonuclease activity, the DNA product is synthesized exclusively by primer extension instead of by nick translation. During the synthesis, one of the four dNTPs is radioactively labeled and incorporated into the new DNA strand being radioactively labeled. By primer extension, it is possible to generate a probe with extremely high specific activity (10^8 to 10^9 cpm/μg), as more than 70% of the labeled dNTP can be incorporated into the new DNA strand. In addition, the DNA template can be as low as 25 ng, as the template remains intact during the reaction instead of being nicked at some points, as seen in nick translation, which usually produces a shorter probe as compared with random-primer labeling.

1. Add 1 μl of DNA template (0.5 to 1.0 μg/μl in ddH$_2$O or TE buffer) to a sterile microcentrifuge tube. Add an appropriate volume (19 to 38 μl ddH$_2$O or TE buffer) to the tube to dilute the DNA template to 25 to 50 ng/μl.

Note: *If the concentration of DNA template is about 50 ng/μl, the dilution of the DNA is not necessary.*

2. Cap the tube and denature the dsDNA into ssDNA (single-strand DNA) by placing the tube in a water bath and boiling for 3 to 6 min. Quickly chill the tube in ice water for 2 to 4 min and briefly spin down.
3. Set up the reaction by adding the following components, in the order shown below, into a fresh microcentrifuge tube on ice.
   - 5X Labeling buffer (10 μl)
   - Denatured DNA template (25 to 50 ng) (1 μl)
   - Three unlabeled dNTP mixtures (500 μ*M* each) (2 μl)
   - Acetylated BSA (10 mg/ml) (optional) (2 μl)

- [α-^{32}P]dCTP or [α-^{32}P]dATP (>3000 Ci/mmol) (2.5 to 4 µl)
- Klenow enzyme (5 units)
- Add ddH$_2$O to a final volume of 50 µl.

Notes: *(1) The standard amount of DNA template used in the reaction should be 25 to 50 ng, since the specific activity of the probe depends on the amount of the template. The lower the amount of DNA used as template, the higher is the resulting specific activity of the probe. (2) Three unlabeled dNTPs can be made by diluting 1 µl of each of the dNTPs (1.5 mM stock solutions) to yield 3 µl of dNTPs mixture at 500 µM for each dNTP. (3) [α-^{32}P]dCTP or [α-^{32}P]dATP (3000 Ci/mmol) should not be >5 µl used in the reaction; otherwise, the background on the hybridization member may be high.*

4. Mix well and incubate the reaction at room temperature for 1 to 1.5 h.
5. Stop the reaction mixture by adding 2 µl of 0.5 M EDTA stop solution and place on ice. Store at −20°C until use.

Notes: *(1) The labeled probe can be directly used for hybridization after being boiled for 5 to 10 min and chilled on ice for 4 min. However, the unpurified probe usually generates a high background (black spots) on the filter. We recommend that the probe be purified from unincorporated [α-^{32}P]dCTP or [α-^{32}P]dATP using a Sephadex G-50 or a Bio-Gel P-60 column. This is described under the protocol for nick translations (**Section I.A**). (2) The specific activity of the probe should be determined by the DE-81 filter-binding assay or the TCA precipitation assay (see the nick-translation protocol in **Section I.A**). The hybridization signal depends on the specific activity of the probe. The higher the specific activity of the probe, the stronger is the signal of the hybridized bands or spots.*

C. Labeling of ssDNA

As compared with labeling of dsDNA, a ssDNA probe only has one complementary strand of a given DNA sequence. The advantage here is the elimination of the potential formation of nonproductive hybrids of the reannealed probe. During the detection of cross-hybridizing sequences in a distantly related species, hybridization between the complementary probes is much more probable and more stable than the hybridization between the probe and its target sequence of interest. This drawback is overcome by the use of ssDNA-probe hybridization. The ssDNA can be prepared from bacteriophage M13 or phagemid or from first-strand cDNA reversely transcribed from mRNA by reverse transcriptase.

Perform the labeling procedure as for random-primer labeling of dsDNA, except that the DNA template should be 0.5 to 1.0 µg in a 25-µl standard reaction mixture.

D. 3'-End Labeling to Fill Recessed 3' Ends of dsDNA

1. Set up a restriction enzyme digestion in order to create 3' ends. In a microcentrifuge tube on ice, add the following components in the order shown:
   - DNA template (1 µg)
   - Appropriate restriction enzyme 10X buffer (2.5 µl)
   - Acetylated BSA (1 mg/ml) (optional) (2.5 µl)
   - Appropriate restriction enzyme (5 units)
   - Add ddH$_2$O to a final volume of 25 µl.

Preparation of Nucleic Acid Probes

Notes: (1) The DNA template can be cDNA, the genomic DNA insert, or plasmid DNA that is dissolved in ddH$_2$O or TE buffer. (2) The restriction enzyme selected should generate recessed 3′ ends in order to be filled with radioactive dNTP. The common restriction enzymes used are EcoR I, Hind III, and BamH I.

2. Incubate the reaction for 1 to 3 h at the appropriate temperature, depending on the particular restriction enzyme.

Notes: It is not necessary to inactivate and remove the enzyme. The digested reaction can be directly used for the labeling reaction.

3. Add 2 to 4 µl of appropriate [α-^{32}P]dNTP (400 to 3000 Ci/mmol) to the digested reaction.

Notes: [α-^{32}P]dNTP used in the reaction depends on the sequence of the protruding 5′ end of the DNA. For EcoR I-digested DNA, the 3′ end can be labeled with [α-^{32}P]dATP. However, for BamH I-digested DNA, the 3′ end should be labeled with [α-^{32}P]dGTP.

dsDNA EcoR I   ...pCpTpTpApAp 5′ [α-^{32}P]dATP   ...pCpTpTpApAp 5′
                     ...pG$_{OH}$ 3′   ...pGp *Ap*A$_{OH}$ 3′

dsDNA BamH I   ...CpCpTpApGp 5′ [α-^{32}P]dGTP   ...CpCpTpApGp 5′
                      ...G$_{OH}$ 3′   ...Gp*G$_{OH}$ 3′

4. Add 2 to 5 units of the Klenow fragment of *E. coli* DNA polymerase I to the reaction and incubate the reaction at room temperature for 15 min.

Note: If the dsDNA to be labeled is predigested and purified, a labeling reaction can be set up as follows (on ice):

- Digested DNA (1 µg)
- Klenow 5X buffer (10 µl)
- Acetylated BSA (1 mg/ml) (optional) (1 µl)
- Three unlabeled dNTPs (2 m*M*) (1 µl)
- Appropriate [α-32P]dNTP (400 to 3000 Ci/mmol) (2µl)
- Klenow DNA polymerase I (2 to 5 units)
- Add ddH$_2$O to a final volume of 50 µl.
- Incubate the reaction at room temperature for 15 min.

5. Heat the reaction at 70°C for 5 min to stop the reaction. Place on ice until use.
6. Separate the labeled DNA from unincorporated dNTP as described for nick-translation labeling.
7. Measure the specific activity of the probe as described in nick-translation labeling (optional).
8. Add one volume of TE-saturated phenol/chloroform to the reaction mixture in step 5. Mix well by vortexing for 1 min and centrifuging at 11,000 g for 5 min at room temperature.
9. Carefully transfer the upper, aqueous layer to a fresh tube and add 0.1 volume of 2 *M* NaCl solution. Mix well. Add 2.5 volumes of chilled 100% ethanol. Allow to precipitate at –20°C for 1 h.
10. Centrifuge at 12,000 g for 5 min, discard the supernatant, and briefly rinse the pellet with 1 ml of 70% ethanol.
11. Dry the pellet under vacuum for 15 min and dissolve the pellet (labeled DNA) in 50 µl of TE buffer. Store at –20°C until use.

E. 3'-End Labeling of ssDNA with Terminal Transferase

1. 3' Tailing of ssDNA primers
   a. Set up a reaction as follows:
      - 5X Terminal transferase buffer (5 µl)
      - ssDNA primers (2 pmol)
      - [α-^{32}P]dATP (800 Ci/mmol) (1.6 µl)
      - Terminal transferase (10 to 20 units/µl) (1 µl)
      - Add ddH$_2$O to a final volume of 25 µl.
   b. Incubate at 37°C for 1 h and stop the reaction by heating at 70°C for 10 min to inactivate the enzyme.

Note: Multiple [α-^{32}P]dATPs were added to the 3' end of the DNA.

   c. Calculate the percentage of incorporation and the specific activity of the probe. The procedures for the TCA precipitation assay and the DE-81 filter-binding assay are described in the section on nick-translation labeling (**Section I.A**).

2. 3'-End labeling of ssDNA with a single [α-^{32}P]cordycepin-5'-triphosphate lacking 3'-OH.
   a. Set up a reaction as follows:
      - 5X Terminal transferase buffer (10 µl)
      - ssDNA primers (10 pmol)
      - [α-^{32}P]Cordycepin-5'-phosphate (3000 Ci/mmol) (7.5 µl)
      - Terminal transferase (10 to 20 units/µl) (2 µl)
      - Add ddH$_2$O to a final volume of 50 µl.
   b. Incubate the reaction at 37°C for 1 h.
   c. Stop the reaction at 70°C for 10 min. The labeled reaction can be directly used in hybridization or purified as described in 3'-end labeling in order to fill the recessed 3' ends of dsDNA.
   d. Calculate the percentage of incorporation and the specific activity of the probe (see 3' tailing of ssDNA, **Section I.A**).

F. 5'-End Labeling Using Bacteriophage T4 Polynucleotide Kinase

1. Set up a dephosphorylation reaction to remove phosphate group from both 5' ends of the linear dsDNA.
   a. Set up a reaction as follows:
      - 10X Calf intestinal alkaline phosphatase (CIAP) buffer (5 µl)
      - 5' Ends DNA (2 to 10 pmol)
      - CIAP diluted in 1X CIAP buffer (0.5 units)
      - Add ddH$_2$O to a final volume of 50 µl.

Notes: *(1) For protruding 5'-termini dephosphorylation, incubate at 37°C for 0.5 h and then add 0.5 units of alkaline phosphatase and incubate for 30 min at 37°C. (2) For recessed 5'-termini or blunt-ends dephosphorylation, incubate at 37°C for 15 min and at 56°C for another 15 min. Add 0.5 units of alkaline phosphatase and incubate at 37°C and 56°C for 15 min, respectively.*

   b. Stop and extract the reaction by adding one volume of TE-saturated phenol/chloroform. Mix well by vortexing for 1 min.
   c. Centrifuge at 11,000 g for 5 min and carefully transfer the top, aqueous phase to a fresh tube.
   d. Add one volume of chloroform:isoamyl alcohol (24:1) to the supernatant. Mix well.

e. Centrifuge as in step 1.c. Carefully transfer the supernatant to a fresh tube.
f. Add 0.1 volume of 2 M NaCl and two volumes of chilled 100% ethanol to the supernatant. Allow to precipitate at −70°C for 30 min.
g. Centrifuge at 12,000 g for 5 min. Discard the supernatant and briefly rinse the pellet with 2 ml of 70% ethanol.
h. Dry the pellet under vacuum for 15 min and dissolve the pellet in 34 µl of forward-exchange 1X buffer.

2. Carry out the kinase reaction as follows:
   a. Add the following to the reaction at step 1.h.
      - Reaction above (34 µl)
      - [α-^{32}P]dATP (3000 Ci/mmol) (15 µl)
      - T4 Polynucleotide kinase (8 to 10 units/µl) (1 µl)
      - Total volume of 50 µl.
   b. Incubate at 37°C for 10 min and stop the reaction by adding 2 µl of 0.5 M EDTA.
   c. Extract and precipitate the labeled DNA as in steps 1.b to 1.h.

Notes: *Ammonium ions are strong inhibitors of the bacteriophage T4 polynucleotide kinase. Do not use any ammonium acetate buffer prior to running the kinase reaction.*

   d. Calculate the percentage of incorporation and specific activity as described for nick-translation labeling (**Section I.A**).

G. Labeling of RNA by *In Vitro* Transcription

Recently, a number of plasmid vectors have been developed for subcloning of cDNA or genomic DNA inserts of interest. These vectors contain polycloning sites downstream from powerful bacteriophage promoters SP6, T7, or T3 in the vector. The cDNA or genomic DNA insert of interest can be cloned at the polycloning site between promoters SP6 and T7 or T3, forming a recombinant plasmid. The cDNA or genomic DNA inserted can be transcribed *in vitro* into single-strand sense RNA or antisense RNA from a linear plasmid DNA with promoter SP6, T7, or T3. During the process of *in vitro* transcription, one of the dNTPs is radioactively labeled and can be incorporated into the RNA strand, which is the labeled RNA. The labeled RNA probe usually has a high specific activity and is much "hotter" than a ssDNA probe. As compared with DNA labeling, the yield of the RNA probe is very high because the template can be repeatedly transcribed. RNA probes can be easily purified from a DNA template merely by the use of *RNase*-free *DNase I* treatment. The greatest advantage of an RNA probe over a DNA probe is that the RNA probe can produce much stronger signals in a variety of different hybridization reactions.

1. Prepare the linear DNA template for *in vitro* transcription. The plasmid is linearized by an appropriate restriction enzyme in order to produce "run-off" transcripts. To make RNA transcripts from the DNA insert, the recombinant plasmid should be digested by an appropriate restriction enzyme that cuts at one site that is very close to one end of the insert.
   a. On ice, set up a plasmid linearization reaction as follows:
      - Recombinant plasmid DNA (µg/µl) (5 µg)
      - Appropriate restriction enzyme 10X buffer (5 µl)
      - Acetylated BSA (1 mg/ml) (optional) (5 µl)
      - Terminal transferase (10 to 20 units/µl) (20 units)
      - Add ddH$_2$O to a final volume of 50 µl.
   b. Incubate at the appropriate temperature for 2 to 3 h.
   c. Extract by adding one volume of TE-saturated phenol/chloroform and mix well by vortexing. Centrifuge at 11,000 g for 5 min at room temperature.
   d. Carefully transfer the top, aqueous phase to a fresh tube and add one volume of chloroform:isoamyl alcohol (24:1) to the supernatant. Mix well and centrifuge as in step c.

e. Transfer the top, aqueous phase to a fresh tube. Add 0.1 volume of 2 M NaCl solution or 0.5 volume of 7.5 M ammonium acetate, and 2 volumes of chilled 100% ethanol to the supernatant. Allow precipitation to occur at −70°C for 30 min or at −20°C for 2 h.

f. Centrifuge at 12,000 g for 5 min, decant the supernatant, and briefly rinse the pellet with 1 ml of 70% ethanol. Dry the pellet for 15 min under vacuum and dissolve the linearized plasmid in 15 µl ddH$_2$O.

g. Take 2 µl of the sample to measure the concentration of the DNA using UV-absorption spectroscopy at 260 and 280 nm. Store the sample at −20°C until use.

2. Blunt the 3′ overhang end using the 3′ to 5′ exonuclease activity of Klenow DNA polymerase. Although this is optional, we recommend that the 3′ protruding end be converted into a blunt end, because some of the RNA sequence is complementary to that of the vector DNA. Enzymes such as *Kpn I*, *Sac I*, *Pst I*, *bgl I*, *Sac II*, *Pvu I*, *Sfi*, and *Sph I* should not be used to linearize plasmid DNA for *in vitro* transcription.

   a. Set up, on ice, an *in vitro* transcription reaction as follows:
      - 5X Transcription buffer (8 µl)
      - 100 mM Dithiothreitol (DTT) (4 µl)
      - RNasin ribonuclease inhibitor (40 units) (Stratagene)
      - Linearized template DNA (0.2 to 1.0 g/µl) (2 µl)
      - Add ddH$_2$O to a final volume of 15.2 µl.
   b. Add Klenow DNA polymerase (5 units/µg DNA) to the reaction and incubate at 22°C for 15 min.
   c. To the reaction, add the following:
      - Mixture of ATP, GTP, CTP, or UTP (2.5 mM each) (8 µl) minus the NTP used for label
      - 120 µM UTP or CTP (4.8 µl) — same as the NTP used for label
      - [α-^{32}P]UTP or [α-^{32}P]CTP (50 µCi at 10 mCi/ml) (10 µl)
      - SP6 or T7 or T3 RNA polymerase (15 to 20 units/µl) (2 µl)
   d. Incubate the reaction at 37 to 40°C for 1 h.

3. Alternatively to step 2, carry out large-scale *in vitro* transcription.
   a. In a microcentrifuge tube on ice, add the following in the order listed below:
      - 5X Transcription buffer (20 µl)
      - 100 mM DTT (8 µl)
      - Ribonuclease inhibitor (100 units)
      - Mixture of ATP, GTP, CTP, or UTP (2.5 mM each) (20 µl)
      - Linearized DNA template (1 to 2.5 µg/µl) (2 µl)
      - [α-^{32}P]UTP or [α-^{32}P]CTP (50 µCi at 10 mCi/ml) (25 µl)
      - SP6, or T7 or T3 RNA polymerase (15 to 20 units/µl) (5 µl)
      - Add ddH$_2$O to a final volume of 100 µl.
   b. Incubate the reaction at 37 to 40°C for 1 to 2 h.

4. Remove the DNA template using *DNase I*.
   a. Add *RNase*-free *DNase I* to a concentration of 1 unit/µg DNA template.
   b. Incubate for 15 min at 37°C.

5. Purify the RNA probe.
   a. Extract the enzyme by adding one volume of TE-saturated phenol/chloroform. Mix well by vortexing for 1 min and centrifuging at 11,000 g for 5 min at room temperature.
   b. Transfer the top, aqueous phase to a fresh tube and add one volume of chloroform:isoamyl alcohol (24:1). Mix well by vortexing and centrifuge at 11,000 g for 5 min.
   c. Carefully transfer the upper, aqueous phase into a fresh tube, and then add 0.5 volume of 7.5 M ammonium acetate solution and 2.5 volumes of chilled 100% ethanol. Allow to precipitate at −70°C for 30 min or at −20°C for 2 h.
   d. Centrifuge at 12,000 g for 5 min. Carefully discard the supernatant and briefly rinse the pellet with 1 ml of 70% ethanol and dry the pellet under vacuum for 15 min.
   e. Dissolve the RNA probe in 20 to 50 µl of TE buffer and store at −20°C until use.

Note: *The quantity and quality of the labeled RNA can be checked by denaturing agarose gel electrophoresis using 4 to 5 µl of the sample (see Northern blotting protocol, **Chapter 6**).*

Preparation of Nucleic Acid Probes

6. The percentage of incorporation and the specific activity of the RNA probe can be determined right after the *in vitro* transcription reaction.
   a. Estimate the cpm used in the transcription reaction. For example, if 50 µCi of NTP was used, the cpm is $50 \times 2.2 \times 10^6$ cpm/µCi = 110×10^6 cpm in 40 µl reaction, or 2.8×10^6 cpm/µl.
   b. Carry out a TCA precipitation assay using 1:10 dilution in ddH$_2$O as described for nick-translation labeling of DNA (**Section I.A**).
   c. Calculate the percentage of incorporation.

   Percent incorporation = (TCA-precipitated cpm/total cpm) × 100%

   d. Calculate the specific activity of the probe. If 1 µl of a 1:10 dilution was used for TCA precipitation, then $10 \times$ cpm precipitated = cpm/µl incorporated. In a 40-µl reaction, $40 \times$ cpm/µl is the total cpm incorporated. If 50 µCi of labeled UTP at 400 µCi/nmol was used, then 50/400 = 0.125 nmol of UTP was added into the reaction. If there were 100% incorporation and UTP represents 25% of the nucleotides in the RNA probe, then $4 \times 0.125 = 0.5$ nmol of nucleotides were incorporated, and 0.5×330 ng/nmol = 165 ng of RNA were synthesized. Then, the total ng RNA probe = % incorporation × 165 ng. For example, if 1:10 dilution of the labeled RNA sample has 2.2×10^5 cpm, then the total cpm incorporated was $10 \times 2.2 \times 10^5$ cpm × 40 µl (total reaction) = 88×10^6 cpm.

Percent incorporation	= $(88 \times 10^6$ cpm$)/[110 \times 10^6$ cpm $(50$ µCi$)]$ = 80%
Total RNA synthesized	= 165 ng × 0.80 = 132 ng RNA
Specific activity of the probe	= $(88 \times 10^6$ cpm$)/(0.132$ µg$)$
	= 6.7×10^8 cpm/µg RNA

Reagents Needed for Radioactive Labeling

[α-^{32}P]dCTP or [α-^{32}P]dATP
Commercially available source
Specific activity of 3000 Ci/mmol

[α-^{32}P]Cordycepin-5'-Phosphate Analog
3000 Ci/mmol

7.5 M Ammonium Acetate Solution
Dissolve 57.8 g ammonium acetate in 50 ml ddH$_2$O and adjust volume to 100 ml.

10X CIAP Buffer
0.5 *M* Tris-HCl, pH 9.0
10 m*M* MgCl$_2$
1 m*M* ZnCl$_2$
10 m*M* Spermidine

Chloroform:isoamyl alcohol (24:1)
Mix and store at room temperature.

DNA Polymerase I
Commercial suppliers

dNTPs: Unlabeled Solutions
2 m*M* of each dNTP

EDTA Solution
    0.5 M EDTA in ddH$_2$O, pH 8.0

10X Forward Exchange Buffer
    0.5 M Tris-HCl, pH 7.5
    0.1 M MgCl$_2$
    50 mM DTT
    1 mM Spermidine

Klenow 5X Buffer
    0.25 M Tris-HCl, pH 7.2
    50 mM MgSO$_4$
    0.5 mM DTT

Klenow DNA Polymerase I
    From commercial source

5X Labeling Buffer
    250 mM Tris-HCl, pH 8.0 (from stock solution)
    25 mM MgCl$_2$
    10 mM DTT
    1 mM HEPES, pH 6.6 (from stock solution)
    26 A$_{260}$ units/ml random hexadeoxyribonucleotides

2 M NaCl Solution
    Dissolve 58.4 g of NaCl in 400 ml ddH$_2$O. Adjust final volume to 500 ml with ddH$_2$O, autoclave, and store at room temperature.

10X Nick-Translation Buffer
    500 mM Tris-HCl, pH 7.5
    100 mM MgSO$_4$
    1 mM DTT
    500 µg/ml BSA (Fraction V, Sigma) (optional)
    Aliquot the stock solution and store at –20°C until use.

NTPs Stock Solutions
    10 mM ATP in ddH$_2$O, pH 7.0
    10 mM GTP in ddH$_2$O, pH 7.0
    10 mM UTP in ddH$_2$O, pH 7.0
    10 mM CTP in ddH$_2$O, pH 7.0

Radioactive dNTP Solution for DNA
    [α-^{32}P]dATP or [α-^{32}P]dCTP (3000 Ci/mmol)

Radioactive NTP Solution for RNA
    [α-^{32}P]UTP or [α-^{32}P]CTP (400 Ci/nmol)

Pancreatic DNase I Solution
    DNase I (1 mg/ml) in a solution containing 0.15 M NaCl and 50% glycerol
    Aliquot and store at –20°C.

RNase-Free DNase I

Preparation of Nucleic Acid Probes

Sephadex G-50 or Bio-Gel P-60 Powder
    Commercial suppliers

ssDNA Primers
    Prepared from bacteriophage M13 or phagemid, or ssDNA isolated from DNA.

TE Buffer
    10 mM Tris-HCl, pH 8.0
    1 mM EDTA

TE-Saturated Phenol/Chloroform
    Thaw phenol crystals at 65°C and mix equal parts of phenol and TE buffer. Mix well and allow the phases to separate at room temperature for 30 min. Take one part of the lower, phenol phase and mix with one part of chloroform:isoamyl alcohol (24:1). Mix well and allow the phases to separate. Store at 4°C until use.

1X TEN Buffer
    10 mM Tris-HCl, pH 8.0
    1 mM EDTA, pH 8.0
    100 mM NaCl

5X Terminal Transferase Buffer
    0.5 M Cacodylate, pH 6.8
    1 mM CoCl$_2$
    0.5 mM DTT
    500 µg/ml BSA

Terminal Transferase
    Available from commercial sources

5X Transcription Buffer
    0.2 M Tris-HCl, pH 7.5
    30 mM MgCl$_2$
    10 mM Spermidine
    50 mM NaCl

II. Nonradioactive Labeling Methods

Recently, nonradioactive labeling techniques have been developed for DNA and RNA probe preparations. These methods have major advantages over traditional radioactive labeling in the following aspects.

    Safer: Radioactive labels (dNTPs) required for radioactive labeling are totally omitted in nonradioactive labeling methods. It is well known that radioactive dNTPs are dangerous to humans.
    Simpler and less expensive: Traditional radioactive labeling is usually complicated, and elaborate steps must be taken to avoid contamination. Since the half-life of the isotope is short, for instance, [α-^{32}P]dATP or [α-^{32}P]dCTP only has a 14-day half-life and is somewhat expensive, the labeled DNA or RNA cannot be kept for a long period of time. In contrast, nonradioactive labeling has simpler

procedures and is relatively inexpensive. The major advantage is that the labeled RNA or DNA can be kept at −20°C for months and can be reused up to five to six times without any significant decrease in the hybridization signals.

Based on the above major advantages, the nonradioactive labeling methods are of much greater interest to researchers and scientists, and thus have the potential of totally replacing traditional radioactive labeling in the near future. However, one major downfall of nonradioactive probes is the problem of high background signals on the membranes. While such background may not be a problem when analyzing Southern blots of miniprep DNA, polymerase chain reaction (PCR) products, or even Northern blots of transcripts produced in high abundance, it is a problem for more sensitive procedures.

There are a number of kits available for the synthesis of nonradioactive probes, such as those from Ambion or Amersham, using random labeling technology and enzyme-conjugated nucleotides for the elongation steps. Molecular Probes (Eugene, OR) also has a variety of kits designed to label DNA and RNA fragments using different fluorophores. Although somewhat costly, these kits come with all the required reagents for the production of high-quality probes. There are also specialized hybridization and stripping reagents that allow the detection of hybridization signals using very low amounts of probe and the repeated use of Northern blots. Such items are often huge time savers if adequate funding is available, and the use of very little probe has the benefit of avoiding the problem of high background signals.

In the following sections, we describe several techniques for labeling nucleic acids using nonradioactive digoxigenin-dUTP as labeled dNTP. The basic principle of the digoxigenin labeling and detection method is that, in a random labeling procedure, random hexanucleotide primers anneal to denatured DNA template. The Klenow enzyme catalyzes the synthesis of a new strand of DNA complementary to the template DNA. During the incorporation of four nucleotides, one of them is labeled by digoxigenin, i.e., DIG-dUTP. After hybridization of the probe with the target DNA sequence that is immobilized onto a membrane filter, an antidigoxigenin antibody conjugated with an alkaline phosphatase will interact with the digoxigenin-dUTP. The detection of the hybridized signal will then be visualized with the chemiluminescent substrate Lumi-Phos 530 (Roche, Indianapolis, IN). When the substrate is hydrolyzed by the alkaline phosphatase conjugated to the digoxigenin antibody, photons will be generated by fluorescence and hit the x-ray film. These will be visible as black spots or bands after the film is developed. The detection can also be visualized with the colorimetric substrates NBT and X-Phosphate, which give a purple/blue color.

A. Random-Primer Labeling of dsDNA

1. Denature the dsDNA template by boiling for 12 min and quickly chill on ice for 4 min. Briefly spin down and place on ice until use.
2. In a microcentrifuge tube on ice, set up the following reaction:
   - Denatured DNA template (25 ng to 3 µg) (10 µl)
   - 10X Hexanucleotide primers mixture (4 µl)
   - 10X dNTPs mixture (4 µl)
   - ddH$_2$O to 38 µl
   - Klenow enzyme (2 units/µl) (2 µl)
   - Total volume of 40 µl
3. Incubate at 37°C for 3 to 12 h.

Notes: *The amount of labeled DNA depends on the amount of DNA templates and on the length of the incubation at 37°C. Based on our experience, the longer the incubation within 12 h, the more DNA is synthesized. However, when incubation is longer than 12 h, there is no significant increase in the amount of DNA labeled.*

4. Add 4 µl of 0.5 M EDTA solution to stop the reaction.
5. Add 0.15 volumes of 3 M sodium acetate buffer (pH 5.2) and 2.5 volumes of chilled 100% ethanol to the reaction. Allow to precipitate at –70°C for 1 h.
6. Centrifuge at 12,000 g for 5 min at room temperature. Carefully discard the supernatant and quickly rinse the pellet with 1 ml of 70% ethanol. Dry the pellet under vacuum for 10 min.
7. Dissolve the labeled DNA in 50 to 100 µl ddH$_2$O or TE buffer. Store at –20°C until use.

*Notes: (1) Labeled DNA (dsDNA) must be denatured before being used for hybridization. This can be done by boiling the DNA for 12 min then quickly chilling on ice for 4 min. (2) After hybridization, the probe contained in the hybridization buffer can be stored at –20°C and reused up to five times. Each time, incubate the used probe solution at 68°C for 5 min prior to reuse. (3) The labeled DNA is recommended, but not required, to estimate the yield by dot blotting of serial dilutions of commercially labeled control DNA and labeled sample DNA on nylon membrane or its equivalent. After hybridization and detection, compare the spot intensities of the control and the sample DNA (see dot blotting in **Chapter 5**).*

B. Nick-Translation Labeling of dsDNA with Digoxigenin-11-dUTP

1. On ice, set up a reaction as follows:
   - dsDNA template, 1 to 2 µl (2 µg)
   - 10X DIG-DNA labeling mixture (4 µl)
   - 10X Reaction buffer (4 µl)
   - *DNase I*/DNA polymerase I (4 µl)
   - Add ddH$_2$O to a final volume of 40 µl.
2. Incubate at 15°C for 40 min and stop the reaction by adding 4 µl of 0.5 M EDTA solution to the reaction and heating to 65°C for 10 to 15 min.
3. Precipitate and dissolve the DNA as described in **Section II.A** for random-primer labeling of dsDNA.

C. 3'-End Labeling with DIG-11-ddUTP

1. On ice, set up a standard reaction as follows:
   - DNA fragments or oligonucleotides (20 to 150 pmol)
   - 5X Reaction buffer (8 µl)
   - CoCl$_2$ solution (8 µl)
   - DIG-11-ddUTP (2 µl)
   - Terminal transferase (2 µl)
   - Add ddH$_2$O to a final volume of 40 µl.
2. Incubate the reaction at 37°C for 30 min and place on ice. Add 2 µl of 0.5 M EDTA buffer.
3. Precipitate and dissolve the labeled DNA in the same manner as described in **Section II.A** for random-primer labeling of dsDNA.

D. 3' Tailing of DNA with DIG-11-dUTP/dATP

Everything in this procedure is the same as for 3'-end labeling (**Section II.C**) except that labeled dNTP is DIG-11-dUTP/dATP instead of DIG-11-ddUTP.

E. 5'-End Labeling with DIG-NHS-Ester

1. Precipitate DNA fragments or oligonucleotides in 50 μl sodium borate solution.

Note: A synthesizer should generate a free NH_2 group at the 5' end of the DNA.

2. Dissolve 1.3 mg DIG-NHS-ester (Roche) in 50 μl dimethylformamide and add it to the 50-μl DNA sample.
3. Incubate at room temperature for 12 to 20 h.
4. Precipitate and dissolve the DNA as described for random labeling of DNA (**Section II.A**).

F. RNA Labeling

1. On ice, set up a reaction as follows:
   - Purified and linearized plasmid DNA template containing insert of interest (1 μg)
   - 10X NTP labeling mixture (4 μl)
   - 10X Transcription buffer (4 μl)
   - RNA polymerase (T7, SP6, or T3) (4 μl)
   - Add DEPC-treated ddH_2O to a final volume of 40 μl.
2. Incubate at 37°C for 2 to 3 h.
3. Add 20 units *DNase I* (*RNase*-free) and incubate at 37°C for 15 min to remove the DNA template. Add 4 μl of 0.5 *M* EDTA solution.
4. Precipitate and dissolve RNA as described for random primer labeling of DNA (**Section II.A**).

Reagents Needed for Nonradioactive Labeling

$CoCl_2$ Solution
   25 m*M* Cobalt chloride

DEPC-Treated Water
   0.1% Diethylpyrocarbonate (DEPC) in ddH_2O
   Incubate at room temperature overnight followed by autoclaving.

Warning: DEPC is highly toxic and should be used only in an exhaust hood.

DIG-11-ddUTP Solution
   1 m*M* Digoxigenin-11-ddUTP (2',3'-dideoxyuridine-5'-triphosphate coupled to digoxigenin via an 11-atom spacer) in ddH_2O

10X DIG DNA Labeling Mixture
   1 m*M* dATP
   1 m*M* dCTP
   1 m*M* dGTP
   0.35 m*M* DIG-11-dUTP
   0.65 m*M* dTTP
   pH 6.5 (+20°C)

Dimethylformamide (100% ACS-grade)

DNase I/DNA Polymerase I
        0.08 milliunits/µl *DNase I*
        0.1 units/µl DNA polymerase I
        50 mM Tris-HCl, pH 7.5
        10 mM MgCl$_2$
        1 mM DTE
        50% (v/v) Glycerol

EDTA Solution
        0.5 M EDTA in ddH$_2$O, pH 8.0

Klenow DNA Polymerase I
        2 units/µl, Labeling grade

10X Labeling dNTP Mixture
        1 mM dATP
        1 mM dCTP
        1 mM dGTP
        0.65 mM dTTP
        0.35 mM DIG-dUTP
        pH 6.5

10X NTP Labeling Mixture for RNA
        100 mM Tris-HCl, pH 7.5
        10 mM ATP
        10 mM CTP
        10 mM GTP
        6.5 mM UTP
        3.5 mM DIG-UTP

10X Random Hexanucleotide Mixture
        0.5 M Tris-HCl, pH 7.2
        0.1 M MgCl$_2$
        1 mM Dithioerythritol (DTE)
        2 mg/ml BSA
        62.5 A$_{260}$ units/ml random hexanucleotides

10X Reaction Buffer for Nick-Translation
        0.5 M Tris-HCl, pH 7.5
        0.1 M MgCl$_2$
        10 mM DTE

5X Reaction Buffer for 3'-End Labeling
        125 mM Tris-HCl, pH 6.6
        1 M Potassium cacodylate
        1.25 mg/ml BSA

RNA Polymerase

    20 units/μl T7, SP6, or T3

3 M *Sodium Acetate Buffer, pH 5.2*

    Dissolve sodium acetate in ddH$_2$O and adjust pH to 5.2 with 3 *M* glacial acetic acid. Autoclave and store at room temperature.

Sodium Borate Solution

    0.1 *M* Sodium borate, pH 8.5

TE Buffer

    10 m*M* Tris-HCl, pH 8.0
    1 m*M* EDTA

Terminal Transferase

    50 units/μl
    0.2 *M* Potassium cacodylate
    1 m*M* EDTA
    0.2 *M* KCl
    0.2 mg/ml BSA
    50% Glycerol, pH 6.5

10X Transcription Buffer

    0.4 *M* Tris-HCl, pH 8.0
    60 *M* MgCl$_2$
    100 m*M* DTT
    20 m*M* Spermidine
    0.1 *M* NaCl
    1 unit *RNase* inhibitor

References

1. **Sambrook, J., Fritsch, E.F., and Maniatis, T.,** *Molecular Cloning, A Laboratory Manual*, 2nd ed., Cold Spring Harbor Press, Cold Spring Harbor, NY, 1989.
2. **Rigby, P.W.J., Dieckmann, M., Rhodes, C., and Berg, P.,** Labeling deoxyribo-nucleic acid to high specific activity *in vivo* by nick translation with DNA polymerase 1, *J. Mol. Biol.*, 113, 237, 1977.
3. **Feinberg, A.P. and Vogelstein, B.,** A technique for radiolabeling DNA restriction endonuclease fragments to high specific activity, *Anal. Biochem.*, 132, 6, 1983.
4. **Young, W.S.,** Simultaneous use of digoxigenin- and radiolabeled-oligodeoxyribonucleotide probes for hybridization histochemistry, *Neuropeptides*, 13, 271, 1989.
5. **Mitsuhashi, M., Cooper, A., Ogura, M., Shinagawa, T., Yano, K., and Hosokawa, T.,** Oligonucleotide probe design — a new approach, *Nature*, 367, 759, 1994.

Chapter 5

Southern Blot Hybridization

Chung-Jui Tsai and Leland J. Cseke

Contents

I. Preparation of DNA Samples 86
    A. Restriction Enzyme Digestion of Genomic DNA 86
    B. Agarose Gel Electrophoresis of DNA 87
II. Blotting DNA onto Nylon Membranes 89
    A. Capillary Transfer Method 89
    B. Alkaline Transfer Method 91
    C. Vacuum Transfer Method 91
III. Preparation of DNA Dot/Slot Blots 92
IV. Hybridization Procedures 94
    A. Hybridization to ^{32}P-Labeled Probes 94
        1. Formamide-Based Hybridization Buffer 95
        2. Aqueous Hybridization Buffer 96
    B. Hybridization to Nonradioactive Probes 98
    C. Stripping DNA Probes 99
        1. Stripping by Alkali 99
        2. Stripping by Boiling 99
V. Troubleshooting Guide 100
    Reagents Needed 101
References 103

Southern[1] first described the detection of specific nucleotide sequences from a pool of digested deoxyribonucleic acid (DNA) fragments separated by electrophoresis and transferred onto a solid support.[2] It was thus termed the Southern blot. In conjunction with filter hybridization, the method relied on the complementary nature of the double-strand DNA and its reversible denaturation/renaturation properties to detect the presence and abundance of target DNA sequences in the fractionated DNA population, using a labeled probe of the DNA of interest. Although the exact procedure has since been modified extensively to achieve a higher degree of sensitivity, the principles remain the same and have had a profound impact on modern molecular biology. With changes, similar procedures are being used to detect levels of gene expression (so-called Northern blot hybridization, see **Chapter 6**) and protein accumulation (so-called Western blot hybridization, see **Chapter 8**). To improve throughput, DNA samples can be spotted, either manually or using commercially available devices, onto membranes directly before hybridization without gel electrophoresis. This modification is termed dot/slot blot hybridization. Unlike Southern blot analysis, where hybridization signals can be resolved according to the sizes of the hybridizing DNA, dot/slot blot analysis is most commonly used for comparison of the "total" signal level among individual samples. It was the forerunner of the present day's DNA microarray technology, which was further modified for high-throughput and global analysis of gene-expression profiles (see **Chapter 15**). Today, Southern hybridization remains as a fundamental method for gene cloning, library screening, DNA mapping, and confirmation of transgene integration.[2-6]

I. Preparation of DNA Samples

A typical Southern blot hybridization involves digestion of the DNA (usually genomic DNA) with appropriate restriction enzyme(s) and fractionation of the digested DNA fragments by agarose gel electrophoresis. The gel containing fractionated DNA is then blotted onto a nylon membrane and immobilized prior to hybridization. For genomic Southern hybridization, only high-molecular-weight genomic DNA should be used. Digestion of high-molecular-weight DNA, however, could be tricky due to the complexity of the genomic DNA and the variable digestion efficiency of different restriction enzymes. Six-base cutters are preferred for genomic DNA digestion, as they generate DNA fragments with a more evenly distributed size range. Four-base cutters should be avoided, since their recognition sites occur more frequently in the genome. Similarly, eight-base cutters are not desirable, as the resultant DNA fragments usually are of higher molecular weight. It is also advised to avoid methylation-sensitive restriction enzymes, as genomic DNA of some eukaryotic organisms (especially plants) is highly methylated.[7] The reaction conditions recommended by the suppliers should be adhered to, as less-than-optimum conditions may contribute to "star activity" (i.e., cleavage at noncanonical sites) of some restriction enzymes.

A. Restriction Enzyme Digestion of Genomic DNA

1. Prepare high-molecular-weight genomic DNA as described in **Chapter 1**. Digest 10 to 20 μg of the genomic DNA to completion using two to three units per μg DNA. Use the desired restriction enzyme(s) with appropriate buffer and supplement, such as bovine serum albumin (BSA), according to the manufacturer's protocols. Incubate at appropriate temperature overnight. If digestion with two or more restriction enzymes is desired, make sure the reaction buffer selected is compatible with all enzymes.
2. In the next morning, check ~0.1 μg of digested DNA in a thin minigel. Too thick a gel would make it difficult to evaluate the DNA fragment distribution pattern, thus requiring more DNA to be loaded. The DNA should appear as a smear with some visible bands when digestion has gone to completion. If high-molecular-weight DNA is still present, or if the DNA smear has a top-heavy distribution, add another one to two units per μg DNA of restriction enzyme(s) for further digestion.

Notes: *(1) It is highly recommended that the digestion efficiency be checked in a minigel prior to the "real" electrophoresis step to avoid any surprises due to incomplete digestion or uneven digestion efficiency among samples. (2) To further improve digestion efficiency, restriction enzymes can be added in three batches, ten units each time, every 2 to 3 h, followed by overnight incubation. However, the total amount of enzymes added should not exceed 10% of the reaction volume.*

3. (Optional) Precipitate the DNA by adding 0.1 volume of 3 M sodium acetate and two volumes of cold 100% ethanol, as described in **Chapter 1**. Wash with 70% ethanol and dry briefly under vacuum. Resuspend the pellet in 30 to 40 μl TE buffer and incubate with lid open at 37 to 50°C for 10 min to evaporate any residual ethanol. Incomplete evaporation of ethanol will lead to sample losses during gel loading. Add DNA loading buffer to a 1X final concentration, mix well, spin down the contents, and place the samples on ice until use. This step is only necessary when the volume of the digestion reaction in step 1 exceeds the capacity of the well to which the sample is to be loaded.

B. Agarose Gel Electrophoresis of DNA

Agarose gel electrophoresis is a standard method used to separate and identify DNA fragments. Agarose extracted from seaweed is a linear polymer consisting of D-galactose and 3,6-anhydro-L-galactose units. It is soluble in hot water (or aqueous buffer) and, when cooled, forms a matrix that serves as a molecular sieve to separate DNA fragments, on the basis of size, under the influence of an electric field. For Southern blot hybridization, we recommend the use of ultrapure-grade agarose to minimize impurities that may affect DNA migration. The concentration of agarose used depends on the range of DNA sizes to be resolved. Table 5.1 can be used as guide for the separation of double-strand (ds) DNA.[8] In general, concentrations of 0.8 to 1% agarose are adequate for fractionation of digested genomic DNA. Electrophoresis buffer also influences migration of DNA in agarose gels. The most commonly used buffers are TAE (Tris-acetate-EDTA) and TBE (Tris-borate-EDTA). The choice of buffer is often a matter of personal preference, since both buffers work adequately well for most purposes. In general, TAE has a lower buffering capacity and is more easily exhausted/overheated during prolonged electrophoresis. TAE buffer has a better resolving power than TBE buffer for high-molecular-weight DNA, but TBE buffer is recommended for resolving smaller DNA fragments.

1. Thoroughly rinse the appropriate gel apparatus with tap water several times, followed by three to five rinses with distilled water. Set up the gel-casting unit with appropriate comb.
2. Add appropriate amounts of agarose and electrophoresis buffer to a clean flask (estimate the volume based on a 0.8- to 1-cm-thick gel). Swirl to mix and heat to melt the agarose in a microwave oven for ~1 min per 100 ml solution (depending on the power of the microwave). Swirl gently to mix and reheat the solution for 30 sec. Repeat until it boils. Mix thoroughly and make sure no gel-like translucent suspensions are present.
3. Slowly cool the agarose solution to 50 to 60°C in a water bath. Alternatively, fast-cool the gel-solution-containing flask by use of running cold tap water applied to the side of the flask and by use of constant swirling. This may take a couple of minutes, depending on the volume of the solution. Add 2 μl of 10-mg·ml^{-1} ethidium bromide (EtBr) per 100 ml of warm agarose gel solution, swirl gently to mix, and pour onto the gel-casting unit. Remove any air bubbles. Allow the gel to solidify for 20 to 30 min at room temperature.

TABLE 5.1
Agarose Concentration for Separation of DNA

Agarose Gel Percentage (w/v)	Linear dsDNA Separation Range (kb)
0.3%	0.5–60
0.6%	1–20
0.7%	0.8–10
0.9%	0.5–7
1.2%	0.4–6
1.5%	0.2–4
2.0%	0.1–3

Caution: *EtBr is a powerful mutagen and a potential carcinogen. Gloves should be worn when working with this compound, especially when preparing a stock solution from the powdered form. Wipe the area with a damp cloth and change the gloves after working with the EtBr powder. All EtBr waste (gel, electrophoresis buffer, etc.) needs to be disposed of properly according to the local safety practices.*

Notes: *EtBr is used to stain DNA molecules by intercalation between the stacked bases of double-strand DNA in agarose gels. It fluoresces orange when illuminated with ultraviolet (UV) light. The merit of adding EtBr to the gel is that DNA bands can be stained and monitored with a UV lamp during electrophoresis. The drawback, however, is that running buffer and gel apparatus are contaminated with EtBr. An alternative is to carry out electrophoresis without EtBr. The gel is then stained with EtBr for 10 to 30 min following electrophoresis.*

4. Place the gel tray in the electrophoresis tank with the well-side toward the cathode (the negative electrode). Cover the gel with 1X electrophoresis buffer to a depth of ~1 mm. Carefully remove the comb by slowly and vertically lifting the comb up away from the gel, making sure not to tear the sample wells and that no air pockets are trapped in the wells. Use a small pipette tip to flush the buffer up and down to remove any bubbles if necessary.
5. Carefully load the commercial DNA standard markers (1 μg per well for genomic blot, or 0.1 μg per well for plasmid DNA/PCR blot) along with the DNA samples, one by one, onto the wells. Leave one well blank between the DNA markers and the samples. For certain applications, the same pipette tip may be used between samples (such as checking miniprep DNAs for proper inserts) if the tip is rinsed in the electrophoresis buffer after each sample. This is not recommended for genomic Southern gels.

Notes: *(1) For electrophoresis of genomic DNA, the final concentration of loading buffer must be at least 1X, or the sample may float out of the well. A smaller amount of loading buffer can be used for electrophoresis of plasmid DNA or PCR (polymerase chain reaction) products. (2) For nonradioactive detection, we recommend that the DNA markers be prelabeled, precipitated, and resuspended, as described in **Chapter 4**, and loaded onto the gel directly without denaturation. The prelabeled DNA standard markers allow easy and accurate size estimation of the hybridization DNA bands after signal detection.*

6. Cover the lid and connect the current. For genomic DNA electrophoresis, run the gel at 1 V per cm of the gel length for 16 h, or until the bromophenol blue (i.e., the "leading") dye reaches to ~2 cm from the end of the gel. A higher voltage, up to 10 V·cm^{-1} of gel length, with a shorter run time can be used for other DNA electrophoresis purposes.
7. Stop the electrophoresis. If the gel contains EtBr, the DNA banding patterns can be visualized directly by placing the gel on a UV transilluminator and photographed with a Polaroid™ camera or gel documentation system. For genomic Southern gels, use a fluorescent ruler alongside the gel for photodocumentation to aid in size estimation later. If the gel does not contain EtBr, stain the gel in diluted EtBr (0.5 μg·ml^{-1} in water) for 10 to 30 min with gentle agitation, and destain in water if necessary, before UV visualization and documentation.

Caution: *Because high voltage is used in gel electrophoresis, care should be taken to avoid electrical shocks. Always wear safety goggles and gloves for protection and minimize exposure to UV light.*

II. Blotting DNA onto Nylon Membranes

Three methods are described here for transfer of fractionated DNA from an agarose gel onto a solid support. The conventional transfer method introduced by Southern[1] relies on a gel-sandwich setup, where a gel is placed over a buffer-soaked wick and overlaid with a piece of transfer membrane. A stack of dry paper towels is then placed on top of the gel sandwich to create a capillary action that "pulls" the buffer from the wick, through the gel and the membrane, and up toward the dry paper towels. Fractionated DNA fragments in the gel are carried upward by capillary buffer flow and retained on the membrane, thus generating an imprint that is identical to the electrophoresis pattern of the gel. The DNA is then permanently fixed onto the membrane by UV cross-linking or baking. Capillary transfer is usually conducted with a neutral, high-salt buffer (such as 10X SSC or 10X SSPE), but an alkaline transfer buffer can be used with nylon membranes with a faster transfer rate. Another rapid-transfer method is vacuum blotting, but it requires the purchase of a vacuum transfer apparatus.

Certain pretreatments of the gel are necessary prior to blotting in order to facilitate DNA transfer. For genomic Southern blotting, the gel is first acid-treated to partially depurinate the DNA. The gel is then soaked in alkali to denature the DNA, followed by neutralization of the gel pH. The depurination step helps to increase the transfer efficiency of DNA fragments larger than 5 kb (kilobase-pairs) by partially fragmenting the DNA, while the denaturation treatment enables hybridization of the blotted DNA with labeled, single-strand probes with complementary sequences. The neutralization step is not necessary for alkaline blotting, as DNA transfer takes place in alkali. Both nitrocellulose and nylon membranes can be used, but a nylon membrane is preferred because nitrocellulose membranes have a lower DNA binding capacity and may not withstand some of the hybridization conditions described in this chapter. Both positively charged and uncharged nylon membranes can be reprobed numerous times, but positively charged nylon membrane tends to give rise to more background. Positively charged nylon membranes are commonly used with alkaline transfer buffer, which facilitates covalent binding of DNA to the membrane.

A. Capillary Transfer Method

1. After photo-documentation of the gel, carefully place the gel in a glass tray upside down and soak the gel in 0.25 N HCl for 10 min with gentle shaking to depurinate the DNA. Omit this step if the sizes of the target fragments (e.g., plasmid DNA or PCR products) are small.

Note: *Do not prolong the incubation in HCl; otherwise, excessive fragmentation will occur, resulting in decreased transfer efficiency because smaller fragments are not effectively retained on the membrane.*

2. Carefully decant the depurination solution. Rinse briefly with distilled, deionized water (ddH$_2$O). Add the alkaline denaturation solution to cover the gel and denature the DNA for 30 min at room temperature with gentle shaking.
3. Pour off the denaturation solution and rinse the gel with ddH$_2$O. Soak the gel in the neutralization solution for 30 min at room temperature with gentle agitation. This step neutralizes the gel without renaturing the DNA.
4. Remove the neutralization solution. Equilibrate the gel in 10X SSC or SSPE transfer buffer for 30 min with shaking.
5. While the gel is being pretreated, cut a piece of nylon membrane and two pieces of 3MM Whatman™ filter paper slightly larger than the gel. Denote the orientation of the membrane by cutting a small piece off from one of the corners, or by marking with a pencil on one of the corners. Wet the membrane in ddH$_2$O and then soak the membrane in transfer buffer until ready.

6. Set up a blotting tray as illustrated in Figure 5.1. The gel casting tray used in electrophoresis can be conveniently used upside-down as a support platform in a glass transfer tray. Add ~500 ml of transfer buffer (10X SSC or 10X SSPE) to the transfer tray. Place a piece of nonwoven wiper (we use TechniCloth™ cleanroom wipers, #18-315C, from Fisher Scientific, Chicago), cut to the width of the support platform, on the platform to serve as a wick. Make sure both ends of the wick are draped in the transfer buffer. Wet the wick thoroughly and roll away any air bubbles with a glass rod or test tube. Add transfer buffer to two-thirds of the height of the support platform.

Notes: *The use of a wick under the gel is not absolutely essential when dealing with procedures such as miniprep DNA restriction digests and PCR products. These procedures usually use such an abundance of DNA that incomplete transfer makes little difference in the desired results.*

7. Assemble the blotting apparatus as follows and as illustrated in Figure 5.1. Place the gel on the wick and remove all bubbles. We have found that used x-ray film (trimmed to an appropriate size) is a good tool for gel transfer. Carefully overlay the gel with the membrane, starting from one side of the gel and slowly proceeding to the other end. Roll away all bubbles using a glass test tube. Record in a notebook the orientation of the membrane (with the reference mark on the corner) in relation to the gel.

Notes: *The membrane should not be handled with bare hands. Wear gloves all the time and use blunt-end forceps to handle membranes.*

8. Cut four strips of SaranWrap™ or Parafilm™ to cover around the edge of the gel and the exposed area of the glass transfer tray. This is to prevent untargeted transfer of the buffer from around the gel. Wet the two pieces of Whatman filter paper in transfer buffer. Carefully lay the wet filter paper, one at a time, on top of the membrane, making sure no bubbles are introduced. Place a stack of paper towels (5 to 10 cm height, cut to appropriate size) over the filter paper. Place a glass plate on top of the paper towels and apply a weight (<0.5 kg) on top.

Note: *The purpose of the weight is to keep close contact between the various layers of the stack. Excessive weight may crush the gel and retard the transfer or lead to fuzzy hybridization bands.*

9. Allow the capillary transfer to occur for 4 h to overnight. Do not allow the apparatus to dry out. Otherwise, a high background may result on the membrane.
10. Remove the paper towel stack and the Whatman filter paper. Without disturbing the remaining setup, use a pencil to mark the wells on the membrane and the positions of the two dyes (xylene cyanole and bromophenol blue) on the edge of the membrane. Remove the membrane and rinse briefly with 2X SSC (or SSPE) to remove gel debris. Place the membrane, DNA side up, on a piece of paper towel to air dry the membrane. Check the blotted gel under UV light to evaluate the transfer efficiency. An efficient transfer should leave no visible EtBr-stained DNA in the gel.
11. Check the membrane under UV light and use a pencil to mark the positions of the DNA standard markers on the edge, if visible. Immobilize DNA onto the membrane by UV cross-linking, using the optimal setting according to the manufacturer's instructions. Alternatively, if a nitrocellulose membrane is used, bake the membrane at 80°C under vacuum for 2 h in a vacuum oven.

Notes: *The nylon membrane should be completely dry before UV cross-linking. We routinely dry the membrane in a 50 to 60°C oven for 1 to 2 h before UV cross-linking with good results (i.e., good signals after several rounds of stripping and reprobing). Depending on the probe and the marker used, the probe can sometimes nonspecifically hybridize to the markers. If this becomes an issue (especially for genomic Southern), simply cut the marker lane off from the membrane.*

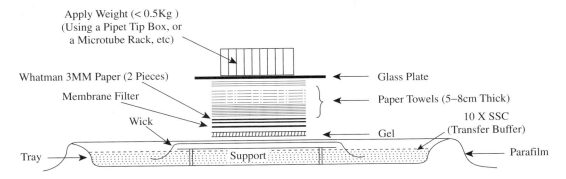

FIGURE 5.1
Standard assembly for capillary transfer of nucleic acids (DNA and RNA) from an agarose gel onto a nitrocellulose or a nylon membrane. (From Kaufman, P.B., Wu, W., Kim, D., and Cseke, L.J., *Handbook of Molecular and Cellular Methods in Biology and Medicine*, 1st ed., CRC Press, Boca Raton, FL, 1995.)

12. Wrap the membrane with SaranWrap and store at room temperature until use.

B. Alkaline Transfer Method

The alkaline transfer was first developed by Reed and Mann[9] in 1985 in order to take advantage of the positively charged nylon membrane so as to covalently bind negatively charged nucleic acids during the capillary transfer. Pretreatment of the gel with depuration solution is still recommended to improve the transfer efficiency of larger DNA fragments, followed by denaturation. Since an alkaline solution is used as the transfer buffer, no neutralization step is required. With the alkaline transfer buffer, DNA is transferred in a fully denatured state with a rate faster than that of the high-salt transfer method (i.e., 2 h vs. overnight). Nitrocellulose membranes are not compatible with the alkaline transfer conditions and should not be used. Uncharged nylon membranes can be used with a lower concentration of alkaline transfer solution.[9] Since DNA is covalently bound to the membrane, no baking or UV cross-linking will be necessary.

1. Pretreat the gel with depurination and denaturation solutions, as described in **Section II.A**.
2. Set up the transfer apparatus, as described in **Section II.A**, using a positively charged nylon membrane and 0.4 N NaOH as the transfer buffer instead of 10X SSC or SSPE. When an uncharged nylon membrane is used, lower the concentration of NaOH to 0.25 N.
3. Allow the transfer to occur for 2 h to overnight.
4. Disassemble the setup as described in **Section II.A**. Rinse the membrane with 2X SSC or SSPE and air dry. The membrane need not be UV cross-linked. It can be stored or used for hybridization immediately.

C. Vacuum Transfer Method

This is a very efficient, vacuum-based transfer method that utilizes commercially available vacuum blotters. Streamlined gel pretreatment and DNA transfer can be completed in about an hour. The following protocol is provided as a guideline. Specific instructions from the manufacturer of the vacuum blotter need to be followed.

1. Cut a nylon membrane to the size of the gel, and cut a silicone mask with an opening smaller (at least 4 mm on all sides) than the size of the gel. Masks can be used multiple times.

Notes: The gel around and under the wells tends to break easily. A break will prevent a vacuum from building up in the subsequent steps. We therefore suggest that the wells be trimmed and that one prepare the membrane and mask sizes accordingly.

2. Place a piece of 3MM Whatman filter on the metal grid of the vacublot apparatus and wet it with distilled water.
3. Place the nylon membrane on the 3MM Whatman filter, followed by the mask. Carefully position and place the gel (well-side facing up) to cover the opening of the mask to form a seal for the entire area. Place the cover of the vacublot into the apparatus to hold the mask in place.
4. Apply the vacuum at 75 mm Hg and leave it on throughout the transfer process.

Notes: A good vacuum seal can be achieved by pulling the mask tightly over the metal grid. The center of the gel (matching the opening area of the underneath mask) will be pulled down slightly as well.

5. Pour 5 to 10 ml of depurination solution to cover the surface of the gel and let it permeate the gel for 5 min.
6. Remove the depurination solution with a pipette and replace it with 20 ml of denaturing solution. Allow the solution to permeate the gel for 10 min.
7. Remove the denaturing solution and replace it with 20 ml of neutralizing solution. Allow the solution to permeate the gel for 10 min.
8. Remove the neutralizing solution and replace it with 20 ml of 20X SSC or 20X SSPE solution. Allow the solution to permeate the gel for 40 min.
9. Remove excess buffer and turn off the vacuum. Remove the gel and check under UV to evaluate the transfer efficiency. Mark the wells (if visible) and the edge of the mask on the membrane with a pencil. Rinse the membrane briefly with 2X SSC or SSPE and proceed to DNA immobilization, as described above in **Section II.A**. Store the wrapped membrane at room temperature.

III. Preparation of DNA Dot/Slot Blots

As compared with standard Southern blots, dot/slot blotting is a simple, fast, and sensitive method used in DNA/DNA, DNA/RNA, and/or RNA/RNA hybridizations (see **Chapter 6** for RNA dot/slot blot applications). Denatured DNA samples can be directly applied to membrane filters without electrophoresis. However, the dot-blot method cannot reveal the sizes of specific bands of interest among different DNA molecules, and it cannot identify possible nonspecific binding of a probe unless appropriate controls are used. Its utility lies in the identification of the presence or absence of given DNA sequences. Therefore, the inclusion of both positively and negatively hybridizing control samples is essential for data interpretation.

1. Denature DNA samples at 95°C for 10 to 15 min and immediately chill on ice. Spin down and add one volume of 20X SSC to the sample. An alternative way is the alkaline denaturing method: add 0.2 volume of 2 N NaOH solution to the sample; leave at room temperature for 15 min; and add 1 volume of neutralization buffer containing 0.5 M Tris-HCl (pH 7.5) and 1.5 M NaCl. Leave at room temperature for 15 min.

Notes: These samples can be directly spotted onto nylon membranes in small volumes using a micropipette. Spot as little volume as possible each time to minimize sample diffusion. Allow sample to dry and repeat until all of sample volume is spotted. The spotting process is repeated for each sample until all DNA is loaded. The membrane can then be treated

Southern Blot Hybridization

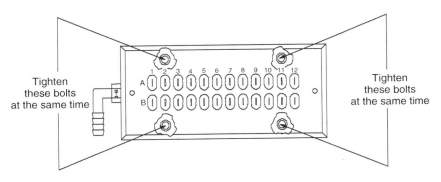

FIGURE 5.2
An example of a DNA/RNA slot-blot vacuum apparatus. The HYBRI-SLOT® Manifold (Whatman Biometra, Cat. No. 21052-014) has typical slot dimensions that are useful for the preparation of membranes used in standard hybridization techniques. Volume = 850 µl; slot dimensions = 0.5 mm × 4.0 mm; slot spacing, center-to-center = 9.0 mm; dimension assembled = 6.3 cm × 14.3 cm × 7.0 cm.

as in step 4 below. However, better consistency (e.g., spot morphology or size) is obtained by using a dot/slot blot filtration apparatus, if available (Figure 5.2).

2. Cut a piece of nylon membrane and 3MM Whatman paper to the size of the membrane support block (or support plate for some brands) of the dot/slot blot vacuum manifold according to the manufacturer's instructions. Wet the membrane in ddH$_2$O followed by 10X SSC, and then assemble it in the manifold with 10X SSC-prewet 3MM Whatman filter paper. Turn on the vacuum and adjust the valve to obtain an appropriate suction (13 to 25 cm Hg, or as specified by the manufacturer). Make sure the assembly produces a tight seal, or cross-contamination between dots/slots may occur. Turn off the vacuum.

Notes: *(1) An appropriate vacuum is required for the blotting process. If the vacuum is too low, diffusion of samples upon loading is usually a problem. However, if the vacuum is pulling too fast (i.e., setting is at too high a level), the efficiency of blotting is reduced. (2) Use a vacuum trap when you connect the manifold to a vacuum pressure gauge to collect waste solution and to prevent contamination of the vacuum pump and/or the gauge.*

3. Apply appropriate sample volume to all the sample wells as quickly as possible. Turn on the vacuum and allow the buffer solution to drain. Use of a multichannel pipette will facilitate the process. A tracking dye, such as bromophenol blue (or common DNA gel-loading dye), can be added to the sample to monitor the process.

Notes: *(1) Some manufacturers recommend that one rehydrate the membrane with TE buffer or ddH$_2$O prior to sample application so as to avoid obtaining halos or weak signals. (2) Apply the samples along one sidewall of the well to obtain an equal distribution of DNA. If the sample volume is less than the manufacturer's recommended level, the sample solutions must be carefully applied to the center of the well.*

4. Add 0.5 to 1 ml (according to the manufacturer) transfer buffer to each well and allow the vacuum to pull the buffer through.
5. While the vacuum is on, disassemble the apparatus by loosening the screws and carefully pulling off the sample template block. Use forceps to lift off the membrane and place the membrane on a piece of paper towel. Turn off the vacuum.
6. Air dry the membrane and immobilize the DNA, with the DNA side up, by UV cross-linking using a UV cross-linker (Fisher Scientific, Chicago, IL) at 120 mjoules·cm^{-2} for 10 sec. Cross-linking can

also be done on a transilluminator (312 nm wavelength is recommended) with the DNA side down for 4 to 6 min.

Notes: *Some protocols call for rinsing the membrane in neutralization buffer, 2X SSC, and/or water. However we have found this to be unnecessary.*

7. Store the wrapped membrane at room temperature, or proceed to hybridization as described in the next section.

IV. Hybridization Procedures

The base-pairing ability of two single-strand molecules of DNA or RNA with complementary sequences is the underlying principle of nucleic acid hybridization. In the case of Southern hybridization, this enables detection of specific DNA sequences denatured and immobilized on membranes using labeled DNA (single-stranded or denatured double-stranded) probes. Several factors affect the detection sensitivity of the hybridization signals. These include membrane type and blotting efficiency, as discussed above; copy number of the target sequences; probe size; labeling efficiency; probe specificity (see **Chapter 4**); and hybridization conditions. Radioactive-labeled (such as ^{32}P) probes can be made to a high specificity, but they are short-lived and pose health concerns. Alternatively, several nonradioactive labeling methods are now available. Nonradioactive probes can be stored for a longer period, and their use reduces radioactive exposure. However, nonradioactive probes usually give rise to a higher background, and their sensitivity is limited and not suitable for detection of single-copy genes.

The conditions under which hybridization takes place also have a profound impact on the probe-target DNA hybridization and, hence, signal detection. This includes hybridization buffer, hybridization stringency (as determined by buffer type and temperature), and washing stringency (as determined by salt concentration and temperature). These conditions also vary between radioactive and nonradioactive methods. In this chapter, we introduce three hybridization protocols suitable for use with ^{32}P-labeled probes. The nonradioactive hybridization method is largely based on the digoxigenin (DIG)-based system that was developed by Roche Applied Science (Indianapolis, IN). DNA transferred and immobilized onto nylon membranes is durable, and the blots can undergo several cycles of hybridization, stripping, and reprobing without reducing the signal level. Stripping methods for removal of labeled probes are also presented.

A. Hybridization to ^{32}P-Labeled Probes

The DNA used for labeling can be a specific gene (full length or partial length) or a fragment or synthetic oligonucleotide with the conserved sequence of interest. It can be derived from the same organism as the target DNA or from a different one. The hybridization and washing stringency depends on the known or estimated homology between the probe DNA and the target DNA. If the target sequence to be detected is identical to that of the probe DNA, a high stringency should be used to avoid nonspecific hybridization and to reduce background. For studies involving the use of a DNA probe derived from a different species, or for detection of different members of the same gene family, a lower stringency is recommended in order to reveal weakly hybridized signals.

Readers are referred to **Chapter 4** for various DNA labeling methods using ^{32}P-labeled nucleotides (e.g., [α-^{32}P]dATP or [α-^{32}P]dCTP). Hybridization that utilizes formamide-based or aqueous hybridization buffers is introduced below.

1. Formamide-Based Hybridization Buffer

The hybridization buffer usually contains a high salt concentration (e.g., 5X SSC, 5X SSPE, or 0.5 M sodium phosphate) in order to promote base-pairing between the target and the probe DNA. The addition of formamide to the hybridization buffer destabilizes the chemical interactions between the two complementary strands of target-probe DNA, thus reducing the melting temperature (T_m) of the DNA duplex. The advantage of using a formamide-containing hybridization buffer is that it allows hybridization to occur at a lower temperature. The lower hybridization temperature helps to retain the target DNA on the membrane, when nitrocellulose membrane is used, after prolonged incubation or repeated hybridization. With the stronger binding strength and capacity of the nylon membrane, however, this is usually not a concern. Disadvantages of using formamide include a reduced hybridization rate and its hazardous nature. Nevertheless, formamide hybridization buffers offer a greater flexibility to manipulate hybridization stringency by adjusting either formamide concentration, temperature, or both.

1. Place the filters in the aqueous 5X SSC or SSPE hybridization buffer (see **Reagents Needed**) and carry out prehybridization for 2 to 4 h with slow rotation at 65°C.

Notes: *(1) We do not recommend using plastic hybridization bags because of the difficulty in getting rid of air bubbles and in obtaining a tight seal. This can cause leaking and radioactive contamination. If a hybridization oven with rotating wheels for hybridization tubes is not available, an appropriate size of plastic tray will be the best alternative to use as a hybridization container. (2) If more than one membrane is to be used for hybridization, place the membranes in sequential order, one at a time, into the hybridization container/tube with prehybridization solution and allow the membrane to thoroughly soak in the buffer before placing another one. This will avoid introducing an air pocket between membrane layers that could cause uneven hybridization. However, it is advised not to exceed three membranes in each hybridization container. (3) The volume of prehybridization solution depends on the container used. Usually less buffer is required when using a hybridization tube. If a plastic bag or container is used, the buffer should be of sufficient volume to cover the entire membrane area.*

2. Prepare the probe, remove unincorporated radioactive nucleotides, and measure the specific activity using a scintillation counter (see **Chapter 4** for details).
3. If a double-stranded DNA probe is used, denature the probe in a microcentrifuge tube placed in boiling water for 10 min and immediately chill on ice. If carrier DNA is to be used, add 100 µg·ml^{-1} carrier DNA (e.g., salmon sperm DNA) to the probe and denature at the same time. Briefly spin down the contents prior to use.
4. While denaturing the probe, decant prehybridization solution and add formamide hybridization solution to the tube. Incubate the tube at the desired hybridization temperature to equilibrate (see hybridization stringency options in steps 6 and 7 below).

Notes: *(1) Probe denaturation is a critical step. If the probes are not completely denatured, a weak hybridization signal will occur. (2) Single-stranded DNA or oligonucleotide probes do not require denaturation. (3) SSPE has a higher buffering capacity than SSC and is preferred for use in the formamide-containing hybridization buffer, as prolonged incubation could cause breakdown of formamide and result in increased hybridization solution pH.[6]*

Caution: *^{32}P-labeled probes must be handled with care and according to local radiation safety procedures. A lab coat and gloves should be worn at all times. Gloves should be changed if any contamination occurs and disposed of as radioactive waste. Waste liquid, pipette* *tips, and papers contaminated with the isotope should be collected in labeled containers. After finishing, a radioactive-contamination survey should be performed and recorded before leaving the area.*

5. Add the denatured probe to the hybridization solution, at a minimum specificity of 2 to 10×10^6 cpm·ml^{-1}. Allow hybridization to proceed overnight for genomic Southern hybridization or for 2 h for dot/slot or plasmid DNA blot hybridization at an appropriate temperature (see below).
6. **For low-stringency hybridization,** use 30 to 40% formamide in the hybridization solution and a hybridization temperature of 42°C or lower.
7. **For high-stringency hybridization,** use 50% formamide in the hybridization solution and a hybridization temperature of 42°C.
8. Discard the hybridization solution as radioactive waste. Rinse the membrane thoroughly with 2X SSC/0.1% SDS washing solution at room temperature to remove excess probes.
9. **For low-stringency wash,** wash the membrane twice, 15 min each, using 2X SSC/0.1% SDS washing solution at 50 to 55°C with slow rotation. Check the signal with a Geiger counter. If the background is still high, continue the wash at a higher temperature (up to 65°C) or wash with 1X SSC/0.1% SDS at 50 to 55°C for one to two times, 15 min each time. Proceed to step 11.
10. **For high stringency wash,** wash the membrane twice, 10 min each, with 2X SSC/0.1% SDS at 65°C followed by two more washes with 0.2X SSC/0.1% SDS, 15 min each, at 65°C with gentle rotation. For best wash results, use a large volume of wash buffer (e.g., 50 ml per hybridization tube) each time.
11. Seal the membrane in a plastic bag or wrap in SaranWrap. Do not allow the membrane to dry if it is to be reprobed. Check the membrane with a Geiger counter to estimate the signal intensity and exposure time.
12. Place the wrapped membrane (DNA-side up) in a cassette that contains an intensifying screen (usually attached to the inner side of the cassette lid) and secure its position with two pieces of 3M Scotch™ tape on the bottom plate of the cassette. For alignment purposes, mark one corner of the wrapped membrane with a Crayola™ glow-in-the-dark marker. In a dark room, with the safe light on, place a piece of x-ray film (slightly larger than the membrane) on top of the membrane and close the cassette lid securely. Place the cassette in a −80°C freezer for a desired period of exposure time. Figure 5.3 shows typical results.

Notes: *The exposure time varies depending on the intensity of the signal. For detection of high-copy-number DNA (such as plasmid or PCR blots), a short exposure at room temperature (even without an intensifying screen) may be adequate. Genomic Southern signals, on the other hand, may take several days to two weeks to obtain. The reader should develop a sense of proper exposure time based on the Geiger counter-surveyed radiation level and experience.*

2. Aqueous Hybridization Buffer

Aqueous hybridization buffers do not contain formamide, thus eliminating the need to handle hazardous chemicals. The solutions are stable and can be stored at room temperature for several months. Since a higher hybridization temperature is used, the aqueous hybridization buffer is not recommended for use with nitrocellulose membranes. The Church hybridization buffer[10] described below is suitable for Southern, Northern, and slot/dot blot hybridizations, whereas the 5X SSC hybridization buffer, as detailed below, is offered as a rapid confirmation and screening method for plasmid or PCR blots.

a. Church Hybridization Buffer. The Church hybridization buffer was first developed by Church and Gilbert in 1984 for hybridization-based genomic sequencing of single-copy genes.[10] It contains a high concentration of SDS (sodium dodecyl sulfate) (7%) as a nonspecific blocking reagent in a phosphate buffer. The solution is stable at room temperature (>6 months) and can be

Southern Blot Hybridization

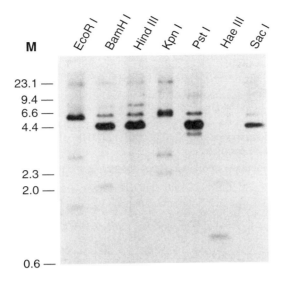

FIGURE 5.3
Southern blot analysis of invertase gene(s) in eukaryotic cells. Genomic DNA was digested with different restriction enzymes followed by electrophoresis on a 0.8% agarose gel. DNA was blotted onto a nylon membrane and hybridized with soluble invertase cDNA used as a probe. The weakly hybridized bands may be other members of the invertase gene family. (From Kaufman, P.B., Wu, W., Kim, D., and Cseke, L.J., *Handbook of Molecular and Cellular Methods in Biology and Medicine*, 1st ed., CRC Press, Boca Raton, FL, 1995.)

used in both prehybridization and hybridization steps, thus simplifying the hybridization protocol. Since no formamide is used, the hybridization stringency can only be adjusted by temperature.

1. Place the membrane in a hybridization tube and prehybridize in ~10 ml of hybridization solution at 65°C for at least 30 min.
2. Prepare the probe. Decant prehybridization solution and add 5 to 10 ml of fresh hybridization solution to the tube. Incubate at 65°C while denaturing the probe. Denature the probe (with 100 µg·ml^{-1} salmon sperm DNA if desired) by boiling for 10 min and quench on ice.
3. Add the probe to the hybridization solution (at ~10^6 cpm·ml^{-1}) and incubate overnight for genomic Southern (i.e., low-copy-number detection) analysis or for 2 h for dot/slot or plasmid DNA blot hybridizations (i.e., high-copy-number detection). Use 65°C for high-stringency hybridization or 50 to 55°C for low-stringency conditions.
4. Discard the hybridization solution as radioactive waste. Rinse the membrane thoroughly with 10 ml of washing solution 2X SSC/0.1% SDS at room temperature.
5. Proceed to low- or high-stringency wash. Then expose the membrane to an x-ray film with an intensifying screen as described in **Section IV.A.1**.

b. 5X SSC Hybridization Buffer. This procedure is almost identical to the formamide-based hybridization procedure except that no formamide is used. It allows rapid screening and confirmation of positive clones during DNA subcloning and PCR analysis. The procedure also assumes that the probe used is identical to the target DNA sequences being hybridized.

1. Prehybridize the membranes in 5X SSC hybridization buffer at 65°C for 15 min to 1 h.
2. Denature the labeled double-stranded DNA probe as described in **Section IV.A.1.3** and add it to the hybridization solution.
3. Allow the hybridization to occur at 65°C for 3 h to overnight. Weaker probes need longer incubation times.
4. Wash the hybridized membranes three times at room temperature with 5X SSC, checking the signal and background with a Geiger counter between washes two and three.
5. Cover the membrane with plastic wrap and proceed to autoradiography as described in **Section IV.A.1**.

B. Hybridization to Nonradioactive Probes

Generally speaking, the nonradioactive hybridization system is less sensitive than the radioactive method, and it is prone to a higher background. Therefore, it is not recommended for detection of single-copy genes in Southern hybridization. The nonradioactive hybridization protocol, however, offers a safer alternative to the radioactive method, especially for student exercises or in laboratories where radioactive material is restricted. Nonradioactive labeled probes can be conveniently made beforehand and stored for a longer period of time. The nonradioactive hybridization method introduced here is based on the DIG hybridization system of Roche Applied Science (Indianapolis, IN). If stripping and reprobing of the blot is desired, we recommend the use of alkaline-labile DIG nucleotides for probe labeling. The DIG-labeled DNA probe is allowed to hybridize with the target DNA mobilized on a nylon membrane, and the hybridized molecules are detected by using an alkaline phosphatase (AP)-conjugated anti-DIG antibody (Roche Applied Science, Indianapolis, IN). A chemiluminescent AP substrate (CDP-Star, Roche Applied Science, Indianapolis, IN) is then added to the membrane, and the membrane is exposed to an x-ray film to record the chemiluminescence. Alternatively, AP-recognized hybridization signals can be detected by a chromogenic assay that utilizes NBT (nitroblue tetrazolium) and BCIP (5-bromo-4-chloro-3-indolylphosphate).

1. Carry out prehybridization, hybridization, and wash steps as described above using either the formamide-based hybridization method or the Church hybridization protocol, except that a nonradioactive DIG-probe is used (see **Chapter 4** for probe preparation).
2. Transfer the hybridized and washed membranes into a clean dish containing maleic acid buffer. Wash for 2 to 4 min with gentle shaking at room temperature.
3. Replace with freshly prepared blocking solution (see **Reagents Needed**) and incubate with shaking for 60 min. Discard the solution.
4. Add diluted anti-DIG-AP antibody (1:10,000 in blocking solution for chemiluminescent detection and 1:5000 for chromogenic detection) to the dish and incubate at room temperature for 40 to 60 min with gentle shaking.
5. Wash the membrane with maleic acid buffer for 20 min with shaking. Repeat once using a fresh washing tray.

Note: *The used antibody solution can be stored at 4°C for up to 2 months and reused for five to six times without any significant decrease in the antibody activity.*

6. Equilibrate the membrane in AP buffer for 1 to 4 min with gentle shaking. Discard the solution. Proceed to step 7 for chemiluminescent detection or step 11 for chromogenic detection.
7. For chemiluminescent signal detection, add 0.5 to 1 ml of CDP-*Star* chemiluminescent substrate dropwise to the membrane (DNA-side up) until the entire surface is covered. Remove excess substrate by lifting the membrane with blunt-ended forceps and wrap the membrane with SaranWrap. Wipe out excess substrate using a paper towel to reduce background. Care should be taken not to introduce air bubbles between the membrane and the top layer of the SaranWrap.
8. Place the wrapped membrane in an exposure cassette with the DNA-side facing up. Repeat if multiple membranes are to be processed.
9. Overlay the wrapped membranes with an x-ray film in a darkroom with the safe light on. Allow the exposure to occur for 2 min to 24 h at room temperature.
10. Develop the x-ray film and adjust the exposure time if necessary.

Notes: *(1) Exposure for more than 4 h may produce a very dark background. Based on our experience, a good hybridization and detection should generate sharp positive signals with 1.5 h of exposure. (2) Repeated exposures can be made up to 2 days, after which time the membrane needs to be washed to remove the old substrate. Repeat steps 6 to 10 if additional exposure is desired.*

11. For chromogenic signal detection, add the chromogenic color development solution (NBT and BCIP in AP buffer) to the membrane and incubate in the dark at room temperature for 30 min to 1 day. Monitor the color development until clear signs are observed. Stop the reaction by rinsing the membrane thoroughly with 10 mM EDTA followed by ddH$_2$O. Store the membrane between two layers of filter paper in the dark.

C. Stripping DNA Probes

After x-ray film exposure and signal development, the hybridized DNA probe can be removed from the target DNA on the membrane so that the membrane can be reused to hybridize with a second probe. DNA immobilized on nylon membranes can be reprobed for up to ten times, but nitrocellulose membranes can be reused for only a couple of times. Stripping methods rely on heat denaturation or alkali treatment to separate the hybridized probe-target DNA duplex. Both protocols work well with nylon membranes. However, nitrocellulose membranes cannot withstand treatment with NaOH, and the boiling method must be used. Special instructions provided by the membrane manufacturer should be followed whenever possible. If chromogenic detection is used to detect nonradioactive signals, the color needs to be removed before stripping. It is very important to note that if reprobing is desired, the membrane cannot be dry during prehybridization, hybridization, washing, and signal detection; otherwise the probe will not be stripped.

1. Stripping by Alkali

This protocol can be used with nylon membranes hybridized with either ^{32}P-labeled or nonradioactive probes. If chromogenic detection is used in the latter case, the color should be removed before stripping.

1. (Optional) When chromogenic detection is used, remove the NBT/BCIP color by incubating the membrane with warm (50 to 60°C) dimethylformamide (DMF) in a fume hood until the blue color has been removed (up to 1 h). Proceed to stripping.
2. Soak membranes in ddH$_2$O for several minutes.
3. Wash the membrane in 0.2 N NaOH and 0.1% SDS at 37°C for 20 min. Discard the stripping solution as radioactive waste. Repeat once.
4. If radioactive probes are used, check the membrane with a Geiger counter. If radioactive signals can still be detected, repeat the stripping step.
5. Rinse the membrane with 2X SSC for 10 min.
6. Proceed to the prehybridization step as described above. If the membrane is not used right away, it can be stored dry between two pieces of filter paper at room temperature.

Notes: *Under some circumstances, it may be necessary to confirm that there is no residual signal on the membrane by autoradiography. In this case, we recommend that the membrane be exposed to an x-ray film for at least one day with an intensifying screen.*

2. Stripping by Boiling

This method can be used with both nylon or nitrocellulose membranes hybridized with ^{32}P-labeled probes.

1. Soak membranes in ddH$_2$O for several minutes.
2. Pour off water and add boiling 0.1% (w/v) SDS solution. Incubate with gentle agitation and allow to cool to room temperature. Repeat this step three or four times until the signal falls to background. The membrane can then be reprobed or stored.

V. Troubleshooting Guide

Symptoms	Solutions
Smearing signal appears for each lane with no clear banding pattern	May be due to poor DNA quality, DNA degradation, or incomplete digestion. Check DNA quality before digestion and monitor the digestion efficiency in a minigel.
	The voltage setting may be too high during electrophoresis.
	Monitor the temperature of the electrophoresis buffer during electrophoresis. Use a buffer-recirculating device, if possible, for a long run, or refresh the buffer during the run.
Bands of the same sizes shift but do so among different lanes	May be due to uneven gel thickness or different salt concentrations among samples.
"Smiling" bands	Electric field not consistent throughout the gel.
	Electrophoresis is too fast. Try to run the gel at a lower voltage setting.
	DNA may be overloaded. Try to prepare the gel with wider lanes.
Unexpected bands	May be due to insufficient blocking or low hybridization and washing stringency. Try to include denatured carrier DNA in both prehybridization and hybridization solutions, or use the Church hybridization buffer.
	Verify the specificity of the probe and make sure the probe does not contain any vector sequences.
Weak signals	May be due to poor probe labeling efficiency and probe specificity. Check the DNA quality or the labeling kit.
	Incomplete transfer. Make sure depurination is carried out properly. Use 20X SSC or SSPE as the transfer buffer.
	Use a lower washing stringency and increase the exposure time.
No signal at all	May be due to incomplete denaturation of target DNA and/or probe, prolonged depurination, or improper UV cross-linking.
High background (radioactive hybridization)	Do not allow the membrane to dry during any stage of the prehybridization, hybridization, and washing.
	Prolong the prehybridization step.
	Increase BSA and SDS concentrations and include carrier DNA in the prehybridization/hybridization solutions.
	Try the Church hybridization protocol.
High (purple) background (nonradioactive hybridization)	Prolong the blocking step by up to 3 h.
	Decrease the probe concentration.
	Do not allow the membrane to dry.
	Increase washing stringency. Prewarm the wash solutions to desired temperature before use.
	Reduce the color-development time by monitoring the color development closely. Stop the reaction by thoroughly rinsing the membrane with EDTA, followed by several changes of ddH$_2$O.
Spotty background	Unincorporated ^{32}P nucleotides are not removed from the probe. Purify the labeled DNA by use of a Sephadex G-50 column.
	Do not use old ^{32}P-labeled nucleotides.
	Rinse the membrane with 2X SSC before immobilization by UV cross-linking or baking.
	For nonradioactive detection, centrifuge the anti-DIG-AP antibody briefly before use, decrease the antibody concentration, or shorten the antibody incubation time.
No signals after reprobing	Stripping of the hybridized probe was not successful, or the filter became dried out during previous hybridization procedure.
	May be due to improper DNA retention. Check the UV cross-linker and perform necessary calibration.

Reagents Needed

Alkaline Phosphatase (AP) Buffer

100 mM Tris-HCl, pH 9.5
100 mM NaCl
50 mM MgCl$_2$

Prepare 1 M Tris (pH 9.5), 5 M NaCl, and 1 M MgCl$_2$ stocks in ddH$_2$O. Autoclave and store at room temperature. Prepare AP buffer by adding appropriate amounts of stock solutions and make up the volume with ddH$_2$O. Store at room temperature.

Alkaline Stripping Solution

0.2 N NaOH
0.1% SDS

Prepare from 10 N NaOH and 10% SDS stocks before use.

Alkaline Transfer Buffer

0.4 N NaOH
1 M NaCl

Dilute from 10 N NaOH and 5 M NaCl stocks before use.

Blocking Solution

Dissolve 10% (w/v) blocking reagent (Roche Applied Science, Indianapolis, IN) in maleic acid buffer by stirring and heating. Autoclave and store at room temperature.

Boiling Stripping Solution

0.1% SDS, heat to boiling

Diluted SDS solution tends to boil vigorously. Close attention should be given when heating up the SDS solution. Alternatively, heat the ddH$_2$O to boiling and carefully add 10% SDS stock to a final concentration of 0.1%.

Color Development Buffer

70 µl BCIP (5-bromo-4-chloro-3-indolylphosphate) (50 mg·ml^{-1})
70 µl NBT (nitroblue tetrazolium) (50 mg·ml^{-1})
10 ml AP buffer

Prepare 50 mg·ml^{-1} NBT in 70% DMF and 50 mg·ml^{-1} BCIP in DMF. Store at 4°C and protect from light. BCIP may precipitate during storage and should be warmed at room temperature to dissolve.

Church Hybridization Buffer

7% (w/v) SDS
1% (w/v) BSA (bovine serum albumin, fraction V)
1 mM EDTA
0.25 M Na-PO$_4$ (pH 7.4)

Prepare 1 M NaH$_2$PO$_4$ and 1 M Na$_2$HPO$_4$ stock solutions. Mix 22.6 ml of 1 M NaH$_2$PO$_4$ and 77.4 ml of 1 M Na$_2$HPO$_4$ to produce 1 M Na-PO$_4$, pH 7.4 stock. For long-term storage, autoclave and store at 4°C.

Prepare the hybridization solution by dissolving 5 g BSA in ~100 ml ddH$_2$O. Add 125 ml of 1 M Na-PO$_4$, 175 ml of 20% SDS, and 1 ml of 0.5 M EDTA. Adjust the volume to 500 ml and store at room temperature.

SDS precipitates out at cool temperature. When this happens, prewarm the hybridization to redissolve before use.

Denaturing Solution
> 1.5 M NaCl
> 0.5 N NaOH
> Dilute from 5 M NaCl and 10 N NaOH stocks and store at room temperature.

Depurination Solution
> 0.25 N HCl
> Dilute from concentrated HCl and store at room temperature.

50X Denhardt's Solution
> 1% (w/v) BSA (bovine serum albumin, fraction V)
> 1% (w/v) Ficoll (Type 400)
> 1% (w/v) PVP (polyvinylpyrrolidone, mol. wt. 40,000)
> Dissolve in ddH$_2$O and adjust the final volume to 500 ml. Filter sterilize and store in 25 to 50 ml aliquots at –20°C.

6X DNA Loading Buffer
> 0.25% Bromophenol blue
> 0.25% Xylene cyanol FF
> 30% Glycerol
> Prepare in ddH$_2$O and store in aliquots at –20°C.

Formamide Hybridization Buffer
> 5X SSC or 5X SSPE
> 5X Denhardt's reagent
> 0.5% SDS
> 50% Formamide
> Formamide concentration may be adjusted for different hybridization stringencies. SSPE has a better buffering capacity and is recommended for use with formamide.

Maleic Acid Buffer
> 100 mM Maleic acid, pH 7.5
> 150 mM NaCl
> Dissolve maleic acid in ddH$_2$O containing NaCl (from 5 M stock). Adjust pH to 7.5 with NaOH. Autoclave and store at room temperature.

Neutralization Solution
> 1.5 M NaCl
> 1 M Tris-HCl, pH 7.5
> 20% SDS
> Dissolve 20 g SDS (sodium dodecyl sulfate) in 100 ml ddH$_2$O and store at room temperature.

Caution: SDS is an irritant and toxic. Handle carefully when weighing the powder. Do not breathe the dust.

3 M Sodium Acetate Solution (pH 5.2)
> Dissolve sodium acetate in ddH$_2$O and adjust pH to 5.2 with 3 M acetic acid. Autoclave and store at room temperature.

5X SSC or 5X SSPE Hybridization Buffer
> 5X SSC or 5X SSPE

5X Denhardt's reagent
0.5% SDS

20X SSC

3 M NaCl
0.3 M Sodium citrate
Dissolve 175.3 g of NaCl and 88.2 g of sodium citrate in ~700 ml ddH$_2$O. Adjust the pH to 7.0 with HCl and the volume to 1 L. Autoclave and store at room temperature.

20X SSPE

3.0 M NaCl
0.2 M NaH$_2$PO$_4$
2 mM EDTA, pH 8.0
Dissolve 175.3 g NaCl in ~500 ml ddH$_2$O. Add 200 ml of 1 M NaH$_2$PO$_4$ stock and 4 ml of 0.5 M EDTA stock. Adjust the pH to 7.4 with NaOH and adjust the volume to 1 L. Autoclave and store at room temperature.

50X TAE

2 M Tris-acetate
50 mM EDTA, pH 8
Dissolve 242 g Tris-base in ~500 ml ddH$_2$O. Add 57.1 ml of glacial acetic acid and 100 ml of 0.5 M EDTA stock. Adjust the volume to 1 L.

5X TBE

225 mM Tris base
225 mM Boric acid
5 mM EDTA, pH 8
Dissolve 54 g Tris-base and 27.5 g boric acid in ~500 ml ddH$_2$O. Add 20 ml of 0.5 M EDTA stock and adjust the volume to 1 L.

TE Buffer

10 mM Tris-HCl, pH 8.0
1 mM EDTA, pH 8.0
Dilute from 1 M Tris (pH 8) and 0.5 M EDTA (pH 8). Autoclave and store at room temperature.

References

1. **Southern, E.M.,** Detection of specific sequences among DNA fragments separated by gel electrophoresis, *J. Mol. Biol.*, 98, 503, 1975.
2. **Danhardt, D.T.,** A membrane-filter technique for the detection of complementary DNA, *Biochem. Biophys. Res. Commun.*, 23, 641, 1966.
3. **Grunstein, M. and Hogness, D.S.,** Colony hybridization: a method for the isolation of cloned DNAs that contain a specific gene, *Proc. Natl. Acad. Sci. U.S.A.*, 72, 3961, 1975.
4. **Lander, E.S.,** Mapping complex genetic traits in humans, in *Genome Analysis, A Practical Approach*, Davies, K.E., Ed., IRL Press, Oxford, 1988, p. 171.
5. **Fourney, R.M. et al.,** Determination of foreign gene copy number in stably transfected cell lines by Southern transfer analysis, in *Gene Transfer and Expression Protocols*, Murray, E.J., Ed., Humana Press, Clifton, NJ, 1991.
6. **Sambrook, J. and Russell, D.W.,** *Molecular Cloning, A Laboratory Manual*, 3rd ed., Cold Spring Harbor Press, Cold Spring Harbor, NY, 2001.

7. **Martienssen, R.A. and Richards, E.J.,** DNA methylation in eukaryotes, *Curr. Opinion Genet. Dev.*, 5, 234, 1995.
8. **Ogden, R.C. and Adams, D.A.,** Electrophoresis in agarose and acrylamide gels, *Methods Enzymol.*, 152, 61, 1987.
9. **Reed, K.C. and Mann, D.A.,** Rapid transfer of DNA from agarose gels to nylon membranes, *Nucleic Acids Res.*, 13, 7207, 1985.
10. **Dyson, N.J.,** Immobilization of nucleic acids and hybridization analysis, in *Essential Molecular Biology, A Practical Approach*, Vol. II, Brown, T.A., Ed., IRL Press, Oxford, 1991, p. 111.
11. **Church, G.M. and Gilbert, W.,** Genomic Sequencing. *Proc. Natl. Acad. Sci. U.S.A.*, 81: 1991, 1984.

Chapter 6

Northern Blot Hybridization

Myeon Haeng Cho, Eun Kyeong Lee, Bin Goo Kang, and Leland J. Cseke

Contents

I.	Electrophoresis of RNA by the Use of Formaldehyde Agarose Gels	106
II.	Blotting RNA onto Nylon or Nitrocellulose Membranes	108
	A. After Electrophoresis	108
	B. Dot/Slot Blots	108
III.	Proper Controls	108
IV.	Hybridization	109
V.	Quantitative Analysis of RNA Following Northern Blot Hybridization	109
VI.	Troubleshooting Guide	110
	Reagents Needed	110
References		111

Northern blot hybridization or ribonucleic acid (RNA) blotting is an RNA analysis procedure in which the size and amount of specific mRNA molecules in total RNAs or poly(A)$^+$ RNAs can be determined. It consists of RNA/DNA (deoxyribonucleic acid) or sense RNA and antisense RNA hybridization. RNA molecules are separated according to their sizes by electrophoresis on a gel under denaturing conditions, transferred to a nitrocelluse or nylon membrane, and fixed to the membrane. The blotted membrane is hybridized with a specific probe, and the mRNA of interest is then visualized by autoradiography or nonisotopic detecting methods.[1–5]

RNA isolation and purification procedures are given in **Chapter 2**. The probe(s) used in Northern hybridization can be specific genomic DNA, cDNA, oligonucleotides, or antisense RNA. The preparation of probe(s) is described in **Chapter 4**. The DNA probe is usually stable but needs longer hybridization.[6,7] The RNA probe, on the other hand, is much hotter, but it is relatively unstable.

Since RNAs, as compared with DNA, are very mobile molecules due to RNA degradation by *RNases*, much care should be taken to maintain the purity and integrity of the RNA. This is very critical for Northern blotting. The most difficult task is to inactivate *RNase* activity. Two common sources of *RNase* contamination are the user's hands and bacteria and fungal molds present on airborne dust particles. To prevent this type of contamination:

1. Gloves should be worn at all times and changed frequently.
2. Whenever possible, disposable plasticware should be autoclaved. Nondisposable glass- and plasticware should be treated with 0.1% diethylpyrocarbonate (DEPC) in distilled, deionized water (ddH_2O) and be autoclaved before use.
3. After treatment, glassware should be baked at 250°C overnight. Gel apparatus should be thoroughly cleaned with detergent and thoroughly rinsed with DEPC-treated water. Apparatus used for RNA electrophoresis, if possible, should be separated from DNA or protein electrophoresis apparatus. Chemicals should be of ultrapure grade and *RNase*-free. Gel mixtures, running buffers, hybridization solutions, and washing solutions should be made with DEPC-treated water.
4. Additional details can be found at the beginning of **Chapter 2**.

I. Electrophoresis of RNA by the Use of Formaldehyde Agarose Gels

Gel electrophoresis is performed using conditions that disrupt RNA secondary structure. This greatly improves resolution and allows for an accurate estimation of the length of the RNA molecule. See **Chapter 2, Section V.A** for additional details.

1. Thoroughly clean an appropriate gel apparatus by washing with detergent. Completely remove the detergent mixture with tap water and rinse with DEPC-treated distilled water three to five times. Allow the apparatus to dry at room temperature.
2. Prepare 100 ml of a 1.5% agarose gel containing 2.2 M formaldehyde. A 1.5% agarose gel is suitable for resolving RNAs in the 0.5 to 8.0 kb (kilobase-pairs) size range. Large RNAs should be separated on 1% or 1.2% agarose gels.
   a. Add 1.5 g agarose to 72 ml of sterile H_2O.
   b. Dissolve the agarose by boiling in a microwave oven. Gently mix and place at room temperature to cool to 50 to 60°C.
   c. While the gel mixture is being cooled, seal the air-dried gel tray at the two open ends with a tape or gasket and put the comb in place.
   d. Add 10 ml of 10X MOPS electrophoresis buffer and 18 ml of deionized formaldehyde.
   e. Gently mix after each addition and slowly pour the mixture into the assembled gel tray placed in a fume hood. Allow the gel to harden for 20 to 30 min at room temperature.

Caution: *Formaldehyde vapors are toxic. DEPC and EtBr are carcinogenic. These chemicals should be handled with care. Gloves should be worn when working with these materials. Gel running buffer containing EtBr should be collected in a special container. Formaldehyde serves to denature the secondary structures of RNA. EtBr is used to stain RNA molecules, which interlace in the regions of secondary structure of RNA and fluoresce orange when illuminated with UV (ultraviolet) light.*

3. While the gel is hardening, prepare the sample(s) in a sterile tube as follows:
   a. For each sample, add the following components:
      - Total RNA (10 to 35 μg per lane) or poly(A)$^+$ RNA (0.2 to 2 μg per lane), 6.0 μl
      - 10X MOPS buffer, 3.5 μl
      - Formaldehyde (37%, 12.3 M), 6.2 μl
      - Formamide (50% v/v), 17.5 μl
      - Ethidium bromide (200 μg/ml), 1.0 μl
      - Add DEPC-treated ddH_2O to a final volume of 35.0 μl.
   b. For RNA standard markers, add the following components:
      - RNA markers with wide range (0.5 to 1.5 μg), 2.0 μl
      - 10X MOPS buffer, 1.0 μl
      - Formaldehyde (37%, 12.3 M), 1.8 μl
      - Formamide (50% v/v), 5.0 μl
      - Add DEPC-treated ddH_2O to a final volume of 10 μl.

Northern Blot Hybridization

    c. Heat the tubes at 65°C for 15 min and immediately chill on ice to denature the RNA sample. Briefly spin down afterward.

    d. Add 3.5 µl and 1 µl of DEPC-treated loading buffer to the sample and RNA standard markers, respectively.

4. Electrophoresis:

    a. Carefully remove the comb and sealing tape or gasket from the gel. Place the gel tray in the electrophoresis tank and add 1X MOPS buffer to the tank right up to the upper edge of the gel or until the gel is covered to a depth of 1.5 to 2 mm above the gel.

Notes: *(1) The comb should be slowly and vertically removed from the gel. Any crack inside the wells of the gel will cause leaking when the samples are loaded. (2) The well end of the gel must be placed at the negative pole end because the negatively charged RNA will migrate toward the positive pole. (3) Based on our experience, we recommend that the 1X MOPS running buffer not cover the gel in order to protect the formaldehyde in the gel from diffusing into the buffer, thus affecting the denaturing efficiency of RNA. However, if the gel is covered with running buffer, each well should be flushed with the buffer using a small pipette tip to cause flow of buffer up and down inside the well several times. The purpose of doing this is to remove any potential bubbles that will adversely influence the loading of samples and electrophoresis. (4) For nonisotopic detection, we recommend that the RNA markers be prelabeled, precipitated, and directly loaded into the gel. The labeling methods are described in* **Chapter 4**. *The prelabeled RNA standard markers allow easy and accurate identification of the sizes of detected bands in the DNA samples after being exposed to x-ray film or after being color-developed. For isotopic detection, however, we do not recommend prelabeled DNA markers due to potentially massive contamination during various steps. If necessary, the hybridized filter can be stripped of the hybridized probe for the RNA samples and reprobed with labeled RNA standard markers, and the marker positions can be compared with those of the detected bands of interest (***Chapter 5, Section IV.C***).*

    b. Prerun the gel at a constant voltage (5 to 10 V/cm) for 10 min and immediately load the RNA samples into each well of the gel. Leave one well blank between the RNA standard markers and the RNA samples.

Note: *Do not insert the pipette tip to the bottom of the well; otherwise, it may break the well and cause sample leaking.*

    c. Measure the gel length between the two electrodes and apply current to 5 to 10 V/cm (constant voltage). Cover the tank, and carry out electrophoresis for 1 h. Cover the gel, which is not submerged with the running buffer, with SaranWrap™ to prevent evaporation of the formaldehyde. Allow the gel to continue to run at a constant voltage for 2 to 4 h or until the first blue dye band reaches a position of 2 cm from the end of the gel.

Notes: *In the middle of electrophoresis, it is recommended that the gel tray be lifted and the buffer at the two ends of the tank be mixed and the electrophoresis then allowed to continue. Overnight running of the gel is not recommended.*

    d. After electrophoresis is complete, photograph the visualized RNA bands under UV light and directly proceed to the blotting procedure.

Caution: *Remember to wear safety glasses and gloves for protection from UV light.*

Notes: *Pictures as a rule are recommended to be taken at relatively longer exposure in order to visualize the smear in each well. After successful electrophoresis of RNA, a long smear with two sharp bands should be visible from each well for total RNA, or a long smear only should be visible for purified poly(A)⁺ RNA (see Figure 2.2). The two sharp bands represent rRNAs: one is 28S rRNA in animals or 25S rRNA in plants, the other is 18S rRNA. If the range of the smear is very limited or the rRNA bands are very weak in total RNA, the RNA samples are likely degraded.*

II. Blotting RNA onto Nylon or Nitrocellulose Membranes

A. After Electrophoresis

The general procedures for Northern blotting are the same as those described under Southern blotting procedures (see **Chapter 5**), except that all solutions and water should be DEPC-treated. The subsequent steps include the same treatment of the gel, set-up of the transfer, cross-linking or baking of the membrane, hybridization, washing, and detection.

Notes: *1) There are a variety of Southern blotting techniques described in **Chapter 5**. It will be necessary for readers to determine which procedure works best for them. 2) In our experience, capillary transfer to the membrane (**Chapter 5, Section II.A**) followed by the Church method of hybridization (**Chapter 5, Section IV.A.2**) works very well. 3) An alternative approach to the transfer step is to rinse the gel with several changes of DEPC-treated ddH$_2$O to remove the formaldehyde. Then both the membrane and the gel are soaked in 30 mM NaOAc, and the transfer is set up in 30 mM NaOAc.*

B. Dot/Slot Blots

Compared with Northern blots, dot blotting is a simple, fast, and sensitive method. The principle of the method is that a known amount of sample RNA is spotted onto an inert support, such as nitrocellulose, and the amount of a specific RNA in the sample is determined by hybridization with a suitable radioactively labeled probe. Denatured RNA samples can be directly applied to membrane filters without electrophoresis. However, the dot-blot method cannot reveal the sizes of specific bands of interest among different RNA molecules. The basic procedures are given below for dot blotting of RNA.

1. Denature RNA samples at 65°C for 10 min in four volumes of denaturing solution containing 66% (v/v) formamide, 21% (v/v) formaldehyde (37%), and 13% (v/v) 10X MOPS buffer.
2. After denaturation, quickly chill the sample on ice for 4 min and briefly spin down.
3. Add one volume of 20X SSC solution to the sample and spot it on the membrane filter as with DNA dot blotting.
4. Fixing of RNA to the membrane, baking, prehybridization, hybridization, washing, and detection are the same as for Northern blotting.

III. Proper Controls

In order to compare the amount of an RNA species in different samples, it is necessary to ensure that equal amounts of RNA are loaded in each lane. For accurate quantitation of a particular RNA species and to verify equal loading and transfer, it is best to strip and reprobe the membrane with a constitutively expressed housekeeping gene such as cyclophilin, actin, or glyceraldehyde-3-phosphate dehydrogenase (GAPDH). These genes are uniformly expressed at moderately abundant levels (~0.1% of poly(A)⁺ RNA or 0.003% of total cellular RNA).

IV. Hybridization

The general procedures for prehybridization, hybridization, washing, detection, autoradiography, and reprobing are the same as described for Southern blotting (see **Chapter 5**), except that all solutions should be DEPC-treated.

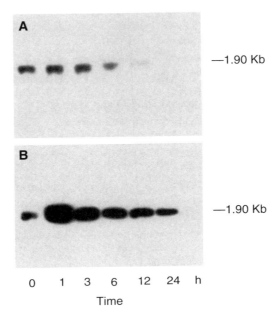

FIGURE 6.1
Northern blot analysis of invertase gene expression in oats (*Avena sativa*). Oat stem segments were gravistimulated for different periods of time. Poly(A)$^+$ RNA was purified from total RNA that was isolated from tissues A and B, respectively. mRNA (2 μg per lane) was electrophoresed in 1% agarose gel containing formaldehyde and blotted onto a nylon membrane. Invertase cDNA was radioactively labeled and used for hybridization. (From Wu, L.-L. et al., *J. Plant Physiol.* 142, 179, 1993.[8] With permission of © Gustav Fischer Verlag; Kaufman, P.B., Wu, W., Kim, D., and Cseke, L.J., *Handbook of Molecular and Cellular Methods in Biology and Medicine*, 1st ed., CRC Press, Boca Raton, FL, 1995.).

V. Quantitative Analysis of RNA Following Northern Blot Hybridization

The intensity of an autoradiographic signal is a measure of the concentration of the specific RNA. Quantitation can be achieved through scanning densitomety of the autoradiographics or the use of a phosphoimager. Data are transferred to a computer software package that generally offers many useful options. The available software for use in gel imaging includes Image Quant™ v3.3 (Molecular Dynamics), Kepler (Large Scale Biology Corp.), PDQUEST (Protein and DNA Imageware Systems) and Quantity One® (Bio-Rad, Hercules, CA). A fact of primary importance is definition of background density. This varies with exposure time, x-ray film type, film age and background radiation on the sample support (gel or membrane). Measurement of a couple of autoradiographs of the same samples, but at different exposure times, allows a double-check of the results. Such issues are not a problem when using a phosphoimager.

A more reliable method for quantitation uses direct counting of the radioactive emissions from the bound probe. This can be achieved easily and without destroying the membrane by using a purpose-built machine. If such a machine is not available, then it is necessary to cut out the individual slots and count them in a scintillation counter. Compare the counts per minute (cpm) obtained for each of the serial dilutions of the samples.

VI. Troubleshooting Guide

Symptoms	Solutions
A smeared signal appears under each lane, with no sharp bands	Electrophoresis is too slow or too fast. Try to use 5 to 10 V/cm at a constant level.
Bands with the same sizes shift but do not occur at the same positions among the different lanes	The thickness of the gel is not even, so the electrophoresis is not uniform.
Many unexpected bands appear	Nonspecific cross-hybridization problems occur. Verify the specificity of the probe and try to use high-stringency conditions of hybridization.
Black background occurs on the filter (for radioisotopic probe hybridization)	Filter became partially dry or blocking efficiency is low, or the quality of the filter is not good. Try to avoid air drying of the filter; increase the percentage of BSA and denatured salmon sperm DNA; and try to use a fresh neutral nylon membrane filter.
Purple background occurs on the filter (for nonisotopic probe hybridization)	Color development was too long. Try to stop the color reaction as soon as desired signals appear.
Unexpected black spots occur on the filter (for probe radioisotopic hybridization)	Unincorporated ^{32}PdCTP was not efficiently removed. Try to use a G-50 Sephadex column to purify the labeled DNA.
Unexpected purple bands occur on the filter (for nonisotopic probe hybridization)	The specificity of the antibody used is not specific. Try to test the quality of the antibody and increase the blocking reagents.
Signals are weak on the filters	The efficiency of labeling is low or the x-ray film exposure time is not long enough.
No signal is seen at all in the filter	Sample and/or probe is not denatured.
No signal occurs at all during subsequent reprobing	The stripping of the hybridized probe was not successful, or the filter became dried out during previous probing.

Reagents Needed

DEPC Water

    0.1% DEPC in ddH$_2$O
    Place at 37°C overnight with a stir bar to inactivate any potential RNase.
    Autoclave to remove the DEPC.

Ethidium Bromide (EtBr)

    10 mg/ml in DEPC-treated ddH$_2$O
    Dissolve well and keep in the dark or in a brown bottle at 4°C.

10X Loading Buffer

    50% Glycerol
    10 mM EDTA
    0.25% (w/v) Bromophenol blue
    0.25% (w/v) Xylene cyanol FF
    Dissolve well in DEPC-treated ddH$_2$O and store at 4°C.

10X MOPS Buffer

    0.2 M 3-(N-Morpholino)propanesulfonic acid (MOPS)
    80 mM Sodium acetate
    10 mM EDTA (pH 8.0)
    Dissolve well after each addition in DEPC-treated ddH$_2$O.
    Adjust the pH to 7.0 with 2 N NaOH.
    Sterile-filter and store at room temperature.

Nylon Membranes or Nitrocellulose Membranes

Heat-sealable Bags

20X SSC Solution

    3 M NaCl
    0.3 M Sodium citrate
- Dissolve sodium acetate in ddH$_2$O and adjust pH to 5.2 with 3 M acetic acid. Autoclave and store at room temperature.

20X SSC Solution

    3 M NaCl
    0.3 M Sodium citrate
- Dissolve 175.3 g of NaCl and 88.2 g of sodium citrate in ~700 ml ddH$_2$O. Adjust the pH to 7.0 with HCl and the volume to 1 L. Autoclave and store at room temperature.

References

1. **Sambrook, J. and Russell, D.W.**, *Molecular Cloning: A Laboratory Manual*, 3rd ed., Cold Spring Harbor Press, Cold Spring Harbor, NY, 2001.
2. **Thomas, P.S.**, Hybridization of denatured RNA and small DNA fragments transferred to nitrocellulose, *Proc. Natl. Acad. Sci. U.S.A.*, 77, 5201, 1980.
3. **Krumlauf, R.**, Northern blot analysis of gene expression, in *Gene Transfer and Expression Protocols*, Murray, E.J., Ed., Humana Press, Clifton, NJ, 1991.
4. **Hames, B.D. and Higgins, S.J.**, *Gene Transcription: A Practical Approach*, Oxford University Press, NY, 1993.
5. **Farrell, R.E.**, *RNA Methodologies: A Laboratory Guide for Isolation and Characterization*, Academic Press, San Diego, CA, 1993.
6. **Jones, P., Qiu, J., and Rickwood, D.**, *RNA Isolation and Analysis*, BIOS Scientific Publishers, St. Thomas House, Oxford, U.K., 1994.
7. **Lasky, R.A.**, *Radioisotopes in Biology: A Practical Approach*, IRL Press, Oxford, U.K., 1989.
8. **Wu, L.-L. et al.**, Molecular basis of the increase in invertase activity elicited by gravistimulation of oat-shoot pulvini, *Plant Physiol.*, 142, 179, 1993.

Chapter 7

Preparation of Monoclonal and Polyclonal Antibodies against Specific Protein(s)

Leland J. Cseke and William Wu

Contents

I. Production of Monoclonal Antibodies 114
    A. Purification of Antigen for Immunization 114
    B. *In Vivo* Immunization of Mice with the Purified Antigen....................................114
    C. Determination of Mouse Immunoglobulin Concentration by ELISA 116
    D. Preparation of Peritoneal Exudate Cells under Sterile Conditions 117
    E. Preparation of Spleen Cells and Myeloma Cells for Fusion 118
    F. Fusion of Spleen Cells and Myeloma Cells under Sterile Conditions 119
    G. Selection and Propagation of Hybridoma Cells 119
    H. Screening of Hybridoma Supernatant by ELISA and Harvesting
       of Monoclonal Antibodies[5] 120
    I. Isotyping of Monoclonal Antibodies 120
    J. Verification of Antibodies by Western Blot Analysis..121
    K. Purification of Monoclonal Antibodies Using Affinity Chromatography[4]............121
    L. Freezing and Thawing Cell Lines under Sterile Conditions.................................122
       1. Freezing Cell Lines 122
       2. Thawing Frozen Cells 122
II. Preparation of Polyclonal Antibodies 122
    A. Production of Antigens of Interest..123
    B. Purification of the Antigen for Immunization 123
    C. *In Vivo* Immunization of Female Rabbits..123
    D. Purification of the Final Polyclonal Antibodies 124
    Reagents Needed 125
References 126

Antibodies are small proteins produced in animals and have long been used as molecular probes in several aspects:[1] (1) detection, measurement, and purification of biological molecules of interest;

(2) *in situ* immunocytochemical localization of specific protein/enzyme in cells and tissues; (3) immunoscreening of expressional cDNA (complementary deoxyribonucleic acid) libraries to identify cDNAs of interest; and (4) treatment of some human diseases. The preparation of antibodies is based on the fact that the immune system of an animal has specific immune responses to and produces antibodies against foreign substances (antigens) such as carbohydrates, nucleic acids, and proteins. The antibodies produced are then secreted into the serum by lymphoid cells in several organs including bone marrow, spleen, and lymph nodes.[2] The serum can be obtained by bleeding an immunized animal. There are two typical types of antibodies:[1-5] (1) monoclonal antibodies raised in mice, and (2) polyclonal antibodies that are usually raised in rabbits. Monoclonal antibodies contain a single antibody specificity, a single affinity, and a single immunoglobulin isotype. However, a preparation of polyclonal antibodies has a mixture of antibody molecules directed against the antigen as well as antibodies that do not react with the antigen of interest. These differences account for the fact that monoclonal antibodies are more specific as compared with polyclonal antibodies. This chapter describes the preparation of both monoclonal and polyclonal antibodies against a protein/enzyme of interest. However, these procedures are not designed for the production of antibodies for clinical use.

I. Production of Monoclonal Antibodies

Monoclonal antibodies are produced by a monoclonal population of cells derived from a single cloned cell, so that all the molecules are identical to each other. The general procedures include: (1) purification of the protein of interest as an antigen;[1,2] (2) *in vivo* immunization of mice by injecting the antigen; (3) removal of plasma cells in the spleen from the immunized mice and hybridization of plasma cells with a myeloma cell line; (4) cloning of cell colonies; (5) screening and selection of a monoclonal population of cells; and (6) harvesting and purification of the monoclonal antibodies secreted into the medium (Figure 7.1).

A. Purification of Antigen for Immunization

The general procedures are described in **Chapter 3**. It is recommended that the protein of interest used as an antigen be as pure as possible. Pure antigens can make the procedure of screening and selection of monoclonal colonies of cells much easier. For glycoproteins, deglycosylation should be carried out because the animal can raise antibodies against the sugar residues, which will reduce the specificity of the monoclonal antibodies and may result in potential cross-reactions between the antibodies and a variety of proteins.

B. *In Vivo* Immunization of Mice with the Purified Antigen

1. Resuspend the purified antigen in PBS (phosphate-buffered saline) in a sterile microcentrifuge tube to a concentration of 1 to 5 µg/µl and use 50 to 500 µg per injection, depending on the particular antigen to be used.
2. Combine 100 µl (100 to 150 µg) of the antigen sample with an equal volume of complete Freund's adjuvant to a final volume of 200 µl for the first injection per mouse. Mix thoroughly to obtain an emulsion using a syringe or a pipette. Slowly take up the emulsified mixture with a 1-ml disposable syringe equipped with an 18- or 20-gauge needle and remove air bubbles.
3. Take a blood sample (0.5 to 1.0 ml) as a preimmune serum from the vein behind the eye of a mouse before the first injection. Allow the blood to clot by placing the sample at 4°C for approximately

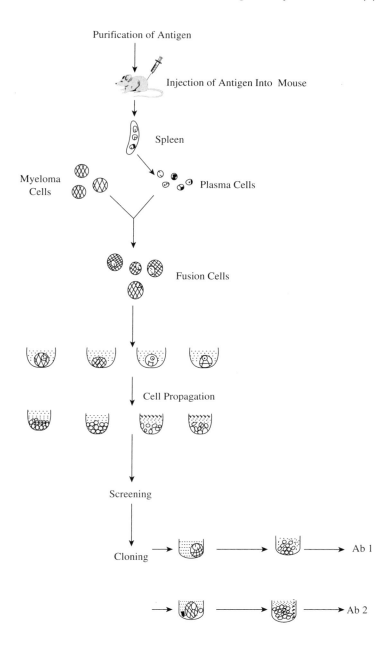

FIGURE 7.1
Schematic representation of procedures used to produce monoclonal antibodies. (From Kaufman, P.B., Wu, W., Kim, D., and Cseke, L.J., *Handbook of Molecular and Cellular Methods in Biology and Medicine,* 1st ed., CRC Press, Boca Raton, FL, 1995.)

30 min, and centrifuge at 7000 g for 10 min. Transfer the serum to a fresh tube, dilute it at 1:100 to 1:500 with PBS, and store at –20°C until use.

4. Carry out immunization of two to four mice (BALB/c strain, 6 to 8 weeks old) by injecting the antigen mixture (200 µl per mouse) into the mice. It is recommended that one should have a helper to hold the mouse for injection or use an appropriate restraining device if no helper is available. Briefly clean the area for injection with a cotton swab wetted with 70 to 75% ethanol and inject the mixture into the mouse using one of the following methods:

TABLE 7.1
Items To Be Included for Each Aliquot of Ig Supernatant Tested

Components	Ig Supernatant	Control 1 (negative)	Control 2 (positive)	Control 3 (positive)	Blank
Antigen (Ag)	+	+	+	+	−
First antibody (1° Ab) [a]	+	−	−	−	−
Mouse Ig standard	−	−	−	+	−
Preimmune serum	−	+	−	−	−
Second antibody (2° Ab)	+	+	−	+	−
Substrate (S)	+	+	+	+	−

[a] Immune serum or Ig supernatant.

    a. Intravenous injection into the tail vein; this is the best method for stimulation of the spleen.
    b. Intra-peritoneal injection (do not push the needle too deep so as to avoid injecting the needle into stomach).
    c. Subcutaneous or intramuscular injection into the thigh muscle.
5. Place the mice back in their cage(s). If the mice were injected with the same antigen, they can be caged together. If different mice were immunized by different antigens, they must be labeled properly and kept in different cages.
6. Mix 100 μl (100 to 150 μg) of the antigen with 100 μl of incomplete Freund's adjuvant to obtain an emulsion and then make the first booster injection 4 weeks after the first injection (primary injection).
7. Repeat booster injections two to three times at 4-week intervals as in step 6.
8. Take a blood sample from the same source of mice as immune serum in order to check the antibody titre by the use of enzyme-linked immunosorbent assay (ELISA) using the preimmune serum as a negative control and the purified mouse immunoglobulin standard as a positive control.
9. Boost the mice with another 100 μl (100 μg) of antigen mixed with 100 μl of Freund's incomplete adjuvant 3 to 4 days prior to doing the cell fusion.

C. Determination of Mouse Immunoglobulin Concentration by ELISA

1. Design a 96-well ELISA plate such that the items in Table 7.1 should be included for each aliquot of immunoglobulin (Ig) supernatant to be tested:

Notes: *(1) Five concentrations (0.001 to 1 μg/μl) of mouse immunoglobulin standard should be used in the same plate or another plate under the same conditions in order to generate a standard curve. (2) For each of the above controls — including the Ig supernatant to be assayed, negative controls, blank control, and five different concentrations of mouse Ig standard — two to four wells of replicates should be set up in order to obtain an average assay value. A 96-well ELISA plate has 12 columns named in the order from 1 to 12, and 8 rows from A to H. All 96 wells can be used for multiple samples, or 48 wells can be used for a few samples, leaving the other 48 wells used as blanks.*

2. Dilute the antigen sample to 50 μg/ml in PBS (pH 7.2). According to the above design, coat appropriate wells of a disposable ELISA plate by adding the antigen to the wells (50 μl per well). Cover the plate and incubate at room temperature for 2 h or at 4°C overnight.
3. Decant the antigen solution from wells by dumping the plate upside down on a paper towel two to three times. Fill each well (100 to 200 μl per well) with 1 to 3% (w/v) bovine serum albumin (BSA) in PBS with 0.05% Tween-20 to block any remaining binding sites in the wells. Incubate at room temperature for 30 to 60 min.

4. Discard the blocking solution as in step 3 and add 50 µl of the first antibody solution, including preimmune serum, immune serum or Ig supernatant, and different concentrations of mouse Ig standard into appropriate wells according to the design in step 1. Allow to incubate at room temperature for 60 min.
5. Carefully remove the first antibody solution from appropriate wells and immediately wash the plate by filling up each well with washing buffer followed by discarding the buffer after 2 min. Repeat washing twice.
6. According to the design in step 1, add 50 µl per well of diluted (1:2000 to 1:5000 in PBS with 1% BSA) second antibody to the appropriate wells. The second antibody is goat antimouse immunoglobulin-peroxidase-conjugate or goat antimouse immunoglobulin-alkaline phosphatase conjugate (Sigma-Aldrich, St. Louis, MO). Incubate the plate at room temperature for 60 min.
7. Wash the plate as in step 5.
8. As quickly as possible, add 50 µl per well of freshly prepared substrate solution to the appropriate wells according to the design in step 1. Allow the enzyme reaction to proceed for 15 min at room temperature in the dark. A color change will occur as the reaction takes place between the enzyme and its substrate. The color developed indicates a positive reaction.
9. Stop the reaction by adding 50 µl per well of 12.5% (v/v) H_2SO_4.
10. Place the plate in a cassette in an autoreader (e.g., Bio-Tek, E1310) and read the absorbance or the optical density at 410 or 490 nm according to the manufacturer's instructions. Draw a standard curve on graph paper using the concentrations as the ordinate and the optical density or absorbance as the abscissa. To determine the concentration of antibodies in the sample, find its density and draw a horizontal line crossing to the standard curve. Starting from the cross-point on the standard curve, draw a vertical line crossing to the ordinate, at which point the concentration of mouse Ig standard is equal to the concentration of antibodies in the sample.

D. Preparation of Peritoneal Exudate Cells under Sterile Conditions

Peritoneal exudate cells (PECs) or endothelial cell growth supplement (ECGS) derived from bovine hypothalamus have been demonstrated to be necessary as the feeder cells for culture of hybridoma cells.

1. Thoroughly clean a laminar-flow cabinet as follows:
   a. Turn on the air flow.
   b. Thoroughly clean the cabinet using 95% ethanol.
   c. Thoroughly clean the cabinet 10 min after step b using 70% ethanol.
   d. Thoroughly clean the cabinet 10 min after step c using 70% ethanol.
2. Sacrifice the nonimmunized mouse (the same strain as the immunized mouse) two days before cell fusion by cervical dislocation or CO_2 asphyxiation (by placing the mouse in a closed container with a few pieces of dry ice). Soak the dead mouse briefly in a beaker of 95% ethanol prior to placing it in the laminar-flow cabinet.
3. Make a small incision in the skin covering the abdominal region and split the skin to expose the peritoneal cavity using sterile scissors and forceps.
4. Inject 3 to 5 ml of serum-free medium (DMEM) into the peritoneal cavity using a disposable syringe with a 20- or 23-gauge needle and gently squeeze the abdomen for 2 to 3 min using the fingers.
5. Slowly withdraw the mixed fluid containing the peritoneal exudate cells and place in a disposable petri dish and dilute the cells to 10 ml with serum-free DMEM.
6. Count the cells as follows:
   a. Dilute 0.1 ml of the cell suspension to 1 ml in a microcentrifuge tube containing 0.9 ml of staining solution (0.1 ml of 2 to 10% [w/v] trypan blue in distilled water is mixed with 0.8 ml of serum-free medium [DMEM]).
   b. Place a 0.4-mm dry coverslip on a dry hemaocytometer slide and place a drop of the diluted cell suspension on the groove at the edge of the slide. The drop should be drawn by capillary action onto the slide and it should form an even film to the edges without overflowing. This indicates the correct volume that is used for counting.

c. Focus on the lines under a microscope and the cells should be visible using a 10X objective lens.
d. Count the number of cells within the four corner areas.
e. Calculate the cell numbers using the following equation:

$$\text{Cells per milliliter} = \frac{\text{total cells counted}}{\text{number of corners counted}} \times 10^4 \times 10 \text{ (dilution factor)}$$

For example,

Corner	Cells counted
1	80
2	100
3	120
4	100
Total	400

Cells per milliliter = $400/4 \times 10^4 \times 10$ (1:10 dilution) = 1×10^7
Total cells of original cell suspension = $1 \times 10^7 \times$ total volume.
Usually one mouse can yield 1×10^6 to 1×10^7 cells.

7. Dilute the cells to 4×10^5 cells/ml with serum-free DMEM and add one drop per well (approximately 50 µl per well, 2×10^3 to 2×10^4 cells per well) of the cell suspension into a 96-well plate using a sterile 5-ml pipette and cover the plate.
8. Place the plate containing the PECs in a cell culture incubator at 37°C with 5% CO_2. Allow cells to incubate for 2 days and check for possible contamination prior to their being used for hybridoma culture.

E. Preparation of Spleen Cells and Myeloma Cells for Fusion

1. Thoroughly clean the laminar-flow hood as described (**Section I.D**).
2. Sacrifice the immunized mouse 2 days before cell fusion by cervical dislocation or CO_2 asphyxiation (by placing the mouse in a closed container with a few pieces of dry ice). Soak the dead mouse briefly in a beaker of 95% ethanol prior to placing it on the sterile bench in the laminar-flow cabinet.
3. Lift a little skin of the abdominal region with one hand using sterile forceps and make a small incision in the skin with the other hand with sterile scissors. Tear the skin to expose the abdominal wall and make a small incision in the abdominal wall. Carefully remove the spleen (maroon color) using forceps and place it in a disposable petri dish containing 10 ml of serum-free medium (DMEM). Carefully remove the connective tissue such as fatty tissue from the spleen by use of a bent needle and a pair of forceps.
4. Inject 2 to 5 ml of serum-free medium (DMEM) into the spleen with one hand using a disposable syringe fitted with a 20- or 23-gauge needle and hold the spleen with the other hand using a pair of forceps. This causes the spleen to swell.
5. Gently tease the tissue apart using sterile forceps and a disposable syringe needle. The spleen cells will be released into the dish.
6. Remove the spleen cell clumps as much as possible and tilt the petri dish so as to cause the cells to flow to one side of the dish. Carefully transfer the cells to a sterile 15- to 30-ml centrifuge tube with a V-shaped bottom (avoid transfer of cell clumps to the tube).
7. Centrifuge at 50 g for 5 to 10 min at room temperature. While the cells are being spun down, prepare working Geys's hemolytic medium, which can reduce any damage to the lympoid cells. The working Geys's hemolytic medium consists of: 8 ml of Geys's solution A; 29 ml distilled, deionized water (ddH_2O); and 2 ml of Geys's solution B. The pH is adjusted to 7.2 using pH indicator and sterile-filtered into a sterile container. The medium should be used within 30 min (20 ml per spleen).
8. Carefully remove the supernatant from the cell pellet as much as possible, and with the fingers, flick the bottom of the tube in order to loosen the cell pellet. Add 4 ml of freshly prepared Geys's hemolytic

medium to the cells and gently suspend the cells using a sterile Pasteur pipette followed by adding 16 ml of the Geys's hemolytic medium. Mix well and allow to stand for exactly 5 min at room temperature.

9. While the spleen cells are being hemolyzed, collect myeloma cells, grown to mid-log phase in T-flasks (mouse myeloma cell lines are commercially available), into a sterile centrifuge tube with a V-shaped bottom by scraping the cells with a sterile rubber policeman.
10. Centrifuge both spleen cells and myeloma cells as in step 7 and resuspend the cells in 10 ml of serum-free DMEM at room temperature. Repeat centrifugation for myeloma cells and resuspend in 10 ml of serum-free DMEM.

F. Fusion of Spleen Cells and Myeloma Cells under Sterile Conditions

1. Count the cells as described in **Section I.D**. The viability should be >80%, and only a few red cells will be visible.
2. Using all of the spleen cells, make a mixture of spleen and myeloma cells (the ratio of spleen cells to myeloma cells is 5:1 or 10:1) in a sterile centrifuge tube with a V-shaped bottom.
3. Prepare aliquots of 50% (w/v) polyethylene glycerol (PEG, 6000) in serum-free DMEM medium and place in a 37°C water bath.
4. Centrifuge the spleen/myeloma cell mixture at 120 g at room temperature for 5 min and completely remove the supernatant.
5. Tap the bottom of the tube to loosen the cell pellet and slowly add 1 ml of 50% (w/v) PEG in serum-free medium to cells at a density of 10^7 to 10^8 cells/ml. The addition should be made dropwise (one drop) with constant swirling. Rapidly mix the cells in the viscous solution by pipetting up and down. After 1 to 2 min, start the dilution in the next step.
6. Add 3 ml (three times the volume of PEG solution) of the warmed, serum-free DMEM to the mixture dropwise (1 drop/1.5 s) over a 1.5-min period with gentle swirling in a 37°C water bath.
7. Add 12 ml of warmed serum-free DMEM as in step 6 over 3 min.
8. Centrifuge as in step 4, decant the supernatant, and resuspend the "fusion" cells in 10 ml of prewarmed complete DMEM containing 20% (v/v) fetal calf serum (FCS). Count the cells and dilute them to a final density of 10^5 to 10^6 cells/ml in complete DMEM.
9. Add 50 µl per well of the diluted "fusion" cell suspension into disposable 96-well plate(s). Add one drop per well (approximately 50 µl per well) of the feeder cell suspension (PECs) prepared in **Section I.D** in 96-well plate(s). Cover the plate(s) and incubate in an incubator (e.g., Queue Cell Culture Incubator) at 37°C with 5% CO_2 for 10 to 24 h prior to the next procedure.

G. Selection and Propagation of Hybridoma Cells

The spleen cells contain genes coding for antibodies against the antigen of interest. The myeloma cells contain genes that are able to multiply indefinitely but cannot code for their own immunoglobulin (Ig). However, they can secrete Ig coded by spleen cell genes after fusion. The myeloma cell line is usually hypoxanthine, aminopterin, and thymidine (HAT) selective, in which aminopterin blocks the main biosynthetic pathway for nucleic acids. The myeloma cells are also usually mutants that do not contain the gene coding for hypoxanthine guanine phosphoribosyl transferase (HGRPT) that is required for the salvage pathway selection. The basic principle of generating hybridized cells is that, after fusion, the mutant parent myeloma cells can proliferate in the absence of aminopterin, but they do not survive in HAT medium. The normal parent spleen cells can survive in the HAT medium via the salvage pathway, but they die out due to their sensitivity to ouabain or amphotericin B methylester in the HAT medium. Only the fusion products or hybrid cells, which contain the HGPRT gene from the parent spleen cells and the ouabain-resistance gene from the parent myeloma cells, can multiply indefinitely and secrete antibodies against the antigen of interest.

1. Add 50 µl per well of prewarmed (37°C) 2X HAT in complete DMEM containing 20% FCS to the growing cells in a sterile laminar-flow hood and return the plate to the incubator.
2. Feed the cells by adding 50 µl per well prewarmed 1X HAT in complete DMEM containing 20% (v/v) FCS once every 3 days.
3. Check for possible contamination and monitor the formation of colonies once every 2 to 4 days under a microscope set near the incubator, and quickly return the plate to the incubator.
4. When the hybridoma colonies (usually a heterogeneous mixture of hybridomas) are visible, start isolation and propagation of monoclonal hybridomas by the limiting dilution method.
   a. Carefully transfer the cells from the well to a sterile tube and add 4 ml of prewarmed 1X HAT in complete DMEM containing 20% (v/v) FCS supplemented with 5% ECGS.
   b. Count the cells as described previously in **Section I.D**.
   c. Dilute the cells to eight, four, two, and one cell(s) in 50 µl in a series of tubes using prewarmed 1X HAT in complete DMEM containing 20% (v/v) FCS supplemented with 10% PECs or 10% ECGS.
   d. Add 50 µl per well containing approximately one cell to the 96-well plate and retain in the incubator for 10 to 14 days or until colonies become visible.

H. Screening of Hybridoma Supernatant by ELISA and Harvesting of Monoclonal Antibodies[5]

1. When the supernatant in the well containing healthy and uncontaminated hybridomas turns from bright red to orange or yellow, it indicates that the supernatant is acid due to the respiration of the hybridoma cells. The supernatant should be carefully removed from the well, and prewarmed, fresh 1X HAT in complete DMEM containing 20% FCS and 10% PECs should be added to the well, which is then returned to the incubator.
2. Carry out ELISA to identify positive supernatants as described in **Section I.C**. A positive supernatant from a specific well contains antibodies against the antigen that was used for the immunization of the mouse.
3. Reclone the hybridoma cells secreting positive supernatant using the limiting dilution method, as the first single-limiting dilution usually does not ensure monoclonality. Normally, by two or three successive clonings, the monoclonal antibody-producing cells can be obtained. The medium should now be changed to HT in complete DMEM containing 10% (v/v) FCS for the subclonings.
4. Carry out ELISA to verify the positive supernatant as in step 2.
5. The healthy and uncontaminated cell lines that produce monoclonal antibodies should be transferred to 24-well plates or T-25 or T-75 disposable culture flasks containing 30 to 60 ml of complete DMEM with 10% FCS for large-quantity production of monoclonal antibodies.

Note: *If one uses T-flasks, the flasks should not be overfilled with medium, and they should be capped tightly.*

6. When the supernatant changes from red to yellow, harvest the supernatant by transferring the culture to a centrifuge tube and centrifuging at 120 g for 10 min at room temperature. In a sterile laminar-flow hood, transfer the supernatant to a fresh tube and adjust the pH to 7.2. After adding sodium azide to a final concentration of 0.01 mM, the supernatant can be stored at –20°C until use. The cell pellet can be resuspended in the complete DMEM for further culture or for being frozen.

I. Isotyping of Monoclonal Antibodies

Antibodies or immunoglobulin (Ig) contain an enzyme-like specificity for certain structural epitopes of the antigen. Antibodies can be classified into five subclasses: IgA, IgD, IgE, IgG, and IgM. IgG can be divided into IgG1, IgG2a, IgG2b, and IgG3. An antibody molecule contains two heavy chains and two light chains, which are held together by disulfide bonds. The monoclonal antibodies obtained can be isotyped as follows:

1. Dilute the antigen sample to 50 µg/ml in PBS (pH 7.2). According to the above design, coat appropriate wells of a disposable 96-well polystyrene ELISA plate by adding the antigen to the wells (50 µl per well). Cover the plate and incubate at room temperature for 2 h or at 4°C overnight.
2. Decant the antigen solution from wells by dumping the plate upside-down on a paper towel two to three times. Fill each well (100 to 200 µl per well) with 1 to 3% (w/v) bovine serum albumin (BSA) in PBS with 0.05% Tween-20 to block any remaining binding sites in the wells. Incubate at room temperature for 30 to 60 min.
3. Discard the blocking solution as in step 2 and add 50 µl per well of monoclonal antibody supernatant as with the first antibodies. Allow to incubate at room temperature for 60 min.
4. Carefully remove the first antibody solution from appropriate wells and immediately wash the plate by filling up each well with washing buffer followed by discarding the buffer after 2 min. Repeat the washing twice.
5. Add 50 µl per well of appropriate subclass-specific rabbit antimouse Ig such as IgG, IgG2a, IgG2b, IgG3, IgA, IgD, and IgM to the appropriate wells and incubate at 37°C for 60 min.
6. Wash the plate as in step 4.
7. Add 50 µl per well of horseradish peroxidase-labeled, affinity-purified goat antirabbit antibody diluted at 1:50 with washing buffer to each well. Allow to incubate at 37°C for 60 min.
8. Wash the plate as step 4.
9. Add 100 µl of freshly prepared peroxidase substrate solution to each well. The solution contains 2,2-azino-di(3-ethylbenzthiazoline sulfonic acid) (ABTS) that is diluted at 1:50 in 100 mM citrate phosphate buffer (pH 4.2) and 0.03% (v/v) hydrogen peroxide. Incubate at room temperature for 30 min. A positive reaction should be bluish green in color.
10. Place the plate in a cassette in an autoreader (e.g., Bio-Tek, E1310) for quantitative measurement and read the absorbance or the optical density at wavelengths of 405 to 415 nm according to the manufacturer's instructions.

J. Verification of Antibodies by Western Blot Analysis

In addition to the use of ELISA to test for positive reactions of the antibodies obtained, the specificity of the antibody produced should also be verified by Western blotting using a protein mixture containing the antigen for immunization. Putative monoclonal antibodies should detect the expected protein of known size in the protein mixture. The detailed procedures are described under the Western blotting protocols in **Chapter 8**.

K. Purification of Monoclonal Antibodies Using Affinity Chromatography[4]

1. Prepare an appropriate size of column by equilibrating Sephadex or equivalent beds conjugated with goat antimouse IgG or IgM antibody with a buffer (pH 7.2) containing 10 mM sodium phosphate and 100 to 200 mM NaCl. Allow the column to stand in a cold room for 1 h and drain away the liquid.
2. Load the antibody supernatant onto the column, collect the eluate, and reload onto the column two or three times in order to obtain optimal binding between the column adsorbant matrix and antibody.
3. Wash the column twice with ten volumes of the total column volume using the equilibrating buffer.
4. Elute the bound antibodies with elution buffer (pH 2.3) containing 100 mM glycine-HCl and 150 mM NaCl.
5. Neutralize the acidic eluate with 1 M Tris/150 mM NaCl to pH 7.2 to 7.4. The purified antibody can be directly stored at −20°C until use. If it is too dilute, the antibody can be precipitated with 50% (w/v) solid ammonium sulfate, centrifugation, resuspension, and dialysis. Alternatively, the eluate can be concentrated by inserting an appropriate concentration tube and centrifuging at 2000 to 3000 g for an appropriate period of time.

L. Freezing and Thawing Cell Lines under Sterile Conditions

Once the monoclonal antibody-producing cell lines are obtained, it is necessary to freeze some of the cells for long-term storage. The frozen cells can be thawed for future culture in order to prepare fresh antibody.

1. Freezing Cell Lines

1. Transfer healthy, uncontaminated, and mid-log grown cells to a centrifuge tube with a V-shaped bottom by scraping the cells with a sterile rubber policeman.
2. Centrifuge at 120 g for 5 min at room temperature and remove the supernatant. Tap the bottom of the tube to loosen the cell pellet and resuspend the cells in 1 ml of HB 101-DMEM (serum free).
3. Count the cells as described previously (**Section I.D**) and dilute the cells to 10^6 cells/ml with HB 101-DMEM.
4. Add an equal volume of cold (4°C) 20% (v/v) dimethyl sulfoxide (DMSO) in HB 101-DMEM to the cell suspension with gentle swirling. Both HB 101-DMEM and 10% DMSO will permit the cells to survive during freezing.
5. Aliquot the cell suspension (1 to 1.5 ml per vial) into 2-ml cryogenic vials set in a bucket of ice in a sterile laminar-flow cabinet. Tightly cap the vials, allow the vials to set on ice for 10 to 15 min, and place the vials inside a styrofoam-insulated cooler. Place the cooler at −80°C for 2 to 3 days to slowly freeze the cells.

Notes: *The vials should be clearly labeled with the cell line name. The monoclonal antibody-producing cell line can be named progressively according to the order of subcloning procedure for specific cell lines. For example, if a primary hybridoma is cultured in the A4-well of the 96-well plate, then transferred to the B2-well that produces a positive supernatant, then subcloned into the B4-well of a 24-well plate and further transferred into T-flask no. 5, which produces monoclonal antibodies, the name for this cell line can be XA4B2B4T5, where X refers to the antigen or protein name. Alternatively, the cell line can be named according to antigen name and be given a number such as X2 or X101.*

6. Transfer the frozen cells to a liquid-nitrogen tank for extended storage. Check and keep the liquid nitrogen above the safety level.

2. Thawing Frozen Cells

1. Remove the cell line of interest from the liquid-nitrogen tank and quickly immerse the vial in a 37°C water bath. Gently swirl the vial and stop thawing when the last ice crystals disappear. The cell suspension should still be cold.
2. Clean the outside of the vial with 75% ethanol prior to opening, and then transfer the cell suspension into a 50-ml centrifuge tube containing 20 ml of complete DMEM with 20% FCS. Mix well.
3. Centrifuge at 120 g for 10 min at room temperature and remove the supernatant. Tap the bottom of the tube to loosen the cell pellet and resuspend the cells in 1 to 2 ml complete DMEM containing 20% FCS.
4. Transfer the cell suspension to one or two wells of a culture plate and culture in an incubator at 37°C with 5% CO_2 as described previously (**Section I.D**).

II. Preparation of Polyclonal Antibodies

The primary difference between monoclonal and polyclonal antibodies is that polyclonal antibodies are a mixture of perhaps hundreds of different antibody molecules that are derived from and specific to different sections of the antigen. There is an array of methods for the preparation of polyclonal antibodies, depending on the specific antigen. These differences mostly pertain to differences in the purification of the antigen. The standard procedure is much more simple than the preparation of monoclonal antibodies and includes the following steps: (1) production of antigens of interest usually through the expression of protein fragments in *E. coli*, (2) purification of the antigen for immunization in rabbits, (3) injection of rabbits with the purified antigen or a homogenized mixture containing a substantial amount of the antigen, (4) collection of blood samples from the rabbits prior to final bleeding to assess the number of booster injections that may be needed to result in a good immune response, and (5) purification of the polyclonal antibodies from the blood serum after the final bleeding. Some of these general procedures are described in **Chapter 3** and this chapter.

A. Production of Antigens of Interest

Produce at least 2 mg of antigen. This will allow enough starting material for the repeated injections that are necessary to induce a strong immune response. The most common methods for this are the induction of expression of a cloned protein cDNA fragment in an *E. coli* strain (e.g., BL21 cell lines) harboring an IPTG inducible expression vector (e.g., pET 22b(+) from Novagen) with the cDNA in the appropriate open reading frame.

Note: *It is often the case (in our experience) that full-length cDNAs do not express well in E. coli depending on the size of the product. Consequently, the induction of a fragment (~500 to 800 bp [base-pairs]) specific to the gene of interest is highly recommended. If chosen carefully, this also may help to produce polyclonal antibodies that do not cross react with related gene products.*

B. Purification of the Antigen for Immunization

Purify the antigen for immunization. It is recommended that the antigen of interest (usually a protein) be as pure as possible. Pure antigen will make the resulting polyclonal antibodies much more reliable and specific. While antigen may be purified using SDS-PAGE (sodium dodecyl sulfate-polyacrylamide gel electrophoresis), there are a number of available kits that may aid in the purification process, such as HIS-tag kits designed to purify cloned and expressed proteins on nickel columns. Since the HIS-tag is only present on the expressed protein, such kits can often yield only one protein after the purification process.

Note: *Unless specifically necessary, glycoproteins should be de-glycosylation because animals can raise antibodies against the sugar residues, which may reduce the specificity of the antibodies and may result in potential cross-reactions between the antibody and a variety of proteins.*

C. In Vivo Immunization of Female Rabbits

1. Carry out *in vivo* immunization of female rabbits with the purified antigen. We recommend the use of at least two rabbits per antigen, since the procedure is time consuming. If one rabbit becomes ill, then it is always best to have a backup.
    a. Resuspend the purified antigen in PBS in a sterile microcentrifuge tube to a concentration of 1 to 5 µg/µl, and use 50 to 200 µg per injection, depending on the particular antigen to be used.

b. Combine 50 to 200 µg of the antigen sample with an equal volume of complete Freund's adjuvant for the first injection per rabbit. If the antigen was originally contained in polyacrylamide gel fragments, then a gel slice containing approximately 200 µg can be homogenized in Freund's adjuvant prior to injection. Mix thoroughly to obtain an emulsion using a sterile syringe or a pipette. Slowly take up the emulsified mixture in a 1-ml disposable syringe fitted with an 18- or 20-gauge needle and remove any air bubbles.
c. Take a blood sample (0.5 to 1.0 ml) as preimmune serum from the ear vein before the first injection. Allow the blood to clot by placing the sample at 4°C for approximately 30 min, and centrifuge at 8000 g for 10 min. Transfer the serum to a fresh tube, dilute it at 1:100 to 1:500 with PBS and store at −20°C until use.
d. Place the rabbit (New Zealand White) in a cage, and briefly clean the areas on the back or the butt area of the rabbit with 70% ethanol. Slowly inject the antigen mixture into the muscle in two to four areas. Return the rabbit to the cage.
e. Boost the rabbit by injecting the same amount of antigen as for the first injection. The antigen is mixed with incomplete Freund's adjuvant as above and injected two to four times at 21-day intervals.
2. Collect blood from the immunized rabbit. Two or four days after the third or fourth injection, take blood from the ear vein, up to 20 to 30 ml per rabbit. This can be done by placing the immunized rabbit in a cage, removing the hair from the ear vein area, and cleaning the area with 95% ethanol. To achieve a good flow of blood, spray the area with some xylene and make a cut in the ear vein using a disposable knife. Lower the ear and collect the blood in a sterile tube. Stop the bleeding with a dry cotton swab by pressing it on the cut area using the fingers until the bleeding stops. Return the rabbit to the cage.
3. Harvest the serum containing polyclonal antibodies against the antigen. Place the blood sample at 4°C for 30 to 90 min and centrifuge the clotted sample at 8000 g for 10 min at 4°C. Carefully transfer the serum containing polyclonal antibodies to a fresh tube, aliquot, and store at −20°C until use.
4. (**Optional**) Purify the polyclonal antibodies.[3] The serum may be purified with an appropriate Sephadex 4B or agarose column conjugated with purified antigen. Load the serum onto the column, collect the eluate, and reload the column several times. The bound antigen and antibody complex can be eluted by using the appropriate buffer. Store the purified antibodies at −20°C until use.
5. Check the quality of the antibodies by ELISA as described previously (**Section I.C**). It is also highly recommended that the antibodies be tested using Western blots and samples of the purified antigen as compared with the preimmune serum (see **Chapter 8** for procedures).

Notes: *(1) It is usually necessary to test a range of antibody and preimmune dilutions to determine the best antibody dilution for Western analysis. We recommend the following range of primary antibody dilutions be used to test blotted membrane strips containing pure antigen or* E. coli *extracts with induced antigen: 1:100, 1:500, 1:1500, 1:3000. Low dilutions usually result in high background, while high dilutions result in little signal. (2) The secondary antibody can usually be used at a dilution of 1:20,000 or 1:30,000.*

D. Purification of the Final Polyclonal Antibodies

Once it is determined that the rabbits are responding properly to the injections of antigen, then the final booster shot can be administered. Often the rabbits are sacrificed and a final large-scale bleeding is done to collect several hundred ml of serum. Purification of the serum can be done as above (step 4), and the final antibodies are ready for use and storage at −20°C.

Notes: *We also recommend that aliquots of the final antibody be made and stored at −80°C for long-term storage.*

Reagents Needed

Complete Freund's Adjuvant
Sigma-Aldrich, St. Louis, MO

Complete Medium (DMEM)
Add 5 to 20% (v/v) sterile, commercial fetal calf serum (FCS) to serum-free medium DMEM, depending on the percentage of FCS required for the particular medium. Thoroughly clean the outside of the FCS bottle prior to opening with 70% ethanol in a laminar-flow cabinet under sterile conditions. Warm the complete DMEM to 37°C before use.

Geys's Solution A (500 ml)
NH_4Cl (17.5 g)
KCl (0.93 g)
$Na_2HPO_4 \cdot 12H_2O$ (0.75 g) (disodium hydrogen orthophosphate)
KH_2PO_4 (0.06 g) (potassium dihydrogen orthophosphate)
D-Glucose (2.5 g)
Phenol red (0.03 g)
Gelatin (12.5 g)
Dissolve well after each addition in 400 ml of ddH_2O and add ddH_2O to a final volume of 500 ml. Aliquot into 20 to 50 ml per vial and autoclave at 15 psi for 15 min. Store at room temperature for up to 1 year.

Geys's Solution B (200 ml)
$MgCl_2 \cdot 6H_2O$ (8.4 g)
$MgSO_4 \cdot 7H_2O$ (2.8 g)
$CaCl_2$ (6.8 g)
Dissolve well after each addition in 150 ml ddH_2O and add ddH_2O to a final volume of 200 ml. Aliquot 10 ml per vial and autoclave at 10 psi for 10 min. Store at room temperature for up to 1 year.

100X HAT Medium (100 ml)
Hypoxanthine (136 mg)
Aminopterin (1.9 mg)
Thymidine (38.8 mg)
Dissolve well after each addition in 100 ml distilled water and autoclave. Aliquot and store at −20°C.

2X HAT Medium
Dilute 100X HAT medium in complete DMEM containing 20% (v/v) FCS and 10% ECGS (endothelial cell growth supplement).

100X HT Medium
The same as 100X HAT medium except no aminopterin is included.

Incomplete Freund's Adjuvant
Sigma-Aldrich, St. Louis, MO

Phosphate-Buffered Saline (PBS)

KH_2PO_4 (anhydrous) (0.92 g)
$K_2HPO_4 \cdot 3H_2O$ (6.39 g)
NaCl (28.6 g)
Add H_2O to a final volume of 3.5 L. The pH should be 7.2 to 7.4.
Autoclave.

Serum-Free Medium (DMEM)

Liquid media are commercially available. For commercial powder, slowly empty one bottle of Dulbecco's modified Eagle medium (DMEM) powder (Sigma-Aldrich, St. Louis, MO) into a clean beaker containing 800 ml distilled, deionized water (ddH_2O) and a stirring bar. After completely dissolving, add 3.7 g $NaHCO_3$. Adjust the pH to 7.2 with 1 N HCl. Add 10 ml of antibiotic-antimycotic (ABAM, usually contains penicillin or streptomycin) stored at –20°C to the medium and add ddH_2O to a final volume of 1000 ml. In a sterile laminar-flow cabinet, immediately filter the sterilized medium into a sterile bottle using a disposable Nalgene filtration unit with 0.22-μm filter under vacuum. Cover the bottle tightly and store at 4°C until use. Prewarm the medium to approximately 37°C and thoroughly clean the outside surface with 70% ethanol before opening.

Substrate Solution for Goat Antimouse Ig-Peroxidase-Conjugate

0.2% (w/v) *o*-Phenylenediamine (OPD)
0.03% (v/v) Hydrogen peroxide
In a buffer (pH 6.0) containing 17 mM citric acid and 65 mM sodium phosphate (dibasic)

Substrate Solution for Goat Antimouse Ig-Alkaline Phosphatase-Conjugate

0.1% (w/v) *p*-Nitrophenyl phosphate (NPP)
10 mM $MgCl_2$
In 50 mM sodium carbonate buffer (pH 9.8 adjusted with sodium bicarbonate)

Washing Buffer or Blotto

5% (w/v) Nonfat powdered milk in PBS or 1.5% BSA in PBS with 0.05% (v/v) Tween-20 detergent

References

1. **Harlow, E. and Lane, D.**, *Antibodies: A Laboratory Manual*, Cold Spring Harbor Press, New York, 1988.
2. **Zola, H.**, *Monoclonal Antibodies: A Manual of Techniques*, CRC Press, Boca Raton, FL, 1984.
3. **Diano, M., Le Bivic, A., and Hirn, M.**, A method for the production of highly specific polyclonal antibodies, *Anal. Biochem.*, 166, 224, 1987.
4. **Olmstead, J.B.**, Affinity purification of antibodies from diazotized paper blots of heterogenous protein samples, *J. Biol. Chem.*, 256, 11955, 1981.
5. **Gaastra, W.**, Enzyme-linked immunosorbent assay (ELISA), in *Methods in Molecular Biology*, Vol. 1, Walker, J.M., Ed., Humana Press, Clifton, NJ, 1984, chap. 38.

Chapter 8

Western Blot Hybridization

Leland J. Cseke, William Wu, June Seung Lee, and Soo Chul Chang

Contents

I. Separation of Proteins by SDS-PAGE and the Use of Two-Dimensional Gels 128
    A. SDS-PAGE..128
        1. Preparation of the Separating Gel...128
        2. Preparation of the Stacking Gel...129
        3. Loading the Samples and Protein Standard Markers onto the Gel 129
        4. Electrophoresis 130
    B. Two-Dimensional Gel Electrophoresis 131
        1. Preparation of the Focusing Gel (Range of pH 4 to 6) 131
        2. Loading the Samples onto the Gel 132
        3. Electrophoresis 133
        4. Postfocusing Procedures 133
II. Staining and Destaining of the Gel 134
    A. Coomassie Blue Staining and Destaining Method...134
    B. Silver Staining Method 135
III. Transfer of Proteins from the Gel to Membranes 136
    A. Wet Blotting 136
    B. Semidry Blotting 137
IV. Immunodetection of Specific Protein(s) 138
    A. Alkaline Phosphatase 138
    B. Chemiluminescence Detection..139
V. Quantitative Analysis of Proteins after Western Blot Hybridization 140
VI. Troubleshooting Guide...140
    Reagents Needed 141
References 145

Western blot analysis is based on a protein/protein hybridization technique that is used for immunodetection of specific antigen(s) of interest in a complex mixture of proteins. This is a simple, sensitive, and effective technology that has been used in immunology, molecular and cellular biology, and protein chemistry. The principle of Western blotting is as follows: (1) A protein mixture is first separated according to molecular size using sodium dodecyl sulfate-polyacrylamide gel electrophoresis (SDS-PAGE). (2) The separated protein molecules are then immobilized onto a nitrocellulose or polyvinylidene difluoride (PVDF) membrane. (3) The specific protein band of interest is identified by use of a specific antibody raised against a specific antigen (protein), which can specifically bind to the antigen (protein) of interest in the protein mixture that is immobilized on the membrane. (4) The antibody–antigen complex is then detected by an enzyme linked to a second antibody and substrate, or by the use of ^{125}I-labeled protein A or its equivalent.[1-4] The antigen can also be directly detected with a fluorescence-labeled antibody, which directly binds to the antigen in a protein mixture. After washing away the nonbound antibody, the antigen-labeled antibody can then be visualized under a microscope fitted with an ultraviolet (UV) lamp that can excite the fluorescent tag using a specific wavelength of light. However, the direct method needs a relatively large amount of the labeled antibody to obtain good detection, and the labeling of an antibody of particular interest is expensive. Therefore, the indirect immunodetection method is widely used today. The following detailed protocols are based on modifications in the method of Towbin et al.,[2] who first developed the technique.

I. Separation of Proteins by SDS-PAGE and the Use of Two-Dimensional Gels

A. SDS-PAGE

SDS is an anionic detergent that denatures proteins and makes them negatively charged by wrapping around the polypeptides. This binding results in equal charge densities per unit length of protein, thereby eliminating the ionic charges of individual amino acids. Consequently, SDS-PAGE can separate and determine the molecular weights (MW) of proteins using standardized protein-size markers. There is a linear relationship between the log of the MW of a polypeptide and its R_f, which is the ratio of the distance from the top of the gel to the polypeptide divided by the distance from the top of the gel to the dye front. A standard curve can then be generated by plotting the R_f of each standard polypeptide marker as the abscissa and the $\log_{10}$ of its MW as the ordinate. The MW of an unknown protein can then be determined by finding the R_f that vertically crosses on the standard curve and reading the $\log_{10}$ MW that horizontally crosses to the ordinate. The antilog of the $\log_{10}$ MW is the actual MW of the protein.

1. Preparation of the Separating Gel

The following protocols are directed toward the use of larger gels. However, minigels are often more common in the analysis of proteins due to their ease of preparation and faster speeds of analysis. If minigels are to be used, the volumes of the reagents contained below can simply be proportionally reduced to match the volumes required for the particular apparatus to be used.

1. Thoroughly clean glass plates and spacers with detergent, wash with tap water, rinse with distilled water several times, and air dry.
2. While wearing gloves, wipe dry the glass plates and spacers with 100% ethanol. Assemble a vertical slab gel unit, such as the Hoefer™ SE 600 Vertical Slab Gel Unit (Amersham Biosciences, Piscataway, NJ), in the casting mode according to the instructions. Place one spacer (1.0 to 1.5 mm thick) at each of the two sides between the two glass plates and fix in place with clamps, forming a sandwich. Repeat for the second sandwich. The size of the separating gel varies with individuals. A standard size is 120 × 140 × 1.5 mm.

Note: *If desired, a little grease oil can be sprayed on the spacer areas at both the top and bottom ends of the sandwiches to prevent potential leaking.*

3. Place the unit on a level surface, and check for potential leaks by pipetting some distilled water into the sandwich. Then drain away the water by inverting the unit.
4. Prepare the separating gel solution on ice in a clean 100-ml beaker or a 125-ml flask in the order shown below (for two standard-sized gels):
   - 20 ml Monomer solution
   - 15 ml Running gel buffer
   - 0.6 ml 10% SDS
   - 24.1 ml distilled, deionized water (ddH_2O)
   - Mix after each addition.
5. Degas the solution to remove any bubbles by use of vacuum or a sonicator for 5 min (optional). Add 0.25 ml of freshly prepared 10% ammonium persulfate (AP) and 20 µl tetramethylethylenediamine (TEMED) and mix by gently swirling. Do not make air bubbles.
6. Immediately pipette the mixture with a 50-ml syringe into the assembled sandwich up to a level that is 4 cm from the top.
7. Take up 0.8 ml of ddH_2O in a 1-ml syringe equipped with a 2-in.(5.08 cm) 22-gauge needle or its equivalent and very slowly load 0.3 ml of the water, starting from one top corner next to the spacer, onto the surface of the acrylamide gel solution. Repeat on the other side of the slab next to the other spacer. The water layer will evenly flow across the surface of the gel mixture. The purpose of applying water is to make the surface of the gel very even. A very sharp gel–water interface can be visible after the gel has polymerized.
8. Drain away the water layer by gently tilting the casting unit and rinse once with 2 ml of overlay solution. Add 1 ml of overlay solution on top of the gel and allow the gel to sit for 2 h or overnight by covering the top of the gel with a piece of Parafilm™ or SaranWrap™ to prevent evaporation.

2. Preparation of the Stacking Gel

1. Prepare the stacking gel solution in a clean 50-ml beaker or a 50-ml flask in the order shown below (for two pieces of standard-sized gels):
   - 2.66 ml Monomer solution
   - 5 ml Stacking gel buffer
   - 0.2 ml 10% SDS
   - 12.2 ml ddH_2O
   - Mix well after each addition.
2. Drain the overlay solution on the separating gel by tilting the casting unit and rinse with 2 ml of the stacking gel mixture. Drain away the stacking gel mixture and insert the comb (1.0 to 1.5 mm thick) into the glass sandwich according to the instructions for the apparatus.

Notes: The stacking gel solution may be filled prior to inserting the comb into the sandwich. However, air bubbles are easily generated and trapped around the teeth of the comb. We recommend that the comb be inserted into the sandwich prior to filling the stacking gel mixture.

3. Add 100 µl of freshly prepared 10% AP and 10 µl TEMED to the stacking gel solution and mix by gently swirling. Do not make air bubbles.
4. Immediately and slowly pipette the solution, from both sides next to the spacers, into the sandwich up to the top. Allow the gel to polymerize and sit for 30 min.

Note: The leftover stacking gel solution can be stored at 4°C to slow the polymerization process. If the stacking gel leaks a little, more solution can then be added to the top of the gel to generate good wells.

3. Loading the Samples and Protein Standard Markers onto the Gel

1. While the stacking gel is polymerizing, prepare samples and protein standard markers. Add one volume of 2X denaturing loading buffer to each of the samples and to the protein standard markers. Cap the tubes and place them in boiling water for 2 to 4 min. Briefly spin down before use.

Notes: At this stage, the proteins and markers are denatured. This is important for efficient electrophoresis. The isolation and purification of proteins are described in **Chapter 3**. The amount of proteins loaded into one well should be 10 to 35 µg in a total of 5 to 15 µl for Coomassie Blue staining and Western blotting, and it should be 5 to 15 µg in 5 to 10 µl for highly sensitive silver staining. The amount of protein standard markers for one well should be 2 to 10 µg.

2. Slowly and vertically pull the comb straight up from the gel. Rinse each well with Tris-glycine running buffer using a syringe or pipette to add the Tris-glycine running buffer. Carefully invert the casting stand to drain the wells and repeat twice.
3. Position the casting stand upright and fill each well with 10 µl of Tris-glycine running buffer. Take up sample or markers into a syringe equipped with a 2-in. (5.08 cm) 22-gauge needle or a pipette with an equivalent tip attached. Carefully insert the needle into the Tris-glycine running buffer in each well and underlay the sample or standard marker in the well. Repeat until all of the samples are loaded.

Notes: The volume loaded into each well should be less than 40 µl for a standard-sized gel or less than 15 µl for a minigel. Overloading may cause samples to float out and cause contamination among wells. We recommend that the markers be loaded into the front left or right well, or both wells. The volume and order of each sample loaded in the two sheets of gels should be identical. One gel will be used for staining in the determination of the MW of the proteins. The other gel will be used in Western blotting. Otherwise, it is also possible to use prestained standard markers. There are various kinds of markers that are commercially available. In this case, one gel may be adequate for Western blot analysis.

4. Carefully overlay the wells with Tris-glycine running buffer, using a syringe or pipette, until the top of the gel is covered.

Notes: The samples and markers containing blue dye should be visible at the bottom of each well. Mark the orientation of samples and record the order and orientation.

4. Electrophoresis

1. Carefully put the upper buffer chamber in place according to the instructions. Remove lower clamps and clamp the sandwiches to the bottom of the upper buffer chamber. **Do not disturb the wells.** Check for any potential leaking by adding 10 to 20 ml of Tris-glycine running buffer into the chamber.
2. Carefully transfer the assembled unit into the lower buffer chamber according to the instructions. Place the entire tank on a magnetic stirrer. Slowly fill the lower chamber with 3 L of Tris-glycine running buffer and add 1 L to the upper chamber, or use an appropriate volume depending on the sizes of the chambers.

Notes: The volume of bottom-chamber buffer should be sufficient to cover two-thirds of the slabs; otherwise, the heat generated during electrophoresis will not be distributed evenly, thus causing distortion of the band patterns on the gel. For the upper chamber, gently fill with the Tris-glycine running buffer, adding it from one corner into the chamber. Do not pour the buffer into the well areas to avoid washing the samples out.

3. Put a stir bar in the lower chamber and stir the buffer for 1 min to remove any air bubbles trapped under the ends of the sandwiches.
4. Add a few drops of dye solution such as phenol red or equivalent into the upper-chamber buffer if the dye in the samples is not sufficient.
5. Place the lid, or its equivalent, on the unit and connect the power supply. The cathode (negative pole) should be connected to the upper buffer chamber and the anode (positive pole) should be connected

to the bottom buffer chamber. Proteins that become negatively charged by SDS will migrate from the cathode to the anode and will separate according to their MW.

6. Set the power supply at a constant power or current and turn it on. Adjust the current to 25 to 30 mA per 1.5-mm thickness of standard-sized gel. The voltage will increase during the running process due to the increase in heat. Allow the gel to run electrophoretically for several hours until the dye reaches the bottom of the gel.
7. Turn off the power supply and disconnect the power cables. Loosen and remove the clamps at both sides of the sandwiches. Place the gel sandwiches on the table or bench. Slowly remove the spacers and use an extra spacer to carefully separate the two glass plates, starting from one corner. Remove the top plate; the gel should be on the bottom glass plate. Make a small cut at the upper left or right corner of the gel to record the orientation by using a razor blade. Use one gel for staining if needed or to determine the MW of different proteins. The other gel can be used for Western blotting.

B. Two-Dimensional Gel Electrophoresis

The use of one-dimensional gels can routinely be used to separate a mixture of proteins on the basis of individual molecular weights. However, under some circumstances, more information may be required about individual proteins. In these cases, two-dimensional (2-D) gel electrophoresis may be used to better separate proteins based on additional criteria.

Two-dimensional gel electrophoresis consists of a first-dimensional gel, which is an isoelectric focusing (IEF) gel, and a second-dimensional gel, SDS-PAGE. IEF separates proteins based on each protein's individual isoelectric point (pI) values, while SDS-PAGE separates them based on their molecular weights. When done correctly, the results of two-dimensional gel electrophoresis can give high resolution of proteins. Classical methodology for the two-dimensional gel electrophoresis is well established by O'Farrell.[5] In order to separate proteins according to their own pIs, pH gradients should be made in a gel. This can be achieved by using carrier ampholytes that have high buffer capacities at their pIs. Upon application of an electric field after mixing these ampholytes in a gel, negatively charged and positively charged ampholytes move toward the anode and cathode, respectively. This movement results in alignment of the ampholytes between the cathode and the anode according to their pIs. However, ampholytes are not isoelectric at times, and loss of basic ampholytes can occur during electrophoresis. This can affect the pH gradient and its stability. Such problems can be overcome using immobilines that are copolymerized with acrylamide and *bis*-acrylamide, resulting in stable pH gradients during electrophoresis. Thus, sufficient focusing can be performed for problematic proteins to attain their isoelectric points.[6] IEF, using immobilized pH gradients, is well-described elsewhere.[7] While there are many different protocols for IEF, the following protocols are a good starting point in the use of two-dimensional gel electrophoresis of most protein extracts.

1. Preparation of the Focusing Gel (Range of pH 4 to 6)

The preparation of the focusing gel is similar to the preparation of the gel for SDS-PAGE, except no stacking gel is required. The same sample is run in several lanes, which are subsequently separated into individual strips for analysis of the pH gradient, staining, and for use on the two-dimensional gel.

1. Thoroughly clean two glass plates and spacers with detergent, wash with tap water, rinse with distilled water several times, and air dry.
2. Wear gloves and wipe dry the glass plates and spacers with 100% ethanol. Assemble the vertical slab gel unit, such as the SE 600 Vertical Slab Gel Unit (Amersham Biosciences), in the casting mode according to the instructions. Place one spacer (1.0 to 1.5 mm thick) at each of the two sides between the two glass plates and fix in place with clamps, forming a sandwich. Repeat for a second sandwich, if required. A standard size is $120 \times 140 \times 1.5$ mm.

Note: If desired, a little grease oil can be sprayed on the spacer areas at both the top and bottom ends of the sandwiches to prevent potential leaking.

3. Check for potential leaks by pipetting some distilled water into the sandwich and then drain away the water by inverting the unit. Set the unit on an even and level surface.
4. Prepare the focusing gel solution in a clean 100-ml beaker or a 125-ml flask in the order shown below (for one standard-sized gel):
   - 8.0 ml Monomer solution
   - 192 µl Ampholyte solution, pH 3 to 10
   - 960 µl Ampholyte solution, pH 4 to 6
   - 21.6 ml ddH$_2$O
   - Ultrapure urea (24.0 g)
   - Mix well after each addition.

Note: If the solution is warmed slightly, the urea will dissolve more rapidly.

5. Degas the solution to remove any bubbles by use of vacuum or a sonicator for 5 min (optional). Add 0.10 ml of freshly prepared 10% AP and 20 µl TEMED and mix by gently swirling. Do not make air bubbles.
6. Immediately pipette the mixture with a 50-ml syringe into the assembled sandwich up to the top and allow the gel to sit for 2 h or overnight by covering the top of the gel with a piece of Parafilm or SaranWrap to prevent evaporation.

Notes: The focusing gel solution may be filled before inserting the comb into the sandwich. However, air bubbles are easily generated and trapped around the teeth of the comb. We recommend that the comb be inserted into the sandwich before filling the focusing gel mixture.

2. Loading the Samples onto the Gel

1. While the focusing gel is polymerizing, prepare the samples. Add one volume of 2X denaturing loading buffer (for IEF) to each of the samples. Cap the tubes and briefly spin down and place on ice until use.

Notes: At this stage, the proteins are denatured, which is important for efficient electrophoresis. The isolation and purification of proteins are described in **Chapter 3**. The amount of proteins loaded into one well should be 10 to 35 µg in a total of 5 to 15 µl. The amount of protein standard markers for one well should be 2 to 10 µg.

2. Slowly and vertically pull the comb straight up from the gel. Rinse each well with distilled water using a syringe or pipette to add the Tris-glycine running buffer. Carefully invert the casting stand to drain the wells and repeat twice.
3. Take up each sample into a syringe equipped with a 2-in. 22-gauge needle or a pipette with an equivalent tip attached. Carefully insert the needle into the Tris-glycine running buffer in each well and underlay the sample or standard marker in the well. Repeat until all of the samples are loaded.

Notes: The volume loaded into each well should be less than 40 µl for a standard-sized gel or less than 15 µl for a minigel. Overloading may cause samples to float out and cause contamination among wells. The volume and order of each sample loaded in the gel should be recorded.

4. Carefully overlay the wells with Tris-glycine running buffer, using a syringe or pipette, until the top of the gel is covered.

Notes: The samples containing blue dye should be visible at the bottom of each well. Mark the orientation of samples and record in a lab notebook.

3. Electrophoresis

1. Carefully put the upper buffer chamber in place according to the instructions. Remove lower clamps and clamp the sandwiches to the bottom of the upper buffer chamber. **Do not disturb the wells.** Check for any potential leaking by adding 10 to 20 ml of Tris-glycine running buffer into the chamber.
2. Carefully transfer the assembled unit into the lower buffer chamber according to the instructions. Place the entire tank on a magnetic stirrer. Slowly fill the lower chamber with 3 L of anolyte solution and add 1 L of catholyte solution to the upper chamber, or use an appropriate volume depending on the sizes of the chambers.

Notes: The volume of bottom-chamber buffer should be sufficient to cover two-thirds of the slabs; otherwise, the heat generated during electrophoresis will not be distributed evenly, thus causing distortion of the band patterns on the gel. For the upper chamber, gently fill with the Tris-glycine running buffer, adding it from one corner into the chamber. Do not pour the buffer into the well areas to avoid washing the samples out.

3. Put a stir bar in the lower chamber and stir the buffer for 1 min to remove any air bubbles trapped under the ends of the sandwiches.
4. Add a few drops of dye solution such as phenol red or equivalent into the upper-chamber buffer if the dye in the samples is not sufficient.
5. Place the lid, or its equivalent, on the unit and connect the power supply. The cathode (negative pole) should be connected to the upper buffer chamber, and the anode (positive pole) should be connected to the bottom buffer chamber. Proteins will migrate from the cathode to the anode and separate according to their individual pIs.
6. Set the power supply at a constant voltage. Adjust the voltage to 70 V and run until the dye front enters the focusing gel. Then set the voltage at 150 V and run until the dye front reaches the bottom of the gel. The current will decrease from 10 mA during the running process.

4. Postfocusing Procedures

1. Turn off the power supply and disconnect the power cables or their equivalent. Loosen and remove the clamps at both sides of the sandwiches. Place the gel sandwich on the table or bench. Slowly remove the spacers and use an extra spacer to carefully separate the two glass plates, starting from one corner. Remove the top plate; the gel should be on the bottom glass plate. Make a small cut at the upper left or right corner of the gel to record the orientation by using a razor blade.
2. In order to determine the pH gradient of the gel, cut a strip of gel into 1-cm slices and suspend each slice in 1 ml of 10 mM KCl for 30 min. Read the pH of the KCl solution.
3. Cut the gel lane-by-lane, leaving one lane attached to the marker lane. This section can be used for staining to visualize and estimate the pIs of the proteins once compared with the pH at different sections of the gel. Indicate where the cathode- or anode-attached ends of the gel strips are located by making appropriate cuts in the gel.
4. Prepare standard SDS-PAGE gels with both resolving gel and stacking gels as described in the SDS-PAGE section of this chapter (**Section I.A**), with the exception that no comb is added to the top of the stacking gel. Instead, the gel is allowed to polymerize, forming a continuous and smooth upper gel.
5. Place the gel strips in equilibration buffer for 15 to 30 min.
6. Insert one gel strip at the top of the stacking gel for SDS-PAGE with or without agarose solution to make the top of the gel uniform.

Note: If the gel strips are not used immediately, store them between two sheets of plastic film at $-80°C$.

7. Perform SDS-PAGE as described in **Section I.A**.

II. Staining and Destaining of the Gel

A. Coomassie Blue Staining and Destaining Method

1. Carefully transfer the gel to be stained to a glass or plastic tray containing 100 to 200 ml of Coomassie Blue (CB) staining solution and stain for 4 to 8 h at room temperature with slow shaking at 60 rpm. Alternatively, cover the tray tightly and stain for 20 to 30 min in a 50°C water bath.

Note: Minigels will stain and destain much more quickly.

2. Remove the staining solution and replace it with destaining solution I. Allow destaining to take place for 1 to 1.5 h.
3. Gently transfer the stained gel and carefully place it between two sheets of supporting matrix. Gently roll the sandwich and insert it into a cylinder holder and cover the two ends of the holder. An alternative way is to keep the gel in the tray and replace destaining solution I with destaining solution II. Destain the gel with shaking at 60 rpm until a clear background is obtained.
4. Place the cylinder into the SE 530 Destainer (Amersham Biosciences) or its equivalent filled with destaining solution II. Put a stir bar into the tank and place the tank on a magnetic stirrer. Turn on the stir bar and allow the gel to destain overnight or until a clear background is obtained.

Notes: (1) Photography of the destained gel to keep as a record is strongly recommended. The photograph may be made after the gel has been dried. However, the gel is sometimes damaged during the drying process. (2) A fast destaining method may be used to destain the gel, using 0.5 to 1% Clorox® (containing sodium hypochlorite) in distilled water. Constantly monitor the destaining of the background. When clear bands are just visible, immediately rinse the gel with a large volume of distilled water several times until the blue color is stable. The disadvantage of this method is that if overdestaining occurs, some bands may disappear. Never use >1.5% Clorox®.

5. Carefully place the gel between a water-prewetted thin film (commercially available for gel drying), remove any air bubbles inside the sandwich to prevent any cracks from developing in the gel, and dry the gel at 50 to 70°C under vacuum for 1 h (an example is shown in Figure 8.1).

Note: There is a variety of ways to dry protein gels, and procedures will vary with different approaches.

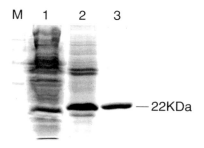

FIGURE 8.1
Coomassie Blue-stained gel showing a protein purification scheme for a peptide of interest. Lane M: marker lane; Lane 1: crude extract from IPTG-induced *E. coli* harboring the expression construct; Lane 2: sample from an intermediate urea-purification step; Lane 3: pure peptide after HIS-tag purification. This fragment was used to prepare specific polyclonal antibodies, as described in **Chapter 7**.

Western Blot Hybridization 135

6. Measure the distance from the top of the gel to each of the bands in the protein standard marker lane, and do the same thing for every band of interest in the sample lanes. Calculate the R_f for each band by dividing the distance from the top of the gel to the specific band by the distance from the top of the gel to the dye front. Generate a standard curve by plotting the R_f of each standard polypeptide marker as the abscissa and the $\log_{10}$ of its MW as the ordinate. The MW of an unknown protein can then be determined with ease by finding its R_f that vertically crosses the standard curve and reading the $\log_{10}$ MW horizontally across to the ordinate. The antilog of the $\log_{10}$ MW is the actual MW of the protein. Repeat until the MWs of all of the visible bands are determined.

B. Silver Staining Method

Silver staining is a very sensitive technique compared with Coomassie Blue staining. Some very weak bands that do not become visible using Coomassie Blue staining become very sharp after silver staining. The disadvantage is that the silver staining procedure is more complicated, and the background using silver staining may be high. In our experience, the following protocols work well.

Notes: *Silver staining is sensitive enough to stain fingerprints on the gel; gloves should be worn when handling the gel. Distilled, deionized water (ddH_2O) should be used in the procedure.*

1. Carefully transfer the slab gel into a clean container containing 400 ml of fixation solution. Allow fixation to occur for 30 to 60 min with slow shaking at 60 rpm.
2. Remove the fixation solution and wash the gel with 500 ml of 30% ethanol for 30 min and then 500 ml of water for 30 min with shaking at 60 rpm. Repeat the ethanol/water washing three more times.

Note: *The gel should shrink in 30% ethanol and swell in water. This helps in the washing of the gel.*

3. Stain the gel in 200 ml of silver nitrate ($AgNO_3$) staining solution (or alternative silver staining solution) for 20 to 30 min with shaking at 60 rpm.
4. Immediately and quickly rinse the gel in 400 ml water three times.
5. Develop the gel in 200 ml of silver developer solution (or alternative silver developer solution) for 10 min with occasional slow shaking.

Note: *The formaldehyde in these solutions is toxic and should not be inhaled. Wear gloves during these procedures.*

6. Stop the development process in 400 ml of silver stopping solution for 10 min with shaking at 60 rpm.
7. Wash the gel in 500 ml water five times at 10-min intervals with shaking at 70 rpm.
8. Clear the bands in Farmer's reducer solution for just 1 to 2 min and quickly rinse the gel with 500 ml water. Repeat washing at least five times at 10-min intervals in distilled water to completely remove Farmer's reducer solution until a clear background on the gel is obtained.
9. If required, repeat the staining, rinsing, developing, stopping, and washing as described in steps 3 through 7.
10. Optional: Stained background can be cleared slightly by incubating the gel with 200 ml of a 1:5 dilution of Farmer's reducer solution for 2 to 5 min, stop, and wash the gel as described in step 7.
11. Dry the gel and calculate the MWs of the different proteins as described in the Coomassie Blue staining method (**Section II.A**).

III. Transfer of Proteins from the Gel to Membranes

A. Wet Blotting

1. Soak the gel, the same size of nitrocellulose membrane or its equivalent, and four pieces of 3MM Whatman™ filter paper (relatively bigger than the gel) in 500 ml of blotting buffer for 15 to 20 min. If polyvinylidene difluoride (PVDF) is used, the membrane should be soaked in 100% methanol briefly (30 sec) and then soaked in blotting buffer.
2. Fill a tray that is large enough to hold the cassette with blotting buffer to a depth of 2.5 to 5 cm.

Notes: *The cassette should be loaded under the buffer in order to avoid trapping any air bubbles between the different layers. Gloves should be worn when assembling the electroblotting apparatus.*

3. Place one-half of the cassette in the tray with the hook facing up, put one dacron sponge on the cassette half, and press on the sponge several times to force out any air bubbles.
4. Place two pieces of the soaked 3MM Whatman filter paper onto the sponge and lay the soaked nitrocellulose membrane or its equivalent on the filter papers. Avoid any air bubbles between the layers.
5. Carefully place the soaked gel on the membrane and lay two soaked pieces of 3MM Whatman filter paper on the gel.

Note: *Gently press on the filter papers to force out any trapped air bubbles that will block the local transfer of proteins to the membrane.*

6. Cover the filter papers with another dacron sponge and place the second half of the cassette on the top of the stack so that the hook is down and faces the hole near the edge of the bottom half.
7. Press the two halves together and slide them toward each other so that the hooks are engaged with the opposite half.
8. Insert the assembled cassette into the blotting chamber according to the instructions. Fill the chamber with 4°C chilled blotting buffer sufficient to cover the cassette. Place a stir bar at the bottom of the chamber.

Notes: *(1) Make sure to insert the cassette in the right orientation so that the membrane is between the gel and the anode. (2) Because heat will be produced quickly due to the current, the current should be monitored so as not to be higher than 1.5 A, which can burn the apparatus. Chilled blotting buffer will help to keep the voltage from increasing too rapidly during the transfer. Some blotting units have a container for ice or an ice block that can also be used to keep the solutions cool. The temperature should be controlled so that it is less than 60°C.*

9. Place the whole chamber on a magnetic stirrer plate, turn on the stir bar at low speed, connect to the power supply, and turn on the current starting at 0.8 to 1 A.
10. Allow the blotting to take place for 60 to 75 min at 1 to 1.5 A.
11. When the blotting is complete, quickly mark the orientation of the membrane and immediately rinse it with 200 ml of phosphate-buffered saline (PBS).

Notes: *If using nitrocellulose membranes, do not let the membrane become dry at this stage, as this usually brings about a higher background during immunodetection. If one's time is limited, the membrane at this stage can be wrapped wet with SaranWrap and stored at 4°C until use. If using PVDF membranes, the membrane should be dried completely after transfer to enhance the protein binding. To determine if the transfer was successful, PVDF* membranes can also be wetted with 20% methanol after drying. The bound proteins will appear as clear areas against an opaque membrane.

12. Incubate the membrane in 400 ml of blocking solution at room temperature for 45 min with shaking at 60 rpm. This step serves to block the remaining binding sites on the membrane. Proceed to immunodetection.

Notes: *The membrane at this stage may be dried and wrapped with filter papers or equivalent, and stored at 4°C until use. To check for the efficiency of blotting, the blotted gel can be stained with Coomassie Blue and then destained. A successful transfer should have no visible bands left.*

B. Semidry Blotting

1. After performing SDS-PAGE, soak the gel in 200 ml of blotting buffer for 10 min.
2. Cut a sheet of nitrocellulose membrane or its equivalent and ten pieces of 3MM Whatman filter paper to the same size as the gel and soak them in blotting buffer before blotting. If PVDF is used, the membrane should be soaked in 100% methanol briefly (30 sec); then, it can be soaked in blotting buffer.
3. Wash the anode (bottom) plate of a semidry blotting apparatus with distilled water and lay five pieces of the soaked 3MM Whatman filter paper on the plate one by one.

Note: *Remove any bubbles trapped between the papers by rolling a pipette over the filter papers.*

4. Overlay the soaked membrane on the top of the 3MM Whatman filter papers and remove any bubbles trapped.
5. Carefully place the soaked gel on the membrane and gently lay five soaked pieces of 3MM Whatman filter paper on the gel.

Note: *Gently press on the filter papers to force out any trapped air bubbles that will block the local transfer of proteins to the membrane.*

6. Put the cathode (top) plate in distilled water and place the plate carefully on the assembled unit.

Note: *Do not disturb the assembled apparatus or bubbles may be produced.*

7. Put anything that weighs approximately 0.5 kg, such as a bottle or beaker that contains 0.5 L of water, on the top plate. This can serve as a weight without crushing the gel.
8. Connect the assembled apparatus to a power supply (bottom plate to positive, red pole). Run the power supply at 1.2 mA·cm^{-2} of membrane for 1 h at room temperature.

Notes: *(1) Two or three gels can be used for blotting by inserting five pieces of 3 MM Whatman paper between each gel and membrane. (2) Minigels usually require less current. At typical 8×10 cm^2 gel can be run at 50 mA for 40 min.*

9. When the blotting is complete, quickly mark the orientation of the membrane and immediately rinse it with 200 ml of phosphate-buffered saline (PBS).

Notes: *Do not let the membrane become dry at this stage, as this usually brings about a higher background during immunodetection. If one's time is limited, the membrane at this stage can be wrapped wet with SaranWrap and stored at 4°C until use. If using PVDF* membranes, the membrane should be dried completely after transfer to enhance the protein binding. To determine if the transfer was successful, PVDF membranes can also be wetted with 20% methanol after drying. The bound proteins will appear as clear areas against an opaque membrane.

10. Incubate the membrane in 400 ml of blocking solution at room temperature for 45 min with shaking at 60 rpm. This step serves to block the remaining binding sites on the membrane. Proceed to immunodetection.

Notes: The membrane at this stage can be dried and wrapped with filter papers or equivalent, and stored at 4°C until use. To check for the efficiency of blotting, the blotted gel can be stained with Coomassie Blue and then destained. A successful transfer should have no visible bands left.

IV. Immunodetection of Specific Protein(s)

A. Alkaline Phosphatase

1. Incubate the unblocked membrane in 400 ml of blocking solution for 45 min with shaking at 60 rpm in order to block the nonbinding sites on the membrane. If the membrane was previously blocked and dried, soak the membrane in 200 ml of PBS solution containing 0.05% (v/v) Tween-20 for 20 min.

Notes: (1) Extensive optimization is required for each antibody produced. To this end, protein dot blots containing antigen and nonantigen samples or gels containing repeated loadings of the same antigen sample can be prepared. After transfer, the membrane can be cut into a number of strips for analysis with different dilutions of both preimmune and immune serum (i.e., 1:400; 1:800; 1:1200; 1:2400, or 1:4000). The different controls given in Table 8.1 are also strongly recommended, since they will assess the quality of the immune serum compared with the preimmune serum. (2) Gloves should be worn when handling the membrane.

2. Transfer the membrane filter to an appropriately sized tray or place the membrane strips in an assay plate using a pair of forceps. Incubate with first or primary antibody (monoclonal or polyclonal antibodies) at the determined optimal concentration diluted in PBS/BSA/T solution (1% [w/v] BSA or equivalent and 0.05% [v/v] Tween-20 in PBS buffer). Allow incubation to take place at room temperature for 1 to 3 h with slow shaking at 60 rpm.

TABLE 8.1
Recommended Controls for the Membrane Strips

Control Groups	Membrane Strip Number (duplicate for each)					
	1	2	3	4	5	6
Antigen	+	+	+	+	+	+
First antibody	−	+	−	−	−	+
Second antibody	−	−	+	−	+	+
Preimmune serum	−	−	−	+	+	−
Color substrate	+	+	+	+	+	+
Color developer	+	+	+	+	+	+

Notes: (1) Monoclonal antibodies are more specific than polyclonal antibodies, which can generate nonspecific band(s). (2) The concentration of antibody used in the incubation varies with different antibodies. (3) The volume of antibody solution should be enough to cover the membrane. However, too much solution should be avoided. (4) More than two membranes or strips in one tray is not recommended for best results.

3. Decant the antibody solution, which can be reused for up to three times, and wash the membrane with four volumes of PBS/BSA/T for 5 min. Repeat the washing three times at 5-min intervals with shaking at 60 rpm.
4. Incubate the membranes in the secondary antibody solution diluted 500 to 3000 times in PBS/BSA/T solution at room temperature for 1 to 1.5 h with shaking at 60 rpm. If the first antibody is a monoclonal one from mice, the secondary antibody should be commercial goat antimouse IgA, IgG, or IgM conjugated with alkaline phosphatase. If the first antibody is a polyclonal one from rabbits, the secondary antibody should be goat antirabbit IgG conjugated with alkaline phosphatase or alkaline peroxidase.
5. Decant the antibody solution, which can be reused for up to three times, and wash the membrane with 4 volumes of PBS/BSA/T for 5 min. Repeat washing three times with shaking at 60 rpm.
6. Incubate the membrane in 0.2 M Tris-HCl (pH 9.2) for 2 min.
7. Detect band(s) by developing the membrane in alkaline phosphatase developer solution for 5 to 40 min until desired band(s) is (are) visible, and quickly stop the development by rinsing the membrane with large volumes of tap water.
8. Air dry the membranes and photograph for a permanent record. The MW of the detected band(s) can be determined by comparison with prestained markers or by matching the membrane with the Coomassie Blue- or silver-stained gel, whose MWs are determined as described in **Section II.A** and **Section II.B**, respectively.

Note: If the optimization procedure is being done, no bands should be visible from membrane strips that were treated with only antigen, only primary antibody, only secondary antibody, or with preimmune antiserum.

B. Chemiluminescence Detection

1. Perform steps 1 through 5 as above (**Section IV.A**) except the secondary antibody may be diluted as much as 1:30,000, depending on the distributor.
2. Place a sheet of SaranWrap on a table and add 0.8 ml per filter (12 × 14 cm^2) of enhanced chemiluminescence (ECL) detection solution (Amersham Life Science, Arlington Heights, IL) or equivalent to the center of the wrap.
3. To remove excess washing solution, dampen the filter briefly and wet the protein-binding side thoroughly by lifting and overlaying the filter with the solution several times. Wear gloves or cross-contamination may occur.
4. Wrap the filter with SaranWrap film, leaving two ends of the film unfolded. Place the wrapped filter on a paper towel and carefully press the wrapped filter, using another piece of paper towel, to force out excess detection solution through the unfolded ends of the SaranWrap film. Excess detection solution can cause a high background if it is not removed.
5. Completely wrap the filter and place it in an exposure cassette with the protein-binding side facing up. Tape the four corners of the filter.
6. In a darkroom with a safelight on, overlay the filter with an x-ray film and close the cassette. Allow exposure to proceed at room temperature for 10 sec to 4 h, depending on the intensity of the signal to be detected and background.
7. In a darkroom, process the film in an appropriate developer and fixer. If an automatic x-ray processor is available, development, fixation, washing, and drying of the film can be completed in 2 min. If a hybridized signal is detected, it appears as a black band on the film (an example from several different tissues is shown in Figure 8.2).

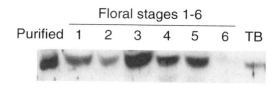

FIGURE 8.2
A Western blot of proteins extracted from six aspen tree floral stages as well as terminal buds. The specific polyclonal antibodies prepared from the purified peptide from Figure 8.1 were used to analyze the specific protein in different floral stages. The differences in band intensities correspond to the differences in the protein concentration in the crude extract (see **Chapter 3**). A small amount of purified *E. coli*-expressed protein fragment was used as a positive control.

Note: *Multiple films may have to be exposed and processed until the signals are desirable. Exposure for more than 4 h may generate a high black background. In our experience, a good hybridization and detection procedure should display sharp bands within 30 min, but one should start with shorter exposures. In addition, the film should be slightly overexposed to obtain a visible background that will help identify the sizes of the band(s) as compared with the marker bands.*

V. Quantitative Analysis of Proteins after Western Blot Hybridization

Quantitation of a protein of interest can be performed after Western blotting by comparing the antigen signal with those derived from a standard curve of increasing concentrations of purified antigen run on the same gel. Typically, four different concentrations of purified antigen will suffice to prepare the standard curve. However, since it is virtually impossible to standardize the signals resulting from the same protein on different gels, a separate standard curve must be run on each and every gel if comparisons are to be made between gels. Careful quantitation and loading of total protein per lane is also necessary for subsequent back-calculations of protein concentrations. Alkaline phosphatase results on membranes or chemiluminescence-detection results on film can be scanned into a quantitation program (such as Gel-Pro® Analyzer version 2.0.1; Media Cybernetics, Silver Spring, MD) or used for densitometry. A better alternative is to develop the Western blot results using a molecular-imager system or phospho-imager equipped with a chemiluminescence screen. These devices have a much broader range of detectable signal intensities and are thus much more accurate. They also come with software required for quantitation, but protocols vary significantly between different companies. Once the standard curve is calculated, it can easily be used to calculate the amount of protein in the band of interest. Such information can then be used to back-calculate the concentration of the antigen of interest in microgram-per-gram fresh weight of tissue.[8]

VI. Troubleshooting Guide

Symptom	Solutions
No signal at all	Reagents were omitted or added incorrectly. Try to use all reagents in the correct order.
	Insufficient antigen (protein) was loaded onto the gel. Try to increase the amount of proteins for one well.
	The primary antibody may not be good. The activity of the antibody may need to be higher.
	The secondary antibody or conjugated alkaline phosphatase loses activity. Try to test the quality of the secondary antibody.

(continued)

Symptom	Solutions
No signal at all *(continued)*	Transfer of proteins to the membrane was not successful. This may be due to placing the membrane between the gel and negative pole. If that happens, all of the proteins may be transferred into the blotting buffer in the chamber. The gel should be stained after transfer to determine if the proteins remain in the gel. The use of multicolor molecular-weight markers will easily provide evidence of good transfer of proteins. To determine if the transfer was successful, PVDF membranes can also be wetted with 20% methanol after drying. The bound proteins will appear as clear areas against an opaque membrane.
High background	Reagents may be too concentrated or the amount of blocking is insufficient. Try to use the right concentrations of solutions and increase the amount of blocking reagent; also, incubate the membrane for a longer time in the blocking solution.
	The amount of primary or secondary antibody is too high. Try optimizing the dilutions of antibodies.
Very weak signal	The concentration of the antigen or the antibodies or the developer solution was too low, or the incubation time was too short. Try to use the optimized concentration of each (presented above) and increase the incubation times. Also, while methanol is often added to the transfer buffer to enhance transfer efficiency, in some cases the use of high levels of methanol in the transfer buffer can degrade the SDS, causing poor transfer. If this is suspected, the amount of methanol in the transfer buffer can reduced to as low as 10%.
Nonspecific bands	The primary antibody raised from the antigen is not specific, resulting in cross-reactions. Try to verify the specificity of the antibody before repeating, and increase the blocking reagents to 5% (w/v) BSA or its equivalent. It may be necessary to purify the antibody against purified antigen.
Negative bands	During chemiluminescence detection, negative bands may appear on the film. This occurs when the amount of antigen on the membrane is too high, resulting in the rapid use of chemiluminescence reagents. Sometimes, if too much protein is loaded on the gel, the gel may even glow green. If this happens, do not waste your film. Run new gels with less protein loaded.

Reagents Needed

0.5% Agarose Solution

Agarose (0.5 g)
Add ddH$_2$O to a final volume of 100 ml and boil.
Cool down to room temperature before using.

Alkaline Phosphatase Developer Solution

15 ml 0.2 *M* Tris-HCl, pH 9.2
0.1 ml Nitroblue tetrazolium (NBT) stock solution (1.0 mg NBT/ml Tris-HCl)
0.05 ml 5-Bromo-4-chloro-3-indolylphosphate (BCIP) stock solution (5 mg BCIP/ml dimethyl formamide)
8 µl 2.0 *M* MgCl$_2$

Alternative Silver Developer Solution

2.5 ml 1% Citric acid
0.26 ml 36% (w/v) Formaldehyde
Add ddH$_2$O to a volume of 500 ml.

Alternative Silver Staining Solution

420 ml 0.36% (w/v) NaOH
28 ml 35% (w/v) Ammonia
Add 80 ml of 20% (w/v) silver nitrate dropwise with stirring.

10% Ammonium Persulfate (AP)
AP (0.2 g)
Dissolve well in 2.0 ml ddH$_2$O.
Store at 4°C for up to 10 days.

Ampholyte Solution, pH 3 to 10
Fluka Brand Cat. #10043 40% in water for electrophoresis (see also Sigma 2002–2003 catalogue, p. 167)

Ampholyte Solution, pH 4 to 6
Fluka Brand Cat. #10038 40% in water for electrophoresis (see also Sigma 2002–2003 catalogue, p. 167)

Anolyte Solution
1.37 ml of phosphoric acid solution (85%, d = 1.685 g/ml)
Add ddH$_2$O to a final volume of 1 L.

Blocking Solution
2 to 5% (w/v) Bovine serum albumin (BSA) or 5% (w/v) dry nonfat milk or equivalent in PBS solution.

Blotting Buffer
Tris (15.2 g)
Dissolve in 1 L ddH$_2$O
Glycine (72.1 g)
Methanol (1 L)
Add ddH$_2$O to 5 L. (The pH is about 8.3.)

Catholyte Solution
Sodium hydroxide (0.8 g)
Dissolve well after each addition in 100 ml ddH$_2$O.
Add ddH$_2$O to a final volume of 1 L.
This solution should be freshly prepared just before electrophoresis.

CB Staining Solution
1% CB solution (125 ml)
Methanol (500 ml)
Acetic acid (100 ml)
Add ddH$_2$O to a final volume of 1000 ml.

1% Coomassie Blue (CB) Solution
Coomassie Blue R-250 (2 g)
Dissolve well in 200 ml ddH$_2$O.
Filter.

2X Denaturing Loading Buffer (for SDS-PAGE)
Stacking gel buffer (5 ml)
10% SDS solution (8 ml)
Glycerol (4 ml)

2-Mercaptoethanol (2 ml)
Add ddH$_2$O to a final volume of 20 ml.
Bromophenol blue (0.02 g)
Divide in aliquots and store at –20°C.

2X Denaturing Loading Buffer (for IEF)

Urea (8 M) (12.0 g), high-purity urea should be used.
100 µl Ampholyte solution, pH 3.5 to 10
500 µl Ampholyte solution, pH 4 to 6
2.5 ml 20% Triton X-100
250 µl 2-Mercaptoethanol
Add ddH$_2$O to a final volume of 15 ml.
Bromophenol blue (0.01 g)
Divide in aliquots and store at –20°C.

Destaining Solution I

Methanol (500 ml)
Acetic acid (100 ml)
Add ddH$_2$O to a final volume of 1000 ml.

Destaining Solution II

Acetic acid (245 ml)
Methanol (175 ml)
Add ddH$_2$O to a final volume of 3.5 L.

ECL Detection Solution

(Amersham Life Science)

Equilibration Buffer

Stacking gel buffer (13.5 ml)
10% SDS solution (23 ml)
Glycerol (10 ml)
2-Mercaptoethanol (5 ml)
Add ddH$_2$O to a final volume of 100 ml.
Bromophenol blue (0.02 g)
Divide in aliquots and store at –20°C.

Farmer's Reducer Solution (Fresh)

0.5% (w/v) Potassium ferricyanide
1% (w/v) Sodium thiosulfate
Dissolve in and add ddH$_2$O to a volume of 400 ml.

Fixation Solution

30% Ethanol (v/v)
10% Acetic acid (v/v)
Add ddH$_2$O to a final volume of 1000 ml.

Monomer Solution

Acrylamide (116.8 g)
N,N-Methylene-*bis*-acrylamide (3.2 g)
Dissolve well after each addition in 300 ml ddH$_2$O.

Add ddH$_2$O to a final volume of 400 ml.
Wrap the bottle with aluminum foil and store at 4°C in the dark.

Caution: *Acrylamide is neurotoxic. Gloves should be worn when handling this chemical.*

Overlay Solution

Running gel buffer (25 ml)
10% SDS solution (1 ml)
Add ddH$_2$O to 100 ml.
Store at 4°C.

PBS/BSA/T Solution

1% (w/v) BSA
0.05% (v/v) Tween-20 in PBS solution

Phosphate-Buffered Saline (PBS) Solution

10 mM NaH$_2$PO$_4$
150 mM NaCl
Adjust the pH to 7.2 with 2 N NaOH.
Make 4 L.

10 mM KCl Solution

Running Gel Buffer

1.5 M Tris (72.6 g)
Dissolve well in 200 ml ddH$_2$O.
Adjust pH to 8.8 with 2 N HCl.
Add ddH$_2$O to a final volume of 400 ml.
Store at 4°C.

10% SDS

SDS (10 g)
Dissolve well in 100 ml warm ddH$_2$O.
Store at room temperature.

Silver Developer Solution (Fresh)

3% (w/v) Sodium carbonate in 0.02% formaldehyde in ddH$_2$O (0.5 ml formaldehyde in 2 L ddH$_2$O)
Make up to 500 ml.

0.1% AgNO$_3$ Staining Solution

0.75 ml 80% AgNO$_3$ (w/v) in 600 ml ddH$_2$O

Silver Stopping Solution

1% Acetic acid in ddH$_2$O

Stacking Gel Buffer
>0.5 M Tris (6 g)
>Dissolve well in 50 ml ddH$_2$O.
>Adjust pH to 6.8 with 2 N HCl.
>Add ddH$_2$O to a final volume of 100 ml.
>Store at 4°C.

Tris-Glycine Running Buffer
>0.25 M Tris (12 g)
>Glycine (57.6 g)
>10% SDS solution (40 ml)
>Add ddH$_2$O to a final volume of 4 L.

References

1. **Knudsen, K.A.,** Proteins transferred to nitrocellulose for use as immunogens, *Anal. Biochem.,* 147, 285, 1985.
2. **Towbin, J., Staehlin, T., and Gordon, J.,** Electrophoretic transfer of proteins from polyacrylamide gels to nitrocellulose sheets: procedure and some applications, *Proc. Natl. Acad. Sci. U.S.A.,* 76, 4350, 1979.
3. **Johnson, D.A., Gautsch, J.W., Sportsman, J.R., and Elder, J.H.,** Improved method for utilizing nonfat dry milk for analysis of proteins and nucleic acids transferred to nitrocellulose, *Gene Anal. Technol.,* 1, 3, 1984.
4. **Kyhse-Anderson, J.,** Electroblotting of multiple gels: a simple apparatus without buffer tank for rapid transfer of proteins from polyacrylamide to nitrocellulose, *J. Biochem. Biophys. Methods,* 10, 203, 1984.
5. **O'Farrell, P.H.,** High resolution two-dimensional electrophoresis of proteins, *J. Biol. Chem.,* 250, 4007, 1975.
6. **Righetti, P.G.,** Immobilised pH gradients: theory and methodology, in *Laboratory Techniques in Biochemistry and Molecular Biology,* Burdon, R.H. and van Knippenberg, P.H., Eds., Elsevier, Amsterdam, 1990.
7. **Berkelman, T. and Stenstedt, T.,** *2-D Electrophoresis Using Immobilized pH Gradients,* Amersham Pharmacia Biotech, Piscataway, NJ, 1998.
8. **Dudareva, N., Cseke, L., Blanc, V.M., and Pichersky, E.,** Evolution of floral scent in *Clarkia*: novel patterns of S-linalool synthase gene expression in the *C. breweri* flower, *Plant Cell,* 8, 1137, 1996.

Chapter 9

cDNA Libraries

Leland J. Cseke, William Wu, and Peter B. Kaufman

Contents

I.	Principles and Strategies for Construction of a cDNA Library	148
II.	Construction and Screening of a cDNA Library Using Lambda DNA as a Vector	152
	A. Vectors Used for cDNA Library Construction	152
	B. Protocols for Construction of a cDNA Library	153
	1. Synthesis of the First-Strand cDNA	153
	2. Synthesis of the Second-Strand DNA	154
	3. Trichloroacetic Acid Assay and Yield Calculations	155
	4. Alkaline Agarose Gel Electrophoresis of cDNAs	156
	5. Troubleshooting Guide Prior to cDNA Packaging	156
	6. Ligation of *EcoR* I Linkers/Adapters to the Double-Strand, Blunt-End cDNAs	158
	7. *In Vitro* Packaging of Ligated DNA	161
	C. Titering and Amplification of the Packaged Phage	161
	D. Large-Scale Ligation and *In Vitro* Packaging	162
	E. Immunoscreening of a Lambda Expression cDNA Library	163
	F. Troubleshooting Guide for Immunoscreening	166
	G. Screening a cDNA Library Using Labeled DNA Probes	166
	1. Screening of a cDNA Library Using a ^{32}P-Labeled Probe	167
	2. Screening of a cDNA Library Using a Nonisotope-Labeled Probe	167
	Reagents Needed	167
III.	Construction of a cDNA Library by Subtractive Hybridization Techniques	171
	A. Synthesis of the First-Strand cDNA	171
	B. Removal of the mRNA Template	172
	C. Hybridization of cDNA to mRNA	172
	D. Separation of cDNA/mRNA Hybrids from Single-Strand cDNA by Hydroxyapatite Chromatography	173
	E. Making a cDNA Library from the Subtracted First-Strand cDNA	173
	Reagents Needed	174

IV.	Construction of Fractional cDNA Libraries Using *Xenopus* Oocytes as an Expression System	175
	Reagents Needed	180
V.	cDNA Cloning and Analysis by the Polymerase Chain Reaction	180
	A. Selection of Oligonucleotides	181
	B. General Amplification of Double-Strand DNA by PCR	181
	C. cDNA Cloning by RT-PCR	184
	D. Analysis of Gene Expression by Semiquantitative PCR	186
	Reagents Needed	189
References		190

I. Principles and Strategies for Construction of a cDNA Library

Construction of a complementary deoxyribonucleic acid (cDNA) library is a highly sophisticated technology that is used in molecular biology studies. The quality and integrity of the cDNA library is directly related to the success or failure of cDNAs that are of interest to investigators.[1-3] In order to obtain a good library, the person working on the library should have a strong molecular biology background and extensive laboratory experience in molecular biology. Fortunately, there are many methods that have been well developed, and multiple commercial kits are available for cDNA cloning. This chapter describes the detailed strategies used in cDNA synthesis, in construction and screening of cDNA libraries, and in the isolation of putative clones. The step-by-step protocols presented will help experienced workers, as well as beginners, to achieve success.

cDNA cloning is a complex series of enzymatic procedures. The general principle is that messenger ribonucleic acid (mRNA) is copied into first-stranded DNA, which is called complementary DNA or cDNA, based on nucleotide bases complementary to each other. This step is driven by avian myeloblastosis virus (AMV) reverse transcriptase using an oligo(dT) primer or random primers. The second-stranded DNA is copied from the first-stranded DNA using DNA polymerase I, generating a double-stranded cDNA. The double-strand cDNA is subsequently ligated to an adapter for preparing the termini for vector ligation. The recombinant construct will then be packaged *in vitro* and cloned in a specific host, thus constructing a cDNA library. cDNA libraries preserve as much of the original cDNAs as possible and allow one to "fish" out any possible cDNA(s) expressed in a given tissue by screening the cDNA library. To accomplish successful screening, a probe having significant sequence similarity must be available.

The classical cDNA cloning method takes advantage of the 3' hairpins generated by AMV reverse transcriptase during first-strand synthesis. These hairpins are then used to prime second-strand cDNA catalyzed by Klenow DNA polymerase and reverse transcriptase. *S1* nuclease is added to cleave the hairpin loop. However, this digestion is difficult to control, causing low cloning efficiencies and the loss of a significant amount of sequence information corresponding to the 5' end of the mRNA. An improved strategy utilizes 4 m*M* sodium pyrophosphate, which greatly suppresses the formation of hairpins during the synthesis of first-stranded cDNA. Second-strand synthesis is then carried out by *RNase* H to create nicks and gaps in the hybridized mRNA template, generating 3'-OH priming sites for DNA synthesis and repair by DNA polymerase I. After treatment with T4 DNA polymerase to remove any remaining 3' protruding ends, the blunt-ended, double-strand cDNAs are ready for adapter or linker ligation (Figure 9.1). *S1* digestion is avoided in this method, and the cloning efficiency is much higher than in the classic method. In addition, the sequence information can be optimal.

A variety of improved methods have been developed for cloning cDNA molecules. The basic strategies can be grouped into two classes. One is random cloning; the other is orientation-specific

cDNA Libraries

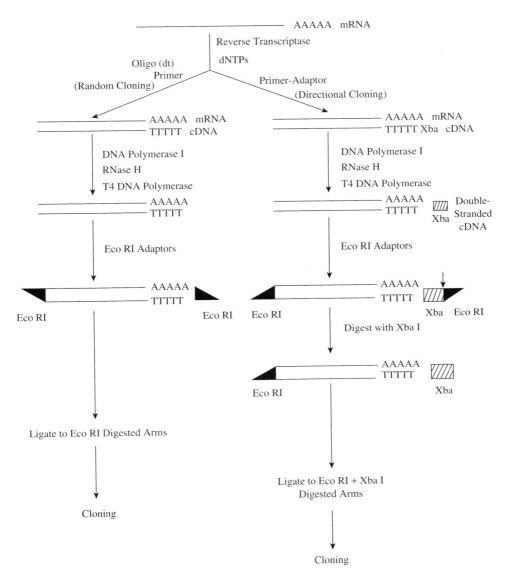

FIGURE 9.1
Diagram of cDNA synthesis scheme showing two cloning strategies: random cloning (left) and directional cloning (right) (From Kaufman, P.B., Wu, W., Kim, D., and Cseke, L.J., *Handbook of Molecular and Cellular Methods in Biology and Medicine*, 1st ed., CRC Press, Boca Raton, FL, 1995.)

cloning. The former is much easier than the latter with respect to the techniques involved. Figure 9.2 shows the scheme for cDNA cloning using a random primer. The random or classical cloning of cDNA uses oligo(dT) as a primer and λgt10 or λgt11 as cloning vectors for the cDNA library. The cDNAs are cloned into a single *EcoR* I site via the addition of *EcoR* I adapters (or linkers) to each end of the cDNA molecules. Because of the single *EcoR* I site cloning, the cDNAs are cloned in random orientations, including both sense and antisense orientations. If the cDNA library is constructed using the nonexpressional vector λgt10 (Figure 9.3), it can be screened by using DNA or RNA as a probe. However, if the expression vector λgt11 is used for the cDNA library, the library can be screened using a specific antibody as well as DNA or RNA as a probe.[4] The disadvantage of using an antibody as a probe to screen the randomly cloned cDNA library is that

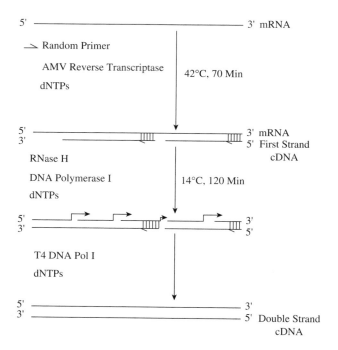

FIGURE 9.2
Scheme for cDNA cloning using a random primer. (From Kaufman, P.B., Wu, W., Kim, D., and Cseke, L.J., *Handbook of Molecular and Cellular Methods in Biology and Medicine*, 1st ed., CRC Press, Boca Raton, FL, 1995.)

the possibility of obtaining the positive clones is at least 50% less. This is because approximately 50% of cloned cDNAs are expressed as antisense RNA, which may interfere with sense RNA by inhibiting the translation of the sense RNA into proteins.

Because of the shortcomings of random cloning, the state-of-the-art strategy involves the efficient orientation-specific or directional cloning of cDNA using a primer-adapter to prime the synthesis of the first-strand cDNA (Figure 9.1). The primer-adapter consists of oligo(dT) adjacent to a unique restriction site (*Xba* I or *Not* I). The subsequent steps are carried out as in random cloning, except that the final double-strand cDNAs with *EcoR* I adapters attached are digested with either *Xba* I or *Not* I. The digested cDNA molecule contains one *EcoR* I and one *Xba* I or *Not* I terminus, which can be ligated with a vector containing the same restriction enzyme termini. As compared with random cloning protocol, directional cloning is a more powerful and valuable technique. In expression vectors such as λgt11 *Sfi-Not* I, the likelihood of expressing the cDNA insert as the correct polypeptide is increased significantly. This is because of the absence of possible antisense RNA interference.

In transcription vectors, such as the λGEM-2 and λGEM-4 vectors (Figure 9.4), all of the cDNA inserts are cloned in the same direction downstream from the promoters for T7 and SP6 RNA polymerase. Total sense or antisense RNA probes can be obtained from the cDNA library to represent all of the sequences in the library. These RNAs can be used for subtraction hybridization to isolate and analyze some rare genes expressed in one organism but not in another, or in different tissues in the same organism. The positive cDNA inserts purified from the directionally cloned library can drive further directional subcloning procedures, using an appropriate plasmid vector, such as pGEM-11 (Promega Corp., Madison, WI), that is digested with the same enzymes, *EcoR* I and *Xba* I or *Not* I. The cDNA insert is cloned in the same direction downstream from the promoters for T7 and SP6 RNA polymerase. At this point, the poly(A)-tail strand cDNA can be easily identified and sequenced using T7 and SP6 primers. In addition, transcription of sense RNA or antisense

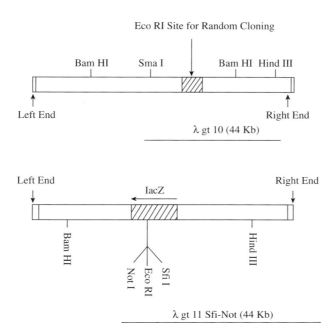

FIGURE 9.3
Structural maps of vectors: λgt10 (top) and λgt11 *Sfi-Not* I (bottom). (From Kaufman, P.B., Wu, W., Kim, D., and Cseke, L.J., *Handbook of Molecular and Cellular Methods in Biology and Medicine,* 1st ed., CRC Press, Boca Raton, FL, 1995.)

RNA, under the control the T7 or SP6 RNA polymerase promoters, can be utilized as RNA probes to screen cDNA or genomic DNA libraries as well as to perform Northern blot analyses.

cDNA cloning is one of the techniques used in state-of-the-art molecular biology. Based on the authors' experience, there are three steps that are critical for success or failure in cDNA library construction. The first is the purity and integrity of the mRNA used for the synthesis of the first-strand cDNA. Any degradation or absence of specific mRNAs will result in partial-length cDNAs or complete loss of some cDNAs, especially for some rarely expressed genes. The second important step is to achieve good synthesis of full-length cDNAs. If this procedure is not carried out properly, even if you have a very good mRNA source, the cDNA library to be constructed will be not good, and you may "fish" out only partial-length cDNAs or no positive clones at all. As long as double-strand cDNAs are obtained, they will be much more stable than mRNAs. However, there is a third essential step in cDNA cloning. This is the ligation of cDNAs with adapters to the vector. If the ligation fails or is not efficient, *in vitro* packaging of recombinant λDNAs cannot be carried out effectively. As a result, the number of plaque-forming units (pfu) will be very low.

In order to construct an exceptionally good cDNA library, the following precautions should be taken, in addition to carefully following the protocols given in this chapter. The risk of *RNase* contamination must be eliminated whenever possible. Specific strategies for the isolation and purification of high quality and quantity of mRNAs as templates are described in **Chapter 2**. The following practices are helpful in creating an *RNase*-free laboratory environment for mRNA and cDNA synthesis:

1. Disposable plastic test tubes, micropipette tips, and microcentrifuge tubes should be sterilized.
2. Gloves should be worn at all times and changed often to avoid finger-derived *RNase* contamination.
3. All glassware and the electrophoresis apparatus used for cDNA cloning should be separated from other labware. Glassware should be treated with 0.1% diethylpyrocarbonate (DEPC) solution, autoclaved to remove the DEPC, and baked overnight at 250°C before being used. Alternatively, glassware

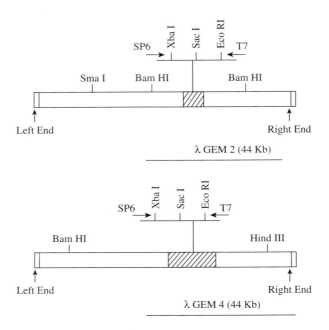

FIGURE 9.4
Structural maps of vectors: λGEM-2 (top) and λGEM-4 (bottom). (From Kaufman, P.B., Wu, W., Kim, D., and Cseke, L.J., *Handbook of Molecular and Cellular Methods in Biology and Medicine*, 1st ed., CRC Press, Boca Raton, FL, 1995.)

and the gel apparatus can be treated overnight with 2% Absolve™ Glassware Cleaner (NEN™ Life Science Products, Boston, MA).

4. Solutions to be used with mRNA and cDNA synthesis should be treated with 0.1% DEPC (v/v) to inhibit *RNase* by acylation. Add DEPC to the solutions and stir vigorously for 20 min, followed by autoclaving or heating the solution at 70°C for 1 h to remove the DEPC.
5. Because DEPC reacts with amines and sulfhydryl groups, reagents such as Tris and DTT (dithiothreitol) cannot be treated directly. These solutions can be made by using DEPC-treated and sterile water. If solutions such as DTT cannot be autoclaved, the solution can be filter sterilized.
6. Other items such as lab surfaces and pipetters can be treated with solutions such as *RNase* AWAY® (Invitrogen™, Carlsbad, CA) to removed residual *RNases*.

Caution: *DEPC is a powerful and toxic acylating agent and should be handled in a fume hood. Never add DEPC into solutions containing ammonia because that will cause the formation of ethyl carbamate, a potent carcinogen.*

II. Construction and Screening of a cDNA Library Using Lambda DNA as a Vector

A. Vectors Used for cDNA Library Construction

The selection of a vector for cDNA cloning depends on the screening method and on the number of pfu to ensure the presence of the desired cDNA. In order to detect some rare cDNAs from low-abundance mRNAs from mammalian cells, approximately 2×10^5 pfu of recombinants must be screened using nucleic acid probes. Bacteriophage lambda vectors, as compared with plasmid vectors, are usually recommended when large numbers of recombinants are needed. This is because of the high cloning efficiency of and subsequent infection by the phage particles. Phage cDNA libraries can be efficiently screened at different pfu densities with either nucleic acid or antibody probes. However, the screening is much more difficult when antibodies are used as probes. The advantage of using plasmids as vectors is that a commercial plasmid contains multiple cloning sites, and the plasmids can be used as tailed vectors and for vector–primer cloning strategies. In addition, the isolation and manipulation of the plasmid clones are relatively easy. However, cloning efficiencies are usually 10- to 50-fold lower using plasmid vectors. The number of colonies screened must be small because, at high densities, the colonies will overlap each other, causing difficulty in screening.

Recently, the advanced features of plasmids have been incorporated into lambda DNA vectors. For example, the λGEM-2 and λGEM-4 vectors (Promega Corp.) were constructed by inserting a 535-bp (base-pair) pGEM-1 vector fragment (λGEM-2) and the entire pGEM-1 plasmid (λGEM-4) into the original cloning site of λgt10. These improved vectors allow directional cloning via *EcoR* I and *Xba* I cloning sites and simplify subcloning procedures. The T7 and SP6 promoters make it possible to synthesize RNA probes, which can be used for subtraction cDNA library construction.

λgt10, λGEM-2, and λGEM-4 are usually the recommended vectors used for cDNA libraries to be screened with DNA or RNA probes. If an antibody is used as a screening probe, the expression vector λgt11 is the better one to use (Figure 9.3). This vector can express the inserted cDNA as part of a β-galactosidase fusion protein. However, since this vector contains a single *EcoR* I cloning site, cDNAs can only be cloned according to a random cloning strategy using oligo(dT) as a primer and *EcoR* I adapters or linkers. The possibility of detecting the desired cDNA clones is very small because of possible antisense RNA interference. On the other hand, another vector, λgt11 *Sfi-Not* I, is used for directional cloning of cDNA (Figure 9.3). All the cDNAs can be cloned in the same direction, thus increasing the likelihood of expression of cDNA inserts as fusion polypeptides. The cDNA library can be screened using an antibody as a probe as well as nucleic acid probes. This cloning strategy effectively increases the possibility of obtaining the cDNA clones of interest.

B. Protocols for Construction of a cDNA Library

There are a number of different procedures available for cDNA production, depending on the enzyme and buffer used. The reader should pay close attention to subtle differences in the procedures accompanying the enzyme used. The following example works well in our laboratories.

1. Synthesis of the First-Strand cDNA

The general scheme of this procedure is outlined in Figure 9.1. It is recommended that two reactions be set up and carried out separately. The benefit of doing this is that if one reaction is somehow stopped by accident during the experiments, the other reaction can be used as a backup. Otherwise, the procedure may have to be started over, thus wasting time and money. The standard reaction given below has a total volume of 25 μl using up to 2 μg of mRNA. For each additional microgram of mRNA, increase the reaction volume by 10 μl. A control reaction should be carried out whenever possible under the same conditions.

1. Anneal 2 μg of mRNA template with 1 μg of primer of oligo(dT) or with 0.7 μg of primer-adapter of oligo(dT) *Not* I/*Xba* I in a sterile *RNase*-free microcentrifuge tube. Add nuclease-free distilled, deionized water (ddH_2O) to a total volume of 15 μl. Heat the tube at 70°C for 5 min and allow it to slowly cool to room temperature to finish the annealing. Briefly spin down the mixture to the bottom of the microcentrifuge tube.
2. To the annealed primer/template, add the following in the order shown. To prevent precipitation of sodium pyrophosphate when it is added to the buffer components, the buffer should be preheated at 40 to 42°C for 4 min before the addition of sodium pyrophosphate and AMV reverse transcriptase. Gently mix well after each addition.

- First-strand 5X buffer (5 μl)
- rRNasin ribonuclease inhibitor (25 units/μg mRNA) (50 units)
- 40 mM Sodium pyrophosphate (2.5 μl)
- AMV reverse transcriptase (15 units/μg mRNA) (30 units)
- Add nuclease-free ddH_2O to a total volume of 25 μl.

Functions: The 5X reaction buffer contains components required for cDNA synthesis. Ribonuclease inhibitor inhibits *RNase* activity and protects the mRNA template. Sodium pyrophosphate suppresses the formation of hairpins that are normally produced in "classical" cloning methods, thus avoiding the S1 digestion step, which is more difficult to carry out. AMV reverse transcriptase catalyzes the synthesis of the first-strand cDNA from mRNA template based on the rule of complementary base pairing.

3. Set up a tracer reaction by transferring 5 μl of the mixture to a fresh tube containing 1 μl of 4 μCi of [α-^{32}P]dCTP (>400 Ci/mmol, less than one week old). The first-strand synthesis will be measured by trichloroacetic acid (TCA) precipitation and alkaline agarose gel electrophoresis using the tracer reaction.

Caution: *[α-^{32}P]dCTP is a dangerous isotope. Gloves should be worn when carrying out the tracer reaction, TCA assay, and gel electrophoresis. Waste materials such as contaminated gloves, pipette tips, solutions, and filter papers should be put in special containers for waste radioactive materials.*

4. Incubate both reactions at 42°C for 1.5 h and place on ice. At this stage, the synthesis of the first-strand cDNA has been completed. To the tracer reaction, add 50 mM EDTA up to a total volume of 100 μl and store on ice for TCA incorporation assays and alkaline agarose gel electrophoresis analysis after extraction. The unlabeled reaction will be directly used for second-strand DNA synthesis.
5. Inactivate the reactions by heating at 70°C for 10 min, add 20 μl ddH_2O to the unlabeled reaction, and store at −20°C until amplification by polymerase chain reaction (PCR).

2. Synthesis of the Second-Strand DNA

1. To the unlabeled first-strand cDNA tube on ice, add the following components in the order given below. Gently mix well after each addition.
   - First-strand reaction mixture (20 μl)
   - Second-strand 10X buffer (10 μl)
   - *E. coli* DNA polymerase I (24 units)
   - *E. coli RNase* H (1 unit)
   - Add nuclease-free ddH_2O to a final volume of 100 μl.

   Functions: *E. coli RNase* H makes nicks and gaps in the hybridized mRNA template, creating 3'-OH priming sites. *E. coli* polymerase I serves to synthesize DNA and repair the gaps from the priming sites, producing the second-strand cDNA.
2. Set up a tracer reaction for the second-strand cDNA by transferring 10 μl of the mixture to a fresh tube containing 1 ml 4 μCi [α-^{32}P]dCTP (>400 Ci/mmol). The tracer reaction will be used for TCA assay and alkaline agarose gel electrophoresis to monitor the quantity and quality of the synthesis of the second-strand DNA.
3. Incubate both reactions at 14°C for 3.5 h. Add 90 μl of 50 mM EDTA to the 10-μl tracer reaction and store on ice.
4. Heat the unlabeled double-strand cDNA sample at 70°C for 10 min to stop the reaction. **At this point, the double-strand cDNAs should be generated.** Briefly spin down to collect the contents at the bottom of the tube and place on ice.
5. Add 4 units of T4 DNA polymerase (2 units/μg input mRNA) to the mixture at step 4 and incubate the tube at 37°C for 10 min. This step functions to make blunt-end, double-strand cDNA by T4 polymerase. The blunt ends are required for later adapter ligations.
6. Stop the T4 polymerase reaction by adding 10 μl of 0.2 M EDTA and place on ice.
7. Extract the cDNAs with one volume of TE-saturated phenol/chloroform. Mix well and centrifuge in a microcentrifuge at top speed for 4 min at room temperature.

cDNA Libraries

8. Carefully transfer the top, aqueous phase to a fresh tube. **Do not take any white materials at the interphase between the two phases.** To the supernatant, add 0.5 volume of 7.5 M ammonium acetate or 0.15 volume of 3 M sodium acetate (pH 5.2), mix well and add 2.5 volumes of chilled (−20°C) 100% ethanol. Gently mix and allow precipitation to occur at −20°C for 2 h.
9. Centrifuge in a microcentrifuge at the top speed for 5 min. Carefully remove the supernatant, briefly rinse the pellet with 1 ml of cold 70% ethanol, and gently drain away the ethanol.
10. Dry the cDNA pellet under vacuum for 15 min. Dissolve the cDNA pellet in 25 μl of TE buffer. It is important to take 2 to 4 μl of the sample to measure the concentration of cDNAs prior to the next reaction. Store the sample at −20°C until use.

Note: *Before adapter ligation, it is strongly recommended that the quantity and quality of both first- and second-strand cDNAs be checked by TCA precipitation and gel electrophoresis as described in* **Section II.B.3** *and* **Section II.B4**, *respectively.*

3. Trichloroacetic Acid Assay and Yield Calculations

1. Spot 4 μl of the first-strand tracer reaction sample and 4 μl of the second-strand reaction sample on glass-fiber filters and air dry. These will represent the total counts per minute (cpm) in the samples.
2. Add another 4 μl of the same reaction samples to fresh tubes containing 100 μl of carrier DNA solution (1 mg/ml) and mix well. Add 0.5 ml of 5% TCA and mix by vortexing. Allow precipitation to occur on ice for 30 min.
3. Filter the precipitated samples through glass-fiber filters. Wash the filters with 6 ml cold 5% TCA four times and rinse once with 6 ml of acetone or ethanol. Allow the filters to dry at room temperature. The samples represent incorporated cpm in the reactions.
4. Place the filters in individual vials and add 10 to 15 ml of scintillation fluid to cover each filter. Count both total and incorporated cpm samples according to the instructions for the scintillation counter. The cpm can also be counted by Cerenkov radiation (without scintillant).
5. Calculate yield of first-strand cDNA as follows:

$$\text{Percent incorporated} = \frac{\text{incorporated cpm}}{\text{total cpm}} \times 100$$

$$\text{dNTP incorporated (nmol)} = [(4 \text{ nmol dNTP}/\mu l) \times \text{reaction volume } (\mu l) \times \text{percent incorporation}]/100$$

$$\text{cDNA synthesized (ng)} = \text{dNTP incorporated (nmol)} \times 330 \text{ ng/nmol}$$

$$\text{Percent mRNA converted to cDNA} = \frac{\text{cDNA synthesized (ng)}}{\text{mRNA in reaction (ng)}}$$

6. Calculate the yield of second-strand cDNA as follows:

$$\text{Percent second-strand DNA incorporated} = \frac{\text{incorporated cpm}}{\text{total cpm}} \times 100$$

$$\text{dNTP incorporated (nmol)} = [0.8 \text{ nmol dNTP}/\mu l \times \text{reaction volume } (\mu l) -$$
$$\text{dNTP (nmol) incorporated in first-strand reaction}] \times \text{percent second-strand incorporation}/100$$

$$\text{Second-strand cDNA synthesized (ng)} = \text{dNTP incorporated (nmol)} \times 330 \text{ ng/nmol}$$

$$\text{Percent converted to double-stranded cDNA} = \frac{\text{second-strand cDNA synthesized (ng)}}{\text{first-strand cDNA synthesized (ng)}} \times 100$$

Note: Reactions yielding 15 to 30% first-strand conversion and 80 to 200% second-strand conversion values can be used to construct a successful cDNA library.

4. Alkaline Agarose Gel Electrophoresis of cDNAs

The quality and the size range of cDNAs synthesized in the first- and second-strand reactions should be checked by 1.4% alkaline agarose gel electrophoresis using tracer reaction samples.

1. Extract the first- and second-strand cDNAs with phenol and precipitate with ethanol. Designate these as the unlabeled reactions.
2. Label λ*Hind* III fragments with ^{32}P in a fill-in reaction. These are used as DNA markers to estimate the sizes of cDNAs. In a sterile eppendorf tube, add the following components in the order shown below.
   - *Hind* III 10X buffer (2 µl)
   - dATP (0.2 mM)
   - dGTP (0.2 mM)
   - [α-^{32}P]dCTP (2 µCi)
   - λ*Hind* III markers (1 µg)
   - Klenow DNA polymerase (1 unit)
   - Add ddH$_2$O to a final volume of 20 µl.
   - Mix well after each addition and incubate the reaction for 15 min at room temperature. Add 2 µl of 0.2 M EDTA to stop the reaction. Transfer 6 µl of the sample directly to 6 µl of 2X alkaline buffer and store the remainder at –20°C.
3. Dissolve 1.4% (w/v) agarose in 50 mM NaCl, 1 mM EDTA solution and melt in a microwave for a few minutes. Cool the gel mixture to about 50°C and pour the gel into a minielectrophoresis apparatus. Allow the gel to harden and equilibrate the gel in alkaline gel running buffer for 1 h prior to electrophoresis.
4. Transfer the same amount of each sample (50,000 cpm) to separate tubes and add an equal volume of TE buffer to each tube. Mix and add one volume of 2X alkaline buffer to each tube. Total volume should be approximately 30 µl. It is recommended that the same numbers of incorporated cpm be loaded for each strand to compare the density of signals on the same autoradiograph.
5. Carefully load the samples into wells and immediately run the gel at 7 V/cm until the dye has migrated to about 2 cm from the end.
6. Stop the electrophoresis and soak the gel in five volumes of 7% TCA at room temperature until the dye changes from blue to yellow. Dry the gel on a piece of 3MM Whatman™ paper on a gel dryer or under a weighted stack of paper towels for 6 h.
7. Wrap the dried gel with SaranWrap™ and expose to x-ray film at room temperature or at –70°C with an intensifying screen for 1 to 4 h (Figure 9.5).

Note: For successful synthesis of cDNAs, the signal should range from 0.15 to 8 kb (kilobasepairs) with a sharp size range of 1.5 to 4 kb (Figure 9.5).

5. Troubleshooting Guide Prior to cDNA Packaging

Obtaining good double-strand cDNA is half the battle in making a cDNA library. Therefore, the synthesis of first- and second-strand cDNAs is extremely important for successful cDNA library construction. Before proceeding further, it is a wise idea to make sure that the double-strand cDNAs synthesized are good enough for the construction of a successful cDNA library. Based on our experience, the following items should be checked before proceeding any further:

1. If the yield of the first-strand cDNA conversion is less than 8%, and if the intensity of the different cDNA sizes shown on the x-ray film is very weak, or if the size range is very small and of lower molecular weight, the potential causes for these problems are due to poor mRNA template or the failure of reverse transcriptase to act, or to the presence of some inhibitors.

cDNA Libraries

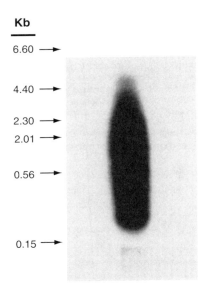

FIGURE 9.5
Eukaryotic cDNAs that were synthesized by using an adapter-primer. The cDNAs show a size range from 0.25 to 5.0 kb. (From Kaufman, P.B., Wu, W., Kim, D., and Cseke, L.J., *Handbook of Molecular and Cellular Methods in Biology and Medicine*, 1st ed., CRC Press, Boca Raton, FL, 1995.)

The quality of the mRNA template is the most important factor for successful cDNA synthesis. If the poly(A)$^+$ RNA is degraded during its isolation and purification procedure, the first-strand cDNA synthesized from the template is only partial-length cDNA. To avoid this problem, it is important to check the quality of the mRNA template prior to the synthesis of first-strand cDNA. This can be done by simply analyzing 2 μg mRNA in 1% agarose–formaldehyde denaturing gel containing ethidium bromide (EtBr). After electrophoresis, take a photograph of the gel under ultraviolet (UV) light. If the smear range is from 0.65 up to 8.5 kb, the integrity of the mRNA is very good for cDNA synthesis. Another possible cause is that AMV reverse transcriptase functions at the beginning of the synthesis of first-strand cDNA, but fails in the middle of strand synthesis. In this case, the cDNA produced is likely to be of only partial length. To ensure that this potential problem does not occur, always use positive mRNA template as a control to check the activity of reverse transcriptase. If the mRNA template and reverse transcriptase are very good, then use the control RNA to check for the presence of inhibitors such as SDS, EDTA, and salts. It is also possible that the ratio of the amount of primer:mRNA is not optimal, which may adversely affect first-strand cDNA synthesis.

2. If the yield of the second-strand DNA conversion is less than 40%, or the size range is very small, the problem is likely to be due to either *RNase* H or DNA polymerase I not functioning well. To check the activity of *RNase* H, treat control RNA with this enzyme and assay by electrophoresis. If the *RNase* H works well, the RNA is nicked into fragments. This will be shown by gel electrophoresis as compared with untreated RNA. On the other hand, the DNA polymerase I may not function well; as a result, the synthesis of second-strand DNA or the repair of gaps will not be completed, thus generating partial-length cDNAs. In this case, set up another reaction using a control experiment, or use fresh DNA polymerase I. If low incorporation is obtained even with the positive control, it is likely that the radioisotope used as a label is not good. Check the specificity and half-life of the isotope and use fresh [α-^{32}P]dCTP as a control. If you have obtained very good double-strand cDNAs, and if the cDNA of interest is included in the expected size range, you have had good success and may go ahead with adapter ligation.

6. Ligation of *EcoR* I Linkers/Adapters to the Double-Strand, Blunt-End cDNAs

The *EcoR* I linker or *EcoR* I adapter ligation system is designed to generate *EcoR* I sites at the termini of blunt-ended cDNA, and it is commercially available. If *EcoR* I linkers are used, digestion with *EcoR* I is required after ligation. However, there is no need to digest with *EcoR* I if one uses the *EcoR* I adapter that has one sticky end. The ligation of *EcoR* I linker or adapter allows cDNAs to be cloned into unique *EcoR* I sites of vectors λgt 10 or λgt 11. If cDNAs are synthesized with an *Xba* I or *Not* I primer-adapter, they can be directionally cloned between the *EcoR* I and *Xba* I or *Not* I sites of λgt 11 *Sfi-Not* I vector or vectors such as λGEM-2 and λGEM-4.

1. Size fractionation of double-strand cDNAs: It is strongly recommended that cDNA samples be size-fractionated prior to ligation with linkers to eliminate <200-bp cDNAs. This can easily be done with Sephacryl S-400 spin columns (Promega Corp.) that exclude double-strand DNA (dsDNA) fragments >270 bp or with gel filtration.
    a. To a vertical column, slowly add approximately 1 to 2 ml of Sephacryl S-400 slurry and drain the buffer completely. The final height of the gel bed should reach the neck of the column at the lower "ring" marking. Rehydrate the column with buffer, cap the top, and seal the bottom. Store at 4°C until use.
    b. Right before use, drain away the buffer and place the column in the wash tube that is provided or a microcentrifuge tube. Briefly centrifuge in a swinging bucket for 4 min at 800 g to remove excess buffer.
    c. Place the column in a fresh collection tube and slowly load the cDNA samples (30 to 60 µl), dropwise, to the center of the gel bed. Centrifuge in a swinging bucket at 800 g for 5 min and collect the eluate. The eluted cDNAs can be directly used or precipitated with ethanol and dissolved in TE buffer.
2. Protection of the internal *EcoR* I sites of cDNAs by methylation:
    a. Set up the *EcoR* I methylation reaction as follows:
        - 1 mM s-Adenosyl-L-methionine (2 µl)
        - *EcoR* I methylase 10X buffer (2 µl)
        - BSA (1 mg/ml) (optional) (2 µl)
        - DNA (1 µg/µl) (2 µl)
        - *EcoR* I methylase (20 units)
        - Add ddH$_2$O to a final volume of 20 µl.
        - Mix well after each addition.
    b. Incubate the tube at 37°C for 20 min and inactivate the enzyme by heating at 70°C for 10 min.
    c. Extract the enzyme with 20 µl of TE-saturated phenol/chloroform, mix well by vortexing, and centrifuge at 11,000 g for 4 min.
    d. Carefully transfer the top, aqueous phase to a fresh tube and add an equal volume of chloroform:isoamyl alcohol (24:1). Vortex to mix and centrifuge as in step c.
    e. Carefully transfer the top, aqueous phase to a fresh tube. **Do not take any white materials.** To the supernatant, add 0.15 volume of 3 M sodium acetate buffer, pH 5.2, and 2.5 volumes of chilled 100% ethanol. Allow precipitation to occur at –70°C for 30 min or at –20°C for 2 h.
    f. Centrifuge at 12,000 g for 10 min and carefully pour off the supernatant. Briefly rinse the DNA pellet with 1 ml of 70% chilled ethanol and dry the pellet under vacuum for 10 min. Dissolve the methylated cDNAs in 15 µl of TE buffer.
3. Ligation of *EcoR* I linkers to the methylated cDNAs
    a. Set up the reaction as follows and mix after each addition.
        - cDNA (0.1 µg/µl) (2 µl)
        - Ligase 10X buffer (2 µl)
        - BSA (1 mg/ml) (optional) (2 µl)
        - Phosphorylated *EcoR* I linkers (60-fold molar excess) (diluted stock linkers, 0.001 A$_{260}$ unit of the linker = 10 pmol)
        - T4 DNA ligase (5 Weiss units)
        - Add ddH$_2$O to a final volume of 20 µl
        - **Always add enzyme last**

b. Incubate at 15°C overnight and stop the reaction by heating at 70°C for 10 min.
c. Cool the tube on ice for 2 min and carry out *EcoR* I digestion as follows:
   - Ligated cDNAs (20 μl)
   - *EcoR* I 10X buffer (5 μl)
   - *EcoR* I (1 unit/pmol of linker used in step a)
   - Add ddH$_2$O to a final volume of 50 μl.
d. Incubate the reaction at 37°C for 2 h and add 5 μl of 0.2 *M* EDTA to stop the reaction.
e. Extract enzymes with 55 μl of TE-saturated phenol/chloroform and mix by vortexing for 1 min.
f. Centrifuge at 11,000 g for 4 min and carefully transfer the upper, aqueous phase to a fresh tube.
   Steps g to j are designed for directional cloning of cDNAs that are synthesized with an *Xba* I or *Not* I primer-adapter. The ligated cDNAs with linkers must be digested with *Xba* I or *Not* I following the *EcoR* I digestion. Alternatively, a double enzyme digestion with *EcoR* I and *Xba* I or *Not* I can be carried out in the same reaction using 10X digestion buffer for *Xba* I or *Not* I in the reaction. However, double digestions are usually of lower efficiency.
g. Set up the *Xba* I or *Not* I digestion reaction as follows:
   - cDNAs from step f (40 μl)
   - *Xba* I or *Not* I 10X buffer (8 μl)
   - *Xba* I or *Not* I (1 unit/pmol of linker used in step a)
   - Add ddH$_2$O to a final volume of 80 μl.
h. Incubate the reaction at 37°C for 2 h and add 8 μl of 0.2 *M* EDTA to stop the reaction.
i. Extract enzymes with 88 μl of TE-saturated phenol/chloroform and mix by vortexing for 1 min.
j. Centrifuge at 11,000 g for 4 min and carefully transfer the upper, aqueous phase to a fresh tube.
k. Remove unligated linkers with a spin column as follows: Place a spin column in the collection tube and slowly load the restriction enzyme-digested supernatant to the top center of the gel bed. Centrifuge in swinging bucket at 800 g for 5 min and collect the eluate containing cDNA linkers.
l. Add 0.5 volume of 7.5 *M* ammonium acetate to the eluate, mix, and add two volumes of chilled 100% ethanol. Allow precipitation at −20°C for 2 h.
m. Centrifuge at 12,000 g for 10 min and briefly rinse the pellet with 1 ml of cold 70% ethanol. Dry the cDNA linker under vacuum for 10 min and resuspend the pellet in 20 μl of TE buffer. Store the sample at −20°C until use.
n. To check the efficiency of ligation, load 4 μl (0.3 μg) of the ligated sample on a 0.8% agarose gel containing EtBr and an equal amount of the unligated, blunt-ended cDNAs next to the ligated sample. After electrophoresis, take a photograph under UV light. Efficient ligation will cause a shift in the migration rate because of the ligation of linker concatemers, which run more slowly than the unligated sample.

4. Ligation of *EcoR* I adapters to cDNAs: The *EcoR* I adapter is a duplex DNA molecule with one *EcoR* I sticky end and one blunt end for ligation to cDNA. Adapters eliminate *EcoR* I methylation. The adapter cDNA does not need to be digested with *EcoR* I after ligation, which is necessary for *EcoR* I linker ligation methodology. However, a phosphorylation reaction should be carried out if the cloning vector is dephosphorylated.
   a. Adapter ligation
      i. Set up *EcoR* I adapters to cDNAs in a microcentrifuge tube as follows:
         - cDNA (0.1 μg/μl) (0.2 μg)
         - Ligase 10X buffer (3 μl)
         - BSA (1 mg/ml) (optional) (3 μl)
         - *EcoR* I adapters (25-fold molar excess)
         - T4 DNA ligase (Weiss units) (8 units)
         - Add ddH$_2$O to a final volume of 30 μl.
      ii. Incubate the reaction at 15°C overnight and stop the reaction at 70°C for 10 min.
   b. Kinase reaction and *Xba* I or *Not* I digestion. This reaction is required for the dephosphorylated vector, but it is not necessary for the phosphorylated vector. For directional cDNA cloning methodology, *Xba* I or *Not* I digestion must be carried out. This can be done simultaneously with the kinase reaction using restriction enzyme 10X buffer instead of the kinase 10X buffer.
      i. From the above step (a.ii) after ligation, place the tube on ice for 1 min and add the following in the order shown:

For kinase reaction only:
- Kinase 10X buffer (5 µl)
- 0.1 mM ATP (diluted from stock) (2.5 µl)
- T4 polynucleotide kinase (10 units)
- Add ddH$_2$O to a final volume of 50 µl.

For both kinase reaction and *Xba* I or *Not* I digestion:
- *Xba* I or *Not* I 10X buffer (5 µl)
- 0.1 mM ATP (diluted from stock) (2.5 µl)
- T4 polynucleotide kinase (10 units)
- Either *Xba* I or *Not* I (8 units)
- Add ddH$_2$O to a final volume of 50 µl.

  ii. Incubate at 37°C for 60 min and extract once by adding one volume of TE-saturated phenol/chloroform. Vortex for 1 min.
  iii. Centrifuge at 11,000 g for 5 min and carefully transfer the top, aqueous phase to a fresh tube.
  iv. Remove unligated adapters as follows: Place a spin column in the collection tube and slowly load the supernatant onto the top center of the gel bed. Centrifuge in swinging bucket at 800 g for 5 min and collect the eluate containing the cDNA-adapter.
  v. Add 0.5 volume of 7.5 M ammonium acetate to the eluate, mix, and add two volumes of chilled 100% ethanol. Allow precipitation to occur at –20°C for 2 h.
  vi. Centrifuge at 12,000 g for 10 min and briefly rinse the pellet with 1 ml of cold 70% ethanol. Dry the cDNA linker under vacuum for 10 min and resuspend the pellet in 20 ml of TE buffer. Store the sample at –20°C until use.
  vii. To check the efficiency of ligation, load 4 µl (0.3 µg) of the ligated sample onto a 0.8% agarose gel containing EtBr and an equal amount of the unligated, blunt-ended cDNAs next to the ligated sample. After electrophoresis, take a photograph under UV light. Efficient ligation will cause a shift in the migration rate because of the ligation of linker concatemers, which run more slowly than the unligated sample.

5. Ligation of cDNA into lambda vectors: For optimal ligation of vectors and cDNA inserts, before performing a large-scale ligation, several ligations of vector arms and cDNA inserts should be carried out to determine their molar ratio and to determine the optimal ligation conditions. The molar concentration of the vectors should remain constant while adjusting the amount of DNA inserts. The molar ratio range of 43-kb λ arms to an average 1.4-kb cDNA inserts is usually from 1:1 to 1:0.5. Two control ligations should be set up. One is the ligation of vectors and positive DNA inserts to check the efficiency, the other is ligation of the vector arms alone to determine the background level of religated arms.
   a. Set up the ligation reactions as follows and carry them out separately:
      i. Sample ligations:

Components	Microcentrifuge tubes			
	A	B	C	D
Ligase 10X buffer (µl)	0.5	0.5	0.5	0.5
λDNA vectors (0.5 µg/µl; 1 µg = 0.04 pmol) (µg)	1	1	1	1
Linker/adapter–cDNA inserts (0.1 µg/µl; 10 ng = 0.01 pmol) (ng)	10	20	30	40
T4 DNA ligase (Weiss unit) (units)	2	2	2	2
Add ddH$_2$O to a final volume of (µl)	5	5	5	5

      ii. Control ligation 1:
         - Ligase 10X buffer (0.5 µl)
         - λDNA vector (0.5 µg/µl) (1.0 µg)
         - Positive-control DNA insert (0.1 µg)
         - T4 DNA ligase (2.0 Weiss units)
         - Add ddH$_2$O to a final volume of 5 µl.

cDNA Libraries

iii. Control ligation 2:
- Ligase 10X buffer (0.5 µl)
- λDNA vector (0.5 µg/µl) (1.0 µg)
- T4 DNA ligase (2.0 Weiss units)
- Add ddH$_2$O to a final volume of 5 µl.

b. Incubate the ligation reactions at room temperature (22 to 24°C) for 2 h and proceed to packaging.

7. *In Vitro* Packaging of Ligated DNA

In vitro packaging is performed in order to use a phage-infected *E. coli* cell extract to supply the mixture of proteins and precursors required for encapsulating the recombinant λDNA. The packaging system is commercially available with the specific bacterial strain host and control DNA.

1. Thaw three Packagene™ extracts (50 µl per extract) on ice.

Note: *Do not thaw the extract at room temperature or 37°C, and do not freeze the extract once it has thawed.*

2. When the extracts have thawed, immediately divide each extract into two tubes on ice. Each tube contains 25 µl of the extract. Add each of the 5 µl of the ligation reaction mixture to a 25-µl extract and mix gently.
3. Incubate at 22 to 24°C for 4 h and add 250 µl phage buffer and 10 µl of chloroform. Gently mix well and allow the chloroform to settle to the bottom of the tube. Store the packaged phage at 4°C for up to 5 weeks even though the titer may drop.

C. Titering and Amplification of the Packaged Phage

1. Partially thaw the specific bacterial strain such as *E. coli* LE 392, Y 1090, Y1089, or KW 251 on ice in a sterile laminar-flow hood. The bacterial strain is usually kept in 15% glycerol and stored at −70°C until use. Pick up a small amount of the frozen bacteria using a sterile wire transfer loop and immediately inoculate the bacteria by gently drawing several lines on the surface of an LB (Luria-Bertaini) plate. The LB plate should be freshly prepared and dried at room temperature for a couple of days prior to use. Invert the inoculated LB plate and incubate it in an incubator at 37°C overnight. Multiple bacterial colonies will grow.
2. Prepare fresh bacterial culture by picking up a single colony from the freshly streaked LB plate using a sterile wire transfer loop and inoculate a culture tube containing 5 ml of LB medium supplemented with 50 µl of 20% maltose and 50 µl of 1 *M* MgSO$_4$ solution. Shake at 160 rpm at 37°C for 6 to 9 h or until the OD$_{600}$ has reached 0.6. Store the culture at 4°C until use.
3. Dilute each of the packaged recombinant phage samples 1000, 5000, or 10,000 times with phage buffer.
4. Add 20 µl of 1 *M* MgSO$_4$ solution and 2.8 ml of melted top agar to 36 sterile glass test tubes in a sterile laminar-flow hood. Cap the tubes and immediately place the tubes in a 50°C water bath for at least 30 min to keep the agar melted.
5. Mix 0.1 ml of the diluted phage with 0.1 ml of fresh bacterial cells in a microcentrifuge tube. Cap the tube and allow the phage to adsorb to the bacteria in an incubator at 37°C for 30 min.

Notes: *If there are four package samples and two package controls, each one should be diluted at 1000, 5000, and 10,000 times in phage buffer. Each diluted phage should be duplicated at this step, and there will be a total of 36 plates to be set up.*

6. Add the incubated phage/bacterial mixture into specified tubes in the water bath. Vortex gently and immediately pour onto the centers of LB plates. Quickly spread the mixture all over the surface of

the LB plates by gently tilting them. Cover the plates and allow the top agar to harden for 15 min in the flow hood. Invert the plates and incubate them in an incubator at 37°C overnight or for 15 h.

Notes: *The top agar mixture should be evenly distributed over the surface of the LB plate. Otherwise, the growth of bacteriophage will be affected and may decrease the pfu number.*

7. Count the number of plaques for each plate and calculate the titer of the phage (pfu) for each sample and for the control.

    Plaque forming units (pfu) per milliliter = (number of plaques/plate) × dilution times × 10

    The last 10 of the calculation refers to the 0.1 ml, per milliliter basis, of the packaging extract used for one plate. For example, if there are 100 plaques on a plate made from a 1/5000 dilution, the pfu per milliliter of the original packaging extract = $100 \times 5000 \times 10 = 5 \times 10^6$.
8. Compare the titers (pfu/ml) from the different samples; determine the optimal ratio of vector arms and cDNA inserts and prepare a large-scale ligation and packaging reaction using optimum ratio conditions.

D. Large-Scale Ligation and *In Vitro* Packaging

1. It is best to set up two sample reactions and carry them out separately. In case one sample fails due to an accident, the other one can serve as a backup. The following is designed for one reaction. It is based on the optimal ratio of vectors and cDNA inserts used in our lab.
    - Ligase 10X buffer (2 µl)
    - λDNA vector (43 kb; 0.5 µg/µl) (2 µg) (Make sure that the vector is phosphorylated.)
    - Linker/adapter ligated cDNAs (average 1.5 to 1.8 kb; 0.2 µg/µl) (0.2 µg)
    - T4 DNA ligase (5 Weiss units)
    - Add ddH$_2$O to a final volume of 20 µl.
2. Incubate the ligation reactions at room temperature (22 to 24°C) for 2 h.
3. Check the efficiency of ligation by loading 0.1 µg of the ligated sample onto a 0.8% agarose gel containing EtBr and an equal amount of the unligated DNAs and unligated vector arms next to the ligated sample. After electrophoresis, take a photograph under UV light. Efficient ligation will cause a shift in the migration rate due to the ligation of vector concatemers, which run more slowly than the unligated DNAs.

Notes: *This is a very important step to check because the ligation is dependent on optimal conditions. Low efficiency of ligation of the vector will cause trouble in packaging. Thus, the pfu/ml for the library will be less than 1×10^4, which is considered to be a poor library. If the efficiency of vector ligation is very high, the recombinant DNAs can be directly used for packaging, as described in steps 8 to 10, or for being precipitated from steps 4 to 7.*

4. For long-term storage of the recombinant DNAs, which actually constitutes the cDNA library, extract the enzyme with one volume of TE-saturated phenol/chloroform, mix well by vortexing, and centrifuge at 11,000 g for 4 min.
5. Carefully transfer the top, aqueous phase to a fresh tube and add an equal volume of chloroform:isoamyl alcohol (24:1). Vortex to mix and centrifuge as in step 4.
6. Carefully transfer the top, aqueous phase to a fresh tube. **Do not take any white materials.** To the supernatant, add 0.15 volume of 3 *M* sodium acetate buffer, pH 5.2, and 2.5 volumes of chilled 100% ethanol. Allow precipitation to occur at –70°C for 30 min or at –20°C for 2 h.

cDNA Libraries   163

7. Centrifuge at 12,000 g for 10 min and carefully pour off the supernatant. Briefly rinse the DNA pellet with 1 ml of 70% chilled ethanol and dry the pellet under vacuum for 10 min. Dissolve the methylated cDNAs in 10 μl TE buffer. Store the recombinant DNAs at −20°C until use.
8. Thaw two packaging extracts (50 ml each) on ice.

Note: *Do not thaw the extract at room temperature or 37°C, and do not freeze the extract once it has thawed.*

9. When the extract has thawed, immediately add 9 μl of the ligation mixture without precipitation or 0.5 μg stored vector cDNA to 50 μl of the extract and mix gently.

Note: *Do not add more than 10 μl of the ligation mixture per 50 μl of packaging extract.*

10. Incubate at 22 to 24°C for 4 h and add 500 μl phage buffer and 25 μl of chloroform. Gently mix well and allow the chloroform to settle to the bottom of the tube. Store the packaged phage at 4°C for up to 5 weeks even though the titer may drop. At this stage, the packaged sample can be used for titering and screening the cDNA library (see the next sections). In order to store the packaged phage for a long time, S buffer, but not phage buffer, can be added to the mixture and extracted with one volume of chloroform (**not phenol**). Add dimethylsulfoxide (DMSO) to final 7% to the top, aqueous phase and store at −70°C.

Notes: *Based on our experience, the cDNA library stored for long periods of time in this way may drop in pfu number several-fold. For long-term storage, we recommend storing the precipitated recombinant DNAs instead of their being packaged at −70°C. When it is time to screen the library, we suggest that a freshly packaged bacteriophage be produced using the stored recombinant DNAs. This will increase the odds of success.*

E.   Immunoscreening of a Lambda Expression cDNA Library

If the cloning vector used is expression vector λgt11 *Sfi-Not* I, the cDNA inserts cloned into the *lac* Z gene of the vector can be expressed as a part of a β-galactosidase fusion protein. The expression cDNA library can be screened with a specific antibody used as a probe. The primary positive clones of interest can be "fished" out from the library. After several rounds of retitering and rescreening of the primary positive clones, putative clones will be isolated. The overall schematic of the procedures is as follows:

1. Prepare an LB medium, pH 7.5, by adding 15 g of Bacto-agar to 1 L of the LB medium and autoclave. Allow the mixture to cool to about 50°C and add ampicillin (100 μg/ml) and tetracycline (15 μg/ml). Mix well and pour 30 to 40 ml of the mixture into each of the 85- or 100-mm petri dishes in a sterile laminar-flow hood. Remove any bubbles with a pipette tip and let the plates cool for 5 min prior to being covered. Allow the agar to harden for 1 h and store the plates at room temperature for up 10 days or at 4°C in a bag for 1 month. The cold plates should be placed at room temperature for 1 to 2 days before use, or this can be done prior to cooling.
2. Partially thaw the specific bacterial strain, such as *E. coli* Y 1090, on ice. Pick up a small amount of the bacteria using a sterile wire transfer loop and immediately streak out *E. coli* Y1090 on LB plates by gently drawing several lines on the surface of an LB plate. Invert the inoculated LB plate and incubate the plate in an incubator at 37°C overnight. Multiple bacterial colonies will grow.
3. Prepare fresh bacterial cultures by picking up a single colony from the freshly streaked LB plate using a sterile wire loop and inoculate into a culture tube containing 50 ml of LB medium supplemented

with 500 μl of 20% maltose and 500 μl of 1 M MgSO$_4$ solution. Shake at 160 rpm at 37°C overnight or until the OD$_{600}$ has reached 0.6. Store the culture at 4°C for up to two days until use.

4. Add 20 μl of 1 M MgSO$_4$ solution and 2.8 ml of melted LB top agar to each of the sterile glass test tubes in a sterile laminar-flow hood. Cap the tubes and immediately place the tubes in a 50°C water bath for at least 30 min.
5. Based on the titration data, dilute the λ cDNA library with phage buffer. Set up 20 to 25 plates for primary screening of the cDNA library. For each 100-mm plate, mix 0.1 ml of the diluted phage containing 2×10^5 pfu of the library with 0.15 ml of fresh bacterial cells in a microcentrifuge tube. Cap the tube and allow the phage to adsorb to the bacteria in an incubator at 37°C for 30 min.

Notes: *The number of clones needed to detect a given clone is the probability that a low-abundance mRNA is converted into a cDNA in the library:*

$$N = [\ln (1-P)] / [\ln (1-1/n)]$$

where N is the number of clones needed, P is the possibility given (0.99), and 1/n is the fractional portion of the total mRNA, which is represented by a single low-abundance mRNA. For example, if P = 0.99, 1/n = 1/37,000, and N = 1.7 × 10^5.

6. Add the incubated phage/bacterial mixture to the tubes from the water bath. Vortex gently and immediately pour onto the centers of LB plates. Quickly spread the mixture over the entire surface of the LB plates by gently tilting them. Cover the plates and allow the top agar to harden for 15 min in a laminar-flow hood. Invert the plates and incubate in an incubator at 42°C for 4 h.

Note: *The top agar mixture should be evenly distributed over the surface of a LB plate. Otherwise, the growth of the bacteriophage will be affected.*

7. Saturate nitrocellulose filter disks in 10 mM IPTG (isopropyl β-D-thiogalactopyranoside) in water for 30 min and air dry the filters at room temperature for 40 min.
8. Carefully overlay each plate with a dried nitrocellulose filter disk from one end and slowly lower it to the other end of the plate, avoiding any air bubbles underneath the filter. Quickly mark the top side and position of each filter in a triangular fashion by punching the filter through the bottom agar using a 20-gauge needle containing a small amount of India ink. Cover the plates and incubate at 37°C for 4 to 6 h. If a duplicate is needed, a second filter may be overlaid on the plate and incubated for another 5 h at 37°C; however, the signal may be relatively weaker. At this stage, IPTG induces the expression of the cDNA library, and the expressed proteins are transferred onto the facing side of the filter.
9. Place the plates at 4°C for 30 min to chill the top agar and to prevent the top agar from sticking to the filter. Move the plates to room temperature and carefully remove the filters using forceps. Label and wrap each plate with Parafilm™, and store at 4°C until use. The filters may be rinsed briefly in TBST buffer to remove any agar.

Notes: *Due to the diffusion feature of proteins, it is strongly recommended that the filters be processed according to the following steps. Storing at 4°C for a couple of days before processing may cause difficulty in identifying the positive plaques on the plate. The filters must not be allowed to dry out during any of the subsequent steps, which are carried out at room temperature. Based on our experience, high background and strange results may appear if even partial drying occurs. If time is limited, the damp filters can be wrapped in SaranWrap, then in aluminum foil, and stored at 4°C for up to 12 h. The filters should be processed individually during the following steps to obtain an optimal detection signal.*

10. Incubate a filter in 10 ml of TBST buffer containing one of the following blocking reagents; (1) 2 to 3% BSA (bovine serum albumin), (2) 1% gelatin, (3) 5% nonfatty milk, or (4) 20% calf serum. This will block nonspecific protein-binding sites on the filter. Treat the filter for 60 min with slow shaking at 50 rpm.

Notes: *It is recommended that one dish be used for one filter only and that the facing side of the filter be down. Each filter should be incubated with about 10 ml of blocking buffer. Less than 5 ml will cause the filter to dry, but more than 15 ml for an 82-mm filter is not necessary.*

11. Carefully transfer the filter to a fresh dish containing 10 ml of TBST buffer with primary antibody.

Notes: *The antibody should be diluted 200X to 10,000X, depending on the concentration of the antibody prepared. The antibody purified with an immunoglobulin G (IgG) fraction or with an affinity column usually produces better signals. The diluted primary antibody can be reused several times if stored at 4°C.*

12. Wash the filter in 20 ml of TBST buffer containing 0.5% BSA for 10 min with shaking at 50 rpm. Repeat washing two times.
13. Transfer the filter to a fresh dish containing 10 ml of TBST buffer with the second antibody–alkaline phosphatase conjugate at 1:5000 to 1:10,000 dilution. Again, this dilution depends on the source of the secondary antibody. Incubate for 40 min with shaking at 50 rpm.
14. Wash the filter as in step 12.
15. Blot off excess liquid on a filter paper and place the filter into 10 ml of freshly prepared AP color-development substrate solution. Allow the color to develop for 0.5 to 5 h or overnight if a relatively high background is acceptable. Positive clones should appear as purple circles/spots on the white filter.
16. Stop the color development when the color has developed to the desired intensity using 15 ml of stopping solution. The filter can be stored in the solution or stored dry. The color will fade after drying, but it can be restored with water.
17. To locate the positive plaques, match the filter to the original plate by placing the filter, facing-side down, underneath the plate with the help of the marks previously made. This can be done by placing a glass plate over a lamp and putting the matched filter and plate on the glass plate. Turn on the light; by doing so, the positive clones can easily be identified. Remove individual positive plugs containing phage particles from the plate using a sterile pipette with the tip cut off or the large end of a Pasteur pipette if a larger plug is desired. Expel the plug into a microcentrifuge tube containing 1 ml of elution buffer. Allow elution to occur for 4 h at room temperature with occasional shaking.
18. Transfer the eluate supernatant to a fresh tube and add 20 µl of chloroform. Store at 4°C for up to 5 weeks.
19. Determine the pfu of the eluate, as previously described. Replate the phage and repeat the screening procedure with the antibody probe several times until 100% of the plaques on the screened plate are positive (Figure 9.6).

Notes: *During the rescreening process, the plaque number used for one plate should be gradually reduced. In our experience, the plaque density for one 100-mm plate in the rescreening procedure is generally decreased from 1000 to 500, to 300, and to 100 pfu.*

20. Amplify the putative cDNA clones and isolate the recombinant λ DNAs by either plate or liquid methods (see the λ DNA isolation section in **Chapter 1**). The purified DNA can be used for subcloning of the cDNA inserts.

F. Troubleshooting Guide for Immunoscreening

Symptoms	Solutions
Plaques are small in one area and large in other areas	This is due to uneven distribution of the top agar mixture. Make sure that the top agar mixture is spread evenly.
Too many positive clones in the primary screening	The specificity of the primary antibody is low, or the concentration is too high, or the blocking efficiency is low. Try to use IgG or an affinity column-purified antibody, carrying out different dilutions of the antibody or increasing the percentage of the blocking reagent in TBST buffer.
Purple background on the filter	Color development is too long. Try to stop the color reaction as soon as desired signals appear.
Unexpected larger purple spots	Air bubbles may be produced when overlaying the filter on the LB plate. Make sure no air bubbles appear underneath the filter.
No signal appears at all in any of the primary screening plates	The pfu numbers used for each plate are too low. For rare proteins, 2×10^6 pfu should be used in the primary screening of the library.
Signals are weak on the filters of subsequent screening	The antibody has low activity, or color development was stopped too early. Try to use an immunoassay to check the quality of the antibody, or increase the color developing time, or try to use high-quality nitrocellulose filters.

G. Screening a cDNA Library Using Labeled DNA Probes

In addition to the immunoscreening methodology described above, a cDNA library constructed in nonexpression or expression vectors such as λgt10 and λgt11 *Sfi-Not* I can be screened with labeled DNA as a probe, which can be labeled by isotopic or nonisotopic methods (see **Chapter 4** for detailed procedures). The library is screened by *in situ* plaque hybridization to the probe. Identification of cDNAs of a potential multigene family can be accomplished under low-stringency hybridization conditions. Based on our experience, reducing the percentage of formamide to 40 or

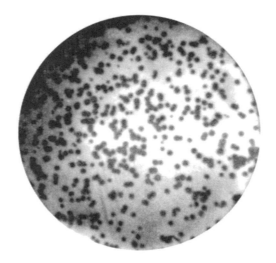

FIGURE 9.6
Final step in progressive screening of a cDNA library showing that 100% of plaques are positive. (From Kaufman, P.B., Wu, W., Kim, D., and Cseke, L.J., *Handbook of Molecular and Cellular Methods in Biology and Medicine*, 1st ed., CRC Press, Boca Raton, FL, 1995.)

cDNA Libraries

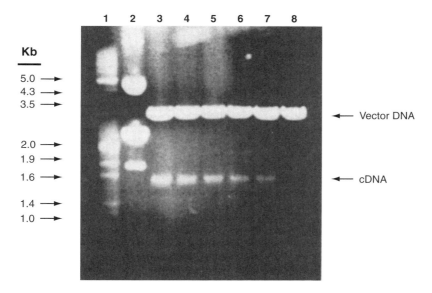

FIGURE 9.7
A successful subcloning of a cDNA insert in a plasmid vector. Plasmid DNA was isolated from different transformant colonies by use of a miniprep. The DNA was digested with *EcoR* I and *Hind* III followed by electrophoresis on a 1% agarose gel containing ethidium bromide. Lane 1: DNA standard molecular-weight markers; Lane 2: undigested plasmid DNA showing three bands because of different levels of supercoiling; Lanes 3–8: DNA from different transformant bacterial colonies.

30% in the hybridization solution and reducing the washing temperatures to room temperature followed by 50 to 56°C will provide the best results. Also, keeping the temperature of hybridization constant and the salt in the wash solution constant at 2X SSC in 0.1% SDS will allow consistent results. If a gene-specific probe is available and its specific clone is desired, nonspecific hybridization should be prevented using high-stringency conditions. Positive plaques will then be isolated to homogeneity by successive rounds of phage titering and rescreening (Figure 9.6). The resulting phage lysates can then be used in large-scale preparations of phage DNAs (see procedures for isolation and purification of phage DNA in **Chapter 1**).

1. Screening of a cDNA Library Using a ^{32}P-Labeled Probe
The general procedures are the same as described under the screening of genomic libraries in **Chapter 10**.

2. Screening of a cDNA Library Using a Nonisotope-Labeled Probe
The general procedures are the same as described under the section on the screening of genomic libraries in **Chapter 10**.

After obtaining positive clones that contain the cDNA of interest, the cDNA should be subcloned for further characterization (Figure 9.7), including restriction mapping, sequencing, glycerol stock and large-scale DNA preps. The detailed procedures are described in **Chapter 10**.

Reagents Needed

Alkaline Gel Running Buffer (Fresh)
        30 mM NaOH
        1 mM EDTA

2X Alkaline Buffer
>20 mM NaOH
>20% Glycerol
>0.025% Bromophenol blue (use fresh each time)

[α-^{32}P]dCTP (>400 Ci/mmol)

7.5 M Ammonium Acetate
>Dissolve 57.8 g ammonium acetate in 50 ml ddH$_2$O and adjust volume to 100 ml.

AP Buffer
>100 mM Tris-HCl, pH 9.5
>100 mM NaCl
>5 mM MgCl$_2$

Blocking Solution
>3% BSA, or 1% gelatin, or 5% nonfatty milk, or 20% calf serum in TBST buffer

1 mg/ml Carrier DNA (e.g., salmon sperm)

Chloroform:Isoamyyl Alcohol (24:1)
>48 ml Chloroform
>2 ml Isoamyl alcohol
>Mix and store at room temperature.

Color Development Substrate Solution
>50 ml AP buffer
>0.33 ml Nitroblue tetrazolium (NBT) stock solution
>0.165 ml 5-Bromo-4-chloro-3-indolylphosphate (BCIP) stock solution
>Mix well after each addition and protect the solution from strong light. Warm the substrate solution to room temperature to prevent precipitation.

EcoR I 10X Buffer
>0.9 M Tris-HCl, pH 7.5
>0.5 M NaCl
>0.1 M MgCl$_2$

EcoR I Methylase 10X Buffer
>1 M Tris-HCl, pH 8.0
>0.1 M EDTA

0.2 M EDTA
>Dissolve 37.2 g Na$_2$EDTA·2H$_2$O in 300 ml ddH$_3$O. Adjust pH to 8.0 with 2 N NaOH solution. Adjust final volume to 500 ml with ddH$_2$O and store at room temperature.

Elution Buffer
>10 mM Tris-HCl, pH 7.5
>10 mM MgCl$_2$

IPTG Solution

10 mM Isopropyl β-D-thiogalactopyranoside in ddH$_2$O.

Kinase 10X Buffer

700 mM Tris-HCl, pH 7.5
100 mM MgCl$_2$
50 mM DTT

LB (Luria-Bertaini) Medium (per liter)

Bacto-tryptone (10 g)
Bacto-yeast extract (5 g)
NaCl (5 g)
Adjust to pH 7.5 with 0.2 N NaOH and autoclave.

LB Plates

Add 15 g of Bacto-agar to 1 L of freshly prepared LB medium and autoclave. Allow the mixture to cool to about 60°C and pour 30 ml of the mixture into 85- or 100-mm petri dishes in a sterile laminar-flow hood with filtered air flowing. Remove any bubbles with a pipette tip and let the plates cool for 5 min prior to their being covered. Allow the agar to harden for 1 h and store the plates at room temperature for up to 10 days or at 4°C in a bag for 1 month. The cold plates should be placed at room temperature for 1 to 2 days before use.

LB Plates Containing Ampicillin and Tetracycline

Add 15 g of Bacto-agar to 1 L LB medium. Autoclave. When the medium cools to 50°C, add ampicillin (100 μg/ml) and tetracycline (15 μg/ml). Pour 30 to 40 ml into each of the 100-mm petri dishes in a laminar-flow hood with filtered air flowing. Remove any bubbles with a pipette tip and let the plates cool for 5 min prior to being covered. Allow the agar to harden for 1 h and store the plates at room temperature for up to 10 days or at 4°C in a bag for 1 month. The cold plates should be placed at room temperature for 1 to 2 days before use.

LB Top Agar

Add 4.0 g agar to 500 ml freshly prepared LB medium. Autoclave and store at 4°C until use. For plating, melt the agar in a microwave. When the solution has cooled to 60°C, add 0.1 ml of 1 M MgSO$_4$ to 10 ml of the mixture. If color-selection methodology is used to select recombinants in conjunction with bacterial-host strain Y1090, 0.1 ml IPTG (20 mg/ml in water, filter-sterilized) and 0.1 ml X-Gal (50 mg/ml in dimethylformamide) should be added to the cooled (55°C) 10 ml of the top agar mixture.

Ligase 10X Buffer

300 mM Tris-HCl, pH 7.8
100 mM MgCl$_2$
100 mM DTT
10 mM ATP

1 M MgSO$_4$

20% (v/v) Maltose

Not I or Xba I 10X Buffer

0.1 M Tris-HCl, pH 7.5
1.5 M NaCl
60 mM MgCl$_2$
10 mM DTT

Phage Buffer

20 mM Tris-HCl, pH 7.4
100 mM NaCl
10 mM MgSO$_4$

S Buffer

Phage buffer + 2% (w/v) gelatin

Sample Buffer

50% Glycerol
0.2% SDS
0.1% Bromophenol blue
10 mM EDTA

First-Strand 5X Buffer

250 mM Tris-HCl, pH 8.3 (at 42°C)
50 mM MgCl$_2$
250 mM KCl
2.5 mM Spermidine
50 mM DTT
5 mM each of dATP, dCTP, dGTP, and dTTP

Second-Strand 10X Buffer

0.5 M Tris-HCl, pH 7.2
0.9 M KCl
30 mM MgCl$_2$
30 mM DTT
0.5 mg/ml Bovine serum albumin (BSA)

2 M Sodium Chloride (NaCl)

Dissolve 58.4 g of NaCl in 400 ml ddH$_2$O. Adjust final volume to 500 ml with ddH$_2$O, autoclave, and store at room temperature.

Stop Solution

20 mM Tris-HCl, pH 8.0
5 mM EDTA

TBST Buffer

10 mM Tris-HCl pH 8.0
150 mM NaCl
0.05% Tween-20

TE Buffer

        10 mM Tris-HCl, pH 8.0
        1 mM EDTA

TE-Saturated Phenol/Chloroform

        Thaw phenol crystals at 65°C and mix in equal parts of phenol and TE buffer. Mix well and allow the phases to separate at room temperature for 30 min. Take one part of the lower, phenol phase and mix with one part of chloroform:isoamyl alcohol (24:1). Mix well, allow phases to separate, and store at 4°C until use.

Trichloroacetic Acid (5% and 7%)

Xba I or Not I 10X Buffer

        300 mM Tris-HCl, pH 7.8
        100 mM MgCl$_2$
        500 mM NaCl
        100 mM DTT

III. Construction of a cDNA Library by Subtractive Hybridization Techniques

Subtractive hybridization is a powerful technology used to identify specific cDNA(s) derived from specific mRNAs expressed by the right gene(s) in one tissue or cell type/line but not in another. The first step is to make the first-stranded cDNAs representing all of the mRNA sequences expressed in each of two tissues or cell types/lines. Traditionally, the single-strand cDNAs from two different tissues or cell types/lines can be labeled and used, without subtractive procedures, to screen a conventional cDNA library in order to identify the sequences differentially expressed in one tissue or cell type/line but not in the other. The drawback of "differential screening" methods not making use of the subtractive technique is that significant hybridization background generally occurs. In that case, only abundantly expressed sequences can be identified. The rarely expressed low-abundance sequences are usually missed. In order to solve this problem, subtractive technology was developed to hybridize the single-stranded cDNAs from one tissue or cell type/line to an excess of mRNAs from a second tissue or cell type/line. Any cDNAs from tissue or cell type/line A that represent sequences expressed in both lines should form DNA/RNA hybrids with the mRNAs from tissue or cell type/line B. The unhybridized cDNAs are specific for tissue or cell type/line A, but not B, and they remain single-stranded. Both single-stranded and hybridized cDNAs can be separated by chromatography on hydroxyapatite (HAP) columns. The unhybridized cDNAs can then be used to construct a subtracted cDNA library in which the sequences specific to one tissue or cell type/line are greatly enriched. Alternatively, the cDNAs can be used to make subtractive probes to screen conventional cDNA libraries to isolate the specific clones. The subtracted cDNA libraries include an enriched population of the cDNA clones, which simplify the screening process due to the much smaller numbers of pfu in the primary screening. This also speeds up the purification process of the cDNA of interest.

A. Synthesis of the First-Strand cDNA

The protocol for this is identical to the one described for cDNA Library Construction. The detailed protocol is described in **Section II.B.1**.

B. Removal of the mRNA Template

1. Add one volume (40 μl) of alkaline denaturing buffer to the above 40 μl of the unlabeled reaction mixture.
2. Incubate at 37°C for 30 min to separate the first-strand cDNA and mRNA template.
3. Add one volume of neutralization buffer to the mixture in step 2.
4. Add 0.1 volume of 3 M sodium acetate buffer, pH 5.2, and 2.5 volumes of chilled 100% ethanol. Place at –80°C for 30 min to precipitate the nucleic acids.
5. Centrifuge at 12,000 g for 5 min and carefully decant the supernatant. Briefly rinse the pellet with 1 ml of 70% cold ethanol and dry the pellet under vacuum for 15 min.
6. Dissolve the pellet in 50 μl of TE buffer.
7. Add *RNase* A (*DNase*-free) to a final concentration of 20 μg/ml and incubate at 37°C for 30 min.

Note: *The function of this step is to hydrolyze the mRNA templates.*

8. Extract the enzyme with one volume of TE-saturated phenol/chloroform, mix well by vortexing, and centrifuge at 11,000 g for 4 min.
9. Carefully transfer the top, aqueous phase to a fresh tube and add an equal volume of chloroform:isoamyl alcohol (24:1). Vortex to mix and centrifuge as in step 8.
10. Carefully transfer the top, aqueous phase to a fresh tube. **Do not take any white material.** To the supernatant, add 0.15 volume of 3 M sodium acetate buffer, pH 5.2, and 2.5 volumes of chilled 100% ethanol. Allow precipitation to occur at –70°C for 30 min or at –20°C for 2 h.
11. Centrifuge at 12,000 g for 10 min and carefully pour off the supernatant. Briefly rinse the cDNA pellet with 1 ml of 70% chilled ethanol and dry the pellet under vacuum for 10 min. Dissolve the cDNAs in 15 μl ddH$_2$O. At this stage, the cDNA should be single-strand DNA, which can be quantified as described previously.

Notes: *The separation of the first-strand cDNAs from hybridized mRNA is extremely important for subtraction hybridization. In order to check that the cDNA is of single-strand or double-strand form with mRNA, the purified cDNAs can be subjected to S1-nuclease assay. The resulting mixture is then subjected to 1.4% agarose gel electrophoresis using undigested cDNA as a control. If the cDNA is single-stranded, and if the S1 digestion is complete, no band should be visible as compared with control cDNA.*

C. Hybridization of cDNA to mRNA

1. Carry out a hybridization reaction in a microcentrifuge tube as follows:
   - 2 to 5 μg first-strand cDNA from tissue or cell type/line A
   - 15 μl 2X hybridization buffer
   - 25 μg Poly(A)$^+$ RNA from tissue or cell type/line B
   - Add ddH$_2$O to a final volume of 30 μl.

   or

   - 2 to 5 μg first-strand cDNA from tissue or cell type/line B
   - 15 μl 2X hybridization buffer
   - 25 μg Poly(A)$^+$ RNA from tissue or cell type/line A
   - Add ddH$_2$O to a final volume of 30 μl.
2. Tightly cover the tubes and wrap the top of the tube with Parafilm to prevent evaporation. Incubate the tubes at 65 to 68°C overnight to 16 h.

Note: *An RNase-free environment should be maintained for the hybridization.*

cDNA Libraries

D. Separation of cDNA/mRNA Hybrids from Single-Strand cDNA by Hydroxyapatite Chromatography

1. Prepare the hydroxyapatite (HAP) column: Add 1g of HAP (DNA Grade Bio-Gel, BioRad, Hercules, CA) in 5 ml of 0.1 M phosphate buffer (PB), pH 6.8, and mix well to form a slurry. Heat in boiling water for 5 min. Place a 5-ml plastic syringe closed at the bottom in a water bath equilibrated to 60°C. Place a cellulose acetate filter cut to size with a cork borer in the bottom of the syringe and wet the filter with 0.1 M PB. Alternatively, place some sterile glass wool at the bottom of the syringe. Slowly add 0.5 to 1 ml of HAP slurry to the bottom-closed column and allow the slurry to settle for 5 min prior to opening the column. Wash the column with 5 volumes of 0.1 M PB (60°C) two times.
2. Load the hybridized sample onto the prepared HAP column: Dilute the sample ten-fold in prewarmed (60°C) 0.1 M PB containing 0.15 M NaCl. Gently load the sample onto the column with the bottom closed. Gently stir the mixture in the column with a needle (**avoid bubbles**) and let it set for 10 min.
3. Open the column from the bottom and collect the effluent containing the single-strand cDNAs. Wash the column twice with 0.5 ml of 0.1 M PB containing 0.15 M NaCl and collect the effluent. The cDNA/mRNA hybrids in the column can be eluted with 0.5 M PB.
4. Pool the effluents together and load onto a fresh HAP column as in steps 2 to 3. The function of this step is to maximally remove any cDNA/mRNA hybrids that potentially remain in the effluent.
5. Pool the effluents together and dialyze against 500 ml ddH$_2$O overnight to remove the PB salts that can interfere with ethanol precipitation.
6. Add 0.15 volume of 3 M sodium acetate buffer, pH 5.2, and 2.5 volumes of chilled 100% ethanol to the dialyzed sample and place at –80°C for 30 min.
7. Centrifuge at 12,000 g for 5 min and briefly rinse the cDNA pellet with 2 ml of 70% cold ethanol. Dry the pellet under vacuum for 10 min and dissolve the cDNAs in 50 μl TE buffer. Take 4 μl of the sample to measure the concentration of cDNA. Store the sample at –20°C until use. At this stage, the single-strand cDNA should represent the sequences specifically expressed in tissue or cell type/line A or B but not in the other. These cDNAs can be labeled as probes to screen a conventional library or they can be used to construct a subtractive cDNA library.

Notes: (1) *In order to ensure that the cDNA is single-strand DNA, take an appropriate amount of the cDNA sample and carry out an S1 digestion assay as referred to in* **Section III.B**. (2) *The mRNA left in the sample from the above procedure will not interfere with subsequent steps.*

E. Making a cDNA Library from the Subtracted First-Strand cDNA

1. Synthesize the second-strand cDNA as follows, using a poly(A) primer.
   a. Set up the following reaction:
      - Subtractive first-strand cDNA (1 μg)
      - Second-strand 10X buffer (50 μl)
      - Poly(A)12-20 primer (0.2 μg)
      - *E. coli* DNA polymerase I (20 units)
      - Add nuclease-free ddH$_2$O to a final volume of 50 μl.
   b. Set up a tracer reaction for the second-strand DNA by removing 5 μl of the mixture to a fresh tube containing 1 μl 4 μCi [α-^{32}P]dCTP (>400 Ci/mmol). The tracer reaction will be used for TCA assay and alkaline agarose gel electrophoresis to monitor the quantity and quality of the synthesis of the second-strand DNA.
   c. Incubate both reactions at 14°C for 3.5 h. Add 95 μl of 50 mM EDTA to the 5-μl tracer reaction and store on ice.
   d. Heat the unlabeled double-strand DNA sample at 70°C for 10 min to stop the reaction. **At this point, the double-strand cDNAs should be generated.** Briefly spin down to collect the contents at the bottom of the tube and place on ice.

e. Add 4 units of T4 DNA polymerase (2 units/μg input cDNA) to the mixture and incubate the tube at 37°C for 10 min. This step functions to make the blunt-end, double-strand cDNA by T4 polymerase. The blunt ends are required for later adapter ligations.
f. Stop the T4 polymerase reaction by adding 10 μl of 0.2 M EDTA and place on ice.
g. Extract the cDNAs with one volume of TE-saturated phenol/chloroform. Mix well and centrifuge in a microcentrifuge at top speed for 4 min at room temperature.
h. Carefully transfer the top, aqueous phase to a fresh tube. To the supernatant, add 0.5 volume of 7.5 M ammonium acetate or 0.15 volume of 3 M sodium acetate (pH 5.2), mix well, and add 2.5 volumes of chilled (−20°C) 100% ethanol. Gently mix and allow precipitation to occur at −20°C for 2 h.
i. Centrifuge in a microcentrifuge at the top speed for 5 min. Carefully remove the supernatant, briefly rinse the pellet with 1 ml of cold 70% ethanol, and gently drain away the ethanol.
j. Dry the cDNA pellet under vacuum for 15 min. Dissolve the cDNA pellet in 25 μl of TE buffer. It is important to take 2 to 4 μl of the sample to measure the concentration of cDNAs prior to the next reaction. Store the sample at −20°C until use.

Note: Before adapter ligation, it is strongly recommended that the quantity and quality of the double-strand (ds) cDNAs be checked by TCA precipitation and gel electrophoresis as described previously.

2. Carry out the ligation of linkers or adapters, *Xba* I or *Not* I digestion for directional cloning, ligation to the λ DNA vector arms, and *in vitro*. The detailed protocols for these procedures are described in **Section II** of this chapter.
3. Screen the library to identify any clones of interest. To clone the DNA sequences present in tissue or cell type/line A and not B, a library made from A can be screened with either subtracted or nonsubtracted probe from both A and B. The desired phage clones can be identified by hybridization to the probe from A but not from B. Because it is a subtractive cDNA library with enriching cDNAs of interest, the phage library should not be plated densely. This is so that single plaques can be distinguished. Normally, 50 to 2000 plaques on a 100-mm plate are enough. The detailed screening and purification procedures are given in **Section II** of this chapter.
4. The specificity of the cDNA inserts in A but not B or in B but not A can be verified by slot-blot screening. A high-titer phage lysate is recommended for this purpose, since a single plaque contains about 10^6 phages, while a high-titer phage lysate can have 10^9 phages per milliliter.
   For one slot:
   - 100 μl Phage lysate
   - 20 μl 1 M Tris-HCl, pH 8.0
   - 4 μl 0.5 M EDTA
   - 124 μl 100% Formamide

   Heat for 10 min at 68°C and chill on ice. Load the sample onto the well of a slot-blotter (Schleicher and Schuell) and turn on the vacuum when all samples are loaded. Wash slots with 200 μl 2X SSC. Dry the membrane at room temperature for 10 min and bake at 80°C under vacuum for 2 h or UV cross-link at an optimal setting. A more detailed procedure for slot-blotting is described in **Chapter 5**. The procedures for prehybridization, hybridization, washing, and exposure are similar to Southern blotting. After exposure, bands should be specific for lysate A or B.

Reagents Needed

Alkaline Denaturing Buffer
0.5 M NaOH
1.5 M NaCl

[α - ^{32}P]dCTP (>400 Ci/mmol)

Chloroform:Isoamyl Alcohol (24:1)

0.2 M EDTA

Dissolve 37.2 g $Na_2EDTA \cdot 2H_2O$ in 300 ml ddH_2O. Adjust pH to 8.0 with 2N NaOH solution. Adjust final volume to 500 ml with ddH_2O and store at room temperature.

First-Strand 5X Buffer

250 mM Tris-HCl, pH 8.3 (at 42°C)
50 mM $MgCl_2$
250 mM KCl
2.5 mM Spermidine
50 mM DTT
5 mM each of dATP, dCTP, dGTP, and dTTP

Hybridization Buffer (2X)

40 mM Tris-HCl, pH 7.7
1.2 M NaCl
4 mM EDTA
0.4% SDS
1 µg/µl Carrier yeast tRNA (transfer RNA) (optional)

Neutralization Buffer

1.5 M NaCl
0.5 M Tris-HCl, pH 7.4

Phosphate Buffer (Stock)

0.5 M Monobasic sodium phosphate
Adjust the pH with 0.5 M dibasic sodium phosphate to 6.8.

3 M Sodium Acetate Buffer, pH 5.2

Dissolve sodium acetate in ddH_2O and adjust pH to 5.2 with 3 M glacial acetic acid. Autoclave and store at room temperature.

Trichloroacetic Acid (5% and 7%)

IV. Construction of Fractional cDNA Libraries Using *Xenopus* Oocytes as an Expression System

In addition to the subtractive technology described above, another powerful and state-of-the-art method of enriching a population of cDNA sequences of interest is to construct a fraction or a portion of the whole cDNA library.[5-7] This technology is especially useful when no specific nucleic acid or protein probe is available but when there is a known physiological function such as enzyme activity. Total mRNA is first fractionated using sucrose gradients to concentrate the particular mRNA in a specific fraction, which is then individually microinjected into *Xenopus* oocytes to check the expression of the interesting mRNA using a specific assay method. The expressed fraction of mRNA will then be used for cDNA synthesis and cloning, thus greatly increasing the possibility of identifying cDNA clones of interest. This is useful for rarely expressed mRNAs. However, the major drawback of the technique is the requirement for specialized and expensive microinjection equipment. The *Xenopus* oocyte is a remarkable giant cell that can be used to test the function of injected DNA, mRNA, and postmodified proteins. It is a very good system that has become a well-established tool for gene expression in eukaryotes.

Protocol

1. Isolate the total RNA and purify the poly(A)$^+$ RNA from the organism of interest as described previously (see the RNA section in **Chapter 2**).
2. Fractionate the mRNA by sucrose gradients containing methylmercuric hydroxide.
   a. Prepare sucrose gradients (10 to 30% w/v) containing 10 mM methylmercuric hydroxide in ultracentrifuge tubes (Beckman SW41 or equivalent). Dissolve the sucrose in sterile water and treat the solutions overnight with 0.1% DEPC at 37°C followed by heating to 100°C for 15 min to remove the DEPC. When the solutions have cooled to room temperature, add 1 M Tris-HCl (pH 7.4) and 0.5 M EDTA (pH 7.4) to final concentrations of 10 mM Tris-HCl and 1 mM EDTA. Finally, add methylmercuric hydroxide to a final concentration of 10 mM. The gradients can be made by (1) a gradient maker, (2) by adding the solutions, one by one, in decreasing density of the solutions one over another, or (3) by adding the solutions, one by one, in increasing density of the solutions under each other to a tube. The most convenient way is to place the tube on dry ice and to add a given density of sucrose solution. After this solution is frozen, add another density of solution. Finally, place the tube at 4°C overnight, gradually allowing it to thaw and diffuse, thus generating a gradient.
   b. Add methylmercuric hydroxide to a final concentration of 20 mM to 100-μl RNA sample (1 μg/μl) and carefully load the RNA sample onto the gradients.
   c. Centrifuge the gradients at 34,000 rpm for 15 to 18 h at 4°C in a Beckman SW41 rotor or its equivalent.
   d. Collect 0.2- to 0.3-ml fractions through a hypodermic needle inserted into the bottom of the centrifuge tube. Dilute each of the fractions with one volume of sterile, DEPC-treated water containing 5 mM 2-mercaptoethanol. Add 60 μl of 3 M sodium acetate buffer and place the tubes at 0°C for 1 h.
   e. Centrifuge at 12,000 g for 15 min at 0°C and carefully decant the supernatants to a fresh tube. Wash the RNA sample with 2 ml of 70% ethanol and place the pellets to dry at room temperature.
   f. Resuspend the RNA in 20 μl of water and add 0.1 volume of 3 M sodium acetate buffer and 3 volumes of chilled 100% ethanol. Mix well and leave the preparation at –70°C for 30 min.
   g. Centrifuge at 12,000 g for 10 min at 4°C. Dry the RNA under vacuum for 15 min. Dissolve the RNA in 20 μl of TE buffer. Take 4 μl of the sample to measure the concentration and quality of the RNA (see **Chapter 2**). Store the sample at –20°C until use.
3. Prepare the *Xenopus* oocytes for microinjection.

   The *Xenopus* oocyte is an egg-forming cell located in the ovary. It is surrounded by thousands of follicle cells. An oocyte contains a nucleus, an animal hemisphere (the relatively darker one), and a vegetal hemisphere. The development of oocytes includes several stages with the largest or stage VI arrested at meiotic prophase. Stage VI oocytes are usually used for microinjection. Oocytes are ideal for microinjection studies because of their large size (1 to 1.2 mm in diameter), which makes it easy to microinject either mRNA into the cytoplasm or DNA into the nucleus. In addition, the oocyte contains sufficient components needed for the expression of injected genes or RNAs, such as RNA polymerases I, II, and III, histones, and ribonucleotide triphosphates. Each oocyte contains more than 200,000 ribosomes, more than 10,000 tRNA molecules, and all of the enzymes required for translation of injected RNA and posttranslational modification of proteins.

   Xenopus laevis adult females are available from Carolina Biological Supply Co. (Burlington, NC) and Nasco (Ft. Atkinson, WI). The frogs can be kept in large plastic tanks with 5 L of dechlorinated water per frog at a depth of 10 cm with air holes at the tops of the tanks at room temperature. The frogs can be fed twice per week with chopped beef heart, Nasco frog brittle, or Purina trout chow. Newly purchased frogs may not eat for the first 1 to 2 weeks. A normal adult female *Xenopus* contains approximately 30,000 oocytes.
   a. For isolating the oocytes, a female frog can be anesthetized by immersing it in 0.5% ethyl *m*-aminobenzoate for 25 min. Alternatively, the frog can be anesthetized by immersing it in a bowl of ice for 30 to 45 min to lower its body temperature.

Caution: Gloves should be worn, since ethyl m-aminobenzoate is a potential carcinogen in humans.

b. Rinse the frog with water once it is immobilized, place it on a dissecting tray, and use a sharp scalpel to dissect the frog and remove the oocytes.
c. If the frog is to be reused, sterile instruments and techniques should be used. Place the frog with the ventral side up on ice to keep the back, but not the head, on ice. Gently swab the lower abdomen of the frog with a cotton ball soaked in alcohol. Lift and hold the skin with a pair of forceps and make a 1- to 2-cm cut in the skin using a pair of dissecting scissors. Make a similar cut through the underlying muscle. Pull out a section of ovary using forceps and cut it off with a pair of scissors.
d. Transfer the ovary to a petri dish containing modified Barth's medium (MBM). Then, put the remaining ovary back into the frog. Sew up the muscle with dissolving suture material and stitch the skin with silk sutures. Place the frog on its ventral side in a container with a small amount of water with the head elevated by wet paper towels. The frog can recover completely within a few hours.
e. Remove individual oocytes from the ovarian clumps by gently pulling the oocyte off at its base using two pairs of fine forceps. Select the largest or stage VI oocytes (1 to 1.2 mm in diameter) and transfer the oocytes with a wide-mouth Pasteur pipette to a fresh petri dish containing MBM. Alternatively, the oocytes can be released by the enzymatic method. Treat the clumps of oocytes in a solution containing 2 mg/ml collagenase (Sigma Type II) in MBM for 2 to 4 h at room temperature with slow shaking (60 rpm). The released oocytes should be washed thoroughly in MBM.
f. Repeat the enzymatic treatment and washing once more to remove the remaining follicular cells. Well-prepared oocytes should have no red tissue on the surface. The oocytes can be cultured for up to 2 weeks with daily changes of medium and removal of dead cells. The vegetal hemisphere of dead cells is very pale as compared with that of live oocytes.
4. Microinject the fractionated mRNA into the healthy oocytes.
a. Select microinjection equipment that is relatively simple. It should include a stereomicroscope (10× to 40×) magnification, a light source, micropipettes, and a microinjection system that is composed of a micromanipulator and microsyringe. The microsyringe should be capable of accurately delivering volumes in the nanoliter range. These systems are commercially available: the Pico-Injector (PLI-100, Medical Systems Corp., Greenvale, NY) and the Eppendorf Micro-injector (Model 5247, Fremont, CA). Assemble the system according to the instructions and test for nanoliter microinjection capability.
b. Incubate the mature oocytes overnight in MBM prior to injection. In order to immobilize the oocytes during microinjection, a piece of polyethylene mesh or its equivalent should be placed at the bottom of a petri dish. The size of the mesh wells should be about the right size to secure the oocyte. Add MBM to cover the mesh.
c. Transfer the oocytes to the support mesh with a sterile Pasteur pipette and orient the oocytes with a hair loop or equivalent.
d. Rinse the needle with sterile water and dry with a piece of clean paper. Slowly inject a small volume of mineral oil into the microsyringe with long, fine, plastic tubing attached to a 1-ml sterile syringe. **Avoid any bubbles.**

Note: *Details on making needles are located in* **Chapter 20**.

e. Briefly spin down the specific fraction of mRNA prepared previously to prevent potential particles from blocking the microinjector. Place 2 µl of the mRNA (0.5 to 1 µg/µl) as a droplet onto a sheet of Parafilm near the needle for ease of filling the needle. Position the micromanipulator close to the mRNA droplet and slowly suck in the droplet. **Avoid any bubbles.**
f. Transfer the oocytes (10 to 20) onto the microinjection support and adjust the fine focus. Slowly adjust the needle to approach individual oocyte and focus again.
g. Under the microscope, orient the oocyte and adjust the needle so that it just reaches the surface of the oocyte in the vegetal region or near the "equator" of the oocyte[9,12] at about a 20 to 30° angle (Figure 9.8). Under the microscope, gently lower the needle toward the inside of the oocyte using the micromanipulator. The local surface of the oocyte gradually becomes depressed at first

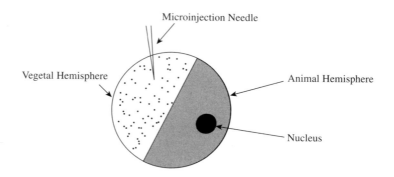

FIGURE 9.8
Diagram showing microinjection of RNA into an oocyte of *Xenopus*. The animal hemisphere is shown as a relatively dark region because of the heavy pigmentation. (From Kaufman, P.B., Wu, W., Kim, D., and Cseke, L.J., *Handbook of Molecular and Cellular Methods in Biology and Medicine*, 1st ed., CRC Press, Boca Raton, FL, 1995.)

and then returns to its original shape when the needle punches inside the oocyte. Immediately stop lowering the needle at this time.

h. Slowly inject 30 to 50 nl of the mRNA sample into one oocyte and carefully withdraw the needle. Repeat the injection of the same volume to each of the other oocytes.

Notes: *The mRNA amount for each oocyte should be determined. It usually ranges from 4 to 400 ng. The microinjection should be carried out carefully so that the contents inside the oocyte, such as yolk platelets, cannot float out.*

i. Carefully transfer the injected oocytes to a petri dish containing MBS and allow the mRNA to be translated for 12 h to 3 days at 20°C prior to analyzing the activity of enzyme or protein product.

Note: *Temperatures of 25°C or higher may decrease the oocytes' chances of survival.*

j. If labeling of the protein is desired, ^{35}S-methionine (1 to 5 mCi/ml of medium) can be added to the medium and incubated for 12 h. Alternatively, the radioactive amino acid can be added to the mRNA sample and be coinjected into the oocytes.

5. Extract the soluble protein to determine the expression of mRNA of interest.
   a. Homogenize the oocytes on ice in the homogenizing buffer (50 μl per oocyte) and centrifuge the homogenate at 10,000 g at 4°C for 5 min.
   b. Carefully transfer the supernatant to a fresh tube. **Avoid any yolk material.** The supernatant can be used for enzyme assay, immunoprecipitation, or electrophysiological response to a specific substrate based on the particular interest and method available.

6. Synthesize the cDNA from the fraction of mRNA of interest.

   The procedures for the synthesis of first- and second-strand cDNAs using *Xba* I primer-adapter with poly(dT) — for the ligation of *EcoR* I adapters to the double-strand cDNAs and for the digestion of the adapter-cDNA with *EcoR* I followed by *Xba* I — are the same as described in **Section II** of this chapter. At this stage, the recombinant cDNAs have one *EcoR* I sticky end and one *Xba* I sticky end, both of which are useful for directional cloning as given below.

7. Carry out ligation of cDNAs to the expressional vector, *in vitro* packaging, and titration.

   The general procedure of ligation of cDNA inserts to appropriate expression vectors, *in vitro* packaging, and titration have been previously described in this chapter (**Section II.B.6, Section II.B.7, Section II.C.,** and **Section II.D.**).

cDNA Libraries

8. Carry out *in vitro* transcription of the cloned cDNA.
   a. Isolation of recombinant phage or plasmid DNA (3×10^4 clones) containing cDNA inserts has been previously described (see **Section II** in this chapter).
   b. Digest the DNA with an appropriate restriction enzyme for making sense RNA. If antisense RNA is desired, the DNA should be digested with another appropriate enzyme, depending on the particular expression vector used.
      - Recombinant DNA (5 μg)
      - Appropriate enzyme 10X buffer (5 μl)
      - Appropriate restriction enzyme (20 units)
      - Add ddH$_2$O to a final volume of 50 μl.
   c. Incubate at the appropriate temperature for 2 to 3 h and add 2 μl of 500 mM EDTA to stop the reaction.
   d. Extract the enzyme with 50 μl of TE-saturated phenol/chloroform, mix well by vortexing, and centrifuge at 11,000 g for 4 min.
   e. Carefully transfer the top, aqueous phase to a fresh tube and add an equal volume of chloroform:isoamyl alcohol (24:1). Vortex to mix and centrifuge as in step d.
   f. Carefully transfer the top, aqueous phase to a fresh tube. To the supernatant, add 0.15 volume of 3 M sodium acetate buffer (pH 5.2) and 2.5 volumes of chilled 100% ethanol. Allow precipitation to occur at –70°C for 30 min or at –20°C for 2 h.
   g. Centrifuge at 12,000 g for 10 min and carefully pour off the supernatant. Briefly rinse the DNA pellet with 1 ml of 70% chilled ethanol and dry the pellet under vacuum for 10 min. Dissolve the DNA in 20 μl TE buffer.
   h. Set up *in vitro* transcription in the presence of the cap analogue GppppG using the linearized DNA templates and T7 RNA polymerase at room temperature, mix well after each addition, and incubate the reaction at 37°C for 2 h.
      - 5X Transcription buffer (16 μl)
      - 0.1 M DTT (8 μl)
      - rRNasin ribonuclease inhibitor (80 units)
      - Nucleotides mixture (2.5 mM each of ATP, GTP, CTP, and 1 mM UTP) (16 μl)
      - 5 mM m^{7}G (5')ppp(5')G (5 μl)
      - Linearized DNA template (1μg/μl) (5 μl)
      - T7 RNA polymerase (80 units)
      - Add ddH$_2$O to a final volume of 80 μl.

   Function: Cleavage of the recombinant phage DNA is to stop the subsequent *in vitro* transcription of cDNAs cloned in the vector immediately after the poly(A) tail, leaving the mRNA of interest intact.
   i. Add *RNase*-free *DNase* I to a final concentration of 1 unit/μg template DNA, which will be hydrolyzed by the enzyme after being incubated for 30 min at 37°C.
   j. Extract the enzyme with 80 μl of TE-saturated phenol/chloroform, mix well by vortexing, and centrifuge at 11,000 g for 4 min.
   k. Carefully transfer the top, aqueous phase to a fresh tube and add an equal volume of chloroform:isoamyl alcohol (24:1). Vortex to mix and centrifuge as in step j.
   l. Carefully transfer the top, aqueous phase to a fresh tube. To the supernatant, add 0.5 volume of 7.5 M ammonium acetate and 2.5 volumes of chilled 100% ethanol. Allow precipitation to occur at –70°C for 30 min or at –20°C for 2 h.
   m. Centrifuge at 12,000 g for 10 min and carefully pour off the supernatant. Briefly rinse the DNA pellet with 1 ml of 70% chilled ethanol and dry the pellet under vacuum for 10 min. Dissolve the DNA in 20 μl TE buffer. Measure the concentration of the RNA (see **Chapter 2**) and store the sample at –70°C until use.
9. Microinject the mRNA transcribed *in vitro* into *Xenopus* oocytes and measure the electrophysiological response with specific substrate as described in step 4.
10. Repeat steps 8 and 9 several times with the cDNA library with stepwise fractionations from 2000 clones to 1000, 500, 200, and 100 until a single, positive clone is obtained.

Reagents Needed

Homogenization Buffer
0.1 M NaCl
1% Triton-X-100
1 mM Phenylmethylsulfonyl fluoride
20 mM Tris-HCl, pH 7.6

10X Ligation Buffer
300 mM Tris-HCl, pH 7.8
100 mM MgCl$_2$
100 mM DTT
10 mM ATP

Modified Barth's Medium (MBM)
88 mM NaCl
1 mM KCl
2.4 mM NaHCO$_3$
15 mM Hepes-NaOH, pH 7.6
0.82 mM MgSO$_4 \cdot$ 7H$_2$O
0.33 mM CaNO$_3 \cdot$ 4H$_2$O
0.41 mM CaCl$_2 \cdot$ 6H$_2$O

5X Transcription Buffer
200 mM Tris-HCl, pH 7.5
30 mM MgCl$_2$
10 mM Spermidine
50 mM NaCl

V. cDNA Cloning and Analysis by the Polymerase Chain Reaction

The polymerase chain reaction (PCR) is a powerful technique used for *in vitro* amplification of specific DNA sequences using appropriate primers.[9-13] PCR is a rapid, sensitive, and inexpensive procedure to amplify a specific DNA of interest, and it is a major breakthrough technology in the analysis of DNA and RNA, DNA cloning, genetic diagnosis, detection of mutations, and genetic engineering.[15-21] In the process, *Taq* DNA polymerase purified from *Thermus aquaticus* is a heat-stable enzyme and carries out the synthesis of a complementary strand of DNA from 5' to 3' direction by the primer extension reaction. The primers are designed so that primer A directs the synthesis of DNA toward the other, primer B, and vice versa, resulting in the synthesis of the region of DNA flanked by the two primers. Because the *Taq* DNA polymerase is stable at high temperatures (94°C), it allows the target sequences of interest to be amplified during many cycles using excess primers and a commercial thermocycler apparatus. The present section describes examples of procedures for PCR-directed dsDNA amplification, reverse-transcription PCR (RT-PCR) cDNA cloning, and semiquantitative PCR for analysis of gene expression.

cDNA Libraries

A. Selection of Oligonucleotides

The specificity of primers is dependent on how well they match the complementary sequence of DNA as well as their length. PCR primers are typically 16 to 30 bp in length, with longer primers often giving better results. We recommend that the following criteria be considered while designing 16- to 30-bp primers. First, highly accurate DNA sequence is essential to ensure that the primer sequence exactly matches the sequence of the template DNA. The region chosen for the oligonucleotide should have a G/C content between 40 and 60%, with an annealing temperature (Tm) of at least 50°C. The Tm can be estimated using the following formula.

$$Tm = 4(G/C) + 2(A/T) - 12°C$$

We prefer the last six nucleotides at the 3' end of the primer to be exactly 50% G/C, with the very last base being a G or a C. Stretches of G or C at the ends of the primers should be avoided, as they often cause problems. In addition, many researchers analyze their primers for the possibility of forming primer dimers or hairpin loops. There are a variety of programs for doing this, but we have not found them to be required.

B. General Amplification of Double-Strand DNA by PCR

The following procedure can be applied to any dsDNA template. However, for the purposes of this chapter, it applies to isolated λDNA, which can then be used to clone specific cDNAs.

1. Using procedures detailed in **Chapter 1**, isolate λDNA from one of the above-prepared cDNA libraries.
2. Carry out PCR amplification.
   a. Design oligonucleotide primers that are complementary to two different regions of the DNA molecule or DNA fragment of interest. For example, two oligonucleotide primers can be designed as follows for the amplification of actin cDNA.

      Primer 1:   5' ATGGATGACGATATCGCTG 3'

      Primer 2:   5' ATGAGGTAGTCTGTCAGGT 3'

   Note: *These primers can generate a product of 568 bp.*

   b. In a 0.2-ml thin-walled PCR tube on ice, set up a standard reaction by adding the following in the order listed:
      - 10X Amplification buffer (5 µl)
      - ddH$_2$O (10 µl)
      - Mixture of four dNTPs (1.25 mM each) (8.5 µl)
      - Primer 1 (50 to 100 ng) in ddH$_2$O (5 µl)
      - Primer 2 (50 to 100 ng) in ddH$_2$O (5 µl)
      - dsDNA in TE buffer (100 to 250 ng) (2 µl)
      - *Taq* DNA polymerase (5 units/µl) (0.5 µl)
      - Add ddH$_2$O to a final volume of 50 µl.

Notes: (1) Thin-walled PCR tubes are highly recommended to allow better heat transfer between the thermocycler block and the sample. (2) Depending on the number of samples (including positive and negative controls), a cocktail containing all reagents with the exception of the DNA template should be made to establish uniform conditions for all reactions. (3) While the amplification buffer that we describe here works well in many cases, reaction buffers vary between different Taq polymerases. We recommend that the buffer supplied with the Taq polymerase be used along with the recommended amount of $MgCl_2$ or $Mg(OAc)_2$.

- c. Overlay the mixture with 50 μl of light mineral oil (Sigma or equivalent) to prevent evaporation of the sample if a thermocycler with a heated lid is not available.
- d. Carry out 25 to 35 cycles of PCR amplification in a PCR machine. The conditions of PCR, including primer annealing temperatures, extension times, and number of cycles, will have to be optimized. However, a good starting example is as follows:

Cycle	Denaturation	Annealing	Polymerization
First cycle	2 min at 94°C	—	—
Subsequent cycles (24–34)	15 sec at 94°C	30 sec at 50°C	3.5 min at 72°C
Last cycle	15 sec at 94°C	30 sec at 50°C	10 min at 72°C

Finally, hold at 4°C until the sample is removed.

Notes: (1) When using thin-walled PCR tubes, it is generally not necessary to denature the DNA template for longer than 15 sec during each cycle. In fact, repeated exposure to high temperatures tends to ruin the integrity of the template. (2) Primer annealing temperatures will have to be optimized. However, a temperature 3°C below the calculated Tm of the primer having the lowest Tm is a good starting point. (3) Other conditions that must be optimized are the amount of $MgCl_2$ used, the extension times, and the number of cycles used. The amount of $MgCl_2$ controls the specificity of the primer. Shorter extension times should be used for the amplification of shorter fragments. We recommend using as few cycles as possible to amplify the DNA fragment of interest. Taq polymerases have higher mutation rates, and each subsequent cycle increases the chance of introducing error. This can be very problematic, especially if the cloned DNA is to be used to generate protein products. We recommend using Taq polymerases that possess proof-reading ability. These enzymes often come with a bound antibody that prevents DNA synthesis until the first heat cycle. The resulting "hot start" greatly reduces nonspecific fragment synthesis; however, the proofreading enzymes also tend to remove the T residues at the end of the synthesized fragments. Consequently, the fragments must be cloned immediately after amplification.

- e. If using mineral oil, remove the reaction mixture from the mineral oil using a pipette with a relatively long tip. Slowly insert the tip into the bottom of the tube and then carefully take up the sample, leaving the oil phase behind. Withdraw the tip from the tube, wipe the outside of the tip with a clean paper towel, and transfer the sample into a fresh tube.
2. Check the purity of the amplified cDNA on an agarose gel and elute the fragment of interest if required.
    - a. Add DNA loading buffer to 10 μl of the amplified cDNA sample and load the sample into 1% agarose gel, which contains EtBr for staining, including positive and negative controls as well as DNA standard markers. Carry out electrophoresis at 120 V.
    - b. Under UV light, inspect the resulting bands to see if an appropriate fragment was amplified.

cDNA Libraries

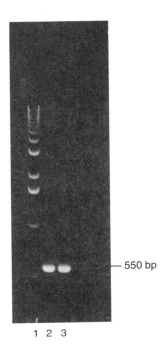

1 2 3

FIGURE 9.9
Amplification of a partial-length cDNA (550 bp) after 30 cycles of PCR. The products were subjected to 1% agarose electrophoresis, stained with ethidium bromide, and photographed. (From Kaufman, P.B., Wu, W., Kim, D., and Cseke, L.J., *Handbook of Molecular and Cellular Methods in Biology and Medicine,* 1st ed., CRC Press, Boca Raton, FL, 1995.)

    c. If a single, clearly resolved band is visualized in the sample of interest, then this DNA can be cloned directly from the PCR reaction using 1 to 7 µl of the PCR reaction and a variety of PCR cloning kits, such as the pGEM Teasy vector (Promega) or the TOPO TA Cloning kits (Invitrogen) (Figure 9.9). We highly recommend this procedure; however, it must be performed immediately after the PCR reactions are finished.

    d. If too many bands are resolved in the sample of interest, then the band of interest will have to be eluted from the gel. Run an appropriate amount of the remaining PCR reaction on a gel, and slice the individual sharp band(s) using a clean razor blade. Trim away excess agarose gel as much as possible.

    e. Procedures for the elution of the bands are detailed in **Chapter 1**.

3. Ligate the fragment to an appropriate TA cloning vector, and transform an appropriate *E. coli* host in preparation for minipreps.
    a. Set up the ligation reaction as follows:
        - Ligase 10X buffer (2 µl)
        - Vector (0.1 µg)
        - cDNA inserts (0.5 µg)
        - T4 DNA ligase (10 Weiss units)
        - Add ddH$_2$O to a final volume of 20 µl.
    b. Incubate the ligation reactions at room temperature (22 to 24°C) for 4 h and proceed to subcloning, as described in **Chapter 10**. (See **Chapter 14, Section IV.C** for further details.)
4. Identify correct clones and verify the sequence.
    a. Prepare miniprep DNA of at least 20 colonies after *E. coli* transformation. Cut these preps with flanking restriction enzymes and run fragments on a 1% agarose gel to identify clones with the correct size fragment. A Southern blot can be prepared using this gel if an appropriate probe can be generated.

Notes: When performing Southern blots on cloned PCR cDNA fragments, it is important to design a probe that lies between but does not include the primers used to create the products. This will prevent the hybridization of primer sequences that may flank nonspecific amplified DNA fragments. This is especially important if the fragment was directly cloned out of the PCR reaction without gel elution due to the higher probability of the presence of nonspecific fragments.

    b. Carry out sequencing of the putative clones having the amplified cDNA as described in **Chapter 11**.
    c. Compare the nucleotide sequence and the deduced amino acid sequence with known protein/enzyme from which the primers were designed.
    d. Label the putative cDNA as a probe that can be used for Southern blot or Northern blot and screen the cDNA library to obtain full-length cDNA.

C. cDNA Cloning by RT-PCR

Reverse-transcription PCR (RT-PCR) is a powerful technique used for cDNA cloning when specific primers are available. It is a relatively fast, simple, and inexpensive procedure as compared with other strategies described above. Total RNA or mRNA is first reverse-transcribed into cDNA, which is then amplified by PCR — the RT-PCR technique. Briefly, specific primers are designed according to conserved nucleic acid motifs in two different regions of known genes or their protein products. The primers are annealed to λDNA isolated from a constructed library or to first-strand cDNA synthesized from mRNA template. Double-strand cDNAs can be generated and amplified using *Taq* DNA polymerase. There are, however, two major disadvantages of applying the PCR cloning strategy: (1) the cDNA obtained is usually of partial length ranging from 200 to 1200 bp, and (2) the annealing of the primer to template is not always specific and may amplify nonspecific sequences referred to as "artifacts." Therefore, we strongly recommend that (1) the PCR products be verified by dot-blot or Southern blot hybridization using the targeting sequence as probe and (2) the positive PCR products be sequenced and compared with other known sequences.

    1. Carry out isolation of total RNA and purification of poly(A)+ RNA from tissue or cell lines of interest (see **Chapter 2**).
    2. Design oligonucleic primers based on conserved amino acid sequences or conserved nucleic acid sequences (if available) in two different regions of known proteins, enzymes, or gene sequences. Each primer can be designed with one specific restriction enzyme site for subcloning of the forthcoming double-strand cDNAs, or a TA cloning-type vector can be used, as described in **Section V.B** of this chapter. Since the use of degenerate primers is slightly more tricky, we use this as an example in this section. However, if primers with specific sequences can be designed, this will provide a much more direct method of obtaining the desired gene.

        In our example, two amino acid-sequence regions, NDPNG and DPCEW, of invertase are conserved from prokaryotes to eukaryotes. Both sequences have been characterized and published in professional journals. The first sequence is close to the N′-terminal; the other is toward the C-terminal. Two degenerate oligonucleotides can be designed according to the two conserved amino acid regions. In order to subclone the forthcoming double-strand cDNAs into a specific vector for sequencing, a *BamH* I restriction site is designed at the N-terminal of the first primer, and a *Hind* III site follows the C-terminal of the second primer. The design is as follows:

Primer 1: 5′ ATC<u>GGATCC</u>AAC(T)GAT(C)CCIAA(C)TGGI 3′ for NDPNG

Primer 2: 3′ GGTGAGCGTCCCTAG<u>TTCGA</u>AGTT 5′ for DPCEW

cDNA Libraries

The sequences underlined are the *BamH* I site at the 5' end and the *Hind* III site at the 3' end of the forthcoming double-strand cDNAs. The "I" in Primer 1 stands for the third position of the codon, which can be any combination of T, C, A, or G. The synthesis of the primers can be done with a DNA synthesizer, or primers can be ordered from a commercial source.

Alternatively, primer 2 (at the 3' end of the gene) can be replaced with oligo(dT)$_{18}$. This takes advantage of the polyT tails at the end of most mRNAs. The length of the forthcoming double-strand cDNA will be longer than that made by the first choice of two specific primers. However, the reduction of specificity may increase the number of nonspecific fragments amplified. We recommend setting up both reactions to increase the odds of success.

Notes: *(1) Degenerate primers should be designed to have as little degeneracy as possible. Therefore, amino acids having four or more codons should be avoided if possible. (2) The small tail at the end of each restriction site is designed to match the sequence of the gene of interest. A tail between three and eight nucleotides is required to allow the restriction enzyme to bind and make the cut. (3) The "I" in the above examples stands for the third position of the codon, which can be any of T, C, A, or G. (4) The primers in this example can generate a 554-bp fragment.*

3. Synthesize the first-strand cDNA followed by the second-strand cDNA as described in **Section II** of this chapter. The second strand can also be synthesized, but PCR amplification can be immediately carried out directly after the first-strand cDNA.
4. Carry out PCR amplification of single-strand cDNA or double-strand cDNA.
   a. In a thin-walled PCR tube on ice, add the following in the order listed:
      - 10X Amplification buffer (10 µl)
      - ddH$_2$O (20 µl)
      - Mixture of four dNTPs (1.25 m*M* each) (17 µl)
      - Primer 1 (100 to 110 pmol) in ddH$_2$O (5 µl)
      - Primer 2 (100 to 110 pmol) in ddH$_2$O (5 µl)
      - Single-strand cDNA or double-strand cDNA in TE buffer (1 to 2 µg)
      - *Taq* DNA polymerase (5 units/µl) (1 µl)
      - Add ddH$_2$O to a final volume of 100 µl.

Notes: *(1) Please see note in Section V.B, "General Amplification of Double-Strand DNA by PCR," in this chapter. (2) The use of excess degenerate primers is recommended due to the lower abundance of each working primer within the degenerate mixture. (3) The volume of this PCR reaction has been doubled from that of Section V.B due to the fact that most degenerate PCR results in many fragments. The desired product will have to be gel eluted, and the larger volume will provide more starting material for cloning.*

   b. Overlay the mixture with 50 µl of light mineral oil (Sigma or equivalent) to prevent evaporation of the sample if a thermocycler with a heated lid is not available.
   c. Carry out PCR amplification in a PCR machine, which is programmed as follows:

Cycle	Denaturation	Annealing	Polymerization
First cycle	2 min at 94°C	—	—
Touchdown cycles (10 cycles)	15 sec at 94°C	30 sec at 55°C	3.5 min at 72°C
Subsequent cycles (24 cycles)	15 sec at 94°C	30 sec at 52°C	3.5 min + 15 sec/cycle at 72°C
Last cycle	15 sec at 94°C	30 sec at 52°C	10 min at 70°C

Finally, hold at 4°C until the sample is removed.

Notes: (1) Please see note in **Section V.B**, "General Amplification of Double-Strand DNA by PCR." (2) The initial ten PCR reactions or "touchdown" cycles work to provide a more stringent environment in which the primer will anneal. This tends to enrich the reaction with the template of interest and provides much more specific end results than traditional cycling methods.

   d. Carefully remove the reaction mixture from the mineral oil as described above in the "general amplification" protocol (**Section V.B.2**) (only if a heated lid was not used).
5. Check the purity of the amplified cDNA on an agarose gel and elute the desired fragment if required. These procedures are described above in the "general amplification" protocol (**Section V.B.2**).
6. Ligate the cDNA to an appropriate vector. While the eluted fragment(s) can be cloned directly into a TA cloning-type vector, as described in **Section V.B.2**, our current example makes use of restriction enzyme sites designed into the primers.
   a. Digest the cDNA and vector, such as pGEM-11Zf(+) (Promega Corp.) or equivalent, with *BamH* I and *Hind* III to generate sticky ends for subcloning.
      - cDNA or vector (2 to 5 µg)
      - *Hind* III 10X buffer (5 µl)
      - BSA (optional) (2.5 µl)
      - *Hind* III (20 units)
      - *BamH* I (20 units)
      - Add ddH$_2$O to a final volume of 50 µl.
   b. Incubate at 37°C for 2.5 h and extract once by adding one volume of TE-saturated phenol/chloroform. Vortex for 1 min.
   c. Centrifuge at 11,000 g for 5 min and carefully remove the top, aqueous phase to a fresh tube.
   d. Extract the supernatant once with one volume of chloroform:isoamyl alcohol (24:1) and centrifuge as in step c.
   e. Add 0.5 volume of 7.5 M ammonium acetate to the supernatant, mix, and add two volumes of chilled 100% ethanol. Allow precipitation at –20°C for 2 h.
   f. Centrifuge at 12,000 g for 10 min and briefly rinse the pellet with 1 ml of cold 70% ethanol. Dry the cDNA linker under vacuum for 10 min and resuspend the pellet in 20 µl of TE buffer. Store the sample at –20°C until use.
   g. Finish setting up the ligation as described above in the (general amplification) procedure.
   h. Analyze putative clones as described in the (general amplification) protocol in **Section V.B.2**.

D. Analysis of Gene Expression by Semiquantitative PCR

This section is not strictly specific to the cloning of cDNA fragments. However, it is absolutely essential for any cDNA cloning (by PCR or by library construction) that a tissue strongly expressing the gene of interest be selected. In preparation for PCR cloning of a cDNA of interest, it is useful to have a quick method to assess which tissues within the organism of interest express the gene of interest. Northern blot hybridization and *in situ* hybridization of mRNA are techniques that are commonly used for analysis of gene expression. The disadvantage of these approaches, however, is the difficulty in obtaining the desired result(s) when analyzing nonconstitutively expressed cell- or tissue-specific genes having low-abundance mRNAs or when limiting amounts of mRNA are available due to degradation. This drawback is overcome by quantitative PCR. The principle of this technology is that total RNAs or mRNAs are isolated from different samples of cell types, tissue types, or developmental stages — including chemically treated (+), untreated (–), and time-course treatments — to study the kinetic expression of specific genes. The same amount of total RNAs or mRNAs from each sample is reverse transcribed into cDNAs under the same conditions. The particular cDNA of interest in each of the samples is then amplified by PCR using specific primers. The same amount of the amplified cDNA from each sample is then analyzed by dot blotting or Southern blotting using the target sequence between primers as a probe. In this case, different amounts of mRNA, expressed by specific genes under different conditions, will generate different amounts of cDNA. By comparing the measurements of the hybridized signal intensities of different samples, the different levels of gene expression can be detected with ease. At the same time, a control mRNA (usually an appropriate housekeeping gene transcript) such as actin mRNA is reverse transcribed and amplified by PCR, using specific primers under exactly the same conditions, to monitor equivalent reverse transcription to cDNA and equivalent amplification in PCR.

1. Isolate total RNAs or purify mRNAs from (1) different cell or tissue types of interest, (2) the same cell or tissue type at different development stages, or (3) the same cell or tissue types under different treatments. The detailed procedures are described in **Chapter 2**.
2. Synthesize the first-strand cDNAs from the isolated total RNAs or mRNAs using reverse transcriptase and oligo(dT) as a primer.
    a. Anneal 1 μg total RNA or 50 ng mRNA template with 2 μg oligo(dT) primer in a sterile *RNase*-free microcentrifuge tube. Add nuclease-free ddH₂O to a total volume of 15 μl. Heat the tube at 70°C for 5 min and allow it to slowly cool to room temperature to finish annealing. Briefly spin down the mixture to the bottom.
    b. Complete the first-strand synthesis as described in **Section II.B.1**.

Notes: *(1) Small amounts of total RNA (1 to 1.5 μg) or mRNA (20 to 50 ng) should be used for semiquantitative analysis of specific mRNA. (2) The conditions for reverse transcription should be the same for each of the comparison samples and actin control, including all the components (volume, temperature, and reaction time).*

3. Amplify the specific cDNA of interest by PCR using specific primers.
    a. Design oligonucleic primers based on conserved amino acid sequences in two different regions of the specific gene product (protein) of interest. For example, two oligonucleotide primers can be designed as follows for the two amino acid-sequence regions, NDPNG and DPCEW, of invertase.

        Primer 1:   5′ AAC(T)GAT(C)CCIAA(C)TGGI 3′ for NDPNG

        Primer 2:   3′ GGTGAGCGTCCCTAG 5′ for DPCEW

Notes: *(1) Degenerate primers should be designed to have as little degeneracy as possible. Therefore, amino acids having four or more codons should be avoided if possible. (2) The "I" in Primer 1 stands for the third position of the codon, which can be any of T, C, A, or G. (3) The primers in this example can generate a 554-bp fragment.*

        For amplification of actin cDNA, the primers can be designed as follows:

        Primer 1:   5′ ATGGATGACGATATCGCTG 3′

        Primer 2:   5′ ATGAGGTAGTCTGTCAGGT 3′

Note: *These primers can generate a product of 568 bp.*

    b. Carry out PCR amplification.
        i. In a thin-walled PCR tube on ice, add the following in the listed order for amplification of one sample:
           - 10X Amplification buffer (10 μl)
           - ddH₂O (20 μl)
           - Mixture of four dNTPs (1.25 m*M* each) (17 μl)
           - Primer 1 (100 to 110 pmol) in ddH₂O (4 μl)

- Primer 2 (100 to 110 pmol) in ddH$_2$O (4 µl)
- cDNA synthesized previously (0.1 to 1 µg) (6 µl)
- Add ddH$_2$O to a final volume of 100 µl.

For no cDNA control:
- 10X Amplification buffer (10 µl)
- ddH$_2$O (20 µl)
- Mixture of four dNTPs (1.25 mM each) (17 µl)
- Primer 1 (100 to 110 pmol) in ddH$_2$O (4 µl)
- Primer 2 (100 to 110 pmol) in ddH$_2$O (4 µl)
- Add ddH$_2$O to a final volume of 100 µl.

Notes: (1) Please see Notes contained in **Section V.B**, "*General Amplification of Double-Strand DNA by PCR.*" (2) *The use of excess degenerate primers is recommended due to the lower abundance of each working primer within the degenerate mixture.*

  ii. Repeat until all samples, including the control actin cDNA, have been set up. Add 2.5 units of *Taq* DNA polymerase (5 units/µl, Perkin Elmer Cetus) to each of the tubes. Gently mix well.

  iii. Overlay the mixture with 50 µl of light mineral oil (Sigma or equivalent) to prevent evaporation of the sample.

  iv. Carry out PCR amplification for 10 to 35 cycles in a PCR cycler, which is programmed as follows:

Cycle	Denaturation	Annealing	Polymerization
First cycle	2 min at 94°C	—	—
Subsequent cycles (35 cycles)	15 sec at 94°C	30 sec at 50°C	4 min at 72°C
Last cycle	15 sec at 94°C	30 sec at 50°C	10 min at 72°C

Notes: (1) *The conditions for amplification should be exactly the same for every sample, whether with actin cDNA or no cDNA.* (2) *The number of cycles in semiquantitative PCR must be kept to a minimum to avoid saturating the PCR reaction with product or the amplification of nonspecific fragments. While this may result in difficultly in visualizing the products, Southern blot analysis is a sensitive procedure and will resolve the products. Since the optimal number of cycles in not known, the following procedure should be used.*

  v. Starting after ten cycles, carefully push a pipette tip through the oil and remove 15 µl of the reaction from each of the samples and from the controls after every five cycles at the end of the appropriate cycle (at 72°C extension phase). Place tubes containing the 15-µl reactions at 4°C until use.

Note: *Sampling of all reactions should be done within 2 min. There should be six samplings for each sample in 35 cycles. Oil should be avoided as much as possible.*

  vi. Use the 15 µl of amplified cDNA from each of the samples and controls to carry out Southern blots or dot-blot hybridization using the appropriate ^{32}P-labeled internal oligonucleotides or the target sequence (e.g., cDNA fragment) between the two primers as a probe. The detailed procedures for Southern blot and dot-blot hybridizations are described in **Chapter 5**.

  vii. After film development, measure the signal intensities of each sample with a densitometer and analyze the data. For cDNA negative controls, no signal should be visible. For actin cDNA, the signal intensities should be increased as amplified cycles, but no differences should be seen between any cell or tissue types or between any developmental stages. In

cDNA Libraries

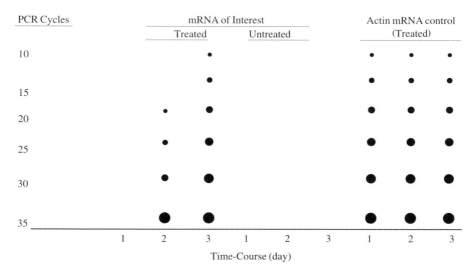

FIGURE 9.10
Diagram showing an analysis of a given mRNA of interest and actin mRNA used as control obtained by the use of semiquantitative PCR. (From Kaufman, P.B., Wu, W., Kim, D., and Cseke, L.J., *Handbook of Molecular and Cellular Methods in Biology and Medicine*, 1st ed., CRC Press, Boca Raton, FL, 1995.)

contrast, if the expression of a specific gene is indeed nonconstitutive, is cell- or tissue-specific, or is chemical-treatment induced, clear signal patterns of amplified cDNA will be seen. The signal for each sample should be stronger with amplified cycles (Figure 9.10). For example, if the signal of amplified invertase cDNA in gibberellin-treated tissue is much stronger than that of untreated tissue — and if the signal is increased from cycle 10, to 15, 20, 25, 30, and 35 — the expression of invertase mRNA, which was used for the synthesis of the cDNA blotted, is likely induced by gibberellin treatment.[8]

viii. To determine the size of PCR products after dot blotting, repeat amplification of 6 µl cDNA synthesized previously and carry out 1.0 to 1.4% agarose gel electrophoresis. This is not necessary if Southern blot hybridization was performed instead for the dot-blot procedure (see Southern blotting in **Chapter 5**).

Reagents Needed

[α-^{32}P]dCTP (>400 Ci/mmol)

7.5 M Ammonium Acetate
Dissolve 57.8 g ammonium acetate in 50 ml ddH$_2$O and adjust volume to 100 ml.

10X Amplification Buffer
100 mM Tris-HCl, pH 8.3
500 mM KCl
15 mM MgCl$_2$
0.1% BSA

AMV Reverse Transcriptase

Mixture of Four dNTPs (1.25 mM each)

First-Strand 5X Buffer

    250 mM Tris-HCl, pH 8.3 (42°C)
    50 mM MgCl$_2$
    250 mM KCl
    2.5 mM Spermidine
    50 mM DTT
    5 mM Each of dATP, dCTP, dGTP, and dTTP

Hind III 10X Buffer

    300 mM Tris-HCl, pH 7.8
    100 mM MgCl$_2$
    500 mM NaCl
    100 mM DTT

Ligase 10X Buffer

    300 mM Tris-HCl, pH 7.8
    100 mM MgCl$_2$
    100 mM DTT
    10 mM ATP

Nuclease-Free ddH$_2$O

rRNasin Ribonuclease Inhibitor

200 mM Sodium Pyrophosphate

TE Buffer

    10 mM Tris-HCl, pH 8.0
    1 mM EDTA

TE-Saturated Phenol/Chloroform

    Thaw crystals of phenol at 65°C and mix with an equal volume of TE buffer. Mix well and allow the phases to separate at room temperature for 30 min. Take one part of the lower, phenol phase and mix with one part of chloroform:isoamyl alcohol (24:1). Mix well, allow the phases to separate, and store at 4°C until use.

Trichloroacetic Acid (5% and 7%)

References

1. **Okayama, H. and Berg, P.**, High-efficiency cloning of full-length cDNA, *Mol. Cell Biol.*, 2, 161, 1982.
2. **Wu, L.-L., Song, I., Karuppiah, N., and Kaufman, P.B.**, Kinetic induction of oat shoot pulvinus invertase mRNA by gravistimulation and partial cDNA cloning by the polymerase chain reaction, *Plant Mol. Biol.*, 21, 1175, 1993.
3. **Sambrook, J., Fritsch, E.F., and Maniatis, T.**, *Molecular Cloning: A Laboratory Manual*, 2nd ed., Cold Spring Harbor Press, Cold Spring Harbor, NY, 1989.
4. **Young, R.A. and Davis, R.W.**, Efficient isolation of genes by using antibody probes, *Proc. Natl. Acad. Sci. U.S.A.*, 80, 1194, 1983.

5. **Heikkila, J.J.,** Expression of cloned genes and translation of messenger RNA in microinjected *Xenopus* oocytes, *Intl. J. Biochem.*, 22, 1223, 1990.
6. **Coleman, A.,** Translation of eukaryotic messenger RNA in *Xenopus* oocytes, in *Transcription and Translation: A Practical Approach,* Hames, B.D. and Higgins, S.J., Eds., IRL Press, Washington, DC, 1984.
7. **Hitchcock, M.J.M., Ginns, E.L., and Marcus-Sekura, C.J.,** Microinjection into *Xenopus* oocytes: equipment, in *Methods in Enzymology,* Vol. 152, Berger, S.L. and Kimmel, A.R., Eds., Academic Press, New York, 1987.
8. **Wu, L.-L., Mitchell, J.P., Cohn, N.S., and Kaufman, P.B.,** Gibberellin (GA_3) enhances cell wall invertase activity and mRNA levels in elongating dwarf pea *(Pisum sativum)* shoots, *Intl. J. Plant Sci.*, 154, 280, 1993.
9. **Parimoo, S., Patanjali, S.R., Shukla, H., Chaplin, D.D., and Weissman, S.M.,** cDNA selection: efficient PCR approach for the selection of cDNAs encoded in large chromosomal DNA fragments, *Proc. Natl. Acad. Sci. U.S.A.*, 88, 9623, 1991.
10. **Frohman, M.A., Dush, M.K., and Martin, G.R.,** Rapid production of full-length cDNAs from rare transcripts: amplification using a single gene-specific oligonucleotide primer, *Proc. Natl. Acad. Sci. U.S.A.*, 85, 8998, 1988.
11. **Winer, M.P.,** Directional cloning of blunt-ended PCR products, *BioTechniques*, 15, 502, 1993.
12. **Erlich, H.A.,** *PCR Technology: Principles and Applications for DNA Amplification,* Stockton Press, New York, 1989.
13. **Innis, M.A., Gelfand, D.H., Sninsky, J.J., and White, T.J.,** *PCR Protocols: A Guide to Methods and Applications,* Academic Press, New York, 1989.
14. **Wang, A.M., Doyle, M.V., and Mark, D.F.,** Quantitation of mRNA by the polymerase chain reaction, *Proc. Natl. Acad. Sci. U.S.A.*, 86, 9717, 1989.
15. **Wu, L.-L., Song, I., Karuppiah, N., and Kaufman, P.B.,** Kinetic induction of oat shoot pulvinus invertase mRNA by gravistimulation and partial cDNA cloning by the polymerase chain reaction, *Plant Mol. Biol.*, 21, 1175, 1993.
16. **Arnold, C. and Hodgson, I.J.,** Vectorette PCR: a novel approach to genomic walking, *PCR Methods Appl.*, 1, 39, 1991.
17. **Barany, F.,** Genetic disease detection and DNA amplification using cloned thermostable ligase, *Proc. Natl. Acad. Sci. U.S.A.*, 88, 189, 1991.
18. **Lu, W., Han, D.-S., Yuan, J., and Andrieu, J.-M.,** Multitarget PCR analysis by capillary electrophoresis and laser-induced fluorescence, *Nature*, 368, 269, 1994.
19. **Bloomquist, B.T., Johnson, R.C., and Mains, R.E.,** Rapid isolation of flanking genomic DNA using biotin-RAGE: a variation of single-sided polymerase chain reaction, *DNA Cell Biol.*, 10, 791, 1992.
20. **Edwards, J.B., Delort, P.M., and Mallet, J.,** Oligodeoxyribonucleotide ligation to single-stranded cDNAs: a new tool for cloning 5′ ends of mRNAs and for constructing cDNA libraries by *in vitro* amplification, *Nucleic Acids Res.*, 19, 5227, 1991.

Chapter 10

Genomic DNA Libraries

Leland J. Cseke and Peter B. Kaufman

Contents

I.	General Strategies and Applications	194
II.	Bacteriophage Lambda Library	195
	A. Construction of a Genomic DNA Library	195
	1. Optimization of Partial Digestion of Genomic DNA with *Sau3A* I	196
	2. Large-Scale Preparation of Partially Digested Genomic DNA	199
	3. Partial Fill-In of Recessed 3′ Termini of Genomic DNA Fragments	200
	4. Small-Scale Ligation of Partially Filled-In Genomic DNA Fragments and Partially Filled-In λGEM-12 Arms	201
	5. *In Vitro* Packaging of Ligated DNA	201
	6. Titering of Packaged Phage on LB Plates	202
	7. Large-Scale Ligation of Partially Filled-In Vector Arms and Partially Filled-In DNA Fragments	203
	B. Screening of the Genomic DNA Library	203
	1. Screening of the Genomic Library with a ^{32}P-Labeled Probe	203
	2. Screening of a Genomic Library Using a Nonradioactive Probe	206
	C. Construction of Partial Genomic Libraries	207
	1. Southern Blotting of Genomic DNA and Identification of Desired Fragments	207
	2. Isolation of Desired Fragments	207
	3. Ligation of Linkers	207
	4. Construction and Screening of the Library	208
	D. Restriction Mapping of Positive Recombinant Bacteriophage DNA Clones	208
	E. Subcloning of the DNA Fragment of Interest	209
	1. Restriction Enzyme Digestion of Vector and DNA Insert for Subcloning	210
	2. Ligation of Plasmid Vector and DNA Insert	212
	3. *E. coli* Transformation	214
	4. Transformation Using Chemically Competent Cells	215
	5. Transformation by Electroporation	216
	6. Selection of Transformants Containing Recombinant Plasmids	216
Reagents Needed		217

III. YAC Libraries...221
    A. Preparation of Cells or Tissues for Isolation and Purification
       of High-Molecular-Weight DNA 222
    B. Isolation of High-Molecular-Weight DNA 223
    C. Isolation of Intact Yeast DNA...223
    D. Isolation of Yeast DNA for PCR Screening 224
    E. Partial Restriction Enzyme Digestion of DNA in Agarose............................224
    F. Preparation of YAC Vectors for Cloning 225
    G. Ligation of Partially Digested Genomic DNA Insert to pYAC4 Vector.......225
    H. Size Fractionation of DNA by CHEF Gel or Other PFGE............................225
    I. Hydrolysis of Agarose by Agarase 226
    J. Preparation of Spheroplasts for Transformation..226
    K. Transformation of Spheroplasts with Recombinant YAC/DNA Insert 226
    L. Verification of YAC Transformants...227
    M. Amplification and Storage of the YAC Library..227
    N. Screening of a YAC Library..227
    Reagents Needed..228
IV. Troubleshooting Guide..232
V. Genomic Cloning Using PCR..232
    A. Selection of Oligonucleotides 232
    B. PCR Amplification of Genomic DNA and Genomic Libraries 233
    C. Isolation of Flanking Sequences by Inverse PCR 235
References 236

I. General Strategies and Applications

Genomic deoxyribonucleic acid (DNA) cloning is a technology that plays an important role in state-of-the-art molecular biology studies. Many of the genes that have been characterized originally come from genomic DNA cloning. A genomic clone contains much more information than a complementary DNA (cDNA) clone because it possesses regulatory regions such as promoter and terminator regions, intron sequences that may also be involved in regulation and overall gene structure, and in some cases methylation information that may be responsible for chromatin structure. If one wishes to identify, characterize, and regulate the expression of a full-length unknown genomic gene, he or she may have to start from molecular cloning of genomic DNA, which is described in detail in this chapter.

The quality and integrity of a genomic DNA library are directly correlated with the success or failure of identifying a gene of interest. A very good library is supposed to contain all DNA sequences of the entire genome. The probability of "fishing out" a DNA sequence of interest depends on the size of a library, which in turn relies on the sizes of DNA fragments selected for cloning. The larger the DNA fragments, the smaller the number of clones in the library. The probability of having an interesting DNA sequence in the library can be calculated by the following equation:[1]

$$N = \ln(1 - P)/\ln(1 - f)$$

where N is the number of recombinants required, P is the desired probability of "fishing out" a DNA sequence, and f is the fractional proportion of the genome in a single recombinant. For example, given a 99% probability of "fishing out" an interesting DNA sequence in a library cloned by 18-kb (kilobase-pairs) fragments of a 4×10^9-bp (base-pairs) genome, the required recombinants are as follows:

$$N = [\ln (1 - 0.99)]/\{\ln [1 - (1.8 \times 10^4/4 \times 10^9)]\} = 1 \times 10^6$$

In general, the average size of DNA fragments selected for cloning is directly related to the cloning vectors that are used to construct genomic DNA libraries. There are three types of vectors that are commonly used for the construction of genomic DNA libraries: (1) bacteriophage lambda vectors, (2) cosmids, and (3) yeast artificial chromosomes (YACs). Lambda vectors can accept 14- to 25-kb DNA fragments. DNA fragments are ligated with lambda vector arms, forming concatemers that are then packaged into bacteriophage lambda (λ) particles. Recombinants numbering 3×10^6 to 3×10^7 are usually necessary to achieve a 99% probability of isolating a specific clone of interest, and the screening procedure of the library is relatively complicated. Cosmids, on the other hand, can accept about 35- to 45-kb DNA fragments. Only approximately 3×10^5 recombinant cosmids can achieve a 99% probability of identifying a particular single-copy sequence of interest. However, it is difficult to construct and maintain a genomic DNA library in cosmids as compared with bacteriophage λ vectors. YAC vectors are very useful to clone 50- to 10,000-kb DNA fragments. This is powerful when one wishes to clone and isolate extra-large genes (e.g., human factor VIII gene, 180 kb in length; the dystrophin gene, 1800 kb in length) in a single recombinant YAC of a smaller size library. This chapter describes and emphasizes the construction of bacteriophage λ libraries and YAC libraries.[2-7] At the end of the chapter, we also describe a good starting place for the use of polymerase chain reaction (PCR) to clone genomic fragments.

II. Bacteriophage Lambda Library

A. Construction of a Genomic DNA Library

Several years ago, the commonly used bacteriophage λ DNA vectors were EMBL3 and EMBL4 (Figure 10.1). Both contain a left arm (20 kb), a right arm (9 kb), and a central stuffer (14 kb), which can be removed and replaced with a foreign DNA insert (9 to 23 kb). The only difference between two λ DNA vectors is the orientations of the polylinker regions.

More recently, new λ DNA vectors have been developed such as λGEM-11 and λGEM-12 (Figure 10.2). These vectors are modified from EMBL3/EMBL4. The sizes of arms and capacity of cloning foreign DNA inserts are the same as EMBL3 and EMBL4. However, there are several advantages over EMBL3 and EMBL4:

1. More restriction enzyme sites are designed in the polylinker region.
2. There are two promoters, T7 and SP6, at the polylinker ends in opposite orientations. These promoters make it possible to directly sequence the DNA insert by using T7 or SP6 primer without subcloning. They also allow one to express the DNA insert into sense RNA (ribonucleic acid) or antisense RNA synthesized by T7 or SP6 RNA polymerase.
3. These vectors are designed to be optimized for the highest recombinant efficiencies and lowest nonrecombinant background.

Because of these merits of λGEM-11 and λGEM-12 — and based on our successful experience with them — we recommend that these two vectors be used for genomic DNA cloning.

The overall procedure of constructing genomic DNA libraries is that high-molecular-weight genomic DNA is partially digested with *Sau3A* into 14- to 24-kb fragments. These fragments are ligated with dephosphorylated λGEM-11 or λGEM-12 arms by T4 DNA ligase. The concatemers are then packaged *in vitro* into bacteriophage λ particles, generating a genomic library. The library is screened using a specific probe, and positive clones will then be isolated. After several rounds of rescreening, the putative DNA clones of interest can be purified.

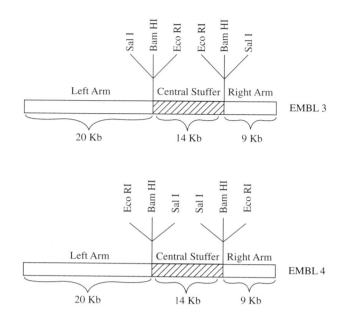

FIGURE 10.1
Structural maps of vectors: EMBL 3 (top) and EMBL 4 (bottom). (From Kaufman, P.B., Wu, W., Kim, D., and Cseke, L.J., *Handbook of Molecular and Cellular Methods in Biology and Medicine*, 1st ed., CRC Press, Boca Raton, FL, 1995.)

1. Optimization of Partial Digestion of Genomic DNA with *Sau*3A I

In order to determine the amount of enzyme that can digest high-molecular-weight DNA into 14- to 23-kb fragments for cloning into λ vectors, small-scale reactions, called the pilot experiments, should be carried out.

1. Prepare 1X *Sau*3A I buffer on ice:
   - 10X *Sau*3A I buffer (0.2 ml)
   - Acetylated bovine serum albumin (BSA) (1 mg/ml, optional) (0.2 ml)
   - Add distilled, deionized water (ddH$_2$O) to a final volume of 2.0 ml.
2. Prepare *Sau*3A I dilutions in ten individual microcentrifuge tubes on ice, as shown in Table 10.1.
3. Set up ten individual small-scale digestion reactions on ice in the order shown in Table 10.2.

Notes: (1) Genomic DNA used in the reactions should be high molecular weight (>150 to 200 kb) with a ratio of 1.85 to 1.95 of A_{260}/A_{280}. The purification of high-molecular-weight genomic DNA is described in **Chapter 1** and results of DNA isolation are shown in Figure 10.3. An impure DNA sample with a ratio <1.75 of A_{260}/A_{280} should not be used for construction of a genomic DNA library. (2) *Sau*3A I dilutions (5 µl) used in the DNA digestion reactions should always be added last from the right dilution tube to the right DNA digestion tube. (3) The final *Sau*3A I concentration used in tubes one to ten should be 1, 0.1, 0.05, 0.025, 0.015, 0.0125, 0.01, 0.0085, 0.005, and 0.0035 unit/µg DNA, respectively.

4. Incubate the ten digestion reactions at the same time at 37°C in a water bath for 30 min. Place the tubes on ice and add 2 µl of 0.2 M EDTA buffer (pH 8.0) to stop the reaction.
5. While the reactions are being carried out, prepare a large size of 0.4% agarose gel in 1X TBE buffer. Melt the agarose in a microwave oven and allow it to cool to about 50°C. Add 10 µl of 10-mg/ml EtBr solution to every 100 ml of gel mixture, mix well, and pour into a prepared gel tray. Allow the gel to harden at room temperature for about 30 min. Place the gel in the apparatus and add 0.5X

Genomic DNA Libraries

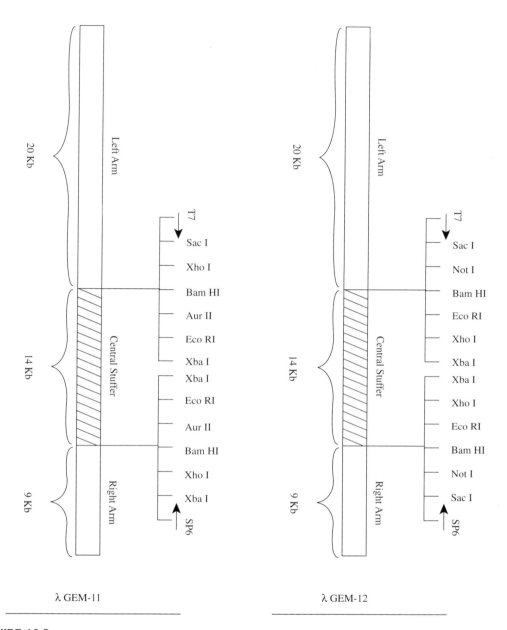

FIGURE 10.2
Structural maps of vectors: λGEM-11 (left) and λGEM-12 (right). (From Kaufman, P.B., Wu, W., Kim, D., and Cseke, L.J., *Handbook of Molecular and Cellular Methods in Biology and Medicine,* 1st ed., CRC Press, Boca Raton, FL, 1995.)

  TBE buffer to cover the gel (1 to 2 mm depth above the gel surface). A minigel should not be used for this purpose.
6. Add 10 µl of 5X DNA loading buffer to each of the ten tubes of digested DNA prepared in step 4.
7. Carefully load 30 µl of each sample onto the wells in the order of numbers one to ten. Load DNA markers (e.g., λ DNA *Hind* III markers) to the left or the right well of the sample wells to estimate the sizes of digested DNA with ease.
8. Electrophorese the gel at 2 V/cm until the bromophenol blue reaches the bottom of the gel. It usually takes about 10 to 12 h.

TABLE 10.1
Preparation of Sau3A I Dilutions

Tube No.	Preparation of Sau3A I (3 units/µl) Dilution	Amount of Dilution
1	2 µl Sau3A I + 28 ml 1X Sau3A I buffer	1/15
2	10 µl of 1/15 dilution + 90 µl 1X Sau3A I buffer	1/150
3	10 µl of 1/150 dilution + 10 µl 1X Sau3A I buffer	1/300
4	10 µl of 1/150 dilution + 30 µl 1X Sau3A I buffer	1/600
5	10 µl of 1/150 dilution + 50 µl 1X Sau3A I buffer	1/900
6	10 µl of 1/150 dilution + 70 µl 1X Sau3A I buffer	1/1200
7	10 µl of 1/150 dilution + 90 µl 1X Sau3A I buffer	1/1500
8	10 µl of 1/150 dilution + 110 µl 1X Sau3A I buffer	1/1800
9	10 µl of 1/150 dilution + 190 µl 1X Sau3A I buffer	1/3000
10	10 µl of 1/150 dilution + 290 µl 1X Sau3A I buffer	1/4500

Source: From Kaufman, P.B., Wu, W., Kim, D., and Cseke, L.J., *Handbook of Molecular and Cellular Methods in Biology and Medicine,* 1st ed., CRC Press, Boca Raton, FL, 1995.

TABLE 10.2
Small-Scale Digestion Reactions, µl

Components	Tube Number									
	1	2	3	4	5	6	7	8	9	10
Genomic DNA (1 µg/µl)	1	1	1	1	1	1	1	1	1	1 µl
10X Sau3A I buffer	5	5	5	5	5	5	5	5	5	5 µl
Acetylated BSA (1 mg/ml, optional)	5	5	5	5	5	5	5	5	5	5 µl
ddH$_2$O	34	34	34	34	34	34	34	34	34	34 µl
Sau3A I dilution in the same order as in Table 10.1	5	5	5	5	5	5	5	5	5	5 µl
Final volume	50	50	50	50	50	50	50	50	50	50 µl

Source: From Kaufman, P.B., Wu, W., Kim, D., and Cseke, L.J., *Handbook of Molecular and Cellular Methods in Biology and Medicine,* 1st ed., CRC Press, Boca Raton, FL, 1995.

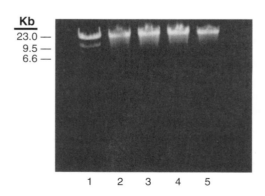

FIGURE 10.3
Isolation of high-molecular-weight genomic DNA from a eukaryotic organism for the construction of a genomic DNA library. Lane 1: DNA standard-molecular-weight markers. Lanes 2 to 5: different DNA samples. (From Kaufman, P.B., Wu, W., Kim, D., and Cseke, L.J., *Handbook of Molecular and Cellular Methods in Biology and Medicine,* 1st ed., CRC Press, Boca Raton, FL, 1995.)

Genomic DNA Libraries

1 2 3 4 5 6

FIGURE 10.4
Partial digestion of genomic DNA shown in Figure 10.3 with *Sau3A* I at different concentrations. Lanes 1 to 6 show progressively diluted *Sau3A* I. The digested DNA was electrophoresed on a 0.4% agarose gel and stained with ethidium bromide. (From Kaufman, P.B., Wu, W., Kim, D., and Cseke, L.J., *Handbook of Molecular and Cellular Methods in Biology and Medicine*, 1st ed., CRC Press, Boca Raton, FL, 1995.)

9. Photograph the gel under ultraviolet (UV) light and find the well that shows the maximum intensity of fluorescence in the desired DNA size range of 14 to 23 kb. The intensity of fluorescence is directly related to the mass distribution of DNA. The amount of *Sau3A* I used to get the maximum intensity of fluorescence from 14- to 23-kb DNA fragments is the optimal concentration that can guide large-scale digestion of DNA for the construction of genomic DNA libraries (Figure 10.4).

2. Large-Scale Preparation of Partially Digested Genomic DNA

1. Guided by the optimized conditions established in step 1.9, carry out a large-scale digestion of 50 µg of high-molecular-weight genomic DNA using half the number of units of *Sau3A* I per microgram of DNA that produced the maximum intensity of fluorescence in the DNA size range of 14 to 23 kb.

Notes: The DNA concentration, time, and temperature should be the same as those used for the small-scale digestion. To ensure success, we recommend that a duplicate large-scale digestion be set up. For example, if tube no. 7 (0.01 unit of Sau3A I per microgram DNA) in the small-scale digestion of the DNA displays a maximum intensity of fluorescence in the size range of 14 to 23 kb, the large-scale digestion of the same DNA can be carried out on ice as shown in Table 10.3.

Incubate the reactions at 37°C in a water bath for 30 min. Then stop the reactions by adding 0.1 ml of 0.2 *M* EDTA buffer (pH 8.0) and place on ice until use.

2. To check the size range of digested DNA, take 30 µl of the partially digested DNA from each sample, add 7 µl of 5X loading buffer, and load the mixture onto 0.4% agarose gel in 1X TBE buffer. Electrophorese and photograph as described for the small-scale digestion.

TABLE 10.3
Large-Scale Digestion Reactions

Components	Tube 1	Tube 2 (duplicate)
Genomic DNA (1 µg/µl)	50 µl	50 µl
10X *Sau*3A I buffer	250 µl	250 µl
1 mg/ml Acetylated BSA (optional)	250 µl	250 µl
ddH$_2$O	1.7 ml	1.7 ml
Diluted *Sau*3A I (0.005 units/mg) prepared as in Table 10.1, tube 9	250 µl	250 µl
Final volume	2.5 ml	2.5 ml

Source: From Kaufman, P.B., Wu, W., Kim, D., and Cseke, L.J., *Handbook of Molecular and Cellular Methods in Biology and Medicine*, 1st ed., CRC Press, Boca Raton, FL, 1995.

Notes: (1) If the digestion is adequate, proceed to step 3. (2) If the digestion is not desirable, repeat large-scale digestion with an appropriate amount of enzyme until a size range of 14 to 23 kb is achieved. If the DNA molecules are under- or over-digested, the undesired DNA fragments will cause failure in packaging in vitro, resulting in a bad library.

3. Extract with one volume of TE-saturated phenol/chloroform. Mix well by inverting for 1 min and centrifuge at 11,000 g for 5 min at room temperature.

Caution: Phenol is toxic. Care should be taken when handling this chemical. Waste phenol should be collected in a special container. Gloves should be worn when dealing with this chemical.

4. Carefully transfer the top, aqueous phase to a fresh tube and repeat step 3.
5. Carefully transfer the upper, aqueous phase to a fresh tube and add one volume of chloroform:isoamyl alcohol (24:1). Mix well and centrifuge as in step 3.
6. Transfer the top, aqueous phase to a fresh tube and add 0.1 volume of 3 *M* sodium acetate buffer (pH 5.2) or 0.5 volume of 7.5 *M* ammonium acetate. Mix and add 2 to 2.5 volumes of chilled 100% ethanol. Allow to precipitate at –70°C for 30 min.
7. Centrifuge at 12,000 g for 10 min at room temperature. Carefully decant the supernatant and briefly rinse the DNA pellet with 5 ml of 70% ethanol. Dry the pellet under vacuum for 10 min. Dissolve the DNA in 200 to 500 µl of TE buffer. Take 5 to 10 µl of the sample to measure the concentration at 260 nm and store the DNA sample at –20°C until use.

3. Partial Fill-In of Recessed 3′ Termini of Genomic DNA Fragments

Since partially filled-in *Xho* I sites of λ vectors are commercially available, the partially digested DNA fragments can be partially filled-in in order to allow ligation to the vectors. The merit of partially filled-in reactions is to prevent DNA fragments from ligating to each other, thus eliminating the need for size-fractionation of genomic DNA fragments.

1. On ice, set up a standard reaction as follows:
   - Partially digested genomic DNA (20 µg)
   - 10X Fill-in buffer containing dGTP and dATP (10 µl)
   - Klenow fragment of *E. coli* DNA polymerase I (5 units/µl) (4 µl)
   - Add ddH$_2$O to a final volume of 100 µl.
2. Incubate the reaction at 37°C for 30 to 60 min.
3. Extract and precipitate as described in steps 2.3 to 2.7, except that the DNA pellet should be dissolved in 20 to 30 µl of ddH$_2$O. Take 2 to 4 µl of the sample to measure the concentration of DNA. Store at –20°C until use.

Genomic DNA Libraries

4. Small-Scale Ligation of Partially Filled-In Genomic DNA Fragments and Partially Filled-In λGEM-12 Arms

The purpose of setting up a small-scale ligation is to optimize the conditions for large-scale ligation.

1. Set up, on ice, sample and control reactions as follows:
   a. Sample reactions:

Components	Tube number				
	1	2	3	4	5
DNA insert (0.3 μg/μl, 0.1 μg = 0.01 pmol) (μl)	0	4	3	2	0.5
Vector DNA (0.5 μg/μl, 1 μg = 0.035 pmol) (μl)	2	2	2	2	2
10X Ligase buffer (μl)	1	1	1	1	1
ddH$_2$O (μl)	6	2	3	4	5.5
T4 DNA ligase (10 to 15 Weiss units/μl) (μl)	1	1	1	1	1
Total volume (μl)	10	10	10	10	10

   b. Positive control reaction:

λGEM-12 Xho I half-site arms (0.5 μg)	2 (μl)
Positive control insert (0.5 μg)	2 (μl)
10X Ligase buffer	1 (μl)
ddH$_2$O	4 (μl)
T4 DNA ligase (10 to 15 Weiss units/μl)	1 (μl)
Total volume	10 (μl)

2. Incubate the ligation reactions overnight at 4°C for *Xho* I half-site arms.

Notes: (1) Setting up a positive control is necessary in order to check the efficiency of ligation, packaging, and titration. If the efficiency is very low, such that plaque-forming units per milliliter (pfu/ml) < 10^4, then either the vectors do not work well or some procedures were carried out incorrectly. Normally, the positive control is pretested by the company, and pfu/ml is usually 10^6 to 10^8. (2) Setting up a ligation of vector arms only is done in order to check the background induced by the vector arms. Normally background pfu/ml < 10^2 to 10^3. If a higher pfu number, such as 10^4 to 10^5, is observed, then the vectors' ligation is very high, which lowers the efficiency of ligation between the vector arms and the DNA insert. In those circumstances, we recommend the use of new vector arms that have a very low self-ligation rate.

5. In Vitro Packaging of Ligated DNA

There are many commercial companies that supply packaging kits used to harbor the newly constructed λ phage. The example below focuses on the use of Packagene™ extracts (Promega, Madison, WI).

1. Thaw three commercial Packagene extracts (50 μl per extract) on ice. Do not thaw the extract at 37°C.
2. Quickly divide each extract into two tubes (25 μl per tube) on ice.
3. Add 4 μl of each of the ligated mixture in step 4.2 into appropriate tubes, each containing 25 μl of Packagene extract. For testing the quality of the Packagene extract, add 0.5 μg of the provided positive packaging DNA into a 25-μl packaging extract.

4. Incubate at 22 to 24°C (room temperature) for 3 to 4 h.

Note: *Longer packaging (6 to 10 h) is acceptable.*

5. Add phage buffer to each packaged tube up to 250 µl and add 10 to 12 µl of chloroform. Mix well by inversion and allow the chloroform to settle to the bottom of the tube. Store the packaged phages at 4°C until use or for up to 4 weeks until a consequent several-fold dropping of the titer is observed.

Note: *When the packaged phage solution is used for titration or screening, do not mix the chloroform at the bottom of the tube into the solution; otherwise, the chloroform may kill or inhibit the growth of the bacterial host.*

6. Titering of Packaged Phage on LB Plates

1. Partially thaw the specific bacterial strain such as *E. coli* LE 392 or KW 251 on ice in a sterile laminar-flow hood. The bacterial strain is usually kept in 20% glycerol and stored at −70°C until use. Pick up a tiny bit of the bacteria using a sterile wire transfer loop, and immediately inoculate the bacteria by gently drawing several lines on the surface of an LB (Luria-Bertaini) plate. The LB plate should be freshly prepared and dried at room temperature for a couple of days prior to use. Invert the inoculated LB plate and incubate the plate in an incubator at 37°C overnight. Multiple bacterial colonies will grow.
2. Prepare a fresh bacterial culture by picking a single colony from the freshly streaked LB plate using a sterile wire transfer loop and inoculate into a culture tube containing 5 ml of LB medium supplemented with 50 µl of 20% maltose and 50 µl of 1 M $MgSO_4$ solution. Shake at 160 rpm at 37°C for 6 to 9 h or until the OD_{600} has reached 0.6. The culture may be stored at 4°C for a day or two.
3. Dilute each of the packaged recombinant phage samples 1000, 5000, or 10,000 times with phage buffer.
4. Add 20 µl of 1 M $MgSO_4$ solution and 2.8 ml of melted top agar to 36 sterile glass test tubes in a sterile laminar-flow hood. Cap the tubes and immediately place the tubes in a 50°C water bath for at least 30 min to keep the agar melted.
5. Mix 0.1 ml of the diluted phage with 0.1 ml of fresh bacterial cells in a microcentrifuge tube. Cap the tube and allow the phage to adhere to the bacteria in an incubator at 37°C for 30 min.
6. Add the incubated phage/bacterial mixture into specific tubes in the water bath. Vortex gently and immediately pour onto the centers of LB plates; then quickly spread the mixture all over the surface of the LB plates by gently tilting them. Cover the plates and allow the top agar to harden for 15 min in the sterile laminar-flow hood. Invert the plates and incubate them in an incubator at 37°C overnight or for 15 h.

Note: *The top agar mixture should be evenly distributed over the surface of an LB plate. Otherwise, the growth of bacteriophage will be affected, thus decreasing the pfu number.*

7. Count the number of plaques for each plate and calculate the titer of the phage (pfu) for each sample and the control.

Plaque-forming units (pfu) per milliliter = number of plaques per plate × dilution × 10

The last denoted 10 of the calculation refers to the 0.1 ml, on a per-milliliter basis, of the packaging extract used for one plate. For example, if there are 100 plaques on a plate made from a 1/5000 dilution, the pfu per milliliter of the original packaging extract = $100 \times 5000 \times 10 = 5 \times 10^6$.

8. Compare the titers (pfu/ml) from different samples, determine the optimal ratio of vector arms and genomic DNA inserts, and prepare a large-scale ligation and packaging reaction using the optimum ratio of conditions.

Genomic DNA Libraries

7. Large-Scale Ligation of Partially Filled-In Vector Arms and Partially Filled-In DNA Fragments

1. Based on the pfu/ml of the small-scale ligations, choose the optimal conditions for large-scale ligation. For example, if tube no. 4 in the small-scale ligation shows the maximum pfu/ml, then a large-scale ligation can be set up as follows:

Partially filled-in DNA inserts (0.3 µg/µl, 0.1 µg = 0.01 pmol)	10 (µl)
Partially filled-in vector arms (*Xho* I half-site arms, 0.5 µg/µl, 1 µg = 0.035 pmol)	10 (µl)
10X Ligase buffer	5 (µl)
ddH$_2$O	20 (µl)
T4 DNA ligase (10 to 15 Weiss units/µl)	5 (µl)
Total volume	50 (µl)

2. Incubate at 4°C for 12 to 24 h.
3. Carry out *in vitro* packaging of ligated DNA using Packagene extracts thawed on ice. Add 9 µl of the ligated mixture to each of three extracts (50 µl per extract).
4. Incubate at 22 to 24°C for 5 to 8 h.
5. Add phage buffer to each tube up to 0.5 ml. Add 25 µl of chloroform to each tube, mix well, and store at 4°C until use or for up to 4 weeks. At this stage, a genomic DNA library has been established.
6. Perform a titration as in steps 6.1 to 6.8 above.

Note: *A good genomic library should have a pfu/ml up to 10^7 to 10^8. If the pfu < 10^4, do not use for screening*

B. Screening of the Genomic DNA Library

1. Screening of the Genomic Library with a ^{32}P-Labeled Probe

1. Prepare LB plates by adding 15 g of Bacto-agar to 1 L of LB medium and autoclave. Allow the mixture to cool to about 50°C and pour 30 to 40 ml of the mixture into each 85- or 100-mm petri dish in a sterile laminar-flow hood with filtered air flowing. Remove any air bubbles with a pipette tip and let the plates cool for 5 min prior to their being covered. Allow the agar to harden for 1 h and store the plates at room temperature up to 10 days or at 4°C in a bag for 1 month. The cold plates should be placed at room temperature for 1 to 2 days before use.
2. Partially thaw the specific bacterial strain, *E. coli* LE392, on ice in a sterile laminar-flow hood. Pick up a tiny amount of the bacteria using a sterile wire transfer loop and immediately streak out the *E. coli* on LB plates by gently drawing several lines on the surface of each LB plate. Invert the inoculated LB plate and incubate the plate in an incubator at 37°C overnight. Multiple bacterial colonies will grow.
3. Prepare a fresh bacterial culture by picking up a single colony from the freshly streaked LB plate using a sterile wire transfer loop and inoculate into a culture tube containing 50 ml of LB medium supplemented with 500 µl of 20% maltose and 500 µl of 1 *M* MgSO$_4$ solution. Shake at 160 rpm at 37°C overnight or until the OD$_{600}$ has reached 0.6. Store the culture at 4°C until use.
4. Add 20 µl of 1 *M* MgSO$_4$ solution and 2.8 ml of melted LB top agar to each of the sterile glass test tubes in a sterile laminar-flow hood. Cap the tubes and immediately place them in a 50°C water bath for at least 30 min.
5. Based on the titration data, dilute the λ DNA library with phage buffer. Set up 20 to 25 plates for primary screening of the genomic DNA library. For each of the 100-mm diameter plates, mix 0.1 ml of the diluted phage containing 2×10^5 pfu of the genomic library with 0.2 ml of fresh bacterial cells in a microcentrifuge tube. Cap the tube and allow the phage to adhere to the bacteria in an incubator at 37°C for 30 min.

6. Add the incubated phage/bacterial mixture into the tubes from the water bath. Vortex gently and immediately pour onto the center of each LB plate. Quickly spread the mixture over the entire surface of each LB plate by gently tilting the plate. Cover the plates and allow the top agar to harden for 15 min in a laminar-flow hood. Invert the plates and incubate in an incubator at 37°C overnight.

Note: *The top agar mixture should be evenly distributed over the surface of the LB plate. Otherwise, the growth of the bacteriophage will be uneven.*

7. Chill the plates at 4°C for 1.5 h.
8. Move the plates to room temperature. Carefully overlay each plate with a dry nitrocellulose filter disk or a nylon membrane disk (prewetting treatment is not necessary) from one side of the plate slowly to the other side, carefully preventing any air bubbles from developing underneath the filter. Quickly mark the top side and position of each filter in a triangular pattern by punching the filter through the bottom agar layer with a 20-gauge needle containing a small amount of India ink. Allow the phage DNA to mobilize onto the facing side of the membrane filters for 1 to 2 min. If a duplicate is needed, a second filter may be overlaid on the plate for 2 to 3 min.

Notes: *We recommend the use of positively charged nylon membrane disks because they tightly bind the negative phosphate groups of the DNA. Nylon membranes are not easily broken, which is usually the case with nitrocellulose membranes.*

9. Carefully remove the filters using forceps and individually place the filters, plaque-side up, on a piece of wet 3MM Whatman™ filter paper saturated with denaturing solution for 4 min at room temperature. This step serves to denature the double-strand DNA for hybridization with a probe.
10. Transfer the filters, plaque-side up, on another piece of 3MM Whatman paper saturated with neutralization solution at room temperature for 4 min. This step functions to neutralize the filters for hybridization.
11. Transfer the filters, plaque-side up, on another piece of 3MM Whatman filter paper saturated with 5X SSC at room temperature for 2 min.
12. Label and wrap each plate with Parafilm™, and store the plates at 4°C until further use. Air dry the filters at room temperature for 15 min. Then wrap the filters with dry 3MM Whatman paper and bake the filters in a vacuum oven at 80°C for 2 h or UV cross-link the membranes using an optimal setting (usually about 1 min). Wrap the filters with aluminum foil and store at 4°C until prehybridization is carried out. This step covalently links the DNA to the filters.
13. Immerse the filters in 5X SSC for 5 min at room temperature to equilibrate the filters.

Note: *Do not let the filters dry during subsequent steps. Otherwise, a high background and/or anomalous results will show up.*

14. Place the filters in the prehybridization solution and carry out prehybridization for 2 to 4 h with slow shaking at 60 rpm. More details on hybridization techniques can be found in **Chapter 5**.

Notes: *(1) We strongly recommend not using plastic hybridization bags because it is usually not easy to get rid of air bubbles nor can the plastic bags be sealed well. This can cause leaking and contamination. An appropriate size of plastic beaker or tray is the best type of hybridization container to use for this purpose. (2) The prehybridization temperature depends on the prehybridization buffer. The temperature should be set at 42°C if the buffer contains 30 to 40% (for low-stringency conditions) or 50% formamide (for high-stringency conditions). If the buffer, on the other hand, does not contain formamide, the temperature is set at 65°C. Low-stringency conditions will help identify cDNAs of a potential multigene family. High-stringency conditions help prevent nonspecific cross-linking hybridization. (3) If many filters are to be used for hybridization, one container should not contain more than three filter disks. Too many filters in one beaker may cause weak hybridization to*

Genomic DNA Libraries

occur. (4) *The volume of prehybridization solution may vary, but we use 15 ml per 100-cm² filter disk.*

15. Denature the labeled double-strand DNA probe contained in a microcentrifuge tube in heat block set at 100°C for 10 min and immediately chill on ice for 5 min to denature the probe for hybridization. Briefly spin down prior to use with a microcentrifuge.

Notes: *(1) This is a critical step. If the probes are not completely denatured, a weak or no hybridization signal may occur. Single-strand oligonucleotide probes, however, usually do not require denaturation. (2) The DNA used for labeling can be a specific gene (usually a conserved partial-length fragment), an oligonucleotide (where synthesis is based on the conserved regions of known DNA), or a specific cDNA (partial or full length) from other organisms. The DNA is usually labeled with [α-³²P]dCTP and is ready for hybridization (see DNA-labeling protocols in* **Chapter 4***). (3) The labeled probe should be separated from the unincorporated nucleotides by use of a G-50 column (see* **Chapter 4** *for details). Otherwise, nonspecific black spots will appear on the filter, causing one to "fish" out false positive plaques. (4) It is recommended, but not required, to calculate the cpm of the labeled probe prior to hybridization. We use a final concentration of probe at 2×10^6 counts per ml.*

Caution: *[α-³²P]dCTP is a dangerous isotope. A lab coat and gloves should be worn when working with this isotope. Gloves should be changed often and put in a special container. Waste liquid, pipette tips, and papers contaminated with the isotope should be collected in labeled containers. After finishing, a radioactive contamination survey should be performed and recorded.*

16. Dilute the purified probe with 1 ml of hybridization solution and add the probe at 2 to 10×10^6 cpm/ml to the hybridization buffer. Mix well and carefully transfer the prehybridized filters to the hybridization solution. Allow hybridization to proceed overnight or up to 19 h.

Notes: *For hybridization, notes are the same as for prehybridization.*

17. Wash the hybridized filters according to the following conditions:
    a. High-stringency conditions
        i. Wash the filters in a solution (50 ml per filter) containing 2X SSC and 0.1% SDS (w/v) for 15 min at room temperature with slow shaking. Repeat once.
        ii. If the background signal remains high when tested with a survey meter, wash the filters in a fresh solution (50 ml per filter) containing 2X SSC and 0.1% SDS (w/v) for 20 min at 65°C with slow shaking. Repeat two to four times.
        iii. Blot the filters dry on a piece of 3MM paper and proceed with autoradiography.
    b. Low-stringency conditions
        i. Wash the filters in a solution (50 ml per filter) containing 2X SSC and 0.1% SDS (w/v) for 10 min at room temperature with slow shaking. Repeat once.
        ii. If the background signal remains high when tested with a survey meter, wash the filters in a fresh solution (50 ml per filter) containing 2X SSC and 0.1% SDS (w/v) for 15 min at 50 to 55°C with slow shaking. Repeat once.
        iii. Blot the filters dry on a piece of 3MM paper and proceed with autoradiography.
18. Wrap the filters, one by one, with SaranWrap™ and place in an exposure cassette. In a dark room with the safe light on, cover the filters with a piece of x-ray film and place the cassette with an intensifying screen at –80°C for 2 to 24 h prior to their being developed.

Notes: *The film should be slightly overexposed in order to obtain a relatively even background. This will help to identify the marks made previously.*

19. In order to locate the positive plaques, match the developed film with the original plate by placing the film underneath the plate with the help of the previously made marks. This can be done by placing a glass plate over a lamp and putting the matched film and plate on the glass plate. Turn on the light so that the positive clones can be easily identified. Make sure that the plaque-facing side exposed on the film faces down to identify the actual positive plaques. Any mismatch will cause failure in picking the appropriate plaque. Remove individual positive plugs containing phage particles from the plate using a sterile pipette with the tip cut off or the large end of a Pasteur pipette. Expel the plug into a microcentrifuge tube containing 1 ml of elution buffer. Allow elution to occur for 4 h at room temperature with occasional shaking.
20. Transfer the supernatant eluate into a fresh tube and add 20 μl of chloroform. Store at 4°C for up to 5 weeks.
21. Determine the pfu of the eluate as described previously. Replate the phage and repeat the screening procedure with the same isotopic probe several times until 100% of the plaques on the plate are positive. This usually requires three screening procedures.

Notes: During the rescreening process, the plaque number used for one plate should be gradually reduced. In our hands, the plaque density for one 100-mm plate in the rescreening procedures is decreased from 1000 to 500, to 300, and to 100.

22. Amplify the putative DNA clones and isolate the recombinant λ DNAs by either plating or liquid methods (see the DNA-isolation section in **Chapter 1**). The purified DNA can be used for subcloning of DNA inserts.

2. Screening of a Genomic Library Using a Nonradioactive Probe

1. Carry out steps 1.1 to 1.13 as described above in the screening by using a ^{32}P-labeled probe.
2. Perform prehybridization and hybridization as described above, except that a nonradioactive probe and appropriate buffers are used (see **Reagents Needed**).
3. Transfer the hybridized, washed filters into a clean dish containing buffer A (15 ml per filter) for 2 to 4 min.
4. Transfer the filters in buffer B (15 ml per filter) for 60 min.
5. Incubate the filters with an antibody solution, which is the anti-DIG–alkaline phosphatase (Roche Molecular Biochemicals, Indianapolis, IN) diluted at 1:10,000 in buffer B at room temperature for 40 to 60 min using 10 ml per filter.
6. Wash the filters (100 ml per filter) in buffer A for 20 min and repeat once using a fresh washing tray.

Note: The used antibody solution can be stored at 4°C for up to 2 months and reused five to six times without any significant decrease in the antibody activity.

7. Equilibrate the filters in buffer C for 1 to 4 min.
8. Detect the hybridized band(s) by one of the following two methods.
   a. Using Lumi-Phos 530 (Roche Molecular Biochemicals) as a substrate:
      i. Add 0.5 ml of the Lumi-Phos 530 to the center of a clean dish, which should be prewarmed at room temperature for 1 h prior to use.
      ii. Briefly damp a filter using a forceps and completely wet the plaque-facing side of the filter and/or both sides of the filter by slowly laying the filter down on the solution several times.
      iii. Wrap the filter with SaranWrap and wipe out the excess Lumi-Phos solution using a paper towel to reduce the black background.
      iv. Place the wrapped filter in an exposure cassette with the plaque side facing up.
      v. Repeat steps ii, iii, and iv until all the filters are done.
      vi. Overlay the wrapped filters with an x-ray film in a darkroom with a safe light on and allow the exposure to occur by placing the closed cassette at room temperature for 2 min to 24 h.
      vii. Develop the film and proceed to positive clone identification as described previously.

Notes: (1) *Exposure for more than 4 h may produce a dark background. Based on our experience, a good hybridization and detection should generate sharp positive spots with 1.5 h of exposure.* (2) *The film should be exposed long enough to resolve some background signal. This helps to identify the marks made previously.*

    b. Using the NBT and BCIP detection method:
        i. Add 40 µl of NBT solution and 30 µl of BCIP solution in 10 ml buffer C for one filter. NBT and BCIP are available commercially.
        ii. Place the filter in the mixture made in step i and put in the dark for color development at room temperature for 30 min to 1 day in order to obtain a clean background.
        iii. Air dry the filters and proceed to identify the positive plaques on the original LB plates as described in steps 19 to 22 in the radioactive screening method described in **Section II.B.1**.

C. Construction of Partial Genomic Libraries

Under some circumstances, the desired clone within a genomic library is found in extremely low abundance, making it highly unlikely to be isolated using the standard libraries described above. In situations where the screening process is not yielding a genomic clone of interest, it may be useful to create a partial genomic library that is enriched in the clone of interest. This can be done by selectively isolating digested genomic DNA fragments from a gel slice that corresponds to the region of a Southern blot showing signal for the desired clone.

1. Southern Blotting of Genomic DNA and Identification of Desired Fragments

1. Prepare good-quality Southern blots that have a clear hybridization signal for the genomic DNA of interest following procedures found in **Chapter 5**.
2. Once it is determined which restriction enzyme produces the best band signal, carefully determine the size of the hybridizing band (in kb) by comparison with the gel photo that was taken prior to blotting and hybridization. The DNA markers positioned next to the digested genomic DNA in the photo can be used to estimate the size of the hybridizing band on the x-ray film.

2. Isolation of Desired Fragments

1. Once the hybridizing band size is known. Cut genomic DNA with the restriction enzyme that resulted in this band, and run a second gel with the same DNA markers. Detailed procedures are found in **Chapter 5**. We recommend that several lanes of digested genomic DNA be run to provide adequate material for the procedures that follow.
2. After staining and visualizing the gel, cut out the sections of the gel that correspond to the size of the hybridizing band seen in the Southern blots.
3. Perform gel extraction on these gel slices using commercially available kits that result in high yields of DNA. Procedures in **Chapter 1** can also be used.
4. Test the isolated DNA for proper purity as described in **Chapter 1** and run a small aliquot on a gel to confirm the presence of fragments of the appropriate size.

3. Ligation of Linkers

Most λ phage packaging supplies require the presence of a specific sequence at the ends of the DNA fragments to be positioned between the phage arms. For this purpose, either commercially available or custom-designed linkers can be ligated to the ends of the isolated DNA fragments prior to the construction of the library. The required linker will depend on the packaging system being used. Procedures for ligations are found in **Chapter 9** as well as earlier in **Section II.A.4** and **Section II.A.7** in this chapter.

TABLE 10.4
Single- and Double-Enzyme Digestions

Components	Tube Number					
	1	2	3	4	5	6
Recombinant positive phage DNA (µg)	20	20	20	20	20	20
Appropriate 10X restriction enzyme buffer (µl)	4	4	4	4	4	4
1 mg/ml Acetylated BSA (optional) (µl)	4	4	4	4	4	4
EcoR I (10 units/l) (µl)	6.7	0	0	6.7	6.7	0
BamH I (10 units/l) (µl)	0	6.7	0	6.7	0	6.7
Xho I (10 units/l) (µl)	0	0	6.7	0	6.7	6.7
Add ddH$_2$O to final volume of (µl)	40	40	40	40	40	40
Enzyme digestion	*EcoR* I	*BamH* I	*Xho* I	*EcoR* I + *BamH* I	*EcoR* I + *Xho* I	*BamH* I + *Xho* I

Source: From Kaufman, P.B., Wu, W., Kim, D., and Cseke, L.J., *Handbook of Molecular and Cellular Methods in Biology and Medicine*, 1st ed., CRC Press, Boca Raton, FL, 1995.

Alternatively, some packaging systems, such as the λGEM-12 *Xho* I half-site arms, allow the use of partially filled-in genomic DNA fragments. In such cases, procedures such as those found in **Section II.A.3**, "Partial Fill-In of Recessed 3′ Termini of Genomic DNA Fragments," can be used to prepare the DNA fragments for library construction.

4. Construction and Screening of the Library

Once the termini of the DNA fragments are prepared, library construction and screening are done as described in **Section II.A.4** through **Section II.B**.

D. Restriction Mapping of Positive Recombinant Bacteriophage DNA Clones

After putative clones are isolated, the next logical step is to locate the gene of interest within the insert. This will tell you how big the gene is and where it is located in the inserted DNA. It is necessary to have this information in order to subclone or directly sequence the gene. For this purpose, restriction mapping should be carried out using three to four restriction enzymes and combinations of these enzymes. The selection of enzymes depends on the restriction enzyme sites contained in the vectors. When λGEM-11 or λGEM-12 (Promega Corp.) is used as the cloning vector, the mapping procedure is as follows:

1. Set up, on ice, a series of single- and double-enzyme digestions as shown in Table 10.4.
2. Incubate the tubes at an appropriate temperature for 2 to 3 h.
3. After restriction enzyme digestion, carry out electrophoresis on a 0.9% agarose gel (Figure 10.5), blot onto a nylon membrane, and perform prehybridization and hybridization using the same probe for screening of the genomic library, washing, and detection as described in detail under Southern blotting in **Chapter 5**.

Genomic DNA Libraries

4. To ensure that the correct results are obtained, we recommend that steps 1 to 3 be repeated. Both primary and repeated Southern blots should have identical hybridization patterns. An example of such results is given below:

Restriction Enzyme Digestion	Bands Observed on Gel (kb) [a]
EcoR I	20, 9, <u>10</u>, 6, 4
BamH I	20, 9, <u>10</u>, 6, 4
Xho I	20, 9, <u>12</u>, 6, 2
EcoR I + BamH I	20, 9, <u>6</u>, <u>4</u>, 2
EcorR I + Xho I	20, 9, <u>6</u>, <u>4</u>, 2
BamH I + Xho I	20, 9, <u>6</u>, 4, <u>2</u>

[a] Hybridized bands are indicated by underlined numbers. Doublets or triplets are possible.

5. Identify and locate the gene in the insert according to the Southern blot hybridization results and draw a restriction map (Figure 10.6). Check out the fragments based on the map as follows:

Restriction Enzyme Digestion	Bands Observed on Gel (kb) [a]
EcoR I	20, 9, <u>10</u>, 6, 4
BamH I	20, 9, <u>10</u>, <u>6</u>, 4
Xho I	20, 9, <u>12</u>, <u>6</u>, 2
EcoR I + BamH I	20, 9, <u>6</u>, <u>4</u>, 4, 4, 2
EcorR I + Xho I	20, 9, <u>6</u>, 6, <u>4</u>, 2, 2
BamH I + Xho I	20, 9, <u>6</u>, <u>6</u>, 4, <u>2</u>, 2

[a] Hybridized bands are indicated by underlined numbers. Doublets or triplets are possible.

In conclusion, the gene identified is 8 kb in size and is located in the middle of the insert (Figure 10.6).

6. After identifying the location of the gene in the insert, carry out the purification of the gene from recombinant λGEM-11 or λGEM-12. Based on the map in step 5, the gene can be cut out with EcoR I digestion of the recombinant λGEM-11 or λGEM-12 followed by electrophoresis on a 0.9% low-melting-point agarose gel. When the electrophoresis is completed, a 10-kb band containing the 8-kb gene can be eluted out of the agarose gel (see **Chapter 1**). The eluted fragment will then be used for subcloning using an expression plasmid vector for sequencing.

Notes: The gene can be directly sequenced within λGEM-11 or λGEM-12, together with the rest of the insert fragment, using SP6 or T7 primer or both. However, it takes longer and is more expensive to sequence the entire 20-kb insert. Therefore, we recommend that the gene be separated from any other insert sequence as much as possible.

E. Subcloning of the DNA Fragment of Interest

There are at least four advantages to subcloning a DNA fragment of interest. First, the DNA fragment can be amplified up to 300-fold using a high copy number of plasmids that replicate in short-life-cycle *E. coli*. Second, plasmids used for subcloning are designed to contain SP6, T7, or T3 promoters upstream from the polycloning sites, thus allowing one to prepare sense RNA or antisense RNA of the insert for analysis. Third, the merit of designing appropriate primer corresponding to SP6, T7, or T3 allows one to sequence the ends of the insert of interest in opposite directions. Finally, a known DNA sequence, such as cDNA or a genomic gene, can be ligated to appropriate vectors for gene transfer and expression analysis. For these purposes, the recombinant vectors (usually plasmids) may need to be subcloned. Therefore, subcloning is currently an essential technique used in molecular biology studies. This section describes one detailed protocol for subcloning a DNA fragment, gene, or cDNA of interest and then selecting transformants. All procedures have been tested successfully in our laboratories.

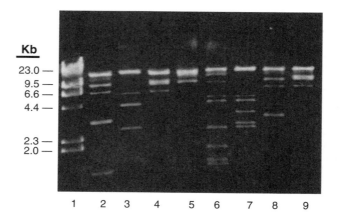

FIGURE 10.5
Digestion of a DNA fragment containing the gene of interest with different restriction enzymes for restriction mapping. The digested DNA was subjected to 0.9% agarose gel electrophoresis, stained with ethidium bromide, and photographed prior to being blotted onto a nylon membrane filter. Lane 1: DNA markers. Lanes 2 to 9: DNA digested with different restriction enzymes, including single- and double-enzyme digestions. (From Kaufman, P.B., Wu, W., Kim, D., and Cseke, L.J., *Handbook of Molecular and Cellular Methods in Biology and Medicine*, 1st ed., CRC Press, Boca Raton, FL, 1995.)

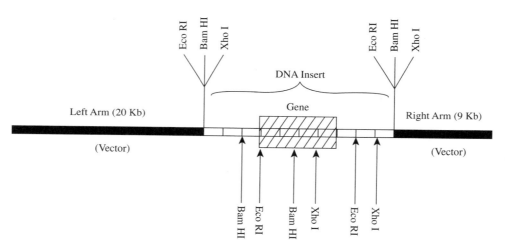

FIGURE 10.6
Diagram showing a restriction map and the location of a given gene of interest. The gene is 8 kb in size and is located in the middle of the genomic DNA fragment that has been cloned in an EMBL 4 vector. (From Kaufman, P.B., Wu, W., Kim, D., and Cseke, L.J., *Handbook of Molecular and Cellular Methods in Biology and Medicine*, 1st ed., CRC Press, Boca Raton, FL, 1995.)

1. Restriction Enzyme Digestion of Vector and DNA Insert for Subcloning

Commercial plasmids such as pGEM series and pBluescript-SK II are available for cloning. Selection of a particular plasmid vector depends on the preferences of the individual investigator. Generally speaking, a standard plasmid for cloning should have the following necessary characteristics: (1) a polycloning site for the insertion of the foreign DNA of interest; (2) SP6, T7, T3, or equivalent promoters upstream from the polylinker site located in opposite directions in order to express the DNA insert for sense RNA, antisense RNA, or protein analysis; (3) the origin of replication for the duplication of the recombinant plasmid in the host cell; (4) a selectable marker gene such as Amp^r for antibiotic selections of transformants; and (5) a selectable marker gene such

Genomic DNA Libraries

as the *lac Z* gene containing the polycloning site for color (e.g., blue/white) screening of interesting bacterial colonies that contain the recombinant plasmids.

a. Preparation of Vectors

1. Set up, on ice, a standard single-restriction enzyme digestion as follows:
   - Plasmid DNA (10 µg)
   - 10X Appropriate restriction enzyme buffer (10 µl)
   - 1 mg/ml Acetylated BSA (optional) (10 µl)
   - Appropriate restriction enzyme (3.3 units/µg DNA)
   - Add ddH$_2$O to a final volume of 100 µl.

Notes: *(1) The restriction enzyme used for vector and DNA-insert digestions should be the same in order to ensure optimal ligation. Some enzyme combinations, however, produce compatible ends. (2) For directional cloning, the plasmid and the DNA insert should be digested using two different restriction enzymes. The double-enzyme digestion of DNA may be set up as a single reaction at the same time or be carried out as two single-enzyme digestions at different times. (3) For double-restriction enzyme digestions, the appropriate 10X buffer containing a higher NaCl concentration than the other buffer may be chosen for the double-enzyme-digestion buffer.*

   Set up, on ice, the double-restriction enzyme reaction as follows:
   - Plasmid DNA (10 µg)
   - 10X Appropriate restriction enzyme buffer (10 µl)
   - 1 mg/ml Acetylated BSA (optional) (10 µl)
   - Appropriate restriction enzyme A (3.3 units/µg DNA)
   - Appropriate restriction enzyme B (3.3 units/µg DNA)
   - Add ddH$_2$O to a final volume of 100 µl.

2. Incubate at an appropriate temperature (e.g., 37°C) for 2 to 3 h. For single-enzyme-digested DNA, proceed to step 3. For double-enzyme-digested DNA, proceed to step 5.

Notes: *(1) To ensure that an optimal ligation to the DNA insert occurs, the vector plasmid should be completely digested. The digestion efficiency can be checked by loading 1 µg of the digested DNA (10 µl) with loading buffer to a 1% agarose minigel. Undigested plasmid DNA (1 µg) and standard DNA markers should be loaded in the adjacent wells as controls. After electrophoresis, the undigested plasmid DNA may reveal multiple bands because of different levels of supercoiled plasmids. However, one band will be visible for a complete single-enzyme digestion; one major band and one tiny band (<70 bp) may be visible after digestion with two different restriction enzymes. (2) Double-restriction enzyme digestion is usually more difficult than a single-enzyme digestion since DNA can be cut by one of two enzymes due to salt concentration in the digestion buffer or other factor(s). To ensure that the vector is digested by two restriction enzymes, the digested vector should be purified and subject to religation by using T4 DNA ligase. The efficiency of digestion is then checked by loading the religated vector DNA onto a 1% agarose gel followed by electrophoresis. When the vector DNA is digested with two restriction enzymes and purified, it cannot be religated. Therefore, only one band appears, which is the linearized DNA. On the other hand, when the DNA is digested with one of two restriction enzymes and purified, it will be religated during the ligation reaction. When that occurs, multiple bands may be visible due to different levels of supercoiled plasmids. (3) After completion of the restriction enzyme digestion of the vector, calf intestinal alkaline phosphatase (CIAP) treatment should be carried out for the above single-restriction enzyme digestion. This treatment removes 5'-phosphate groups, thus preventing recircularization of the vector during ligation. Otherwise, the vector will close on itself during ligation, and the efficiency of ligation* between the vector and the DNA insert will be very low. For double-restriction enzyme-digested vectors, the CIAP treatment is not necessary.

3. Carry out the CIAP treatment by adding the following directly to the single-enzyme-digested DNA sample (90 µl).
   - 10X CIAP buffer (15 µl)
   - CIAP diluted in 10X CIAP buffer (0.01 unit/pmol ends)
   - Add ddH$_2$O to a final volume of 150 µl.

Notes: (1) CIAP and 10X CIAP should be kept at 4°C. CIAP treatment should be set up at 0°C. (2) Calculation of the amount of ends is as follows: There are 9 µg digested DNA left after the removal of 1 µg (of the 10 µg digested DNA) for checking on agarose gel. If the vector is 3.2 kb, the amount of ends can be calculated by the formula below:

$$\text{pmol ends} = [\text{amount DNA}/(\text{base pairs} \times 660/\text{base pair})] \times 2$$
$$= 9/3.2 \times 1000 \times 660 \times 2$$
$$= 4.2 \times 10^{-6} \times 2$$
$$= 8.4 \times 10^{-6} \, \mu M$$
$$8.4 \times 10^{-6} \times 10^{-6} = 8.4 \text{ pmol ends}$$

4. Incubate at 37°C for 1 h and add 2 µl of 0.5 M EDTA buffer (pH 8.0) to stop the reaction.
5. Extract with one volume of TE-saturated phenol/chloroform. Mix well by vortexing for 1 min and centrifuge at 11,000 g for 5 min at room temperature.
6. Carefully transfer the top, aqueous phase to a fresh tube and add one volume of chloroform:isoamyl alcohol (24:1) to the supernatant. Mix well and centrifuge as in step 5.
7. Carefully transfer the upper, aqueous phase to a fresh tube and add 0.1 volume of 3 M sodium acetate buffer (pH 5.2) or 0.5 volume of 7.5 M ammonium acetate to the supernatant. Briefly mix and add 2 to 2.5 volumes of chilled 100% ethanol to the supernatant. Allow to precipitate at –70°C for 1 h or at –20°C for 2 h.
8. Centrifuge at 12,000 g for 10 min and carefully decant the supernatant. Briefly rinse the DNA pellet with 1 ml of 70% ethanol and dry the pellet under vacuum for 20 min. Dissolve the DNA pellet in 20 to 40 µl ddH$_2$O. Take 4 ml of the sample to measure the concentration of the DNA at 260 nm. Store the sample at –20°C until use.

Note: Adding 0.5 volume of 7.5 M ammonium acetate to the supernatant at step 7 yields a higher amount of DNA precipitation than by adding 0.1 volume of 3 M sodium acetate buffer (pH 5.2).

b. Preparation of DNA Insert

1. Purify DNA insert from an agarose gel as described under the DNA elution section in **Chapter 1**.

Note: Inserting DNA whose size is <4 kb is easier for successful subcloning than DNA whose size is 4 to 12 kb.

2. Carry out restriction enzyme digestion, purification, and precipitation the same as for vector DNA (see steps 1 to 8 in **Section II.E.1.a**).

2. Ligation of Plasmid Vector and DNA Insert

To achieve optimal ligation, the ratio of vector to DNA insert (1:1, 1:2, 1:3, and 3:1 molar ratios) should be optimized by using small-scale reactions. The following reaction protocol is standard for the ligation of a 3.2-kb plasmid vector and a 2.533-kb DNA insert.

Genomic DNA Libraries

TABLE 10.5
Calculation of Molar Ratios

Vector DNA:Insert DNA	Amount of DNA (µg)	
	Vector	Insert
1:1	1	0.792
1:2	1	1.584
1:3	1	2.376
3:1	1	0.264

Source: From Kaufman, P.B., Wu, W., Kim, D., and Cseke, L.J., *Handbook of Molecular and Cellular Methods in Biology and Medicine*, 1st ed., CRC Press, Boca Raton, FL, 1995.

TABLE 10.6
Components of Ligation Reactions

Components	Ligation Reactions			
	1 (1:1)	2 (1:2)	3 (1:3)	4 (3:1)
Plasmid DNA as vector (µg)	1	1	1	1
Insert-DNA (µg)	0.792	1.584	2.376	0.244
10X Ligase buffer (µl)	1	1	1	1
T4 DNA ligase (Weiss units)	4	4	4	4
Add ddH$_2$O (µl)	10	10	10	10

Source: From Kaufman, P.B., Wu, W., Kim, D., and Cseke, L.J., *Handbook of Molecular and Cellular Methods in Biology and Medicine*, 1st ed., CRC Press, Boca Raton, FL, 1995.

1. Calculate the molar weights of vector and DNA insert:
   1 M Plasmid vector-DNA = $3.2 \times 1000 \times 660 = 2.112 \times 10^6$
   1 M DNA insert = $2.533 \times 1000 \times 660 = 1.672 \times 10^6$
   where 660 refers to the average molecular weight of a nucleotide.
2. Calculate the molar ratio of vector to DNA insert using Table 10.5.
3. Set up the ligations as in Table 10.6 on ice.

Note: *The restriction enzyme-digested plasmid (vector) DNA and DNA insert should be dissolved in ddH$_2$O (nuclease free) at 0.5 to 1.0 µg/µl. If the DNA is less than 0.4 µg/µl, the DNA should be precipitated so as to dissolve at about 1 µg/µl.*

4. Incubate the reactions at 4°C for 12 to 24 h, or at 16°C for 4 to 6 h, or at room temperature (22 to 25°C) for 1 to 2 h.

Note: *After the ligations are completed at the above temperatures, the mixture can be stored at 4°C until use.*

5. Check the efficiency of the ligations by 1% agarose electrophoresis. When the electrophoresis is complete, photograph the gel stained with EtBr under UV light. As compared with unligated vector or DNA insert, high-efficiency ligation should make it possible to visualize less than approximately 10% unligated vector and DNA insert by estimating the intensity of fluorescence. Approximately 90% of the vector and DNA insert are ligated to each other and show strong band(s) with molecular-weight shifts compared with the vector and DNA-insert sizes. By comparing the efficiency of ligations using different molar ratios, the optimal conditions can be determined with ease. These can be used as a guide for large-scale ligation.

Note: The small-scale ligation above is optional, but it is strongly recommended that it be carried out.

6. The large-scale ligation of vector and DNA insert is based on the optimal conditions determined by small-scale ligations. For example, if one uses a 1:2 molar ratio of plasmid-DNA:insert-DNA as the optimal ligation condition, a large-scale ligation can be carried out as follows:
   - Plasmid DNA as vector (3 µg)
   - Insert-DNA (4.75 µg)
   - 10X Ligase buffer (3 µl)
   - T4 DNA ligase (Weiss units) (15 to 50 units)
   - Add ddH$_2$O to a final volume of 30 µl.

   Incubate the ligation mixture at 4°C for 12 to 24 h, or at 16°C for 4 to 6 h, or at room temperature (22 to 25°C) for 1 to 2 h. Store at 4°C until use. Proceed to carrying out the transformation.

3. E. coli Transformation

1. Prepare the LB medium and LB plates as described in **Reagents Needed**. This should be done well before ligation.
2. Prepare competent cells using one of the following methods, depending on the procedure to be used for transformation. This should be completed well before ligation.

a. Preparation of Chemically Competent Cells

There are many methods for the preparation of chemically competent cells, and there are also commercially available competent cells. The protocol that follows works equally well as commercially available cells in our laboratories.

1. Streak the appropriate *E. coli* strain (e.g., DH5αF′ or JM109 for color screening) directly from a small amount of frozen glycerol stock stored at −70°C onto the surface of an LB plate using a sterile platinum wire loop. Invert the plate and incubate in a 37°C incubator for 12 to 16 h. Bacterial colonies will become visible.

Notes: *(1) It is not necessary to thaw the frozen bacteria at room temperature or at 4°C. A small amount of bacteria adhering to the wire loop is sufficient for inoculation. Repeated freeze/thaw cycles will ruin the culture. (2) If using one of the bacterial strains above, LB plates should contain thiamine-HCl for the selection of F′ in the cell, which is necessary for color screening.*

2. Inoculate a well-isolated colony from the plate at step 1 into 10 ml of 2X LB medium. The medium may be supplemented with 0.1 ml of 20% maltose and 0.1 ml of 1 M MgSO$_4$ solution if desired, but we find that this is not necessary in most cases. Incubate at 37°C overnight with shaking at 200 rpm.
3. Next day, prewarm two 1-l flasks containing 150 ml of 2X LB medium to 37°C. Add 1.5 ml of cells from step 2 to the LB medium. Again, 1.5 ml of 20% maltose and 1.5 ml of 1 M MgSO$_4$ solution may be added to each flask if desired. Incubate at 37°C with shaking at 200 rpm. Measure the OD$_{600}$ or A$_{600}$ every 20 to 30 min until the A$_{600}$ reaches 0.45 to 0.55. It usually takes 2 to 4 h.
4. Chill the flasks in ice water for 2 h and centrifuge the cells at 3000 g for 15 min at 4°C.

Note: *From this point on, it is absolutely necessary to keep the cells ice cold throughout the rest of the procedure.*

5. Drain off the supernatant and resuspend the cells in 5 ml of ice-cold CaCl$_2$/MnCl$_2$ salt solution. Dilute this suspension to 70 ml with the same buffer.
6. Incubate the cells on ice for 45 to 60 min.
7. Centrifuge at 2000 g for 10 min at 4°C and gently resuspend the cells in a total of 15 ml (for both flasks) of ice-cold CaCl$_2$/MnCl$_2$ salt solution.

8. Add 3.45 ml of sterile 80% glycerol dropwise with gentle swirling to the cell solution. Aliquot the cells at 0.2 ml per tube on ice, freeze all tubes on dry ice or in liquid nitrogen, and then store at –70°C until use.

b. Preparation of Competent Cells for Electroporation
1. Carry out step 1 to step 4 in the protocol for **Section II.E.3.a**, "Preparation of Chemically Competent Cells."
2. Wash the cells by resuspending the pellet with 100 ml distilled water or low-salt buffer to reduce the ionic strength of the cell suspension.
3. Centrifuge at 2000 g for 10 min at 4°C and carefully decant the supernatant.
4. Repeat steps 2 and 3 twice.
5. Resuspend the cells in 200 ml of low-salt buffer or distilled water. Add glycerol dropwise with gentle swirling to 10% (v/v). Dispense the cell suspension into 20 µl per tube aliquots at approximately 3×10^9 cells/ml. Freeze on dry ice or in liquid nitrogen, and then store at –70°C until use.

4. Transformation Using Chemically Competent Cells
1. Thaw the desired number of 0.2 ml aliquots of frozen chemically competent cells on ice. Each transformation requires only 50 µl of cells. So, one tube is adequate for four transformations.
2. Add 3 µl of frozen-thawed DMSO to every 0.2 ml aliquot, mix, and add 50 µl of competent cells to the recombinant plasmid DNA, as shown in Table 10.7.

Note: *DMSO may help to make the cell permeable to the DNA; however, addition of DMSO is not absolutely necessary.*

3. Incubate on ice for 30 min.
4. Heat shock at 42°C for 2 min and place on ice for 1 min (optional).
5. Transfer the cell suspension to sterile culture tubes and add 1 ml of LB medium containing 20 µl of 20% maltose and 20 µl of 1 M MgSO$_4$ solution. Incubate at 37°C for 1 to 2 h with shaking at 140 rpm to recover the cells.

Note: *(1) Addition of maltose and MgSO$_4$ is only required for certain strains of bacteria and can be omitted in most cases. (2) Addition of more LB medium and further incubation is not necessary when using antibiotics such as ampicillin or ticarcillin due to their method of action. In these cases, a good time saver is to plate the cells immediately after heat shock.*

6. Add 50 to 150 µl of the culture per plate to the centers of LB plates containing the appropriate antibiotic (e.g., 50 µg/ml ampicillin), 0.5 mM IPTG, and 40 µg/ml X-Gal. Quickly spread the cells over the entire surface of the LB plates using a flame-sterile, bent glass rod.

Note: *If many transformations are to be done, we recommend using different amounts of DNA at step 2 and plating at the same volume in order to determine the optimal conditions.*

TABLE 10.7
Aliquot Components

Components	Aliquot Number		
	1	2	3
Cells (µl)	20	20	20
DMSO (µl)	3	3	3
DNA (µl)	1 (15 ng)	1 (100 ng)	1 (200 ng)

Source: From Kaufman, P.B., Wu, W., Kim, D., and Cseke, L.J., *Handbook of Molecular and Cellular Methods in Biology and Medicine*, 1st ed., CRC Press, Boca Raton, FL, 1995.

5. Transformation by Electroporation

1. Thaw three aliquots of 20 μl of frozen, electroporation competent cells on ice. Place recombinant DNA sample on ice.
2. Chill three disposable microelectroporation chambers (Invitrogen) on ice.
3. Connect the power cable to an Eppendorf 2510 Electroporator (Eppendorf, AG, Hamburg, Germany).
4. Set up the pulse control as follows:
   - Power: charge
   - Capacitance (μF): 330
   - High ohm/low ohm: low ohm
   - Charge rate: fast
   - Set the voltage booster at 4 kV for *E. coli*.
5. Add ice water to the chamber safe up to four-fifths of the volume, and place the chamber rack in the chamber safe.
6. Add 1, 2, and 3 μl of recombinant plasmid DNA (e.g., pGEM, pBKS II, 0.5 to 1 μg/μl) to three aliquots of 20 μl of ice-thawed cells, respectively. Gently mix and immediately place on ice until use.

Note: *Avoid any air bubbles during mixing. The total volume of the transformation mixture should be <25 μl.*

7. Use a pipette to transfer one aliquot and carefully place the mixture drop between the electrode poles in the microelectroporation chamber. Gently cover the chamber.

Note: *The liquid should not be allowed to drop into the bottom of the chamber, and there should be no air bubbles within the drop.*

8. Gently place the chamber into the cell in the safe rack, cover the chamber safe, and turn the electric-shock pointer toward the cell containing the bacterial cell chamber.
9. Connect the power from the voltage booster to the chamber safe, and turn on power for both the pulse control and the voltage booster units.
10. Press the charge button on the pulse control unit up to 365. When the dc voltage goes down to 345 to 350 V, turn the button from "charge" to "arm." Quickly push the trigger button for 1 s. The voltage goes down to <10 and the voltage booster should read 1.9 to 2.0 kV.
11. Turn off the power and carefully remove the chamber from the cell. The transformed cell suspension should still be between the positive and negative electrode poles.
12. Quickly transfer the transformed cell suspension into a test tube containing 2 ml LB medium without ampicillin, as the cells are quite weak. Mix well and place at room temperature for no longer than 15 min.
13. Repeat steps 7 to 12 until all the samples are transformed.
14. Incubate at 37°C for 1 to 2 h with shaking at 150 rpm to recover the cells.
15. Use a sterile, bent glass rod to spread 20, 50, 100, and 200 μl of each of the three recovered transformant cells over the entire surface of LB plates containing 50 μg/ml ampicillin, 0.5 mM IPTG, and 40 μg/ml X-Gal.

6. Selection of Transformants Containing Recombinant Plasmids

1. Invert all the plates prepared at step 1 (**Section II.E.4**) and step 1 to step 15 (**Section II.E.5**) and incubate in a 37°C incubator for 12 to 16 h until colonies are visible.
2. Chill the plates at 4°C for 1 h to maximally expose the blue colonies that may not be obvious when they are first taken from the incubator.

Notes: *Blue colonies contain nonrecombinant plasmid. β-galactosidase expressed by the lac Z gene hydrolyzes X-Gal, forming a blue color. White colonies are supposed to bear recombinant plasmids in which foreign DNA was inserted at the polycloning site in the lac Z*

Genomic DNA Libraries

gene. The interrupted lac Z gene cannot express β-galactosidase activity. Therefore, the colonies remain very white. (2) In our experience, small cloned fragments are not always capable of disrupting the lac Z gene. In these cases, the colonies with the correct clones may appear as pale blue.

3. Inoculate individual white colonies into 5 ml of LB medium. Incubate at 37°C overnight with shaking at 160 rpm.

Note: To verify white colonies, at least 20 individual colonies should be analyzed.

4. Isolate plasmids as described in **Chapter 1**.
5. Digest the plasmids with the same restriction enzyme(s) as used for subcloning of the DNA insert of interest.
6. Carry out electrophoresis as previously described.
7. Photograph and verify the sizes of the vector and DNA insert as compared with unligated vector and DNA insert as controls.
8. This gel can also be used for Southern blotting (see **Chapter 5**) using unligated insert-DNA as a probe. Hybridized bands indicate that the DNA insert of interest has been successfully subcloned in the plasmids.
9. Inoculate 0.5 ml of the verified white colony cells into 100 ml of LB medium containing 50 µg/ml ampicillin and 1 ml of 1 M MgSO$_4$ solution. Incubate at 37°C overnight while shaking at 160 rpm.
10. Aliquot 1 ml of the culture to eppendorf tubes and add glycerol dropwise to 15% (w/v). Freeze on dry ice or in liquid nitrogen and store at −70°C for further use.

Note: 7% DMSO works equally well for storage of cells as 15% glycerol.

11. Use the remainder of the culture to carry out large-scale isolation and purification of the recombinant plasmids as described in **Chapter 1**. This plasmid DNA can be used for sequencing or RNA analysis of the insert of interest.

Reagents Needed

7.5 M Ammonium Acetate Solution
Dissolve 57.8 g ammonium acetate in 50 ml ddH$_2$O and adjust volume to 100 ml.

Anti-DIG-Alkaline Phosphatase
Antidigoxigenin conjugated to alkaline phosphatase (Roche Molecular Biochemicals)

BCIP Solution
50 mg/ml 5-Bromo-4-chloro-3-indolylphosphate (X-phosphate), in 100% dimethylformamide

Buffer A
100 mM Tris-HCl
150 mM NaCl, pH 7.5

Buffer B
2% (w/v) Blocking reagent (Roche Molecular Biochemicals) or equivalent, such as BSA, nonfat milk, or gelatin in buffer A
Dissolve well by stirring with a magnetic stir bar.

Buffer C
    100 mM Tris-HCl, pH 9.5
    100 mM NaCl
    50 mM MgCl$_2$

10X CIAP Buffer
    0.5 M Tris-HCl, pH 9.0
    10 mM MgCl$_2$
    1 mM ZnCl$_2$
    10 mM Spermidine

Denaturing Solution
    1.5 M NaCl
    0.5 M NaOH
    Autoclave.

50X Denhardt's Solution
    1% (w/v) BSA (bovine serum albumin)
    1% (w/v) Ficoll (Type 400, Pharmacia)
    1% (w/v) PVP (polyvinylpyrrolidone)
    Dissolve well after each addition. Adjust to the final volume to 500 ml with distilled water and sterile filter. Divide the solution into 50 ml aliquots and store at −20°C. Dilute tenfold into prehybridization and hybridization buffers.

Elution Buffer
    10 mM Tris-HCl, pH 7.5
    10 mM MgCl$_2$

10X Fill-In Buffer
    0.5 M Tris-HCl, pH 7.2
    0.1 M MgSO$_4$
    1 mM DTT
    0.5 g/ml Acetylated BSA
    10 mM dATP
    10 mM dGTP

Hybridization Buffer for Nonradioactive Probes
    Add DIG-dUTP-labeled probe to the appropriate volume of fresh prehybridization buffer.

Hybridization Buffer for Radioactive Probes
    5X SSC
    0.5% SDS
    5X Denhardt's reagent
    0.2% Denatured salmon sperm DNA
    [a-^{32}P]-Labeled DNA probe

0.1 M IPTG Solution
    1.2 g IPTG in 50 ml ddH$_2$O
    Filter-sterilize and store at 4°C.

LB (Luria-Bertaini) Medium (per liter)

Bacto-tryptone (10 g)
Bacto-yeast extract (5 g)
NaCl (5 g)
Adjust the pH to 7.5 with 2 N NaOH solution and autoclave. When it has cooled, store at 4°C until use. Make 4 L.

2X LB Medium (per liter)

Bacto-tryptone (20 g)
Bacto-yeast extract (10 g)
NaCl (10 g)
Adjust the pH to 7.5 with 2 N NaOH solution and autoclave. When it has cooled, store at 4°C until use. Make 4 L.

LB Plates

Add 15 g of Bacto-agar to 1 L of freshly prepared LB medium and autoclave. Allow the mixture to cool to about 60°C and pour 30 ml of the mixture into each 85- or 100-mm-diameter petri dish in a sterile laminar-flow hood. Remove any air bubbles with a pipette tip and let the plates cool for 5 min prior to their being covered. Allow the agar to harden for 1 h and store the plates at room temperature up to 10 days or at 4°C in a bag for 1 month. The cold plates should be placed at room temperature for 1 to 2 days before use. If antibiotics are required, plate can be supplemented with 50 µg/ml ampicillin after autoclaving when it has cooled to 50 to 55°C. Then 0.5 mM IPTG, and 40 µg/ml X-Gal can be added directly to the surface of LB plates and allowed to soak in to the medium for about 60 min.

LB Top Agar

Add 4.0 g agar to 500 ml freshly prepared LB medium and autoclave. Store at 4°C until use. For plating, melt the agar in a microwave oven. When the solution has cooled to 60°C, add 0.1 ml of 1 M MgSO$_4$ per 10 ml of the mixture.

10X Ligase Buffer

0.3 M Tris-HCl, pH 7.8
0.1 M MgCl$_2$
0.1 M DTT
10 mM ATP

5X Loading Buffer

38% (w/v) Sucrose
0.1% Bromophenol blue
67 mM EDTA

Low-Salt Buffer (200 ml)

CaCl$_2$-2H$_2$O (2.94 g)
MnCl$_2$-4H$_2$O (2.77 g)
NaOAc (0.65 g) (anhydrous)
Bring close to 200 ml; adjust the pH to 5.5 with acetic acid; adjust volume to 200 ml; filter sterilize and chill for storage.

20% (v/v) Maltose

1 M MgSO$_4$

NBT Solution
75 mg/ml Nitroblue tetrazolium salt in 70% (v/v) dimethylformamide

Neutralizing Solution
1.5 M NaCl
0.5 M Tris-HCl, pH 7.4
Autoclave.

Phage Buffer
20 mM Tris-HCl, pH 7.5
100 mM NaCl
10 mM MgSO$_4$

Prehybridization Buffer for Nonradioactive Probes
5X SSC
0.1% N-Lauroylsarcosine
0.02% Sodium dodecyl sulfate (SDS)
1% Blocking reagent
Dissolve well on a heating and stirring plate at 65°C after each addition. Add sterile water to final volume or add 50% formamide to the mixture if using the formamide method.

Prehybridization Buffer for Radioactive Probes
5X SSC
0.5% SDS
5X Denhardt's reagent
0.2% Denatured salmon sperm DNA

Note: *This is only one example of an effective hybridization solution widely used during genomic library screening. Other well-established protocols and recipes are detailed in* **Chapter 5**. *We have also had very good success with the Church method of hybridization.*

10X Sau3A I Buffer
0.1 M Tris-HCl, pH 7.5
1 M NaCl
70 mM MgCl$_2$

S Buffer
Phage buffer + 2% (w/v) gelatin

3 M Sodium Acetate Buffer, pH 5.2
Dissolve sodium acetate in ddH$_2$O and adjust pH to 5.2 with 3 M glacial acetic acid. Autoclave and store at room temperature.

20X SSC Solution (1 L)
3 M NaCl
0.3 M Na$_3$ citrate (trisodium citric acid)
Autoclave.

5X SSC Solution
Dilute 20X SSC solution four times with sterile water.

5X TBE Buffer
Tris base (54 g)
Boric acid (27.5 g)
20 ml of 0.5 M EDTA, pH 8.0

TE Buffer
10 mM Tris-HCl, pH 8.0
1 mM EDTA, pH 8.0

TE-Saturated Phenol/Chloroform
Completely melt the phenol crystals in a 65°C water bath. Mix equal parts of phenol and TE buffer and allow the phases to separate. Mix one part of the lower, phenol phase with one part of chloroform:isoamyl alcohol (24:1). Allow the phases to separate and store at 4°C in an aluminum-foil-wrapped or dark bottle.

Thiamine-HCl Plates
Na_2HPO_4 (6 g)
KH_2PO_4 (3 g)
NaCl (0.5 g)
NH_4Cl (1 g)
Agarose (15 g)
Add ddH_2O to 1 L and autoclave.
Cool to 50°C and add:
  2 ml of 1 M $MgSO_4$
  0.1 ml of 1 M $CaCl_2$
  10 ml of 20% Glucose
  1 ml of 1 M Thiamine-HCl

Trituration Buffer
0.1 M $CaCl_2$
70 mM $MgCl_2$
40 mM Sodium acetate, pH 5.5
Freshly prepare and sterile-filter.

X-Gal Stock Solution
50 mg/ml Stock in N,N'-dimethylformamide

III. YAC Libraries

In bacteriophage λ libraries, genomic DNA is digested by a four-base "cutter," *Sau*3A I, and fragments of 14 to 23 kb are ligated to λ DNA vector arms. The flaw in this strategy is that large genes (>23 kb) are usually divided into multiple fragments contained in different clones. The gene sequence is located in multiple overlapping fragments. Therefore, it is usually impossible to identify a large gene or operon (>23 kb) in a single clone. To overcome this disadvantage, the technique of constructing yeast artificial chromosome (YAC) libraries has been developed. In a YAC library, extra-large genomic DNA molecules with average size of 800 to 1000 kb can be cloned with ease using YAC vectors. Theoretically, any size of gene can be readily isolated from a single positive clone in a standard YAC library. This breakthrough in gene cloning has made it possible to accomplish the human genome project, various plant genome projects, and other genome projects by genome mapping, physical mapping, and chromosomal walking.

YAC vectors used in cloning are artificially designed to carry necessary sequences such as the centromere (CEN4), an autonomous replicating sequence (ARS), the telomeres (TEL), selectable marker genes, and the cloning site for insertion of the genomic DNA of interest. The YAC vectors, such as pYAC4, are propagated as bacterial plasmids, linearized by an appropriate restriction enzyme (e.g., *EcoR* I), and ligated to genomic insert-DNA. The recombinant, linear YAC vectors are then transformed into an appropriate yeast host strain, such as AB1380 (*MAta ade 2-1 ura3 can1-100 lys2-1 trp1 his 5*).

A. Preparation of Cells or Tissues for Isolation and Purification of High-Molecular-Weight DNA

Construction of a YAC library requires that the molecular weight (MW) of genomic DNA be as high as possible. The purity and integrity of the isolated DNA is crucial for pulsed-field gel electrophoresis (PFGE) and cloning. To obtain intact genomic DNA, traditional shearing and solvent (e.g., phenol/chloroform) extraction should be avoided. Instead, cells or protoplasts are lysed *in situ* in an agarose plug, digested with appropriate restriction enzyme, and checked by PFGE or field-inversion gel electrophoresis (FIGE).

1. For cultured cells:
    a. Wash the cells with five volumes of ice-cold phosphate-buffered saline (PBS) and centrifuge at 100 g for 5 min at room temperature. Carefully decant the supernatant and repeat the washing twice.
    b. Resuspend the cells in ice-cold cell suspension buffer at approximately 5×10^7 cells/ml.
2. For fresh animal tissues:
    a. Slice the tissues into 1- to 2-mm pieces using a clean razor blade or equivalent and transfer to an ice-cold glass homogenizer with a tight-fitting pestle.
    b. Add four volumes of ice-cold PBS to the homogenizer, briefly suspend the tissue slices, and homogenize for 2 to 4 min on ice.
    c. Filter the homogenate through two layers of cheesecloth to remove the cell fragments.
    d. Centrifuge the cells at 1000 g for 5 min at room temperature. Carefully decant the supernatant and wash the cells three times with five volumes of ice-cold PBS.
    e. Resuspend the cells in ice-cold cell suspension buffer at approximately 5×10^7 cells/ml.
3. For frozen animal tissues:
    a. Grind frozen tissues in liquid nitrogen to a fine powder using a chilled mortar and pestle.
    b. Transfer the powder to a centrifuge tube containing five volumes of ice-cold PBS, suspend the powder into PBS, and centrifuge at 1000 g for 5 min at room temperature.
    c. Decant the supernatant and wash the cells three times in five volumes of ice-cold PBS.
    d. Resuspend the cells at 5×10^7 cells/ml in ice-cold cell suspension buffer.
4. For fresh plant tissues:
    For plant tissues, we recommend that one prepare protoplasts instead of the entire cells, whose cell walls are difficult to digest.
    a. Remove six to eight of the youngest, fully expanded leaves from plants grown in the greenhouse or under sterile conditions. Peel off the lower side of the epidermis using a pair of jeweler's forceps and place the leaf tissue in a petri dish.
    b. Surface-sterilize the leaves from greenhouse-grown plants by immersing the leaves in 5 to 10% Clorox® solution (sodium hypochlorite) for 5 to 10 min followed by thorough rinsing with 40 ml sterile distilled water four to five times to remove the Clorox.

c. Add 5 to 10 ml of sterile enzyme medium, mix, and incubate in the dark at room temperature (24 to 25°C) for 18 to 20 h without shaking.

Note: At this stage, the cell walls are hydrolyzed by enzymes such as cellulase. This can be monitored by looking at the cells under a microscope. Cells with hydrolyzed cell walls will appear as perfectly spherical protoplasts.

d. Add 15 ml of washing medium and gently shake to loosen the protoplasts from undigested leaf materials.
e. Filter through a nylon mesh (50 μm pore diameter) to remove undigested materials. The protoplasts are in the filtrate solution.
f. Centrifuge the protoplasts at 1000 g for 5 min at room temperature and carefully decant the supernatant.
g. Resuspend the protoplasts in 4 ml of washing medium and centrifuge as in step f.
h. Resuspend the protoplasts in 1 ml of washing medium, add 1 ml of 18% sucrose, which will become an underlayer in the protoplast suspension, and centrifuge at 120 g for 5 min at room temperature.
i. Carefully transfer the protoplasts from the interface using a wide-bore Pasteur pipette to a clean centrifuge tube and add 1 ml protoplast suspension buffer.
j. Count the protoplasts using a microscope and a haemacytometer.
k. Centrifuge at 1000 g for 5 min at room temperature and resuspend the protoplasts at approximately 5×10^7 protoplasts/ml in cell suspension buffer.

B. Isolation of High-Molecular-Weight DNA

1. Prepare an equal volume of 1% (w/v) low-melting temperature agarose in cell suspension buffer. Melt the agarose in a microwave and allow to cool to 42°C.
2. Warm an equal volume of cell suspension or protoplast suspension (5×10^7 cells/ml) to 42°C and add to the agarose gel mixture (42°C). Mix well to ensure that the cells or protoplasts are evenly dispersed throughout the agarose.
3. Add the melted agarose-cell/protoplast mixture to an ice-cold plug former or to preformed Plexiglas™ molds (50 to 100 μl) or equivalent tubes using a 1-ml pipette that has the tip cut off.
4. Allow the plugs to harden for 30 min on ice and carefully remove the plugs by pushing them out of the mold. Cut the cylindrical plugs into smaller blocks, if necessary.
5. Place the plugs or blocks in 50 volumes of lysis buffer and incubate for 24 h at 50°C with shaking at 60 rpm. Replace the old lysis buffer with fresh lysis buffer and continue to incubate at 50°C for 24 h with shaking at 60 rpm.

Notes: Cells or protoplasts are lysed and large DNA molecules are released, which remain trapped within the agarose matrix and are protected from mechanical shearing. The degraded cell materials diffuse out of the agarose matrix. After lysis, the plugs or blocks can be stored at 4°C in fresh lysis buffer for years.

6. Rinse the plugs four times in $TE_{10.5}$ buffer and store at 4°C until use.

C. Isolation of Intact Yeast DNA

1. Harvest yeast cells from a liquid culture by centrifugation at 1000 g for 5 min at 4°C and decant the supernatant.
2. Wash the cell pellet by resuspending the cells in five volumes of ddH$_2$O and centrifuge as in step 1. Repeat washing once.
3. Resuspend the cells in 50 mM EDTA buffer (pH 8.0) at approximately 4×10^9 cells/ml on ice.

4. Prepare an equal volume of 1% (w/v) low-melting temperature agarose gel in ddH$_2$O. Melt in a microwave and allow to cool to 42°C.
5. Warm an equal volume of yeast cell suspension to 42°C and add to an equal volume of agarose gel mixture (42°C). Mix well and pour the mixture into a plug mold on ice. Allow to harden at 0°C for 30 min.
6. Carefully transfer the plugs into ten volumes of SCEM buffer containing 1 unit/ml of Zymolyase 20-T, and incubate at 37°C for 5 h.
7. Replace SCEM buffer with ten volumes of DLS buffer and incubate at 50°C for 3 h. Replace the old DLS buffer with fresh DLS buffer and incubate for another 3 h.
8. Rinse the plugs four times with four volumes of TE$_{10.5}$ buffer and store at 4°C until use.

D. Isolation of Yeast DNA for PCR Screening

1. Carry out steps 1 to 8 in **Section III.C**.
2. Dilute the plugs with ten volumes of ddH$_2$O and boil for 5 min and quickly centrifuge at 10,000 g for 4 min at room temperature.
3. Transfer the supernatant to a fresh tube, measure the DNA concentration (about 2 to 3 ng/µl), and have it ready for PCR screening.

E. Partial Restriction Enzyme Digestion of DNA in Agarose

The DNA purified in agarose plugs or blocks is almost intact and should be partially digested with a rare cutting enzyme used for YAC cloning or be completely digested with an appropriate enzyme for PFGE analysis.

1. Complete the restriction digestion for PFGE:
   a. Incubate agarose plugs in 50 volumes of TE$_{10.1}$ buffer (pH 7.6) at room temperature for 30 min.
   b. Transfer the plugs to individual microcentrifuge tubes and add ten volumes of appropriate 1X restriction enzyme buffer to each tube. Incubate the tubes for 30 min at 4°C.
   c. Remove the buffer and add two volumes of the same fresh 1X restriction enzyme buffer. Add 40 to 50 units of the appropriate restriction enzyme (e.g., *EcoR* I) to each tube and incubate at the optimal temperature for the enzyme overnight.
   d. Soak the plugs in 50 volumes of cold TE$_{10.1}$ buffer (pH 7.6) at 4°C for 1 h in order to diffuse out any salt in the restriction buffer from the plugs. The plugs can be individually loaded onto PFGE.
2. Partial *EcoR* I restriction enzyme digestion of genomic DNA for YAC cloning:
   Due to the STAR activity of *EcoR* I, special care must be taken not to incubate the digestion for too long of a period. The shorter incubation period prevents the enzyme from nonspecifically shortening the DNA fragments.
   a. Rinse plugs containing DNA three times in 50 volumes of 1X restriction enzyme buffer lacking Mg^{2+}.
   b. Remove the buffer and add one volume of 1X restriction buffer lacking Mg^{2+} with 4 units/µg of *EcoR* I at 4°C for 1 h.
   c. Add Mg^{2+} from a stock solution of 100 mM MgCl$_2$ to the desired concentration to initiate *EcoR* I digestion.
   d. Immediately incubate at 37°C for 1 h.
   e. Stop the reaction by removing the restriction buffer and adding ten volumes of cold TE$_{10.5}$ buffer. Store at 4°C until use.

Genomic DNA Libraries

F. Preparation of YAC Vectors for Cloning

YAC4 is the most widely used yeast artificial chromosome vector that is propagated as bacterial plasmids pYAC4 in *E. coli*.

1. Prepare an *E. coli* culture containing pYAC4. The procedure for incubation of *E. coli* is described in **Chapter 1**.
2. Linearize pYAC4 with the restriction enzyme *BamH* I in order to release a HIS3 spacer fragment between the telomeres. The *BamH* I site is then dephosphorylated to prevent ligation between the telomeres and the HIS3 spacer. The procedures for *BamH* I digestion and dephosphorylation with CIAP are described in the section on subcloning of insert DNA in a plasmid vector (**Section II.E**).
3. Extract the linearized and dephosphorylated pYAC4 with phenol/chloroform, chloroform:isoamyl alcohol (24:1); precipitate in ethanol; and dissolve the DNA in ddH$_2$O as described for plasmid DNA isolation in **Chapter 1**.
4. Carry out restriction enzyme digestion with *EcoR* I to open the cloning site in the intron of the *SUP*4tRNA gene as described previously.
5. Extract, precipitate, and resuspend the DNA in ddH$_2$O as in step 3. Store at −20°C until use.

G. Ligation of Partially Digested Genomic DNA Insert to pYAC4 Vector

1. Briefly rinse the agarose plug containing partially digested genomic DNA twice with 20 volumes of ddH$_2$O followed by 10 volumes of 1X ligation buffer.
2. Discard the buffer and add the prepared pYAC4 vector to the plug at a vector-DNA:insert-DNA molar ratio of 40:1. The mass of vector is approximately equal to the mass of insert.
3. Melt the plugs in a 68°C water bath for 5 min and transfer to 37°C.
4. Preheat 2X ligation buffer containing 4000 units/ml DNA T4 to 37°C for 2 min and add one volume of the buffer to the gel mixture prepared in step 3. Gently mix and allow the ligation reaction to incubate at 37°C for 2 to 3 h.
5. Transfer the reaction mixture to room temperature and continue to incubate the reaction overnight.

Note: Longer incubation at room temperature is acceptable.

H. Size Fractionation of DNA by CHEF Gel or Other PFGE

It is important that the ligated reaction be size-fractionated prior to transformation. In 1% (w/v) low-melting-point agarose CHEF gel, apply switching conditions to retain fragment above a particular size in a compression zone. By use of switching times of 15 sec on the CHEF apparatus, DNA fragments <300 kb are allowed to migrate as a function of their sizes. However, fragments >300 kb migrate more slowly in a compression zone without resolution. Electrophoresis is carried out using 0.5X TBE at 10°C.

Notes: After electrophoresis and staining, multiple DNA bands should be visible. They are, from the bottom to the top, HIS3 spacer fragment, unligated left or right YAC4 arms, ligated right-to-right arms, ligated left-to-right arms, ligated left-to-left arms, <300-kb zone, and 300- to 1500-kb compression zone from which the ligated DNA can be recovered.

I. Hydrolysis of Agarose by Agarase

Agarase treatment can be directly carried out inside the agarose matrix to hydrolyze agarose and to release the recombinant YAC/DNA insert. The solution containing the ligated YAC/DNA and oligosaccharide can be directly transformed into yeast spheroplasts without purification.

1. Equilibrate the agarose slice or block in 30 mM NaCl in ddH_2O.
2. Discard the solution and place in a 68°C water bath until the agarose has melted.
3. Transfer to 37°C and add agarase (40 to 80 units per gram of agarose) to the melted agarose mixture.
4. Incubate the reaction mixture at 37°C for 2 to 3 h. Store at room temperature or 4°C until use for transformation.

J. Preparation of Spheroplasts for Transformation

1. Inoculate a single AB1380 colony (yeast host) into 200 ml of YPD medium and incubate at 30°C until the culture has reached mid-log growth phase.
2. Harvest the cells by centrifuging at 900 g for 5 min at room temperature.
3. Discard the supernatant and resuspend the cells in 20 ml ddH_2O. Centrifuge as in step 2. Repeat rinsing once.
4. Resuspend the cells in 10 ml of 1 M sorbitol solution, count the cells under the microscope in the hemocytometer, and centrifuge as in step 2.
5. Resuspend the cells in 5 ml of SCEM and add Zymolyase 20-T (5 to 20 units/1.5 × 10^9 cells). Incubate at 30°C for 15 min with gentle shaking at 60 rpm.
6. Centrifuge at 500 g and wash the pellet once in 10 ml of 1 M sorbitol solution.
7. Centrifuge as in step 6 and resuspend the cells in YPD medium containing 1 M sorbitol. Allow the cells to recover for 30 min at room temperature.
8. Add 5 ml of STC and centrifuge as in step 6. Wash the cells in 10 ml of STC buffer, centrifuge, and resuspend the cells in 5 ml of STC buffer.
9. Check the spheroplasts with a phase-contrast microscope. A good preparation should have <5% of lysed cells in the STC buffer, but 100% lysed cells in water added to the slide. The cells at this point are stable and ready for transformation.

K. Transformation of Spheroplasts with Recombinant YAC/DNA Insert

1. Add 0.5 to 1.0 volume of 2 M sorbitol solution to the size-fractionated, agarase-treated liquid mixture (**Section I**) and aliquot into 10-ml tubes (15 µl per aliquot).
2. Add 0.1 ml of the cell suspension (see **Section J**) to each of four aliquots, gently mix, and incubate at room temperature for 15 min.
3. Add 1 ml of PEG solution to each tube and incubate at room temperature for 15 min.
4. Centrifuge at 500 g for 10 min at room temperature and carefully discard the supernatant.
5. Resuspend the cell pellet in 0.15 ml of SOS per tube and incubate at 30°C for 45 to 60 min.
6. Add 3 ml of TOP lacking uracil, which is prewarmed to 40°C, to each tube and transfer to petri dishes containing SORB without uracil.
7. Incubate at 30°C for 3 to 4 days until transformants appear.

Note: *YAC cloning in yeast has a low transformation efficiency, which is usually 3 × 10^3 YAC transformants per microgram DNA. This low transformation efficiency is due to the large size of the transforming DNA as well as the particular morphology of yeast.*

Genomic DNA Libraries

L. Verification of YAC Transformants

1. Transfer individual, primary transformants onto SD medium without both uracil and tryptophan.
2. Incubate at 30°C with shaking. Only YAC positive clones can grow on SD medium because of the expression of the tryptophan gene and because they have the red phenotype feature due to the interrupted *SUP*4tRNA gene.
3. Verify the YAC clones by PFGE and by Southern blot hybridization using genomic DNA insert fragments as probe(s).

M. Amplification and Storage of the YAC Library

1. Triplicate storage of arrayed YAC libraries used for large-scale physical mapping is carried out as follows:
   a. Incubate single colonies in 0.6 ml of YPD medium in Micronics™ racks (96 × 1 ml) at 30°C with agitation for 36 to 40 h. Incubate the cultures in SD medium for colony filters and PCR pools.
   b. Add 0.2 ml of 80% glycerol to each culture and mix well.
   c. Transfer 0.2 ml of the cell suspension to each of three microtiter plates. One is used as the master library, and the other two are used as working libraries. Wrap the plates with SaranWrap and quickly store at −80°C.
   d. Cap the Micronics and store at −80°C.
2. Nonorganized storage of a pooled YAC library used for identifying one or a few of YAC clones of interest is carried out as follows:
   a. Scrape off and pool approximately 500 colonies grown on SD medium lacking uracil and tryptophan. Incubate these cells in 500 ml of YPD medium at 30°C for 6 to 10 h with agitation.
   b. Remove about 7×10^8 yeast cells in duplicate to make approximately 20 µg total DNA in 2 ml for 500 to 1000 PCR reactions.
   c. Centrifuge the remaining part of the culture at 900 g for 10 min at room temperature.
   d. Carefully discard the supernatant and resuspend the cell pellet in 20% (v/v) glycerol in YPD medium and store the pooled yeast in aliquots at −80°C.
   e. Incubate approximately 2500 colonies from an aliquot on SD medium lacking uracil at 30°C with agitation for 2 days.
   f. Make five replicates on nylon filters. Two of the colony filters are stored on 3MM Whatman paper soaked with 20% (v/v) glycerol at −80°C. Three colony filters are to be used for colony screening.

N. Screening of a YAC Library

1. PCR screening of pooled YACs for exclusion of a large part of the library is carried out as follows:
   a. Incubate superpools of 1920 colonies in 20 of 96-well microplates at 30°C for 36 h.
   b. Isolate DNA from yeast cells as described in **Section III.D**.
   c. Carry out PCR reactions using approximately 20 to 40 ng DNA from a yeast pool in a final volume of 20 µl overlayered with a drop of light mineral oil unless a heated lid is available. The PCR procedure is described in detail in **Chapter 9**. The PCR cycles depend on the primers and the size of the PCR products. The conditions need to be optimized.
   d. Analyze the PCR products on 1.0 to 1.5% (w/v) agarose gels or Seckem gels. Positive pools should have the expected PCR band as compared with the positive control lane and the marker lane.
   e. Repeat steps a to d by screening 96 colonies included in the positive pool.
   f. Store the positive colony pool in 20% (v/v) glycerol at −80°C.
2. Colony screening on a nylon membrane is carried out as follows:
   a. Thaw the glycerol stocks from the positive pool of the PCR screening and incubate in 0.15 ml SD medium without uracil in microtitre plates at 30°C for 3 days.

b. Inoculate a charged nylon membrane with 96 or 384 YAC clones from cultures at step a.
c. Place the inoculated membranes on SD medium without uracil and trytophan on 22.5 × 22.5 cm plates and incubate at 30°C for 3 days or until colonies are approximately 2 mm in diameter.

Note: *Avoid any air bubbles under the membranes in these procedures.*

d. Individually transfer the membranes onto the 22 × 22 cm 3MM Whatman paper plates, saturated in SCEM containing 0.5 unit/ml Zymolyase 20-T for at least 30 min. Seal the plates with Parafilm and incubate at 30°C overnight.
e. Incubate the membrane at room temperature on 3MM Whatman paper saturated as follows:

Saturated with	Incubation Time
10% (w/v) SDS	5 min
0.5 N NaOH	10 min
Transfer onto dry 3MM Whatman paper	5 min
0.2 M Tris-HCl, pH 7.5, 2X SSC	3 × 5 min

f. Air dry the membranes for 2 h or under vacuum for 1 to 2 h at 80°C. Store the membrane in aluminum foil or in prehybridization buffer until use.
g. Prehybridize the membranes at 65°C for 3 to 6 h in 20 ml per filter prehybridization solution with 7% (w/v) PEG 8000, 10% (w/v) SDS, and 100 µg/ml of sonicated denatured salmon sperm DNA.
h. Hybridize the membrane at 65°C using fresh prehybridization buffer with at least 3×10^5 cpm/ml probe.
i. Carry out washing and exposing the membranes as described for Southern blotting in **Chapter 5**. Positive clone(s) should be visualized as black spot(s).
j. Store individual positive clones in 20% (v/v) glycerol at –80°C.
3. Verification of positive YAC clones is carried out as follows:
   a. Streak out individual positive clones from the glycerol stock on SD medium lacking uracil and incubate at 30°C for 3 days.
   b. Inoculate individual colonies (usually two to four) in 5 ml of YPD medium and incubate at 30°C overnight with agitation.
   c. Harvest the cells and extract the DNA as previously described.
   d. Carry out PFGE using two to four of the agarose plugs prepared, followed by Southern blot hybridization.
   e. Check the insert size of the genomic DNA in the positive YAC clones.
   f. Prepare 20% (v/v) glycerol stocks of the putative clones and store at –80°C. In the future, the stocks can be used for genomic and physical mapping.

Reagents Needed

Agarase Stock Solution

2000 units/ml in ddH$_2$O with 50% glycerol
Store at –20°C.

Amino Acids

AAs (Sigma)TRP⁻	URA⁻ (600 mg/l)	URA⁻ (520 mg/l)
Ade A 9795	1	1
Arg A 3909	4	—
His H 9511	2	2
Iso I 7383	6	6
Leu L 1512	6	6
Llys L 1262	5	5
Met M 2893	2	2
Phe P 5030	5	5
Thr T 1645	20	20
Trp T 0271	4	—
Tyr T 1020	5	5
Total	60	52

Cell Suspension Buffer

10 mM Tris-HCl, pH 7.6
100 mM EDTA, pH 8.0
20 mM NaCl

Clorox Solution

5 to 10% (v/v) Clorox (sodium hypochlorite solution or commercial bleach) in ddH_2O

CPW-Salt Solution (1 L)

KH_2PO_4 (27.2 g)
KI (0.16 mg)
$CuSO_4 \cdot 5H_2O$ (0.025 mg)
KNO_3 (0.101 g)
$MgSO_4 \cdot 7H_2O$ (0.246 g)

DLS Buffer

1% (w/v) dodecyl lithium sulfate (DLS)
50 mM NaCl
10 mM Tris-HCl, pH 7.8

10X EcoR I Buffer

100 mM NaCl
10 mM Tris-HCl, pH 7.9

Enzyme Medium

9% (w/v) Mannitol
3 mM 2-(N-Morpholino)-ethane-sulfonic acid (MES)-KOH, pH 5.8
1% (w/v) Cellulase
0.2% (w/v) Macerozyme
Make up in CPW-salt solution.

10X Ligase Buffer

300 mM Tris-HCl, pH 7.8
100 mM MgCl$_2$
100 mM DTT
10 mM ATP

Lysis Buffer

0.5 M EDTA, pH 8.0
1% (w/v) N-Laurylsarcosine
1 mg/ml Proteinase K

PEG Buffer

20% (w/v) PEG 8000
10 mM Tris-HCl, pH 8.0
10 mM CaCl$_2$
Filter-sterilize.

Phosphate-Buffered Saline (PBS)

NaCl (8 g)
KCl (0.2 g)
Na$_2$HPO$_4$ (1.44 g)
KH$_2$PO$_4$ (0.24 g)
Dissolve well after each addition in 800 ml ddH$_2$O.
Adjust the pH to 7.4 with 2 N HCl and add ddH$_2$O to 1 L.
Autoclave and store at room temperature.

SCEM Buffer

1 M Sorbitol
10 mM EDTA, pH 8.0
100 mM Sodium acetate, pH 5.8
Autoclave, cool to 30°C, and add 30 mM 2-mercaptoethanol.

SD Medium (Yeast Synthetic Drop-Out Medium Supplement)

Order from Sigma Chemical Co.: Cat. No. Y1876 is medium supplement without tryptophan, and Cat. No. Y 1501 is medium supplement without uracil.

SORB Buffer

0.9 M Sorbitol
3% (w/v) D-Glucose
0.67% (w/v) Yeast nitrogen base lacking amino acids
1.5 to 2% (w/v) Bactoagar
Adjust the pH to 5.8. Autoclave.
Cool to 37°C and add the amino acids required.

2 M Sorbitol Solution

2 M Sorbitol in ddH$_2$O
Autoclave.

SOS Buffer

1 M Sorbitol
6.5 mM CaCl$_2$
0.25% (w/v) Yeast extract (Difco)
0.5% (w/v) Bactopeptone (Difco)
20 μg/ml Uracil and tryptophan
Adjust the pH to 5.8. Filter-sterilize.

STC Buffer

1 M Sorbitol
10 mM Tris-HCl, pH 8.0
10 mM CaCl$_2$
Autoclave.

Sucrose Solution

18% (w/v) Sucrose
3 mM MES-KOH, pH 5.8
Make up in CPW-salt solution. Autoclave.

TE$_{10.5}$ Buffer

10 mM Tris-HCl, pH 7.8
5 mM EDTA, pH 8.0

TOP

1 M Sorbitol
2% (w/v) D-Glucose
0.67% (w/v) Yeast nitrogen base lacking amino acids
1% (w/v) Bactoagar
Adjust the pH to 5.8. Autoclave.
Cool to 37°C and add the amino acids required.

Washing Medium

3 mM MES-KOH, pH 5.8
2% (w/v) KCl
Make up in CPW-salt solution. Autoclave.

YPD Medium

1% (w/v) Yeast extract (Difco)
2% (w/v) Bactopeptone (Difco)
2% (w/v) D-Glucose
1.5 to 2% (w/v) Bactoagar (Difco)
Adjust the pH to 5.8. Autoclave.

Zymolyase 20-Ton Stock Solution

1000 units/ml in ddH$_2$O
Filter-sterilize and store at 4°C.

IV. Troubleshooting Guide

Symptoms	Solutions
Plaques are small on one side and large on other side	This is due to uneven distribution of the top agar mixture. Make sure that the top agar mixture is spread evenly.
Too many positive clones occur in the primary screening	A nonspecific cross-linking problem occurs. Try to use high-stringency conditions of hybridization.
Black background on the filter	Filter was partially dried, or blocking efficiency is low, or the quality of the filter is not good. Try to avoid air drying of the filter; increase the percentage of BSA and denatured salmon sperm DNA, and try to use a fresh, neutral, nylon membrane filter.
Unexpected black spots on the filter	Unincorporated ^{32}PdCTP filter was not efficiently removed. Try to use a G-50 gel column to purify the labeled DNA.
No signal occurs at all in any of the plates used for the primary screening procedure	The pfu numbers used for each plate are too low. A threshold of 2×10^5 pfu should be used in the primary screening of the library.
Signals are weak on the filters	The efficiency of labeling is low, or the x-ray film exposure time is not long enough.
No signals occur at all on the later rescreening of the plaque picked up from the primary screening	False positive clones were most likely picked up due to mismatch of the exposed film with the original LB plate.

V. Genomic Cloning Using PCR

Under some circumstances, one may be able to take advantage of genomic PCR cloning techniques to isolate new genomic clones. This is usually the case when a good amount of sequence information is known about the cDNA of interest, such as nucleotide sequence, relatedness of the cDNA sequence to other genes, and possible intron positions as determined by comparison with related gene sequences. There is a wide variety of PCR techniques that address this issue. In this section, we suggest two possibilities that have worked well in our laboratories.

A. Selection of Oligonucleotides

The same rules apply to oligonucleotide primer design as detailed in **Chapter 9** and **Chapter 11**. However, when dealing with the isolation of genomic clones, one must make some special considerations.

1. If possible, any new cDNA sequence should be compared with other related genes whose intron positions have already been determined. This will help the researcher avoid possible intron junctions that could interrupt the primer sequence and cause the primer not to bind in that region. These regions should be avoided during primer design.
2. Based on the sequence of the cDNA, selection of primer sets at the very beginning of the gene of interest and at the very end may allow cloning in only one PCR amplification. However, since genomic clones in animals and especially in plants often have large introns, it is wise to create additional primer sets that can be used to amplify the 5′ and 3′ ends of the genomic sequence separately. If an overlapping region is included, then these two regions can later be assembled using unique restriction sites or PCR to extend the ends of the clones in reactions containing a mixture of the 5′ and 3′ clones. We recommend setting up reactions for the 5′ end, the 3′ end, and the entire clone at the same time using several sets of primers at each location. This gives better odds of successful amplification.

3. Due to the low abundance of each specific genomic sequence, especially for single-copy genes, the initial cycles in the PCR reactions are critical. To provide the best results, primers should carefully be designed to have almost identical annealing temperatures. A melting temperature of between 60 and 70°C usually works well. A discussion of the calculation and design of primer melting temperatures can be found in **Chapter 9**.

B. PCR Amplification of Genomic DNA and Genomic Libraries

Some of the most important factors in performing successful genomic PCR are the selection of primers, the quality of the DNA being used, the amount of DNA used, and the quality of the *Taq* polymerase used. Primers that are too far apart due to large introns or those that fall on intron junctions will not work. Only highly pure DNA works well in the PCR process, and too much DNA often inhibits amplification. Procedures for the preparation of high-quality genomic DNA are found in **Chapter 1**. Finally, the enzyme used is critical. It must be capable of producing large DNA fragments up to 10 kb in length. We prefer to use either the Advantage™ 2 Polymerase Mix or the Advantage Genomic Polymerase Mix from Clontech, Palo Alto, CA. Both enzymes are a mixture of a robust *Taq* polymerase and a proofreading enzyme that helps to provide longer and more accurate amplifications. The Advantage Genomic Polymerase Mix also comes with a bound antibody that prevents activity until the first heat cycle, which gives it an intrinsic "hot start." Other companies have similar high-quality enzymes.

Due to the low abundance of most genomic sequences and the number of cycles required in this method, even very slight contamination with the cDNA clone can cause difficulties in the amplification of the genomic clone. Usually, optimization of the PCR reactions is necessary before the reactions are highly efficient. The following procedure covers these criteria and can be applied to any genomic DNA template or λDNA isolated from a prepared genomic library.

1. Using procedures detailed in **Chapter 1**, isolate genomic DNA or λDNA from one of the above prepared genomic DNA libraries.
2. Carry out PCR amplification.
   a. Design oligonucleotide primers that are specific and complementary to two different regions of the gene of interest following the criteria detailed above. Degenerate primers designed from protein sequences can also be used, but many nonspecific bands are usually amplified.
   b. In a 0.2-ml thin-walled PCR tube on ice, set up the reactions by adding the following in the order listed below. It is also important to set up negative controls using no DNA template and only one primer at a time.
      - 10X Advantage Genomic Polymerase Buffer (Clonetech) (5 µl)
      - ddH$_2$O (10 µl)
      - Mixture of four dNTPs (1.25 mM each) (8.5 µl)
      - Primer 1 (50 to 100 ng) in ddH$_2$O (5 µl)
      - Primer 2 (50 to 100 ng) in ddH$_2$O (5 µl)
      - Genomic DNA or λDNA in TE buffer (0.2 to 2 µg) (2 µl)
      - Advantage Genomic Polymerase Mix (0.5 µl)
      - Add ddH$_2$O to a final volume of 50 µl.

Notes: *(1) Thin-walled PCR tubes are highly recommended to allow better heat transfer between the thermocycler block and the sample. (2) Depending on the number of samples (including negative controls), a cocktail containing all reagents with the exception of the DNA template should be made to establish uniform conditions between all reactions.*

   c. Overlay the mixture with 50 µl of light mineral oil (Sigma or equivalent) to prevent evaporation of the sample if a thermocycler having a heated lid is not available.

d. Carry out "touchdown" PCR amplification in a PCR machine. The conditions of PCR, including primer annealing temperatures, extension times, and number of cycles, will have to be optimized. However, a good starting example is as follows:

Cycle	Denaturation	Annealing	Polymerization	
First:				
T1	3 sec at 94°C	15 sec at 70°C	5 min at 70°C	2 cycles
T2	3 sec at 94°C	15 sec at 68°C	5 min at 70°C	2 cycles
T3	3 sec at 94°C	15 sec at 66°C	5 min at 70°C	2 cycles
T4	3 sec at 94°C	15 sec at 64°C	5 min at 70°C	2 cycles
T5	3 sec at 94°C	15 sec at 62°C	5 min at 70°C	2 cycles
Subsequent	3 sec at 94°C	15 sec at 60°C	5 min at 70°C	29 cycles
Last	3 sec at 94°C	15 sec at 60°C	10 min at 72°C	

Finally, hold at 4°C until the sample is removed.

Notes: *(1) When using thin-walled PCR tubes, it is generally not necessary to denature the genomic DNA template for longer than 3 sec during each cycle. In fact, exposure to high temperatures tends to ruin the integrity of the template. An initial heat-denature step is also not necessary. (2) The initial ten cycles are at higher temperatures, making the primer annealing more stringent. This enriches the reaction with the template of interest prior to the final 30 cycles. Final primer annealing temperatures will have to be optimized. However, a temperature 3°C below the calculated annealing temperature (Tm) of the primer having the lowest Tm is a good starting point. (3) Other conditions that must be optimized are the extension times and the number of cycles used. Shorter extension times should be used for the amplification of shorter genes. We recommend using as few cycles as possible to amplify the DNA fragment of interest. Taq polymerases (even with a proofreading enzyme) have higher mutation rates, and each subsequent cycle increases the chance of introducing error. (4) Proofreading enzymes also tend to remove the T residues at the end of the synthesized fragments. Consequently, the fragments must be cloned immediately after amplification.*

  e. If using mineral oil, remove the reaction mixture from the mineral oil using a pipette with a relatively long tip. Slowly insert the tip into the bottom of the tube and then carefully take up the sample, leaving the oil phase behind. Withdraw the tip from the tube, wipe the outside of the tip with a clean paper towel, and transfer the sample into a fresh tube.

2. Check the purity of the amplified cDNA on an agarose gel and elute the fragment of interest if required.
   a. Add DNA loading buffer to 10 μl of the amplified cDNA sample, and load the sample into 1% agarose gel, which contains EtBr for staining, including positive and negative controls as well as DNA standard markers. Carry out electrophoresis at 120 V.
   b. Under UV light, inspect the resulting bands to see if an appropriate fragment was amplified.
   c. If a single, clearly resolved band is visualized in the sample of interest, then this DNA can be cloned directly from the PCR reaction using 1 to 7 μl of the PCR reaction and a variety of PCR cloning kits, such as the pGEM Teasy vector (Promega) or the TOPO TA cloning kits (Invitrogen, Carlsbad, CA). (See Figure 9.9 in **Chapter 9**.) We highly recommend this procedure; however, it must be performed immediately after the PCR reactions are finished.
   d. If too many bands are resolved in the sample of interest, then the band of interest will have to be eluted from the gel. Run an appropriate amount of the remaining PCR reaction on a gel, and slice the individual sharp band(s) using a clean razor blade. Trim away excess agarose gel as much as possible.
   e. Procedures for the elution of the bands are detailed in **Chapter 1**.

3. Ligate the fragment to an appropriate TA cloning vector, and transform an appropriate *E. coli* host in preparation for minipreps.
   a. Set up the ligation reaction as follows:
      - Ligase 10X buffer (2 µl)
      - Vectors (0.1 µg)
      - Amplified genomic DNA fragment (0.5 µg)
      - T4 DNA ligase (10 Weiss units)
      - Add ddH$_2$O to a final volume of 20 µl.
   b. Incubate the ligation reactions at room temperature (22 to 24°C) for 4 h and proceed to subcloning, as described in **Section II.E**.
4. Identify correct clones and verify the sequence.
   a. Prepare miniprep DNA of at least 20 colonies after *E. coli* transformation. Cut these preps with flanking restriction enzymes, and run fragments on a 1% agarose gel to identify clones with the correct size fragment. A Southern blot can be prepared using this gel if an appropriate probe can be generated, and restriction maps can be prepared using the miniprep DNA from potential positive clones.

Notes: When performing Southern blots on cloned PCR fragments, it is important to design a probe that lies between but does not include the primers used to create the products. This prevents the hybridization of primer sequences that may flank nonspecific amplified DNA fragments. This is especially important if the fragment was directly cloned out of the PCR reaction without gel elution due to the higher probability of the presence of nonspecific fragments.

   b. Carry out sequencing of the putative clones having the amplified cDNA as described in **Chapter 11**.
   c. Compare the nucleotide sequence and the deduced amino acid sequence with the known genes from which the primers were designed.
   d. If it is determined that a given clone is correct, then glycerol stocks and large-scale DNA preps should be prepared and stored for future use.

C. Isolation of Flanking Sequences by Inverse PCR

In some cases, only a small portion of a gene sequence is known. In these situations, the researcher needs to make efforts to obtain the sequence of the regions of the gene that flank the known sequence. There are a wide variety of approaches to this problem, many of which include PCR. In this section, we describe one technique that has worked well in our laboratory.

Inverse PCR makes use of primers that are designed in the opposite orientation from the direction that they are normally prepared. In other words, the primers face away from each other and are used to amplify surrounding genomic DNA that was cut with restriction enzymes and religated to form a collection (or library) of circular DNAs. The resulting fragments can be as large as 6 kb. The protocol is as follows:

1. Using procedures detailed in **Chapter 1**, isolate genomic DNA from the organism of interest. This DNA must be of high quality.
2. Perform single-enzyme digestions, each with 10 µg of genomic DNA, using several different enzymes. Since unknown sequences are being dealt with, it is not known which enzyme will provide the best results. Four- and six-cutter enzymes will result in different size fragments.
3. Run a small portion of each digest on a 1% agarose gel to confirm that the cutting was complete.
4. If the digestion was successful, extract with one volume of TE-saturated phenol/chloroform. Mix well by vortexing and centrifuge at 11,000 g for 5 min at room temperature.

5. Carefully transfer the upper, aqueous phase to a fresh tube and add one volume of chloroform:isoamyl alcohol (24:1). Mix well and centrifuge as in step 4.
6. Transfer the top, aqueous phase to a fresh tube and add 0.1 volume of 3 M sodium acetate buffer (pH 5.2) or 0.5 volume of 7.5 M ammonium acetate. Mix and add 2 to 2.5 volumes of chilled 100% ethanol. Allow to precipitate at −70°C for 30 min.
7. Centrifuge at 12,000 g for 10 min at room temperature. Carefully decant the supernatant and briefly rinse the DNA pellet with 1 ml of 70% ethanol. Dry the pellets under vacuum for 10 min. Dissolve the DNA in a small amount of ddH$_2$O in preparation for ligation.
8. Carry out ligations as described previously for large-scale ligations in **Section II.A.7**. This will create a collection of circular DNA molecules upon which some will have the partial known sequence of the gene of interest.
9. Repeat steps 4 to 7 to prepare the DNA for PCR. If a unique restriction site is available within the region of known sequence and between the two primers, the circular DNA can be linearized prior to PCR. However, while this may prevent the polymerase from looping around the circular DNAs, we have found that it is not strictly necessary.
10. Design oligonucleotide primers that are specific but in opposite orientations to each other within the region of known sequence following the criteria detailed earlier in this chapter. This should be done well in advance.
11. Perform PCR as described in **Section IV.B**, "PCR Amplification of Genomic DNA and Genomic Libraries," using the prepared primers in each of the ligated samples of cut genomic DNA. Analysis is also done as described in that section. The same primers can be used to sequence the region lying between the primers.

References

1. **Sambrook, J., Fritsch, E.F., and Maniatis, T.,** *Molecular Cloning: A Laboratory Manual*, 2nd ed., Cold Spring Harbor Press, Cold Spring Harbor, NY, 1989.
2. **Tonegawa, S., Brock, C., Hozumi, N., and Schuller, R.,** Cloning of an immunoglobin variable region gene from mouse embryo, *Proc. Natl. Acad. Sci. U.S.A.*, 74, 3518, 1977.
3. **Saiki, R.K., Gelfand, D.H., Stoffel, S., Scharf, S., Higuchi, R., Horn, G.T., Mullis, K.B., and Erlich, H.A.,** Primer-directed enzymatic amplification of DNA with a thermostable DNA polymerase, *Science*, 239, 487, 1988.
4. **Wu, L.-L., Song, I., Kim, D., and Kaufman, P.B.,** Molecular basis of the increase in invertase activity elicited by gravistimulation of oat-shoot pulvini, *J. Plant Physiol.*, 142, 179, 1993.
5. **Erlich, H.A.,** *PCR Technology: Principles and Applications for DNA Amplification*, Stockton Press, New York, 1989.
6. **Innis, M.A., Gelfand, D.H., Sninsky, J.J., and White, T.J.,** *PCR Protocols: A Guide to Methods and Applications*, Academic Press, New York, 1989.
7. **Jones, D.H. and Winistorfer, S.C.,** A method for the amplification of unknown flanking DNA: targeted inverted repeat amplification, *BioTechniques*, 15, 894, 1993.
8. **Sukharev, S.I., Blount, P., Martinac, B., Blattner, F.R., and Kung, C.,** A large-conductance mechanosensitive channel in *E. coli* encoded by *mscL* gene, *Nature*, 368, 265, 1994.

Chapter 11

DNA Sequencing and Analysis

Leland J. Cseke, William Wu, and Peter B. Kaufman

Contents

I.	Preparation of DNA for Sequencing 238	
	A. Preparation of Double-Strand Plasmid DNA 238	
	B. Preparation of Single-Strand Template DNA 239	
	1. Preparation of Single-Strand M13 DNA 239	
	2. Preparation of Single-Strand DNA Using Helper Phage R408 239	
	C. Purification of Double-Strand λDNA 240	
	D. Preparation of Double-Strand DNA Fragments 240	
	E. Preparation of Unidirectional Deletions of the Target DNA 240	
	Reagents Needed 241	
II.	DNA Sequencing by Dideoxynucleotides Chain Termination 243	
	A. General Considerations and Strategies 243	
	1. DNA Polymerase 243	
	2. Radioactively Labeled dNTP 243	
	3. Compressions 243	
	B. Sequencing Reactions 244	
	1. Selection of Oligonucleotides 244	
	2. Sequencing of Double-Strand Plasmid DNA 244	
	3. Sequencing of Single-Strand DNA 246	
	4. Sequencing of Bacteriophage Lambda Linear Duplex DNA 246	
	Reagents Needed 247	
	C. Denaturing Polyacrylamide Gel Electrophoresis 250	
	1. Preparation of Sequencing Gel 250	
	2. Electrophoresis 253	
	3. Autoradiography 255	
	4. Use of Formamide Gels for Sequencing of G-C-Rich Templates 255	
	Reagents Needed 257	
	D. Extending Sequence Far from the Primer 259	
III.	Cycle Sequencing and Automated Sequencing 259	
	A. Selection of Oligonucleotides 260	

	B.	DNA Synthesis and Termination Reactions by PCR 261
		1. Protocol..261
	Reagents Needed..261
	C.	DNA Sequencing Using an Automated Sequencer 262
		1. Sequencing Reactions 263
		2. Sample Preparation for Injection..................................264
		3. Analysis of Sequence Chromatograms.........................265
	Reagents Needed..266
IV.	Troubleshooting Guide...266
	A.	Manual Sequencing: Problems, Possible Causes, and Solutions........266
	B.	Automated Sequencing: Problems, Possible Causes, and Solutions ...268
References...269

Deoxyribonucleic acid (DNA) sequencing is a necessary technique for almost all molecular biology studies, including DNA cloning, characterization, mutagenesis, DNA recombination, and regulation of gene expression.[1-3] There are several well-established methods for nucleic acid sequencing, and in recent years, advances in automated sequencing have made DNA sequence analysis an easy and routine process. The present chapter describes in detail the preparation of DNA for sequencing,[4,5] DNA sequencing by dideoxynucleotide chain termination,[2,3] and cycle sequencing.[1,4,5,6] These techniques are the foundation of most sequencing applications. However, with the advent of automated sequencing and the use of web-based search engines, we also briefly describe how the newer machines and sources have greatly increased the speed of sequence analysis. The following techniques are routinely used in our labs.

I. Preparation of DNA for Sequencing

Two of the most critical factors to obtaining high-quality sequence from any sequencing technique is (1) preparation of proper sequencing oligonucleotides and (2) obtaining highly pure and intact DNA free of contaminating agents. A variety of DNA templates can be used during sequencing, including single-strand DNA, double-strand DNA, λDNA, and DNA fragments isolated from agarose gels. The following sections detail the preparation of each type of template. Considerations for oligonucleotides are described in **Section II**.

A. Preparation of Double-Strand Plasmid DNA

Double-strand plasmids containing DNA inserts of interest can be directly sequenced. Both strands of the DNA insert can be simultaneously sequenced in opposite directions using two different primers, which are annealed to the appropriate sites, in two separate sequencing reactions. This simultaneous sequencing method greatly speeds up the manual sequencing procedure; it is especially useful for sequencing large DNA templates. Cesium chloride gradients, polyethylene glycol (PEG) precipitation, adsorption to glass, columns, and other common DNA purification methods all produce suitable DNA. The cleaner the template is, the higher the success rate will be. We have found that the CsCl gradient method produces excellent results, but it is time-consuming and relatively expensive. The large-scale preparation method, on the other hand, is simple, cheaper, and works equally well in our laboratory if care is taken during phenol/chloroform purification steps. The detailed protocols for isolation and purification of plasmid DNA are described in **Chapter 1**.

DNA Sequencing and Analysis

B. Preparation of Single-Strand Template DNA

Single-strand DNA (ssDNA) used as a template usually reveals excellent nucleotide sequencing results. There are several published methods for preparing single-strand DNA from clones in M13 vectors and hybrid plasmid-phage (phagemid) vectors.[7,8] The following protocol works well for purification of ssDNA from plasmid-phage (phagemid) vectors. Phagemid vectors include M13mp9, M13mp12, M13mp13, M13mp18, and M13mp19.

1. Preparation of Single-Strand M13 DNA

1. Dilute freshly cultured host cells (e.g., JM101) 1:100 in 1.5 ml of 2X YT medium in 4 microcentrifuge tubes.
2. Infect each tube with a purified M13 plaque.
3. Incubate at 37°C with shaking at 160 rpm for 5 to 9 h.
4. Centrifuge at 500 to 1500 g for 10 min at room temperature.
5. Carefully transfer the supernatant containing the phage to fresh tubes and add $^1/_9$ volume of 40% (w/v) PEG-8000 and $^1/_9$ volume of 5 M sodium acetate (pH 7.0) to each tube.
6. Allow the DNA to precipitate at 4°C for 20 to 30 min.
7. Centrifuge at 12,000 g for 15 min and carefully decant the supernatant. Wipe off the inside of the tube to remove PEG as much as possible using a clean Kimwipe™.
8. Suspend the pellet in each tube in 0.2 ml of TE buffer (pH 8.0) and add 0.2 ml of 50 mM Tris-HCl-saturated (pH 7.5) phenol:chloroform (3:1). Mix well by vortexing.
9. Centrifuge at 10,000 g for 5 min and transfer the upper, aqueous phase to fresh tubes.
10. Repeat step 9 once.
11. Extract the supernatant with one volume of chloroform:isoamyl alcohol (24:1). Centrifuge as in step 9. Transfer the supernatant to a fresh tube.
12. To the supernatant, add 0.1 volume of 3 M sodium acetate (pH 5.2) and 2 to 4 volumes of chilled 100% ethanol. Mix well and place at –70°C for 20 to 30 min.
13. Centrifuge at 12,000 g for 15 min, decant the supernatant, and briefly rinse the pellet with 1 ml of 70% ethanol. Dry the DNA pellet under vacuum for 30 min.
14. Dissolve the DNA in 50 to 100 µl of TE buffer. Measure the concentration of the DNA sample and check the quality and quantity by agarose gel electrophoresis. Store the sample at –20°C until sequencing is performed.

2. Preparation of Single-Strand DNA Using Helper Phage R408

An alternative method to the use of M13 helper phage is the use of R408 helper phage. Both phages yield good-quality DNA for use in sequencing.

1. Plate out the transformants of an appropriate *E.coli* strain that have the putative phagemids containing the DNA insert of interest on an LB plate. Invert the plate and incubate at 37°C in order to obtain single colonies.
2. Inoculate a single colony in 5 ml of LB medium containing 75 µg/ml ampicillin and 12 µg/ml tetracycline or appropriate antibiotics. Incubate at 37°C overnight with shaking at 150 rpm.
3. Add 0.3 ml of the overnight culture to 3 ml of superbroth in a 50-ml conical tube. Incubate at 37°C with shaking at 150 rpm for 2 to 3 h.
4. Add 8 µl of helper phage R408 (pfu = 1×10^9, available from Stratagene, LaJolla, CA) to the culture at step 3 and continue to incubate for 8 to 10 h.
5. Transfer 1.5 ml of the culture to each of two microcentrifuge tubes and centrifuge at 11,000 g for 2 min.
6. Transfer 1.2 ml of the supernatant from each tube into a fresh tube and add 0.3 ml of PEG precipitation buffer to the supernatant. Vortex for 1 min and leave at room temperature for 20 min.
7. Centrifuge at 12,000 g for 15 min and decant the supernatant completely.
8. Resuspend the PEG pellet in 0.3 ml of TE buffer (pH 8.0) and extract it with one volume of phenol:chloroform:isoamyl alcohol (25:24:1). Mix by vortexing and centrifuge at 11,000 g for 5 min.

9. Transfer the top, aqueous phase to a fresh tube and repeat extraction as in step 8 twice.
10. Precipitate the single-strand DNA by adding 0.5 volume of 7.5 M ammonium acetate (pH 7.5) and 2.5 volumes of chilled 100% ethanol to the supernatant. Place at –70°C for 30 min.
11. Centrifuge at 12,000 g for 20 min at 4°C, decant the supernatant, and briefly rinse the DNA pellet with 1 ml of 70% ethanol. Dry the pellet under vacuum for 30 min and dissolve the DNA in 10 μl of TE buffer (pH 8.0). Combine the two DNA samples into one tube and take 2 to 4 μl to measure the concentration of the single-strand DNA at 260 and 280 nm. Store the sample at –20°C until use.

C. Purification of Double-Strand λDNA

Double-strand bacteriophage and other linear double-strand DNA templates can be directly sequenced without subcloning into plasmid vectors. However, the double-strand DNA should be treated with T7 gene 6 exonuclease (United States Biochemical Corp., Cleveland, OH) in order to generate a single-strand DNA template prior to doing the sequencing reaction. The detailed methods for purification of λDNA are described in **Chapter 1**. The procedure for subsequent DNA treatment is detailed below:

1. Carry out a restriction enzyme digestion of the bacteriophage linear duplex DNA.
   This is strongly recommended since the ends of lambda vectors are usually protected or modified such that the efficiency of T7 gene 6 exonulease degradation is not optimal. However, if the DNA is pretreated with an appropriate restriction enzyme, the T7 gene 6 exonuclease works well. The restriction enzyme chosen should be unique and close to the cloning site of the DNA insert of interest.
   a. Set up a restriction enzyme reaction in a microcentrifuge tube on ice as follows:
      - Purified lambda duplex DNA (2 to 3 μg/μl) (15 μg)
      - 10X restriction enzyme buffer (1.5 μl)
      - Appropriate restriction enzyme (45 to 50 units)
      - Add distilled, deionized water (ddH$_2$O) to a final volume of 15 μl.
   b. Incubate at the appropriate temperature (e.g., 37°C) for 60 min and directly proceed to step 2.
2. Digest the double-strand DNA fragments by the use of T7 gene 6 exonuclease.
   a. Add 75 units of T7 gene 6 exonuclease diluted in TE buffer to 15 μl of the restriction enzyme-digested DNA sample.
   b. Incubate at 37°C for 20 to 30 min.
   c. Stop the reaction by heating at 80°C for 10 to 15 min.
3. Proceed immediately with sequencing reactions as detailed below.

D. Preparation of Double-Strand DNA Fragments

Double-strand DNA fragments obtained from restriction-endonuclease digestions or PCR (polymerase chain reaction) reactions can also be used as templates for sequencing. It is critical in these cases to take steps to remove the restriction-digestion buffers or any excess nucleotides. This can be effectively done using agarose gel purification, as detailed in **Chapter 1**. However, there are a large number of available kits for both gel purification and PCR reaction cleanup. Such kits often give highly pure DNA fragments.

E. Preparation of Unidirectional Deletions of the Target DNA

In case of large sizes of DNA, it takes a longer time to sequence the entire DNA using stepwise sequencing methods based on the design of a series of sequencing primers. However, such large DNA fragments can often undergo simple deletion, giving all the same primers access to different

DNA Sequencing and Analysis

stretches of DNA. Restriction digests — using fortuitive sites or unidirectional deletions using exonuclease III — can be used to generate a series of shorter fragments with overlapping ends. These progressively deleted fragments can then be simultaneously sequenced in a short time. The general principles and procedures are outlined below, and detailed protocols are found in **Chapter 12**.

The principle behind unidirectional deletion is as follows:

1. A recombinant plasmid phagemid, or bacteriophage M13 replicative-form DNA, that contains the cloned DNA of interest is linearized with two appropriate restriction enzymes. Both enzymes should cut the recombinant DNA between one end of the target DNA and the binding site for the universal sequencing primer. One enzyme should cleave near the target DNA and must generate either a recessed 3' end (or 5' protruding end) or a blunt end. The other enzyme should cleave near the binding site for the universal sequencing primer and must generate a four-base protruding 3' end or be filled in with α-phosphorothioate dNTPs.
2. The linearized DNA is progressively deleted with exonuclease III, which only digests the DNA from the blunt or 5' protruding terminus, leaving the 3' protruding (overhang) or α-phosphorothioate-filled end intact. The digestion proceeds unidirectionally from the site of cleavage to the target DNA sequence. The digestion is terminated by removing appropriate amounts of the samples at appropriate time intervals, generating a series and progressive deletions of shorter fragments.
3. The exposed single strands are then cleaved by nuclease S1 or mung-bean nuclease, producing blunt termini at both ends of the DNA fragments.
4. The shortened DNA is then recircularized by using T4 DNA ligase, which is transformed into an appropriate bacterial host. Transformants can be selected with appropriate antibiotics in the culture medium.
5. The recombinant plasmids are purified and subjected to nucleotide sequencing.

Reagents Needed

7.5 M *Ammonium Acetate (pH 7.5)*

Dissolve appropriate amount of ammonium acetate in 50 ml ddH$_2$O. Adjust pH to 7.5 and increase volume to 100 ml.

Chloroform:Isoamyl Alcohol (24:1)

48 ml Chloroform
2 ml Isoamyl alcohol
Mix and store at room temperature.

LB (Luria-Bertaini) Medium

Tryptone (10 g)
Yeast extract (5 g)
NaCl (5 g)
Dissolve in ddH$_2$O and adjust the volume to 1 L. Adjust pH to 7.5 with 2 *N* NaOH. Autoclave.

LB Plates

Add 15 g of agar to 1 L of LB medium and autoclave. When the medium cools to 50 to 60°C, add appropriate antibiotics, mix well, and pour 20 to 25 ml of medium into 100-mm petri dishes. Allow the media to harden in a laminar-flow hood. Store at room temperature for 10 days or at 4°C for up to 2 months.

PEG Precipitation Buffer

3.5 *M* Ammonium acetate, pH 7.5
20% (w/v) PEG, 8000

40% (w/v) PEG-8000

Dissolve 40% (w/v) PEG (mol. wt. 8000) in ddH$_2$O and store at 4°C.

Buffered Phenol

We strongly recommend purchase of saturated phenol (pH 8.0) from a commercial vendor. Most buffered phenol comes with a separate bottle of equilibration buffer to adjust the phenol phase to pH 8.0. To use, add the entire bottle of equilibration buffer to the phenol bottle and 0.1% (w/v) hydroxyquinoline (as an antioxidant). Stir the mixture for 15 min and allow the phases to separate (requires several hours) before use. Store at 4°C.

Caution: *Phenol is highly corrosive and can be absorbed through skin. Wear gloves and work in a fume hood when handling phenol.*

Phenol:Chloroform:Isoamyl Alcohol

Mix one part of buffered phenol with one part of chloroform:isoamyl alcohol (24:1). Top the organic phase with TE buffer (about 1 cm height) and allow the phases to separate. Store at 4°C in a light-tight bottle.

3 M Sodium Acetate Solution (pH 5.2)

Dissolve sodium acetate in ddH$_2$O and adjust pH to 5.2 with 3 M glacial acetic acid. Autoclave and store at room temperature.

5 M Sodium Acetate Solution (pH 7.0)

Prepare as above with appropriate amount and pH 7.0.

Superbroth Medium

Bacto-tryptone (12 g)
Bacto-yeast extract (24 g)
0.4% Glycerol (v/v)
Dissolve in a total volume of 900 ml in ddH$_2$O and autoclave. Cool to about 50°C and add 100 ml of phosphate buffer containing 170 mM KH$_2$PO$_4$ and 720 mM K$_2$HPO$_4$. Autoclave again.

TE Buffer (pH 8.0)

10 mM Tris, pH 8.0
1 mM EDTA, pH 8.0

1 M Tris (pH 7.5)

Dissolve 121.1 g Tris-HCl in 800 ml ddH$_2$O. Adjust pH with 2 N HCl to desired pH. Adjust final volume to 1 l, autoclave, and store at room temperature.

50 mM Tris-HCl-Saturated (pH 7.5) Phenol:Chloroform (3:1)

Mix phenol and chloroform in a ratio of 3:1 and mix with Tris pH 7.5 that has been diluted to 50 mM.

2X YT Medium (1 L)

Bactotryptone (10 g)
Yeast extract (10 g)
NaCl (5 g)
Autoclave. Store at 4°C and warm up to 37°C prior to use.

II. DNA Sequencing by Dideoxynucleotides Chain Termination

Since Sanger et al.[2] developed the dideoxynucleotides chain termination method of DNA sequencing, it has been modified and well established by the use of a superior enzyme and superior DNA cloning vectors. The general principles for this method include:

1. A synthesized oligonucleotide primer anneals to the 3' end of the DNA template to be sequenced.
2. A DNA polymerase catalyzes *in vitro* the synthesis (5' → 3') of a new DNA strand that is complementary to the template starting from the primer site using deoxynucleoside 5'-triphosphates (dNTPs); one of the dNTPs is usually α-^{35}SdATP, α-^{35}SdCTP, α-^{32}PdATP, or α-^{32}PdCTP.
3. After the synthesis reaction of the new-strand DNA has been carried out for a random time period, it is terminated by the incorporation of a nucleotide analog that is an appropriate 2',3'-dideoxynucleotide 5'-triphosphate (ddNTP).

All four ddNTPs lack the 3'-OH group, which is required for DNA chain elongation. Based on the nucleotide bases of the DNA template, one of the four ddNTPs in each reaction is used, and the enzyme-catalyzed polymerization is then terminated at each site where the ddNTP is incorporated, generating a population of chains with different sizes. Therefore, by setting up four separate reactions, each with a different ddNTP, complete nucleotide sequence information for the DNA strand will be revealed after electrophoresis and autoradiography.

A. General Considerations and Strategies

1. DNA Polymerase

We strongly recommend the use of the latest version of Sequenase DNA polymerase (United States Biochemical, or USB). Sequenase Version 2.0 DNA polymerase is a superior enzyme that is genetically modified from wild-type T7 DNA polymerase. It has no 3' → 5' exonuclease activity and has the properties of high processivity and high purity. It is fast and simple, and it efficiently uses nucleotide analogs for sequencing (e.g., ddNTPs, alpha-thio dNTPs, dITP). Also, it has less radioactivity background compared with that observed by using avian myeloblastosis virus (AMV) reverse transcriptase or the large fragment of *E. coli* DNA polymerase I (Klenow enzyme).

2. Radioactively Labeled dNTP

^{32}P-labeled dNTP (dATP or dCTP) has a high energy level and a short half-life (usually 14 days). A major disadvantage of using ^{32}P is that it can give diffuse bands on autoradiographic x-ray film, which can limit the readable information of DNA sequence at the top of the sequencing gel. In contrast, [^{35}S]dATP has a lower energy level and a longer half-life (usually 84 to 90 days), and it can improve autoradiographic resolution. In this chapter, we focus on the use of [^{35}S]dATP for DNA sequencing.

3. Compressions

Compressions represent a common problem in DNA sequencing, primarily due to dG- and dC-rich regions, which cannot be fully denatured during electrophoresis. This usually causes interruption of the normal pattern of migration of DNA fragments. The bands are usually spaced closer than usual (compressed together) or occur further apart than usual, resulting in a significant loss of sequence information. In order to solve this problem, we use dITP or 7-deaza-dGTP (USB) to replace the nucleotide dGTP, which forms a weaker secondary structure that can be readily denatured during electrophoresis. We have found that, with this modification, the bands are sharper, and compressions were eliminated by the use of dITP. In some cases, some bands appear to be weak using both dGTP and dITP. This limitation can be eliminated by using pyrophosphatase (USB) in the presence of Mg^{2+}, Mn^{2+}, or both. The manganese can improve band uniformity and sequence information close to the priming site.

B. Sequencing Reactions

1. Selection of Oligonucleotides

Sequencing primers must be selected carefully for proper location and specificity. For sequencing a region of the DNA template close to the cloning site of the vector, the primers are complementary to the specific vector sequence and are generally identified on vector maps. These primers are most often commercially available from USB, Promega, (Madison, WI) Biosciences Clontech (Palo Alto, CA), and other companies that supply the vectors. For the sequence beyond 250 to 500 nucleotides from the cloning site, new primers should be designed based on the nucleotide sequence information obtained from the last sequencing of the DNA template. The designed primers, also called extending primers, can be synthesized with a DNA synthesizer or, more commonly, they are ordered from one of many companies that specialize in oligonucleotide synthesis. Such primers should be designed approximately 100 bp (base-pairs) upstream of the start of the region of desired sequence data, because small sequencing fragments are difficult to resolve on sequencing gels. The specificity of primers is dependent on the initial quality of the DNA sequence and the length of the primers. Sequencing primers are typically 16 to 30 bp in length, with longer primers often giving better results. Also, when sequencing genomic DNA, one must take into consideration the possible locations of introns that may interrupt the sequence of primers based on cDNAs.

We recommend that the following criteria be considered while designing 16- to 30-bp sequencing primers. First, highly accurate DNA sequence is essential to ensure that the primer sequence exactly matches the sequence of the template DNA. The region chosen for the oligonucleotide should have a G/C content between 40 and 60%, with an annealing temperature (Tm) of at least 50°C. The Tm can be estimated using the following formula.

$$Tm = 4(G/C) + 2(A/T) - 12°C$$

We prefer the last six nucleotides at the 3' end of the primer to be exactly 50% G/C, with the very last base being a G or a C.

2. Sequencing of Double-Strand Plasmid DNA

1. Denature plasmid DNA by the alkaline-denaturation method, which provides good results. Transfer an appropriate volume of the purified plasmid sample (approximately 1 μg/μl) to a microcentrifuge tube and add one volume of freshly prepared denaturation solution containing 0.4 M NaOH and 0.4 mM EDTA (pH 8.0) to the sample. Incubate at 37°C for 30 to 40 min.

Notes: *The amount of DNA to be denatured should be in excess to the amount of DNA to be sequenced. For example, if using 3 to 5 μg DNA for one primer sequencing reaction, the amount of DNA for two primer sequencing reactions for both strands at opposite directions will be 6 to 7 μg. If the yield of DNA following denaturation, neutralization, and precipitation is 80%, the amount of DNA used for denaturation should be 7.2 to 8.4 μg. We routinely use double the amount of DNA for denaturation and measure the DNA concentration after it is precipitated. This ensures that one has a sufficient amount of DNA for sequencing reactions.*

DNA Sequencing and Analysis

2. Add 0.1 volume of 3 M sodium acetate buffer (pH 5.2) to the denatured sample to neutralize the mixture.
3. Add two to four volumes of 100% ethanol to the mixture and allow precipitation to occur at –70°C for 20 min.
4. Centrifuge at 12,000 g for 15 min, decant the supernatant, and briefly rinse the DNA pellet with 1 ml of chilled 70% ethanol to wash away the salt. Dry the DNA under vacuum for 10 min to completely evaporate the ethanol.
5. Dissolve the DNA in 15 μl ddH$_2$O and place the tube on ice. Quickly measure the concentration of DNA using 2 to 3 μl of the sample at 260 and 280 nm. Immediately proceed to primer annealing.

a. **Setup of the Template-Primer Annealing Reaction**
   1. Transfer 3 to 5 μg freshly denatured plasmid DNA to each of two microcentrifuge tubes on ice for the two opposite primers. Add ddH$_2$O to a total volume of 7 μl in each tube.
   2. To each tube, add 2 μl of 5X reaction buffer and 1 μl of appropriate primer. This gives approximately 1:1 (template:primer) molar stoichiometry. Each tube contains a total of 10 μl mixture.
   3. Place the tubes in a plastic rack or equivalent and heat at 65°C in a water bath for 2 to 3 min. Quickly transfer the rack together with the tubes to a beaker or tray containing an appropriate volume of 60 to 63°C water, which allows one to slowly cool the sample to <30°C or room temperature over 20 to 30 min. If one uses a heating block, the tubes can be heated at 65°C for 2 min followed by slow cooling to room temperature by turning off the heat. The heating block will slowly cool down. When the temperature drops to <30°C, the annealing is completed. The real annealing temperature is 50 to 52°C. Some laboratories prefer to anneal at 50 to 55°C for 15 to 30 min followed by slow cooling.

b. **Preparation for Labeling and Termination**
   1. While the annealing mixture is being cooled, remove the necessary materials from a commercial sequencing kit, which is usually stored at –20°C, and remove [^{35}S]dATP from the freezer (–80°C). Thaw the materials on ice.
   2. Label four microcentrifuge tubes for each template-primer mixture, which are A, G, T, and C that, respectively, represent ddATP, ddGTP, ddTTP, and ddCTP.
   3. Add 2.5 μl of termination mixture of ddATP, ddGTP, ddTTP, and ddCTP to the labeled tubes A, G, T, and C, respectively. Cap each tube and keep at room temperature until use.
   4. Dilute the 5X labeling mixture five-fold as a working concentration and store on ice until use. For example, 2 μl of labeling mixture is diluted to total 10 μl with ddH$_2$O.

Notes: *(1) Mix the mixture in each stock tube well by pipetting it up and down prior to removal of an appropriate amount of mixture from each tube. (2) There are two sets of labeling and termination mixtures in the sequencing kit (USB). For regular noncompression sequencing, the one labeled as dGTP should be used. However, if compression appears due to G–C-rich sequences, the one labeled as dITP is strongly recommended for use. dITP replaces the nucleotide analog dGTP. We have found that it significantly improves the compressed bands (see Figure 11.1).*

   5. Dilute Sequenase Version 2.0 T7 DNA polymerase (1:8 dilution) as follows:
      - Ice-cold enzyme dilution buffer (6.5 μl)
      - Pyrophosphatase (0.5 μl)
      - DNA polymerase stored at –20°C (1 μl)

Notes: *The enzyme should not be diluted in glycerol enzyme dilution buffer if one uses TBE in the gel and in the running buffer. The diluted enzyme should be stored on ice and used within 50 min.*

c. **Carrying Out the Labeling Reaction When the Template-Primer Annealing Is Completed**
   1. Briefly spin down the annealed mixture and store on ice.
   2. Add the following to the tube containing 10 μl of annealed mixture in the order shown below:
      - DTT (1 μl)

- Diluted labeling mixture (2 μl)
- [^{35}S]dATP (1000 to 1500 Ci/mmol) (1 μl)
- Diluted Sequenase DNA polymerase (2 μl)

Caution: *[^{35}S]dATP is dangerous. It should be handled carefully using an appropriate Plexiglas™ protector shell. Gloves should also be worn.*

Note: *Adding 1 μl of Mn buffer may enhance the bands close to the primer.*

3. Gently mix well to avoid any air bubbles and incubate at room temperature for 3 to 5 min.
4. While labeling, place the previously labeled tubes containing termination mixture (step b.3) in a 37°C heating block for at least 2 min.

d. **Carrying Out the Termination Reactions**
 1. Carefully and quickly transfer 3.5 μl of the labeled mixture to each termination tube (A, G, T, and C) warmed at 37°C, mix, and quickly return to the 37°C heating block.
 2. Continue to incubate the tube at 37°C for 5 to 6 min.
 3. Add 4 μl of stop solution to each tube, mix, and cap the tubes. Store at 4°C for immediate use or at −20°C for later use.

Notes: *The samples should undergo electrophoresis within 4 days, even when stored at −20°C. Denature the samples at 75 to 80°C for 2 to 3 min prior to loading them into the sequencing gel.*

3. Sequencing of Single-Strand DNA

The procedures are very similar to those described in Sequencing of Double-Strand DNA (**Section II.B.2**) except for following:

1. No double-strand DNA denaturation is necessary. These steps are simply deleted in the above procedure, and the appropriate primer and template DNA can be directly annealed in an annealing reaction.
2. The temperature for primer-template annealing step can be from 50 to 55°C in a heating block for 15 to 25 min followed by slow cooling by turning off the heat block.
3. All other steps are identical to what was previously discussed.

4. Sequencing of Bacteriophage Lambda Linear Duplex DNA

For the purposes of sequencing, double-strand bacteriophage λDNA has been denatured by heating or by using alkaline (NaOH) conditions. The drawback is, unlike denatured plasmid DNA, that the denatured λDNA readily renatures during the process of primer-template annealing, competing with or preventing the primer from being annealed to the DNA template. This usually causes the failure of complete sequencing. DNA inserts cloned into bacteriophage λDNA vectors can now be directly sequenced without being subcloned into appropriate plasmid vectors. This can be achieved by using bacteriophage T7 gene 6 exonuclease (USB). The basic principle is that the enzyme only degrades double-strand DNA from 5′ → 3′. It simultaneously removes deoxynucleotides from both strands of DNA in opposite directions until the middle point of the DNA molecule is reached, generating two half-length, single-strand DNA fragments. The two half-length DNA fragments represent the sense and antisense orientations of the DNA cloned in the vectors. The DNA insert in the single-strand products can then be sequenced using appropriate primers annealed to the cloning sites of the vectors in both directions. The sequence of new-strand DNA synthesized from the antisense-strand fragment is actually the sequence of the sense-strand DNA of interest. Therefore, the sequence generated from the sense-strand fragment needs to be converted to the comple-

DNA Sequencing and Analysis

mentary sequence in order to be combined with the sequence obtained from the antisense-strand fragment at their 3' ends.

Once the λDNA is isolated, as detailed in **Chapter 1**, and treated as detailed in **Section I** ("Preparation of DNA for Sequencing" in this chapter) sequencing reactions are carried out as in **Section II.B.3** for single-strand DNA sequencing.

Reagents Needed

Denaturation Solution
0.4 M NaOH
0.4 mM EDTA (pH 8.0)

DTT Solution
100 mM Dithiothreitol

Enzyme Dilution Buffer
10 mM Tris-HCl, pH 7.5
5 mM DTT
0.5 mg/ml BSA (bovine serum albumin)

Glycerol Enzyme Dilution Buffer
20 mM Tris-HCl, pH 7.5
2 mM DTT
0.1 mM EDTA
50% Glycerol

5X Labeling Mixture for dGTP
7.5 μM dGTP
7.5 μM dTTP
7.5 μM dCTP

5X Labeling Mixture for dITP
7.5 μM dITP
7.5 μM dTTP
7.5 μM dCTP

Mn Buffer (only for dGTP)
150 mM sodium isocitrate
100 mM $MnCl_2$

Pyrophosphatase
5 units/ml in buffer containing the following:
10 mM Tris-HCl, pH 7.5
0.1 mM EDTA, pH 7.5
50% Glycerol

Radioactively Labeled dNTP
[^{35}S]dATP (1000 to 1500 Ci/mmol)

5X Reaction Buffer
0.2 M Tris-HCl, pH 7.5
0.1 M MgCl$_2$
0.25 M NaCl

Sequence-Extending Mixture for dGTP
180 μM dATP
180 μM dGTP
180 μM dCTP
180 μM dTTP
50 mM NaCl

Sequence-Extending Mixture for dITP
180 μM dATP
360 μM dITP
180 μM dCTP
180 μM dTTP
50 mM NaCl

Sequenase Version 2.0 T7 DNA Polymerase
13 units/μl in buffer containing the following:
20 mM KPO$_4$, pH 7.4
1 mM DTT
0.1 mM EDTA, pH 7.4
50% Glycerol

3 M Sodium Acetate Solution (pH 5.2)
Dissolve sodium acetate in ddH$_2$O and adjust pH to 5.2 with 3 M glacial acetic acid. Autoclave and store at room temperature.

Stop Solution
20 mM EDTA, pH 8.0
95% (v/v) Formamide
0.05% Bromophenol blue
0.05% Xylene cyanol FF

ddATP Termination Mixture for dGTP
80 μM dATP
80 μM dGTP
80 μM dCTP
80 μM dTTP
8 μM ddATP
50 mM NaCl

ddCTP Termination Mixture for dGTP
80 μM dATP
80 μM dGTP
80 μM dCTP
80 μM dTTP
8 μM ddCTP
50 mM NaCl

DNA Sequencing and Analysis

ddGTP Termination Mixture for dGTP
    80 μM dATP
    80 μM dGTP
    80 μM dCTP
    80 μM dTTP
    8 μM ddGTP
    50 mM NaCl

ddTTP Termination Mixture for dGTP
    80 μM dATP
    80 μM dGTP
    80 μM dCTP
    80 μM dTTP
    8 μM ddTTP
    50 mM NaCl

ddATP Termination Mixture for dITP
    80 μM dATP
    80 μM dITP
    80 μM dCTP
    80 μM dTTP
    8 μM ddATP
    50 mM NaCl

ddCTP Termination Mixture for dITP
    80 μM dATP
    80 μM dITP
    80 μM dCTP
    80 μM dTTP
    8 μM ddCTP
    50 mM NaCl

ddGTP Termination Mixture for dITP
    80 μM dATP
    160 μM dITP
    80 μM dCTP
    80 μM dTTP
    2 μM ddGTP
    50 mM NaCl

ddTTP Termination Mixture for dITP
    80 μM dATP
    80 μM dITP
    80 μM dCTP
    80 μM dTTP
    8 μM ddTTP
    50 mM NaCl

Note: All materials should be stored at $-20°C$ except $[^{35}S]dATP$, which should be stored at $-70°C$ and thawed on ice prior to use.

C. Denaturing Polyacrylamide Gel Electrophoresis

1. Preparation of Sequencing Gel

Gel preparation is one of the most important steps requisite for successful DNA sequencing. The DNA sequencing gel is a very thin gel (0.2 to 0.4 mm thick). Its quality and integrity directly influence the resolution and maximum reading information of a DNA sequence. There are two common problems in preparing a sequencing gel. One is the trapping of air bubble(s) in the gel, while the other is the leaking of the gel mixture from the bottom or the side-edges of the gel apparatus. Each of these problems can cause failure or a poor quality of the gel to occur. Both can result in false data for the DNA sequence. This, of course, wastes time and money. The detailed protocol given below can solve these problems and is successfully and routinely used in our laboratory.

1. Prepare a gel mixture as follows:
   a. Dissolve urea in distilled, deionized water (ddH$_2$O), in the amounts shown in Table 11.1 or Table 11.2, in a clean beaker using a stir bar.

Caution: *The Long Ranger™ mixture (JT Baker Chemicals, Phillipsburg, NJ) and the acrylamide/bis-acrylamide are neurotoxins. Gloves should be worn when handling the gel mixture.*

   b. Warm at 40 to 50°C to dissolve the urea with stirring. Do NOT boil the mixture or polymerization may occur.
   c. Add the components shown in Table 11.3, Table 11.4, or Table 11.5 to the mixture with stirring at room temperature.
   d. Optional: Filter and degas under vacuum for 2 to 3 min and transfer the gel mixture to a plastic squeeze bottle or leave it in the beaker if using a syringe to fill the gel apparatus with the liquid.
2. While the gel mixture is being cooled (25 to 30°C) and mixed with a stirring bar, begin to clean the glass plates. Gently place the glass plates, one by one, in a sink. Thoroughly clean the glass plates with detergent using a sponge. Thoroughly wash away the soap residue with running tap water followed by distilled water. Spray (95 or 100%) ethanol onto the plates to completely remove the soap and oily residues. Finally, wipe the plates dry using clean paper towels with no dust (e.g., bleached single-fold towels, James River Corp., Norwalk, CT).
3. Lay one glass plate down on the lab bench and coat its surface with 0.5 to 1.0 ml of Sigmacote® (Sigma-Aldrich, St. Louis, MO) to form a thin film. This can prevent the gel from sticking to the glass plates when they are separated from each other after electrophoresis. Mark the coated plate on

TABLE 11.1
Recipe if Using the Long Ranger Gel Mixture

Components	Small-Sized Gel (50 ml)	Large-Sized Gel (100 ml)
Ultrapure urea (7 M)	21 g	42 g
ddH$_2$O	12 ml	24 ml

TABLE 11.2
Recipe for Regular Acrylamide Gel

Components	Small-Sized Gel (50 ml)	Large-Sized Gel (100 ml)
Ultrapure urea (7–8.3 M)	21–25 g	42–50 g
ddH$_2$O	12 ml	24 ml

the back with a marker pen for later identification. The other plate should not be coated because it serves as the bottom plate when the plates are separated from each other.
4. Assemble a glass-plate sandwich by placing a spacer (0.2 or 0.4 mm thick) on each of the two side edges of one glass plate and cover the plate with the other plate. There are different ways to assemble the glass plates and to pour the gel mixture into the sandwich, depending on specific manufacturer's instructions.

a. **Pouring the Gel Mixture Horizontally**

1. Place two Styrofoam™ supporters (approximately 5 to 10 cm thick, one for the top and the other for the bottom of the sandwich) or two plastic pipette-tip boxes with an even surface on a flat bench. Put the glass sandwich on the two supports and check the flatness with a level. It is not necessary to use tape to seal the edges of the glass sandwich especially when two raised edges (about 0.03 to 0.5 cm above the glass surface) on each side of the bottom plate are used to hold the top plate in place. Clamping the side edges of the plates with manufacturer's clamps is optional. The glass sandwich should not be placed directly on the hard surface of the lab bench because later tapping on the top plate while pouring the gel may cause damage to the glass plate (see step 2).
2. Immediately prior to pouring the gel, add N,N,N'N'-tetramethylethylenediamine (TEMED) and freshly prepared 10% ammomium persulfate solution (APS) to the gel mixture. (For a 100-ml gel mixture, the volume for TEMED and APS should be 48 μl and 490 μl, respectively.) Quickly and gently mix the mixture by swirling the squeeze bottle to avoid air bubbles, and immediately begin to pour the gel mixture into the sandwich. Starting at the middle region at the top of the sandwich, slowly and continuously squeeze the bottle with one hand to cause outward flow of the mixture. Simultaneously use a roll of Scotch™ masking tape (1 to 2 cm wide and 10 to 12 cm diameter) to tap the top glass plate along the front of the flowing solution with the other hand until the sandwich is completely

TABLE 11.3
Recipe if Using the Long Ranger Gel Mixture

Components	Small-Sized Gel (50 ml)	Large-Sized Gel (100 ml)
5X TBE	12 ml	24 ml
Long Ranger mixture	5 ml	10 ml
Add ddH$_2$O up to	50 ml	100 ml

TABLE 11.4
Recipe for Regular Acrylamide Gel Using TBE Buffer

Components	Small-Sized Gel (50 ml)	Large-Sized Gel (100 ml)
5X TBE	10 ml	20 ml
6% Acrylamide/*bis*-acrylamide	2.8 g/0.15 g	5.7 g/0.3 g
8% Acrylamide/*bis*-acrylamide	3.8 g/0.2 g	7.6 g/0.4 g
Add ddH$_2$O up to	50 ml	100 ml

TABLE 11.5
Recipe for Regular Acrylamide Gel Using Glycerol Gel Buffer (for Sequenase Verison 2.0 DNA Polymerase Diluted with Glycerol Enzyme Dilution Buffer)

Components	Small-Sized Gel (50 ml)	Large-Sized Gel (100 ml)
10X TBE	5 ml	10 ml
6% Acrylamide/*bis*-acrylamide	2.8 g/0.15 g	5.7 g/0.3 g
8% Acrylamide/*bis*-acrylamide	3.8 g/0.2 g	7.6 g/0.4 g
Add ddH$_2$O up to	50 ml	100 ml

filled. The gel mixture distributes into the sandwich by capillary suction, and the tapping helps to cause even flow of the gel mixture. If squeezing of the bottle is stopped at any time, make sure to squeeze the bottle to remove any air bubbles before allowing the mixture to flow into the sandwich again. It is strongly recommended that two people work together while pouring the gel. One person can focus on squeezing the bottle to pour the gel, while the other is responsible for tapping the top glass plate to help the flow and prevent bubble formation. The gel mixture can also be loaded into the sandwich by the use of a syringe or a pipette, depending on personal perference.

3. Immediately and slowly insert the comb (0.2 or 0.4 mm thick, depending the thickness of the spacers), upside down, into the gel mixture to make a straight, even edge along the top of the gel. Avoid any bubbles underneath the comb. To prevent the gel from forming between the comb and the glass plates, clamp the comb in place together with the glass plates with 2 to 3 appropriately sized clamps. It is recommend that a 24-well comb be used for a small gel and a 48- or 60-well comb for a large gel. The color of the comb should be off-white or white so that each well can be clearly seen while loading the DNA samples.

4. Allow the gel to polymerize for 0.5 to 1.5 h. The gel can be subjected to electrophoresis directly by removing the comb and mounting the gel cassette onto the sequencing apparatus according to the manufacturer's instructions. For multiple loadings at approximately 2-h intervals, we recommend that the gel be poured late in the day or at night, and that electrophoresis be carried out early the next morning. If this is the option, the polymerized gel should be covered at the top and bottom edges of the cassette with clean paper towels wetted with distilled water. Wrap the paper towels with SaranWrap™ to keep the gel from drying. Leave it overnight so that the electrophoresis can be performed the next morning.

b. Pouring the Gel Mixture at an Angle

1. Tightly seal the two side edges of the sandwich using tape. Clamping each side of the sandwich also works well.

2. Seal the bottom of the sandwich in one of the following ways: (1) Preseal the bottom to ensure no leaking. This can be done by transferring approximately 5 ml of the gel mixture to a tray and adding 5 µl of TEMED and 45 µl of freshly prepared 10% APS. Mix well and immediately immerse the bottom of the sandwich, which is held vertically. The mixture containing a high concentration of APS will quickly polymerize and seal the bottom of the sandwich. After the bottom is sealed, the rest of the gel mixture can be poured. (2) Leave the bottom unsealed and attach the special bottom tray (provided by the manufacturer) to the bottom of the sandwich. Some kinds of commercial apparatus have disposable bottom spacers available, which can be inserted into the sandwich at the bottom and can then be sealed by tape.

3. Immediately prior to pouring the gel, add TEMED and freshly prepared 10% APS to the gel mixture contained in a clean squeeze bottle. For 100 ml of gel mixture, the volume for TEMED and APS should be 48 μl and 490 μl, respectively. Quickly and gently mix the solutions by swirling the squeeze bottle to avoid bubbles, and immediately begin to pour the gel mixture into the sandwich. Raise the top of the sandwich to a 45° angle with one hand. Using the other hand, slowly squeeze the bottle to cause flow of the gel mixture at the top corner along one side of the sandwich. The angle of the plate and the rate of flowing should be adjusted to avoid air bubbles. When the sandwich is half-full, place the top of the plate on a small box or support on the lab bench and the bottom of the plate on the bench, forming an angle of 25 to 30°. Continuously add the gel mixture into the sandwich and simultaneously use a roll of Scotch masking tape (1 to 2 cm thick and 10 to 12 cm diameter) or its equivalent to tap the top glass plate along the front of flowing gel with the other hand in order to avoid formation of any air bubbles until the sandwich is completely filled. The gel mixture can also be loaded into the sandwich with a syringe or pipette, depending on personal preference.

4. Immediately and slowly insert the comb (0.2 or 0.4 mm thick, depending on the thickness of the spacer), upside down, into the gel mixture to make a straight, even top front of the gel. Avoid formation of any air bubbles underneath the comb. To prevent gel from forming between the comb and glass plates, clamp the comb in place between the glass plates with 2 to 3 appropriately sized clamps. It is recommend that a 24-well comb be used for a small gel and a 48- or 60-well comb for a large gel. The color of the comb should be off-white or white so that each well can be seen clearly while loading the DNA sample.

DNA Sequencing and Analysis

TABLE 11.6
Examples with Constant Power

Gel Volume	Constant Power	Electrophoresis Temperature
35 cm³	35–38.5 W	45–55°C
70 cm³	70–77 W	45–55°C

5. Slowly lay the plates on a flat surface and allow the gel to polymerize for 0.5 to 1.5 h. The gel can be directly subjected to electrophoresis by removing the comb and mounting the gel cassette onto the sequencing apparatus according to the manufacturer's instructions. For multiple loadings (e.g., four loadings at 2-h intervals), we recommend that the gel be poured late in the day or at night, and that the electrophoresis be carried out early the next morning. If this is the option, the polymerized gel should be covered at the top and bottom edges of the cassette with clean paper towels wetted with distilled water. Wrap the paper towels with SaranWrap to keep the gel from drying. Leave it overnight so that the electrophoresis can be performed the next morning.

2. Electrophoresis

1. Remove the comb and carefully insert the cassette into the gel apparatus according to the manufacturer's instructions. Many systems make use of a metal plate mounted to one side of the cassette to help radiate excess heat that builds during the running of the gel.
2. Add a sufficient quantity of 0.6X TBE running buffer (diluted from 5X TBE stock buffer) to both top (cathode) and bottom (anode) chambers.
3. Gently wash the top surface of the gel several times by pipetting the running buffer up and down.
4. Vertically hold the left and right sides of the comb using two hands and carefully insert the comb into the top of the sandwich to form the wells. The insertion of the comb should be slow and even until all the teeth just touch the surface of the gel. Remove any air bubbles trapped in the wells by using a pipette.

Notes: *It is acceptable if the teeth of the comb protrude a little bit deeply into the gel (<0.5 mm from the surface of the gel). However, if one inserts the teeth too deeply into the gel or repeatedly pulls and inserts the comb several times, the surface of the gel will be badly damaged and cause leaking while loading samples, resulting in contamination of the wells and inaccurate sequence data.*

5. Calculate the volume of the gel (length × width × thickness of the spacer) and set up the power supply. We strongly recommend that the power supply be set at a constant power (watts) using 1.0 to 1.1 W/cm³. If constant power is not available from the power supply, 0.5 to 0.8 mA/cm³ constant current should be set up. Examples are given in Table 11.6.

Note: *If one sets the power supply at a nonconstant level (e.g., high voltage or high current), the gel may burn or melt during the electrophoresis due to the heat generated.*

6. Connect the power-supply unit to the gel apparatus with the cathode located at the top of the gel and anode at the bottom of the gel so that the DNA samples bearing a negative charge will migrate downward.
7. Turn on the power and prerun the gel for 10 to 12 min.
8. Denature the labeled DNA samples at 75°C for 2.5 to 3 min prior to loading them in the wells. The tubes containing the samples should be capped tightly while heating in order to prevent evaporation.

Note: *For heating, it is recommended that one place the tubes in a heating block in the order of A, T, G, C, or A, G, T, C. This can help prevent a potential mix-up while loading the samples into the gel.*

9. Turn off the power, wash the wells briefly to remove urea, and quickly and carefully load each of the samples to appropriate wells (3.5 to 4 μl per well).

Notes: *(1) There should be no leaking between wells; otherwise, the DNA samples will be contaminated and result in mix-up of the nucleotide sequence. (2) There are special sample-loading pipette tips available for DNA sequencing. It is recommended that multiple tips be used for loading multiple wells to prevent the possibility of contamination. If one uses a 0.4-mm spacer, we recommend the use of normal 100- to 200-μl pipette tips for loading the wells. This is faster, and no air bubbles develop as compared with using the commercial sequencing loading tips, which are long and flat. (3) After taking samples from each tube, the tubes should immediately be capped. All tubes should be briefly spun down and stored at 4 or –20°C until the next loading. (4) It is very important that the samples loaded into the respective wells be recorded in a notebook in the order of DNA samples loaded (e.g., A, G, T, C, or A, T, G, C, depending on the particular loading sequence). This will help in reading the DNA sequence following autoradiography.*

10. After all the samples are loaded, turn on the power, which is set at an appropriate constant voltage, and allow to electrophorese. The running time depends on the size of DNA. Normally, the run time for 250, 400 to 450, and >500 bp are 2, 4, and 6 to 8 h, respectively. For multiple loadings (e.g., two to five times loading, depending on the size of the gel and volume of the DNA sample), monitor the migration of the first blue dye (bromophenol blue or BPB), which migrates at 40 bp. When the BPB reaches approximately 2 to 3 cm from the bottom, carry out the next loading so that there are overlapping sequences between the adjacent loads. This is very useful for the correct reading of DNA sequences after autoradiography.

Notes: *The power supply must be turned off, and the wells to be loaded should be washed with the running buffer using a pipette to remove bubbles. The DNA samples can be denatured again at 75°C for 2 to 3 min prior to loading. The loading order of the samples should be the same as for the previous loading in order to avoid misreading the DNA sequences.*

Warning: *The electrophoresis is carried out at a high voltage, so care must be taken. Prior to each loading of the samples, make sure that the power is turned off. Monitor the running for a few minutes until stable conditions are established, especially when a new apparatus is used.*

11. After the electrophoresis has been completed, disconnect the power supply and decant the upper-chamber buffer to a ^{35}S-waste container. Lay the cassette containing the gel on a table or bench and allow it to cool for 15 to 25 min prior to separating the glass plates. Dump the bottom-chamber buffer into a waste container. While the gel is being cooled, cut a piece of 3MM Whatman™ paper that is larger than the gel to prepare for gel drying (e.g., dry-ice bucket or freeze-trap, pump, and dry apparatus).

Note: *If the glass plates are separated immediately following electrophoresis, the gel may be uneven, broken, or it may cause difficulty in separating the glass plates.*

12. Remove the clamps and/or the heat-cooling plate from the glass plates. Starting at one corner of the spacer site, carefully insert a spatula into the sandwich and slowly lift the top glass plate (the Sigma-coated plate) with one hand while holding the bottom plate with the other hand until the top glass plate is completely separated from the gel. Remove the top plate; the gel should stick to the bottom plate.

Note: As long as the top plate is loosened, continue to lift it until it is completely separated from the other plate. Do not allow the top plate to come back into contact with the gel in the middle of this apparatus or it may damage the gel.

13. Carry out fixation. If using the Long Ranger gel mixture, the electrophoresed gel does not need to be fixed. If, instead, regular acrylamide gel mixture is used, fixing of the gel should be performed. Immerse the gel together with the bottom plate in a large tray containing a sufficient volume of fixing solution containing 15 to 20% of ethanol or methanol and 5 to 10% acetic acid for 15 to 20 min. Remove the gel together with the plate from the tray, drain off excess fixing solution, and gently place bleached, clean paper towels on the gel to remove excess solution.
14. Gently and slowly lay the prepared 3MM Whatman paper on the gel from the bottom side or the upper side of the gel until it covers the entire gel. Gently and firmly press the paper thoroughly onto the gel surface using a Styrofoam™ block (approximately 50 cm long, 5 cm wide, and 5 cm thick). Starting from one corner, slowly peel the 3MM Whatman paper with the gel on it from the bottom plate. Place it on a flat surface with gel-side up and carefully cover the gel with SaranWrap without any air bubbles or wrinkles forming between the SaranWrap and the gel.
15. Assemble the gel on the drying apparatus according to the manufacturer's instructions and dry the gel at 70 to 80°C under vacuum for 30 to 55 min, depending on the strength of the vacuum pump. Turn off the heat and allow to cool for another 20 to 30 min under vacuum. Proceed to autoradiography.

Notes: (1) The position of the water trap should be lower than the gel dryer for effective suction of the water. If no frozen or refrigerated trap is available, place a water-trap flask or bottle in a bucket containing some dry ice and ethanol. Connect the trap to a vacuum pump and to the gel dryer. (2) If the gel is removed right after the heat is turned off, the gel may crack and/or become uneven. It is recommended that the gel be allowed to cool while under vacuum before being removed from the gel dryer.

3. Autoradiography

1. Peel off the SaranWrap from the gel and place the gel on an appropriate exposure cassette. In a dark room under a safety light, place an appropriately sized piece of x-ray film (Kodak XAR-film or Amersham x-ray film) on the surface of the gel and close the cassette.
2. Allow exposure to occur at room temperature (for ^{35}S) by placing the cassette in a dark place for 1 to 4 days, depending on the intensity of the signal. It is recommended that one develop the film after exposure for 24 h and decide the length of further exposure (Figure 11.1).
3. Develop the exposed film using an automatic developing machine or its equivalent in a dark room under a safety light. Proceed with DNA-sequence reading by recording the sequence of the bands in each of the four lanes, working from the bottom to the top of the photo (Figure 11.2).

4. Use of Formamide Gels for Sequencing of G-C-Rich Templates

Compressions are a common phenomenon that usually result in unreadability on a developed sequencing film. The problem is primarily due to intramolecular base pairing in a primer extension that is G-C rich. The local folded structures or hairpin loops migrate faster through the gel matrix than unfolded structures, and they persist during the electrophoresis, resulting in bands running very close together with a gap or increased band spacing in the region above. One way to solve this problem is to increase the denaturing condition in the gel matrix using urea and formamide. The procedures for preparing a formamide gel (**Section II.C.1**), electrophoresis (**Section II.C.2**), and autoradiography (**Section II.C.3**) are similar except for the following steps.

1. The gel mixture is made as shown in Table 11.7, Table 11.8, or Table 11.9. Dissolve the urea by placing the container in a 55 to 60°C water bath with shaking. The mixture should be cooled to 25 to 30°C prior to adding TEMED (0.12%, v/v) and 10% APS (0.78%, v/v).

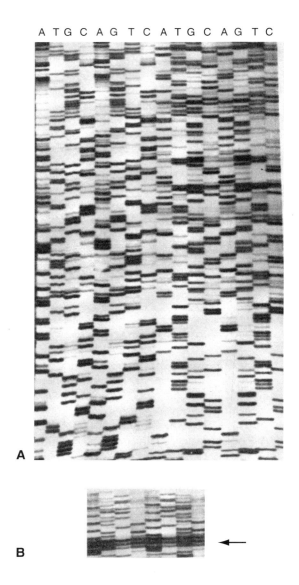

FIGURE 11.1
Autoradiograph showing a portion of DNA sequences obtained by the dideoxynucleotide chain termination method: (A) normal sequencing gel, (B) sequencing-gel compressions as denoted by the arrow. (From Kaufman, P.B., Wu, W., Kim, D., and Cseke, L.J., *Handbook of Molecular and Cellular Methods in Biology and Medicine*, 1st ed., CRC Press, Boca Raton, FL, 1995.)

2. Use 1X TBE running buffer instead of 0.6X TBE for both the regular acrylamide gel mixture and the Long Ranger gel mixture (JT Baker Chemicals). If the DNA polymerase is diluted with the glycerol-enzyme dilution buffer, the running gel buffer should contain 1X glycerol-tolerant gel buffer instead of TBE buffer.
3. Double the electrophoresis time, since the formamide slows the migration of DNA by about 50%.
4. After the electrophoresis, fix the gel in 20% ethanol or methanol and 10% acetic acid for 15 min.

DNA Sequencing and Analysis

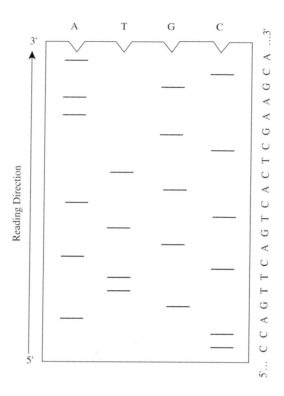

FIGURE 11.2
Diagram illustrating the reading of an autoradiograph obtained by dideoxy DNA sequencing. (From Kaufman, P.B., Wu, W., Kim, D., and Cseke, L.J., *Handbook of Molecular and Cellular Methods in Biology and Medicine,* 1st ed., CRC Press, Boca Raton, FL, 1995.)

TABLE 11.7
Using the Long Ranger Gel Mixture

Components	Small-Sized Gel (50 ml)	Large-Sized Gel (100 ml)
Ultrapure urea	21 g	42 g
Ultrapure formamide	20 ml	40 ml
Long Ranger mixture	8 ml	16 ml
5X TBE	10 ml	20 ml
Add ddH$_2$O up to	50 ml	100 ml

Reagents Needed

Acrylamide/bis-acrylamide
See Table 11.4 and Table 11.5

10% Ammonium Persulfate Solution (APS)
APS (0.1 g) in 1 ml ddH$_2$O (make fresh)

TABLE 11.8
Recipe for a Regular Acrylamide Gel Mixture (if Sequenase Version 2.0 DNA Polymerase is Diluted with a Buffer without Glycerol)

Components	Small-Sized Gel (50 ml)	Large-Sized Gel (100 ml)
Ultrapure urea (7 M)	21 g	42 g
Ultrapure formamide	15–20 ml	30–40 ml
6% Acrylamide/*bis*-acrylamide	2.8 g/0.15 g	5.7 g/0.3 g
8% Acrylamide/*bis*-acrylamide	3.8 g/0.2 g	7.6 g/0.4 g
5X TBE	5 ml	10 ml
Add ddH$_2$O up to	50 ml	100 ml

TABLE 11.9
Recipe for a Regular Acrylamide Gel Mixture (if the Sequenase Version 2.0 DNA Polymerase is Diluted in Glycerol Enzyme Dilution Buffer)

Components	Small-Sized Gel (50 ml)	Large-Sized Gel (100 ml)
Ultrapure urea (7 M)	21 g	42 g
Ultrapure formamide	15–20 ml	30–40 ml
6% Acrylamide/*bis*-acrylamide	2.8 g/0.15 g	5.7 g/0.3 g
8% Acrylamide/*bis*-acrylamide	3.8 g/0.2 g	7.6 g/0.4 g
10X Glycerol-tolerant gel buffer	5 ml	10 ml
Add ddH$_2$O up to	50 ml	100 ml

Formamide
    Ultrapure grade

10X Glycerol-Tolerant Gel Buffer
    Tris base (108 g)
    Taurine (36 g)
    Na$_2$EDTA·2H$_2$O buffer (pH 8.0) (4.65 g)
    Dissolve well after each addition in 700 ml ddH$_2$O. Then add ddH$_2$O to a final volume of 1000 ml. Autoclave. Store at room temperature.

High-Power Supply
    It should have an upper limit of 2500 to 3000 V.

Long Ranger Gel Mixture
    Long Ranger gel solution is available from many scientific supply companies and is a JT Baker reagent (JT Baker: Cat. #4730-02) (50% concentrate).

5X TBE Buffer
    Tris base (54 g)
    Boric acid (27.5 g)
    0.5 M EDTA buffer (pH 8.0) (20 ml)
    Dissolve well after each addition in 700 ml ddH$_2$O. Then add ddH$_2$O to a final volume of 1000 ml. Autoclave. Store at room temperature.

TEMED
N,N,N′N′-tetramethylethylenediamine (Sigma-Aldrich: item T 9281)

Urea
Ultrapure grade

D. Extending Sequence Far from the Primer

We routinely use the sequence-extending mixture (USB) to obtain sequences beyond 600 to 800 bases from the primer with high resolution. Using multiple loadings, more than 1500 to 2000 bases from the primer can be extended. The procedures for extending sequencing are similar to those described in **Section II**, with the following exceptions:

1. The labeling mixture should be undiluted or diluted for two- to three-fold instead of five-fold dilution.
2. Increase the amount of [^{35}S]dATP (1000 to 1500 Ci/mmol) from the regular volume of 1 µl to 1.5–2 µl.
3. Extend the labeling reaction from the regular time of 2–5 to 4–7 min at room temperature.
4. For the termination reaction, instead of adding 2.5 µl of ddNTPs to the appropriate tubes (A,G,T, and C), add 2.5 µl of a mixture of the appropriate sequence-extending mixture (USB) and appropriate termination mixture (USB) at ratios of 2:0.5, 2:1.0, and 1.5:1 (v/v), depending on the particular DNA sequence, to appropriate tubes.
5. Use a relatively long gel (80 to 100 cm) and extend electrophoresis for 12 to 16 h using a high-quality sequencing apparatus.

III. Cycle Sequencing and Automated Sequencing

DNA molecules or fragments (single-strand, double-strand, and plasmid DNA) can now be directly sequenced by combining PCR technology and the dideoxynucleotide chain termination method. The use of thermostable DNA polymerases allows sequencing reactions to be cycled through alternating periods of thermal denaturation, primer annealing, and extension/termination to increase the signal levels generated from template DNA.[9–13] This amplification process employs a single primer, so the amount of product increases linearly with the number of cycles. The analysis of the subsequent products can be accomplished either on standard "manual" sequencing gels or by newer automated machines that greatly enhance the speed and ease with which sequence data can be manipulated. In recent years, this automation process has been expanding into new and exciting methods of sequencing.[14] Chip-based DNA sequencing devices, for example, miniaturize the process of electrophoresis onto a small glass chip. Such devices can sequence a fragment of DNA as the DNA is passed through microchannels etched onto the surface of the glass chip. When a sample is injected into the chip's channels, an electric current separates the DNA molecules based on their size, just like other methods of sequencing. However, such methods currently require specialized and expensive equipment that are not practical for most labs. Consequently, this section will focus on one example for both manual-cycle sequencing and automated-cycle sequencing.

The preparation of a DNA template is the single most important factor that the user can change to affect the quality of data produced. There is a very persuasive correlation between cleanliness and consistency of template DNA and the quality of sequence data produced. While any type of DNA can be used for these techniques, PCR products are typically very good sequencing substrates for either manual- or automated-cycle sequencing systems for several reasons: They are relatively free of inhibitory contaminants; they are often of relatively uniform concentrations; and they typically do not include large amounts of unsequenced template DNA that can interfere with injection. We do recommend, however, that PCR primers and nucleotides be removed.

A. Selection of Oligonucleotides

Special attention must be paid to good primer design and annealing temperatures for cycle sequencing, since even minor nonspecific annealing will amplify during PCR reactions. Often, commercially available oligonucleotides can be used with great success. Since the popular multiple cloning sites are all derived from similar sequences, vector primers, such as the control 40 M13 forward primer, can serve for the sequencing primer in most of the common vectors. Among the vectors compatible with this primer are: M13mp8, M13mp9, M13mp12, M13mp13, M13mp18, M13mp19, pUC18, pUC19, and virtually any vector featuring blue/white screening with β-galactosidase activity.

If it is necessary to design new sequencing primers, the same criteria as those detailed in **Section II.B.1** should be followed. The length of the primer and its sequence will determine the melting temperature and specificity. For the cycling temperatures normally used, the primer should be about 18 to 25 nucleotides long. It is also a good idea to check the sequence of the primer for possible self-annealing (dimer formation could result) and for potential "hairpin" formation, especially those involving the 3′ end of the primer. Finally, check for possible sites of false priming in the vector or other known sequence if possible.

For most applications, 5 pmoles of primer should be used for each sequencing reaction. Primers should initially be quantified at 260 nm (OD_{260}). If the primer has N bases, the approximate concentration (pmol/μl) is given by the following formula:

$$\text{Concentration (pmol/}\mu\text{l)} = OD_{260}/(0.01 \times N)$$

where N is the number of bases.

The following list can also be used to estimate the number of ng in one pmol of a given size primer:

Amount of Primer per pmole
15 mer or 15 bases (5 ng)
16 mer or 16 bases (5.3 ng)
17 mer or 17 bases (5.7 ng)
18 mer or 18 bases (6.0 ng)
19 mer or 19 bases (6.3 ng)
20 mer or 20 bases (6.7 ng)
21 mer or 21 bases (7.0 ng)
22 mer or 22 bases (7.3 ng)
23 mer or 23 bases (7.6 ng)
24 mer or 24 bases (8.0 ng)
25 mer or 25 bases (8.3 ng)
26 mer or 26 bases (8.6 ng)
27 mer or 27 bases (9.0 ng)
28 mer or 28 bases (9.3 ng)
29 mer or 29 bases (9.6 ng)
30 mer or 30 bases (10.0 ng)

DNA Sequencing and Analysis

B. DNA Synthesis and Termination Reactions by PCR

1. Protocol

The following protocol is a good method for preparing the four terminating-sequence reactions using PCR to be subsequently run on gels.

1. Label four 0.5-μl microcentrifuge tubes (A, G, T, and C) for each set of sequencing reactions for each primer. A, G, T, and C represent ddATP, ddGTP, ddTTP, and ddCTP, respectively.
2. Add 0.5 ml of 2X stock mixture of dNTPs/ddATP, dNTPs/ddGTP, dNTPs/ddTTP, and dNTPs/ddCTP to the labeled tubes A, G, T, and C, respectively. Add 0.5 μl ddH$_2$O to each tube, generating 1X working mixture solution. Cap the tubes and store on ice until use.
3. Prepare the following mixture for each set of four sequencing reactions for each primer in a microcentrifuge tube (0.5 ml) on ice.
   - Appropriate primer (16 mer to 27 mer, 10 to 30 ng/μl), 2 to 5 pmol (15 to 27 ng), depending on the size of primer
   - DNA template (0.4 to 7 kb [kilobase-pairs], 10 to 100 ng/μl), 100 to 1000 ng, depending on the size of the template
   - [α-^{35}S]dATP (>1000 Ci/mmol), 1 to 1.2 μl or [α-^{32}P]dATP (800 Ci/mmol), 0.5 μl
   - 5X Sequencing buffer, 4 μl
   - Add ddH$_2$O to a final volume of 17 μl.

4. Add 1 μl (5 units/ml) of sequencing-grade *Taq* DNA polymerase (Promega Corp.) to the mixture at step 3. Gently mix by pipetting up and down, and store on ice.
5. Remove 4 μl of the primer-template-enzyme mixture at step 4 to the bottom of each tube containing 1 μl of dNTPs and the appropriate ddNTP prepared in step 2.
6. If a thermal cycler with a heated lid is not available, overlay the mixture in each tube with approximately 20 μl of mineral oil to prevent evaporation during the PCR amplification.
7. Place the tubes in a thermal cycler preheated to 95°C and begin 30 cycles following the cycling profiles given in Table 11.10, depending on the size of primer and the size of DNA template.

TABLE 11.10
Thermal Cycling Profiles

Profile	Predenaturation	Cycling			Last
		Denaturation	Annealing	Extension	
Template[a] (4 kb)	94°C, 2 min	94°C, 1 min	50°C, 1 min	70°C, 1.5 min	4°C
Template[b] (>4 kb)	95°C, 2 min	95°C, 1 min	60°C, 1 min	72°C, 2 min	4°C

[a] Primer is <24 bases or with G-C content <40%.
[b] Primer is >24 mer or <24 bases with G-C content >50%.

8. After the PCR cycling is completed, carefully remove the mineral oil from each tube using pipette tips. Add 3.5 μl of stop solution to inactivate the enzyme activity.
9. The reaction mixture can be directly subjected to electrophoresis following denaturation at 75°C for 2 min, or it can be stored at 4°C until use. Load 3 μl per well and avoid inclusion of mineral oil. The procedures for electrophoresis and autoradiography are described in **Section II.C.2** and **Section II.C.3** of this chapter.

Reagents Needed

2X dNTPs/ddATP Mixture
   dATP (80 μM)
   dTTP (80 μM)

       dCTP (80 μM)
       7-Deaza dGTP (80 μM)
       ddATP (1.4 mM)

2X dNTPs/ddCTP Mixture
       dATP (80 μM)
       dTTP (80 μM)
       dCTP (80 μM)
       7-Deaza dGTP (80 μM)
       ddCTP (800 μM)

2X dNTPs/ddGTP Mixture
       dATP (80 μM)
       dTTP (80 μM)
       dCTP (80 μM)
       7-Deaza dGTP (80 μM)
       ddGTP (120 μM)

2X dNTPs/ddTTP Mixture
       dATP (80 μM)
       dTTP (80 μM)
       dCTP (80 μM)
       7-Deaza dGTP (80 μM)
       ddTTP (2.4 mM)

5X Sequencing Buffer
       0.25 M Tris-HCl, pH 9.0 at room temperature
       10 mM MgCl$_2$

Stop Solution
       10 mM NaOH
       95% Formamide
       0.05% Bromophenol Blue
       0.05% Xylene cyanole

Taq DNA Polymerase
       Sequencing grade (Sigma, Promega, or Clontech)

C.   DNA Sequencing Using an Automated Sequencer

New methods in automated sequencing have greatly reduced the amount of work required for sequence analysis. We have made very good use of the DYEnamic™ ET terminator cycle sequencing kit (Amersham Biosciences, Piscataway, NJ). To run a sequence with this kit, a reaction premix provided with the kit is combined with your template DNA and primer. The single reaction mixture is thermally cycled. After cycling, the reaction products are precipitated with salt and ethanol, or they are passed through a gel filtration resin and concentrated to remove unincorporated dye-labeled terminators. The reaction products are resuspended in MegaBACE™ (Amersham Biosciences) loading solution and run on an ABI Prism 310 capillary DNA sequencing instrument.

The technology makes use of dye-labeled dideoxynucleotides and a new DNA polymerase. Each dideoxy terminator is labeled with two dyes. One of these dyes, fluorescein, has a large

DNA Sequencing and Analysis

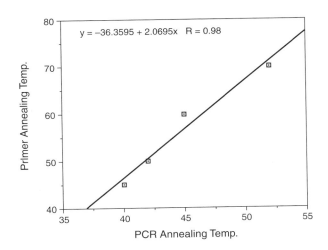

FIGURE 11.3
Sequencing PCR temperatures as determined by linear curve-fitting of data points from sequencing reactions having given sequencing read lengths of over 700 bp. y-axis: calculated primer melting temperature; x-axis: PCR temperature determined to be best for sequencing.

extinction coefficient at the wavelength (488 nm) of the argon ion laser in the instrument. The fluorescein donor dye absorbs light energy from incident laser light and transfers the collected energy to an "acceptor" dye. Each of the four chain terminators — ddG, ddA, ddT or ddC — have a different acceptor dye coupled with the fluorescein donor. The acceptor dyes then emit light at their characteristic wavelengths. The fluorescence is detected by the instrument allowing identification of which nucleotide caused the termination event. The acceptor dyes are the same standard rhodamine dyes used in DYEnamic™ ET primers (Amersham Biosciences): rhodamine 110, rhodamine-6-G, tetramethyl rhodamine, and rhodamine X. The kit is formulated with dITP. Use of this nucleotide instead of dGTP affords resolution of even very strong compression artifacts for better accuracy.

1. Sequencing Reactions

The melting temperature of the primer should be kept in mind when choosing cycle temperatures. We have found that cycle annealing temperatures well below those of the calculated annealing temperature provide more reliable sequencing results. The control primer included in the kit, for example, is moderately long (23 bases) with a melting temperature of about 75°C under sequencing reaction conditions. The best sequencing results come from 50°C cycling annealing temperatures. If one needs to use many primers with different annealing temperatures, we recommend the preparation of a graph, such as the one depicted in Figure 11.3, which comprises data points from successful sequencing reactions. Please note that the duration of the annealing and denaturation steps does not seem to be critical, and even brief pauses at these temperatures seem to be effective.

1. DNA template: the recommended quantities of DNA template are 0.1 to 1 µg (40 to 400 fmol) of single-strand DNA or 0.2 to 2 µg (80 to 800 fmol) of plasmid (double-strand) DNA. The DNA should be diluted in deionized water to a total volume of not more than 11 µl.

Notes: *(1) Whenever possible, tubes should be kept capped and on ice to minimize evaporation of the small volumes employed. Additions should be made with disposable-tip micropipettes, and care should be taken not to contaminate stock solutions. (2) Do not use buffers with more than 0.1 mM EDTA because these could significantly reduce the effective metal cofactor concentration in the reactions. Similarly, primer should be diluted in water or a buffer with no more than 0.1 mM EDTA.*

2. For each reaction, combine the proper amount of the sequencing reagent premix with the diluted DNA template in the following manner:
   - DYEnamic ET reagent premix, 8 μl
   - Primer (5 μM), 1 μl
   - DNA template, 11 μl
   - Total, 20 μl

Note: *The reagent premix contains Thermo Sequenase II™ DNA polymerase, a thermostable DNA polymerase specifically engineered for cycle sequencing (Amersham Biosciences).*[7,13]

3. When all reagents have been added, treat the samples appropriately for your thermal cycler. Cap the tubes or use a plate sealer where appropriate. Mix the contents well. A brief centrifugation will bring the solution to the bottom of the tube/plate. Place the plate or tubes into the thermal cycler.

Notes: *(1) We highly recommend the use of thin-walled 200-μl PCR tubes. These tubes provide better results than 500-μl tubes because the thin walls transfer heat more rapidly. (2) The specific cycling parameters used will depend on the primer sequence and the amount and purity of the template DNA.*

4. Start the cycling program. For most primers and the recommended amount of purified DNA, the following cycle parameters are appropriate:
   - 95°C, 20 sec
   - 50°C, 15 sec
   - 60°C, 2 min
   - 27 to 32 cycles

Notes: *(1) Initial denaturation steps at 95 to 98°C should be avoided if possible, since the enzyme rapidly thermally denatures. (2) The number of cycles required will primarily depend on the amount of template DNA (in fmol) used for sequencing. It will also depend on the purity of the DNA and the sensitivity of fluorescent detection of the equipment. Performing many cycles (greater than 40) can give rise to artifacts arising from the formation of PCR products with the sequencing primer and template or from the presence of contaminating DNA. If it is desired to do many cycles, ensure that the annealing step is sufficiently stringent to prevent false priming.*

2. Sample Preparation for Injection

1. To each reaction tube add 2 μl of sodium acetate/EDTA buffer.
2. Add 2.5 volumes (55 μl) of 100% ethanol to each reaction or 60 μl of 95% ethanol. The final concentration of ethanol should be 70%. Mix well.

Notes: *(1) Final ethanol concentrations of 65% produce sequences with weak signals, whereas final ethanol concentrations of 75% produce sequences with "blob" artifacts from unincorporated dye terminator. (2) It is unnecessary to use cold ethanol for precipitation, and it is unnecessary to incubate the samples at low temperature for precipitation.*

3. For microcentrifuge tubes, centrifuge in a microcentrifuge (~12,000 rpm) for 15 min at room temperature. For 96-well plate precipitations, we recommend at least 2500 g for 30 min. We have found this parameter to be important. If you choose to spin the plates at a lower centrifugal force, increase the time. We routinely use 3100 g for 30 min.

DNA Sequencing and Analysis

4. Draw off supernatant by aspiration or by performing a brief inverted spin, removing as much liquid as possible.
5. Add 500 μl of 70% ethanol to wash the pellet. Centrifuge briefly.
6. Draw off supernatant or repeat inverted spin and vacuum dry or air dry the pellets. (Do not overdry.)
7. Resuspend each pellet in 10 μl of MegaBASE loading solution (Amersham Biosciences). Extreme care should be taken to redissolve the DNA completely at this step so that the longest possible sequences can be obtained. If the ethanol precipitation and wash steps were centrifuged in a fixed-angle rotor, the DNA pellet will be on the side of the microcentrifuge tube. This material must be washed to the bottom of the tube to assure that the entire reaction product is loaded on the gel. Vortex vigorously (10 to 20 sec) to ensure complete resuspension. Briefly centrifuge to collect the sample at the bottom of the tube or plate and to remove bubbles.

Note: *As an alternative to ethanol precipitation, spin columns or gel filtration plates can be used according to the manufacturer's instructions. We have had good success with the Millipore Multiscreen™ system (Millipore Corp., Billerica, MA) using Sephadex G-50. Wash the gel filtration resin at least once with water to remove residual salt and ions that can interfere with injection. The final material from spin plates is typically in 20 μl of water.*

8. Set up the instrument and run parameters following the manufacturer's recommendations, and run the samples as directed.

Note: *Most often, each reaction must be transferred into a tube specifically designed for the instrument being used. This prevents electrode damage during the injection process.*

3. Analysis of Sequence Chromatograms

While automated sequencing has the powerful advantages of speed, accuracy, and the convenience of automatically prepared sequence data, we highly recommend that each base-pair be manually double-checked from the peak data on the final chromatograms. Minor problems, such as suboptimal peak determination files (mobility files and/or matrix files) or "blips" caused by impurities in the reagents or water used, can generate sequencing artifacts that result in single base-pair errors throughout the sequence. This is especially true at the beginning of the runs when the peaks are crowded or at the end of the runs when the peaks become more broad. Such artifacts can be avoided by simply looking over the chromatograms (see Figure 11.4 for an example). This task is made easy by choosing four distinct colors, one for each ddNTP. Detected errors can be edited in any word-processing program.

Once a sequence is determined to be error free, it can be copied and pasted into a wide variety of sequence analysis programs and web sites for an array of analysis techniques. Programs such as Vector NTI Suite™ (InforMax, Frederick, MD), Gene Construction Kit™ (SciQuest, Research Triangle Park, NC), and DNAsis™ (Hitachi Sofware, South San Francisco, CA) are still quite

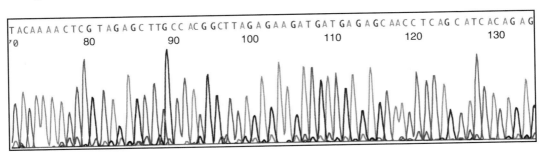

FIGURE 11.4
An example of a DNA sequence chromatogram resulting from the completion of automated DNA sequencing.

popular. Some of our current favorite web sites include NCBI blast searches for sequence identification (http://www.ncbi.nlm.nih.gov/BLAST/), PSORT web server (http://psort.ims.u-tokyo.ac.jp) for the identifcation of sequence motifs, Multialign interface (http://prodes.toulouse.inra.fr/multalin/multalin.html) for sequence alignments, and the Boxshade server (http://www.ch.embnet.org/software/BOX_form.html) to generate eye-pleasing representations of the alignments.

Reagents Needed

Electrophoresis Polymer for Capillary Sequencing

We use POP-6™ (Performance Optimized Polymer 6) along with 10X running buffer with EDTA supplied by Applied Biosystems (Foster City, CA). Other combinations can be used, depending upon the read lengths required. Capillary polymer should be used according to the instructions provided with the instrument.

Gel Reagents

Sequencing gels should be prepared according to the instructions provided with the instrument.

MegaBASE Loading Solution (Amersham Biosciences)

Sequencing Reagent Premix (Amersham Biosciences)

This premix is supplied with the DYEnamic ET terminator cycle sequencing kit.

Sodium Acetate/EDTA buffer

1.5 M Sodium acetate
250 mM EDTA

IV. Troubleshooting Guide

A. Manual Sequencing: Problems, Possible Causes, and Solutions

1. Gel pulls away from the comb during electrophoresis.
   Cause: Too much heat is built up.
   Solution: Be sure to set the power supply at a constant power or a constant current. Do not set the voltage at a particular level, since the current undergoes changes during electrophoresis, generating temperatures high enough to melt the surface of the gel. When that happens, multiple loadings of the samples are impossible.
2. No bands occur at all on the developed x-ray film.
   Possible causes:
   a. The primer does not work.
   b. Double-strand DNA is not denatured well, such that the primer fails to anneal to the template.
   c. Some component(s) is (are) missed during the labeling reaction.
   d. Sequenase Version 2.0 T7 DNA polymerase has lost its activity.
   Solution: Be sure to denature the DNA template completely and add all components necessary for the labeling reaction. Try to use the control DNA and primer provided.
3. Bands are fuzzy.
   Possible causes:
   a. Urea is not washed from the wells prior to loading the samples.
   b. Labeled samples are overheated during denaturation.
   c. It takes too long to finish loading all the samples, resulting in some annealing of DNA.

Solutions:
   a. Be sure to wash the surface of the gel prior to drying the prerun and repeat washing after the prerun before loading the samples into the wells. The washing can be done with the running buffer using a pipette tip.
   b. Control the time for denaturing of the labeled samples to between 2 to 3 min, and immediately load the sample into the wells. For many samples, the loading should be carried out quickly so that all of the samples should be loaded within 2 min.
4. No clear bands appear except a smear in each lane.
   Possible causes:
   a. Preparation of DNA template is bad.
   b. Labeling reaction goes too long and results in very old labeling.
   c. Labeled DNA samples are not denatured well at 75°C prior to their being loaded into the gel.
   d. Gel polymerizes too rapidly (10 to 15 min) due to excess 10% APS added.
   e. The gel is electrophoresed at a too-cold or too-hot temperature.
   f. The gel is dried at too-warm a temperature.
   Solutions:
   a. Make sure the DNA template is very pure without any nicks. This can be checked by spectrophotometer measurement with a ratio of A_{260}/A_{280} of 1.9 to 2.0. Furthermore, the purified DNA should be checked by agarose gel electrophoresis.
   b. Use 0.5 ml of freshly prepared APS per 100 ml of gel mixture and make sure the gel mixture is cooled to room temperature prior to its being poured into the glass sandwich.
   c. Keep the labeling reaction time to 2 to 5 min for regular sequencing and 4 to 7 min for extended sequencing.
   d. Be sure to denature the labeled samples at 75 to 80°C for 2 to 3 min prior to their being loaded into the gel.
   e. Dry the gel at 75 to 80°C under vacuum but not above 80°C.
5. All the bands are weak.
   Possible causes:
   a. Primer concentration is too low or the annealing of primer and template does not work well.
   b. Double-strand linear DNA and double-strand plasmid DNA are too large due to the presence of a large DNA insert, resulting in difficulty in denaturation.
   c. Labeling reaction goes on for too long.
   d. Labeled DNA samples may not be completely denatured before loading into the gel.
   Solutions:
   a. Heat the primer and double-strand DNA template at 65°C for 3 to 4 min and slowly cool down to room temperature over 20 to 35 min.
   b. Use the alkaline-denaturing method to denature large-sized DNA templates. If this still does not work well, try to fragment the DNA insert to be sequenced and subclone for further sequencing.
   c. Make sure the labeling reaction is carried out properly and denature the labeled sample at 75 to 80°C for 3 to 4 min prior to their being loaded into the gel.
6. Bands occur across all four lanes in some areas called compressions.
   Possible cause: Occurrence of nucleotide sequences with strong secondary structure.
   Solution: Use an appropriate amount of dITP to replace dGTP and an appropriate amount of pyrophosphatase in the labeling reaction.
7. Bands are faint near the primer.
   Possible cause: Insufficient DNA template or insufficient primer.
   Solutions:
   a. Use 1 to 1.5 µg of single-strand M13 DNA or 3 to 5 µg plasmid DNA per reaction.
   b. Increase the molar ratio of primer:DNA template from 1:1 to 1:4–5.
   c. Use 1 µl of Mn buffer per regular labeling reaction.
8. Bands are faint or blank in one or two lanes.
   Possible cause: Some components may have been missed or added improperly.
   Solution: Be sure that all the components are added properly as for other lanes revealing bands with good resolution.

B. Automated Sequencing: Problems, Possible Causes, and Solutions

Prior to diagnosing problems with the sequencing reaction chemistry, it is important that the sequencing system be determined to be in optimal working condition. This can be accomplished by injecting sequencing standards and performing electrophoresis according to the following protocol. It is based on information from the Amersham Biosciences DYEnamic ET Terminator Cycle Sequencing kit instruction manual. If the overall read length is short, then this may indicate that the instrument is in need of maintenance such as capillary cleaning and focusing.

1. Weak signals
   a. DNA preparation may be bad. Try control DNA.
   b. Poor quality template. Contaminants carried over with template preparations can precipitate and interfere with efficiency of electrokinetic injection. There is a strong correlation between the quality of template DNA and the success of sequencing experiments.
   c. Insufficient template DNA or insufficient number of cycles. Try using more DNA or more cycles.
   d. Incorrect temperatures for primers used. Try a lower annealing temperature for cycling.
   e. Too little primer used. The recommended amount of primer is 5 pmol.
   f. Primer is bad. Some primers form dimers, hairpins, etc., interfering with annealing with the template. Try a different primer.
   g. Salt in samples. If products were prepared by spin columns, confirm that they were eluted in water. Some preparations of size-exclusion chromatography media are preswollen in a salt-containing buffer. These must be washed several times in water to remove salt. Some methods recommend several washes of dry media to remove residual ions that interfere with injection.
   h. Too much ethanol during precipitation. Use the recommended volumes or calculate such that the precipitation contains 70% ethanol. Excess ethanol will precipitate salts, buffers, and contaminants in template DNA, all of which will compete for the sequencing products and reduce the effective signal.
   i. Wrong precipitation salt/wrong volume used. The protocol works best with ammonium acetate, not sodium acetate.
   j. Excess EDTA in primer or template. Resuspend primer and template in water or dilute buffer containing less than 0.1 mM EDTA.
   k. Old loading buffer. Old solutions can increase the ionic strength of the buffer and reduce the efficiency of injection.
   l. Try changing injection conditions. Try reducing AND increasing the injection time to find better conditions. With capillary sequencing, sometimes injecting less gives more signal.
2. Extensions appear short (read length limited to less than 350 bases)
   a. There is too much template DNA. In some cases, the use of too much DNA, especially PCR product DNA, can exhaust the supply of ddNTPs and cause the sequence to suddenly fade prior to reaching 350 bases. This is especially prevalent when also using too much primer. Use less than 1 pmol of template DNA and 5 pmol primer for each sequence. Use of less template has the added benefit of reducing the concentration of potential contaminants.
   b. Increase the time-of-extension step in cycles to 2 to 4 min.
3. New symptom (sequences with late starts and peaks that are broad, and poorly resolved)
   a. Resuspend product in a larger volume of loading buffer. Try 20, 50, or 100 μl.
   b. The run voltage is too high. Limit run voltage to no greater than 9 kV.
   c. Try changing injection conditions. Try reducing AND increasing the injection time by three-fold to find better conditions.
   d. Try reducing injection voltage. Reduce voltage to 2 kV or 3 kV.

Note: *Overloaded samples frequently have low signals, since the peaks are broad and diffuse. It is very common to misdiagnose overloading as not enough signal. Under optimal conditions, all samples should begin being detected within a few minutes of each other. Samples with late starts and broad peaks are overloaded.*

4. Localized broad peaks or very tall early peaks — terminator blobs.
   Cause: Insufficient removal of unused terminators.
   Solution: Carefully follow the protocol for cleanup using 70% ethanol and the ammonium acetate supplied with the kit.
5. Peak spacing changes during the run (the "accordion" effect).
   Cause: Samples are "almost" overloaded.
   Solution: Follow suggestions for overloading in 3a to 3d above.
6. Noisy sequences or double sequences
   Cause: Annealing temperature too low.
   Solution: Raise the annealing temperature or eliminate it entirely to perform cycling between 95°C and 60°C. We find that the effective annealing temperature of primers is higher with this terminator product than with other terminator products. If problems persist, please contact Amersham Biosciences technical service for assistance.

References

1. **Cullmann, G., Hubscher, U., and Berchtold, M.W.**, A reliable protocol for dsDNA and PCR product sequencing, *BioTechniques*, 15, 578, 1993.
2. **Sanger, F., Nicken, S., and Coulson, A.R.**, DNA sequencing with chain termination inhibitors, *Proc. Natl. Acad. Sci. U.S.A.*, 74, 5463, 1977.
3. **Church, G.M. and Gilbert, W.**, Genomic sequencing, *Proc. Natl. Acad. Sci. U.S.A.*, 81, 1991, 1984.
4. **Wiemann, H.V., Grothues, D., Sensen, C., Zimmermann, C.S., Stegemann, H.E., Rupp, T., and Ansorge, W.**, Automated low-redundancy large-scale DNA sequencing by primer walking, *BioTechniques*, 15, 714, 1993.
5. **Reynolds, T.R., Uliana, S.R.B., Floeter-Winter, L.M., and Buck, G.A.**, Optimization of coupled PCR amplification and cycle sequencing of cloned and genomic DNA, *BioTechniques*, 15, 462, 1993.
6. **Bishop, M.J. and Rawlings, C.J.**, *Nucleic Acid and Protein Sequence Analysis: A Practical Approach*, IRL Press, Oxford, 1987.
7. **Tabor, S. and Richardson, C.C.**, DNA sequence analysis with a modified bacteriophage T7 DNA polymerase, *Proc. Natl. Acad. Sci. U.S.A.*, 84, 4767, 1987.
8. **Sheen, J.Y. and Seed, B.**, Electrolyte gradient gels for DNA sequencing, *BioTechniques*, 6, 942, 1988.
9. **Murray, V.**, Improved double-stranded DNA sequencing using the linear polymerase chain reaction, *Nucleic Acids Res.*, 17, 8889, 1989.
10. **Carothers, A.M., Urlaub, G., Mucha, J., Grunberger, D., and Chasin, L.A.**, Point mutation analysis in a mammalian gene: rapid preparation of total RNA, PCR amplification of cDNA, and Taq sequencing by a novel method, *BioTechniques*, 7, 494, 1989.
11. **McMahon, G., Davis, E., and Wogan, G.N.**, Characterization of c-Ki-ras oncogene alleles by direct sequencing of enzymatically amplified DNA from carcinogen-induced tumors, *Proc. Natl. Acad. Sci. U.S.A.*, 84, 4974, 1987.
12. **Levedakou, E.N., Landegren, U., and Hood, L.E.**, A strategy to study gene polymorphism by direct sequence analysis of cosmid clones and amplified genomic DNA, *BioTechniques*, 7, 438, 1989.
13. **Tabor, S. and Richardson, C.C.**, Selective inactivation of the exonuclease activity of bacteriophage T7 DNA polymerase by *in vitro* mutagenesis, *J. Biol. Chem.*, 264, 6447, 1989.
14. **Noble, D.**, DNA sequencing on a chip, *Anal. Chem.*, 67, 201, 1995.

Chapter 12

DNA Site-Directed and Deletion Mutagenesis

Leland J. Cseke and William Wu

Contents

I. Mutagenesis by Deletion 272
    A. Performing Exonuclease III Deletions 273
        1. Subcloning of the DNA Insert 273
        2. Purification of Recombinant Plasmids for Exonuclease Deletions 273
        3. Double-Restriction Enzyme Digestions of Circular Plasmids 275
        4. Performing Series Deletions of the Linearized DNA with
           Exonuclease III 276
        5. Transformation Using Recircularized Plasmids and Estimation
           of the Sizes of Deletion Subclones 277
    B. Performing *Bal* 31 Deletions 277
        1. Linearizing Recombinant Plasmids or DNA Fragments Using
           Appropriate Endonuclease(s) 279
        2. Performing *Bal* 31 Deletions 279
        3. Filling in the 5′ Overhangs of the Deleted Plasmids 280
        4. Digestion with the Second Restriction Enzyme 280
        5. Vector Preparation 280
        6. Recircularization of the Deleted Fragments with the New Vector 281
        7. Transformation of Recircularized Plasmids and Estimation of the Sizes
           of Deletion Subclones 281
II. Oligonucleotide Site-Directed Mutagenesis 281
    A. PCR-Based Mutation of DNA 282
    B. Generation of Single-Strand Phagemid DNA 284
    C. Performing Site-Directed Mutagenesis on Single-Strand Phagemid DNA 284
    Reagents Needed 285
References 289

DNA mutagenesis is generally defined as any process through which an altered deoxyribonucleic acid (DNA) sequence is produced. It is a technology that plays an important role in determining the functions of different gene domains; in the identification of transcriptional regulatory sequences such as promoters, enhancers and silencer elements; and in the production of DNA constructs.[1-6] While there are many variations on mutagenesis techniques, the present chapter describes several well-established methods, including DNA deletions and oligonucleotide site-directed mutagenesis. However, it does not address the use of mutagenic reagents or the use of radiation to generate random mutations.

The general principles and procedures discussed in this chapter are as follows:

1. The DNA of interest is mutated *in vitro* by an appropriate method followed by ligation to a vector.
2. The recombinant DNA construct ligations are then transferred into an appropriate cell line or host organism using selective media to allow only the growth of cells containing the constructs of interest. This is followed by identification of correct constructs using miniprep DNA preparations and/or polymerase chain reaction (PCR) methods, as detailed in **Chapter 1, Chapter 4,** and **Chapter 9**.
3. Once the new constructs are introduced into the study organism, the effects or functions of the mutated gene or DNA fragment can then be analyzed in transfected cells or from transgenic animals or plants.[3-10]

I. Mutagenesis by Deletion

Deletion is a simple and commonly used method of mutagenesis. It has found use in the generation of truncated DNA fragments for sequence analysis, where much of the sequence is unknown, or for analysis of promoter-region function. If enough sequence information is known, and appropriate restriction sites are present within the DNA of interest, deletion constructs can be made simply by digesting the DNA to remove various regions. If blunt ends or compatible cohesive termini are available after restriction enzyme digestion, the truncated DNA and vector can be directly ligated to each other using DNA ligase. If, however, the DNA and vector ends are incompatible, the Klenow fragment of *E. coli* DNA polymerase I, bacteriophage T4 DNA polymerase, mung-bean nuclease, or nuclease S1 can be used to generate blunt ends so that the fragments can be ligated and recircularized with DNA ligase. It is important to remember that different restriction endonucleases may generate compatible overhanging ends. For example, a *BamH* I cohesive end is compatible with ends resulting from digestion with *Bgl* II, *Bcl* I, *Dpn* II, and *BstY* I. It is always best to consult compatibility tables for this type of information. The methods necessary for restriction enzyme digestion, ligation, DNA isolation, and identification of correct clones have already been detailed in **Chapter 1, Chapter 9,** and **Chapter 10**.

More commonly, however, such fortuitive sites are not present. In these cases, nested series of deletions can be produced, using exonuclease III or *Bal* 31, by progressively digesting sequences from the ends of the target DNA. If the recombinant DNA is a plasmid, the plasmid should first be linearized with an appropriate restriction enzyme that cleaves at one end of the target sequence based on known DNA sequences. The linearized DNA is subjected to time-course digestions by the exonuclease, and the overhanging ends undergo treatment to generate blunt termini for efficient ligation. The truncated DNA can then be ligated to an appropriate vector with T4 DNA ligase and transformed into *E. coli*. Subsequent analysis of miniprep DNA will identify colonies containing distinct deletion constructs.

A. Performing Exonuclease III Deletions

Exonuclease III is an enzyme that catalyzes the stepwise removal of mononucleotides from 3'-hydroxyl termini of duplex DNA. A limited number of nucleotides are removed during each binding event, resulting in progressive deletions within the population of DNA molecules.[11] The preferred substrates are blunt or recessed 3'-termini, although the enzyme also acts at nicks in duplex DNA to produce single-strand gaps. The enzyme is not active on single-strand DNA, and thus 3'-protruding termini are resistant to cleavage. The degree of resistance depends on the length of the extension, with extensions of four bases or longer showing no cleavage. This property can be exploited to produce unidirectional deletions from a linear molecule with one resistant (3'-overhang) and one susceptible (blunt or 5'-overhang) terminus.[11] The overall procedure is depicted in Figure 12.1.

1. Subcloning of the DNA Insert

The DNA of interest should first be subcloned into an appropriate plasmid vector, such as one of the pGEM series (Promega Corp., Madison, WI), whose multiple cloning sites contain unique restriction sites located at the end of the DNA insert requiring deletion. One of the two unique restriction sites should be near the end of the insert and must generate a blunt or 5' overhang end that is necessary for exonuclease III activity. For unidirectional deletions, the other enzyme must produce a 3' overhang end to protect the rest of the plasmid from exonuclease III activity. The unique restriction enzymes are listed in Table 12.1 and **Table 12.2**, respectively. The general procedures for subcloning of the DNA of interest into plasmids are described in detail in **Chapter 10**.

2. Purification of Recombinant Plasmids for Exonuclease Deletions

The quality and integrity of isolated plasmids are critical for successful deletions with exonuclease III. Any nicks in the plasmids may cause exonuclease III digestion to occur from the nick points, resulting in single-strand gaps. Therefore, uniform, circular double-strand plasmids should be prepared and used for the enzyme deletion. We recommend the following two protocols:

a. Purification of Circular Plasmid DNA by Equilibrium Sedimentation in CsCl-Ethidium Bromide Gradients

CsCl gradients, while a bit more complicated than alkaline lysis methods, are known to give high yields of intact, nick-free plasmids. The detailed procedure for this type of isolation is described under the section on purification of plasmid DNA in **Chapter 1**.

b. Purification of Supercoiled Plasmid DNA by the Acid-Phenol Extraction Method

1. Isolate plasmids using standard alkaline lysis methods as described in **Chapter 1**. The method used should contain an *RNase* digestion step, and the DNA should be dissolved in distilled, deionized water (ddH$_2$O) (1 to 2 μg/μl).
2. To every 0.1 ml DNA sample, add 5 μl of 1 M sodium acetate buffer (pH 4.0) and 3.8 μl of 2 M NaCl solution. Mix well by pipetting.
3. Remove the nicked and linear DNA by adding one volume of acid-phenol to the mixture. Mix well by vortexing.
4. Centrifuge at 11,000 g for 5 min. Then, carefully transfer the top, aqueous phase (supernatant) to a fresh tube.
5. Repeat steps 3 to 4 twice.
6. Add 0.1 volume of 0.5 M Tris-HCl (pH 8.6) and extract once with 1 volume of chloroform:isoamyl alcohol (24:1). Mix well.
7. Centrifuge at 11,000 g for 5 min and carefully transfer the upper, aqueous phase to a fresh tube.
8. Add 0.1 volume of 2 M NaCl solution and 2 to 4 volumes of 100% ethanol. Allow preparation to incubate at –70°C for 30 min.

9. Centrifuge at 12,000 g for 15 min, decant the supernatant, and briefly rinse the pellet with 1 ml of 70% ethanol. Dry the DNA pellet under vacuum for several minutes.
10. Dissolve the DNA in 10 to 50 µl of TE buffer, measure the concentration of the sample, and store at −20°C until use.

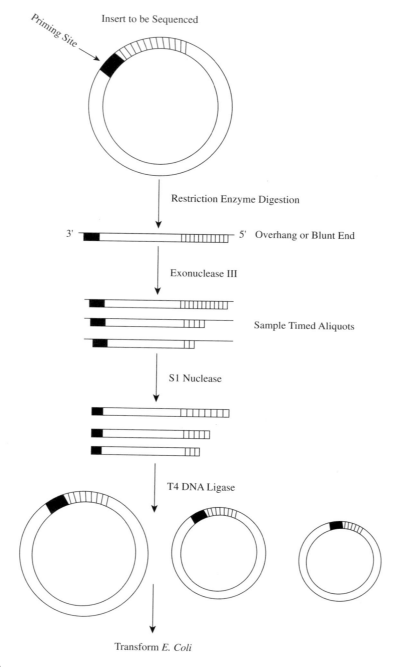

FIGURE 12.1
Diagram showing progressive deletions of a cloned DNA fragment resulting from the exonuclease III protocol. (From Kaufman, P.B., Wu, W., Kim, D., and Cseke, L.J., *Handbook of Molecular and Cellular Methods in Biology and Medicine*, 1st ed., CRC Press, Boca Raton, FL, 1995.)

TABLE 12.1
Examples of Unique Enzymes for Generating 5′ Protruding or Blunt Ends

Enzyme	Recognition Sequence	Enzyme	Recognition Sequence
Not I	5′..GC*GGCC GC..3′ 3′..CG CCGG*CG..5′	Sma I	5′..CCC*GGG..3′ 3′..GGG*CCC..5′
Xba I	5′..T*CTAG A..3′ 3′..A GATC*T..5′	Xho I	5′..CT*CGA G..3′ 3′..GA GCT*C..5′
Sal I	5′..G*TCGA C..3′ 3′..C AGCT*G..5′		

Note: The asterisks indicate the restriction enzymerestriction enzyme cutting site.

TABLE 12.2
Examples of Unique Enzymes That Produce Exonuclease III-Resistant 3′ Protruding (Overhang) Termini

Enzyme	Recognition Sequence	Enzyme	Recognition Sequence
Sph I	5′..G CATG*C..3′ 3′..C*GTAC G..5′	Pvu I	5′..CG AT*CG..3′ 3′..GC*TA GC..5′
Sac I	5′..G AGCT*C..3′ 3′..C*TCGA G..5′	Kpn I	5′..G GTAC*C..3′ 3′..C*CATG G..5′
Aat II	5′..G ACGT*C..3′ 3′..C*TGCA G..5′	Bgl I	5′..GCCN NNN*NGGC..3′ 3′..CGGN*NNN NCCG..5′

Note: The asterisks indicate the restriction enzyme cutting site.

3. Double-Restriction Enzyme Digestions of Circular Plasmids

This is an essential step for successful unidirectional deletion of the DNA insert, which will be carried out later. According to the restriction enzyme recognition sequence information that is generated during the subcloning process above, two unique and different restriction enzymes should be located near the end of the DNA insert to be deleted. One enzyme cleaves close to the target DNA and must produce blunt or 5′ overhang ends that will allow exonuclease III to unidirectionally digest the DNA. The other enzyme must generate a 3′ overhang end that is resistant to exonuclease III digestion in order to protect the rest of the plasmid from degradation (see Table 12.1 and Table 12.2 for examples).

1. Set up a standard **double-restriction enzyme digestion** as follows:
   - Plasmid DNA or DNA to be inserted (20 µg)
   - 10X appropriate restriction enzyme buffer (10 µl)
   - 1 mg/ml acetylated bovine serum albumin (BSA) (required only with specific enzymes) (1 µl)
   - Appropriate restriction enzyme A (3.4 units/µg DNA)
   - Appropriate restriction enzyme B (3.4 units/µg DNA)
   - Add ddH$_2$O to a final volume of 100 µl.

Note: *For double-restriction enzyme digestions, the appropriate 10X buffer containing a higher NaCl concentration than the other buffer may be chosen for the double-enzyme digestion buffer. There are also charts available from each company that give recommended buffers for a wide variety of restriction enzyme combinations.*

2. Incubate at the appropriate temperature, depending on the restriction enzyme (e.g., 37°C) for 2 to 3 h.

Note: The digestion efficiency can be checked by loading 2 µl of the digestion mixture with loading buffer onto a 1% agarose minigel. When running such gels, undigested constructs and standard DNA markers should also be loaded in the adjacent wells

3. Extract with one volume of phenol:chloroform:isoamyl alcohol (25:24:1). Mix well by vortexing for 1 min and centrifuge at 11,000 g for 5 min at room temperature.
4. Carefully transfer the top, aqueous phase to a fresh tube and add one volume of chloroform:isoamyl alcohol (24:1) to the supernatant. Mix well and centrifuge as in step 3.
5. Carefully transfer the upper, aqueous phase to a fresh tube and add 0.1 volume of 3 M sodium acetate buffer (pH 5.2) or 0.5 volume of 7.5 M ammonium acetate to the supernatant. Briefly mix and add 2 to 2.5 volumes of chilled 100% ethanol to the supernatant.
6. Centrifuge at 12,000 g for 10 min and carefully decant the supernatant. Briefly rinse the DNA pellet with 1 ml of 70% ethanol and dry the pellet briefly under vacuum. Dissolve the DNA pellet in 40 µl ddH$_2$O. Take 2 µl of the sample to measure the concentration of the DNA at 260 nm. Store the sample at –20°C until use.

Note: Addition of 0.5 volume of 7.5 M ammonium acetate to the supernatant at step 5 yields a higher amount of precipitated DNA than by adding 0.1 volume of 3 M sodium acetate buffer (pH 5.2).

4. Performing Series Deletions of the Linearized DNA with Exonuclease III

1. Label 15 to 30 microcentrifuge tubes (0.5 ml), depending on the size of insert to be deleted, and add 7.5 to 8.0 µl of fresh nuclease S1 mixture to each tube. Store on ice until use.
2. Dissolve or dilute 5 to 6 µg linearized DNA in a total 60 µl of 1X exonuclease III buffer.
3. Warm the sample at an appropriate temperature, start digestion by adding 250 to 550 units of exonuclease III, and mix as quickly as possible. Transfer 2.5 to 3.0 µl of the reaction mixture at 0.5-min intervals to the labeled tubes containing nuclease S1 (step 1). Quickly mix by pipetting up and down several times and place on ice until use.

Notes: (1) The amount of DNA, the volume of the reaction mixture, and the amount of exonuclease III should be optimized experimentally, depending on the size of insert to be deleted. (2) The digestion rate depends on the reaction mixture temperature, as shown in Table 12.3. (3) One does not need to change buffer between exonuclease III and nuclease S1, since the S1 buffer contains zinc cations and has a low pH, which can inactivate the exonuclease III. A small amount of exonuclease III buffer will not inhibit the activity of nuclease S1.

4. After all of the samples have been taken, transfer all of the tubes to room temperature and carry out nuclease S1 digestion of single DNA strands by incubating at room temperature for 30 to 35 min.
5. Terminate the reaction by adding 1 µl of S1 stopping buffer and heat at 70°C for 10 min.
6. Check the efficiency of digestion by preparing a 1% agarose gel using 2 to 3 µl of sample from each time point.

**TABLE 12.3
Dependence of Digestion Rate on Temperature**

Temperature (°C)	Digestion Rate (bp·min^{-1})
4	25–30
22	80–85
25	90–100
30	200–220
37	455–465
45	600–630

7. Carry out blunting of termini by adding 1 µl of Klenow mixture and 1 µl of dNTP mixture to each tube and incubate at 37°C for 10 min.
8. Recircularize the DNA by adding 40 µl of T4 DNA ligase mixture to each tube, mixing, and then incubating at room temperature for 60 to 70 min.

5. Transformation Using Recircularized Plasmids and Estimation of the Sizes of Deletion Subclones

The newly recircularized DNA constructs must now be introduced into an appropriate host and plated for single colonies. The general procedures for transformation and selection of transformants are described in detail under the section on subcloning in **Chapter 10**.

The approximate size of the various deletions at each time point can be determined using a miniprep of DNA treated with restriction enzymes that cut on either side of the truncated DNA inserts. We recommend that at least five or six minipreps be done on each time point in order to give a modest representation of the fragments resulting from each digestion. Once the DNA is cut and run on a gel with appropriate DNA markers, constructs containing the desired size fragments can be cultured further and stored for future use. The miniprep DNA can also be used for DNA sequencing (**Chapter 11**), subcloning (**Chapter 10**), or other analysis. In some cases, such as when restriction sites are not available, it may be desirable to perform PCR on each miniprep to determine the size of the fragments. Such reactions require the presence of PCR primers that flank the region of the deletion, and useful methods for these procedures can be found in **Chapter 9** and **Chapter 10**.

Note: *The resulting PCR fragments can often be sequenced directly.*

If a simple test for approximate sizes of digested inserts is desired, instead of going through the process of full minipreps, the following **fast screening method** may be of use.[12]

1. Mark five to ten microcentrifuge tubes (0.5 ml) for each time point of deletions and the transformant colonies to be analyzed in the same order.
2. Individually pick a portion (approximately two-thirds) of each of five to ten transformant colonies for each time point of the deletions using sterile toothpicks or pipette tips, and transfer each to the bottom of the appropriate tubes labeled in step 1. Wrap the plates containing partial colonies of transformants and store at 4°C until future use.
3. Add 50 µl of 10 mM EDTA (pH 8.0) to each tube and suspend the cells by vortexing for 1 min.
4. Add 50 µl of 2X lysis buffer to each tube, vortex for 1 min, and incubate at 70°C for 5 min. Cool to room temperature.
5. Add 3 µl of 2 M KCl and 0.5 µl of 0.2% bromophenol blue and vortex. Incubate on ice for 5 min.
6. Centrifuge at 9000 g for 4 min at 4°C. Transfer the supernatant to fresh tubes.
7. Load appropriate supercoiled DNA markers and 50 to 60 µl of each supernatant onto a 0.7% agarose gel containing ethidium bromide (10 µl of 10 mg/ml ethidium bromide per 100 ml of gel mixture). Allow electrophoresis to occur until the dye reaches three-quarters the length of the gel.
8. Culture appropriate colonies from the plates stored at 4°C, which contain the expected sizes after deletions, and isolate plasmids as described in **Chapter 1**.

B. Performing *Bal* 31 Deletions

Bal 31 nuclease has two types of activity. It acts as an exonuclease, degrading double-strand DNA and RNA from both 5′ and 3′ ends. It also possesses single-strand endonuclease activity that can act at sites of nicks within the DNA.[13] Consequently, while the strategy for the use of *Bal* 31 to make deletions is essentially the same as that of exonuclease III, one must keep in mind that the subsequent digestions are not unidirectional. Still, useful deletions can be obtained if appropriate steps are incorporated into the procedure. Figure 12.2 depicts the overall strategy.

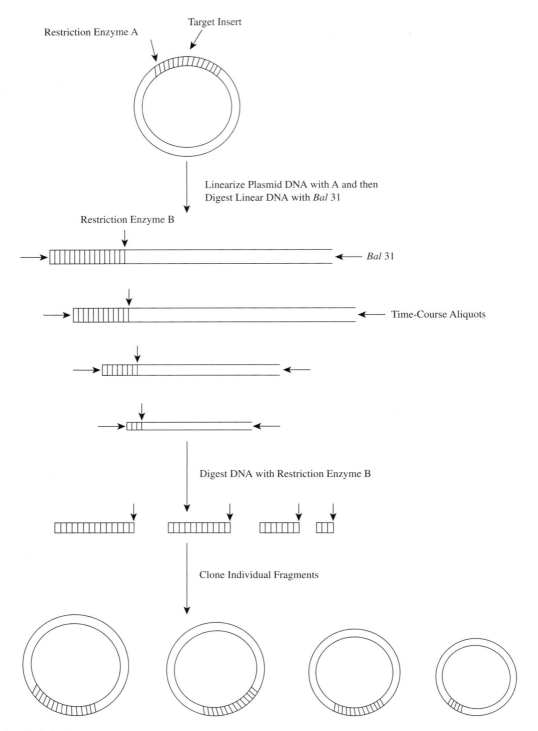

FIGURE 12.2
Mutagensis strategy by nested deletions with *Bal* 31. (From Kaufman, P.B., Wu, W., Kim, D., and Cseke, L.J., *Handbook of Molecular and Cellular Methods in Biology and Medicine,* 1st ed., CRC Press, Boca Raton, FL, 1995.)

DNA Site-Directed and Deletion Mutagenesis

As described above for exonuclease III, some sequence information and restriction-endonuclease locations must be known. The following procedure makes use of two unique restriction-endonuclease sites. One is located near the end of the DNA region to be deleted and will linearize the entire plasmid. The other site is located within the DNA insert of interest and will eventually act as a site that will make one end of each bidirectional deletion uniform in all fragments, releasing the rest of the truncated plasmid vector (Figure 12.2). Just like with exonuclease III, it is essential to prepare nick-free DNA constructs using either CsCl gradient purification or the acid-phenol protocol detailed in **Section I.A.2.a** and **Section I.A.2.b**, respectively.

1. Linearizing Recombinant Plasmids or DNA Fragments Using Appropriate Endonuclease(s)

A nick-free plasmid or a DNA fragment of interest is first linearized by using a restriction-endonuclease site at one end of the DNA fragment of interest.[14] In this protocol, we prefer to make use of restriction enzymes that yield 5' overhangs (Table 12.1).

1. Set up a standard **single-restriction enzyme digestion** as follows:
   - Plasmid DNA (20 µg)
   - 10X appropriate restriction enzyme buffer (10 µl)
   - 1 mg/ml acetylated BSA (add only if required) (1 µl)
   - Appropriate restriction enzyme (3.4 units/µg DNA)
   - Add ddH$_2$O to a final volume of 100 µl.

   In some cases, it may be necessary to digest the DNA with more than one enzyme. This is often the case when a nonvector DNA fragment is to undergo deletions followed by ligation to an appropriate vector.

2. Incubate at the appropriate temperature, depending on the restriction enzyme (e.g., 37°C), for 2 to 3 h, and complete the purification steps detailed above in steps 3 through 6 in **Section I.A.3** for the double-restriction digestions used in preparation for exonuclease III deletions.

2. Performing *Bal* 31 Deletions

1. Add 10 µg of the linearized recombinant plasmid (diluted to 90 µl with ddH$_2$O) to a microcentrifuge tube containing 90 µl of 2X *Bal* 31 buffer, mix well, and incubate at 30 or 37°C. Lower temperatures will result in slower nuclease activity.
2. Label six microcentrifuge tubes (numbered one to six) at room temperature and add 70 µl of phenol:chloroform:isoamyl alcohol (25:24:1) to each tube.
3. Transfer 30 µl of the DNA mixture from step 1 into tube 1 and mix by vortexing. This represents the zero time-point for the enzyme degradation. Meanwhile, add 3 µl of diluted *Bal* 31 solution (0.2 unit/µl) to the remaining DNA mixture in step 1.
4. Transfer 30 µl of the *Bal* 31/DNA mixture from step 1 into tubes numbered two to six at intervals of 2 to 5 min, respectively. Immediately vortex the tubes to stop the *Bal* 31 activity.
5. Centrifuge all tubes at 11,000 g for 5 min at room temperature.
6. Carefully transfer the top, aqueous phase to fresh tubes and add one volume of chloroform:isoamyl alcohol (24:1) to the supernatant. Mix well and centrifuge as in step 5.
7. Carefully transfer the upper, aqueous phase to fresh tubes and add 0.1 volume of 3 *M* sodium acetate buffer (pH 5.2) or 0.5 volume of 7.5 *M* ammonium acetate to each supernatant. Briefly mix and add 2 to 2.5 volumes of chilled 100% ethanol to the supernatant.
8. Centrifuge at 12,000 g for 10 min and carefully decant the supernatant. Briefly rinse the DNA pellet with 1 ml of 70% ethanol and dry the pellet under vacuum for several minutes. Dissolve the DNA pellet in 20 µl ddH$_2$O. Store the samples at −20°C until use.
9. Take 2 µl from each sample corresponding to each of the time points and analyze the extent of deletion by agarose gel electrophoresis as described previously.

3. Filling in the 5′ Overhangs of the Deleted Plasmids

At this point, each tube will contain a mixture of various sizes of deleted fragments of interest as well as the original plasmid. For the purpose of subcloning the truncated fragments of interest, we make use of a second restriction site located within the DNA fragment of interest (Figure 12.2). Ideally, this site will be found in the new vector that will be prepared for ligation below. If so, then the vector can be prepared with one blunt end and one specific end, thus allowing directional cloning. If this is not the case, then both ends should be blunted, and the reader should first do the steps detailed in the next section, "Digestion with the Second Restriction Enzyme," before doing these procedures.

1. Set up the 5′ overhang reaction mixture for each deletion in a microcentrifuge tube on ice as follows:
   - Linearized and deleted plasmid (10 µl)
   - 10X 5′ overhang buffer (10 µl)
   - 2 mM dNTP mixture (10 µl)
   - DNA polymerase I (Klenow fragment) (10 units)
   - Add ddH$_2$O to a final volume of 100 µl.
2. Incubate at room temperature for 20 min and heat at 70°C for 5 min to stop the reaction.
3. Add one volume of TE buffer (pH 7.5) and carry out phenol:chloroform extraction and precipitation as described in **Section I.A.3**. Dissolve the DNA in 40 µl of ddH$_2$O.

4. Digestion with the Second Restriction Enzyme

This step creates a protruding cohesive end that, if planned properly, will allow for directional cloning of the fragment of interest as well as a uniform end for each of the timed digestions above. If the second restriction site does not exist in the cloning vector into which the fragments will be cloned, or if it will generate constructs with fragments in the wrong direction, this portion of the protocol should be performed prior to blunting the ends of the fragments.

1. Using all 40 µl of DNA prepared just above, set up a single-restriction digest as described above for linearizing the plasmid.
2. Incubate at the appropriate temperature depending on the restriction enzyme (e.g., 37°C) for 2 to 3 h, and complete the purification steps detailed in steps 3 through 6 in **Section I.A.3** for the double-restriction digestions used in preparation for exonuclease III deletions.

Note: *The deleted fragments can be isolated from deleted plasmid fragments using a variety of methods (see* **Chapter 1***); however, this is not absolutely necessary for this procedure.*

5. Vector Preparation

A new vector also needs to be prepared for ligation of the truncated DNAs. If the correct sites are available, then this can be done simply by performing a double-restriction enzyme digestion as detailed in preparation for exonuclease III digestions (**Section I.A.3**). One enzyme must create a blunt end, and the other will be specific to the chosen internal site in the fragment of interest. However, if such a site is not available, then a single blunt-end digestion can be done, or the ends of the vector cut with an enzyme generating a 5′ protruding cohesive end can be blunted as detailed in **Section I.B.3**.[14] After completion of restriction enzyme digestion, treatment with shrimp or calf intestinal alkaline phosphatase (CIAP) is highly recommended. This treatment removes 5′-phosphate groups, thus preventing recirculization of the vector during ligation.

1. Carry out CIAP treatment by adding the following directly to the single-enzyme-digested or blunted-ended plasmid vector DNA (20 µl).
   - 10X CIAP buffer (15 µl)
   - CIAP diluted in 10X CIAP buffer (0.01 unit/pmol ends)
   - Add ddH$_2$O to a final volume of 150 µl.

DNA Site-Directed and Deletion Mutagenesis

Notes: Calculation of the amount of ends in pmol is as follows: There should be 19.6 µg digested DNA left after taking 0.4 µg of 20 µg digested DNA for checking on an agarose gel. If the vector is 3.2 kb (kilobase-pairs) in size, the amount of ends can be calculated by the formula below:

$$\text{pmol of ends} = \frac{\text{amount of DNA}}{\text{(base pairs} \times 660/\text{base pairs)}} \times 2$$

$$= \frac{19.6}{3.2 \times 1000 \times 660} \times 2$$

$$= 1.86 \times 10^{-5} \ \mu M$$

$$= 18.6 \ \text{pmol of ends}$$

2. Incubate at 37°C for 1 h and add 2 µl of 0.5 M EDTA (pH 8.0) to stop the reaction.
3. Extract with one volume of phenol:chloroform:isoamyl alcohol (25:24:1). Mix well by vortexing for 1 min and centrifuge at 11,000 g for 5 min at room temperature.
4. Carefully transfer the top, aqueous phase to a fresh tube and add one volume of chloroform:isoamyl alcohol (24:1) to the supernatant. Mix well and centrifuge as in step 5.
5. Carefully transfer the upper, aqueous phase to a fresh tube and add 0.1 volume of 3 M sodium acetate buffer (pH 5.2) or 0.5 volume of 7.5 M ammonium acetate to the supernatant. Briefly mix and add 2 to 2.5 volumes of chilled 100% ethanol to the supernatant.
6. Centrifuge at 12,000 g for 10 min and carefully decant the supernatant. Briefly rinse the DNA pellet with 1 ml of 70% ethanol and dry the pellet briefly under vacuum. Dissolve the DNA pellet in 40 µl ddH$_2$O. Take 2 µl of the sample to measure the concentration of the DNA spectrophotometrically at 260 nm. Store the sample at –20°C until use.

6. Recircularization of the Deleted Fragments with the New Vector

1. On ice, set up the following ligations for each of the desired deletions:
   - Linearized, deleted DNA fragments (500 ng)
   - Prepared vector (50 ng)
   - 10X T4 DNA ligase buffer (3 µl)
   - T4 DNA ligase (10 Weiss units)
   - Add ddH$_2$O to a final volume of 30 µl.
2. Incubate the reactions at 4°C for 12 to 24 h, or at 16°C for 4 to 6 h, or at room temperature (22 to 25°C) for 1 to 2 h.

Note: After ligation is finished at the above temperatures, the mixture can be stored at 4°C until later use.

7. Transformation of Recircularized Plasmids and Estimation of the Sizes of Deletion Subclones

Desired deletion subclones can now be identified using the methods described in the same section for exonuclease III (**Section I.A.5**). It is also recommended that linearized, full length plasmid be run as a control along with the digested miniprep DNA to identify any constructs resulting from the truncated vector.

II. Oligonucleotide Site-Directed Mutagenesis

Oligonucleotide site-directed mutagenesis is a powerful technique that has been widely used to generate site-specific mutations in any cloned gene or complementary DNA (cDNA) of interest.

The basic principle is that a mismatched oligonucleotide is synthesized, annealed to a specific region of single-strand target DNA, and elongated by DNA polymerase, generating a mutant second strand of DNA containing the modified sequence in the region of interest.

One of the most common types of mutagenesis done today makes use of standard PCR techniques to introduce sequence modifications such as new restriction-endonuclease sites at the ends of fragments. Such practices are essential to the development of a wide variety of DNA constructs, allowing DNA fragments of interest to be inserted into vector sequences in a highly specific and accurate manner. However, single-strand DNA can also be prepared upon which the synthetic primer can be used to alter the sequence and subsequently generate the changes on both strands (Figure 12.3). Both of these approaches are discussed here.

A. PCR-Based Mutation of DNA

The most important issue in the use of PCR to alter DNA sequences is proper planning and primer design. For this process to be successful, we recommend the following steps.

1. Obtain highly accurate DNA sequence where the sequence data on both strands of the DNA of interest exactly match each other.
2. Identify the location of the desired mutation(s). If genomic DNA is to be used, make sure intron positions are taken into account.
3. Design oligonucleotides using the following criteria:
   a. Place the restriction enzyme site in the desired location.
   b. Choose approximately 22 base-pairs (bp) toward the 3' end of the sequence that exactly match the sequence of the original DNA. In this region of the oligonucleotide, the G/C content should be between 40 and 60%, with an annealing temperature (Tm) of at least 50°C. The Tm can be estimated using the following formula.

$$Tm = 4(G/C) + 2(A/T) - 12°C$$

   We prefer the last six nucleotides at the 3' end of the primer to be exactly 50% G/C, with the very last base being a G or a C.
   c. Add an additional eight nucleotides to the 5' end of the restriction site that exactly match the original DNA sequence. This will not only help the PCR reaction to accommodate the mismatched bases, but it will also provide a tail of DNA, which will allow most restriction digestions to be done on the resulting PCR products with added restriction sites. The length of the necessary tail, however, does vary in some cases. This information can be obtained from the manufacturer.
4. Perform PCR reactions on a small amount of template DNA using standard PCR techniques (see **Chapter 9** and **Chapter 10**). Reaction conditions will vary according to the type of DNA used and the Tm of the primers along the ~22-bp stretch that matches the original sequence.
5. The resulting PCR products can be subcloned into an appropriate vector that makes use of the overhanging A base that is added to each end of each PCR fragment. PCR products with added restriction sites can also be subcloned directly using restriction digests and standard ligations to appropriate vectors/constructs (see **Chapter 10**).

Note: *Taq polymerases having proofreading functions are able to remove the overhanging A base during extended periods of time. If such proofreading polymerases are required, we strongly recommend that the PCR product be cloned immediately after the PCR reaction is done.*

DNA Site-Directed and Deletion Mutagenesis

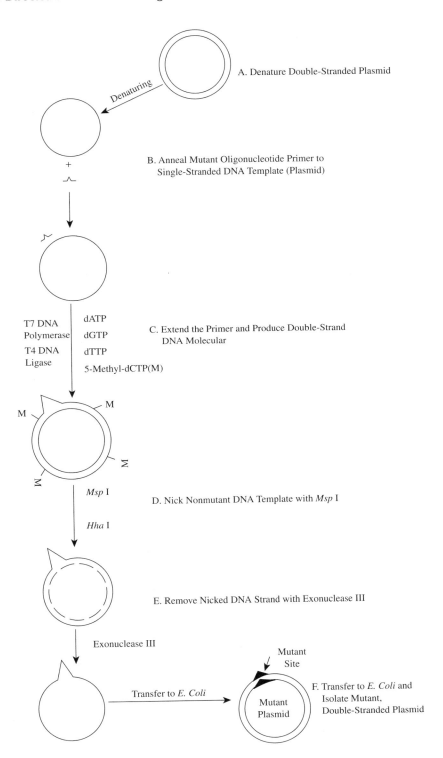

FIGURE 12.3
Strategy for generating site-specific mutations by oligonucleotide-directed *in vitro* mutagenesis. (From Kaufman, P.B., Wu, W., Kim, D., and Cseke, L.J., *Handbook of Molecular and Cellular Methods in Biology and Medicine*, 1st ed., CRC Press, Boca Raton, FL, 1995.)

B. Generation of Single-Strand Phagemid DNA

Longer stretches of altered DNA sequences can be prepared when mutagenic oligonucleotides are annealed to single-strand DNA. The target, single-strand DNA can be prepared by subcloning the target DNA fragment into a single-strand phage vector such as M13. An alternative way is to use a commercial Bluescript-based series of pCAT vectors containing the origin of replication of phage, called phagemids. The plasmid can generate single-strand F1-packaged phage with the aid of helper phage following transformation into *E. coli* strains harboring an F1 episome (Figure 12.3).

1. Plate out the transformants of *E. coli* that have the recombinant phagemids and the specific region of the DNA of interest on an LB plate. Invert the plate and incubate at 37°C to obtain single colonies.
2. Inoculate a single colony in 5 ml of LB medium containing 50 µg/ml ampicillin and 12 µg/ml tetracycline. Incubate at 37°C overnight with shaking at 150 rpm.
3. Add 0.3 ml of the overnight culture to 3 ml of superbroth in a 50-ml conical tube. Incubate at 37°C with shaking at 150 rpm for 2 to 3 h.
4. Add 8 µl of helper phage R408 (plaque-forming units [pfu] = 1×10^{11}, available from Stratagene, LaJolla, CA and Cedar Creek, TX) to the culture at step 3 and continue to incubate for 8 to 10 h.
5. Transfer 1.5 ml of the culture to each of two microcentrifuge tubes and centrifuge at 11,000 g for 2 min.
6. Transfer 1.2 ml of the supernatant from each tube to a fresh tube and add 0.3 ml of PEG precipitation buffer to the supernatant. Vortex for 1 min and leave at room temperature for 20 min.
7. Centrifuge at 12,000 g for 15 min and decant the supernatant completely.
8. Resuspend the PEG pellet in 0.3 ml of TE buffer (pH 8.0) and extract it with one volume of phenol:chloroform:isoamyl alcohol (25:24:1). Mix by vortexing and centrifuge at 11,000 g for 5 min.
9. Transfer the top, aqueous phase to a fresh tube and repeat extraction as in step 8 twice.
10. Precipitate the single-strand DNA by adding 0.5 volume of 7.5 M ammonium acetate (pH 7.5) and 2.5 volumes of chilled 100% ethanol to the supernatant. Place at –70°C for 30 min.
11. Centrifuge at 12,000 g for 20 min at 4°C, decant the supernatant, and briefly rinse the DNA pellet with 1 ml of 70% ethanol. Dry the pellet under vacuum for 30 min and dissolve the DNA in 10 µl of TE buffer (pH 7.6). Combine the two DNA samples into one tube and take 2 to 4 µl to measure the concentration of the single-strand DNA at 260 and 280 nm. Store the sample at –20°C until use.

C. Performing Site-Directed Mutagenesis on Single-Strand Phagemid DNA

After the single-strand DNA is obtained, the DNA is phosphorylated by T4 polynucleotide kinase. The synthetic primer containing the mutated, mismatched sequence is annealed to the single-strand DNA as a template. The second strand of DNA can be synthesized using T7 DNA polymerase and T4 ligase in the presence of 5-methyl-deoxycytosine triphosphate, generating a methylated new strand of DNA containing methylcytosine residues. However, the DNA template is not methylated and can be nicked at multiple points with the methylation-sensitive restriction-endonuclease *Msp* I. Incubation with exonuclease III results in the degradation of the nicked strand. The mutant, unnicked strand DNA can then be used to transform an appropriate nonrestrictive *E. coli* host (e.g., SDM cells). For efficient site-directed mutagenesis, a T7 *in vitro* mutagenesis kit is strongly recommended.

1. Set up the phosphorylation reaction of the mutant, mismatched oligonucleotides as follows in a microcentrifuge tube on ice:
    - Mutant oligonucleotides (200 pmol)
    - 10X T4 polynucleotides kinase buffer (2 µl)
    - 10 mM ATP (4 µl)

DNA Site-Directed and Deletion Mutagenesis

- T4 polynucleotide kinase (1 unit)
- Add ddH$_2$O to a final volume of 20 µl.

2. Incubate at 37°C for 40 min and heat at 70°C for 10 min to inactivate the enzyme.
3. Set up the annealing reaction in a microcentrifuge tube as follows:
    - Single-strand DNA template (6 µg)
    - Phosphorylated mutant oligonucleotide (15 pmol)
    - 5X annealing buffer (1.5 µl)
    - Add ddH$_2$O to a final volume of 20 µl.
4. Incubate at 65°C for 5 min and allow to cool slowly to room temperature. This can be done by placing the rack containing the tube in a beaker filled with 65°C water. The cooling procedure takes about 30 min.
5. Synthesize the mutant, methylated new strand by setting up the following reaction:
    - Annealed template and oligonucleotide primer (10 µl)
    - 10X synthesis mixture (5 µl)
    - T7 DNA polymerase (12.5 units)
    - T4 DNA ligase (25 units)
    - Add ddH$_2$O to a final volume of 50 µl.
6. Incubate at 4°C for 60 to 80 min and heat at 70°C for 10 min to terminate the reaction.
7. Carry out the nicking reaction of the DNA template (unmethylated strand) by adding 20 units of *Msp* I and 20 units of *Hha* I to the mixture and incubate at 37°C for 50 min.
8. Remove the nicked strand by adding 150 units of exonuclease III and incubate at 37°C for 50 min.
9. Stop the reaction by heating at 70°C for 10 min.
10. Take 1 µl of the reaction mixture to transform competent SDM *E. coli* cells using the electroporation method (see the section on subcloning in **Chapter 10**). It is recommended that one carry out at least five samples of transformation.
11. Pick up at least 15 colonies for preparation of single-strand phagemid DNA and carry out sequencing by the dideoxy method (see **Chapter 11**). Colonies determined to have the desired mutation(s) can be propagated and subcloned using standard techniques.

Reagents Needed

Acid-Phenol (pH 4.0)

Melt phenol crystals at 55 to 65°C in a water bath.
Add 250 ml of 50 m*M* sodium acetate buffer (pH 4.0) diluted from 1 *M* sodium acetate buffer to 250 ml of the melted phenol with stirring for 2 to 3 h at room temperature.
Stop the stirring and allow the phases to separate for 30 to 45 min.
Remove the top aqueous phase by aspiration. Remove a small aliquot to determine the pH. Repeat the second to fourth steps two or three times until the pH reaches 4.0. Store at 4°C until use.

7.5 M Ammonium Acetate

Dissolve 57.8 g ammonium acetate in 50 ml ddH$_2$O and adjust volume to 100 ml.

5X Annealing Buffer

200 m*M* Tris-HCl, pH 7.5
100 m*M* MgCl$_2$
250 m*M* NaCl

10 mM ATP

Dissolve in ddH$_2$O and store at −20°C. It is best to make this fresh.

2X Bal 31 Buffer

  40 mM Tris-HCl, pH 8.0
  24 mM CaCl$_2$
  24 mM MgCl$_2$
  400 mM NaCl
  2 mM EDTA, pH 8.0

Bal 31 Dilution Buffer

  20 mM Tris-HCl, pH 6.8
  0.1 M NaCl
  5 mM CaCl$_2$
  5 mM MgCl$_2$

Diluted Bal 31 Solution

  0.2 unit/µl in *Bal* 31 dilution buffer

Note: The diluted enzyme should be stored at 4°C instead of –20°C.

Calf Intestine Alkaline Phosphatase (CIAP)

10X CIAP Buffer

  0.5 M Tris-HCl, pH 9.0
  10 mM MgCl$_2$
  1 mM ZnCl$_2$
  10 mM Spermidine

Chloroform:Isoamyl Alcohol (24:1)

  48 ml Chloroform
  2 ml Isoamylalcohol
  Mix and store at room temperature.

dNTP Mixture

  0.13 mM dATP
  0.13 mM dTTP
  0.13 mM dGTP
  0.13 mM dCTP

2 mM dNTP Mixture

  2 mM each of dATP, dCTP, dGTP, and dTTP in 20 mM Tris-HCl, pH 7.5

0.5 M EDTA

  Dissolve 93.06 g Na$_2$EDTA·2H$_2$O in 300 ml ddH$_2$O. Adjust pH to 8.0 with 2 N NaOH solution. Adjust final volume to 500 ml with ddH$_2$O and store at room temperature.

10X Exonuclease III Buffer

  0.66 M Tris-HCl, pH 8.0
  6.6 mM MgCl$_2$

1X Exonuclease III Buffer

  Dilute 10X exonuclease III buffer in ddH$_2$O.

DNA Site-Directed and Deletion Mutagenesis

Klenow Buffer
    20 mM Tris-HCl, pH 8.0
    100 mM MgCl$_2$

Klenow Mixture
    6 to 12 units Klenow DNA polymerase
    60 µl Klenow buffer

LB (Luria-Bertaini) Medium
    Tryptone (10 g)
    Yeast extract (5 g)
    NaCl (5 g)
    Dissolve in ddH$_2$O and adjust the volume to 1 L. Adjust pH to 7.5 with 2 N NaOH. Autoclave.

2X Lysis Buffer
    8% (v/v) 2.5 M NaOH
    5% (v/v) of 10% SDS
    20% (w/v) Sucrose

Msp I, Hha I, and Exonuclease III (Promega)

Nuclease S1 Mixture (Fresh)
    54 µl of 10X S1 buffer
    0.344 ml ddH$_2$O
    120 units Nuclease S1

10X 5′ Overhang Buffer
    500 mM Tris-HCl, pH 7.2
    100 mM MgSO$_4$
    1 mM DTT (dithiothreitol)
    500 µg/ml BSA

PEG Precipitation Buffer
    3.5 M Ammonium acetate, pH 7.5
    30% (w/v) PEG, mol. wt. 8000

Phenol:Chloroform:Isoamyl Alcohol
    Mix one part of buffered phenol with one part of chloroform:isoamyl alcohol (24:1). Top the organic phase with TE buffer (about 1-cm height) and allow the phases to separate. Store at 4°C in a light-tight bottle.

2 M Potassium Chloride Solution
    Dissolve potassium chloride in ddH$_2$O. Autoclave and store at room temperature.

S1 Buffer
    2.5 M NaCl
    300 mM Potassium acetate, pH 4.6
    10 mM ZnSO$_4$
    50% Glycerol

S1 Stopping Buffer
>300 mM Tris base
>50 mM EDTA, pH 8.0

1 M Sodium Acetate Solution (pH 4.0)
>Dissolve sodium acetate in ddH$_2$O and adjust pH to 4.0 with 3 M glacial acetic acid. Autoclave and store at room temperature.

3 M Sodium Acetate Solution (pH 5.2)
>Dissolve sodium acetate in ddH$_2$O and adjust pH to 5.2 with 3 M glacial acetic acid. Autoclave and store at room temperature.

2 M Sodium Chloride Solution
>Dissolve sodium chloride in ddH$_2$O. Autoclave and store at room temperature.

Superbroth Medium
>Bacto-tryptone (12 g)
>Bacto-yeast extract (24 g)
>0.4% (v/v) Glycerol
>Dissolve in a total volume of 900 ml of ddH$_2$O and autoclave. Cool to about 50°C and add 100 ml of phosphate buffer containing 170 mM KH$_2$PO$_4$ and 720 mM K$_2$HPO$_4$. Autoclave again.

10X Synthesis Mixture
>0.1 M Tris-HCl, pH 7.5
>5 mM dATP
>5 mM dGTP
>5 mM dTTP
>5 mM 5-Methyl-dCTP
>20 mM DTT
>10 mM ATP

T4 DNA Ligase (Promega)

10X T4 DNA Ligase Buffer
>0.5 M Tris-HCl, pH 7.6
>0.1 M MgCl$_2$
>10 mM ATP

T4 DNA Ligase Mixture
>0.1 ml 10X Ligase buffer
>0.79 ml ddH$_2$O
>0.1 ml 50% PEG
>10 µl of 0.1 M DTT
>5 units T4 DNA ligase

T4 Polynucleotide Kinase
>1 unit/µl

10X T4 Polynucleotide Kinase Buffer
500 mM Tris-HCl, pH 7.5
100 mM MgCl$_2$
50 mM DTT
1 mM spermidine
1 mM EDTA

T7 DNA Polymerase I Klenow Fragment (Promega)
1 unit/µl

TE Buffer
10 mM Tris, pH 8.0
1 mM EDTA, pH 8.0

1 M Tris (pH 7.5 or 8.6)
Dissolve 121.1 g Tris-HCl in 800 ml ddH$_2$O. Adjust pH with 2 N HCl to desired pH. Adjust final volume to 1 l, autoclave, and store at room temperature.

References

1. **Sambrook, J., Fritsch, E.F., and Maniatis, T.,** *Molecular Cloning: A Laboratory Manual,* 2nd ed., Cold Spring Harbor Press, Cold Spring Harbor, NY, 1989.
2. **Dag, A.G., Bejarano, E.R., Buck, K.W., Burrell, M., and Lichtenstein, C.P.,** Expression of an antisense viral gene in transgenic tobacco confers resistance to the DNA virus tomato golden mosaic virus, *Proc. Natl. Acad. Sci. U.S.A.,* 88, 6721, 1991.
3. **Della-Cioppa, G., Bauer, S.C., Taylor, M.L., Rochester, D.E., Klein, B.K., Haughn, G.W., Smith, J., Mazur, B.J., and Somerville, C.,** Transformation with a mutant *Arabidopsis* acetolactate synthase gene renders tobacco resistant to sulfonylurea herbicides, *Mol. Gen. Genet.,* 211, 266, 1988.
4. **Przibilla, E., Heiss, S., Johanningmeier, U., and Trebst, A.,** Site-specific mutagenesis of the D1 subunit of photosystem II in wild type *Chlamydomonas, Plant Cell,* 3, 169, 1991.
5. **Fobert, P.R., Miki, B.L., and Iyer, V.N.,** Detection of gene regulatory signals in plants revealed by T-DNA-mediated fusions, *Plant Mol. Biol.,* 17, 837, 1991.
6. **Morrison, H.G. and Desrosiers, R.C.,** A PCR-based strategy for extensive mutagenesis of a target DNA sequencing, *BioTechniques,* 15, 454, 1993.
7. **Goldschmidt-Clermont, M.,** Transgenic expression of aminoglycoside adenine transferase in the chloroplast: a selectable marker for site-directed transformation of *Chlamydomonas, Nucleic Acids Res.,* 15, 4083, 1991.
8. **Bloch, C.A. and Ausubel, F.M.,** Paraquat-mediated selection for mutations in the manganese-superoxide dismutase gene *sod*A, *J. Bacteriol.,* 168, 795, 1986.
9. **Watkins, B.A., Davis, A.E., Cocchi, F., and Reitz, M.S.,** A rapid method for site-specific mutagenesis using larger plasmids as templates, *BioTechniques,* 15, 700, 1993.
10. **Klein, R., Silos-Santiago, I., Smeyne, R.J., Lira, S.A., Brambilla, R., Bryant, S., Zhang, L., Snider, W.D., and Barbacid, M.,** Disruption of the neurotrophin-3 receptor gene *trk*C eliminates Ia muscle afferents and results in abnormal movements, *Nature,* 368, 249, 1994.
11. **Henikoff, S.,** Unidirectional digestion with exonuclease III creates targeted breakpoints for DNA sequencing, *Gene,* 28, 351, 1984.
12. **Vandeyar, M.A., Weiner, M.P., Hutton, C.J., and Batt, C.A.,** A simple and rapid method for the selection of oligodeoxynucleotide-directed mutants, *Gene,* 65, 129, 1988.

13. **Bencen, G.H., Wei, C.F., Robberson, D.L., and Gray, H.B., Jr.,** Terminally directed hydrolysis of duplex ribonucleic acid catalyzed by a species of the BAL31 nuclease from *Alteromonas espejiana*, *J. Biol. Chem.*, 259, 13584, 1984.
14. **Hauser, C.R. and Gray, H.B., Jr.,** Vectors containing infrequently cleaved restriction sites for use in BAL31 nuclease-assisted and end-label-mediated analysis of cloned DNA fragments, *Genet. Anal. Tech. Appl.*, 8, 139–147, 1991.

Chapter 13

DNA Footprinting and Gel-Retardation Assay

William Wu and Leland J. Cseke

Contents

I. Footprinting Protocols 293
   A. Preparation of Single-End (3′ or 5′ End) Labeled DNA Probe 293
   B. Preparation of Denaturing Polyacrylamide Gels 293
   C. *DNase* I Protection Assay 293
   D. Electrophoresis 294
   E. Troubleshooting Guide 296
   Reagents Needed 296
II. Gel Retardation Assay Protocol 298
   A. Preparation of Single-End (3′ or 5′ end) Labeled DNA Probe 298
   B. Preparation of Protein Sample for the Assay 298
   C. Preparation of Nondenaturing Polyacrylamide Gels 298
   D. Formation of DNA–Protein Complexes and Electrophoresis 299
   E. Autoradiography of the Gel 299
   F. Troubleshooting Guide 301
   Reagents Needed 301
III. Footprinting Following the Gel Retardation Assay Using Copper–Phenanthroline 302
References 303

In prokaryotes and eukaryotes, the binding of proteins to specific deoxyribonucleic acid (DNA) sequences is important to the regulation of many cellular processes, such as replication, recombination, repair, and transcription. Gene expression in eukaryotic cells is regulated by *cis*-acting regulatory elements and *trans*-acting regulatory factors that control transcription initiation of the genes. Over the years, many DNA sequences responsible for this regulation have been identified, and special proteins binding to these regulatory sequences have been classified.

The most direct information about DNA sequences bound by specific proteins such as transcriptional factors or equivalent come from a powerful technique called DNA footprinting or *DNase* I protection analysis. This assay was developed as a qualitative technique to locate protein-binding sites on DNA.[1] In parallel to the isolation and purification of the DNA-binding proteins, this analysis was used as a quantitative assay of the binding affinities and strength of proteins to their specific binding sites.[2]

The basis of DNA footprinting is (1) the protection of the phosphodiester backbone of DNA from *DNase* I-catalyzed hydrolysis afforded by bound proteins, followed by (2) the separation of the hydrolysis products on denaturing DNA-sequencing gels, and (3) visualization of the binding sites by autoradiography.[3,4] First, a particular double-strand DNA (dsDNA) is labeled at the 5' end with ^{32}P or its equivalent and allowed to interact with particular proteins of interest.[5,6] The DNA–protein complex is then subjected to a DNA endonuclease digestion such as *DNase* I. However, the nicking by *DNase* I is so brief that no DNA molecule gets more than one single strand cut. The specific DNA fragment bound by protein should be resistant to *DNase* I nicking. If the DNA contains n base-pairs in which there are y base-pairs bound by the protein, then n different sizes of DNA pieces would exist in nonprotein-interacted DNA molecules as a control.[6-8] Since the protein binding prevents the specific DNA fragment from endonuclease digestion, only $n - y$ different sizes of DNA fragments are present in protein-binding DNA molecules.

These fragments from both control and DNA–protein-interacted samples can be separated by gel electrophoresis, and their sizes can be determined by the distance from the labeled 5' end to the nicks. The protein-binding region of the DNA can then be determined with ease. Furthermore, the actual points of the bound region of the DNA fragment can be identified by dimethyl sulfate protection techniques. Dimethyl sulfate methylates only A and G but not C and T. The methylation should not occur in the actual protein-contact region of the DNA fragment. The actual size of the regions of contact by the protein can then be identified by the positions of nonmethylated A and G in the entire bound region of the DNA fragment. The DNA–protein complex can be isolated and denatured for DNA and protein sequencing in order to identify the specific sequence bound by the protein.

The gel retardation assay or the gel mobility shift assay, on the other hand, are methods used to detect the specific protein or protein fraction that binds to a specific DNA sequence.[9] The gel retardation assay typically involves the addition of protein to defined end-labeled fragments of double-strand DNA, separation of protein-DNA complex from naked DNA by electrophoresis, and visualization of the DNA by autoradiography.[10,11] The DNA molecule is highly negatively charged and migrates quickly toward a positive electrode in an electric field. During a native or nondenaturing polyacrylamide gel electrophoresis, DNA molecules are separated based on their sizes. Smaller molecules migrate faster than larger ones. When a DNA molecule or DNA fragment is bound by protein(s), it moves more slowly through the gel. The larger the bound protein, the slower the DNA–protein complex migrates. Based on this principle of the gel retardation assay, DNA molecules or fragments of interest (usually its length and sequence are known) are radioactively or nonradioactively labeled and mixed with a specific protein or protein fraction for a specified period of time. The mixture is loaded onto a native polyacrylamide gel and subjected to electrophoresis followed by autoradiography or its equivalent. As compared with control DNA fragments without the protein mixture, the free DNA molecules or fragments will run rapidly close to the bottom of the gel, but those molecules or fragments bound by protein(s) are retarded above the free-DNA bands. All these bands are revealed on the developed x-ray film or an equivalent membrane. Factors that influence the electrophoretic mobility of DNA–protein complexes include the molecular weight of the protein and the DNA,[12] the ionic strength and the pH of the electrophoresis buffer, the concentration of the gel matrix, and temperature.[13]

The gel retardation assay has several advantages for characterization of the DNA–protein complexes.[3,4,14] First, femptomole or picomole quantities of sequence-specific binding proteins can be detected by their effect on the mobility of a radiolabeled target DNA, even when other DNA-binding proteins are present, as, for example, in a crude cell extract. The gel retardation assay can be completed within a few hours and is therefore ideal for monitoring the purification of *trans*-acting transcription factors and other sequence-specific binding proteins. Target sequences for a DNA-binding protein can be identified in a mixed population of DNA fragments. DNA–protein complexes can be resolved into several components that reflect alternative stoichiometries,

DNA Footprinting and Gel-Retardation Assay

interaction between several different proteins, or multiple proteins competing for the same site. The assay can easily be adapted for quantitative determination of association rate constants, dissociation rate constants, abundance, cooperativity, and specificity.

I. Footprinting Protocols

A. Preparation of Single-End (3' or 5' End) Labeled DNA Probe

DNA purification is described in **Chapter 1**. Radioactively labeled DNA fragments of interest are prepared by single-end labeling in order to achieve high specific activity. The labeling procedures are described in **Chapter 4**. The probe DNA is purified by extraction and precipitation with ethanol. The DNA is dissolved in an appropriate volume of TE buffer, and then the cpm (counts per minute) of the probe is measured. The sample is stored at –20°C until use.

B. Preparation of Denaturing Polyacrylamide Gels

1. Thoroughly clean an appropriate vertical gel electrophoresis apparatus, including glass plates (15 × 60 cm or equivalent), spacers (0.4 mm), tanks, and comb (0.4 mm). Siliconize one of the glass plates with 1 to 1.5 ml of Sigmacote™ (Sigma Chemicals, St. Louis, MO), wash it with ethanol, then wipe dry. Assemble the sandwich according to the instructions for the apparatus.
2. In a clean beaker, prepare 100 ml of the gel mixture as shown in Table 13.1.
3. Gently heat and stir the gel mixture until the urea is dissolved and filter or degas (optional). Cool to room temperature.
4. Add 250 µl of 10% ammonium persulfate and 40 µl of $N,N,N'N'$-tetramethylethylenediamine (TEMED) to the gel mixture, gently mix well, and then immediately pour the gel. Insert the comb and allow the gel to polymerize for 1 to 1.5 h. See **Chapter 11** for more details on pouring polyacrylamide gels.

Notes: The glass sandwich can be laid on a flat table or bench without sealing the bottom edge. The gel mixture can be poured horizontally by capillary action. If the gel is poured vertically or at an angle, the side and bottom edges must be sealed tightly to prevent leaking. The polymerized gel can be wrapped with wet paper towels at the top and bottom and covered with SaranWrap™ to prevent the gel from drying out. The gel can be left overnight until use.

TABLE 13.1
Preparation of Gel Mixture

Stock Solution	Polyacrylamide Gel Concentration (%)		
	5	6	7
5X TBE solution (ml)	20	20	20
40% Acrylamide (ml)	12.5	15	17.5
Ultrapure urea (g)	48	48	48
ddH$_2$O (ml)	32.5	30	27.5

C. *DNase* I Protection Assay

1. Prepare the DNA mixture as follows:
   - Single-end-labeled DNA fragments (2 to 4 × 10^5 cpm) (5 to 10 µl)
   - 20% Polyvinyl alcohol (10 µl)
   - Nonspecific competitor DNA poly(dI-dC) (1 mg/ml) (1 µl)
   - Add TE buffer to a final volume of 50 µl.

2. Prepare the protein mixture as follows:
   - 2X protein-binding buffer (25 μl)
   - DNA-binding protein (50 ng to 20 μg) (1 to 20 μl)
   - Add distilled, deionized water (ddH$_2$O) to a final volume of 50 μl.

Notes: *The extraction and purification of proteins (cytosolic and nuclear proteins) are described in **Chapter 3**. The protein used for the assay can be a specific protein or protein fraction of a total protein mixture.*

3. Set up a DNA–protein interaction by combining the protein mixture and DNA mixture into 100 μl and incubating the reaction mixture at room temperature for 20 min. Meanwhile, set up a free-DNA sample by mixing another 50 μl of the DNA mixture with 25 μl of 2X binding buffer and 25 μl ddH$_2$O.
4. Add one volume of salt solution to each tube and incubate for 1 to 2 min.
5. Add 10 μl of the diluted *DNase* I solution to each tube and incubate at room temperature for 1 min. Stop the reaction by adding 200 μl *DNase* I stopping solution followed by incubating at 37°C for 20 to 30 min.

Notes: *The DNase I should be prediluted from 1:1000 to 1:10,000 with 2X protein-binding buffer and salt solution. A series of DNase I protection assays should be carried out using the same amount of DNA in order to find the optimal dilution. Use the amount of DNase I that digests about one half of the labeled DNA probe in reaction step 5.*

6. Extract by adding 200 μl or one volume of phenol:chloroform:isoamyl alcohol (24:24:1) to each tube. Mix well by vortexing for 15 sec.
7. Centrifuge at 11,000 g for 5 min and carefully transfer the upper, aqueous phase to a fresh tube and add 2.5 volumes of chilled 100% ethanol to the tube. Allow precipitation to occur at –70°C for 30 min or at –20°C for 2 h.
8. Centrifuge at 12,000 g for 5 to 8 min, decant the supernatant, and briefly rinse the pellet with 1 ml of cold 70% ethanol. Dry the pellet under vacuum for 15 min. Dissolve the pellet in 10 μl of formamide-loading dye and retain until loading of sample into the gel.

Note: *The DNA–protein complex sample and control (free-DNA sample) should be loaded adjacent to each other in order to easily identify the missing bands that were bound by proteins.*

9. Determine the total radioactivity recovered for each sample by counting for 1 min in a scintillation counter (see **Chapter 4**).

D. Electrophoresis

1. Carefully remove the comb from the gel sandwich and rinse the wells with distilled water. Attach the sandwich to the apparatus, fill the tanks with 1X TBE solution, and flush the wells with buffer using a pipette.
2. Prerun the gel for 30 to 70 min at 45 to 70 W constant power, depending on the size of the gel. Circulate the buffer between tanks during the electrophoresis.
3. Stop the prerunning and heat the prepared samples (from step 8 in **Section I.C**) at 95°C for 2 to 4 min and immediately load the same amounts of radioactivity (6000 to 10,000 cpm per lane) onto the gel.
4. Connect the electrophoresis apparatus to the power supply and run the electrophoresis in 1X TBE buffer at the same constant power as for prerunning for 2 to 4 h or until the bromophenol blue is about 2 cm from the bottom of the gel.
5. Stop the electrophoresis, remove the gel plates from the tank, and allow to cool for 10 min at room temperature.
6. Slowly separate the glass plates, starting from one corner near the spacer, with a spatula or its equivalent. Remove the top plate and put it aside.

Notes: *(1) As long as the top glass plate starts to separate from the other plate, keep going and never allow the two glass plates to become attached again; otherwise, the gel may be torn. (2) The gel usually sticks on the bottom plate. If the gel sticks to the top glass plate, carefully turn the plate over and lay it on the table or bench.*

7. Carry out autoradiography as follows:
   a. Lay one piece of 3MM Whatman™ filter paper on the gel and gently press it onto the filter using an appropriate amount of pressure to make the gel adhere to the filter.
   b. Starting from one corner of the gel, carefully peel off the gel, together with the filter paper, and lay it on the table or bench. Cover the gel with SaranWrap and remove any air bubbles between the gel and the SaranWrap.
   c. Dry the gel at 80°C for 30 to 60 min under vacuum.
   d. Remove the SaranWrap from the gel and place it in the exposure cassette. Lay an x-ray film on the gel in the dark room and cover the cassette. Allow exposure to take place at room temperature or at –70°C with an intensifying screen for the desired time.
   e. Develop the x-ray film and analyze the data. When comparing with the control lane, the missing bands in DNA–protein-interacted sample are the regions bound by protein (Figure 13.1).

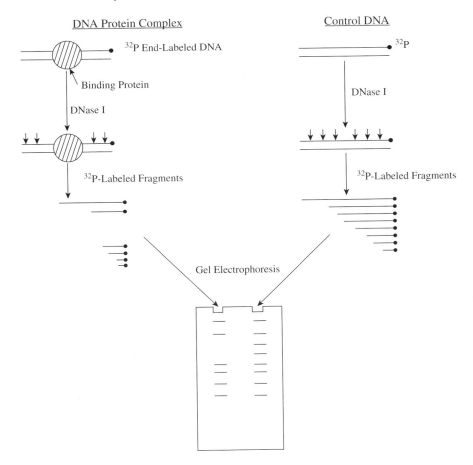

FIGURE 13.1
Diagram showing the nuclease protection assay of DNA–protein interactions. Following gel electrophoresis and autoradiography, the missing bands represent the regions that are bound by specific protein(s). (From Kaufman, P.B., Wu, W., Kim, D., and Cseke, L.J., *Handbook of Molecular and Cellular Methods in Biology and Medicine*, 1st ed., CRC Press, Boca Raton, FL, 1995.)

E. Troubleshooting Guide

1. Single-strand breaks in the end-labeled DNA fragment must be avoided because they give false signals indistinguishable from genuine *DNase* I cleavage and hence can mask an otherwise good footprint. It is therefore advisable to check the fragment on a denaturing gel before use. Always use a freshly labeled fragment (3 to 4 days at the most), because radiochemical nicking will degrade it.
2. Denaturing gels are not denaturing unless run hot. A double-strand form of the DNA fragment is therefore often seen on the autoradiogram, especially at low levels of *DNase* I digestion, and can sometimes be misinterpreted as a hypersensitive cleavage. By running a small quantity of undigested DNA fragment in parallel with the footprint, this error can be avoided.
3. The optimum concentration of magnesium should be determined empirically for each DNA fragment. With each new binding site or DNA fragment, a magnesium titration curve should be set up by varying the concentration of magnesium (0.1 mM to 10 mM) while keeping all other variables constant. Similarly, the optimum temperature (from 0 to 37) of binding and protection should be determined in trial reactions.
4. The concentration of *DNase* I required to produce a convincing ladder of DNA fragments usually varies between 50 and 500 ng/ml. The exact concentration must be determined empirically and can vary from one DNA to another and among different preparations of the same fragment. Cleavage depends on the specific activity of the *DNase* I, the purity of the DNA fragment, and the sequence of the DNA fragment in question. Some investigators find it easier to vary the incubation time or temperature in reaction condition, rather than to vary the amount of *DNase* I added to the reaction.

Reagents Needed

40% Acrylamide Stock Solution
Acrylamide (76 g)
N,N'-Methylene-*bis*-acrylamide (2 g)
Add ddH$_2$O to a final volume of 200 ml.
Heat at 37°C to dissolve each component completely after each addition, sterile-filter through a 2-µm membrane, and store at 4°C in an aluminum-foil-wrapped bottle for a period of up to 5 weeks.

Notes: Acrylamide is a potent neurotoxin. Always wear gloves when working with acrylamide and, when handling the dry powder, a face mask is also essential.

10% Ammonium Persulfate Solution (Fresh)
Ammonium persulfate (0.1 g)
Add ddH$_2$O to a final volume of 1 ml.

DNase I Stopping Solution
0.2 M NaCl
40 mM EDTA
1% SDS
125 µg/ml tRNA
100 µg/ml *Proteinase K* (add prior to use)

Enzyme
DNase I (1 mg/1 ml)
Dissolve the enzyme in 10 mM Tris-Cl (pH 8.0). Store the solution frozen in small aliquots at −20°C.

DNA Footprinting and Gel-Retardation Assay

Formamide-Loading Dye Buffer
   95% Formamide
   4% EDTA, pH 8.0
   0.2% Bromophenol blue
   0.2% Xylene cyanol

Phenol:Chloroform:Isoamyl Alcohol
   24:24:1 (v/v/v)
   Mix one part of buffered phenol with one part of chloroform:isoamyl alcohol (24:1). Top the organic phase with TE buffer (about 1-cm height) and allow the phases to separate. Store at 4°C in a light-tight bottle.

20% Polyvinyl Alcohol (Sigma Chemicals)
   Dissolve the polyvinyl alcohol in ddH$_2$O and store the solution frozen in 100-μl aliquots at −20°C.

2X Protein-Binding Buffer
   40 mM Tris-HCl, pH 7.9
   2 mM EDTA
   20% Glycerol
   0.2 ml Nonidet P-40
   4 mM MgCl$_2$
   2 mM DTT

Salt Solution
   10 mM MgCl$_2$
   5 mM CaCl$_2$

5X TBE Buffer
   Tris base (54 g)
   Boric acid (27.5 g)
   0.5 M EDTA, pH 8.0 (20 ml)
   Add ddH$_2$O to a final volume of 1000 ml.
   Autoclave.

1X TBE Buffer
   Dilute 5X TBE buffer with ddH$_2$O.

TE Buffer
   10 mM Tris-HCl, pH 8.0
   1 mM EDTA

Ultrapure Urea (Sigma Chemicals)

II. Gel Retardation Assay Protocol

A. Preparation of Single-End (3' or 5' end) Labeled DNA Probe

1. Prepare DNA fragments by digesting the DNA molecules of interest with appropriate restriction enzyme(s) that create a 3'- or 5'-end overhang.

Notes: DNA purification is described in **Chapter 1**. DNA containing the binding site of interest can come from complementary DNA (cDNA), genomic DNA, plasmid DNA, or a synthetic oligonucleotide. The sizes and/or sequences of the DNA fragments used for gel retardation assay should be known.

2. Radioactively label the DNA fragments of interest by end labeling (3' or 5' end) to achieve a high specific activity. The labeling procedures are described in **Chapter 4**. Purify the probe DNA by extraction and precipitation with ethanol. Dissolve the DNA in an appropriate volume of TE buffer and measure the cpm of the probe. Store the sample at –20°C until use.

B. Preparation of Protein Sample for the Assay

The extraction and purification of proteins (cytosolic and nucleic proteins) are described in **Chapter 3**. The protein used for the assay can be a specific protein or a protein fraction of a total protein mixture.

Notes: The protein or protein fraction must be nondenatured. The tertiary structure of the protein(s) is important for binding to the DNA fragment. For optimal binding, the amount of proteins should be in excess of the amount of DNA, usually 5 to 15 µg per lane in the gel for protein fraction or 50 ng to 5 µg per lane for a purified protein.

C. Preparation of Nondenaturing Polyacrylamide Gels

1. Thoroughly clean and assemble a vertical electrophoresis apparatus according to the instructions for the apparatus. The gel size is normally 14 × 19, 15 × 20, or 20 × 26 cm using 1- or 1.5-mm spacers. Insert an appropriate comb into the sandwich. Check if leaking occurs by using distilled water.

Notes: The comb can be inserted into the sandwich after pouring the gel mixture. However, air bubbles may be trapped around the teeth; this should be avoided.

2. Prepare the native gel mixture containing 4 to 8% acrylamide in the order listed below (for 100 ml gel, 4% polyacrylamide):
   - 5X TBE buffer (10 ml)
   - 30% (w/v) Acrylamide stock solution (13.3 ml)
   - ddH$_2$O (75 ml)
   - Mix well after each addition and degas (optional).
   - Add 0.9 ml of freshly prepared 10% (w/v) ammonium persulfate (AP) solution and 100 µl TEMED to the gel mixture. Gently mix and immediately pour the solution carefully into the glass sandwich or assembled mold.
3. Allow the gel to polymerize for about 1 h at room temperature and carefully remove the comb. Rinse the wells (5 × 3 × 1 or 1.5 mm) thoroughly with ddH$_2$O several times.
4. Attach the gel sandwich properly to the electrophoresis tanks. If a notched glass plate is used, it should face inward toward the buffer reservoir.

5. Fill the buffer tanks with 0.5X TBE running buffer. Flush out the wells to remove any air bubbles with a Pasteur pipette or its equivalent.
6. Connect to the power supply (positive electrode should be connected to the bottom tank) and prerun the gel for 1.5 h at 200 V at room temperature or at 350 V for 1.5 h at 4°C.

Notes: *Native or nondenaturing polyacrylamide gels should run at 3 to 8 V/cm. Too high a voltage can cause overheating, resulting in "bowing" of the DNA bands. Whenever possible, it is strongly recommended that recirculation of the running buffer between the top and bottom buffer tanks be maintained by using a pump. This will help maintain uniform pH and ionic strength throughout the system during the electrophoresis.*

D. Formation of DNA–Protein Complexes and Electrophoresis

1. While the gel is prerunning, set up the DNA and protein-binding reactions in microcentrifuge tubes as indicated below:
   - 2X DNA–protein-binding buffer (10 μl)
   - Nonspecific competitor DNA (1 mg/ml poly[dI-dC] or poly[dA-dT]) (1.5 μl)
   - DNA-binding protein (5.0 μl)
   - Add ddH$_2$O to a final volume of 20 μl.

Notes: *(1) The nonspecific competitor DNA serves to provide a large excess of low-affinity binding sites that absorb nonspecific DNA-binding proteins, thus increasing the detection of the specific protein–DNA complexes.[15] (2) The normal DNA competitors are synthetic copolymers such as poly(dI-dC) or poly(dA-dT).[16]*

2. Incubate the reaction at 22 to 25°C or desired temperature for 20 to 30 min.
3. Add 1 to 2 μl of labeled DNA fragments (1 to 500 fmole; specific activity >5000 cpm/fmole if ^{32}P-labeled) to the reaction and continue incubation at the same temperature for 10 to 30 min.
4. Add 2 μl of 10X loading dye to the binding reaction mixture and store at room temperature until use.
5. Stop the prerun and carefully load the reaction mixture onto the gel (10 μl per well) — along with the appropriate controls (i.e., free DNA controls).
6. Start the electrophoresis at the same voltage as for pre-electrophoresis for 1 to 2 h or until the bromophenol blue reaches a distance of 1 cm from the bottom of the gel. Maintain the running buffer circulation between tanks during electrophoresis.
7. Stop the electrophoresis, remove the gel plates from the tank, and allow to cool for 10 min at room temperature.
8. Slowly separate the glass plates, starting from one corner near the spacer, with a spatula or its equivalent. Remove the top plate and put it aside.

Notes: *(1) As long as the top glass plate starts to separate from the other plate, keep going and never allow the two glass plates to become attached again; otherwise, the gel may be torn. (2) The gel usually sticks to the bottom plate. If the gel sticks to the top glass plate, carefully turn the plate over and lay it on the table or bench.*

E. Autoradiography of the Gel

1. Direct autoradiography for gel elution. If the bands of the DNA–protein complex are to be recovered from the gel, the gel should not be fixed or dyed.
   a. Cover the gel properly without air bubbles present and wrap it together with the supporting glass plate using SaranWrap.

b. In a dark room with a safety light, place an x-ray film on the gel and stick the edges to the plate with tape. Wrap the whole apparatus with light-tight aluminum foil and invert the gel so that the glass plate on its top serves as a weight. Allow exposure to occur at room temperature or at −70°C with an intensifying screen for several hours or until desired bands are visible.

Note: Never use a metal film cassette because it may break the glass plate and crush the gel.

c. Develop the film and analyze the data. Free DNA fragments should be visualized as black bands down near the bottom of the gel, but DNA–protein-complex bands are retarded in their movement and thus occur at the upper part of the gel or above the free-DNA bands. The more proteins that are bound to the DNA, the stronger is the signal, and the smaller is the amount of free DNA that is present.

2. Dry the gel and carry out autoradiography as follows:
    a. Lay one piece of 3MM Whatman filter paper on the gel and gently press it onto the filter using an appropriate amount of pressure to make the gel stick to the filter.
    b. Starting from one corner of the gel, carefully peel off the gel, together with the filter paper, and lay it on the table or bench. Cover the gel with SaranWrap and remove any air bubbles that occur between the gel and the SaranWrap.
    c. Dry the gel at 80°C for 30 to 60 min under vacuum.
    d. Remove the SaranWrap from the gel and place it in an exposure cassette. Lay an x-ray film on the gel in the dark room and cover the cassette. Allow exposure to take place at room temperature or at −70°C with an intensifying screen for the desired time.
    e. Develop the x-ray film and analyze the data (Figure 13.2).

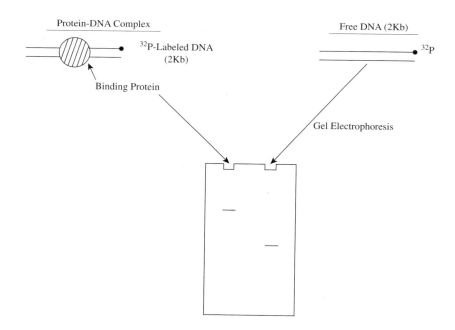

FIGURE 13.2
Schematic representation of gel retardation. DNA bound by specific protein(s) is retarded during gel electrophoresis as compared with the free-DNA control. (From Kaufman, P.B., Wu, W., Kim, D., and Cseke, L.J., *Handbook of Molecular and Cellular Methods in Biology and Medicine*, 1st ed., CRC Press, Boca Raton, FL, 1995.)

F. Troubleshooting Guide

1. The DNA fragment used in the reaction should be longer than 20 bp (base-pairs), with the recognition sequence at least 4 bp away from each end of the fragment. Polymerizing the recognition sequence by ligation of monomer sites or by chemical synthesis may enhance detection of rare or low-affinity binding proteins.
2. Types of retardation gel are variable. A 4 to 5% polyacrylamide gel with a ratio of acrylamide:bisacrylamide of 30:1 is recommended for proteins with a molecular mass of 15 to 500 kDa and DNA fragments 12 to 600 bp in length. High-molecular-weight DNA fragments can be separated on gels with the same percentage acrylamide but with a ratio of acrylamide:bisacrylamide of 50:1 or even 80:1. In general, such gels are rather difficult to handle and give much broader bands than 30:1 gels. Agarose gels are not used unless the mass of the binding protein(s) or the DNA is very large (>1 kb [kilobase-pairs]).[17,18] Because of their large pore sizes, agarose gels cannot detect protein–DNA complexes.
3. Most protein–DNA complexes are stable and carry a slight negative charge at neutral pH, and thus 0.5X TBE is the standard buffer of choice. TBE buffers should be used with caution, since ions can interact with proteins and give artifactual double bands. Retardation buffers should be of low ionic strength, because at high salt concentration the protein–DNA complex carries only a small fraction of the current and does not migrate very far. A high salt concentration may also increase heating during electrophoresis. In the small minority of cases where the protein component of the complex carries a strong negative or positive charge, it may be necessary to use acid (e.g., TAE at pH 6.0) or alkaline buffers (e.g., 50 mM Tris-glycine at pH 9.2).
4. Certain transcription factors, particularly those of prokaryotic origin, require the presence of a specific cofactor for optimum binding. For a given incubation time and temperature, determine the concentration dependence on monovalent (Na^+, K^+) and divalent cations (Mg^{2+}, Zn^{2+}). Cofactors such as cAMP, ATP, and GTP may be required for stable binding to DNA.
5. The size of a protein in a protein–DNA complex can be measured in lysates by cross-linking with ultraviolet (UV) irradiation and then resolving the complexes by sodium dodecyl sulfate-polyacrylamide gel electrophoresis (SDS-PAGE). In many cases, the presence of a previously characterized DNA-binding protein can be confirmed by immunological analysis. An antibody that binds specifically to a particular DNA-binding protein can either prevent formation of a specific protein–DNA complex or further retard ("supershift") the electrophoretic mobility of the complex.

Reagents Needed

30% Acrylamide Stock Solution

Acrylamide (58 g)
N,N'-Methylene-*bis*-acrylamide (1 g)
Add ddH_2O to a final volume of 200 ml.
Heat at 37°C to dissolve each component completely after each addition and sterile-filter through 2 µm membrane. Store at 4°C in an aluminum-foil-wrapped bottle for up to 5 weeks.

10% Ammonium Persulfate Solution (Fresh)

Ammonium persulfate (0.1 g)
Add ddH_2O to a final volume of 1 ml.

2X DNA Protein-Binding Buffer

40 mM Tris-HCl, pH 7.9
100 mM NaCl
20% Glycerol
0.2 mM DTT

10X Loading Dye
> 50% Glycerol
> 0.25% Bromophenol blue
> 0.25% Xylene cyanol

Nonspecific Competitor DNA
> Poly(dI-dC) (1 mg/ml)
> Dissolve an appropriate amount of poly (dI-dC) in ddH$_2$O and store the solution in 100-µl aliquots at −20°C.

0.5X Running Buffer
> Dilute 5X TBE buffer with ddH$_2$O.

5X TBE Buffer
> Tris base (54 g)
> Boric acid (27.5 g)
> 0.5 M EDTA (pH 8.0) (20 ml)
> Add ddH$_2$O to a final volume of 1000 ml.
> Autoclave.

1X TBE or 0.5X TBE Buffer
> Dilute 5X TBE buffer with ddH$_2$O.

III. Footprinting Following the Gel Retardation Assay Using Copper–Phenanthroline

This method combines and simplifies the above two techniques and allows one to obtain detailed nucleotide sequence information. The 1,10-phenanthroline–copper ion is able to cleave DNA when the DNA–protein complex is embedded within the nondenaturing-gel matrix following the gel retardation assay.[19,20] The cleaved DNA is eluted from the gel and resolved by denaturing-gel electrophoresis.

1. Carry out the gel retardation experiment as described previously except that the standard reaction should be scaled up 6- to 12-fold and loaded in several wells in the gel.
2. After electrophoresis, carefully remove the top glass plate and place the gel with its supporting glass plate in a tray or its equivalent containing 200 ml of 10 mM Tris-HCl (pH 8.0) to equilibrate the gel for 10 to 14 min at room temperature.
3. While the gel is equilibrating, prepare the following mixture:
   - 9 mM CuSO$_4$ (1 ml)
   - 40 mM 1,10-phenanthroline (1 ml)
   - Let the mixture stand for 1 to 1.5 min at room temperature to turn blue, and then add 18 ml of ddH$_2$O.
4. Add the 20 ml of the CuSO$_4$–phenanthroline solution to the equilibrated gel and initiate the reaction by adding 0.1 ml of 3-mercaptopropionic acid diluted in 20 ml ddH$_2$O. The solution turns brown.
5. Allow the cleavage reaction to proceed at room temperature for 1 min without any shaking.

Note: *The optimal time of incubation of the cleavage reaction should be predetermined.*

6. Add 10 ml 1.2% 2,9-dimethyl-1,10-phenanthroline in 100% ethanol to stop the reaction for 2 min. The solution should change to a yellow color.

7. Decant the solution and rinse the gel once with distilled water. Partially drain excess water and wrap the gel together with its supporting glass plate with SaranWrap. Carry out autoradiography with the wet gel as described for the gel retardation protocol (**Section II.E**).
8. Slice out the region of the gel containing the DNA–protein complex and the same area of the gel of the free-DNA probe. Place the gel slices into a microcentrifuge tube containing 0.5 ml of gel elution buffer containing 0.2 M NaCl, 20 mM EDTA (pH 8.0), 1% SDS, and 1 mg/ml transfer RNA (tRNA).
9. Crush the gel slices with a clean spatula and allow the elution to proceed at 37°C for at least 6 h or overnight.
10. Briefly spin down to remove the gel pellet. Add one volume of phenol:chloroform:isoamyl alcohol (24:24:1) to the supernatant.
11. Centrifuge at 11,000 g for 5 min and carefully transfer the supernatant to a fresh tube. Add 2.5 volumes of chilled 100% ethanol to the supernatant. Allow to precipitate at –70°C for 30 min.
12. Centrifuge at 12,000 g for 5 min and rinse the pellet with 1 ml of 70% ethanol. Dry the pellet for 15 min under vacuum and dissolve it in 10 to 20 µl of formamide-loading dye.
13. Denature the sample at 90 to 95°C for 4 min and load the same amount of radioactivity (cpm) onto a prerun urea denaturing gel or an appropriate sequencing gel (see prerunning and electrophoresis in the footprinting protocol, **Section I.D**).

Notes: *The free and bound DNA samples should be loaded in the adjacent lanes. An A + G ladder or DNA probe cleaved with guanosine-specific or cytidine- and thymidine-specific reagents, respectively, for chemical sequencing should be loaded in lanes to the left or right of the samples for identification of the sequence in the protein-bound region of the DNA.*

14. Dry the gel and carry out autoradiography to visualize the DNA band profile as described in the footprinting section, **Section I.D**.

References

1. **Galas, D.J. and Schmitz, A.,** DNase footprinting: a simple method for the detection of protein-DNA binding specificity, *Nucleic Acids Res.*, 5, 3157, 1978.
2. **Brenowitz, M., Senear, D.F., Shea, M.A., and Ackers, G.K.,** Quantitative DNase footprint titration: a method for studying protein-DNA interactions, *Methods Enzymol.*, 130, 132, 1986.
3. **Jost, J.P. and Saluz, H.P.,** *A Laboratory Guide to* in vitro *Studies of Protein-DNA Interactions*, Birkhauser Verlag, Basel, Switzerland, 1991.
4. **Kneale, G.C.,** *DNA-Protein Interactions; Principles and Protocols*, Humana Press, Totowa, NJ, 1994.
5. **Cartwright, I.L. and Kelly, S.E.,** Probing the nature of chromosomal DNA-protein contacts by *in vivo* footprinting, *BioTechniques*, 11, 188, 1991.
6. **Saluz, H.P. and Jost, J.P.,** A simple, high-resolution procedure to study DNA methylation and *in vivo* DNA–protein interactions on a single-copy gene level in higher eukaryotes, *Proc. Natl. Acad. Sci. U.S.A.*, 86, 2602, 1989.
7. **Mueller, P.R. and Wold, B.,** Ligation-mediated PCR: applications to genomic footprinting, *Methods Enzymol.*, 2, 20, 1991.
8. **Mueller, P.R. and Wold, B.,** *In vivo* footprinting of a muscle specific enhancer by ligation mediated PCR, *Science*, 246, 780, 1989.
9. **Fried, M. and Crothers, D.M.,** Equilibria and kinetics of lac repressor-operator interactions by polyacrylamide gel electrophoresis, *Nucleic Acids Res.*, 9, 6505, 1981.
10. **Revzin, A.,** Gel electrophoresis assays for DNA-protein interactions, *BioTechniques*, 7, 346, 1989.
11. **Fried, M.G.,** Measurement of protein-DNA interaction parameters by electrophoresis mobility shift assay, *Electrophoresis*, 10, 366, 1989.
12. **Fried, M.G. and Crothers, D.M.,** Kinetics and mechanism in the reaction of gene regulatory proteins with DNA, *J. Mol. Biol.*, 172, 263, 1984.

13. **Eisenberg, S., Civalier, C., and Tye, B.K.,** Specific interaction between a *Saccharomyces cerevisiae* protein and a DNA element associated with certain autonomously replicating sequences, *Proc. Natl. Acad. Sci. U.S.A.*, 85, 743, 1988.
14. **Sambrook, J. and Russel, D.W.,** *Molecular Cloning: A Laboratory Manual*, Cold Spring Harbor Laboratory Press, Cold Spring Harbor, NY, 2001.
15. **Shanblatt, S.H. and Revzin, A.,** Kinetics of RNA polymerase-promoter complex formation: effects of nonspecific DNA-protein interactions, *Nucleic Acids Res.*, 12, 5287, 1984.
16. **Varshavsky, A.,** Electrophoretic assay for DNA-binding proteins, *Methods Enzymol.*, 151, 551, 1987.
17. **Berman, J., Eisenberg, S., and Tye, B.K.,** An agarose gel electrophoresis assay for the detection of DNA-binding activities in yeast cell extracts, *Methods Enzymol.*, 155, 528, 1987.
18. **Topol, J., Ruden, D.M., and Parker, C.S.,** Sequences required for *in vitro* transcriptional activation of a *Drosophila* hsp70 gene, *Cell*, 42, 527, 1985.
19. **Spassky, A. and Sigman, D.S.,** Nuclease activity of 1,10-phenanthroline-copper ion: conformational analysis and footprinting of the lac operon, *Biochemistry*, 24, 8050, 1985.
20. **Sigman, D.S. and Chen, C.H.,** Chemical nucleases: new reagents in molecular biology, *Annu. Rev. Biochem.*, 59, 207, 1990.

Chapter 14

Differential Display

Darren H. Touchell, Yun-Shuh Wang, Scott A. Harding, and Chung-Jui Tsai

Contents

I.	RNA Preparation	306
	A. Total RNA Extraction	307
	B. *DNase* Treatment	307
II.	RT-PCR	308
	A. Reverse Transcription	308
	B. Radioactive PCR Amplification	309
III.	Denaturing Polyacrylamide Gel Electrophoresis	310
	A. Gel Preparation	310
	B. Electrophoresis	310
IV.	Isolation of Differentially Displayed Bands	311
	A. Excision and Elution of DNA Bands	311
	B. PCR Reamplification	313
	C. Cloning	313
	1. T/A Cloning	313
	2. Blunt-End Cloning	314
	D. Analysis of Transformants	314
	1. Colony PCR	314
	2. Miniprep	314
	3. Restriction Digestion and Sequencing	315
V.	Confirmation of Differential Expression	315
	A. Northern Hybridization	315
	B. Reverse Northern Hybridization	315
VI.	Troubleshooting Guide	316
	Reagents Needed	316
References		317

Differential display (DD), as first described by Liang and Pardee[1,2] in 1992, is a reverse-transcription polymerase chain reaction (RT-PCR)-based method for rapid identification of differentially expressed genes. It offers several advantages over conventional methods,[3] such as differential screening of complementary deoxyribonucleic acid (cDNA) libraries and subtractive cDNA library preparation and screening (see **Chapter 9**). First, employing PCR reduces the amounts of starting ribonucleic acid (RNA) needed. At the same time, it enables amplification of transcripts, thus permitting detection of lower-abundance messenger RNA (mRNA) species that are frequently undetectable using hybridization-based methods (e.g., conventional subtractive library screening). Second, the entire DD procedure from RNA extraction to banding-pattern detection can be completed in a few days, whereas hybridization signals from library construction and screening-based approaches routinely take weeks to obtain. Third, DD is not limited to pairwise comparisons, but it can be readily expanded to compare transcript profiles among multiple, closely related samples, thus allowing time-course studies or developmental gradients to be followed. Although it does not offer the high throughput of DNA microarrays (see **Chapter 15**), it is versatile and relatively inexpensive to set up and perform in most molecular biology labs.

In a typical DD experiment, DNA-free total RNA is reverse-transcribed into three subpopulations of cDNAs using three different oligo dT primers, each with a one-base (A, C, or G) anchor at the 3' end. Each cDNA subpopulation is then amplified by radioactive PCR using a combination of the same anchored oligo dT primer and an arbitrary sense primer in the presence of $[\alpha\text{-}^{35}\text{S}]\text{dATP}$. The PCR products are separated in a polyacrylamide denaturing sequencing gel and then visualized by autoradiography. By comparing the banding patterns of two or more samples, differentially up- or down-regulated cDNA fragments representing differentially expressed genes can be identified. Selected cDNA fragments can then be recovered, PCR-amplified, cloned, and sequenced for use as probes to screen cDNA libraries or to confirm gene expression patterns.

Since its advent, DD has been widely used to identify differentially expressed genes in a variety of plant, animal, and human tissues.[3-5] Although the principles remain the same, the precise protocol may vary, depending upon tissue type and the desired outcome of the experiment. The most commonly encountered problem is false positives.[1-6] It is highly recommended that the starting RNA samples be of similar high quality and free of DNA contamination, and that duplicate reactions be performed and displayed side-by-side to avoid false positives. With care and empirical optimization, DD can provide reproducible results for identification of differentially expressed cDNA probes. The protocol presented here has been used in our lab to identify a suite of genes that are differentially expressed across vascular cambium of aspen (*Populus tremuloides* Michx.),[7] and they are currently being employed to study gene expression profiles in leaves of transgenic aspen with variously modified metabolism.

I. RNA Preparation

High-quality RNA is essential for the success of differential display. There are a number of methods available to extract and prepare total RNA from plant or animal tissues (see **Chapter 2**). The method of choice depends largely on the amounts of tissue available and the tissue type. For example, when tissues are not limiting, a moderate- to large-scale RNA preparation method, such as those introduced in **Chapter 2**, is recommended. In this way, the same RNA samples can be used for subsequent confirmation of differential gene expression. On the other hand, commercial RNA extraction kits are very reliable for dealing with limited amounts of starting material (such as apical buds or leaves of plants). However, the same extraction method should be adhered to for different samples of the same tissue type throughout the investigation, because altering procedures may change cDNA banding patterns. The RNA extraction method described here is based on the RNeasy™ Plant Mini Kit (Qiagen, Valencia, CA) for small plant samples. The readers are referred to **Chapter 2** for additional RNA extraction protocols.

A. Total RNA Extraction

1. Collect plant tissues quickly and immediately freeze in liquid N_2. Store tissues at $-80°C$ or under liquid N_2 until required.
2. Grind approximately 50 to 100 mg (fresh wt.) of tissue to a fine powder with a liquid-N_2-precooled mortar and pestle. Be sure not to allow the tissue to thaw.
3. Transfer the frozen powder to a second mortar (at room temperature) containing 375 µl of guanidine isothiocyanate lysis buffer (RLT, Qiagen, Valencia, CA) with 1% (v/v) 2-mercaptoethanol (freshly added) and continue grinding tissues at room temperature to ensure good homogenization and total disruption of cells. The use of a second mortar and pestle is necessary to avoid ice formation.
4. With a large-orifice pipette tip, transfer lysate to a Qiagen shredder spin column placed in a 2-ml collection tube. If the homogenate is too viscous, add more RLT lysis buffer. Centrifuge at 10,000 g for 2 min to collect the flow-through lysate.

Notes: **DO NOT** *add more than 700 µl lysate to the column. If lysate exceeds the volume of the column, load the remaining lysate onto the column successively and centrifuge again.*

5. Carefully transfer the flow-through lysate to a new 2-ml microcentrifuge tube and add 0.5 volume of 100% ethanol. Mix by pipetting.
6. Immediately load sample onto an RNeasy minicolumn and centrifuge at 10,000 g for 15 sec at room temperature. Discard the flow-through. If the sample exceeds 700 µl, add the remaining mixture onto the column successively and centrifuge again.
7. Wash the column with 700 µl wash buffer RW1 (Qiagen, Valencia, CA), centrifuge at 10,000 g for 15 sec, and discard the flow-through and the collection tube.
8. Place the column in a new 2-ml microcentrifuge tube and add 500 µl RPE (Qiagen, Valencia, CA) buffer. Centrifuge at 10,000 g for 15 sec to wash the column. Discard the flow-through and the collection tube. The RPE wash step may be repeated.
9. Place the column in a new 2-ml microcentrifuge tube and centrifuge at top speed for 1 min. Discard the flow-through and the collection tube.
10. Transfer the column to a new 1.5-ml microcentrifuge tube and add 30 µl DEPC-ddH_2O (diethylpyrocarbonate-distilled, deionized water) directly onto the spin column membrane. Centrifuge at 10,000 g for 1 min to elute the total RNA. Repeat if the expected RNA yield is >30 µg.
11. Check 0.5 to 1 µl RNA in a minigel and measure RNA concentration as described in **Chapter 2**. Proceed to *DNase* I digestion and store the rest in small aliquots at $-80°C$.

B. *DNase* Treatment

1. Set up *DNase* I digestion in a 1.5-ml microcentrifuge tube in the following order:
   - Total RNA, 10 to 20 µg
   - 10X *DNase* I buffer, 10 µl
   - *DNase* I (10 units·µl^{-1}), 1 µl
   - ddH_2O, to 100 µl
2. Mix and incubate at 37°C for 30 min.
3. Add 100 µl of DEPC-ddH_2O to the reaction mixture and extract once with 200 µl phenol:chloroform:isoamyl alcohol (25:24:1) and once with chloroform:isoamyl alcohol (24:1). If the initial RNA amounts are low, back-extract the organic phase of phenol:chloroform:isoamyl alcohol with 200 µl of DEPC-ddH_2O, followed by extraction of the aqueous phase with an equal volume of chloroform:isoamyl alcohol.
4. Carefully transfer aqueous phase to a new tube (and pool all fractions together if back extraction is performed) and precipitate RNA with 0.1 volume of 3 M NaOAC and 2 volumes of 100% ethanol at $-80°C$ for 30 min.
5. Centrifuge at 12,000 g for 20 min at 4°C to pellet the RNA. Wash with 70% ethanol and briefly air dry the pellet under vacuum.

6. Dissolve pellet in 50 μL of DEPC-ddH$_2$O.
7. Check 1 μl in a mini-TAE gel and determine the RNA concentration as described in **Chapter 2**.

II. RT-PCR

DD is essentially a RT-PCR method for detection of differences (e.g., presence or abundance) in mRNA species between two or more populations, with two key modifications. First, total RNA is reverse-transcribed into three subpopulations of cDNAs in order to reduce the complexity of banding patterns following PCR amplification, electrophoresis, and autoradiography. Second, radioactive PCR is employed to increase the sensitivity of detection. The highly sensitive nature of the method, however, also makes it prone to false positives.[1–6] We recommend that the following measures be taken to improve the reproducibility of the method: (1) Perform all RT reactions in duplicate and load duplicate reactions side-by-side during electrophoresis. Recover only the bands showing differential expression in both reactions for further study. (2) A master mix should be prepared when working with multiple samples to minimize pipetting errors. (3) Enzymes and reagents from different suppliers, or otherwise prepared differently, represent another variable that should be avoided. PCR reactions with the same primer combination should be performed at the same time, i.e., using the same master mix. (4) Negative controls should be included to aid troubleshooting (see note in **Section II.A**).

A. Reverse Transcription

We routinely use 0.2 to 0.3 μg DNA-free total RNA in a 20-μl reverse transcription (RT) reaction for synthesis of first-strand cDNAs. The cDNAs are enough for 10 PCR reactions, each yielding 50 to 100 resolved bands with a size range of 300 to 1000 nucleotides (nt) after denaturing acrylamide gel electrophoresis and autoradiography. However, the optimum amount of total RNA required to yield satisfactory banding patterns needs to be empirically determined. We recommend a trial experiment using 0.2 to 1.2 μg of total RNA for RT to optimize RNA concentration. The following protocol is based on 0.2 μg total RNA per RT reaction. For each RNA sample, set up three RT reactions, each using one of the three one-base-anchored oligo dT primers ($T_{11}A$, $T_{11}C$, and $T_{11}G$) that will be used again later in radioactive PCR.

1. Prior to starting, dilute a fraction of RNA samples to a final concentration of 0.1 μg·μl^{-1}. Set up three RT reactions per RNA sample in 0.5 ml microcentrifuge tubes using an anchored oligo dT primer ($T_{11}M$, M = A, C, or G) as follows. Prepare each RT reaction in duplicate (e.g., by pipetting each component separately).
   - 2 μ*M* Anchored $T_{11}M$ primer, 2 μl
   - DNA-free total RNA (0.1 μg·μl^{-1}), 2 μl
   - DEPC-ddH$_2$O, 7.25 μl
2. Incubate at 70°C for 10 min, then chill on ice. Spin briefly to collect the contents. Add the following:
   - 5X First-strand buffer, 4 μl
   - 100 m*M* DTT, 2 μl
   - 200 μ*M* dNTP, 2 μl
   - SuperScript II RTase, 0.75 μl (Invitrogen-Life Technologies, Carlsbad, CA)
3. Incubate at 42°C for 1 h and inactivate the enzyme by incubating the reaction at 70°C for 15 min. Store cDNA at –80°C until use.

Notes: (1) If multiple RNA samples are to be analyzed at the same time, a master mix enough for [number of RNA samples $\times 3$ ($T_{11}M$) $\times 2$ (duplicates)]+ 1 reactions should be prepared for step 2 in order to minimize pipetting errors that could have effects on the banding patterns/intensities and lead to false positives. (2) A negative control (without reverse transcriptase) for each RNA sample should be included to assess DNA contamination. DNA bands produced from negative controls would indicate contamination of RNA with genomic DNA. (3) It is a common practice to perform differential display reactions in duplicate as another measure to minimize false positives. Duplicate RT reactions are recommended for each new cDNA template prepared. Once reproducibility of the cDNA templates derived from duplicate RT reactions is confirmed, duplicate RT reactions may be skipped in subsequent experiments when the same cDNA template is used with other primer combinations. However, duplicate radioactive PCR reactions are still recommended. See discussion in the note in **Section II.B**.

B. Radioactive PCR Amplification

1. Prepare a master mix for PCR reactions, excluding cDNA template and anchored primer; both will be added individually. We recommend the RAPD 10-mer kits (20 arbitrary primers per kit) from Qiagen Operon (Alameda, CA). Each 20-μl PCR reaction consists of the following:
   - ddH$_2$O, to 20 μl
   - 10X PCR buffer, 2 μl
   - 20 μM dNTP, 2 μl
   - 2 μM Anchored T$_{11}$M primer, 2 μl
   - 2 μM Arbitrary primer, 2 μl
   - cDNA template, 2 μl
   - ^{35}S-dATP (>1000 Ci·mmole^{-1}), 0.6 μl
   - *Taq* polymerase (1 unit·μl^{-1}), 0.2 μl

Notes: When one arbitrary primer is used, there will be 3 (primer combinations) $\times 2$ (duplicates) = 6 radioactive PCR reactions (excluding negative controls for each starting RNA sample). The number of reactions to be prepared is usually determined by the number of (1) wells available in the sequencing gel, (2) RNA samples to be compared, and (3) arbitrary primers used. For analysis of two RNA samples (such as xylem and phloem RNAs, e.g., see Figure 14.1), PCR products will be loaded in groups of four, each group consisting of two xylem and two phloem duplicate reactions using the same primer combination in a sequencing gel. A 33 $\times$ 42 cm gel with 78-well capacity can be run to analyze up to 12 primer combinations (i.e., 4 arbitrary $\times$ 3 anchored T$_{11}$M primers) with a spacer blank lane between each primer combination. Thus, the total number of PCR reactions to be prepared is 12 $\times$ 4 = 48, and the number of lanes to be used in the gel, including spacer lanes, is 12 $\times$ (4 + 1) = 60.

Caution: Care should be exercised when using [α-^{35}S]dATP. A lab coat and gloves should be worn when working with this isotope. Gloves should be changed often and disposed of as radioactive waste. Waste liquid, pipette tips, and papers contaminated with the isotope should be collected in labeled containers. After finishing, a radioactive contamination survey should be performed and recorded.

2. If the thermal cycler does not have a heated lid, overlay the PCR reaction mixture with a drop of mineral oil (~20 μl).
3. Program the thermal cycler for the following PCR conditions: 94°C for 30 sec, 37°C for 2 min, and 72°C for 40 sec for 40 cycles followed by a final extension of 72°C for 10 min.
4. Add 2 μl loading dye to the PCR products. If mineral oil is used, add one drop (~20 μl) of chloroform:isoamyl alcohol (24:1) to extract the mineral oil into the organic phase at the bottom. The DNA phase, stained blue with the loading dye, is on the top layer. Store the samples on ice.

III. Denaturing Polyacrylamide Gel Electrophoresis

A. Gel Preparation

1. It is important to clean glass plates thoroughly to prevent bubbles from forming in the gel and gels from sticking to the plates. Scrub plates with detergent using a sponge.
2. Wash off excess detergent and dry plates with Kimwipes™. Clean plates with Windex™ using Kimwipes, then wipe with ethanol to remove any residues. Repeat if necessary to ensure that plates are clean.
3. Silanize the notched plate by dispensing ~1 ml Gel Slick™ solution (BMA, East Rutherford, NJ) evenly on the inner side of the notched plate. Use Kimwipes to wipe the solution over the entire plate thoroughly. Rinse with water and wipe clean with ethanol.
4. Clean the spacers and comb by washing with detergent. Wipe with ethanol using Kimwipes.
5. Set up the plates by laying the unnotched plate, inner side up, on a styrofoam rack (such as those in Falcon centrifuge tube packages). Position the spacers on both sides of the plate and carefully place the notched plate, silanized side down, on the other plate. Align the plates and make sure the spacers are still positioned at the edges of the plates. Clamp the plates together, two clamps per side.
6. Seal the bottom of the plates with gel casting tape.
7. Prepare 75 ml of a 5.5% Long Ranger™ denaturing acrylamide gel solution (for a 35 × 43 cm gel) by dissolving 31.5 g urea in 20 ml ddH$_2$O, 18 ml 5X TBE, and 8.25 ml Long Ranger gel solution (50%, BMA, East Rutherford, NJ). Adjust the volume to 75 ml with ddH$_2$O and filter through a Whatman™ No. 1 filter paper. This also degases the gel solution.
8. Add 75 μl of $N,N,N'N'$-tetramethylethylenediamine (TEMED) and 750 μl of 10% ammonium persulfate (APS) to the gel solution and mix gently.
9. Transfer the gel solution to a small plastic beaker. While holding the plates at an angle of ~45°, smoothly and without pausing pour the gel solution into one corner of the sandwiched plates.
10. When the solution approaches the top of the notched plate, lay the plates down on the styrofoam rack. Insert flat side of a sharkstooth comb into the gel until the sharkstooth-end is flush with the edge of the unnotched plate. Make sure no air bubbles are introduced. Use three large binder clips to clamp the plates and comb together. Cover with plastic wrap if desired.
11. Allow the gel to polymerize for at least 1 h before use. The gel can be stored at room temperature for one day; thus, it is recommended that the gel be poured the day before electrophoresis and allowed to polymerize overnight.

B. Electrophoresis

1. Remove clamps and gel sealing tape from the gel sandwich. Carefully remove the comb and rinse the plates with tap water to remove dried polyacrylamide/urea near the top and the bottom of the gel. Wash the sharkstooth comb thoroughly to remove urea residues and Kimwipe dry.
2. Fill the bottom buffer reservoir with 1X TBE. Attach the gel to the electrophoresis apparatus, with the notched plate facing the apparatus, and clamp into position according to the manufacturer's instructions. Fill the upper chamber with 1X TBE.

Differential Display

3. Use a syringe to rinse the top of the gel with 1X TBE, making sure that excess urea and polyacrylamide are flushed away. Carefully reinsert the sharkstooth comb into the slot, teeth down, until the tips of the teeth touch the gel. Rinse wells thoroughly with 1X TBE, using a syringe to remove urea and polyacrylamide.
4. Prerun the gel at 55 W constant power for 30 min until the gel temperature reaches 50 to 60°C. Mark the wells by group (i.e., identifying spacer lanes) with a marker to aid sample loading.
5. Denature 4 to 12 DNA samples from **Section II** above at 80°C for 4 min. Chill on ice (optional).
6. While denaturing the DNA samples, turn off the power supply and flush the wells again with a syringe of 1X TBE to remove urea and polyacrylamide. Immediately load 4 µl of each DNA sample and leave a blank lane between groups of samples of different primer combinations. Rinse the pipette tip thoroughly in the upper buffer reservoir between loadings in order to prevent carryover contamination. Connect the electrodes, turn on the power supply, and run the gel at 55 W constant power for ~5 min, or until the samples migrate into the gel.

Note: *We recommend loading no more than 12 samples at a time (in less than ~5 min) to ensure that an adequate gel temperature (50 to 60°C) is maintained for DNA denaturation.*

7. While running the gel, repeat step 5 to denature another batch of samples. After the first batch of samples has migrated into the gel, turn off the power supply, wash wells, and load samples as described in step 6. Repeat until all samples are loaded.
8. Run the gel at 55 W for 4 h or until the xylene cynol FF (the lagging dye) is close to the bottom of the gel.
9. Turn off the power supply and disconnect the electrodes. Drain electrophoresis buffer from the upper reservoir. Remove the gel from the electrophoresis apparatus.
10. Lay the gel down on a styrofoam rack, notched plate facing up, and allow the gel to cool for 10 to 20 min. Discard the bottom reservoir buffer.

Note: *The electrophoresis buffer should be disposed of as radioactive waste according to local radiation safety procedures.*

11. Remove any remaining gel sealing tape and spacers. Use a metal spatula to pry up the notched plate at one corner and slowly lift the plate off the gel. Carefully lay a piece of 35 × 43 cm Whatman 3MM chromatography paper over the gel to form a firm contact. Slowly peel the filter paper with gel, from one corner, away from the glass plate and transfer the filter paper to a gel dryer, gel side up. Cover the gel with plastic wrap and dry at 75°C under vacuum following the manufacturer's instruction for the gel dryer.
12. Mark the gel with a fluorescent dye or sticker at two opposite corners to aid in orientation later. Expose the gel to x-ray film for 2 to 3 days at room temperature before developing.

Note: *Fluorescent orientation stickers can be homemade using small pieces of laboratory labeling tape marked with a Crayola™ glow-in-the-dark marker, or simply use any glow-in-the-dark stickers available in stores.*

IV. Isolation of Differentially Displayed Bands

A. Excision and Elution of DNA Bands

1. Align the film with gel using fluorescent stickers as guides and secure with tape. Mark differentially expressed bands by piercing the film and the gel assembly with a push pin or a syringe needle. Pierce on both sides of each band (see inset in Figure 14.1 for an illustration).
2. Remove film and excise bands with a single-edged razor blade.

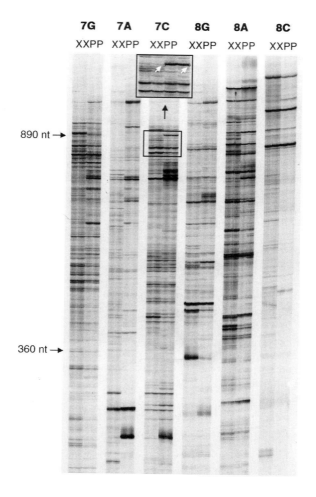

FIGURE 14.1
An autoradiogram of mRNA differential display showing cDNA banding patterns derived from total RNA of aspen xylem (X) and phloem (P) using six primer combinations (two arbitrary × three anchored oligo dT primers).[7] Each reaction was performed in duplicate and loaded side by side. Note that certain primer combinations (e.g., 8C) may not work as well as others. DNA bands with differential xylem or phloem abundance are identified and excised by aligning the film with the gel and punching holes through the margins of the bands with a push pin, as illustrated by arrows in the inset.

Note: *Wipe the razor blade thoroughly with ddH$_2$O followed by ethanol between samples to avoid carryover contamination.*

3. Elute DNA by soaking each excised band (dried gel/filter paper) in 100 μl of ddH$_2$O in a 1.5 μl microcentrifuge tube for overnight at room temperature. Alternatively, the gel/filter paper slices can be soaked in ddH$_2$O for 10 min at room temperature, followed by boiling for 15 min.
4. Centrifuge for 2 min at 12,000 g to pellet the gel and filter paper.
5. Transfer supernatant to a new microcentrifuge tube. Add 10 μl of 3 M sodium acetate, 10 μg of glycogen (Roche Applied Science, Indianapolis, IN) and 300 μl of ethanol to precipitate DNA at −80°C for 30 min.
6. Centrifuge at 4°C for 20 min at 12,000 g. Wash with 70% ethanol and dry briefly under vacuum.
7. Resuspend the pellet in 10 μl ddH$_2$O.

Notes: *(1) It is recommended that only bands with reproducible differential patterns and that are greater than 300 nt be selected. Smaller fragments will provide minimal sequence information. (2) It is quite common to observe bands appearing as doublets or quadruplets in a DD gel. These bands are likely to arise from a single clone due to differential mobility of the two complementary DNA strands and/or the addition of a single 3'-adenosine by the Taq polymerase.*[2]

B. PCR Reamplification

1. Reamplify excised bands in 40-µl reactions as follows using the original primer combinations and PCR conditions:
   - ddH$_2$O, to 40 µl
   - 10X PCR buffer, 4 µl
   - 200 µM dNTP, 4 µl
   - 2 µM Anchored T$_{11}$M primer, 4 µl
   - 2 µM Arbitrary primer, 4 µl
   - DNA template, 4 µl
   - *Taq* polymerase (1 unit·µl^{-1}), 0.4 µl
2. Check 10 µl of the PCR products in a 1% (w/v) agarose minigel. If amplification is high and the size of the PCR product is consistent with the sequencing gel, proceed to T/A cloning (see **Section IV.C**). However, if amplification is low, it may be necessary to reamplify, using the PCR product as a template.

Note: *Failure in PCR reamplification may be a sign of inaccurate band excision (due to poor alignment of the gel and the film). Reexpose the gel to another x-ray film to evaluate the band excision and repeat A and B above if necessary.*

C. Cloning

Taq polymerase-amplified PCR products carrying characteristic single 3'-adenosine overhangs can be readily ligated to cloning vectors with single 3'-thymidine overhangs, a process termed T/A cloning. T vectors can be purchased from commercial vendors such as Invitrogen (Carlsbad, CA) and Promega (Madison, WI), created by restriction enzyme digestion (e.g., *Hph* I, *Mbo* II, *Xcm* I),[5] provided one of these restriction sites resides on (or was engineered into) the multiple cloning sites of the vector, or prepared by 3'-terminal transferase tailing reaction with ddTTP (see **Chapter 4**). Alternatively, the PCR products can be blunt-ended and cloned into blunt-end vectors, following appropriate pretreatment of inserts and vectors.[8] For optimum cloning efficiency, we recommend the use of commercial T/A cloning vectors or cloning kits. Comments on T/A or blunt-end cloning strategies are provided below.

1. T/A Cloning

PCR products can be used directly without further manipulation. If multiple bands or smears are observed, based on agarose gel electrophoresis, gel purification of the target fragments may be necessary in order to avoid false positives. The 3'-A overhangs of PCR products are not stable and degrade easily during storage. Thus, PCR products that have been stored prior to cloning may need to be incubated with a small aliquot of *Taq* polymerase with appropriate buffer at 72°C for 5 min before use. Follow manufacturers' protocols for the T/A cloning procedures.

2. Blunt-End Cloning

PCR products should be blunt-ended to remove the overhangs by T4 DNA polymerase and then phosphorylated by T4 polynucleotide kinase. Both reactions can be performed simultaneously in one single step. The phosphorylated and polished PCR products are then cloned into a dephosphorylated blunt-end vector. The procedures are summarized as follows:

1. Blunt-end treatment of PCR products: Set up a 20-μl reaction containing 100 to 500 ng PCR products, 1X T4 DNA polymerase buffer, 100 μM dNTPs, 1 μg BSA, 1 μM ATP, and 1 to 2 units each of T4 DNA polymerase and T4 polynucleotide kinase. Incubate at 37°C for 30 min. Heat to inactivate enzymes at 75°C for 10 min.
2. Vector preparation: Digest 100 to 500 ng vector of choice with an appropriate enzyme (e.g., *Eco* RV or *Sma* I) and buffer in a 20-μl reaction volume. Incubate at appropriate temperature for 3 h to overnight. Dilute the reaction to 100 μl with 10 μl of 10X alkaline phosphatase buffer, ddH$_2$O, and 1 to 2 units of alkaline phosphatase. Incubate at 37°C for 30 min. If the alkaline phosphatase is also active in the restriction enzyme buffer, perform the two reactions in one. Extract DNA once with phenol:chloroform:isoamyl alcohol (25:24:1) and once with chloroform:isoamyl alcohol (24:1) and precipitate in ethanol. Resuspend DNA in ddH$_2$O.
3. Ligation: Set up a 20-μl ligation reaction with 10 to 20 ng of vector and an appropriate amount of insert (vector:insert = 1:5 molar ratio) in 1X ligation buffer with 1 unit of ligase. Incubate overnight at 16°C.
4. Transformation: Add 10 μl of ligation reaction to an aliquot (100 μl) of competent cells and incubate on ice for 15 to 30 min. Heat shock at 42°C for 2 min and chill on ice. Add 0.5 ml LB medium to the cells and incubate at 37°C with shaking.
5. Plate 50 to 200 μl of the transformation and reagents for blue and white selection, if appropriate, onto selection plates. Incubate in a 37°C oven overnight.

D. Analysis of Transformants

We recommend screening transformants by colony PCR in order to rapidly identify clones with expected insert sizes. Positive colonies are then used for small-scale liquid LB medium inoculation, and the resulting cultures are used for miniprep plasmid DNA extraction. The plasmid DNA can be used for restriction enzyme digestion and sequencing in order to confirm the identity of the cloned fragments.

1. Colony PCR

Number and pick isolated colonies with pipette tips and suspend in 50 μl TE buffer containing 1% Triton X-100. Boil for 5 min to lyse the DNA. Prepare master mix for PCR reaction as described in **Section IV.B** using universal vector primers (such as M13 forward and reverse primers) and 2 μl of colony lysate as the template. Program the thermal cycler according to the melting temperatures of the primers and the expected insert sizes. Analyze the PCR products in an agarose minigel.

2. Miniprep

Bacterial inoculation and minipreparation of plasmid DNA for colony PCR-positive colonies are conducted as described in **Chapter 1**.

3. Restriction Digestion and Sequencing

The size of the cloned fragment can be reconfirmed again by restriction digestion of plasmid DNA (see **Chapter 5**). The identity of the cloned fragments is obtained by sequencing of the plasmid DNA, as described in **Chapter 11**.

V. Confirmation of Differential Expression

One major drawback of DD is the high frequency of false positives. Thus, it is essential to verify the differential expression patterns of the cloned DD fragments by Northern or reverse Northern hybridization analysis before the DD clones are chosen for further study. Gene-expression patterns can be confirmed by standard Northern hybridization, provided a relatively large amount of the initial or equivalent RNA sample is available. At least 10 μg of total RNA per sample is needed for each Northern blot. Although blots may be reprobed several times (see **Chapter 6**) to analyze several DD clones, the throughput is limited, and a probe must be generated each time a DD clone is to be checked. Alternatively, reverse Northern hybridization[6] can be employed to analyze multiple DD clones simultaneously. The cloned DD fragments, either PCR amplified or restriction enzyme digested, are spotted onto duplicate nylon membranes (e.g., slot/dot blots) and hybridized with cDNA probes made from the initial RNA samples. Thus, only two probes are needed to assess the differential expression of numerous DD clones in two RNA samples. In many cases, reverse Northern hybridization is a cost-effective method to screen and eliminate false positives prior to sequencing analysis. The remaining positive DD clones can then be subjected to sequencing analysis or used as probes for screening DNA libraries and standard Northern blot analysis.

A. Northern Hybridization

1. Prepare RNA blots as described in **Chapter 6**. RNA samples used for DD RT-PCR, or extracted from identical tissues, may be used.
2. Digest the DD clones with appropriate restriction enzymes and gel-purify the inserts for labeled-probe preparation as described in **Chapter 4**. Carry out Northern blot hybridization according to methods cited in **Chapter 6**.

B. Reverse Northern Hybridization

1. Prepare slot/dot blots as described in **Chapter 5**.
2. Prepare first-strand cDNA probes, as described in **Section II.A**, using the initial total RNA samples as templates with the following modifications: (1) using total RNA at 5 to 10 μg, (2) using dNTPs at 10 mM each except dATP at 5 mM, and (3) addition of 3 μl of [α-^{32}P]dATP (>3000 Ci·mmole^{-1}, 10 mCi·ml^{-1}).
3. Purify probes and carry out hybridization as described in **Chapter 5**.

Note: *It is highly recommended that for either Northern or reverse Northern verification, cloned DD fragments (from clones obtained in **Section IV.D**) be used instead of the reamplified PCR products (from **Section IV.B**). PCR products may be heterogenic, which could cause false positives to occur during subsequent analysis.*

VI. Troubleshooting Guide

Symptoms	Solutions
Smeared lanes	May be due to poor RNA quality, PCR reactions, gel quality, or electrophoresis conditions.
	Recheck the RNA samples in a minigel.
	Check the PCR reagents with a known gene and primer pair.
	Prepare the gel with freshly made solutions. Do not use TBE buffer with precipitation.
	Make sure that the gel has been prerun adequately and that the wells are washed thoroughly to remove urea and polyacrylamide before loading samples. Make sure that the samples are properly denatured.
	If all lanes are smearing, check all of the above. If only some lanes are smearing, the problems may arise from individual RT or PCR reactions.
Blank lanes or faint bands	May be due to poor RNA quality, RT, or PCR reactions. See notes above.
	Reaction conditions may not be optimized. Try different amounts of cDNA templates, more PCR cycles, or a different *Taq* polymerase.
	Use [^{35}S]-dATP with a higher specific activity (>1000 Ci·mmole^{-1}) or [^{33}P]-dATP.
	Try longer exposure time.
	Bad arbitrary primer. We have noticed that arbitrary primers consisting of G or GC at the 3'-end are most efficient and reliable with plant RNAs. Some primers never worked (may be of sporadic poor quality).
Fuzzy bands	May be due to poor gel quality, insufficient polymerization time, or improper electrophoresis conditions. See notes above.
No PCR reamplification	May be due to improper alignment of the film with the gel.
	Reexpose the gel to a new film and assess the accuracy of band recovery.
High frequency of false positives	Use high-quality starting RNA (with minimal degradation and DNA free).
	Maintain a high level of consistency of the procedures between different samples and reactions. Prepare master mix whenever possible. Use independently extracted RNA samples for Northern or reverse Northern analysis.

Reagents Needed

10% Ammonium Persulfate (APS)

Dissolve 0.5 g of APS in 5 ml ddH$_2$O. Store at 4°C and prepare fresh weekly.

Chloroform:Isoamyl Alcohol = 24:1

Mix and store at room temperature.

10X DNase I Buffer

0.1 *M* Tris, pH 8.3
0.5 *M* KCl
15 m*M* MgCl$_2$
Prepare in DEPC-ddH$_2$O and autoclave. Store at room temperature.

Ethanol, 70% and 100%

50% Long Ranger Stock Solution

Phenol:Chloroform:Isoamyl Alcohol = 25:24:1

Polyacrylamide Gel Loading Dye
    80% deionized formamide
    10 mM EDTA, pH 8
    0.1% Xylene cyanol FF
    0.1% Bromophenol Blue
    Store at 4°C in small aliquots.

3 M Sodium acetate, pH 5.2, DEPC-treated
    Dissolve sodium acetate in ddH$_2$O and adjust pH to 5.2 with 3 M glacial acetic acid.

5X TBE
    0.45 M Tris-base (54 g·l^{-1})
    0.45 M Boric acid (27.5 g·l^{-1})
    10 mM EDTA, pH 8

TEMED

References

1. **Liang, P. and Pardee, A.B.,** Differential display of eukaryotic mRNA by means of polymerase chain reaction, *Science*, 257, 967, 1992.
2. **Liang, P., Averboukh, L., and Pardee, A.B.,** Distribution and cloning of eukaryotic mRNAs by means of differential display: refinements and optimization, *Nucl Acids Res.*, 21, 3269, 1993.
3. **Lievens, S., Goormachtig, S., and Holsters, M.,** A critical evaluation of differential display as a tool to identify genes involved in legume nodulation: looking back and looking forward, *Nucl. Acids Res.*, 29, 3459, 2001.
4. **Matz, M.V. and Lukyanov, S.A.,** Different strategies of differential display: areas of application, *Nucl. Acids Res.*, 26, 5537, 1998.
5. **Zhang, M. et al.,** Expression genetics: a different approach to cancer diagnosis and prognosis, *Trends Biotechnol.*, 16, 66, 1998.
6. **Zhang, H., Zhang, R., and Liang, P.,** Differential screening of gene expression difference enriched by differential display, *Nucl. Acids Res.*, 24, 2454, 1996.
7. **Wang, Y.S.,** Isolation and characterization of cDNAs involved in vascular development of quaking aspen (*Populus tremuloides*), Ph.D. dissertation, Michigan Technological University, Houghton, 2002.
8. **Sambrook, J. and Russell, D.W.,** *Molecular Cloning: A Laboratory Manual*, 3rd ed., Cold Spring Harbor Laboratory Press, New York, 2001.

Chapter 15

Functional Genomics and DNA Microarray Technology

Gopi K. Podila, Chandrashakhar P. Joshi, and Peter B. Kaufman

Contents

I.	What Is This Technology and What Can It Tell Us?	320
II.	Pros and Cons of DNA Macroarrays	321
III.	DNA Microarray Protocols	321
	A. Overview of the Process	321
	B. Growing Bacterial EST (cDNA) Clones	322
	C. Isolation of Plasmid Templates for PCR	323
	Materials, Reagents, and Solutions	324
	D. PCR Amplification of cDNA Inserts	324
	Materials for PCR Amplification	325
	1. Check PCR Products by Agarose Gel Electrophoresis	326
	Materials for PCR Product Analysis by Agarose Gel Electrophoresis	327
	E. Purification and Quantification of PCR Products	327
	1. Fluorometric Determination of PCR DNA Concentration	328
	Materials and Reagents for PCR Product Quantification	328
	F. RNA Preparation from Samples	329
	Reagents and Materials for RNA Purification	330
	G. Preparation of Poly-L-Lysine Coated Slides	330
	Materials and Reagents for Preparing Poly-L-Lysine Coated Slides	331
	H. Printing of Microarrays	331
	1. Postprocessing of Printed Slides	332
	Materials and Reagents for Slide Printing	333
	Materials for 30 Arrays	333
	I. Reverse Transcription and Fluorescent-Labeling of RNA	333
	Materials for Reverse Transcription of RNA and Labeling of cDNA	335
	J. Array Hybridization	335
	K. Array Postprocessing	337
	Materials and Reagents for Hybridization and Postprocessing	337

	L.	Scanning DNA Microarrays	337
		1. Slide Scanning	338
	M.	Image Processing and Data Analysis	339
		1. Data Normalization and Analysis	339
IV.	Troubleshooting Guide		340
V.	Applications of DNA Microarrays in Various Groups of Organisms		341
	A.	Microorganisms	341
		1. Yeast (*Saccharomyces cerevisiae*)	341
		2. Bacteria	341
	B.	Plants	341
	C.	Humans	342
	Reagents Needed		342
References			345

I. What Is This Technology and What Can It Tell Us?

With the advent of high-throughput deoxyribonucleic acid (DNA) sequencing techniques, a score of prokaryotic and eukaryotic genomes have recently been sequenced to completion (e.g., *Escherichia coli*,[5] yeast,[10] human,[26] *Arabidopsis*[25]). This large amount of sequence data can be searched *in silico*, i.e., on the computer, using a number of bioinformatics tools that are developed to infer a large number of open reading frames or protein coding regions that occur in each of these genomes. Eukaryotic genes are more difficult to predict due to their split nature where coding (exon) and noncoding (intron) regions of the gene are interspersed. The gene-prediction algorithms are either generic, based on all the information available about genes and the proteins they encode from various organisms, or they are organism-specific, based on actual sequence information available only in that organism. In either case, such computer-based gene predictions should still be regarded as tentative. Ideally, molecular biologists must ensue further experimental confirmation of the gene function.

These rapid developments in genome sequencing have ushered in a new era of functional genomics, where the function of each newly discovered gene is determined. Again, computer programs are employed to find the similarity between the encoded proteins from these predicted genes with the proteins that exist in the data banks. About 50 to 60% of the genes can thus be tentatively assigned some function that needs further experimental confirmation. The remaining 40 to 50% of the genes belong to a "functions unknown" category. A number of molecular tricks can be performed with all these genes in order to probe their likely function(s), but as a rule of thumb, it is assumed that more expression of a gene at the transcriptional level (ribonucleic acid [RNA]) in a particular process or cell/tissue or treatment indicates its importance in the process. Thus, a gene with high level of expression is considered to be more important in a particular situation than a gene expressed at the low level. It is also assumed that if more transcripts are present, it should translate into more proteins. Both these assumptions have been proved to be wrong in many instances. For example, many transcription factors are expressed at a low level but are very important for transcriptional regulation of a process, and a variety of posttranscriptional processes can control the levels of proteins produced. In spite of these caveats, measuring the level of messenger RNA (mRNA) expression and comparing these levels among various treatments are common practices in molecular biology research. Northern blots are often used for these types of experiments, but doing Northern blots with thousands of genes is practically impossible. Furthermore, a number of experimental constraints make such an elaborate setup impossible.

The newly developed microarray technology solves this problem.[22] Apart from providing data about the expression of a gene, microarrays also provide a composite picture or snapshot of the gene-expression profile in a particular life process. Coordinate expression of genes is extremely important in real-life situations, and understanding which genes are simultaneously expressed, and with what frequency, has a far-reaching impact on our understanding of the biological processes occurring in nature. Although a variety of arrays exist, we will focus here on complementary DNA (cDNA) microarrays that are capable of profiling expression of tens of thousands of cDNAs by first putting them (printing) on a glass microscope slide using a robotic "arrayer" and hybridizing them with fluorescent-labeled DNA probes from two or more types of treatments.[6] This procedure allows comparison of the relative abundance of each of these genes in the context of expression of other genes. Subsequently, the hybridization patterns are digitally recorded as an image. These data are further mined with appropriate image-analysis and clustering software in order to understand the complex expression profile of these newly discovered genes. Thus, microarrays are highly suitable for studies in biology and medicine, since they facilitate our ability to decipher more complex biological phenomena.

II. Pros and Cons of DNA Macroarrays

Another option for studying gene expression of hundreds or a few thousands of genes is the use of macroarrays. They are very similar to microarrays, except the DNA is printed onto a nitrocellulose or nylon membrane, and radioactive tagged targets are used to hybridize to the DNA sequences on the membrane. Because of the size of DNA spots, the area in which these can be printed is a lot larger than that for microarrays. Macroarrays are more suitable where a few hundred to two to three thousand clones need to be analyzed, and the amount of RNA needed to make the labeled targets is limiting. Macroarrays only need a few micrograms of total RNA, whereas microarrays usually need 30 to 100 μg of total RNA to prepare labeled targets.[4,15] The setup costs for preparing and analyzing macroarrays are within the reach of small labs, whereas setup costs for microarrays are much higher. In general microarrays provide higher resolution and sensitivity, do not use radioactivity, and provide much more reproducibility. However, with proper fine-tuning of the protocols, the macroarrays can deliver sensitivity levels very similar to microarrays.[4] Macroarrays may be well suited for small labs with fewer gene sets to work with or for labs focusing on a selected group of genes specific for a biological function.[12,18,27] For macroarrays, image analysis and data processing are usually done by using a phosphorimager and supporting software. The preference of whether to use microarrays or macroarrays depends on the specific needs of the individual laboratory. For many small labs, macroarrays could be a suitable alternative.

III. DNA Microarray Protocols

A. Overview of the Process

Fluorescent microarray technology is useful for making comparisons of the relative abundance of particular messages relative to a designated source of mRNA that serves as a reference point. With the increased commercial support available in recent years, it is now possible for departments or large laboratories to set up a microarray facility. The following protocols are intended to serve as a basic introduction to making and using cDNA microarrays. The microarray experiment consists of three major steps: Step 1 involves the preparation of DNA templates, cDNA amplification by polymerase chain reaction (PCR), and RNA purification. It is assumed that the lab preparing the microarrays has access to plasmid clones that represent the genes whose expression patterns and/or relative message levels are being studied. Plasmid templates are used as PCR substrates to produce DNA representations of cDNA inserts. The PCR products are then analyzed, purified and quantified, and spotted onto poly-L-lysine coated microscope slides. Step 2 entails (1) the preparation of poly-L-lysine-coated slides and (2) the printing of microarrays with cDNAs from Step 1 using RNA isolated from cell or tissue samples to be examined as the substrate for reverse transcription in the presence of fluorescent label-tagged nucleotides. This produces the tagged representations of the mRNA pools of the samples that will be hybridized to the gene-specific cDNA probes immobilized on the glass slide microarray. Step 3 involves hybridization and data recovery and analysis. This covers the steps in which fluorescent label-tagged cDNAs (targets) hybridize to their complements (probes) on the microarray, and the resulting localized concentrations of fluorescent molecules are detected and quantified.

B. Growing Bacterial EST (cDNA) Clones

1. Incubate sealed master plates overnight at 37°C.
2. Prepare sets of standard 96-well round-bottom plates by labeling all plates and placing 100 µl of LB (Luria-Bertaini) broth containing 100 µg·ml^{-1} carbenicillin (or an appropriate antibiotic as needed) in each well. These plates will serve as working copies.

Note: *If you want to preserve the master set of plates, make replicate copies of the master plate to serve as working copies when the master plate is first replicated.*

3. Centrifuge master plates for 2 min at 1000 rpm in a horizontal microtiter plate rotor to remove condensation and droplets from the seals before opening.

Note: *Bacterial culture fluid on the sealers can easily be transferred from one well of the plate to others, thus cross-contaminating the stocks.*

4. Partially fill a container with 100% alcohol. Dip the 96-pin replicating tool in the alcohol. Remove from the alcohol bath and then flame the pins to sterilize them. After the inoculation block is allowed to cool briefly, dip the replicating tool in the master plate, and then into the daughter 96-well plate. Repeat as necessary for each plate that you need to inoculate.

Note: *Before beginning the replication process, it is useful to mark all master and daughter plates with a color marker pen near the A1 well so as to reduce mistakes in relative orientation of the plates, or use notched plates.*

5. Place the inoculated LB plates with lids on into a 1-gal Zip-lock™ bag containing a moistened paper towel and grow overnight at 37°C.

Note: *Many 37°C incubators tend to dry out microtiter plate cultures. Placing the plates in a highly humidified bag avoids this problem.*

6. Fill deep-well plates (R.E.A.L plates from Qiagen, Chatsworth, CA) with 1.3 ml of Superbroth (50 µg·ml^{-1} carbenicillin) per well. These plates will serve as the source of culture for template preparation.
7. Using the replicating tool, inoculate the deep-well plates directly from the freshly grown clone stocks. Cover the openings of the deep-well plates with Qiagen Airpore™ Tape Sheets and place the plastic lid over the sheet. Place the plates in a 37°C shaker incubator at 220 rpm for 22 to 24 h.

Note: *If you would like to store any of the working plates, add 50 µl of 45% (w/v) sterile glycerol to each well of any working plates, mix, and store frozen (–80°C). These plates serve as culture sources for future use.*

C. Isolation of Plasmid Templates for PCR

1. Harvest the bacterial cells in the block by centrifugation for 5 min at 1500 g in a centrifuge with a rotor for 96-well microplates. The block should be covered with adhesive tape during centrifugation. Remove medium by inverting the block. To remove the medium, peel off the tape and quickly invert the block over a waste container. Blot the inverted block firmly on a paper towel to remove any remaining droplets of medium.

Note: Do this step quickly so as to avoid losing any part of the pellet, as the pellet will be loose.

2. Resuspend each bacterial pellet in 0.3 ml Buffer R1. Use an eight-channel pipette with a large fill volume (1 ml per channel) for buffer delivery. Tape the block and mix by vortexing.
3. Add 0.3 ml Buffer R2 to each well, seal the block with new tape, mix gently but thoroughly by inverting ten times, and incubate at room temperature for 5 min.

Note: Buffer R2 should be checked before use for sodium dodecyl sulfate (SDS) precipitation caused by low storage temperatures. If necessary, redissolve SDS by warming. Do not vortex the lysates at this step, as this may cause shearing of the bacterial genomic DNA. Do not incubate for more than 5 min. Additional incubation may result in increased levels of open circular plasmid. At the end of the incubation, the lysate should appear viscous and free of bacterial cell clumps.

4. Add 0.3 ml Buffer R3 to each well, seal the block with new tape, and mix immediately by inverting ten times. Gently inverting the taped block ten times ensures uniform precipitation.

Note: Optionally, place the block in a boiling water bath for 5 min to denature and precipitate proteins and carbohydrates that are not removed by alkaline lysis.

5. Transfer the lysates to the wells of the QIAfilter 96-well plate (Qiagen Inc., Chatsworth, CA) using an eight-channel pipette adjusted to 1-ml fill volume. Apply vacuum (−200 to −300 mbar) until the lysates are completely transferred to the square-well block in the QIAvac base.

Note: Occasionally, precipitate will clog the end of a pipette tip. Lightly tapping the tip on the bottom of the well of the culture block will break through the precipitate and allow the remaining material to be transferred. Any unused wells can be sealed with tape for later use.

6. Take the square-well block containing the cleared lysates from the vacuum manifold. Add 0.7 volume of room-temperature isopropanol to each well (0.63 ml for 0.9 ml of lysate), tape the block, and mix immediately by inverting three times.

Note: If you use an optional boiling method, the lysate needs to be cooled prior to addition of isopropanol. Otherwise volatilization of isopropanol will cause the tape to detach from the wells. When preparing multiple sets of 96 samples, add isopropanol to one block, tape, and mix by inversion before proceeding to the next block. This will minimize separation of the tape from the block before the samples are mixed.

7. Centrifuge the block at 2500 g for 15 min at room temperature to pellet the plasmid DNA. Remove the supernatants by quickly inverting the block over a waste container, then tapping the block, upside down, onto a paper towel.

Note: Mark the orientation of the block before centrifugation so that it can be spun in the same orientation in the ethanol wash step. DNA pellets from isopropanol precipitations have a glassy appearance and may be difficult to see. Handle the block carefully to avoid dislodging the pellets.

8. Wash each DNA pellet with 0.5 ml of 70% ethanol. Centrifuge the block (in the same orientation as before) at 2500 g for 2 min. Remove the wash solutions by inverting the block, then tapping it firmly, upside down, onto a paper towel. Air dry the pellets for 15 min or dry under vacuum for 10 min. Make sure that no alcohol droplets are visible after air drying.

Note: It is important not to overdry the DNA pellets, as this will make them difficult to dissolve. If this occurs, the DNA pellets can be heated at 50°C to completely redissolve the pellets.

9. Redissolve the DNA pellets in 50 to 250 μl 10 mM Tris·HCl, pH 8.5 buffer.

Note: The optimal volume of buffer to use will depend on the copy number of the plasmid and the desired DNA concentration. Avoid repeated pipetting, which can shear the DNA.

Materials, Reagents, and Solutions

96-well R.E.A.L. miniprep kit (Qiagen, Inc., Chatsworth, CA)
LB broth (Biofluids, Rockville, MD)
Superbroth (Biofluids)
96-pin inoculating block (Cat. #VP 4088, V&P Scientific, Inc., San Diego, CA)
Airpore Tape Sheets, (Cat. #19571, Qiagen, Inc.)
Sterile 96-well plate seals, (e.g., Cat. #SEAL-THN-STR, Elkay Products, Inc., Shrewsbury, MA)
96-well U-Bottom Microtiter Plates, Cat. #3799; and 96-well V-Bottom Microtitre Plates, Cat. #3894 (Corning, Inc., Corning, NY)
Centrifuge with a horizontal ("swinging bucket") rotor with a depth capacity of 6.2 cm for spinning microtiter plates and filtration plates (e.g., Sorvall Super T 21, Sorvall, Inc., Newtown, CT)
37°C shaker incubator with holders for deep-well plates
37°C water bath
65°C incubator
Vortex mixer
−80°C freezer
−20°C freezer
Carbenicillin (Invitrogen, Carlsbad, CA)
Ethanol (200 proof USP ethyl alcohol)
Isopropanol
1 M Tris-HCl, pH 8
0.5 M NaEDTA, pH 8

D. PCR Amplification of cDNA Inserts

cDNA inserts from plasmid templates of EST clones need to be amplified, purified, and quantified before being used for printing microarrays. The following protocols describe the general methods to obtain PCR products for printing microarrays. Because of the large numbers of clones to be amplified, it is best to use 96-well formatted PCR plates for amplification, which will also facilitate printing microarrays using robotics.

1. For each 96-well plate to be amplified, prepare a PCR reaction mixture containing the following ingredients:
   - 10X PCR buffer, 1000 μl
   - dATP (100 mM), 20 μl
   - dGTP (100 mM), 20 μl
   - dCTP (100 mM), 20 μl
   - dTTP (100 mM), 20 μl
   - M13F primer (1 mM), 5 μl (See Note following.)
   - M13R primer (1 mM), 5 μl (See Note following.)
   - Ampli-*Taq* polymerase (5 units·μl^{-1}), 100 μl
   - ddH$_2$O, 8800 μl

Note: Primers used for PCR amplification depend on the vector in which the cDNA inserts are located. M13 primers are generally useful for many commonly available cloning vectors. Keep all reagents on ice and return the enzyme tube promptly back to the freezer.

2. Label 96-well PCR plates and aliquot 100 μl of PCR reaction mix to each well. Gently tap plates to ensure that no air bubbles are trapped at the bottom of the wells.
3. Add 1 μl of purified EST plasmid template to each well. Mix well with pipette.

Note: Mark the donor and recipient plates at the corner near the A1 well to facilitate correct orientation during transfer of the template. It is important to watch that the pipette tips are all submerged in the PCR reaction mix when delivering the template. Missing the liquid is easier when multichannel pipettes are used. Always use sterile filtered tips to avoid contamination.

4. Place PCR plate covers on plates and centrifuge the plates at 2700 rpm for 1 min.
5. Place the PCR plates in a thermal cycler (Eppendorf Master Cycler) and run the following cycling program:
   Initial denaturation, 96°C × 2 min
   Denaturation, 94°C × 30 sec
   Primer annealing, 55°C × 30 sec × 30 cycles
   Primer extension, 72°C × 2 min × 30 cycles
   Final extension, 72°C × 5 min

Note: After PCR, plates can be held at 4°C while quality controls are performed on PCR products.

Materials for PCR Amplification

AmpliTaq® DNA Polymerase with GeneAmp® PCR Reaction Kit, Cat. #N808-0156 (Applied Biosystems, Foster City, CA)
10X PCR Buffer II
MgCl$_2$ solution
Platinum *Taq* polymerase
M13 forward and reverse primers (Invitrogen, Carlsbad, CA)
Forward: 5' GTT TTC CCA GTC ACG ACG TTG 3'
Reverse: 5' TGA GCG GAT AAC AAT TTC ACA CAG 3'
dNTP kit (100 mM of each dNTP), Cat. #10297-018 (Invitrogen)
MilliQ water
MicroAmp® optical 96-well reaction plate, Cat. #N801-0560 (Applied Biosystems)
MicroAmp 96-well full plate cover, Cat. #N801-0550 (Applied Biosystems)
Multiscreen® PCR filter plate, Cat. #MANU3050 (Millipore Corp., Bedford, MA)
Cap mat, Cat. #40002-002 (VWR, West Chester, PA)
Falcon Microtest U-bottom 96-well plate, Cat. #353077 (BD Biosciences, Palo Alto, CA)

1. Check PCR Products by Agarose Gel Electrophoresis

If this is the first time the template for these cDNAs is being amplified, then it is necessary to analyze 2 µl of each PCR product on a 2% agarose gel. If amplified products from this template have been previously tested, then quality test the PCR products by analyzing one row of wells from each amplified plate. Gel imaging allows for both a rough quantitation of product (by comparing with known DNA standards) and an excellent characterization of the product. Band sizes, as well as the number of bands observed in the PCR products, are necessary for interpretation of final results of the hybridization. The use of gel well formats suitable for loading from 96-well plates and programmable pipetters makes this form of analysis feasible on a large scale.

1. Cast a 2% agarose gel in 1X TAE (40 mM Tris base, 40 mM acetate, 1 mM EDTA, pH 8.2) with four combs (50 tooth) and submerge in an electrophoresis apparatus with sufficient 1X TAE buffer to cover the surface of the gel.
2. Prepare a reservoir of loading buffer, using 12 wells of a microtiter plate.
3. Program pipetter (Matrix 12-channel pipetter) to sequentially carry out the following steps:
   a. Fill with 2 µl
   b. Fill with 1 µl
   c. Fill with 2 µl
   d. Mix a volume of 5 µl five times
   e. Expel 5 µl
4. Load 2 µl of PCR product from wells A1 to A12 of the PCR plate using sterile filtered tips.
5. Load 2 µl of loading buffer from the reservoir. Place tips in clean wells of disposable mixing tray and allow pipette to mix the sample and loading dye.
6. Position the pipette tip in a well row so that the tip containing the PCR product from well A1 is in the second well of the row, and the other tips are in every other succeeding well.
7. Repeat the process (changing tips for each load), loading PCR plate row B starting in the third well, interleaved with the A row, the C row starting at well 26, and the D row at well 27, interleaved with the C row.
8. Place 5 µl of 100-bp (base-pairs) size standards in wells 1 and 50. Repeat this process, loading samples from rows E, F, G, and H in the second 50-well row of gel wells, loading samples from two 96-well PCR plates per gel, or single-row samples from 16 PCR plates.

Note: *Since it will take time to load all the wells, to reduce diffusion and mixing of samples, apply voltage to the gel for a minute between loading each well strip. This will cause the DNA to enter the gel, and reduce band spreading and sample loss.*

9. Apply a voltage of 200 V to the gel and run until the bromophenol blue (faster band) has nearly migrated to the next set of wells.

Note: *For a gel that is 14 cm in the running dimension and 3 cm between each row of wells, 200 V for 15 min is sufficient.*

10. Use a photodocumentation system or take a digital photo of the gel on a UV table and store image for future reference.

Note: *The gels should show bands of fairly uniform brightness distributed in size between 600 to 2000 base-pairs, depending on the sizes of cDNAs amplified. Further computer analysis of such images can be carried out with image-analysis packages to provide a list of the number and size of bands. Ideally, this information can be made available during analysis of the data from hybridizations involving these PCR products.*

Functional Genomics and DNA Microarray Technology

Materials for PCR Product Analysis by Agarose Gel Electrophoresis

Gel electrophoresis apparatus with capacity for four 50-well combs, (e.g., Cat. #D3, Owl Scientific, Woburn, MA)
50X Tris-acetate electrophoresis buffer (Amersham Biosciences, Piscataway, NJ)
Agarose (Amersham Biosciences)
Dye solution (xylene cyanol/bromophenol blue) (e.g., Cat. #351-081-030, Quality Biological, Inc., Gaithersburg, MD)
Glycerol (enzyme grade)
Ethidium bromide solution (10 mg·ml^{-1})
100 base-pair ladder size standard
Programmable, 12-channel pipetter (e.g., Cat. #2019, Matrix Technologies, Lowell, MA)
Disposable microtiter mixing trays (e.g., Falcon Cat. #353911, Becton Dickinson, Franklin Lakes, NJ)
Electrophoresis power supply (Fisher Biotech, Pittsburgh, PA)
1X TAE buffer
  50X TAE buffer, 40 ml
  Ethidium bromide (10 mg·ml^{-1}), 0.1 ml
  ddH$_2$O, 960 ml
Loading buffer
  Glycerol (enzyme grade), 4.0 ml
  DEPC (diethylpyrocarbonate) water, 0.9 ml
  Dye solution,* 0.1 ml

Note: *This solution is 0.25% (w/v) xylene cyanol and 0.25% (w/v) bromophenol blue.

  100-bp size standards
  DNA ladder (1 mg·ml^{-1}), 50 µl
  1 M Tris-HCl (pH 8.0), 5 µl
  0.5 M EDTA (pH 8.0), 5 µl
  Loading buffer, 440 µl

E. Purification and Quantification of PCR Products

1. Spin down PCR reaction plates and then transfer the PCR products (100 µl) to a Multiscreen filter plate and place the filter on a vacuum manifold filtration system (e.g., Cat. #MAVM0960R, Millipore Corp., Bedford, MA).
2. Apply a vacuum pressure of approximately 250 to 380 mm Hg for 10 min or until plate is dry.

Note: Filter until no more fluid is visible in the well. The filter may appear wet and shiny even when dry, so do not use the appearance of the filter as a guide. It is also important to check all of the wells in the plate before removing from vacuum; some wells filter more slowly than others.

3. Remove plate from manifold filtration system and add 100 µl of MilliQ water to each well. Place filter plate on a shaker and shake vigorously for 20 min to resuspend the DNA.
4. Pipette the purified PCR product to a new Falcon U-bottom 96-well plate. Seal PCR storage plates with a plastic cap mat or adhesive foil lid and store at −20°C until needed for making microarray printing plates.

Note: *For long-term storage after filtration, aliquot equal volumes of purified PCR product into multiple plates. Store one plate with a cap mat at −20°C (for short-term use) and dry down the remaining plates in a speedvac and store at 4°C in a desiccator.*

5. Resuspend dried PCR product in MilliQ water just before preparing microarray printing plates.

1. Fluorometric Determination of PCR DNA Concentration

It is impractical for most labs to determine exact quantification of PCR products, especially if thousands of cDNAs must be prepared. However, it is possible to use a strategy where excess DNA is spotted, so that the exact quantities used do not produce much variation in the observed results. When using this strategy, it is necessary to track the productivity of the PCR reactions. Fluorometry methods using 96-well plates provide a simple way to obtain an approximate concentration of the double-strand PCR product from many thousands of samples.

1. Label 96-well plates for fluorescence assay.
2. Add 200 µl of Fluor buffer to each well. Add 1 µl of PCR product from each well in a row of a PCR plate to a row of the fluorometry plate. Samples can be added to rows A through G of the fluorometry plate.
3. In the final row of the fluorometry plate, add 1 µl of each of the series of double-strand DNA (dsDNA) standards: 0 $\mu g \cdot ml^{-1}$ (TE only), 50, 100, 250, and 500 $\mu g \cdot ml^{-1}$ dsDNA. Repeat this series twice in the final row.
4. Set the fluorometer for excitation at 346 nm and emission at 460 nm. Adjust as necessary to read the plate. If the fluorometer does not support automated analysis, export the data table to Excel.
5. Establish that the response for the standards is linear and reproducible from the range of 0 to 500 $\mu g \cdot ml^{-1}$ of dsDNA.
6. Calculate the concentration of dsDNA in the PCR reactions using the following equation after subtracting the average value obtained from the sample containing 0 $\mu g \cdot ml^{-1}$ from all other sample and control values:

$$[dsDNA\ (\mu g \cdot ml^{-1})] = [(PCR\ sample\ value)/(average\ 100\ \mu g \cdot ml^{-1}\ value)] \times 100$$

Note: *Constantly tracking the yields of the PCRs makes it possible to rapidly detect many of the ways in which PCR can fail or perform poorly.*

Materials and Reagents for PCR Product Quantification

Reference double-strand DNA (0.5 $\mu g \cdot ml^{-1}$) (e.g., Cat. #15612-013, Invitrogen, Carlsbad, CA)
96-well plates for fluorescent detection (e.g., Cat. #7105, Dynex, Chantilly, VA)
Fluorometer (e.g., Cat. #LS50B, Perkin Elmer, Norwalk, CT)
FluoReporter Blue dsDNA Quantitation Kit (Cat. #F-2962, Molecular Probes, Eugene, OR)
TE buffer
12-channel multipipetters
Computer equipped with Microsoft Excel™ software
dsDNA standards: 0 $\mu g \cdot ml^{-1}$, 10 $\mu g \cdot ml^{-1}$, 20 $\mu g \cdot ml^{-1}$, 50 $\mu g \cdot ml^{-1}$, 100 $\mu g \cdot ml^{-1}$, 250 $\mu g \cdot ml^{-1}$, 500 $\mu g \cdot ml^{-1}$
Fluor buffer
Hoechst 33258 solution* (from kit), 25 µl
TNE Buffer** (from kit), 10 ml

Note: ** Hoechst 33258 solution contains the dye at an unspecified concentration in a 1:4 mixture of DMSO:H_2O. ** TNE Buffer is 10 mM Tris-HCl (pH 7.4), 2 M NaCl, 1 mM EDTA.*

Functional Genomics and DNA Microarray Technology

F. RNA Preparation from Samples

This protocol details the methods used to extract RNA from cells and to purify the RNA by a combination of phase extraction and chromatography. Other RNA extraction protocols described in this book can also be used, depending on the type of tissues or source of samples. We have also found that the RNeasy™ kits sold by Qiagen work for plant, animal, or fungal samples and produce very clean RNA suitable for fluorescent labeling by reverse transcription.

1. If starting with cells harvested from tissue culture, wash the cell pellet twice in DPBS.
2. If starting with cells from tissue culture, add 1 ml of Trizol per 2×10^7 cells and mix by shaking. If starting with tissue, add 100 mg of frozen tissue directly to 4 ml of Trizol, and dissociate by homogenization with a rotating-blade tissue homogenizer. If starting from fungal or plant samples, grind the tissue in liquid nitrogen first. If the sample size is small (100 mg or less), grinding can be done in a microfuge tube using sterile polypropylene pestles. Add 2 ml of Trizol reagent per each 100 mg plant or fungal sample and homogenize with a polytron using a narrow probe for small volumes.
3. Add 2/10 volume of chloroform and shake for 15 sec.
4. Let stand for 3 min. Centrifuge at 12,000 g for 15 min at 4°C.
5. Take off the supernatant and add it to a polypropylene tube, recording the volume of the supernatant.
6. Add 0.53 volume of ethanol to the supernatant slowly while vortexing. This will produce a final ethanol concentration of 35%.

Note: *The ethanol should be added drop by drop and allowed to mix completely with the supernatant before more ethanol is added. If a high local concentration of ethanol is produced, the RNA in that vicinity will precipitate.*

7. Add the supernatant from an extraction of 2×10^7 to 1×10^8 cells or up to 250 mg of plant or fungal tissues to an RNeasy maxicolumn, which is seated in a 50 ml centrifuge tube.
8. Centrifuge at 2880 g in a clinical centrifuge with a horizontal rotor at room temperature for 5 min.
9. Pour the flow-through back onto the top of the column and centrifuge again.

Note: *A significant amount of RNA is not captured by the column matrix in the first pass of the RNA-containing solution through the column.*

10. Discard the flow-through and add 15 ml of RW1 buffer to the column. Centrifuge at 2880 g for 5 min.
11. Discard flow-through; then add 10 ml of RPE buffer. Centrifuge at 2880 g for 5 min.
12. Discard flow-through and add another 10 ml of RPE buffer. Centrifuge at 2880 g for 10 min.
13. Put the column in a fresh 50 ml tube and add 1 ml of DEPC-treated water from the kit to the column.
14. Let stand for 1 min.
15. Centrifuge at 2880 g for 5 min.
16. Repeat steps 13, 14, and 15.
17. Aliquot out 400 µl portions of the column eluate to 1.5 ml eppendorf tubes.
18. Add 1/10 volume of 3 M sodium acetate (pH 5.2) and 1 ml of ethanol to each tube and mix.
19. Let stand for 15 min.
20. Centrifuge at 12,000 g at 4°C for 15 min.
21. Wash pellet two times in 75% EtOH, then store at −80°C.

Note: *Many times RNA prepared may have residual impurities that need to be removed before reverse transcription.*

22. Resuspend RNA at approximately $1 \text{ mg} \cdot \text{ml}^{-1}$ in DEPC H_2O.

23. Concentrate to greater than 7 mg·ml^{-1} by centrifugation on a MicroCon™ 100 filter unit and centrifuge at 500 g, checking as necessary to determine the rate of concentration.
24. Determine the concentration of RNA in the concentrated sample by spectrophotometry. Store at –80°C.

Reagents and Materials for RNA Purification

Trizol Reagent (Cat. #15596-018, Invitrogen, Carlsbad, CA)
RNeasy Maxi Kit (Cat. #75162, Qiagen, Chatsworth, CA)
Chloroform
Ethanol (200 proof USP ethyl alcohol)
DPBS (Dulbecco's phosphate buffered saline)
3 M sodium acetate, pH 5.2
DEPC water
RPE buffer (add four volumes of ethanol per volume of RPE concentrate supplied in Qiagen kit)
RW1 buffer (supplied in Qiagen kit)
75% EtOH
Ethanol (100%), 375 ml
MicroCon 100 (Cat. #42412, Millipore Corp., Bedford, MA)
High-speed centrifuge for 15 ml tubes
Clinical centrifuge with horizontal rotor for 50 ml conical tubes
Tissue homogenizer (e.g., Polytron PT1200 with Polytron-Aggregate-Dispergier-und-Mischtechnik 147a Ch6014 #027-30-520-0, Brinkmann Instruments, Inc., Westbury, NY)

G. Preparation of Poly-L-Lysine Coated Slides

Slides coated with poly-L-lysine have a surface that is both hydrophobic and positively charged. The hydrophobic character of the surface minimizes spreading of the printed spots, and the charge helps to position the DNA on the surface in a way that makes cross-linking more efficient.

1. Place slides into slide racks and place racks in glass tanks with 500 ml of cleaning solution. Gold Seal Slides™ are highly recommended, as they have been found to have consistently low levels of autofluorescence.

Note: *It is important to wear powder-free gloves when handling the slides. It is also important to change gloves frequently, as random contact with skin and surfaces transfers grease to the gloves.*

2. Place tanks on platform shaker for two hours at 60 rpm.
3. Pour out cleaning solution and wash in ddH$_2$O for 3 min. Repeat wash four times.
4. Transfer slides to 30-slide plastic racks and place into small plastic boxes for coating.
5. Place slides in 200 ml poly-L-lysine solution per box and make sure that the slides are completely submerged.
6. Place slide boxes on platform shaker for 1 h at 60 rpm.
7. Rinse slides three times with ddH$_2$O.
8. Submerge slides in ddH$_2$O for 1 min.
9. Spin slides in centrifuge for 2 min at 400 g and dry slide boxes used for coating.
10. Place slides back into slide box used for coating and let stand overnight before transferring to new slide box for storage.

Note: *Slides should be completely dry before handling.*

11. Allow slides to age for two weeks on the bench, in a new slide box, before printing on them. The coating dries slowly, becoming more hydrophobic with time.

Functional Genomics and DNA Microarray Technology 331

Note: Slide boxes used for long-term storage should be plastic and free of cork lining. The glue used to affix the cork will leach out over time and give slides stored in these types of boxes a greasy film that has a high degree of autofluorescence. Clean all glassware and racks used for slide cleaning and coating with highly purified H_2O only. Do not use detergent.

Materials and Reagents for Preparing Poly-L-Lysine Coated Slides

Gold Seal microscope slides (Cat. #3011, Becton Dickinson, Franklin Lakes, NJ)
Ethanol (100%)
Poly-L-lysine (Cat. #P8920, Sigma, St. Louis, MO)
50-slide stainless steel rack, Cat. #900401, and 50-slide glass tank, Cat. #900401 (Wheaton Science Products, Millville, NJ)
Sodium hydroxide
Stir plate
Stir bar
Platform shaker
30-slide rack (plastic), Cat. #196, and 30-slide box (plastic), Cat. #195 (Shandon Lipshaw, Pittsburgh, PA)
Sodium chloride
Potassium chloride
Sodium phosphate dibasic heptahydrate
Potassium phosphate monobasic
Autoclave
0.2-μm filter (Nalgene, Inc., Rochester, NY)
Centrifuge: Sorvall T21
Slide box (plastic with no paper or cork liners), (e.g., Cat. #60-6306-02, PGC Scientific, Gaithersburg, MD)
1 L Glass beaker
1 L Graduated cylinder
1 M Sodium borate, pH 8.0
  Dissolve 61.83 g of boric acid in 900 ml of DEPC H_2O. Adjust the pH to 8.0 with 1 N NaOH. Bring volume up to 1 L. Sterilize with a 0.2-μm filter and store at room temperature.
Cleaning solution
  H_2O, 400 ml
  Ethanol, 600 ml
  NaOH, 100 g
  Dissolve NaOH in H_2O. Add ethanol and stir until the solution clears. If the solution does not clear, add H_2O until it does.
Poly-L-lysine solution
  Add poly-L-lysine (0.1% w/v), 35 ml; PBS, 35 ml; H_2O, 280 ml to a total volume of 350 ml.

H. Printing of Microarrays

There are a variety of printers and pins available for transferring PCR products from titer plates to slides. A list of microarray printer makers is given below:

- Beecher Instruments: http://www.beecherinstruments.com
- BioRobotics: http://www.BioRobotics.com/
- Cartesian Technologies: http://www.cartesiantech.com/
- Engineering Services: http://www.ESIT.com/
- Genetic Microsystems: http://www.geneticmicro.com
- Genetix: http://www.genetix.co.uk/
- Gene Machines: http://www.genemachines.com
- Genomic Solutions: http://www.genomicsolutions.com/

- Intelligent Automation Systems: http://www.ias.com
- Packard: http://www.packardinst.com/

The following steps provide a general description of the process and useful tips.

1. Preclean the print pins according to the manufacturer's specifications.
2. Load the printer slide deck with poly-L-lysine-coated slides (see protocol in **Section III.G**, above).
3. Thaw the plates containing the purified PCR products and centrifuge for 2 min at 1000 rpm in a horizontal microtiter plate rotor to remove condensation and droplets from the seals before opening.
4. Transfer 5 to 10 µl of the purified PCR products to a plate that will serve as the source of DNA samples for the printer.

Note: *Printing with quill-type pins usually requires that the volume of fluid in the print source be sufficiently low so that when the pin is lowered to the bottom of the well, it is submerged in the solution to a depth of less than a millimeter. This keeps the pin from carrying a large amount of fluid on the outside of the pin and producing variable, large spots on the first few slides printed. Capillary-action-based pins are useful in delivering small volumes, but these tend to clog and deliver doughnut-shaped spots. The pin-and-loop system provides more uniform spots, but it uses larger volumes of sample and, consequently, lower spot densities.*

5. Run a repetitive test print on the first slide. To do this, the pins are loaded with the DNA solution, and then the pins serially deposit this solution on the first slide in the spotting pattern specified for the print.

Note: *This test is necessary to check the size and shape of the specified spotting pattern and its placement on the slide. It also serves to verify that the pins are loading and spotting, and that a single loading will produce as many spots as are required to deliver material to every slide in the printer.*

6. If one or more of the pins is not performing at the desired level, reclean or substitute another pin and test again.
7. Once it is confirmed that all pins are performing, carry out the full print.

1. Postprocessing of Printed Slides

After all the slides are printed, remove slides from the printer, label with the print identifier and the slide number by writing on the edge of the slide with a diamond scribe, and place in a dust-free slide box to age for 1 week.

Note: *It is useful to etch a line, which outlines the printed area of the slide, onto the first slide. This serves as a guide to locate the area after the slides have been processed and the salt spots are washed off.*

1. Place slides, printed face up, in a casserole dish and cover with SaranWrap™. Expose slides to a 450-mJ dose of ultraviolet (UV) irradiation in the Stratalinker or similar UV crosslinker.
2. Transfer slides to a 30-slide stainless steel rack and place rack into a small glass tank.
3. Dissolve 6.0 g succinic anhydride in 325 ml 1-methyl-2-pyrrolidinone in a glass beaker by stirring with a stir bar.

Note: *Nitrile gloves should be worn and work carried out in a chemical fume hood while handling 1-methyl-2-pyrrolidinone (a known teratogen).*

4. Add 25 ml 1M sodium borate buffer (pH 8.0) to the beaker. Allow the solution to mix for a few seconds, then pour rapidly into glass tank with slides.

Note: Succinic anhydride hydrolyzes quite rapidly once the aqueous buffer solution is added. To obtain quantitative passivation of the poly-L-lysine coating, it is critical that the reactive solution be brought in contact with the slides as quickly as possible.

5. Place the glass tank on a platform shaker in a fume hood for 20 min. Small particulates resulting from precipitation of reaction products will be visible in the fluid.
6. While the slides are incubating on the shaker, prepare a boiling H_2O bath to denature the DNA on the slides.
7. After the slides have been incubated for 20 min, transfer them into the boiling H_2O bath. Immediately turn off the heating element after submerging the slides in the bath. Denature DNA on the slides in the H_2O bath for 2 min.
8. Transfer the slides into a glass tank filled with 100% ethanol and incubate for 4 min.
9. Remove the slides and centrifuge at 400 rpm for 3 min in a horizontal microtiter plate rotor to dry the slides.
10. Transfer slides to a clean, dust-free slide box and let stand overnight before using for hybridization.

Materials and Reagents for Slide Printing

Robotic slide printer

Materials for 30 Arrays

	Quantity needed
Humid chamber (Cat. #H 6644, Sigma-Aldrich, St. Louis, MO)	1
Inverted heat block (70 to 80°C)	1
Diamond scriber (Cat. #52865-005, VWR, West Chester, PA)	1
Slide rack (Cat. #121, Shandon Lipshaw, Pittsburgh, PA)	1
Slide chamber (Cat. #121, Shandon Lipshaw)	2
Succinic anhydride (Cat. #23,969-0, Sigma-Aldrich)	6 g
1-Methyl-2-pyrrolidinone (Cat. #32,863-4, Sigma-Aldrich)	325 ml
Sodium borate (1 M, pH 8) (Use boric acid and adjust pH with NaOH)	15 ml
ddH_2O	~1 L
2 L beaker	1
100% ethanol	350 ml

I. Reverse Transcription and Fluorescent-Labeling of RNA

1. If using an anchored oligo dT primer, anneal the primer to the RNA in the following 17 µl reaction:

Component	For Cy5 labeling	For Cy3 labeling
Total RNA (>7 µg·µl⁻¹)	150 to 200 µg	50 to 80 µg
Anchored primer (2 µg·µl⁻¹)	1 µl	1 µl
DEPC H_2O	to 17 µl	to 17 µl

If using an oligo dT(12 to 18) primer, use 1 µl of primer (1µg·µl⁻¹).

Note: The incorporation rate for Cy5-dUTP is less than that of Cy3-dUTP, so more RNA is labeled to achieve more-equivalent signal from each species. Use 0.2 ml thin-walled PCR tubes to do this reaction.

2. Heat to 65°C for 10 min and cool on ice for 2 min.
3. Add 23 µl of reaction mixture containing either *Cy5*-dUTP or *Cy3*-dUTP nucleotides, mix well by pipetting, and use a brief centrifuge spin to concentrate in the bottom of the tube:
   Reaction mixture:
   - 5X first-strand buffer, 8 µl
   - 10X low T dNTP mix, 4 µl
   - *Cy5* or *Cy3* dUTP (1 mM), 4 µl
   - 0.1 M Dithiothreitol (DTT), 4 µl
   - Rnasin (30 units·µl^{-1}), 1 µl
   - Superscript II (200 units·ul^{-1}), 2 µl
   - Total volume, 23 µl

Note: Superscript reverse transcriptase is very sensitive to denaturation at air/liquid interfaces. So be very careful to suppress foaming in all handling of this reaction.

4. Incubate at 42°C for 30 min. Then add 2 µl Superscript II. Make sure the enzyme is well mixed in the reaction volume and incubate at 42°C for 30 to 60 min. Add 5 µl of 0.5 M EDTA at the end of incubation period to stop the reaction.

Note: You should stop your reaction with EDTA before adding NaOH, since nucleic acids precipitate in alkaline magnesium solutions.

5. Add 10 µl 1 N NaOH and incubate at 65°C for 60 min to hydrolyze residual RNA. Cool to room temperature.

Note: The purity of the sodium hydroxide solution used in this step is critical. Impure NaOH or NaOH stored for long time in a glass container will produce a solution that will degrade the Cy5 dye molecule, turning the solution yellow. It is possible to achieve better results by reducing the time of hydrolysis to 30 min.

6. Neutralize by adding 25 µl of 1 M Tris-HCl (pH 7.5).
7. Desalt the labeled cDNA by adding the neutralized reaction, 400 µl of TE (pH 7.5) and 20 µg of human C0t-1 DNA to a MicroCon 100 cartridge. Pipette to mix, then centrifuge for 10 min at 500 g.
8. Wash again by adding 200 µl TE (pH 7.5) and concentrating to about 20 to 30 µl (approximately 8 to 10 min at 500 g).

Note: A MicroCon 30 cartridge can be used to speed the concentration step. In this case, centrifuge the first wash for approximately 4.5 min at 16,000 g and the second (200 µl wash) for about 2.5 min at 16,000 g.

9. Recover by inverting the concentrator over a clean collection tube and centrifuge for 3 min at 500 g.

Note: If the Cy5-labeled cDNA forms a gelatinous blue precipitate in the concentrated volume, then it indicates the presence of contaminants. If this happens, do not use it for hybridization, as it will bind nonspecifically to all arrayed DNA.

When concentrating by centrifugal filtration, overly long spins can remove nearly all the water from the solution being filtered. This will make it difficult to remove the fluorescent-

tagged nucleic acids from the membrane. If control of volumes proves difficult, the final concentration can be achieved by evaporating liquid in the SpeedVac. Vacuum evaporation does not degrade the performance of the labeled cDNA as long as the sample is not completely dried.

10. Take a 2 to 3 μl aliquot of the Cy5-labeled cDNA for analysis, leaving 18 to 28 μl for hybridization.
11. Analyze the Cy5-labeled cDNA on a 2% agarose gel (6 cm wide × 8.5 cm long, 2 mm wide teeth) in Tris acetate electrophoresis buffer (TAE).

Note: For maximal sensitivity when running samples on a gel for analysis, use loading buffer with minimal dye and do not add ethidium bromide to the gel or running buffer.

12. Scan the gel on a Molecular Dynamics Storm™ fluorescence scanner (setting: red fluorescence, 200 μm resolution, 1000 V on PMT)

Note: Successful labeling produces a dense smear of probe from 400 bp to >1000 bp, with little pileup of low-molecular-weight products. Weak labeling and significant levels of low-molecular-weight material indicate poor labeling. A fraction of the observed low-molecular-weight material is unincorporated fluorescent-tagged nucleotide.

Materials for Reverse Transcription of RNA and Labeling of cDNA

dNTPs, 100 mM each, store frozen, −20°C (Cat. #27-2035-02, Amersham Biosciences, Piscataway, NJ)
pd(T)12 to 18, resuspend at 1 mg·ml^{-1}, and store frozen at −20°C (Cat. #27-7858, Amersham Biosciences)
Anchored oligo primer (anchored; 5′-TTT TTT TTT TTT TTT TTT TTV N-3′), resuspend at 2 mg·ml^{-1}, store frozen at −20°C (e.g., Cat. #3597-006, Sigma-Genosys, The Woodlands, TX)
Cy3-dUTP, 1 mM, and Cy5-dUTP, 1 mM, store frozen at −20°C, light sensitive
Rnasin (RNase inhibitor), store frozen at −20°C (Cat. #N211A, Promega, Madison, WI)
Superscript II RNase H^{-} reverse-transcriptase kit, store frozen at −20°C (Cat. #18064-014, Invitrogen, Carlsbad, CA)
Cot-1 DNA, 1 mg·ml^{-1}, store frozen at −20°C (Cat. #15279-011, Invitrogen)
0.5 M EDTA, pH 8.0
1 N NaOH
1 M Tris-HCL, pH 7.5
TE, pH 7.4
DEPC water
5X Tris-acetate buffer
15 ml round-bottom polypropylene centrifuge tubes (Fisher, Pittsburgh, PA)
50 ml conical polypropylene centrifuge tubes (Fisher)
1.5 ml eppendorf tubes (Brinkmann, Westbury, NY)
0.2 ml thin-wall PCR tube (Perkin Elmer, Norwalk, CT)
5X First-strand buffer (provided with Superscript II)
TAE buffer

J. Array Hybridization

1. Determine the volume of hybridization solution required (you would need 0.033 μl for each mm^2 of slide surface area covered by the coverslip used to cover the array. An array covered by a 24 mm × 50 mm coverslip will require 40 μl of hybridization solution).

Note: The volume of the hybridization solution is critical. When too little solution is used, it is difficult to seat the coverslip without introducing air bubbles over some portion of the arrayed cDNAs, and the coverslip will not sit at a uniform distance from the slide. When too much volume is applied, the coverslip will move easily during handling, leading to misplacement relative to the arrayed cDNAs and to nonhybridization in some areas of the array.

2. For a 40 μl hybridization, pool the Cy3- and Cy5-labeled cDNAs into a single 0.2 ml thin-wall PCR tube and adjust the volume to 30 μl by either adding DEPC H$_2$O or by removing water in a speedvac. If using a vacuum device to remove water, do not use high heat or heat lamps to accelerate evaporation. The fluorescent dyes could be degraded.
3. For a 40 μl hybridization, the following recipe works best:

	High Sample Blocking	High Array Blocking
Cy5 + Cy3 probe	30 μl	28 μl
Poly d(A) (8 mg·ml^{-1})	1 μl	2 μl
Yeast tRNA (4 mg·ml^{-1})	1 μl	2 μl
Human C0t-1 DNA (10 mg·ml^{-1})	1 μl	0 μl
20X SSC	6 μl	6 μl
50X Denhardt's blocking solution	1 μl (optional)	2 μl
Total volume	40 μl	40 μl

Note: Arrays and samples may vary, making it necessary to change the composition of the hybridization cocktail. If there is residual hybridization to control repeat DNA samples on the array, more C0t-1 DNA should be used, as in the high-sample-blocking formulation. If there is a diffuse background or a general haze on all of the array elements, more nonspecific blocker components should be used, as in the high-sample-array-blocking formulation.

4. Mix the components well by pipetting, heat at 98°C for 2 min in a PCR cycler, cool quickly to 25°C, and add 0.6 μl of 10% SDS.
5. Centrifuge for 5 min at 14,000 g. The fluorescent-labeled cDNAs have a tendency to form small aggregates which result in bright background on the array slide. Hard centrifugation will pellet these aggregates, allowing you to avoid introducing them to the array.
6. Apply the labeled cDNA to a 24 mm × 50 mm glass coverslip and then touch with the inverted microarray.

Note: Applying the hybridization mix to the array and placing a coverslip over it requires some dexterity to get the positioning of the coverslip and the exclusion of air bubbles. It is helpful to practice this operation with buffer and plain slides before attempting actual samples. The hybridization solution is added to the coverslip first, since some fluor-aggregates remain in the solution and will bind to the first surface they touch.

7. Place the slide in a microarray hybridization chamber, add 5 μl of 3X SSC in the reservoir, if the chamber provides one, or at the scribed end of the slide, and seal the chamber. Submerge the chamber in a 65°C water bath and allow the slide to hybridize for 16 to 20 h.

Note: There is a wide variety of commercial hybridization chambers. It is worthwhile to prepare a mock hybridization with a blank slide, load it in the chamber, and incubate it to test for leaks or drying of the hybridization fluid, either of which cause severe fluorescent noise on the array.

K. Array Postprocessing

1. Remove the hybridization chamber from the water bath, cool, and carefully dry off. Unseal the chamber and remove the slide.

Note: *As there may be negative pressure in the chamber after cooling, it is necessary to remove water from around the seals so that it is not pulled into the chamber and onto the slide when the seals are loosened.*

2. Place the slide, with the coverslip still affixed, into a Coplin jar filled with 0.5X SSC/0.01% SDS wash buffer. Allow the coverslip to fall from the slide and then remove the coverslip from the jar with a pair of forceps. Allow the slide to wash for 2 to 5 min.
3. Transfer the slide to a fresh Coplin jar filled with 0.06X SSC. Allow the slide to wash for 2 to 5 min.

Note: *The sequence of washes needs to be adjusted so as to allow for noise removal, depending on the source of the sample RNA. Useful variations are to add a first wash, which is 0.5X SSC/0.1% SDS, or to repeat the normal first wash twice.*

4. Transfer the slide to a slide rack and centrifuge at low rpm (700 to 1000) for 3 min in a clinical centrifuge equipped with a horizontal rotor for microtiter plates.

Note: *If the slide is simply air dried, it frequently acquires a fluorescent haze. Centrifuging off the liquids results in a lower fluorescent background. As the rate of drying can be quite rapid, it is suggested that the slide be placed in the centrifuge immediately upon removal from the Coplin jar.*

Materials and Reagents for Hybridization and Postprocessing

20X Saline-sodium citrate (SSC) (Cat. #S-6639, Sigma-Aldrich, St. Louis, MO)
10% Sodium dodecyl sulfate (SDS) (Cat. #15553-035, Invitrogen, Carlsbad, CA)
Bovine serum albumin (BSA) (Cat. #A-9418, Sigma-Aldrich)
Formamide, redistilled (Cat. #15515-081, Invitrogen)
Isopropanol (Cat. #A451-1, Fisher Scientific, Pittsburgh, PA)
Coplin jar (Cat. #25457-200, VWR, West Chester, PA)
Human Cot1-DNA (Cat. #15279-011, Invitrogen)
Mouse Cot1-DNA (Cat. #18440-016, Invitrogen)
Poly(A)-DNA (Cat. #27-7836-01, Amersham Biosciences, Piscataway, NJ)
Microscope cover glass (Cat. #12-545J, Fisher Scientific)
Hybridization chamber (Cat. #2551, Corning Costar, Corning, NY)
1 L of 0.22 μm CA (cellulose acetate) filter system (Cat. #430517, Corning)

L. Scanning DNA Microarrays

One of the key issues in automation of microarray analysis is the ability to integrate the scanning and analysis of the DNA microarrays. There are many scanners and software packages available for obtaining the data points from DNA microarrays and for analysis of the results. It is not possible to discuss all of these instruments and how they work. Most of the scanners have similar functions, and in general, it is preferable to use scanners that can simultaneously scan for both *Cy5* and *Cy3* and also have the ability to detect many varieties of fluorescent tags. A list of scanners and web sites of manufacturers follows.

- Axon: http://www.axon.com
- Beecher Instruments: http://www.beecherinstruments.com
- GSI Lumonics: http://www.gsilumonics.com
- Genetic Microsystems: http://www.geneticmicro.com
- Genomic Solutions: http://www.genomicsolutions.com/
- Molecular Dynamics: http://www.mdyn.com
- Virtek: http://www.virtek.ca/

1. Slide Scanning

We are currently using the GenePix 4000 scanner made by Axon (Axon, Union City, CA). This scanner uses dual lasers to scan at 635 nm and 532 nm to excite *Cy5* and *Cy3*, simultaneously.

1. Turn on the scanner and start the GenePix™ software.
2. Open the scanner door and insert the slide in the chip holder.
3. Set photomultiplier tube (PMT) settings to 600 for both *Cy5* and *Cy3* channels.
4. Perform a low-resolution "prescan" first in order to determine the boundaries of the array and the quality of the hybridization signal.

Note: *If you are using a scanner that scans one channel at a time, Cy5 is more susceptible to photodegradation than Cy3. Thus, it is necessary to first scan the Cy5 channel. However, scanners such as Axon Genpix 4000 can scan both channels simultaneously.*

5. Draw a scan-area marquis.
6. Adjust the PMT based on "prescan" of hybridization signals.

Note: *For gene-expression hybridizations, the ratio over the entire scan area should be 1.0. If it is not, adjust the red and green PMT values to achieve a ratio of 1.0.*

7. Set lines to average as two in the settings before scanning the array for data collection.

Note: *Setting the lines to average to two will allow the machine to scan each pixel twice and average the data counts collected. This will help to reduce the background noise.*

8. Perform a high-resolution data scan. This scan can be represented as a histogram in order to see relative intensities of both channels.

Note: *The histogram represents every pixel from the scan. It is important to manually scrutinize the array to remove any artifacts or dirt spots from the array and to obtain better normalized values. Ideally, pixels should be represented across the entire intensity range. This would also allow one to set up threshold values and to discard values too high or too close to the background.*

9. Once PMT levels have been set, and the intensity ratio is near 1.0, perform a data scan and save the results. Save image of the scan as a multi-image TIFF file.

Note: *For the image file, use the name that includes date prefix. Each image file should be stored as a 16-bit TIFF file.*

10. Once the image is scanned and stored, assign spot identities and calculate the results. Now these results can be put together as an array list file.

Functional Genomics and DNA Microarray Technology

Note: Array list files can be used for further image analysis (relative gene-expression levels) and interpretation of data.

Data from each fluorescence channel are collected and stored as a separate 16-bit TIFF image. These images are analyzed to calculate the relative expression levels of each gene and to identify differentially expressed genes. The analysis process can be divided into two steps: image processing and data analysis. The contrast in this image has been adjusted to allow faint spots to be easily visualized. Important aspects of the hybridization to note are the low level, uniform background, and the good signal-to-noise ratio.

M. Image Processing and Data Analysis

Image processing involves three stages:

1. Spots representing the arrayed genes must be identified from signals that can arise due to precipitated probe or other hybridization artifacts or due to dust on the surface of the slide.

Note: The problem of grid spot location is usually coupled with the difficulty of estimating the fluorescence background. For microarrays, it is important that the background be calculated locally for each spot rather than globally for the entire image, since uneven background can often arise during the hybridization process.

2. Estimate background and calculate background-subtracted hybridization intensities for each spot.

Note: There are currently two approaches to calculate intensities, namely, the use of the median or the mean intensity for each spot. Since array analysis generally uses ratios of measured Cy3 to Cy5 intensities to identify differentially expressed genes, the mean and the integrated intensities are equivalent. In general, the mean intensities are considered to be a better option for calculation of relative gene-expression levels.

3. A grid is created for each spot after the spots are identified from the background. For each spot found in a grid element, local background is calculated and subtracted. Finally, the integrated intensities for both *Cy3* and *Cy5* channels are calculated.

Note: Several image-processing software packages are available, and some are listed below. Measured intensities are entered into the gene-expression database to capture gene-expression data.

BioDiscovery: http://www.biodiscovery.com/
Silicon Genetics: http://www.sigenetics.com/
Spotfire: http://www.spotfire.com/
Stanford University: http://rana.stanford.edu/software/
TIGR: http://www.tigr.org/softlab/
ResGen: http://www.pathways.resgen.com

1. Data Normalization and Analysis

Following image processing, the data generated for the arrayed genes need to be further analyzed in order to identify differentially expressed genes.

1. The first step in this process is the normalization of the relative fluorescence intensities in each of the two scanned channels.

Note: Normalization is necessary to adjust for differences in labeling and detection efficiencies for the Cy5 and Cy3 and for differences in the quantity of starting RNA from the two samples examined in the assay. Both of these can cause a shift in the average ratio of Cy5 to Cy3, and the intensities must be rescaled before analysis. The normalization process is dependent on the assumption that for either the entire collection of arrayed genes — or a subset of the genes, such as housekeeping genes, or a set of controls — the ratio of measured expression averaged over the set should be one.

2. Normalize data by doing the following:
   a. Average total measured fluorescence intensity so that the total integrated intensity across all the spots in the array is equal for both channels.

Note: While the intensity for any one spot may be higher in one channel than the other, when averaged over thousands of spots in the array, these fluctuations should average out.

   b. Perform linear regression analysis for expression of many genes from closely related samples.

Note: For closely related samples, one would expect many of the genes to be expressed at nearly constant levels. Consequently, a scatterplot of the measured Cy5 vs. Cy3 intensities should have a slope of one.

   c. Use a subset of housekeeping genes to calculate a mean value for distribution of transcription levels and the standard deviation. The ratio of measured *Cy5* to *Cy3* ratios for these genes can be modeled and the mean of the ratio adjusted to one.[7]

Note: During the normalization process, care must be taken in handling genes expressed at low levels. Statistical differences in the measured levels can result in variation in the ratios that are calculated, and inefficiencies in labeling for either of the two dyes can cause these low-intensity genes to disappear from the arrays. Rather than using a standard two fold-up or -down regulation for cutoff, the confidence-intervals approach proposed by Chen et al.[7] can be used to identify differentially expressed genes. If it is also important to identify genes that are truly differentially expressed from random changes, three independent microarray assays starting from independent RNA isolations should be performed.

IV. Troubleshooting Guide

As with many protocols dealing with RNA, the quality of microarray data depends a lot on the starting material, namely RNA. So the purity and quality of RNA used to prepare fluorescent-labeled cDNA targets is very critical. Secondly, the cDNA probes printed on glass slides also determine the outcome of data quality. It is critical to maintain the quality and purity of PCR products and the amounts placed on the slides. Variations in the amount of DNA in each spot can lead to variations in data, which are not a true reflection of gene expression. As mentioned in **Section III.E**, "Purification and Quantification of PCR Products," using excess cDNA will compensate for this problem. Thirdly, the quality of poly-L-lysine-coated slides and the uniformity of the coating is important for the quality of the data. Finally, the washing and postprocessing of the slides after hybridization also determine the quality of data.

It is not possible to list an individual troubleshooting guide for the whole process of microarray preparation and analysis. We have listed the troubleshooting tips for each section as Notes. We recommend that you read these notes carefully and check that the procedure and the results are working properly at each step before moving onto the next step. It is also recommended that you print some test slides with fewer ESTs and use good controls to check the whole process before printing high-density microarrays.

V. Applications of DNA Microarrays in Various Groups of Organisms

In the last two years there has been a tremendous increase in the number of publications on the use of DNA microarray analysis in various biological systems and/or how genome-wide gene expression is helping to understand complex biological phenomena. It is beyond the scope of this chapter to discuss or include all of these papers. The following sections briefly describe how microarray analysis has been used in various organisms.

A. Microorganisms

The small genomes of prokaryotes and simple eukaryotes such as yeast are well-suited to DNA microarray-based genomewide gene-expression profiling. In these microorganisms, DNA microarray experiments can be coupled to well-controlled experimental systems, thereby yielding highly reliable and interpretable information to understand the complex cellular processes.

1. Yeast (*Saccharomyces cerevisiae*)

The availability of the complete genome sequence and well-characterized genetics has made yeast (*Saccharomyces cerevisiae*) highly amenable for DNA microarray studies. One of the first reports on the use of DNA microarrays to monitor genomewide gene expression was from yeast.[8,9,29] Similarly, DNA microarray analyses of biological processes leading to sporulation, metabolic shifts between aerobic and fermentation, cell cycle and mitosis, DNA-damaging agents, and drug responses was done using yeast model system.[7,13,16,17,23] These studies have provided more comprehensive information that is being used to identify novel functions for many genes.

2. Bacteria

Microarray and macroarray technology has also been used to study gene expression from prokaryotic systems such as *E. coli*. Genomewide expression analysis of *E. coli* for 4290 genes under various conditions — such as growth in rich medium vs. poor medium or the effect of antibiotics — has identified many useful new drug targets.[19,24] Similarly, new drug targets were identified from microarray analysis of *Mycobacterium tuberculosis*.[28] In the future, these types of studies will facilitate further study of the interaction between bacterial pathogens and mammalian hosts, leading to the identification of virulence genes and host resistance genes.

B. Plants

Schena et al.[21] first demonstrated quantitative monitoring of differential gene expression between *Arabidopsis* root and leaf tissues using a 45 cDNA microarray. By the year 2000, 2375 cDNAs from *Arabidopsis* had been used to study the gene-expression patterns during plant–fungal interaction.[22] In addition, plant response to drought stress and low temperatures, mechanical wounding, and herbivory and nutrient treatments have also been explored using microarrays.[1] The number of cDNAs on the microarray has increased to 11,251 in a recent publication by Schaffer et al.[20] What about nonmodel species? Recently, microarrays have been used to identify genes involved in determining the flavor in strawberry[2] and the wood development in poplars.[14]

C. Humans

One of the fastest growing areas of microarray-based application is in human studies. Proper classification of a cancer type has a tremendous significance in treatment, increasing chances of patient survival. Golub et al.[11] have used microarray-based analysis of human cancers. Human lymphomas have been classified by a variety of means since the original proposal by Thomas Hodgkin in 1832. Diffuse, large B-cell lymphoma, which claims 25,000 cases each year, is difficult to classify due to heterogeneity in classification systems and treatments. Ash et al.[3] used microarrays with 17,856 cDNAS that provided more accurate description of lymphomas in a variety of malignancies. Such method provides an accurate identification of previously undetected and clinically significant subtypes of cancer. Similarly, microarray-based gene-expression profiling is also used to identify host-response genes specific to a viral infection. Zhu et al.[30] examined mRNA changes in primary human fibroblasts 40 min to 24 h postinfection with the human cytomegalovirus (HCMV), and they monitored the response of 6800 human mRNAs. They were able to identify expression of human genes that might affect the replication and pathogenesis of HCMV. In the future, such studies will have a major impact on designing novel drugs to prevent infection by various viral pathogens.

Reagents Needed

50X Denhard Solution

Dissolve 1 g of Ficoll (type 400, Amersham Pharmacia, Piscataway, NJ), 1 g polyvinylpyrrolidone, and 1 g BSA (Fraction V; Sigma) in 75 ml of deionized water. Make up the volume to 100 ml with deionized water. Sterilize the contents using a 0.2-μm filter and store in aliquots.

DEPC Water

Add 1 ml of DEPC (Sigma, St. Louis, MO) to deionized water and stir for 30 min. Autoclave the water.

0.5 M EDTA

Dissolve 93.06 g $Na_2EDTA \cdot 2H_2O$ in 300 ml ddH_2O. Adjust pH to 8.0 with 2 N NaOH solution. Adjust final volume to 500 ml with ddH_2O and store at room temperature.

Ethanol (100% and 70%)

First-Strand 5X Buffer

250 mM Tris-HCl, pH 8.3 (at 42°C)
50 mM $MgCl_2$
250 mM KCl
2.5 mM Spermidine
50 mM DTT
5 mM Each of dATP, dCTP, dGTP, and dTTP

Functional Genomics and DNA Microarray Technology

Hybridization Buffer (2X)
40 mM Tris-HCl, pH 7.7
1.2 M NaCl
4 mM EDTA
0.4% SDS
1 µg/µl Carrier yeast tRNA (transfer RNA) (optional)

IPTG Solution
10 mM Isopropyl β-D-thiogalactopyranoside in distilled deionized water (ddH_2O).

LB (Luria-Bertaini) Medium (per liter)
Bacto-tryptone (10 g)
Bacto-yeast extract (5 g)
NaCl (5 g)
Dissolve in ddH_2O and adjust pH to 7.5 with 0.2 N NaOH. Autoclave.

LB Plates Containing Carbenicillin
Add 15 g of Bacto-agar to 1 L LB medium. Autoclave. When the medium cools to 50°C, add carbenicillin (0.1 g/ml). Pour 30 to 40 ml into each of the 100 mm petri dishes in a laminar-flow hood with filtered air flowing. Remove any bubbles with a pipette tip and let the plates cool for 5 min prior to being covered. Allow the agar to harden for 1 h and store the plates at room temperature up for 10 days or at 4°C in a bag for 1 month. The cold plates should be placed at room temperature for 1 to 2 days before use.

LB Top Agar (500 ml)
Add 4 g agar to 500 ml of LB medium and autoclave.

10X Ligation Buffer
300 mM Tris-HCl, pH 7.8
100 mM $MgCl_2$
100 mM DTT
10 mM ATP

Phosphate Buffer (Stock)
0.5 M Monobasic sodium phosphate
Adjust the pH with 0.5 M dibasic sodium phosphate to 6.8.

Poly-L-Lysine Solution
Add poly-L-lysine (0.1% w/v), 35 ml; PBS, 35 ml; and sterile deionized water, 280 ml, to a total volume of 350 ml.

3 M Sodium Acetate Buffer (pH 5.2)
Dissolve sodium acetate in ddH_2O and adjust pH to 5.2 with 3 M glacial acetic acid. Autoclave and store at room temperature.

1 M Sodium Borate (pH 8.0)
Dissolve 61.83 g of boric acid in 900 ml DEPC water. Adjust the pH with 1 N NaOH to pH 8.0. Bring the volume up to 1 L. Sterilize with a 0.2 µm filter and store at room temperature.

10% Sodium Dodecyl Sulfate Solution

Dissolve 10 g of SDS in 80 ml of deionized water and bring the volume to 100 ml. Autoclave the solution.

20X SSC (Sodium Saline Citrate) Solution

Dissolve 175.3 g of NaCl and 88.2 g of sodium citrate in 800 ml of water. Adjust the pH to 7.0 with a few drops of 10 N solution of NaOH. Adjust the volume to 1 L with water. Dispense into aliquots and autoclave.

Succinic Anhydride Solution

Dissolve 6.0 g succinic anhydride in 325 ml 1-methyl-2-pyrrolidinone in a glass beaker by stirring with stir bar.

Superbroth Medium

Bacto-tryptone (12 g)
Bacto-yeast extract (24 g)
0.4% Glycerol (v/v)

Dissolve in a total volume of 900 ml in ddH$_2$O and autoclave. Cool to about 50°C and add 100 ml of phosphate buffer containing 170 mM KH$_2$PO$_4$ and 720 mM K$_2$HPO$_4$. Autoclave again.

1X TAE Buffer

40 mM Tris-base
40 mM Acetate
1 mM EDTA

5X TBE Buffer

Tris-base (54 g)
Boric acid (27.5 g)
0.5 M EDTA buffer, pH 8.0 (20 ml)

Dissolve well after each addition in 700 ml ddH$_2$O. Then add ddH$_2$O to a final volume of 1000 ml. Autoclave. Store at room temperature.

TE Buffer

10 mM Tris-HCl, pH 8.0
1 mM EDTA, pH 8.0

TNE Buffer

10 mM Tris-HCl, pH 8.0
2 M NaCl
1 mM EDTA, pH 8.0

1 M Tris-HCl, pH 7.5 or 8.0

Dissolve 121.1 g Tris-HCl in 800 ml ddH$_2$O. Adjust pH with 2 N HCl to desired pH. Adjust final volume to 1 L, autoclave and store at room temperature.

References

1. **Aharoni, A. and Vorst, O.,** DNA microarrays for functional genomics, *Plant Mol. Biol.*, 48, 99–118, 2002.
2. **Aharoni, A., Keizer, L.C.P., Bouwmeester, H.J., Sun, Z., Alvarez-Huerta, M., Verhoeven, H.A., Blaas, J., van Houwelingen, A.M.M.L., De Vos, R.C.H., van der Voet, H., Jansen, R.C., Guis, M., Mol, J., Davis, R.W., Schena, M., van Tunen, A.J., and O'Connell, A.P.,** Identification of the SAAT gene involved in strawberry flavor biogenesis by use of DNA microarrays, *Plant Cell*, 12, 647–662, 2000.
3. **Ash, A., Alizadeh, A., Eisen, M.B., Davis, E.R., Ma, C., Lossos, I.S., Rosenwald, A., Boldrick, J.C., Sabet, H., Tran, T., Yu, X., Powell, J.I., Yang, L., Marti, G.E., Moore, T., Hudson, J., Lu, L., Lewis, D.B., Tibshirani, R., Sherlock, G., Chan, W.C., Greiner, T.C., Weisenburger, D.D., Armitage, J.O., Warnke, R., Levy, R., Wilson, W., Grever, M.R., Byrd, J.C., Botstein, D., Brown, P.O., and Staudt, L.M.,** Distinct types of diffuse large β-cell lymphoma identified by gene expression profiling, *Nature*, 403, 503–511, 2002.
4. **Bertucci, F., Beranard, K., Loroid, B., Chang, Y.C., Granjeaud, S., Birnbaum, D., Nguyen, C., Peck, K., and Jordan, B.R.,** Sensitivity issues in DNA array-based expression measurements and performance of nylon microarrays for small samples, *Human Mol. Genet.*, 8, 1715–1722, 1999.
5. **Blattner, F.R., Plunkett, G., III, Bloch, C.A., Perna, N.T., Burland, V., Riley, M., Collado-Vides, J., Glasner, J.D., Rode, C.K., Mayrew, G.F. et al.,** The complete genome sequence of *Escherichia coli* K-12, *Science*, 277, 1453–1474, 1997.
6. **Brown, P.O. and Botstein, D.,** Exploring the new world of the genome with DNA microarrays, *Nat. Genet.*, 21(suppl.), 33–37, 1999.
7. **Chen, Y., Dougherty, E.R., and Bittner, M.L.,** Ratio-based decisions and the qualitative analysis of cDNA mciroarray images, *J. Biomed. Optics*, 24, 364–374, 1997.
8. **Chu, S., DeRisi, J., Eisen, M., Mulholland, J., Botstein, D., Brown, P.O., and Herzkowitz, I.,** The transcriptional program of sporulation in budding yeast, *Science*, 282, 699–705, 1999.
9. **De Risi, J.L., Iyer, V.R., and Brown, P.O.,** Exploring the metabolic and genetic control of gene expression on a genomic scale, *Science*, 278, 680–686, 1997.
10. **Goffeau, A., Aert, R., Agostini-Carbone, M.L., Ahmed, A., Aigle, M., Alberghina, L., Albermann, K., Albers, M., Aldea, M., Alexandraki, D. et al.,** The yeast genome directory, *Nature*, 387(suppl.), 1–105, 1997.
11. **Golub, T.R., Slonim, D.K., Tamayo, P., Huard, C., Gaasenbeek, M., Mesirov-J.P., Coller, H., Loh, M.L., Downing, J.R., Caligiuri, M.A. et al.,** Molecular classification of cancer: class discovery and class prediction by gene expression monitoring, *Science*, 286, 531–537, 1999.
12. **Gonzalez, P., Zigler, S., Epstein, D.L., and Borras, T.,** Identification and isolation of differentially expressed genes from very small tissue samples, *BioTechniques*, 26, 884–892, 1999.
13. **Gray, N.S., Wodicka, L., Thunnissen, A.W.H., Norman, T.C., Kwon, S., Espinoza, F.H., Morgan, D.O., Barnes, G., LeClerc, S., Meijer, L. et al.,** Exploiting chemical libraries, structure, and genomics in the search for kinase inhibitors, *Science*, 281, 533–538, 1998.
14. **Hertzberg, M., Aspeborg, H., Schrader, J., Andersson, A., Erlandsson, R., Blomqvist, K., Bhalerao, R., Uhlé, M., Teeri, T.T., Lundeberg, J., Sundberg, B., Nilsson, P., and Sandberg, G.A.,** Transcriptional roadmap to wood formation, *PNAS U.S.A.*, 98, 14732–14737, 2001.
15. **Jordan, B.R.,** Large-scale expression measurement by hybridization methods: from high-density membranes to "DNA chips," *J. Biochem.*, 124, 251–258, 1998.
16. **Lockhart, D.J, Dong, H., Byrne, M.C., Follettie, M.T., Gallo, M.V., Chee, M.S., Mittmann, M., Wang, C., Kobayashi, M., Horton, H., and Brown, E.L.,** Expression monitoring by hybridization to high-density oligonucleotide arrays, *Nat. Biotechnol.*, 14, 1675–1680, 1996.
17. **Marton, M.J., DeRisi, J.L., Bennet, H.A., Iyer, V.R., Meyer, M.R., Roberts, C.J., Stoughton, R., Burchard, J., Slade, D., Dai, H. et al.,** Drug validation and identification of secondary drug target effects using DNA microarrays, *Nat. Med.*, 4, 1293–1301, 1998.

18. **Podila, G.K., Zheng, J., Balasubramanian, S., Sundaram, S., Hiremath, S., Brand, J., and Hymes, M.,** Molecular interactions in ectomycorrhizas: identification of fungal genes involved in early symbiotic interactions between *Laccaria bicolor* and red pine, *Plant Soil*, 244, 117–128.
19. **Richmond, C.S., Glasner, J.D., Mau, R., Jin, H., and Blattner, F.R.,** Genome-wide expression profiling in *Escherichia coli* K-12, *Nucleic Acids Res.*, 27, 3821–3835, 1999.
20. **Schaffer, R., Landgraf, J., Accerbi, M., Simon, V., Larson, M., and Wisman, E.,** Microarray analysis of diurnal and circadian-regulated genes in *Arabidopsis*, *Plant Cell*, 13, 113–123, 2001.
21. **Schena, M., Shalon, D., Davis, R.W., and Brown P.O.,** Quantitative monitoring of gene expression patterns with a cDNA microarray, *Science*, 270, 467–470, 1995.
22. **Schenk, P.M., Kazan, K., Wilson, I., Anderson, J.P., Richmond, T., Somerville, S.C., and Manners, J.M.,** Coordinated plant defense responses in *Arabidopsis* revealed by microarray analysis, *PNAS*, 97: 11655–11660, 2000.
23. **Spellman, P.T., Sherlock, G., Zhang, M.Q., Iyer, V.R., Anders, K., Eisen, M.B., Brown, P.O., Botstein, D., and Futcher, B.,** Comprehensive identification of cell cycle-regulated genes of the yeast *Saccharomyces cerevisiae* by microarray hybridization, *Mol. Biol. Cell*, 9, 3273–3297, 1998.
24. **Tao, H., Bausch, C., Richmond, C., Blattner, F.R., and Conway, T.,** Functional genomics: expression analysis of *Escherichia coli* growing on minimal and rich media, *J. Bacteriol.*, 181, 6425–6440, 1999.
25. **The *Arabidopsis* Genome Initiative,** Analysis of the genome sequence of the flowering plant *Arabidopsis thaliana*, *Nature*, 408, 796–815, 2000.
26. **Venter, C. et al.,** The sequence of the human genome, *Science*, 291, 1304–1351, 2001.
27. **Voiblet, C., Duplessis, S., Encelot, N., and Martin, F.,** Identification of symbiosis-regulated genes in *Eucalyptus globulus-Pisolithus tinctorius* ectomycorrhiza by differential hybridization of arrayed cDNAs, *Plant J.*, 25, 181–191, 2001.
28. **Wilson, M., DeRisi, J., Kristensen, H.-H., Imboden, P., Rane, S., Brown, P.O., and Schoolnik, G.K.,** Exploring drug-induced alterations in gene expression in *Mycobacterium tuberculosis* by microarray hybridization, *Proc. Natl. Acad. Sci. U.S.A.*, 96, 12833–12838, 1999.
29. **Wodicka, L., Dong, H., Mittmann, M., Ho, M.-H., and Lockhart, D.J.,** Genome-wide expression monitoring in *Saccharomyces cerevisiae*, *Nat. Biotechnol.*, 15, 1359–1367, 1997.
30. **Zhu, H., Cong, J.P., Mamtora, G., Gingeras, T., and Shenk, T.,** Cellular gene expression altered by human cytomegalovirus: global monitoring with oligonucleotide arrays, *Proc. Natl. Acad. Sci. U.S.A.*, 95, 14470–14475, 1998.

Chapter 16

In Vitro Translation of mRNA(s) and Analysis of Protein by Gel Electrophoresis

William Wu and Leland J. Cseke

Contents

I.	Synthesis of Proteins from Purified mRNAs Using Rabbit Reticulocyte Lysates	348
II.	Synthesis of Proteins from Purified mRNAs Using Wheat Germ Extract	348
	A. Preparation of Wheat Germ Extract	348
	B. *In Vitro* Translation of mRNAs Using Wheat Germ Extract	349
III.	Synthesis of Proteins from mRNA Transcribed *In Vitro*	350
	A. *In Vitro* Transcription of mRNA from a Subcloned cDNA Insert	350
	B. *In Vitro* Translation of the *In Vitro* Transcribed mRNA	350
IV.	TCA Assay for Amino Acid Incorporation	350
V.	Analysis of Labeled Proteins by SDS-PAGE	352
	A. Preparation of the Separating Gel	352
	B. Preparation of the Stacking Gel	353
	C. Loading the Samples and Protein Standard Markers onto the Gel	353
	D. Electrophoresis	354
	Reagents Needed	355
References		356

Purified messenger ribonucleic acids (mRNA) from total RNAs or mRNA transcribed *in vitro* from cloned complementary deoxyribonucleic acid (cDNA) inserts can be translated in cell-free extracts to produce protein(s). The synthesized protein(s) can then be analyzed by sodium dodecyl sulfate-polyacrylamide gel electrophoresis (SDS-PAGE), two-dimensional gel electrophoresis, immunoprecipitation, or biological activity assay, depending on the particular research interest.[1-5] The present chapter describes, step by step, the protocols for *in vitro* translation of mRNA(s) using commercial lysates of rabbit reticulocytes and wheat germ extracts, and protocols for analysis of the products by SDS-PAGE. The basic principle is that commercial translational kits contain translational machinery, including all the components such as amino acids, transfer RNA (tRNA), ribosomal RNA (rRNA), and factors (initiation, elongation, and termination) for the *in vitro* synthesis of either radioactively labeled or unlabeled polypeptide(s).

I. Synthesis of Proteins from Purified mRNAs Using Rabbit Reticulocyte Lysates

1. Isolate the total RNAs from cell cultures or tissues of interest (see **Chapter 2**).
2. Purify poly(A)$^+$ RNA from the total RNA (see **Chapter 2**).
3. Thaw the reagents, on ice, from the commercial translation kit stored at –20°C or –70°C.

Note: We recommend that the translation kit components not be thawed at room temperature or at 37°C. This will decrease the efficiency of in vitro translation.

4. Add purified mRNAs (0.1 to 0.2 μg in 2 to 4 μl) in a sterile microcentrifuge tube. Denature the secondary structures of the mRNA by heating at 65 to 67°C for 10 min and quickly chilling the tube on ice for 5 min. Spin down briefly.
5. Set up a standard reaction on ice as follows:
  Sample:
   - Denatured mRNAs (4 μl)
   - Distilled, deionized water (ddH$_2$O) (14 μl)
   - RNasin ribonuclease inhibitor (40 units/μl) (2 μl)
   - Micrococcal nuclease-treated rabbit reticulocyte lysate (70 μl)
   - 1 m*M* Amino acid mixture (minus methionine) (2 μl)
   - [^{35}S]Methionine (1200 Ci/mmol) at 10 mCi/ml (8 μl)
   - Total volume of 100 μl

Caution: [^{35}S]Methionine is radioactive. Care should be taken.

  Background control (no mRNA):
   - ddH$_2$O (9 μl)
   - RNasin ribonuclease inhibitor (40 units/μl) (1 μl)
   - Micrococcal nuclease-treated rabbit reticulocyte lysate (35 μl)
   - 1 m*M* Amino acid mixture (minus methionine) (1 μl)
   - [^{35}S]Methionine (1200 Ci/mmol) at 10 mCi/ml (4 μl)
   - Total volume of 50 μl

Note: After use, the components should be quickly stored at –70 or –20°C.

6. Incubate the reactions at 30°C for 1 h and proceed to analysis of the protein products (see **Section IV** and **Section V**).

II. Synthesis of Proteins from Purified mRNAs Using Wheat Germ Extract

A. Preparation of Wheat Germ Extract

All steps should be carried out on ice or in a cold room.

1. Equilibrate Sephadex G-25 powder overnight in column buffer (~4 g for every 100 ml). Then prepare a Sephadex G-25 column (2 × 30 cm) in a cold room (see **Chapter 3** for general column preparation).

In Vitro Translation of mRNA(s) and Analysis of Protein by Gel Electrophoresis

2. Grind wheat germ (3 to 4 g) in liquid N_2 in a chilled mortar or blender. Add more liquid N_2 to keep the powder from thawing and repeat grinding until a fine powder is obtained.

Note: Gloves should be worn because liquid N_2 can burn the skin.

3. Briefly warm the powder at room temperature for about 2 min and add 10 ml of cold homogenization buffer to the mortar. Repeat grinding for 2 min or until a fine homogenate is visible.
4. Transfer the homogenate, using a clean spatula, into a sterile 30-ml Corex centrifuge tube on ice and centrifuge at 20,000 g for 10 min at 4°C.
5. Carefully transfer the supernatant to a fresh tube.

Note: Be careful to prevent any pellet material from entering the supernatant.

6. Pass the supernatant through the Sephadex G-25 column. Check for the occurrence of precipitation of a few drops of effluent in a test tube containing 10% trichloroacetic acid (TCA). As long as precipitation occurs, collect the effluent in a sterile tube until approximately an equal volume of the supernatant loaded is collected. Add column buffer as needed.
7. Centrifuge the supernatant in a Corex tube at 20,000 g for 10 min at 4°C.
8. Divide the extract into 0.4-ml portions in microcentrifuge tubes on ice. Immediately freeze them in liquid N_2 for 4 min and then store at $-70°C$ until use.

Note: The cell-free extract prepared from the wheat embryos contains tRNA, rRNA, and the soluble factors for initiation, elongation, and termination needed for in vitro translation of mRNA. The endogenous mRNA level is relatively low.

B. *In Vitro* Translation of mRNAs Using Wheat Germ Extract

1. Thaw all reagents on ice.

Note: We recommend that the components not be thawed at room temperature or at 37°C. This will decrease the efficiency of in vitro translation.

2. Add purified mRNAs (0.1 to 0.2 µg in 2 to 4 µl) in a sterile microcentrifuge tube. Denature the secondary structures of the mRNA by heating at 65 to 67°C for 10 min and quickly chilling the tube on ice for 5 min. Briefly spin down.
3. Set up a standard reaction on ice as follows:
   Sample:
   - Denatured mRNAs in ddH$_2$O (4 µl)
   - ddH$_2$O (24 µl)
   - RNasin ribonuclease inhibitor (40 units/µl) (2 µl)
   - Wheat germ extract (60 µl)
   - 1 m*M* Amino acid mixture (minus methionine) (2 µl)
   - [^{35}S]Methionine (1200 Ci/mmol) at 10 mCi/ml (8 µl)
   - Total volume of 100 µl

   Background control (no mRNA):
   - ddH$_2$O (14 µl)
   - RNasin ribonuclease inhibitor (40 units/µl) (1 µl)
   - Wheat germ extract (30 µl)
   - 1 m*M* Amino acid mixture (minus methionine) (1 µl)
   - [^{35}S]Methionine (1200 Ci/mmol) at 10 mCi/ml (4 µl)
   - Total volume of 50 µl
4. Incubate the reactions at 25°C for 1 h and proceed to products analysis (**Section IV** and **Section V**).

III. Synthesis of Proteins from mRNA Transcribed *In Vitro*

A. *In Vitro* Transcription of mRNA from a Subcloned cDNA Insert

A number of plasmid vectors contain polycloning sites downstream from the powerful bacteriophage promoters SP6, T7, or T3. The cDNA or genomic DNA insert of interest can be cloned at the polycloning site between promoters SP6 and T7 or T3, forming a recombinant plasmid. The cDNA or genomic DNA inserted can be then transcribed *in vitro* into single-strand sense RNA from a linear plasmid DNA under the control of the SP6, T7, or T3 promoter.

Note: Care must be taken to make sure that the DNA is cloned in the correct frame for the promoter that will be used for the transcription.

1. In a microcentrifuge tube on ice, add the following in the order listed below:
   - 5X Transcription buffer (20 µl)
   - 0.1 M dithiothreitol (DTT) (8 µl)
   - rRNasin ribonuclease inhibitor (100 units)
   - Mixture of ATP, GTP, CTP, and UTP (2.5 mM each) (20 µl)
   - Linearized DNA template (1 to 2.5 µg/µl) (2 µl)
   - SP6, T7, or T3 RNA polymerase (15 to 20 units/µl) depending on the specific promoter (5 µl)
   - Add ddH$_2$O to a final volume of 100 µl.
2. Incubate the reaction mixture at 37 to 40°C for 1 to 2 h.
3. Add *RNase*-free *DNase* I to a concentration of 1 unit/µg DNA template.
4. Incubate for 15 min at 37°C.
5. Extract the enzyme by adding one volume of phenol/chloroform/isoamyl alcohol (25:24:1). Mix well by vortexing for 1 min and centrifuging at 11,000 g for 5 min at room temperature.
6. Transfer the top, aqueous phase into a fresh tube and add one volume of chloroform:isoamyl alcohol (24:1). Mix well by vortexing and centrifuging at 11,000 g for 5 min.
7. Carefully transfer the upper, aqueous phase to a fresh tube. Add 0.5 volume of 7.5 M ammonium acetate solution and 2.5 volumes of chilled 100% ethanol. Allow to precipitate at –70°C for 30 min or at –20°C for 2 h.
8. Centrifuge at 12,000 g for 5 min. Carefully discard the supernatant and briefly rinse the pellet with 1 ml of 70% ethanol and dry the pellet under vacuum for 15 min.
9. Dissolve the RNA in 20 to 50 µl of ddH$_2$O and store at –20°C until use.

Note: The quantity and quality of the RNA can be checked by denaturing agarose gel electrophoresis using 4 to 5 µl of the sample (see **Chapter 6**, "Northern Blot Hybridization").

B. *In Vitro* Translation of the *In Vitro* Transcribed mRNA

These procedures are described in **Section I** and **Section II** of this chapter.

IV. TCA Assay for Amino Acid Incorporation

1. Remove 2 µl of the translation reaction mixture from both the translation reaction and background control and directly spot the aliquots onto individual glass-fiber filters. Air dry the filters and transfer the filter to a vial containing an appropriate volume of liquid scintillation "cocktail" to cover the filter. Use a liquid scintillation counter to record the total counts present in the translation reaction mixture. These samples represent the total counts per minute (cpm) in each sample.

In Vitro Translation of mRNA(s) and Analysis of Protein by Gel Electrophoresis

Note: Radioactive markers can also be prepared using protocol found in **Chapter 4**.

8. Dry the gel under vacuum at 60 to 80°C and carry out autoradiography by exposing the dried gel on x-ray film for 6 to 24 h or for appropriate period of time.
9. Analyze the results.

Note: All gel materials and solutions are likely contaminated with radioactivity and should be disposed of properly.

Reagents Needed

10% Ammonium Persulfate (AP)
AP (0.2 g)
Dissolve well in 2.0 ml ddH$_2$O.
Store at 4°C for up to 10 days.

Bromophenol Blue Solution
Bromophenol blue (0.05 g)
Sucrose (20 g)
Dissolve well after each addition in 15 ml ddH$_2$O.
Add ddH$_2$O to a final volume of 50 ml, aliquot, and store at −20°C.

Column Buffer
40 mM HEPES/KOH, pH 7.6
0.1 M Potassium acetate
5 mM Magnesium acetate
4 mM DTT

2X Denaturing Buffer
Stacking gel buffer (5 ml)
10% SDS solution (8 ml)
Glycerol (4 ml)
2-Mercaptoethanol (2 ml)
Add ddH$_2$O to a final volume of 20 ml.
Divide in aliquots and store at −20°C.

Homogenization Buffer
40 mM HEPES/KOH, pH 7.6
1 mM Magnesium acetate
0.1 M Potassium acetate
2 mM CaCl$_2$
4 mM DTT

Monomer Solution
Acrylamide (116.8 g)
N,N-Methylene-*bis*-acrylamide (3.2 g)
Dissolve well after each addition in 300 ml ddH$_2$O.
Add ddH$_2$O to a final volume of 400 ml.
Wrap the bottle with aluminum foil and store at 4°C in the dark.

Caution: Acrylamide is neurotoxic. Gloves should be worn when handling this chemical.

Overlay Buffer

>Running gel buffer (25 ml)
>10% SDS solution (1 ml)
>Add ddH$_2$O to 100 ml.
>Store at 4°C.

Running Buffer

>0.25 M Tris (12 g)
>Glycine (57.6 g)
>10% SDS solution (40 ml)
>Add ddH$_2$O to a final volume of 4 L.

Running Gel Buffer

>1.5 M Tris (72.6 g)
>Dissolve well in 200 ml ddH$_2$O.
>Adjust pH to 8.8 with 2 N HCl.
>Add ddH$_2$O to a final volume of 400 ml.
>Store at 4°C.

10% SDS

>SDS (10 g)
>Dissolve well in 100 ml ddH$_2$O.
>Store at room temperature.

Stacking Gel Buffer

>0.5 M Tris (6 g)
>Dissolve well in 50 ml ddH$_2$O.
>Adjust pH to 6.8 with 2 N HCl.
>Add ddH$_2$O to a final volume of 100 ml.
>Store at 4°C.

10% (w/v) TCA Solution

>20 g Trichloroacetic acid in 200 ml ddH$_2$O

References

1. **Sambrook, J., Fritsch, E.F., and Maniatis, T.,** *Molecular Cloning: A Laboratory Manual*, 2nd ed., Cold Spring Harbor Press, Cold Spring Harbor, NY, 1989.
2. **Smith, B.J.,** SDS polyacrylamide gel electrophoresis of proteins, in *Methods in Molecular Biology, Proteins*, Vol. 1, Walker, J.M., Ed., Humana Press, Clifton, NJ, 1984, chap. 6.
3. **Merril, C.R.,** Gel-staining techniques, in *Guide to Protein Purification, Methods in Enzymology*, Vol. 182, Deutscher, M.P., Ed., Academic Press, San Diego, 1990, chap. 36.
4. **Knudsen, K.A.,** Proteins transferred to nitrocellulose for use as immunogens, *Anal. Biochem.*, 147, 285, 1985.
5. **Kyhse-Anderson, J.,** Electroblotting of multiple gels: a simple apparatus without buffer tank for rapid transfer of proteins from polyacrylamide to nitrocellulose, *J. Biochem. Biophys. Methods*, 10, 203, 1984.

Chapter 17

Plant Tissue and Cell Culture

Donghern Kim, Leland J. Cseke, Ara Kirakosyan, and Peter B. Kaufman

Contents

I.	Basic Concept of Totipotency of Plant Cell and Tissue Culture	358
II.	Basic Cell and Tissue Culture Techniques	360
	A. Media Preparation and Media Kits	360
	1. Preparation of Stock Solutions	362
	2. Preparation of Culture Media	363
	B. Inoculating Seeds or Plant Organ Explants into Sterile Media	364
	1. Callus Induction from Rice Embryos and Rice Callus Subculture	364
	2. Callus Induction from Carrot Roots and Carrot Callus Subculture	365
	3. Callus Induction from Tobacco Leaves and Callus Subculture	365
	C. Measurement of Growth of Callus Tissues in Culture	365
	1. Counting Cell Numbers	365
	2. Determining Cell Dry Weight	366
	3. Determining Packed-Cell Volume	366
	D. Induction of Greening of Callus Cultures	366
	1. Photosynthetic and Photoautotropic (Green) Cell Cultures	366
	E. Induction of Shoots and Roots in Callus Cultures	367
	1. Plant Regeneration from Rice Callus	367
	2. Induction of Organogenesis from Cotyledonary-Petioles of Chinese Cabbage	368
	F. Initiation of Cell-Suspension Cultures from Callus Cultures	368
	1. Cell-Suspension Cultures of Rice Cells Derived from Immature Embryos	368
	2. Cell-Suspension Cultures and Somatic Embryogenesis of Carrot	369
	G. Troubleshooting Guide	370
	1. Sterile Techniques To Prevent Microbial Contamination	370
	2. Use of Antibiotics	371
	3. Changing Culture Conditions	371

III.	Applications with Plant Tissue Cultures	371
	A. Induction of Somatic Embryogenesis from Callus Cultures	371
	B. Somatic Cell Hybridization	371
	1. Isolation of Protoplasts from Cell-Suspension Cultures of Wheat (*Triticum aestivum*) for Somatic Cell Hybridization	372
	C. Synthesis of Useful Secondary Metabolites	372
	D. Preservation of Plant Tissue Cultures by Means of Cryopreservation	374
	1. Janus Green B	375
	2. Fluorescein Diacetate	375
	3. Triphenyltetrazolium Chloride (TTC)	375
References		375

Success in plant molecular biology and biotechnology is largely dependent on *in vitro* culture of plant cells and tissues.[1–7,9,10] Obtaining transgenic plants that contain foreign genes has been extensively practiced by using plant tissue culture techniques, since it offers an opportunity to scrutinize the *in vivo* function of particular gene products and to develop new cultivars having novel traits. Numbers of transgenic crops, including herbicide-resistant soybean and Bt-corn, are currently grown in the field for commercial purposes. Potential use of plant cells cultured *in vitro* is also extensively examined in terms of producing commercially valuable products. For the last decade, there has been a large increase in the numbers of plant species successfully cultured, either as cell suspensions or as hairy roots, for the production of secondary metabolites. There are also diverse examples of plant tissue and cell cultures that are used in biotechnology and crop breeding for disease resistance, salt tolerance, and herbicide resistance. This chapter provides several examples of the establishment of photosynthetic cell cultures and the production of secondary metabolites from cultured plant cells.

I. Basic Concept of Totipotency of Plant Cell and Tissue Culture

Plant tissue excised from plants can be cultured *in vitro* and regenerated to whole plants if the culture medium contains suitable nutrients and plant growth hormones. This is due to plants having a unique property called totipotency, the ability of plant cells to develop into whole plants or plant organs. From different parts of plants, dedifferentiated cell masses called callus are formed, usually in the presence of 2,4-dichlorophenoxyacetic acid (2,4-D) at the level of a few mg·l^{-1}. Differentiation of these calli to whole plants is achieved by changing culture conditions, such as levels and compositions of plant growth hormones. Regeneration of plants from callus tissue is usually achieved either by organogenesis or by somatic embryogenesis. In the case of organogenesis, plant organs and tissues, such as shoots, roots, and vascular tissue connecting shoot and root, are formed independently of each other. On the other hand, plant organs regenerated via somatic embryogenesis are thought to originate from a single cell in a callus or from suspension-cultured cells. At the cellular level, cytodifferentiation, such as chloroplast development, is worthwhile to discuss. Since many secondary metabolites are synthesized in plastids, establishing cell cultures with differentiated plastids is necessary for the production of such secondary metabolites. On the other hand, some secondary metabolites can be produced by the addition of precursors to cell cultures.

Since large numbers of plant species have been cultured and various culture methods have been developed, it is most practical to follow previously reported protocols. However, if such protocols are not available for a particular plant species, one should try to use several well-known culture media with different auxin and cytokinin combinations. Auxin and cytokinin are two key components used in plant tissue culture. The ratio of these plant growth regulators in a culture medium greatly affects the course of development of a plant tissue. For example, high auxin/low cytokinin is usually a favorable condition for root development, while low auxin/high cytokinin promotes shoot development. Therefore, an optimal ratio of plant growth regulators should be determined by experiments with various auxin/cytokinin ratios. The quantities held for these plant growth regulators are also important for successful plant tissue culture. For example, 1 mg·l^{-1} 2,4-D is better for callus induction from carrot roots than 5 mg·l^{-1} 2,4-D. On the other hand, callus induction from rice embryos is very poor in the presence of 1 mg·l^{-1} 2,4-D but excellent with 2.5 mg·l^{-1} 2,4-D. Other components in tissue culture media that one should consider are quantities and types of carbon and nitrogen sources. Commonly used nitrogen sources are ammonium and nitrate salts, amino acids, casein hydrolysate, and urea. As a carbon source, sucrose is most widely used. However, other carbon compounds such as D-glucose or even carbon dioxide (CO_2) can replace sucrose in certain types of plant tissue culture. Even though various plant tissue culture methods have been developed, all of these methods are based on general tissue culture techniques. These techniques include the maintenance of sterile conditions, the preparation of culture media, the initiation of plant tissue cultures, and the control of tissue culture growth and development.

Plant cells can be cultured *in vitro* in a liquid medium as single cells or small cell clumps (cell-suspension cultures). Suspension-cultured plant cells are especially attractive materials for bioreactors to produce valuable medicines and other useful metabolites. The advantages of growing the suspension-cultured cells in a bioreactor over culturing callus are that: (1) suspension-cultured cells usually grow more rapidly than callus tissues; (2) suspension-cultured cells are more homogeneous than cells in a callus; and (3) every cell in a liquid medium can come in contact with elicitors or precursors for the metabolite production in the medium directly, while only a small portion of cells in calli can be in contact with the agar medium. Thus, major populations of cells in callus tissue can absorb elicitors or precursors only indirectly. In this chapter, we describe several basic techniques used for the induction and maintenance of plant suspension cultures. These techniques include: (1) the initiation of cell-suspension cultures; (2) the determination of cell growth; (3) differentiation of suspension-cultured cells; and (4) production of secondary metabolites.

Cell-suspension cultures are usually initiated from callus tissue or from differentiated explants such as hypocotyls. Liquid medium (MS, B5, and LS media) containing 2,4-D (1 to 2 mg·l^{-1}) is worth testing first as an initiation medium if no culture medium is established for the suspension culture of a particular plant species. When the cell-suspension culture is to be initiated from callus tissue, one should consider how to break the callus into single cells or small cell clumps. One way to achieve this is to use friable portions of callus tissue. In this case, cells are easily dispersed in the liquid medium by one or two subcultures with continuous shaking at about 100 to 125 rpm. A second way to obtain cell suspensions is to culture hard callus in continuously shaken liquid medium and keep subculturing it by transfer of the callus tissue to fresh medium until it becomes friable. In some cases, the friability of callus tissue increases suddenly after many passages of subculturing. Sometimes, mechanical disruption of callus tissue by the use of a spatula when the callus tissue is subcultured is helpful to speed up the breakage of callus tissue into small cell clumps. Continuous shaking of liquid cultures at about 100 to 125 rpm itself is one way to disrupt callus mechanically due to the shearing forces of swirled liquid media. However, if the shaking speed is too high, cells can be damaged by the excessive shearing force. Another important attribute of continuous shaking of cell-suspension cultures is aeration. Since water contains only limited amounts of oxygen, and oxygen's diffusion from the air into the liquid phase is a slow process, stationary liquid-cultured cells will develop under the stress of anoxia (absence or reduced levels of oxygen), which eventually leads to cell death. The initial density of explants or callus is important for successful induction of cell-suspension cultures. When a cell suspension is initiated from differentiated tissue, a few grams of explants per 100 ml of liquid medium should be inoculated. If cultures are overdiluted (cell density is too low), cell growth becomes very slow. Once a cell-suspension culture is established, it is maintained by regular subculturing once every 1 to 2 weeks. During this maintenance phase, one should keep monitoring cell shape and growth. Cell growth is usually measured by three methods that include counting cell number, measuring dry weight, and measuring packed cell volume (see **Section II.C**).

II. Basic Cell and Tissue Culture Techniques

A. Media Preparation and Media Kits

Plant cell and tissue culture media provide nutrients and growth-regulating substances to ensure proper cell growth and morphogenesis of the tissue in culture. Most of the components found in culture media are required for whole plant growth. For example, the six major elements (N, P, K, Ca, Mg, and S) and eight essential minor elements (Fe, Mn, Zn, B, Cu, Mo, Co, and I; cobalt and iodide are not included in some culture media) are included in the culture media as macronutrients at millimolar (mM) concentrations and as micronutrients at micromolar (μM) concentrations, respectively. The culture media also contain several components that whole plants do not need to be provided from an outside source. Organic carbon, preferably in the form of sucrose, which can be substituted by D-glucose or D-fructose for the purpose of a given tissue culture experiment, is required for plant tissue growth. It is generally an essential component of the culture medium, since most *in vitro* cultured plant cells and tissues are "heterotropic," which means that carbon energy is necessary for tissue or cell growth. On the other hand, green plants are "photoautotropic," since they can produce ATP and organic carbon using light energy. In some cases, green tissue and cells are cultured and are able to do photosynthetic carbon assimilation, but they still need to be fed organic carbon ("photomixotropic").

Besides macronutrients, micronutrients, and organic carbon, other components are also included in the culture media, such as growth regulators, vitamins, amino acids/nitrogen compounds, and agar. Different types of vitamins are currently identified, and their biochemical functions have been characterized. However, not all of the vitamins known today are added to plant culture media. The vitamins most frequently added are thiamine, nicotinic acid, and pyridoxine. These vitamins are usually added to media at concentrations ranging from 0.1 to 5 or 10 mg per liter. Another important substance added together with vitamins is myo-inositol. Phospho-inositol is known to participate in cellular signal transduction. Usually, about 100 mg of myo-inositol is added to 1 L of culture medium. Since amino acids, if culture media contain them, are an immediate nitrogen source for cultured plant cells, more stimulation of cell growth is often observed as compared with cells without amino acid supplements. Frequently used amino acids or organic nitrogen source are L-glycine (2 mg·l^{-1} for MS medium, 3 mg·l^{-1} White medium), L-glutamine, L-asparagine, casein hydrolysate, and adenine sulfate. Since amino acids or organic nitrogen are not an absolute requirement for plant tissue culture, and since organic nitrogen substances may be detrimental to cell culture, one should carefully examine the effect of the addition of organic-N on tissue/cell growth and morphogenesis.

Unlike animal systems, growth regulators for plants are small organic substances. Even though many natural and synthetic growth regulators are available to manipulate plant growth and development, the most frequently used growth regulators are largely auxin, cytokinin, gibberellin, and abscisic acid. The composition and concentrations of the growth regulators in the medium should be optimized for a particular plant tissue and cell culture system. Agar provides solid support for plant tissue culture. When heated culture medium containing agar powder is cooled down, a semisolid agar gel is formed, the firmness of which depends on the concentration of agar. Typical concentrations of agar in solid culture media range from 0.5 to 1% (w/v).

Even though culture media with diverse compositions of nutrients and supplements have been developed, several culture media are manufactured commercially and are widely used in plant tissue culture. In Table 17.1, the compositions of several prominent tissue culture media are listed. Murashige-Skoog (MS) medium is the most prominent one among them. Originally, the medium was developed for callus induction from tobacco pith. It typically contains a high amount of ammonium ion. Gamborg's B5 medium is also the one heavily used in plant tissue culture. Typical features of B5 medium are that its inorganic salt composition is simpler and that it contains a lesser amount of ammonium ion than MS medium. B5 medium was originally developed for soybean suspension cultures. White's medium contains nitrate as a sole inorganic nitrogen source. Its use in plant tisssue culture is usually for organ cultures, such as micropropagation.

TABLE 17.1
Composition and Concentration of Nutrients in Several Plant Tissue Culture Media

	Media			
	MS	B5	N6	White
Inorganic Salts, (mg·l^{-1})				
Macronutrients				
$(NH_4)_2SO_4$	—	134	463	—
KNO_3	1900	2500	2830	80
$Ca(NO_3)_2·4H_2O$	—	—	—	300
NH_4NO_3	1650	—	—	—
$NaH_2PO_4·H_2O$	—	150	—	16.5
KH_2PO_4	170	—	400	—
KCl	—	—	—	65
Na_2SO_4	—	—	—	200
$MgSO_4·7H_2O$	370	250	185	720
$CaCl_2·2H_2O$	440	150	166	—
Micronutrients				
$FeSO_4·7H_2O$	27.8	—	27.8	—
Na_2EDTA	37.3	—	37.3	—
Sequestrene 330Fe	—	28	—	—
$Fe_2(SO_4)_3$	—	—	—	2.5
$MnSO_4·4H_2O$	22.3	13.2	4.4	7
$ZnSO_4·7H_2O$	8.6	2	1.5	3
$CuSO_4·5H_2O$	0.025	0.038	—	—
H_3BO_3	6.2	3	1.6	3
$Na_2MoO_4·2H_2O$	0.25	0.25	—	—
$CoCl_2·6H_2O$	0.025	0.025	—	—
KI	0.83	0.75	0.8	0.75
Organic Substances, (mg·l^{-1})				
Sucrose	30,000	20,000	50,000	20,000
Myo-inositol	100	100	—	—
Thiamine·HCl	0.1	10	1	0.1
Pyridoxin·HCl	0.5	1	0.5	0.1
Nicotinic acid	0.5	1	0.45	0.5
Casein hydrolysate	0.04–10	—	—	—
L-glycine	2	—	2	3
Others				
Agar (%)	0.5–1	0.5–1	0.5–1	0.5–1
pH (prior to autoclaving)	5.7–5.8	5.8	5.8	6.2

1. Preparation of Stock Solutions

Even though commercially premixed powders of many standard tissue culture media are available, culture media can also be made easily from stock solutions. To prepare stock solutions and culture media properly, the solubility, heat stability, and working concentration of each component should be considered. Statements listed below are worth remembering for successful preparation of plant culture media.

1. Fe ion is usually chelated with EDTA, since it is easily precipitated in the stock solution during storage.
2. Calcium and phosphate ions form a water-insoluble product, calcium phosphate. Therefore, calcium chloride and phosphate should be separated from each other when preparing stock solutions.
3. It is better to mix together components whose working concentrations and heat stability are similar each other.
4. Table 17.2 contains a recipe of stock solutions for MS media.
5. Possible uses and stock preparation for plant growth regulators are shown in Table 17.3. Some growth regulators are not stable when heated. In this case, stock solutions should be sterilized by filtration (see **Section II.G**, "Troubleshooting Guide").
6. Storage of stock solutions:
   a. Macro- and micronutrient solutions: Store at 4°C.
   b. FeNaEDTA: Keep in brown bottle and store at 4°C.
   c. Vitamin and growth regulator stocks: aliquot and store in the freezer at –80°C.

TABLE 17.2
Stock Solutions for MS Media

I. Macro-I (10X)

NH_4NO_3	$16.5 \text{ g} \cdot l^{-1}$
KNO_3	$19.0 \text{ g} \cdot l^{-1}$
$MgSO_4 \cdot 7H_2O$	$3.7 \text{ g} \cdot l^{-1}$
KH_2PO_4	$1.7 \text{ g} \cdot l^{-1}$

II. Macro-II (10X)

$CaCl_2 \cdot 2H_2O$	$4.4 \text{ g} \cdot l^{-1}$

III. Micro (1000X)

$MnSO_4 \cdot H_2O$	$17 \text{ g} \cdot l^{-1}$
$ZnSO_4 \cdot 7H_2O$	$8.6 \text{ g} \cdot l^{-1}$
H_3BO_3	$6.2 \text{ g} \cdot l^{-1}$
KI	$830 \text{ mg} \cdot l^{-1}$
$NaMoO_4 \cdot 2H_2O$	$182.5 \text{ mg} \cdot l^{-1}$
$CuSO_4 \cdot 5H_2O$	$25 \text{ mg} \cdot l^{-1}$
$CoCl_2 \cdot 6H_2O$	$26 \text{ mg} \cdot l^{-1}$

IV. Vitamins and amino acid (100X)

Thiamine·HCl	$10 \text{ mg} \cdot l^{-1}$
Pyridoxine·HCl	$50 \text{ mg} \cdot l^{-1}$
Nicotinic acid	$50 \text{ mg} \cdot l^{-1}$
L-glycine	$200 \text{ mg} \cdot l^{-1}$

V. FeNaEDTA[a] (100X)

$Na_2EDTA \cdot 2H_2O$	3.73 g/l
$FeSO_4 \cdot 7H_2O$	2.78 g/l

[a] To prepare FeNaEDTA, autoclave mixture of Na_2EDTA and $FeSO_4$ at 121°C for 1 h and cool down to room temperature. If the chelation is not sufficient, Fe will precipitate during storage at 4°C.

TABLE 17.3
Possible Uses and Stock Preparation for Plant Growth Regulators

Group	Growth Regulator[a]	Potential Use	Stock Conc. (mg/ml)	Stock Preparation
Auxins	IAA NAA 2,4-D IBA Pichloram Dicamba	To initiate root formation To induce embryogenesis To initiate callus formation To inhibit bud formation To inhibit root elongation To promote cell enlargement	0.5–1	Dissolve powder in a small volume of alcohol and adjust volume with H_2O Dissolve powder in 0.1 N NaOH and adjust pH afterward; adjust volume with H_2O
Cytokinins	BAP 2iP Kinetin TDZ Zeatin	To promote shoot formation To inhibit root growth To promote cell division To modulate callus initiation To inhibit shoot elongation To delay senescence	0.5–1	Dissolve powder in a small volume of alcohol and adjust volume with H_2O
Gibberellin	More than 100 GAs are now identified[b]	To stimulate shoot elongation To break dormancy To inhibit root development	0.5–1	Dissolve powder in a small volume of alcohol and adjust volume with H_2O
Abscisic acid		To stimulate embryo maturation To stimulate bulb formation To promote dormancy	0.5–1	Dissolve powder in a small volume of alcohol and adjust volume with H_2O
Polyamines	Putrescine Spermidine	To promote root formation To promote embryogenesis To promote shoot development To delay senescence	0.5–1	Dissolve powder in a small volume of alcohol and adjust volume with H_2O

[a] IAA: indole-3-acetic acid; NAA: β-naphthaleneacetic acid; 2,4-D: 2,4-dichlorophenoxyacetic acid; IBA: indole-3-butyric acid; pichloram: 4-amino-3,5,6-trichloropicolinic acid; dicamba: 3,6-dichloro-o-anisic acid; BAP: 6-benzylaminopurine; 2iP: 6-(γ,γ-dimethylallylamino)purine; kinetin: 6-furfurylaminopurine; TDZ: thidiazuron; zeatin: 6-(4-hydroxy-3-methyl-but-2-enylamino)purine.

[b] Most common one used is GA_3, gibberellic acid.

2. Preparation of Culture Media

In this section, a protocol to prepare a semisolid MS medium from stock solutions is introduced. If another medium is desirable, one should prepare suitable stock solutions and follow the instructions below. Since basal salt mixture is the same for some prominent tissue culture media, it would be convenient to purchase such premixed powdered salts and use them for preparing culture media.

1. To distilled water, add 100 ml of Macro-I and Macro-II, 10 ml of FeNaEDTA, and 1 ml of Micro stocks. Or, dissolve MS basal medium mixture for 1 L in distilled water. Adjust volume to approximately 800 ml.

Note: *The MS salt mixture supplied by Sigma Chemical Company (St. Louis, MO) sometimes leaves undissolved salt particles. On the other hand, other companies' MS salt mixtures can be dissolved completely. Although it is not known whether such undissolved particles severely affect the plant tissue culture, one can purchase a salt mixture suitable for culture purposes.*

2. Add the following chemicals:
   - 30 g sucrose
   - 10 ml vitamin and amino acid stock solution
   - 100 mg myo-inositol
   - Suitable amounts of plant growth regulators (if they are relatively heat stable)

Note: *If a growth regulator is heat labile, filter sterilize the stock solution and add the aliquot to medium after step 9.*

3. Adjust the pH of the solution to 5.8.
4. Adjust the volume to 1 L.
5. Pour the solution into a 2-L flask.
6. Add 8 g phytoagar.
7. Seal the mouth of the flask with two layers of aluminum foil.
8. Autoclave at 121°C for about 20 min.

Note: *Sometimes a given medium contains heat-labile compounds. In this case, media without these compounds should be autoclaved first. After step 9, the medium is supplemented with the heat-labile compounds from filter-sterilized stock solutions.*

9. Cool the solution to about 50°C.
10. Dispense medium into sterile petri dishes (about 20 to 30 ml per 100-mm dish).
11. Wrap dishes with SaranWrap™ and store at room temperature until use.

Note: *If the culture medium contains heat-labile compounds, it is better to use freshly prepared medium or to keep the medium in a refrigerator.*

B. Inoculating Seeds or Plant Organ Explants into Sterile Media

Theoretically, almost all types of plant cells are able to be cultured *in vitro*. However, it is observed that callus tissue is most frequently formed from rapidly dividing cells. Therefore, apical meristems in shoot and root tips and cambial tissue are very good explants for callus induction. In rice seeds, callus is formed from scutellum, which is located at the boundary between rice embryo and starch-filled endosperm. Other points worthy of mentioning here are (1) that optimal conditions for tissue culture should be differentiated according to types of explants and plant species and (2) that the culture efficiencies of a particular type of explants from the same plant species are diversely dependent on cultivars and varieties. Therefore, it is very important to choose the right materials to be cultured. In this section, three well-established tissue culture systems are introduced.

1. Callus Induction from Rice Embryos and Rice Callus Subculture

1. Prepare dehusked rice seeds by removing seed coats.
2. Sterilize seeds chemically as described in **Section II.G**, "Troubleshooting Guide."
3. Place sterile seeds on N6 or MS medium plus 2 mg·l^{-1} 2,4-D in petri dishes.

Note: *Rice seeds are submerged halfway into the medium so that the embryo faces upward and is exposed to the air while the endosperm faces downward. Usually 10 to 16 seeds are placed in each petri dish.*

4. Close the petri dish with its cover and seal the unit with a strip of Parafilm™.
5. Incubate in the dark at 28°C for 2 to 3 weeks and keep watching for the appearance of callus.

Plant Tissue and Cell Culture

Note: *Callus is formed from the scutellum (cotyledon or "seed-leaf") tissue of the rice embryo. After the appearance of callus, you may find endosperm, coleoptile, and callus tissue with or without roots.*

6. Excise calli from other tissues carefully by the use of sterile forceps and transfer them to fresh medium that contains the same composition of nutrients and plant growth hormones.
7. Subculture rice calli regularly every 2 to 3 weeks. When transferred, the callus should be subdivided into smaller pieces (about 0.5 cm^3).

2. Callus Induction from Carrot Roots and Carrot Callus Subculture

1. Remove roots from carrots grown in the greenhouse or in the field, or obtain roots of carrots from a local food market. Collect transverse segments from the middle part of a root by cutting into 5-cm lengths.
2. Wash in a mild soap solution and rinse with distilled water.
3. Sterilize root pieces chemically as described in **Section II.G**, "Troubleshooting Guide."
4. Trim away the end 3 to 5 mm of each explant and slice the remaining tissue into 3- to 5-mm lengths.
5. Remove the epidermis with the use of a scalpel.
6. Cut each slice into four to six pieces and place on agar containing B5 medium plus 1 mg·l^{-1} 2,4-D (six to nine pieces per petri dish) and seal the petri dish with Parafilm.
7. Incubate the explants in the dark at 24 to 26°C and keep watching for the formation of callus tissue.
8. After callus formation has occurred, it should be removed from the explant and transferred to fresh medium.
9. Continue subculturing regularly by transferring callus pieces to fresh medium of the same composition (every 2 to 3 weeks).

3. Callus Induction from Tobacco Leaves and Callus Subculture

1. Put tobacco seeds in a bag of nylon mesh and sterilize the seeds chemically by submerging the bag in a 20% commercial bleach for 10 min followed by extensive washes with sterile water.
2. Place the seeds on agar containing MS medium without any plant growth regulator (MSO medium) and let the seeds germinate under continuous white light.
3. Grow tobacco seedlings until young leaves appear and expand.
4. Remove leaves from sterile tobacco plants and cut into pieces of about 0.5 × 0.5 cm.
5. Place the leaf pieces on the agar containing MS medium supplemented with 1 mg·l^{-1} NAA and 5 mg·l^{-1} BAP (four pieces per petri dish).
6. Seal the petri dish with Parafilm and incubate in the dark at 27°C and watch for the formation of callus from the cut surfaces of the leaf pieces.
7. Collect callus tissue and subculture on fresh medium as described in **Section II.B.2** above.

C. Measurement of Growth of Callus Tissues in Culture

In the case of microbe culture, the growth is usually measured by light absorbance at 600 nm or by the number of colonies after culturing diluted sample suspension. However these methods are not adequate for measuring plant cell or callus growth. In this section, we present three different ways to assess plant cell growth in culture.

1. Counting Cell Numbers

1. Take 1 ml of cell-suspension culture and mix with 4 ml of 12% (w/v) chromium trioxide (CrO_3).
2. Heat at 70°C for about 20 min until the cells are stained; then disrupt cell clumps by vigorous pumping of the culture with a syringe.
3. Spin down the cells by centrifugation at about 3000 rpm for 5 min.

4. Decant supernatant carefully by the use of a Pasteur pipette and resuspend the pellet of cells in 1 ml of water.
5. Count the cell number by the use of a hemocytometer and a light microscope.

2. Determining Cell Dry Weight

1. Take 5 ml from each cell-suspension culture and pass through a preweighed filter paper or glass filter.
2. Wash the filter residue twice with distilled water and dry the residue and the filter overnight at 80°C.
3. After cooling down to room temperature in a desiccator, measure the weight of the dried cells and filter paper and calculate the dried cell weight by subtracting the weight of the filter paper from the weight of the filter paper plus dried cells.

3. Determining Packed-Cell Volume

1. Sample 5 ml of cells from a cell-suspension culture and put this aliquot of cells in a 15-ml graduated conical centrifuge tube.
2. Centrifuge the cells down at 2000 g for 5 min and measure the volume of the packed cells.

Note: During culture, cell growth should be monitored regularly every 3 to 4 days. If the volume of the cell suspension is too small to harvest cells, one way to measure cell growth is to use a flask with a side arm on the side wall of the flask. When the packed-cell volume is to be measured, cells settle down on the bottom of the side arm for a fixed period of time, and one can thus measure the height of the cell mass by the use of a ruler.

D. Induction of Greening of Callus Cultures

Chloroplasts are the intracellular organelles typical of photosynthesizing organisms. Beside being the location of photosynthesis, chloroplasts are also the location where synthesis of various secondary metabolites takes place. Heterotropic callus and cells also contain plastids. However, the number of plastids is very low, and these plastids are not fully differentiated. Greening of cultured cells is inhibited by the presence of 2,4-D, an auxin-type growth regulator, and sucrose. Therefore, culture systems in which 2,4-D and sucrose are omitted should be designed for the successful induction of cell greening.

1. Photosynthetic and Photoautotropic (Green) Cell Cultures

a. Rice

1. Sterilize rice seeds as described (**Section II.G**) and imbibe seeds in sterile water for 2 h.
2. Carefully excise embryos from endosperms and place them on the agar in petri dishes containing MS medium supplemented with 3 mg·l^{-1} NAA, 6 to 8 mg·l^{-1} kinetin, and 0.8% (w/v) phytoagar (about 10 to 16 embryos per petri dish).
3. Culture embryos under continuous light at 28°C. After 2 to 3 weeks, calli are formed from the scutellum (cotyledon) region of embryos.
4. Visually select the green portions of the callus and initiate cell suspension by culturing about 100 green callus tissues in 20 ml of the liquid N6 or MS medium supplemented with 10 g·l^{-1} sucrose, 2 mg·l^{-1} NAA, and 4 mg·l^{-1} kinetin under continuous light with shaking (100 rpm).
5. Subculture regularly (twice a week) by decanting the old medium using a sterile pipette and replacing it with fresh medium of the same composition until fine and green cell-suspension cell lines are established.

b. Soybean

1. Prepare petri dishes containing KT medium for callus induction. KT medium contains MS basal salts, 1% (w/v) sucrose, 1 mg·l^{-1} NAA, 0.2 mg·l^{-1} kinetin, vitamins, and 0.8% phytoagar. Vitamins

Plant Tissue and Cell Culture

are supplied from the stock solution (concentrated 1000 times). 100 ml of the stock solution contains 20 mg nicotinamide, 20 mg pyridoxine·HCl, 10 mg biotin, 10 mg choline chloride, 10 mg calcium panthothenate, 10 mg thiamine·HCl, 5 mg folic acid, 5 mg p-aminobenzoic acid, 5 mg riboflavin, and 0.015 mg cyanocobalamin.

2. Prepare sterile soybean seedlings by growing chemically sterilized seeds on the agar containing MS basal medium in wide-mouth bottles or Magenta™ vessels under continuous light at 25 to 28°C.
3. Excise cotyledons from 5- to 6-day-old seedlings and place them on the KT medium (6 to 10 cotyledons per petri dish). Seal the petri dishes with a strip of Parafilm and incubate them under continuous light at 26 to 28°C.
4. After calli emerge, collect greener parts of the callus tissues by visual selection and initiate cell-suspension cultures in liquid KT medium. For the initiation of cell cultures, put five to six pieces of callus tissue into 10 ml of KT medium in 125-ml flasks. After closing the mouth of each flask with a double layer of sterile aluminum foil, incubate the flasks on a gyratory shaker (100 to 125 rpm) under continuous light. Cell lines established in this way are called SB-M photomixotropic soybean cell lines.
5. After the SB-M cell line is established, transfer about 400 mg of cells to 80 ml of KT^0 medium (KT without sucrose) supplemented with 5 mM HEPES buffer (pH 7.0) in 250-ml flasks and plug flasks with rubber stoppers. In order to supply CO_2, two 16- or 17-gauge needles are placed in rubber stoppers, and the ends of both needles are connected to small membrane filter units (membrane pore size = 0.45 μm). This rubber-stopper–needles filter unit should be autoclavable. Connect one of the two filter units to a gas humidifier and further to a gas tank containing 5% CO_2 by means of polypropylene tube.
6. Adjust the flow of 5% CO_2 at 10 to 15 ml/min per flask while culturing the cells on the gyratory shaker (100 rpm) under continuous light (200 to 300 $\mu E \cdot m^{-2} s^{-1}$).

Note: This protocol is based on experiments performed by Jack Widholm's group at the University of Illinois, Champaign-Urbana, IL.

E. Induction of Shoots and Roots in Callus Cultures

1. Plant Regeneration from Rice Callus

1. Prepare MS 16 medium for rice shoot regeneration from callus tissue.

Note: MS 16 medium is basically the same as MS medium, as described in Table 17.1, except that the medium contains 2 mg·l^{-1} kinetin and 1 mg·l^{-1} NAA as plant growth hormones, and 1.6% (w/v) phytoagar as solid support. The concentration of phytoagar is higher for plant regeneration than for callus induction, since it results in a greater amount of plant regeneration and lower vitrification of regenerated plants.

2. Place five callus clumps (about 1 cm³) on the agar per petri dish.
3. Incubate calli under continuous white light at about 28°C and watch for the appearance of green spots on the surface of the callus and for green shoots.

Note: Regular 40-W fluorescent lamps give sufficient light energy for shoot induction and greening if they are placed about 50 cm above the cultures.

4. Transfer calli containing shoots to MSO medium in a bottle or a culture tube.

Note: MSO medium is MS medium without plant growth hormones. One can prepare this medium by following the steps described in **Section II.A** on media preparation, except that no plant growth regulators are added.

5. Continue to culture calli under continuous white light and watch for the appearance of roots. One may also find that roots appear in step 4.
6. When the size of the regenerated plantlets is about 10 cm, plant them in pots containing sterile vermiculite.
7. While maintaining high humidity by covering the whole unit with a transparent plastic bag, grow plantlets in the culture room. This step is designated as hardening.
8. After hardening for 1 to 2 weeks, transplant regenerants to pots and transfer to the greenhouse.

2. Induction of Organogenesis from Cotyledonary-Petioles of Chinese Cabbage

1. Germinate sterile Chinese cabbage on the agar containing MSO medium under white light (continuous or 16:8 h of light:dark cycles).
2. While waiting for germination and seedling growth, prepare MS medium supplemented with 3 mg·l^{-1} IBA (indole-3-butyric acid) and 10 mg·l^{-1} BAP.
3. Collect cotyledons with petioles or hypocotyls from 4- to 5-day-old Chinese cabbage seedlings and place them on the medium (10 to 16 explants per petri dish).
4. Culture them under continuous white light while watching for the appearance of shoots and roots.
5. Transferring plant regenerants, hardening, and transferring plantlets to the greenhouse are essentially the same as described in the section on rice organogenesis (**Section II.E.1**).

Note: *(1) Shoot regeneration directly from explants is especially important in plant transformation, since prolonged tissue culture may result in unwanted mutation of plant regenerants. Furthermore, direct shoot induction is a faster way to obtain plant regenerants. Usually the entire procedure for the plant tissue culture, including the callus induction, subculture, and the regeneration of plants from callus, takes several months at best. Therefore, many researchers try to induce plant organogenesis directly from explants after transformation by Agrobacterium infection or particle bombardment. (2) Another important consideration in obtaining transgenic plants is the presence of selection pressure. Antibiotics such as kanamycin and hygromycin are commonly used to elicit selection pressure. However, the presence of such antibiotics reduces the culture efficiency.*

F. Initiation of Cell-Suspension Cultures from Callus Cultures

1. Cell-Suspension Cultures of Rice Cells Derived from Immature Embryos

1. At 10 to 20 days after anthesis or shedding of pollen (milk stage, embryos are 0.5 to 1 mm long), remove panicles and husks (palea and lemma) and sterilize chemically.
2. Separate embryos from endosperms carefully. Use of a stereomicroscope is recommended.
3. Place ten embryos on agar containing N6 or MS medium supplemented with 2.5 mg·l^{-1} 2,4-D (per 9-cm petri dish) in which the scutellum (cotyledon) surfaces face upward. Orientation of embryos is important for successful callus induction.
4. Select visually friable embryogenic callus (small clumps whose size is less than 5 mm in diameter). Friable rice callus tissue is yellowish-white with "loose" morphology.
5. Place calli in 10 ml of liquid medium containing N6 medium plus 2 mg·l^{-1} 2,4-D.
6. Incubate callus tissues in the dark at 26°C on the shaker (100 rpm).
7. For the first 2 weeks, subculture once every 2 to 3 days by removing all of the medium and replacing it with fresh medium by the use of a sterile pipette.
8. From 2 weeks to about 2 months, subculture twice a week. During this period, selection of fine cell clumps should be achieved by the use of a wide-bore pipette. Care must be taken not to include large clumps and necrotic (brown or black dead) cells.

2. Cell-Suspension Cultures and Somatic Embryogenesis of Carrot

During somatic embryogenesis toward plantlet regeneration, undifferentiated suspension-cultured cells undergo four developmental stages — globular, heart, torpedo, and cotyledonary stages. These are classified based on the shape of developing embryos. In general, the process takes three to four weeks from embryogenic cells to the mature cotyledonary stage, at which point elongated radicles and hypocotyls are the typical shape of the embryos.

1. Induce carrot callus to form from sterilized roots, hypocotyls, and cotyledons on agar containing Gamborg's B5 medium supplemented with 0.1 to 1 mg·l^{-1} 2,4-D.
2. After 3 weeks, transfer 5 to 10 calli (about 1 g fresh weight) to 20 ml liquid medium of the same composition in a 125-ml flask.
3. Incubate under continuous light at 26°C with shaking at 100 rpm.
4. Subculture regularly (at 2-week intervals) by decanting half of the old medium and replacing it with fresh medium by the use of a sterile pipette. Watch for the formation of a cell suspension.
5. After the carrot cell-suspension culture is established, continue to subculture by taking 5 ml of cell suspension and inoculating it in 45 ml of fresh medium once every week.

Note: *Sieving cells through a nylon mesh increases the degree of uniformity by excluding larger cell clumps. To further increase this uniformity, it may be helpful to centrifuge the sieved cells in a 16% Ficoll solution containing 2% sucrose. Sucrose is added to the medium to adjust the osmotic pressure.*

6. Pass carrot cells through a 125-µm sterile nylon mesh and subsequently through 32-µm nylon mesh. The cell clumps, sized from 32 to 125 µm, are collected for the induction of somatic embryogenesis.
7. Collect cells by centrifugation at about 100 g for 10 min. Then wash the cells twice with liquid MS medium lacking auxin (MSO medium).

Note: *MS medium is not an absolute requirement for the embryogenesis. Gamborg's B5 medium is also a popular growth medium used in the experiment.*

8. Resuspend the cells in 20 ml of liquid MSO medium in a 125-ml flask. Adjust the cell density to 1×10^5 to 1×10^6 cells per ml. If the cell density is lower than this range, the addition of 0.1 mg·l^{-1} zeatin and 0.1 mg·l^{-1} ABA (abscisic acid) is helpful for facilitating embryogenesis.

Note: *Cell culture without growth hormone is the general condition for the induction of somatic embryogenesis. However, supplementing growth hormone (e.g., 1 to 2 mg·l^{-1} of BA) occasionally helps to promote embryogenesis.*

9. Incubate cells under ambient light conditions with continuous shaking at 80 to 100 rpm.
10. Embryos usually appear in about 1 week, and 3 to 4 weeks are sufficient for the maturation of the embryos. Every 3 to 4 days during the somatic embryogenesis, harvest clumps bigger than 500 µm. Filter the culture through a sterile mesh and count the number of somatic embryos at each stage per milliliter of cell culture using a stereomicroscope.
11. Transfer the mature embryos on the agar containing an MSO medium and grow under continuous light. Small plantlets that contain several leaves and roots can be transplanted to sterile vermiculite. Alternatively, the embryos can be "packaged" in alginate beads for enhanced survival when shifting them from suspension culture to soil or vermiculite.

Note: Before these plantlets are moved to a greenhouse, a hardening process is necessary, since they have been grown in flasks with high relative humidity. One way to achieve this is to grow the plantlets under a vinyl or polyethylene film cover with high humidity and then to reduce the humidity successively.

G. Troubleshooting Guide

The most frequently experienced troubles in culture are (1) microbial contamination and (2) failure in or nonproper response of explants to culture medium in terms of cell growth (callus formation) and differentiation (organogenesis, embryogenesis, etc.). In order to avoid microbial contamination, care must be taken to isolate cultures from possible contamination sources. Frequent contamination sources are personnel's hands, the explant itself, the medium, and the equipment. Therefore, everything, including hands, should be sterilized as described below in **Section II.G.1**. If the problem is not solved by sterilization, antibiotics would be another troubleshooting option.

1. Sterile Techniques To Prevent Microbial Contamination

1. Autoclave: For most tissue culture media and equipment that are heat stable. Items are heated by hot steam (121°C) under pressure for 20 to 30 min.
2. Dry oven: Especially for flasks and glass pipettes. Items are heated by dry air at 200°C for 2 to 3 h.
3. Flame sterilization: Metal equipment such as scalpels and forceps are dipped in ethanol and the ethanol on the items is burned with a flame or infrared heater.
4. Filter sterilization: For liquid media and stock solutions. Solutions are filtered through sterile membrane filters (pore size typically 0.2 µm). Pressure or vacuum is applied to force solutions to pass through membrane filters.
5. Laminar flow cabinet: This provides an aseptic workbench area for handling plant tissue to be cultured. Once sterilized, plant tissue must be handled with sterile equipment.
6. Chemical sterilization: For plant materials to be cultured:
   a. Put plant materials in a sterile flask (100 or 250 ml size).
   b. Add about 50 ml of 70% (v/v) ethanol and shake for 30 s. Then discard ethanol using a pipette.
   c. Add 20 to 50% (v/v) commercial bleach (sodium hypochlorite, 3 to 5%) containing a drop of liquid detergent soap sufficient to submerge plant materials completely. Shake for 10 to 20 min on a rotary shaker at about 100 rpm.

Note: The optimum period of time for the bleach treatment is determined by trial and error. However, general guidelines are as follows: (1) seeds and other hard plant tissue can be treated longer than leaves and other soft tissues; (2) the treatment should be stopped when small bleached spots appear on leaves; and (3) the minimum period of chemical treatment to obtain sterility is the optimum period.

   d. Decant the bleach solution using a sterile pipette.
   e. Wash plant tissues with sterile water extensively (more than three times).

Note: (1) Every part of the plant body can be sterilized chemically. However, the best way to obtain sterile plant tissue is to grow plant seedlings aseptically from chemically sterilized seeds. (2) Maintaining aseptic culture conditions is critical for successful tissue culture. In order to achieve this, one should be extremely careful not to allow any possibility of bacterial or fungal contamination. Plant tissues to be cultured should be handled with sterile equipment in an aseptic laminar-flow cabinet and must be placed on a sterile surface. Using antibiotics is often recommended. Hence, the flow hoods are never really sterile.

2. Use of Antibiotics

1. Carbenicillin and ampicillin inhibit microbial cell-wall synthesis. The optimal concentration to eliminate *E. coli* ranges from 75 to 100 µg/ml. Microbes with β-lactamase are resistant.
2. Chloramphenicol inhibits formation of ribosomal 50S subunits. The treatment concentration is 50 µg/ml. Chloramphenicol acetyl transferase detoxifies the antibiotic.
3. Hygromycin inhibits translation by acting on the 70S ribosome. This antibiotic is not a good choice to prevent microbial contamination, since it is toxic to plant tissue (20 µg/ml). Concentration of hygromycin for the selection of resistant microbes is about 100 µg/ml for *E. coli*.
4. Kanamycin/neomycin/G418/deoxystreptamine aminoglycoside: Kanamycin inhibits translation and is frequently used for the selection of resistant plants or microbes expressing nptII genes.

3. Changing Culture Conditions

If explants do not respond to culture treatment, the following points should be considered:

1. Explants: Change tissue types or cultivars to be cultured.
2. Medium: Salt composition and concentration are worth changing. (e.g., from MS to B5).
3. Growth regulator: Different auxins and cytokinins are available. Try other growth regulators with similar effects. Sometimes the production of ethylene during the culture causes trouble. In this case, use silver nitrate or activated charcoal to remove the ethylene.
4. Culture environment: Consider changing the culture room temperature as well as the length and intensity of illumination.

III. Applications with Plant Tissue Cultures

A. Induction of Somatic Embryogenesis from Callus Cultures

Somatic embryogenesis refers to the development of embryoids from somatic cells in suspension cultures. Such embryoids can be produced by transferring cell suspensions to a fresh liquid medium (e.g., MS or modifications of MS like that cited in **Section II.E** and **Section II.F**) devoid of plant hormone, such as 2,4-D. All stages of embryogenesis are then expressed in the cultures over a 7- to 10-day period for cultures maintained at room temperature, illuminated at 200 lux, and shaken at 125 rpm. Classic examples here are carrot (*Daucus carota*), daylily (*Hemerocallis* sp.), and tulip tree (*Liriodendron tulipifera*). Once embryoids are formed in the cell-suspension cultures, they can be "packaged" in alginate beads for use as artificial seeds.

B. Somatic Cell Hybridization

The development of methods for protoplast isolation and culturing with subsequent plant regeneration is critical for the successful somatic hybridization of higher plants. Plant regeneration from protoplasts is an infrequent event. Therefore, at least in the case of cereals, there is little chance of producing transformed plantlike organisms. In this case, since it is impossible to isolate viable protoplasts immediately from plant organs, cell and tissue cultures are used as the protoplast source.

The density, viability, and further development of protoplasts depend on the genotype and age of the plant, the duration of cell culturing, the physiological state of the protoplasts, and on the procedures used for the protoplast enzymatic isolation, purification, and culturing.

1. **Isolation of Protoplasts from Cell-Suspension Cultures of Wheat (*Triticum aestivum*) for Somatic Cell Hybridization**
    1. After 10 to 12 rounds of subculturing, protoplasts are isolated from adapted 6-day-old cultures at the phase of the exponential growth. To this end, cells (1 g) are placed in 10 ml of solution containing 1% cellulase (Onozuka R-10, Serva, Germany), 0.5% pectolase Y-23 (Serva), 0.1% casein hydrolysate (Merck, Germany), 0.6 M mannitol (Serva), and 11 ml of "marine water" (780 mOs·kg^{-1} H$_2$O) (pH 5.8).
    2. Incubation is conducted on a shaker (50 rpm) at 26°C for 3 h. After dilution with an equal volume of marine water, protoplasts are separated from undigested cells and cell aggregates by sequential sieving through a series of metal sieves with mesh apertures of 0.25, 0.1, and 0.05 mm and precipitated by centrifugation at 500 g for 3 min.
    3. The sedimented protoplasts are washed twice with marine water, resuspended in 1 ml of a KM (Kao and Michayluk)[4] nutrient medium (devoid of p-aminobenzoic acid, choline chloride, vitamin D, NAA, xylose, mannose, cellobiose, and all organic acids except sodium pyruvate).
    4. Aliquots of protoplasts are taken for the determination of protoplast density. The density is adjusted to 10^5 protoplasts·ml^{-1} by adding nutrient medium, and 1-ml aliquots are transferred to petri dishes on agar-solidified (0.8% agar) nutrient medium.
    5. The protoplasts are cultured at 27°C; 1 ml of nutrient medium containing 2 mg/l 2,4-D is added every 3 to 5 days. Colonies of protoplasts are formed in 14 to 16 days.
    6. Protoplast fusion can be performed in a PEG-containing medium with alkaline pH (10.5) and a high calcium concentration (70 m*M*). This process is called somatic cell hybridization. It can be performed between different species of wheat (*Triticum* spp.). This protocol also works well for members of the Brassicaceae family (cabbage and its relatives in the genus *Brassica*).

C. Synthesis of Useful Secondary Metabolites

The synthesis of plant secondary metabolites occurs in several types of cultures: (1) shoot cultures, (2) root cultures, (3) cell-suspension cultures, and (4) globular structures in liquid media. For example, synthesis of hyperforin and hypericin secondary metabolites occurs in significant amounts in shoot cultures of selected strains of St. John's wort (*Hypericum perforatum*), often exceeding levels produced in intact plants. For production of secondary metabolites by root cultures, good examples include isoflavonoids in edible legumes such as soybean (*Glycine max*), kudzu (*Pueraria montana*), and fava beans (*Vicia faba*). In these instances, such cultures must be derived from roots per se. A classic example of cell-suspension cultures is that of skikonin production in *Lithospermum erythrorhizon*. In this instance, two kinds of media are used: one is used to stimulate cell multiplication, and the second one is used to favor shikonin biosynthesis. A second example is that of hawthorn (*Crataegus monogyna*), where the production of procyanidins and flavonoids is better in cell-suspension cultures than in callus cultures. The enhancement of secondary-metabolite production in fine-cell-suspension cultures or in liquid-cultivated cell aggregates is different from that of shoots and roots grown in liquid culture.

Cell aggregates, such as compact globular structures, are excellent producers of secondary metabolites. The formation and growth parameters of globular structures derived from callus cultures are helpful in modeling and investigating their use in biotechnology. For example, the globular structures reach a critical mass during the culture process, which does not change during further culture. However, long-term culture further increases the total biomass because there is an increase in the quantity of globules. The biomass yield of globular structures under these conditions is considerably higher than that of suspension cultures, which makes them ideal for use in bioreactors.

The strategy of secondary-metabolite production needs further development. Several approaches can be used to obtain higher yields of the desired products. One of these approaches involves screening and selection of high-productivity cell lines. For some plants, this approach is still unsolved, especially concerning metabolites that are produced in highly differentiated cells and tissues, such as monoterpenes and triterpenoid glycosides that are synthesized in highly specialzed glands and trichomes (epidermal hairs). In these cases, such metabolites are not produced by plant cell and callus cultures because of the absence of highly differentiated structures. There are also stability problems of cell lines derived from differentiated plant tissue. For some cell lines, secondary-metabolite yields reach very low levels after several subcultures. Other cell lines have high metabolite production rates that are achieved by (1) choosing fluorescent cell clones by means of flow cytometry and cell morphology and (2) monitoring cell density by step-gradient centrifugation with ficoll or sucrose (see **Chapter 22**) followed by selection of low-, medium-, and high-density layers in the gradient for separate cultivation *in vitro*.

In all of the above approaches, extensive screening and selection research is needed for all cell lines before industrial applications can be achieved. The only industrial applications achieved so far have been for berberine production in cultures of *Berberis* spp., shikonin from *Lithospermum erythrorhizon*, and vinblastine and vincristine from *Catharanthus roseus*.

Another common and useful approach is elicitation. Some biologically active agents, including cell-wall polysaccharides and glycoproteins, cause an elicitation phenomenon. Elicitor molecules derived from fungal pathogens can act as transcriptional upregulators in leguminous plants. Subsequent studies of plant defense mechanisms have provided a great deal of valuable information. For cell cultures, elicitation is a key strategy for enhancing secondary-metabolite biosynthesis, especially involving the use of new and novel kinds of elicitors.

Production of secondary metabolites in plants is often associated with morphological differentiation. A flower is the most obvious example: the colored secondary metabolites occur frequently in flower petal tissues. Other tissues often connected with secondary metabolites are the roots and bark of trees and ripe fruits. In cell-suspension cultures, there is a continuous selection for rapid growth, which opposes differentiation. Differentiation occurs only at the end of a growth cycle in the stationary phase, often coinciding with secondary-metabolite production.

Given this relationship between synthesis of secondary metabolites and morphological differentiation, the culture of differentiated cells is of interest. Root and shoot cultures have been obtained from various plants that yield secondary metabolites similar to those found in the plant; these compounds are not usually produced in cell-suspension cultures. Cell aggregates have proved to be excellent producers of some secondary metabolites not found in cell-suspension cultures.

Another approach is epigenetic manipulation in order to improve cell growth and productivity. Epigenetic manipulation involves establishment of media that favor optimal growth and secondary-metabolite biosynthesis, the use of biotic and environmental stress treatments, and the induction of metabolite precursors.

Genetic manipulation refers to the acquisition of new genetic information by plant cell and organ cultures. Such genetic manipulation results in (1) greater stability in growth and secondary-metabolite biosynthesis and (2) the ability of cultures to produce new compounds not found in intact plants. Biological agents used to effect such genetic changes in plant cell and organ cultures include the crown gall bacterium, *Agrobacterium tumifasciens*, which produces "shooty" teratomas, and *A. rhizogenes*, which produces "rooty" teratomas. These transgenic cultures are especially useful in producing novel secondary metabolites as well as metabolites that are produced by green-shoot tissues and roots in differentiated plants. Transfer of useful genes derived from microorganisms, insects, and animals is also being achieved to produce human vaccines, interferon, and natural insecticides (such as the Bt gene from the bacterium *Bacillus thuringiensis*). This is achieved by direct insertion of foreign DNA into protoplasts and plant cells in culture or by nononcogenic *Agrobacterium tumifasciens* containing specific vectors. Transgenic plants obtained from such cultures will be especially useful for production of medicinally important secondary metabolites.

D. Preservation of Plant Tissue Cultures by Means of Cryopreservation

Storage of seeds would be the easiest and the most economical way to preserve higher plants. However, this is not possible for vegetatively propagated plants or for plant tissues and cells that are cultured *in vitro*. Preservation at extremely low temperatures or cryopreservation (e.g., −196°C under liquid nitrogen) would be an effective way for the long-term storage of these materials. For the establishment of cryopreservation protocol, one should consider factors affecting the viability of plant tissues after the preservation. For example, the material itself to be preserved should be carefully chosen. Apical meristems prepared aseptically have been widely used in cryopreservation. If cultured plant cells are to be cryopreserved, cells early in the exponential growth phase often survive freezing well. Less-succulent plant cells also have a high chance to survive freezing, since one of the main reasons for the occurrence of cell death during cryopreservation is the formation and growth of ice particles inside the cells during freezing. The addition of suitable cryoprotectants should also be optimized. Dimethyl sulfoxide (DMSO) and glycerol are two examples of cryoprotectants widely used. DMSO rapidly penetrates biological membranes and can be treated up to 10% without causing visible physiological damage to tissues. On the other hand, glycerol is less toxic to plant tissues, but diffuses into plant cells more slowly than DMSO. The third factor affecting the viability of cryopreserved tissue is the rate of cooling. Slow freezing of material by lowering temperature at a rate less than 1°C per min down to about −40°C is necessary to ensure maximal survival of the tissue. After extended periods of storage, frozen tissue is rapidly thawed to prevent reforming of ice crystals, and the viability of recovered tissues is monitored. There are several histochemical methods to visualize tissue viability in that cells or tissue sections (unfixed) can be stained with colored or fluorescent dyes. In this section, a generalized protocol is described for the cryopreservation of apical meristems and cultured cells using DMSO as a cryoprotectant. One should keep in mind that modification of this protocol may be necessary to optimize the conditions specific to a particular type of tissue to be cryopreserved.

1. Prepare micropropagated shoots and excise shoot tip meristems (about 0.5 mm in size). If cultured cells are to be cryopreserved, actively growing cells should be harvested.
2. Preculture meristems or cells in the presence of half-strength cryoprotectants. After meristems are placed on solid mass-propagation media containing 5% (v/v) DMSO, incubate them at 20 to 25°C with 12 to 16 h of photoperiod. For cultured cells, DMSO is slowly added to actively growing culture to the final concentration of 5% (v/v), and cells are returned to shaker. After 48 h of the preculture, meristems or cells are placed on the ice (or in an ice bath) for about 1 to 2 h.
3. Transfer meristems to prechilled liquid MS or B5 medium in a flask and add the same volume of prechilled medium containing 20% (v/v) DMSO slowly in a stepwise fashion for about 30 min to 1 h. Then, stand the flask in an ice bath for 30 min to allow penetration of DMSO into plant tissue and equilibration. In the case of cultured cells, the cells are allowed to settle down in an ice bath, and excess media solution is discarded by pipetting. Then, transfer the cell suspension to a cryogenic glass ampoule and add DMSO slowly to a final concentration of 10%.
4. Transfer meristems in the cryoprotecting solution to an ampoule and seal ampoules using a torch. Put sealed ampoules in a programmable freezing chamber and lower the temperature to −40°C at the rate of 0.5 to 1°C/min.
5. Store frozen ampoule under liquid nitrogen. To examine the viability of cryopreserved tissue, remove ampoules from the liquid nitrogen tank and thaw them at room temperature.
6. The viability of tissue and cells are most directly assessed by culturing and subsequently measuring cell and tissue growth. To count the number of viable cells in the population recovered, cells are stained with color or fluorescent dyes.

1. Janus Green B
   1. Prepare 0.1% (w/v) Janus Green B stock solution in distilled water.
   2. Add 1/100 volume of the dye solution to cell suspension and incubate for 20 min at room temperature.
   3. Using a Pasteur pipette, transfer cells onto a microscope slide and place a coverslip on the cell preparation.
   4. Using a microscope, count the number of viable and dead cells. Mitochondria of viable cells are stained a blue color, while dead cells are completely stained.

2. Fluorescein Diacetate
   1. Dissolve 5 mg fluorescein diacetate in 1 ml of acetone and dilute 50 times with 50 mM phosphate buffer (pH 5.8).
   2. Stain the cells in the dye solution for 5 min at room temperature.
   3. Using a fluorescence microscope and light microscope, count the numbers of viable cells and total cells. Viable cells emit fluorescence when excited by UV illumination, while dead cells do not fluoresce.

3. Triphenyltetrazolium Chloride (TTC)

With this method,[8] cell viability can be assessed by using a spectrophotometer. Viable cells reduce triphenyltetrazolium chloride, which eventually forms a formazan. The amount of formazan is measured by light absorption at 530 nm following the extraction of the pink-colored product with ethanol.

   1. Prepare 0.6% triphenyltetrazolium chloride in 50 mM phosphate buffer (pH 5.8).
   2. Incubate cells (100 mg) in 3 ml of the dye solution for 15 h at 30°C.
   3. Discard the solution using a Pasteur pipette and rinse the cells with distilled water twice.
   4. Collect cells by low-speed centrifugation and add 7 ml of 95% ethanol.
   5. Incubate the tubes for 5 min at 80°C.
   6. After cooling the solution, measure the light absorbance at 530 nm.

References

1. **Bhojwani, S.S. and Razdan, M.K.**, *Plant Tissue Culture: Theory and Practice*, Elsevier Science, New York, 1983.
2. **Debergh, P.C. and Zimmerman, R.H.**, *Micropropagation: Technology and Application*, Kluwer Academic Publishers, Dordrecht, The Netherlands, 1991.
3. **Evans, D.A., Sharp, W.R., and Ammirato, P.V.**, Eds., *Handbook of Plant Cell Culture*, Vol. 4, Techniques and Applications, Macmillan, New York, 1986.
4. **Kao, K.N. and Michayluk, M.R.**, Nutritional requirements for growth of *Vicia hajastana* cells and protoplasts at a very low population density in liquid media, *Planta*, 126, 105–110, 1975.
5. **Kyte, L. and Kleyn, J.**, *Plants from Test Tubes: An Introduction to Micropropagation*, Timber Press, Portland, OR, 1996.
6. **Lindsey, K.**, Ed., *Plant Tissue Culture Manual*, Kluwer Academic Publishers, Dordrecht, The Netherlands, 1991.
7. **Pollard, J.W. and Walker, J.M.**, Eds., *Plant Cell and Tissue Culture: Methods in Molecular Biology*, Vol. 6, Humana Press, Clifton, NJ, 1990.
8. **Reinert, J. and Bajaj, Y.S.**, *Applied and Fundamental Aspects of Plant Cell, Tissue and Organ Culture*, Springer-Verlag, New York, 1977.
9. **Trigiano, R.N. and Gray, D.J.**, Eds., *Plant Tissue Culture Concepts and Laboratory Exercises*, CRC Press, Boca Raton, FL, 1996.
10. **Vasil, I.K.**, Ed., *Cell Culture and Somatic Cell Genetics of Plants*, Vols. 1 to 5, Academic Press, New York, 1984–1988.

Chapter 18

Gene Transfer and Expression in Animals

William Wu, Peter B. Kaufman, and Leland J. Cseke

Contents

I. Gene Transfer and Analysis of Expression in Mammalian Cells 379
   A. Selection of Vectors, Promoters, and Enhancers for Gene Transfer and Expression 379
      1. SV40-Based Vectors 379
      2. Retrovirus Vectors 382
   B. Purification of Plasmid Vectors and DNA Fragments or Genes to Be Used for Transfer 384
      1. Purification of Plasmid Vectors, DNA Fragments, cDNA, or Drug-Selectable Marker Gene by Restriction Enzyme Digestion Followed by Elution from an Electrophoresed Gel 384
      2. Preparation of Blunt-End DNA Fragments 387
   C. Preparation of Double-Strand Oligonucleotides for Cloning 388
   D. Construction of Chimeric Genes for Transfection 389
      1. Ligation of Plasmid Vector and DNA Insert 389
      2. Transformation of an Appropriate Strain of *E. coli* with the Ligated, Recombinant Plasmids and Isolation of the Plasmids for Transfection 390
   E. Transfection of Reporter DNA Constructs into Cultured Cells 391
      1. Calcium Phosphate Transfection 391
      2. Transfection by DEAE-Dextran 392
      3. Transfection by Electroporation 393
   F. Analysis of Expression of Reporter Gene or Inserted Gene of Interest in Transfected Cells 394
      1. Preparation of Cytoplasmic Extract from Transfected Cells for Enzyme Assay 394
      2. Chloramphenicol Acetyl Transferase (CAT) Enzyme Assay 394

		3.	Luciferase Assay 395
		4.	β-Galactosidase Assay... 395
		5.	Isolation of RNA from Transfected Cells for Transcripts Analysis........... 395
		6.	Analysis of RNA by Northern or Dot-Blot Hybridization......................... 396
		7.	Analysis of the Expression of Transfected Gene and Transcription Initiation Site by Primer Extension 396
	G.	Selection of Stable Transformants.. 397
		1.	Resistance to Aminoglycoside Antibiotics 397
		2.	Thymidine Kinase 397
		3.	Hygromycin-β-Phosphotransferase 397
		4.	Tryptophan Synthetase 398
	Reagents Needed 398
II.	Gene Transfer and Analysis of Expression in Mice 402
	A.	Preparation of Chimeric Genes for Microinjection 402
	B.	Preparation of Oocytes 402
	C.	Microinjection of DNA Constructs into the Oocytes 403
	D.	Reimplantation of the Injected Eggs into Recipient Female Mice and Generation of Founder Mice 403
	E.	Indentification of Transgenic Mice 404
		1.	Isolation of DNA from Mouse Tails 404
		2.	Isolation of RNA from the Organs (Embryo, Brain, and Liver) of the Transgenic Mice Using Lithium Chloride/Urea 404
		3.	Southern Blot, Slot Blot, and Northern Blot Hybridizations 405
		4.	Analysis of the Expression of the Reporter Gene by Staining Embryos from Transgenic Mice 405
	F.	Selection of Transgenic Lines by Breeding the Transgenic Mice 405
	Reagents Needed 406
References 407

Gene engineering technology has been well established in mammalian cells and animals (e.g., mice) as a major revolutionary approach to address a vast spectrum of fundamental biological questions. These include activity assays of promoters, enhancers, and silencers of interest; analysis of expression of mutated genes in their coding regions; analysis of the expression of reporter genes; and applications of gene therapy, using an antisense approach for knocking out the expression of cancer genes or introducing normal genes into patients containing specific mutant genes. Therefore, gene transfer[1-7] has a broad range of applications in animal systems. It consists of highly involved techniques that include specific gene isolation and characterization, gene recombination, gene transfer, and analysis of the gene expression in transformed cells or transgenic animals.[2,4,8-12] This technology constitutes one of the major advances in medicine and molecular genetics in recent years.[1,7,13-15] It has allowed the possibility of treating some human diseases resulting from defects in single genes by transferring normal genes into some of the cells of the patients, a technique known as gene therapy. Some genes that are involved in tumor or cancer formation have been isolated and characterized. This chapter describes detailed procedures for gene transfer and analysis of gene expression in mammalian cells and mice. The techniques can also be adapted for other animal systems.

I. Gene Transfer and Analysis of Expression in Mammalian Cells

Two types of mammalian-cell gene-transfer systems have been developed. One is the transient expression system in which exogenous genes are transferred into cells, and the introduced genes are allowed to be expressed for 1 to 3 days. The transfected cells are lysed and the gene products are analyzed. The object of transient transfection is usually to obtain a burst in the expression of the transferred gene(s), but such transfection is not suitable for the selection of transformed, stable cell lines. The other is the stable transformation system in which foreign genes are introduced into cells and stably integrated into the chromosomes or genomes of the host cells. The integrated deoxyribonucleic acid (DNA) replicates efficiently and is maintained during cell division. The gene expression can be analyzed from successive generations of divided cells, establishing genetically stable transformed cell lines. Both gene-transfer systems involve the construction of chimeric genes (e.g., promoter, enhancer, vector, interesting gene to be transferred, and/or reporter gene), selection of cell lines, gene transfer into cells, cell culture, and analysis of the expression of the introduced genes in the transformed cells.

A. Selection of Vectors, Promoters, and Enhancers for Gene Transfer and Expression

Selection of an appropriate vector, including promoters, enhancers, selectable markers, and poly(A) signals, is very important for the success of gene transfer and analysis of gene expression. A number of vectors have been developed, and each vector has its own strengths and weaknesses. We recommend and describe the following vectors that have been well established and widely used in mammalian systems.

1. SV40-Based Vectors

The simian virus 40 (SV40) is a small double-strand circular DNA tumor virus with a 5.2 kb (kilobase-pairs) genome, which has been well studied. This genome contains an early region encoding the tumor (T) antigen, a late region encoding the viral coat proteins, the origin of replication (ori), and enhancer elements near the ori. The ori and early region play essential roles in the expression of genes. Two divergent transcription units are produced from a single complex promoter/replication region. These viral transcripts are referred to as *early* and *late* due to the time of maximal expression during infection. Both transcripts contain introns and are polyadenylated.

DNA manipulation technology has made it possible to mutate the SV40 genome and fuse it with a plasmid such as pBR322, generating a series of valuable vectors that have been used for gene transfer and expression.

a. Expression Vectors for Testing the Activity of the Promoter of Interest

These plasmid vectors are constructed so as to contain an appropriate reporter gene-coding region, the chloramphenicol acetyl transferase (CAT) gene-coding region including the start and stop sites, the SV40 small T antigen region for intron and polyadenylation signal, multiple cloning sites for the insertion of a foreign promoter of interest, β-lactamase Amp^r-coding region, and the origin of replication. In order to test the activity of a promoter of interest, the promoter can be inserted into the multiple cloning sites (MCS) upstream from the reporter gene. To enhance and test putative promoter activity, the SV40 enhancer elements can be inserted. For the selection of eukaryotic

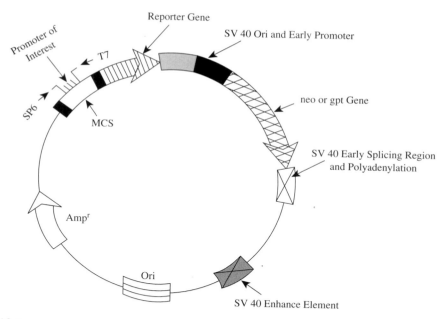

FIGURE 18.1
General map of a desirable expression vector used for insertion and analysis of a given promoter of interest. (From Kaufman, P.B., Wu, W., Kim, D., and Cseke, L.J., *Handbook of Molecular and Cellular Methods in Biology and Medicine*, 1st ed., CRC Press, Boca Raton, FL, 1995.)

transformants, an appropriate antibiotic-resistant gene (e.g., the *E. coli neo^r* gene and the *gpt* gene) should be inserted between (a) SV40 ori and early promoter and (b) SV40 early splicing region and polyadenylation (Figure 18.1).

b. Vectors for Monitoring the Expression of Introduced Genes

A typical expression vector should contain a strong promoter such as SV40 or long terminal repeat (LTR) and other elements as shown in Figure 18.2 and Figure 18.3.

There are three widely used reporter genes as follows:

1. CAT gene: A bacterial gene encoding for CAT has proved to be useful as a reporter gene for monitoring the expression of transferred genes in transformed cells or transgenic animals, since eukaryotic cells contain no endogenous CAT activity. The CAT gene was isolated from the *E. coli* transposon, Tn9, and its coding region was fused to an appropriate promoter. CAT enzyme activity can be assayed readily by incubating the cell extracts with acetyl CoA and ^{14}C-chloramphenicol. The enzyme acylates the chloramphenicol, and products can be separated by thin-layer chromatography (TLC) on silica gel plates, followed by autoradiography.
2. Luciferase gene: The luciferase gene, which encodes for firefly luciferase, has been isolated and widely used as a highly effective reporter gene. As compared with the CAT assay, the assay of luciferase activity is more than 100-fold more sensitive. It is much more simple, rapid, and relatively inexpensive. Luciferase is a small, single polypeptide with a molecular weight of 62 kDa, and it does not need posttranslational modification for the activity. Other advantages of using the luciferase gene as a reporter gene are that mammalian cells do not have endogenous luciferase activity, and that luciferase can produce chemiluminescent light with very high efficiency, which can be readily detected.
3. β-galactosidase gene: *Lac* Z encoding for β-galactosidase is also widely used as a reporter gene. The cell extracts of transformants can be directly assayed for β-galactosidase activity with a spectrophotometric method.

Gene Transfer and Expression in Animals

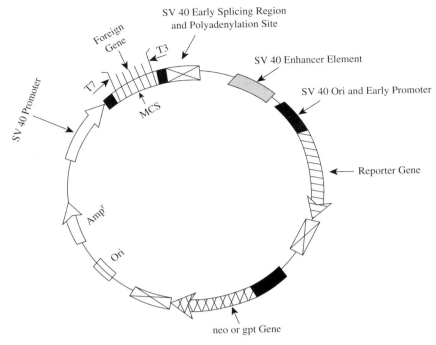

FIGURE 18.2
Structural map of an expression vector with an SV40 promoter that can be used for cloning and analysis of a foreign gene of interest. (From Kaufman, P.B., Wu, W., Kim, D., and Cseke, L.J., *Handbook of Molecular and Cellular Methods in Biology and Medicine*, 1st ed., CRC Press, Boca Raton, FL, 1995.)

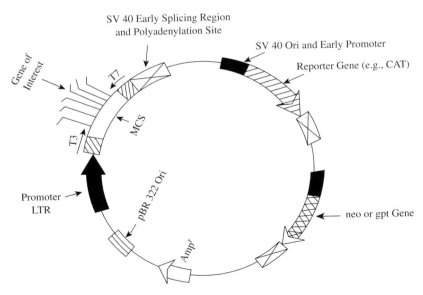

FIGURE 18.3
Structural map of an overexpression vector with a retroviral promoter, LTR, that can be used for cloning and analysis of a foreign gene of interest. (From Kaufman, P.B., Wu, W., Kim, D., and Cseke, L.J., *Handbook of Molecular and Cellular Methods in Biology and Medicine*, 1st ed., CRC Press, Boca Raton, FL, 1995.)

Note: *Some investigators have used uidA or the gusA gene from E. coli, which encodes for β-glucuronidase (GUS), as a reporter gene used for gene transfer. However, a disadvantage here is that mammalian tissues contain endogenous GUS activity, thus making the enzyme assay more difficult. The GUS reporter gene, on the other hand, is an excellent reporter gene widely used in higher plant systems because plants do not contain detectable endogenous GUS activity.*

2. Retrovirus Vectors

It has been demonstrated that retroviruses can be used as effective vectors for gene transfer in mammalian systems. There are several advantages over SV40-based plasmid vectors, which include the following: (1) the retroviral genome can stably integrate into the host chromosomes of infected cells and be passed from generation to generation, thus providing an excellent vector for stable transformation; (2) retroviruses have a broad infectivity and expression-host range for any animal cells via viral particles; (3) integration is site-specific with respect to the viral genome at long terminal repeats (LTRs), which can easily preserve the structure of the gene to remain intact after integration; and (4) the viral genomes are very plastic and manifest a high degree of natural size manipulation. Before we describe the construction of retrovirus-based vectors for mediating gene transfer, it is necessary to briefly elaborate the life cycle of retroviruses.

Retroviruses, such as Rous sarcoma virus (RSV) and Moloney murine leukemia virus (MoMLV), are ribonucleic acid (RNA) viruses that cause a variety of diseases, including tumors in human beings. The viruses contain two transfer RNA (tRNA) primer molecules, two copies of genomic RNA (38S), reverse transcriptase, *RNase* H, and integrase, which are packaged within an envelope. The viral envelope contains glycoproteins that serve to determine the host range of infection. When a virus or virion attaches to a host cell, the viral glycoproteins in the envelope bind to specific receptors in the plasma membrane of the host cell. The bound complex then facilitates the internalization of the virus that loses its coat as it passes through the cytoplasm of the host cell. In the cytoplasm, reverse transcriptase contained in the viral genome catalyzes the formation of a double-strand DNA molecule from the single-strand virion RNA. The DNA molecule circularizes, enters the nucleus, and integrates into the chromosome of the host cell, forming a provirus. Subsequently, the integrated provirus serves as a transcriptional template for both messenger RNAs (mRNA) and virion genomic RNA. Such transcription is catalyzed by the host RNA polymerase II. The mRNAs undergo translation to produce viral proteins and enzymes using the host machinery. All of these components are packaged into viral core particles. The particles move through the cytoplasm, attach to the inner side of the plasma membrane, and bud off. This cycle of infection, reverse transcription, transcription, translation, virion assembly, and budding is repeated over and over again, thus causing infection of new host cells.

The best-understood retrovirus is RSV. The mechanism of synthesis of a double-strand DNA intermediate (provirus) from viral RNA is unique and complex. The nucleotide base sequence of the DNA molecule is different from that of the viral RNA. The sequence U3-R-U5, which is a combination of the 5′ r-u5 segment and the 3′ u3-r segment of the RNA, is present at both 5′ and 3′ ends of the double-strand DNA molecule. The U3-R-U5 is called the long terminal repeat (LTR). The complete scheme can be divided into eight steps.

1. One of the proline tRNA primers anneals to the pbs region in the 5′ r-u5 of the viral genome RNA. Reverse transcriptase catalyzes the extension of the tRNA primer from its 3′-OH end to the 5′ end, generating a DNA fragment called 3′ R′-(U5)′-tRNA.
2. *RNase* H removes the cap and poly(A) tail from the viral RNA, and the viral r-u5 segment in the double-strand region.

3. The 3′ R′-(U5)′-tRNA separates from the pbs region, jumps to the 3′ end of the viral RNA and forms an R′/r duplex.
4. The 3′ R′-(U5)′-tRNA elongates up to the primer binding site (pbs) region of the viral RNA by reverse transcriptase, producing the minus (−) strand of DNA.
5. *RNase* H removes the u3-r from the 3′ end of the viral RNA, followed by synthesis of DNA from the 3′ end of the RNA by reverse transcriptase, forming the first LTR (U3-R-U5) that contains the promoter sequence. This is part of the plus (+) strand DNA.
6. All RNAs including the tRNA primer are removed by *RNase* H.
7. The U3-R-U5 (PBS) is separated from (U3)′R′(U5)′, jumps to the 3′ end of the complementary strand of DNA, and forms a (PBS)/(PBS)′ complex. This is an important process for the virus, since the "act" of jumping brings the promoter in the U3 region from the 3′ end to the 5′ end of the plus strand DNA. The promoter is now upstream from the coding region for *gag-pol-env-src* in the 38S RNA genome.
8. Reverse transcriptase catalyzes the extension of DNA from the 3′ termini of both strands, generating a double-strand DNA (provirus). The LTRs at both ends of the DNA contain promoter and enhanced elements for transcription of the virus genome. The double-strand DNA molecule integrates into the chromosomes of the infected cell.

Because retroviruses can cause tumors in animals and human beings, we obviously do not want the whole virus genome to be used for gene transfer and as an expression vector. Recent DNA recombination technology makes it possible to modify the retrovirus genome so that it is an efficient tool for gene transfer.

The simplest type of gene-transfer system is that in which all or most of the *gag*, *pol*, and *env* genes in the provirus are deleted. However, all the *cis*-active elements such as the 5′ and 3′ LTRs, PBS(+), PBS(−), and psi (ψ) are left intact. A selectable gene such as *neo*, *gpt*, *dhfr*, or *hprt* is inserted at the initiating ATG site for *gag*. The expression of the inserted gene is controlled by the 5′ LTR. This manipulated vector is then fused to a plasmid fragment (e.g., pBR322) containing the origin of replication (ori) and an antibiotic-resistant gene. The recombinant plasmids are propagated in *E. coli*. The vectors are then used as a standard DNA-mediated transfection of suitable recipient cells, and stable transformants can be selected by using an appropriate antibiotic chemical such as G418.

The partial viral RNA transcribed from the transformants can be further packaged into retroviral particles to become an infectious recombinant retrovirus. This can be done by cotransfecting the host cells with a helper virus that can produce *gag* and *env* proteins, which can recognize the psi site on the recombinant transcript and which become packaged to form particles. The disadvantage of this strategy is that the culture supernatant will contain recombinant and wild-type viruses. Some vectors are constructed by deleting the psi region, replacing the 3′ LTR with a SV40 terminator, the poly(A) signal, and fusing it with the backbone of pBR322. Another recombinant expression vector is constructed by deleting *gag* and *env* genes and by inserting a selectable marker gene (e.g., *neo*) or a reporter gene, which is followed by polylinker sites for the insertion of the foreign gene or complementary DNA (cDNA) of interest, at the ATG site for *gag* downstream from the 5′ LTR. The 3′ LTR can be replaced with the SV40 poly(A) signal downstream from the introduced gene. The recombinant vector is then fused with the pBR322 backbone containing the origin of replication and the Amp^r gene. (Delete the CAT gene and insert a recombinant viral vector at the polylinker sites, where a foreign gene can be inserted downstream from the promoter in the 5′ LTR.) The recombinant plasmids are cloned and amplified in *E. coli*. The foreign gene or cDNA of interest can be inserted at the polylinker site downstream from the 5′ LTR of recombinant vectors, and gene transfer can be carried out by a standard plasmid DNA-mediated procedure.

There are also other recombinant viral vectors used for gene transfer, depending on individual research groups. A typical expression vector is shown in Figure 18.3.

B. Purification of Plasmid Vectors and DNA Fragments or Genes to Be Used for Transfer

The DNA to be studied in transfection experiments may be a potential or putative regulatory sequence, such as a promoter or an enhancer element; a cDNA-coding region; a complete gene; or the truncated/mutated, partial-length segment of a gene of interest. Whatever the source, the DNA fragment or gene to be transferred should be partially characterized. This should include having some known information on DNA size, nucleotide sequence, and restriction enzyme sites. The DNA fragment or gene is then separated and purified from the rest of the DNA in order to be ligated to an appropriate vector for transfection. The general procedures and protocols are described below.

1. Purification of Plasmid Vectors, DNA Fragments, cDNA, or Drug-Selectable Marker Gene by Restriction Enzyme Digestion Followed by Elution from an Electrophoresed Gel

a. Preparation of Vectors

Select appropriate plasmid vectors such as pCAT and pGL2 series (Promega Corp., Madison, WI), and other pSV-based vectors, or retrovirus-based vectors, depending on the particular experiments you wish to carry out.

1. For the transfer and over- or under-expression of the gene or cDNA of interest, a standard vector may have: (1) a strong promoter, such as the SV40 promoter or a retroviral LTR; (2) multiple cloning sites downstream from the promoter, which can be used for the insertion of the exogenous gene; (3) SV40 small T intron (early splice) and poly(A) signal downstream from the foreign gene; (4) an enhancer element from SV40 or a retrovirus, which can be upstream from the promoter or upstream from the polylinker sites in either orientation; (5) SV40 ori early promoter followed by the coding region (initiation at AUG) of a reporter gene (e.g., CAT gene or luciferase gene) downstream from the SV40 promoter; (6) SV40 ori early promoter followed by the coding region (initiation at AUG) of selectable marker gene (e.g., neo^r); (7) the Col E1 ori and F1(+) ori of replication for high copy number; and (8) the Amp^r gene for the selection of antibiotic-resistant bacteria.
2. For testing the activity of the promoter of interest, the basic characteristics of the vector are almost the same as described in step 1, except that (1) there is no promoter upstream from the reporter gene, and (2) the polylinker sites are located upstream from the reporter gene for the insertion of the promoter of interest.
3. For testing the activity of enhancer or silencer of interest, the basic backbone of the vector may be similar to that described in step 1, except that there is no enhancer. The enhancer element of interest can be cloned upstream from the promoter or downstream from the reporter gene at the polylinker sites in either orientation. The silencer of interest is usually linked upstream from the promoter.

b. Restriction-Enzyme Digestion of Plasmid Vectors and DNA of Interest for Ligation

The selection of restriction enzymes should be based on what kinds of restriction enzyme sites are present in the polylinker sites of the vector and what restriction enzyme sites are in the DNA to be inserted. Choose a unique restriction enzyme(s) for the vector and for DNA, which can cut the inserted DNA outside of the region of interest (e.g., cDNA-coding region, promoter, enhancer, or mutated and recombinant genes). The general procedures are described as follows:

1. Set up, on ice, a standard restriction enzyme digestion as follows:
   Single-restriction enzyme digestion:
   - Plasmid vectors or DNA to be inserted (10 μg)
   - 10X Appropriate restriction enzyme buffer (10 μl)
   - 1 mg/ml Acetylated bovine serum albumin (BSA) (optional) (10 μl)
   - Appropriate restriction enzyme (3.3 units/μg DNA)
   - Add distilled, deionized water (ddH$_2$O) to a final volume of 100 μl.

Double restriction enzyme digestion:
- Plasmid DNA or DNA to be inserted (10 μg)
- 10X Appropriate restriction enzyme buffer (10 μl)
- 1 mg/ml Acetylated BSA (optional) (10 μl)
- Appropriate restriction enzyme A (3.3 units/μg DNA)
- Appropriate restriction enzyme B (3.3 units/μg DNA)
- Add ddH$_2$O to a final volume of 100 μl.

Note: *(1) The restriction enzyme(s) used for the vector and DNA-insert digestions should be the same to ensure the optimal conditions for ligation. (2) For directional cloning, the plasmid and the DNA insert can be digested using two different restriction enzymes. The double-enzyme digestion of DNA can be set up in a single reaction at the same time or carried out with two single-enzyme digestions performed at different times. (3) For double restriction enzyme digestions, the appropriate 10X buffer containing a higher NaCl concentration than for the other buffer can be chosen as the double-enzyme-digestion buffer.*

2. Incubate at an appropriate temperature, depending on the appropriate restriction enzyme (e.g., 37°C) for 2 to 3 h. For single-enzyme-digested DNA, proceed to step 3. For double-enzyme-digested DNA, proceed to step 5.

Note: *(1) To ensure an optimal ligation to the DNA insert, both vectors and DNA should be completely digested. The digestion efficiency can be checked by loading 1 μg of the digested DNA (10 μl) with loading buffer to a 1% agarose minigel. In the meantime, undigested vectors and DNA (1 μg) and standard DNA markers should be loaded in the adjacent wells. After electrophoresis, the undigested plasmid DNA may reveal multiple bands because of different levels of supercoiling. However, one band should be visible for a complete single-enzyme digestion; one major band and one small band should be visible after digestion with two different restriction enzymes. One should be able to distinguish the digested DNA to be transferred from the undigested DNA. (2) After completion of the restriction enzyme digestion, calf intestinal alkaline phosphatase (CIAP) treatment should be carried out for the above single-restriction enzyme digestion of the plasmid vector. This treatment removes 5′-phosphate groups, thus preventing recircularization of the vector during ligation. Otherwise, the efficiency of ligation between the vector and the DNA insert will be very low. For incompatible double restriction enzyme-digested vectors, the CIAP treatment is not necessary.*

3. Carry out CIAP treatment by adding the following directly to the single-enzyme-digested plasmid vector DNA sample (90 μl).
   - 10X CIAP buffer (15 μl)
   - CIAP diluted in 10X CIAP buffer (0.01 unit/pmol ends)
   - Add ddH$_2$O to a final volume of 150 μl.

Note: *(1) CIAP and 10X CIAP should be kept at 4°C. CIAP treatment should be set up at 4°C. (2) Calculation of the amount of ends is as follows. There are 9 μg digested DNA left after taking 1 μg of 10 μg digested DNA for checking on agarose gel. If the vector is 3.2 kb, the amount of ends can be calculated by the formula below:*

$$\text{pmol ends} = \{\text{amount of DNA}/[\text{base-pairs} \times (660/\text{base-pair})]\} \times 2$$
$$= [9/(3.2 \times 1000 \times 660)] \times 2$$
$$= 4.2 \times 10^{-6} \times 2$$
$$= 8.4 \times 10^{-6} \, \mu M$$
$$= 8.4 \times 10^{-6} \times 10^{-6} = 8.4 \text{ pmol ends}$$

4. Incubate at 37°C for 1 h and add 2 µl of 0.5 M EDTA buffer (pH 8.0) to stop the reaction.
5. Extract with one volume of TE-saturated phenol/chloroform. Mix well by vortexing for 1 min and centrifuge at 11,000 g for 5 min at room temperature.
6. Carefully transfer the top, aqueous phase to a fresh tube and add one volume of chloroform:isoamyl alcohol (24:1) to the supernatant. Mix well and centrifuge as in step 5.
7. Carefully transfer the upper, aqueous phase to a fresh tube and add 0.1 volume of 3 M sodium acetate buffer (pH 5.2) or 0.5 volume of 7.5 M ammonium acetate to the supernatant. Briefly mix and add 2 to 2.5 volumes of chilled 100% ethanol to the supernatant. Allow to precipitate at −70°C for 1 h or at −20°C for 2 h.
8. Centrifuge at 12,000 g for 10 min and carefully decant the supernatant. Briefly rinse the DNA pellet with 1 ml of 70% ethanol and dry the pellet under vacuum for 20 min. Dissolve the DNA pellet in 20 to 40 µl ddH$_2$O. Take 4 µl of the sample to measure the concentration of the DNA at 260 nm. Store the sample at −20°C until use.

Note: *Adding 0.5 volume of 7.5 M ammonium acetate to the supernatant at step 7 yields a higher amount of DNA precipitated than adding 0.1 volume of 3 M sodium acetate buffer (pH 5.2).*

c. **Electrophoresis and Elution of DNA**

Method 1: Elution of DNA by Agarose Gel Slices

1. Electrophorese the digested vectors and DNA to be inserted on 0.8 to 2% agarose, depending on the DNA sizes as described under Southern blotting procedures in **Chapter 5**, except that one should use low-melting-point agarose instead of normal agarose.
2. Transfer the gel onto a long-wavelength ultraviolet (UV) transilluminator (305 to 327 nm) to visualize the DNA bands. Excise the band(s) of interest with a clean razor blade and place the slices into a microcentrifuge tube. The DNA can be eluted from the agarose gel slices by one of two procedures below.

Note: *(1) The gel should not be placed on a short-wavelength (e.g., <270 nm) UV transilluminator, since UV light may cause breakage inside the DNA fragment. This significantly inhibits the subcloning of the insert of interest. (2) UV light is toxic to the human body. Protective glasses, gloves, and a lab coat should be worn when using UV light. (3) The gel slices containing the DNA bands of interest should be trimmed of excess unstained gel area as much as possible.*

3. Carry out elution of DNA by using one of the following methods:
   a. Freezing and thawing method
      i. Add two volumes of TE buffer to the gel slices and completely melt the gel in a 60 to 70°C water bath.

Note: *The gel slices can be directly melted without adding any TE buffer. The DNA concentration is usually high, but total yield of elution of the DNA fragment is much lower than when TE buffer is added.*

   ii. Immediately chill the melted gel solution on dry ice and place the tube at −70°C for at least 20 min.
   iii. Thaw the gel mixture by tapping the tube vigorously. It takes about 5 to 10 min to thaw the gel into a resuspension status.
   iv. Centrifuge at 11,000 g for 5 min at room temperature.
   v. Carefully transfer the liquid phase containing the elution DNA fragment to a fresh tube.
   vi. Extract EtBr with three volumes of water-saturated *n*-butanol.

vii. Precipitate the DNA by adding 0.1 volume of 3 M sodium acetate buffer (pH 5.2) and 2.5 volumes of chilled 100% ethanol into the DNA solution. Allow to precipitate at −70°C for 1 h and centrifuge at 12,000 g for 5 min at room temperature.

viii. Discard the supernatant and briefly rinse the pellet with 1 ml of 70% ethanol. Dry the DNA under vacuum for 15 min and dissolve the DNA in an appropriate volume of ddH$_2$O or TE buffer. Measure the concentration of the eluted DNA and store at −20°C until use.

b. Phenol/chloroform/isoamyl alcohol extraction method

i. Add one volume of Tris buffer-equilibrated phenol to the gel slices, vortex the microcentrifuge tube for 2 min, and place the tube in a dry ice–methanol bath for 20 min.

ii. Centrifuge at 12,000 g at room temperature for 15 min and transfer the aqueous phase to a fresh tube.

iii. Reextract the phenol phase with 0.3 ml ddH$_2$O, mix, and repeat step ii.

iv. Extract the pooled aqueous phases with one volume of phenol:chloroform:isoamyl alcohol (25:24:1) and vortex for 1 min.

v. Repeat step ii and precipitate the DNA as described in step vii of the freezing and thawing method above.

Method 2: Elution of DNA Fragments in Wells of Agarose Gel

1. Carry out DNA digestion and gel electrophoresis as described in step 3 in Method 1, except that one should add running buffer up to the upper edges of the gel instead of covering the gel.
2. During electrophoresis, monitor the separation of the DNA bands stained by EtBr in the gel using a long-wavelength UV lamp. Stop electrophoresis and use a razor blade or a spatula to make a well in front of the band of interest. Add 20 to 60 μl of running buffer to the well.
3. Continue electrophoresis until the band migrates into the well. This is done by monitoring the band that is stained by EtBr.
4. Stop the electrophoresis and transfer the solution containing the DNA of interest from the well to a fresh tube.
5. Extract EtBr with three volumes of water-saturated n-butanol.
6. Precipitate and dissolve the DNA as described in step 3 in Method 1.

2. Preparation of Blunt-End DNA Fragments

If both the vector and the digested DNA fragment have sticky or overhanging ends at the right sites, they can be readily ligated to each other. If not, it is usually necessary to trim the DNA ends prior to their being inserted into a blunt-ended site such as the *Sma* I site of the vector. This can be done either by filling in the 5′ overhangs with DNA polymerase I (Klenow fragment) (see Method 1 below) or by digesting the 3′ overhangs with exonuclease T4 DNA polymerase (see Method 2), or by mung bean nuclease that digests single-strand DNA at both 5′ and 3′ overhangs (see Method 3).

Method 1: Filling In 5′ Overhangs

1. Set up the 5′ overhang reaction in a microcentrifuge tube on ice:
   - DNA fragment or vector DNA (2 to 4 μg in 4 μl)
   - 10X 5′ Overhang buffer (2 μl)
   - 2 mM dNTP mixture (2 μl)
   - DNA polymerase I (Klenow fragment) (2 to 3 units)
   - Add ddH$_2$O to a final volume of 20 μl.
2. Incubate at room temperature for 20 min and heat at 70°C for 5 min to stop the reaction.
3. Add one volume of TE buffer (pH 7.5) and carry out phenol:chloroform:isoamyl alcohol (25:24:1) extraction and precipitation as described in the section on the elution of DNA by agarose gel slices.

Method 2: Removal of 3' Overhangs

1. Set up a standard reaction in a microcentrifuge tube on ice as follows:
   - DNA fragment or vector DNA (2 to 4 µg in 4 µl)
   - 10X 3' Overhang buffer (2 µl)
   - 2 mM dNTP mixture (2 µl)
   - T4 DNA polymerase (2 to 3 units)
   - Add ddH$_2$O to a final volume of 20 µl.
2. Incubate at 12°C for 20 min and heat at 70°C for 5 min to stop the reaction.
3. Add one volume of TE buffer (pH 7.5) and carry out phenol:chloroform:isoamyl alcohol extraction and precipitation as described in the section on the elution of DNA by agarose gel slices.

Method 3: Using Mung Bean Nuclease

1. Set up a standard reaction in a microcentrifuge tube on ice as follows:
   - DNA fragment or vector DNA (2 µg in 2 µl)
   - 10X Mung bean nuclease buffer (2 µl)
   - Mung bean nuclease (5 units)
   - Add ddH$_2$O to a final volume of 20 µl.
2. Incubate at room temperature for 20 min and heat at 70°C for 10 min to stop the reaction.
3. Add one volume of TE buffer (pH 7.5) and carry out phenol:chloroform:isoamyl alcohol extraction and precipitation as described in the section on the elution of DNA by agarose gel slices.

Note: Mung bean nuclease is better than S1 due to the lower intrinsic activity on double-strand DNA.

C. Preparation of Double-Strand Oligonucleotides for Cloning

Procedure:

1. Design two complementary oligonucleotides of interest, which can be synthesized by a DNA synthesizer. The oligonucleotides should be purified afterwards.
2. Add 100 pmol of each of the two complementary oligonucleotides in a total volume of 20 µl in a microcentrifuge tube.
3. Add 1 µl of 20X annealing buffer to the tube, incubate at 90°C for 6 min, followed by slow cooling to room temperature. This can be done by placing the heated tube in a rack in a beaker filled with an appropriate volume of 90°C water. The cooling procedure should take 20 to 30 min for complete annealing.
4. Add phosphate groups to the 5' ends of the oligonucleotides by the kinase reaction.
   a. Set up a reaction as follows in a microcentrifuge tube on ice:
      - The annealed oligonucleotides (15 µl)
      - 10X Polynucleotides kinase buffer (3 µl)
      - 10 mM ATP (1.5 µl)
      - 100 mM DTT (1.5 µl)
      - T4 Polynucleotide kinase (15 units)
      - Add ddH$_2$O to a final volume of 30 µl.
   b. Incubate at 37°C for 40 min and heat at 70°C for 10 min to inactivate the enzyme.
   c. Add one volume of TE buffer (pH 7.5) and carry out extraction and precipitation.

D. Construction of Chimeric Genes for Transfection

After the necessary components for gene transfer are purified as described in **Section I.B**, chimeric gene constructs can be made in either a circular plasmid form or in a linear form. Generally, standard chimeric gene constructs for gene transfer, gene expression, and selection of transformants should include a variety of components: vectors, promoters, enhancers, SV40 introns, poly(A) signals, reporter gene-coding region, cDNA or genomic DNA of interest, and a drug-selection marker gene such as aminoglycoside 3'-phosphotransferase *(aph)* gene, thymidine kinase *(tk)* gene, hygromycin B phosphotransferase gene, and tryptophan synthetase gene. For each gene, cDNA, or drug-selection marker gene to be transferred, its coding region should be downstream from an appropriate promoter, followed by the splicing intron (e.g., SV40 intron) and poly(A) signal. An enhancer element can be inserted upstream from the promoter or downstream from the poly(A) signal. Commercial plasmid vectors are available from Promega (Madison, WI) or other companies. Plasmids contain most of the necessary components, including polylinker sites for insertion of the foreign DNA of interest. However, those plasmid vectors such as the pCAT series and the pGL-2 series do not contain a drug-selection marker gene for the selection of stable transformants. Therefore, an appropriate marker gene should be fused to the recombinant gene constructs for transfection. An alternative way is to cotransfect a plasmid bearing the marker gene and a plasmid containing the DNA of interest into the cells. Nevertheless, we recommend that all necessary components be fused together in a single plasmid DNA for efficient transfection, analysis of the expression of the introduced genes, and selection of stable transformants. However, it has been reported that a chimeric gene containing bacterial DNA could affect the expression of the introduced gene. Therefore, the gene of interest can be purified, and then, one can cotransfect the cells with separate DNA containing a drug-selectable gene, depending on the particular gene-transfer experiment. A typical construction of chimeric genes involves ligation one by one, recircularization of the recombinant plasmid, transfer into an appropriate bacterial host for amplification, selection of transformants, and purification of the recombinant plasmids for transfection.

1. Ligation of Plasmid Vector and DNA Insert

To achieve optimal ligation, it is strongly recommended that the ratio of vector to DNA insert (1:1, 1:2, 1:3, and 3:1 molar ratios) be optimized by using a small-scale reaction series. The following reaction protocol is standard for the ligation of 3.2-kb plasmid vector and 2.533-kb DNA insert.

1. Calculate the molar weight of the vector and the DNA insert as follows:

$$1\ M\ \text{plasmid vector} = 3.2 \times 1000 \times 660 = 2.112 \times 10^6$$

$$1\ M\ \text{insert DNA} = 2.533 \times 1000 \times 660 = 1.672 \times 10^6$$

where 660 refers to the average molecular weight of a nucleotide.

2. Calculate the molar ratio of vector to insert DNA as follows:

Vector DNA:Insert DNA	Amount of DNA (µg)	
	Vector	Insert
1:1	1	0.792
1:2	1	1.584
1:3	1	2.376
3:1	1	0.264

3. Set up the following ligations on ice:

	Ligation Reactions			
Components	1 (1:1)	2 (1:2)	3 (1:3)	4 (3:1)
Plasmid DNA as vector (μg)	1	1	1	1
DNA insert (μg)	0.792	1.584	2.376	0.244
10X Ligase buffer (μl)	1	1	1	1
T4 DNA ligase (Weiss units)	4	4	4	4
Add ddH$_2$O to achieve final volume (μl)	10	10	10	10

Note: *The restriction enzyme-digested plasmid (vector) and DNA insert should be dissolved in ddH$_2$O (nuclease-free) to 0.5 to 1.0 μg/μl. If the DNA is less than 0.4 μg/μl, the DNA should be precipitated so as to dissolve in about 1 μg/μl.*

4. Incubate the reactions at 4°C for 12 to 24 h, or at 16°C for 4 to 6 h, or at room temperature (22 to 25°C) for 1 to 2 h.

Note: *After ligations are finished at the above temperatures, the mixture can be stored at 4°C until use.*

5. Check the efficiency of ligation by 1% agarose electrophoresis. When the electrophoresis is complete, photograph the gel stained with EtBr under UV light. Compared with the unligated vector or DNA-insert wells, a highly efficient ligation should allow one to visualize less than approximately 10% unligated vector and insert DNA by estimating the intensity of fluorescence. Approximately 90% of the vector and insert DNA are ligated to each other and show strong band(s) with molecular-weight shifts compared with the sizes of vector and insert DNA. By comparing the efficiency of ligation using different molar ratios, the optimal conditions can be determined with ease. These results can be used as a guide for large-scale ligations.

Note: *The above small-scale ligation is optional, but it is strongly recommended that one carry it out before attempting large-scale ligations.*

6. Carry out large-scale ligation of vector DNA and insert DNA based on the optimal conditions determined by small-scale ligations. For example, if one uses a 1:2 molar ratio of plasmid DNA:insert-DNA as being the optimal for the ligation, a large-scale ligation can be carried out as follows:
   - Plasmid DNA as vector (3 μg)
   - DNA insert (4.75 μg)
   - 10X Ligase buffer (3 μl)
   - T4 DNA ligase (15 Weiss units)
   - Add ddH$_2$O to a final volume of 30 μl.
7. Incubate the ligation at 4°C for 12 to 24 h, or at 16°C for 4 to 6 h, or at room temperature (22 to 25°C) for 1 to 2 h. Store at 4°C until use. Proceed to transformation.

2. Transformation of an Appropriate Strain of *E. coli* with the Ligated, Recombinant Plasmids and Isolation of the Plasmids for Transfection

1. Prepare the LB (Luria-Bertaini) medium and LB plates as described previously. This should be done before ligation.
2. Prepare the competent cells as previously described. This should be completed before ligation.
3. Prepare competent cells and carry out transformation of bacteria in order to amplify the recombinant plasmids. The competent cells (such as DH5aF′, JM109, and HB 101 — for transformation, selection

6. Place the TLC plate in a tank containing approximately 100 ml of solvent (chloroform:methanol, 95:5). When the solvents migrate up to about 1 cm from the top of the plate, remove the plate and air dry.
7. Expose the plate to x-ray film overnight at room temperature. In order to carry out a quantitative assay, for each sample, cut a square corresponding to the monoacetylated form from the silica plate and place in vial. Add 5 ml of scintillation fluid to the vial and count in a scintillation spectrometer.

3. Luciferase Assay

Luciferase encoded by the reporter luciferase gene is based on an oxidation reaction mediated by luciferyl-CoA. Light is produced as a result of this reaction.

1. For each sample, add 20 to 30 μl of the cell extract as described previously (**Section I.F.1**) to a vial containing 100 to 150 μl of luciferase assay reagent at room temperature.
2. Quickly place the reaction in a luminometer and measure light produced for a period of 10 sec per sample over a 2-min period, depending on the sensitivity. All samples must be done within 2 min because of the decrease in luciferase activity.

Note: *The luciferase activity decreases very rapidly.*

4. β-Galactosidase Assay

1. For each sample prepared in **Section I.F.1**, add the following components into a microcentrifuge tube in the order shown below:

   For sample:
   - Magnesium solution (3 μl)
   - 0.1 M Sodium phosphate buffer (200 μl)
   - CPRC (4 mg/ml) (66 μl)
   - Cell extract prepared (31 μl)
   - Total volume: 300 μl

   For blank:
   - Magnesium solution (3 μl)
   - 0.1 M Sodium phosphate buffer (200 μl)
   - CPRC (4 mg/ml) (66 μl)
   - 0.25 M Tris-HCl, pH 8.0 (31 μl)
   - Total volume: 300 μl

2. Incubate at 37°C for 30 to 60 min and add 0.7 ml of 1 M Na$_2$CO$_3$ to stop the reaction.
3. Measure the absorbance, against blank reference, at A$_{574}$.

5. Isolation of RNA from Transfected Cells for Transcripts Analysis

In order to obtain molecular proof for transformed cells, the expression of transferred genes should be analyzed at the mRNA level in addition to reporter enzyme assays as described in **Section I.F.1–4**.

1. Harvest transfected cells by centrifugation at 200 g for 4 min. Wash the cells and repeat centrifuging three times with PBS buffer.
2. Resuspend the cells in 5 ml of 4 M guanidinium isothiocyanate buffer and vigorously vortex for 0.5 to 1 min to completely disrupt the cells.
3. Add one volume of water-saturated phenol:chloroform. Shake vigorously for 20 sec.
4. Centrifuge at 9000 g for 10 min at 4°C. Carefully transfer the top, aqueous phase to a fresh tube.
5. Repeat phenol:chloroform extraction once as in steps 3 and 4.
6. Precipitate RNA by adding 0.1 volume of 3 M sodium acetate buffer and 2.5 volumes of chilled 100% ethanol. Place the tube at −20°C for 2 h.
7. Centrifuge at 10,000 g for 10 min at 4°C and discard the supernatant. Briefly rinse the pellet with 1 ml of 70% ethanol and dry the pellet under vacuum for 10 min.
8. Resuspend the RNA in 200 μl of TE buffer.

9. Add 200 ul of 8 M LiCl solution to the RNA solution and place at –20°C for 3 to 5 h to precipitate the RNA. Centrifuge at 11,000 g for 15 min at 4°C and carefully decant the supernatant. Briefly rinse the pellet with 2 ml of 70% ethanol and dry the pellet under vacuum for 15 min. Dissolve the RNA in 100 μl of water, 0.1 volume of 3 M sodium acetate buffer, and 3 volumes of chilled ethanol. Place at –20°C for more than 2 h. Repeat step 7. Dissolve the RNA in 50 μl of TE buffer. Take 5 μl of the sample to measure the concentration and quality (see **Chapter 2, Section V**). Store the sample at –20°C until use.

Note: *All solutions for RNA work need to be DEPC treated. See* **Chapter 2** *for more details on working with RNA.*

6. Analysis of RNA by Northern or Dot-Blot Hybridization

Total RNA, isolated as described in **Section I.F.5**, can be used to carry out dot-blot hybridization or Northern blot hybridization to check the presence and size of mRNA expressed by the introduced gene in transfected cells. The probe should be a gene fragment that was used for transfection. If a hybridization signal shows up, then the introduced gene is indeed expressed in the transformants. This is strong evidence of successful transfection. If not, the transfection may have failed, or the introduced gene cannot be expressed due to some adverse influence, such as the presence of an inhibitor. The detailed procedures for Northern blot and dot-blot hybridizations are described in **Chapter 6**.

7. Analysis of the Expression of Transfected Gene and Transcription Initiation Site by Primer Extension

Primer extension is a powerful method used to analyze the expression of a gene as well as the initiation site of transcription.

1. Based on the nucleotide sequence information from the introduced gene, synthesize an oligonucleotide primer that will hybridize close to the 5′ end of the transcript.
2. End-labeling the primer:
   a. Set up, on ice, the following reaction in a microcentrifuge tube:
      - Oligonucleotide primer (40 pmol) (1.5 μl)
      - Polynucleotide kinase buffer (7.5 μl)
      - [γ-^{32}P]ATP (190 mCi) (19 μl)
      - T4 Polynucleotide kinase (30 units) (4 μl)
      - Add ddH$_2$O to a final volume of 75 μl.
   b. Incubate at 37°C for 40 min and add 75 μl of TE buffer, pH 8.0.
   c. Purify the end-labeled probe by the use of Sephadex G50 spin column subjected to centrifugation. Store the eluate containing the probe at –20°C until use.
3. Anneal the labeled primer with specific mRNA in the total RNA isolated previously (**Section I.F.5**).
   a. Set up the annealing reaction as follows:
      - 5X Annealing buffer (4 μl)
      - Total RNA (10 to 15 μg) (6 μl)
      - End-labeled primer (4 μl)
      - Add ddH$_2$O to a final volume of 20 μl.
   b. Cap the tube and heat to 70°C for 3 min in a heating block. Slowly cool the heating block to 35°C and incubate at 35°C overnight.
4. Carry out primer extension:
   a. Set up the following reaction:
      - Annealed primer/RNA mixture (15 μl)
      - Reverse transcriptase buffer (15 μl)
      - 0.1 M DTT (15 μl)
      - 10 mM dNTP (8 μl)

- 1 mM Actinomycin D (8 μl)
- AMV reverse transcriptase (15 units)
- Add ddH$_2$O to a final volume of 150 μl.
    b. Incubate at 42°C for 1 to 1.5 h and stop the reaction by adding 8 μl of formamide dye mixture and heating at 95°C for 5 min.
5. Carry out separation of primer and extended products by electrophoresis on a 6% polyacrylamide sequencing gel. The detailed procedure for nucleotide sequencing is described in **Chapter 11**.
6. Expose the gel to x-ray film at –70°C for an appropriate time period and identify the size of the extended product and the initiation site.

G. Selection of Stable Transformants

Foreign DNA enter the cells after transfection, but only a small portion of the DNA gets transferred into the nucleus, where some of the DNA might be transiently expressed for a few days. An even smaller portion of the DNA, which is introduced in the nuclei of some transfected cells, can integrate into the chromosomes and be expressed in future generations of the cell population, forming stable transformants. The selection of the stable transformants depends on the drug-selectable marker gene fused in the recombinant plasmid, a linear DNA construct, or a separate plasmid that can be cotransfected into the cells with the plasmid containing the foreign gene of interest. Generally, the transfected cells are cultured for 24 to 30 h in the medium lacking specific antibiotics in order to allow these cells to express the selectable marker gene. The cells are then sub-cultured in a 1:5 dilution in a selective medium containing an appropriate drug. The cells are cultured in the selective medium for 2 to 4 weeks with weekly or frequent changes of medium to remove dead cells and cellular debris. Only those cells that bear and express the drug-selectable marker gene can survive in the selective medium. There are a number of selectable markers that have been used for transfection. The common ones are described below.

1. Resistance to Aminoglycoside Antibiotics

If cells are transfected by recombinant gene constructs that contain a bacterial gene (Tn5 or Tn60) for aminoglycoside 3'-phosphotransferase (*aph*), the stable, dominant transformants can confer resistance to aminoglycoside antibiotics such as kanamycin (*kan*), neomycin (*neo*), and geneticin (G418). To select the stable transformants, 0.1 to 0.8 mg/ml of G418 (Sigma Chemicals Co., St. Louis, MO) can be added to complete media. This is the most widely used selection method.

2. Thymidine Kinase

If cell lines lacking the thymidine kinase gene (*tk*$^-$) are transfected by recombinant gene constructs containing the *tk* gene, the stable transformants can be selected by culturing them in a medium containing hypoxanthine, aminopterin, and thymidine (HAT). Nontransfected cells cannot metabolize this lethal analog and will die.

The HAT stock solution can be prepared as follows. Dissolve 15 g hypoxanthine and 1 mg aminopterin in 8 ml of 0.1 N NaOH solution. Adjust the pH to 7.0 with 1 N HCl. Add 5 mg thymidine and add ddH$_2$O to 10 ml. Sterile-filter it. Dilute the stock HAT to 100-fold in the culture medium.

3. Hygromycin-β-Phosphotransferase

Transformants containing the hygromycin-β-phosphotransferase gene can be selected by culturing the transfected cells in a medium containing the antibiotic hygromycin B, which is a protein-synthesis inhibitor. Hygromycin B stock is available from Sigma Chemicals Co. (St. Louis, MO). One uses 10 to 400 μg/ml in the culture medium.

4. Tryptophan Synthetase

If cells are transfected with recombinant gene constructs containing the tryptophan synthetase gene (*trp B*), the stable, dominant transformants can be selected by culturing the cells in a medium lacking tryptophan, an essential amino acid.

Reagents Needed

7.5 M Ammonium acetate
    Dissolve 57.8 g ammonium acetate in 50 ml ddH$_2$O and adjust volume to 100 ml.

5X Annealing Buffer
    2 *M* NaCl
    0.125 *M* piperazone-*N,N'-bis* (2 ethane sulfonic acid) (PIPES)
    5 m*M* EDTA, pH 6.8

20X Annealing Buffer
    200 m*M* Tris-HCl, pH 7.9
    40 m*M* MgCl$_2$
    1 *M* NaCl
    20 m*M* EDTA, pH 8.0

[γ-^{32}P]ATP 10 mCi/μl, 3000 Ci/mmol

2.5 M CaCl$_2$

Calf intestinal alkaline phosphatase (CIAP)

10X CIAP Buffer
    0.5 *M* Tris-HCl, pH 9.0
    10 m*M* MgCl$_2$
    1 m*M* ZnCl$_2$
    10 m*M* Spermidine

Chloroform:isoamyl alcohol (24:1)
    48 ml Chloroform
    2 ml Isoamyl alcohol
    Mix and store at room temperature.

CPRG Solution
    4 mg/ml chlorophenol red-β-D-galactopyranoside in 0.1 *M* sodium phosphate buffer

DEAE-Dextran Solution
    10 mg/ml DEAE-dextran (M$_r$ 5 × 10^5, chloride form [Sigma, Madison, WI]) in ddH$_2$O
    Autoclave. Store at 4°C. Warm up to 37°C prior to use.

DMEM Medium
    Available from Gibco/BRL (Rockville, MD) or prepare it as follows:
    Dissolve one bottle of Dublecco's modified eagle's medium (DMEM) powder in 800 ml ddH$_2$O; add 3.7 g NaHCO$_3$; adjust the pH to 7.2 with 1 *N* HCl; add ddH$_2$O to 1 L.

Sterile-filter the medium (do not autoclave) in a laminar-flow hood. Divide the medium into 200- to 500-ml aliquots in sterile bottles. Store at 4°C until use. Prior to use, add 10 to 20% fetal bovine serum (FBS), depending on the particular cell type, to the medium and warm to room temperature.

T4 DNA Ligase
2 Weiss units/μl

DNA Polymerase I (Klenow Fragment)
1 unit/μl

T4 DNA Polymerase

DNA To Be Transferred
Circular or linear recombinant DNA constructs in 50 μl of TE buffer, pH 7.5

2 mM dNTP Mixture
2 mM Each of dATP, dCTP, dGTP, and dTTP in 20 mM Tris-HCl, pH 7.5

0.5 M EDTA, pH 8.0
Dissolve 93.06 g NA_2 EDTA·$2H_2O$ in 300 ml ddH_2O. Adjust pH to 8.0 with 2 N NaOH solution. Adjust final volume to 500 ml with ddH_2O and store at room temperature.

Formamide Dye Mixture
10 mM NaOH
1 mM EDTA, pH 8.0
0.1% (w/v) Bromophenol blue
0.1% (w/v) Xylene cyanol
80% (v/v) Deionized formamide

4 M Guanidinium Thiocyanate Buffer
4 M Guanidinium thiocyanate
25 mM Sodium citrate, pH 7.0
0.5% Sarcosyl
0.7% (v/v) 2-Mercaptoethanol
Sterile-filter and store in a light-tight bottle at room temperature up to 3 months.

10X HBS Buffer
8.18% (w/v) NaCl
5.94% (w/v) HEPES, pH 7.1
4-(2-Hydroxyethyl)-1-piperazine ethanesulfonic acid
0.2% (w/v) Na_2HPO_4

2X HBS Buffer
Dilute the 10X HBS in ddH_2O.
Adjust the pH to 7.1 with 1 N NaOH solution.

8 M LiCl Solution for RNA Isolation
8 M LiCl in DEPC-treated ddH_2O
Sterile-filter the solution.

5X Ligase Buffer

0.25 M Tris-HCl, pH 7.5
50 mM MgCl$_2$
5 mM ATP
5 mM DTT
25% (v/v) PEG, 8000

Luciferase Assay Reagent

20 mM Tricine
1.07 mM ((MgCO$_3$)$_4$Mg(OH)$_2$)·5H$_2$O
2.67 mM MgSO$_4$
0.1 mM EDTA, pH 7.8
33.3 mM DTT
0.27 mM Coenzyme A
0.47 mM Luciferin
0.53 mM ATP
Adjust the pH to 7.8.

Magnesium Solution

0.1 M MgCl$_2$
5 M 2-Mercaptoethanol

Mung Bean Nuclease

Immediately before use, the enzyme is diluted to 5 units/μl in mung bean nuclease dilution buffer.

10X Mung Bean Nuclease Buffer

300 mM Sodium acetate, pH 5.0
500 mM NaCl
10 mM ZnCl$_2$
50% (v/v) Glycerol

Mung Bean Nuclease Dilution Buffer

10 mM Sodium acetate, pH 5.0
0.1 mM Zinc acetate
1 mM Cysteine
0.1% (v/v) Triton X-100
50% (v/v) Glycerol

10X 3' Overhang Buffer

300 mM Tris-acetate, pH 7.9
660 mM Potassium acetate
100 mM Magnesium acetate
5 mM DTT
1 μg/μl BSA

10X 5' Overhang Buffer

500 mM Tris-HCl, pH 7.2
100 mM MgSO$_4$
1 mM dithiothreitol (DTT)
500 μg/ml BSA

PBS Buffer

0.137 M NaCl
2.7 mM KCl
4.3 mM Na_2HPO_4
1.47 mM KH_2PO_4
The final pH is 7.2 to 7.4.

T4 Polynucleotide Kinase

1 unit/ml

10X Polynucleotide Kinase Buffer

500 mM Tris-HCl, pH 7.5
100 mM $MgCl_2$
50 mM DTT
1 mM Spermidine
1 mM EDTA

AMV Reverse Transcriptase

50 units/μl

Reverse Transcriptase Buffer

1 M Tris-HCl, pH 8.3
500 mM KCl
100 mM $MgCl_2$

3 M Sodium acetate buffer, pH 5.2

Dissolve sodium acetate in ddH_2O and adjust pH to 5.2 with 3 M glacial acetic acid. Autoclave and store at room temperature.

0.1 M Sodium Phosphate Buffer, pH 7.3

TE buffer

10 mM Tris, pH 8.0
1 mM EDTA, pH 8.0

TE-saturated phenol/chloroform

1X Trypsin–EDTA Solution

0.05% (w/v) Trypsin
0.53 mM EDTA, pH 7.6
Dissolve in PBS buffer, pH 7.4 to 7.6.

Note: *All solutions, buffers, and media must be sterilized.*

II. Gene Transfer and Analysis of Expression in Mice

The mouse is a well-established model organism that has been widely used for gene transfer and analysis of gene expression in mammalian systems. As compared with animal-cell culture systems, transgenic mice allow the gene of interest to be integrated into the germ line of the founder mice. Therefore, every cell in future generations of transgenic animals contains the introduced gene(s). Thus, the expression of the transferred gene(s) can be analyzed during different developmental stages of mice as well as in different generations. In addition, the use of a drug-resistant marker gene for selection in cultured cell systems is not necessary in transgenic animals. The drawback of generating transgenic mice, however, is that it requires expensive equipment for oocyte microinjection, is time-consuming, and requires animal-care facilities.

The general procedures include the following: construction of chimeric genes; preparation of oocytes; microinjection; reimplantation of injected eggs; generation of transgenic mice, F1 and future generations; and analysis of the introduced gene in transgenic mice at DNA, mRNA, and protein levels.

A. Preparation of Chimeric Genes for Microinjection

The general procedures are similar to those previously described in **Section I**, except that a drug-selectable marker gene is not necessarily included in the chimeric constructs. The purified gene of interest can be directly transferred into eggs, since a chimeric gene containing bacterial DNA might affect the expression of the introduced gene in transgenic mice.

B. Preparation of Oocytes

1. Four days (94 to 96 h) before harvesting the eggs, induce super-ovulation of 10 to 15 female mice (3 to 4 weeks old) by injecting 5 units per female of pregnant mare's serum (PMS) as a stimulating hormone.
2. Inject 5 units per female of human chorionic gonadotrophin (HCG) as a stimulating hormone 75 to 76 h after the injection of PMS. This injection should be made 19 to 20 h before harvesting the eggs.
3. After the injection of HCG, individually cage the females with stud males overnight by maintaining on a 12-h light:12-h dark cycle.
4. Check the females for vaginal plugs. The mated females should have visible, vaginal plugs.
5. Sacrifice the super-ovulated females that have visible, vaginal plugs by cervical dislocation.
6. Briefly clean the body surface with 70% ethanol, dissect the oviducts, and transfer to disposable petri dishes containing 30 ml of medium A. Carefully transfer the oocytes (the fertilized eggs) from the oviducts to the medium.

Note: *Under a dissecting microscope, the fertilized eggs can be identified as having two nuclei that are called pronuclei. The male pronucleus is bigger than the female pronucleus.*

7. Remove medium A, carefully transfer the oocytes to a fresh petri dish, and incubate the oocytes with hyaluronidase for an appropriate time period to remove the cumulus cells.
8. Extensively wash the oocytes with four changes of medium A.
9. Carefully transfer the oocytes to a fresh petri dish containing 10 to 30 ml of medium B. Place the dish in a 37°C incubator with 5% CO_2/air.

C. Microinjection of DNA Constructs into the Oocytes

The fertilized eggs are incubated in a 37°C incubator. Normally, microinjection of foreign DNA into the oocytes should be carried out during the period of 3 to 8 h after harvesting, as most of the pronuclei can be readily seen. Later, the nuclei may start to break down before the first cleavage, leaving the nuclei unable to be injected. The equipment for microinjection should include:

1. An injection microscope on a vibration-free table. For example, a Diaphot TMD microscope with Nomarski optics is available from Nikon Ltd. (Melville, NY).
2. An inverted microscope for setting up the injecting chamber, pipette holder, and needle. A SMZ-2B binocular microscope is available from Leitz Instruments Co. (Stuttgart, Germany).
3. Two sets of micromanipulators (Leitz Instruments Ltd.)
4. Kopf needle puller model 750 (David Kopf Instruments, Tujunga, CA).
5. A fine needle and pipette holder.
6. A Schott-KL-1500 cold light source (Schott, Auburn, NY).

The procedure is as follows:

1. Thoroughly clean a depression slide with teepol-based detergent and extensively rinse it with running distilled water for 20 min. Rinse it with ethanol and allow to air dry. Add a drop of Medium A on the depression slide followed by a drop of liquid paraffin on the top of the medium drop.
2. Assemble all parts that are necessary for the injection according to the instructions for the microinjection equipment, and carry out pretesting of the procedure prior to injection.
3. Carefully transfer the eggs into the medium drop using a handling pipette that can draw the eggs inside by suction under the microscope.
4. Slowly and carefully fill the injection needle with the DNA sample to be introduced by capillary action.
5. Under the microscope, hold the egg with the holding pipette and use the micromanipulator to carefully insert the needle into the male pronucleus, which is larger than the female pronucleus. Slowly inject the DNA sample into the pronucleus. The pressure is maintained on the DNA sample in the needle by a syringe connected to the needle holder. As soon as the pronucleus swelling occurs, remove the needle. The volume transferred is approximately 1 to 2 pl.
6. Transfer the injected egg to one side and repeat injections until all the eggs have been injected. About 40 to 60 eggs can be injected in 1 to 2 h. Some eggs may lyse as a result of the procedure. It is recommended to inject approximately 300 eggs in a day, and about two-thirds of these may survive for transfer into the oviduct of the female. Twenty to thirty-five pups are expected, and four to six of them will be transgenic mice.
7. Carefully transfer the injected eggs into a fresh dish containing medium B and place it in the incubator.

D. Reimplantation of the Injected Eggs into Recipient Female Mice and Generation of Founder Mice

The injected eggs can be reimplanted into recipient female mice right after injection (one-cell-stage embryo) or after being incubated in the incubator overnight (two-cell-stage embryo). It is recommended that the injected eggs be transferred at the one-cell stage so that the introduced DNA integrates into one of their chromosomes at the same site. Therefore, all the cells in transgenic animals contain the foreign DNA. Usually, the introduced DNA integrates randomly at a single site. In some cases, multiple copies of the DNA are arranged head-to-tail and form, in some, tandem repeats. The detailed mechanism of integration is not well understood.

1. Prepare pseudo-pregnant females by caging four to eight F1 females (6 to 9 weeks old) with vasectomized males (two females per male) on the day they are needed. Thus, the females will be in the correct hormonal state to allow the introduced embryos to implant, but none of their own eggs can form an embryo. The pseudo-pregnant females can be distinguished by inspecting the vaginal area for swelling and moistness.
2. Carefully make a small incision in the body wall of a half-day pseudopregnant female (e.g., C57B1/CBA F1), and gently withdraw the ovary and oviduct from the incision. Hold the oviduct in place using a clip placed on the fat-pad attached to the ovary.
3. Under a binocular dissecting microscope, transfer the injected eggs or embryos into the infundibulum using a sterile glass transfer pipette. Generally, 10 to 20 eggs can be injected into each oviduct.
4. Carefully place the oviduct and ovary back in place and seal the incision.
5. Allow the foster mothers to recover by briefly warming them under an infrared lamp (do not overheat), and house two or three to one cage. Live offspring are usually born 18 days after the surgery. These initial transgenic offspring are called the founder animals or the G0 in terms of genetics.

E. Indentification of Transgenic Mice

In order to identify and prove transgenic mice, at least two experimental analyses must be carried out. One is genomic DNA analysis by the Southern blot or slot blot hybridization to check whether the transferred gene is integrated into the genome of the mouse, using the introduced DNA as a probe. The other is an analysis of the expression of the introduced gene at the mRNA level by Northern blot hybridization, using the DNA as probe. If positive signals are shown by Southern and Northern blot hybridizations, the introduced gene is successfully integrated and expressed in the transgenic mice. For proof of a stably transgenic line of mice, these analyses should be carried out for several generations.

1. Isolation of DNA from Mouse Tails

1. Spray ethyl chloride around the mouse tail to partially freeze the tail.
2. Cut a piece of the tail at approximately 0.5 cm from the end of the tail and place into a microcentrifuge tube containing 700 μl of DNA isolation buffer.
3. Repeat steps 1 and 2 for other mice.
4. Add 25 μl of 10 mg/ml of *Proteinase K* to each tube. Incubate the tubes at 55°C overnight to digest the tail pieces (homogenizing is an optional procedure).
5. Add 700 μl of phenol:chloroform to each tube and shake the tubes on a shaker at 150 rpm for 15 min.
6. Centrifuge at 11,000 g for 10 min and carefully transfer the top, aqueous phase to fresh tubes.
7. Repeat steps 5 and 6 (optional).
8. Extract with 0.7 ml of chloroform, centrifuge at 11,000 g for 5 min, and transfer the upper, aqueous phase to fresh tubes.
9. Add 0.8 volume of isopropanol to each tube and allow to precipitate at room temperature or at –20°C for 30 min.
10. Centrifuge at 11,000 g for 10 min, decant the supernatant, and rinse the pellet with 1 ml of 70% ethanol. Completely dry the DNA pellet under vacuum for 30 min at room temperature.
11. Dissolve the DNA in 50 μl of TE buffer. Take 5 μl of the sample to measure the DNA concentration at 260 and 280 nm. Normally, 60 to 100 μg of DNA can be obtained for each sample.

2. Isolation of RNA from the Organs (Embryo, Brain, and Liver) of the Transgenic Mice Using Lithium Chloride/Urea

Note: RNA can also be isolated from lymphocytes of blood taken from the tails of transgenic mice.

1. Dissect appropriate tissue from transgenic mice.
2. Homogenize the tissue in 8 ml of LiCl/urea isolation solution per gram of tissue to a fine homogenate.
3. Sonicate the sample on ice for 2 min to shear the DNA. Place the sample at 4°C overnight to precipitate RNA.
4. Centrifuge at 9000 g for 20 min, decant the supernatant, and resuspend the RNA pellet in 4 ml of LiCl/urea isolation solution.
5. Repeat step 4.
6. Dissolve the RNA in 3 ml of TE buffer (pH 7.6) containing 0.4% (w/v) SDS. Add one volume of phenol:chloroform to the mixture and mix.
7. Centrifuge at 11,000 g for 5 min and carefully transfer the upper, aqueous phase to a fresh tube.
8. Add 0.1 volume of 3 M sodium acetate buffer (pH 5.2) and two volumes of chilled 100% ethanol to the tube and allow precipitation to occur at –20°C for 60 min.
9. Centrifuge at 11,000 g for 10 min, decant the supernatant and rinse the pellet with 1 ml of 70% ethanol. Completely dry the DNA pellet under vacuum for 30 min at room temperature.
10. Dissolve the RNA in 40 μl of DEPC ddH_2O. Take 5 μl of the sample to measure the RNA concentration at 260 and 280 nm.

Note: All solutions for RNA work need to be DEPC treated. See **Chapter 2** for more details on working with RNA.

3. Southern Blot, Slot Blot, and Northern Blot Hybridizations

Follow the detailed procedures as described in **Chapter 5** and **Chapter 6**, except that the probe should be a partial- or full-length sequence of the gene that has been introduced into the mouse eggs.

4. Analysis of the Expression of the Reporter Gene by Staining Embryos from Transgenic Mice

1. Kill females by cervical dislocation and dissect out the uterus that contains embryos.
2. Transfer the uterus to a fresh petri dish containing 20 ml of PBS. Dissect out embryos using a clean pair of forceps.
3. Wash the embryos with PBS. The washed embryos can be frozen in liquid nitrogen for further RNA isolation or used directly for staining ß-galactosidase expression (as in the following steps).
4. Fix the embryos in cold (4°C) fixative solution for 1 to 1.5 hours.
5. Wash the embryos four times in 40 ml of washing solution at room temperature.
6. Stain the embryos overnight in 20 ml of staining solution at room temperature in the dark. Rinse three times in PBS and store at 4°C in PBS. Stained embryos should be blue in color.

F. Selection of Transgenic Lines by Breeding the Transgenic Mice

The sexes of transgenic founders are usually distinguished at approximately 3 weeks after the mice are born. The mice become sexually mature and can breed at about 6 weeks. To establish a transgenic line, allow the founder mice to breed with nontransgenic mice. Males can mate with females within 24 h of the birth of a litter, and the female can continue to produce litters at 3-week intervals. After running Southern blot and Northern blot hybridization analyses as described previously in **Chapter 5** and **Chapter 6**, respectively, the mice should be allowed to breed again for up to four to five generations. If the introduced gene is stably expressed in several generations, a transgenic mouse line has been established.

Reagents Needed

DNA Isolation Buffer
50 mM Tris-HCl, pH 8.0
100 mM EDTA, pH 8.0
100 mM NaCl
1% (w/v) SDS

Fixative Solution
1% (v/v) Formaldehyde
0.2% (v/v) Glutaraldehyde
0.02% (v/v) NP-40
2 mM MgCl$_2$
5 mM EDTA
Make in PBS

LiCl/Urea Isolation Solution
3 M LiCl
6 M Urea

Medium A
4.78 mM KCl
94.66 mM NaCl
1.71 mM CaCl$_2$
1.19 mM KH$_2$PO$_4$
1.19 mM MgSO$_4$
4.15 mM NaHCO$_3$
23.38 mM Sodium lactate
0.33 mM Sodium pyruvate
5.56 mM Glucose
20.85 mM HEPES, pH 7.4 (adjust with 0.2 M NaOH)
BSA, 4 g/L
Penicillin G (potassium salt), 100,000 units/L
Streptomycin sulfate, 50 mg/L
Phenol red, 10 mg/L

- Dissolve well after each addition.
- Sterile-filter and store at 4°C.
- Warm up to 37°C prior to use.

Medium B
4.78 mM KCl
94.66 mM NaCl
1.71 mM CaCl$_2$
1.19 mM KH$_2$PO$_4$
1.19 mM MgSO$_4$
25 mM NaHCO$_3$
23.38 mM Sodium lactate
0.33 mM Sodium pyruvate
5.56 mM Glucose
BSA, 4 g/L
Penicillin G (potassium salt), 100,000 units/L

Streptomycin sulfate, 50 mg/L
Phenol red, 10 mg/L
- Dissolve well after each addition.
- Sterile-filter and store at 4°C.
- Warm up to 37°C prior to use.

PBS Solution (1 L)

NaCl (7.6 g)
Na_2HPO_4 (3.8 g)
NaH_2PO_4 (0.42 g)

Phenol:Chloroform

One part of TE buffer-saturated (pH 8.0) phenol and one part of chloroform

Staining Solution

5 mM $K_3Fe(CN)_6$
5 mM $K_4Fe(CN)_6 \cdot 3H_2O$
2 mM $MgCl_2$
0.01% (v/v) Sodium deoxycholate
0.02% (v/v) NP-40
1 mg/ml X-gal (5-Bromo-4-chloro-3-indolyl-β-D-galactopyranoside) diluted from stock solution

TE Buffer

10 mM Tris-HCl, pH 7.6
1 mM EDTA, pH 7.6

TE Buffer

10 mM Tris-HCl, pH 8.0
1 mM EDTA, pH 8.0

TE-Saturated Phenol:Chloroform (1:1)

Washing Solution

0.02% (v/v) in PBS

References

1. **Yang, N.S., Burkholder, J., Roberts, B., Martinell, B., and McCabe, D.,** *In vitro* and *in vivo* gene transfer to mammalian somatic cells by particle bombardment, *Proc. Natl. Acad. Sci. U.S.A.*, 87, 9568, 1990.
2. **Klein, T.M., Wolf, E.D., Wu, R., and Sanford, J.C.,** High-velocity microprojectiles for delivering nucleic acids into living cells, *Nature*, 327, 70, 1987.
3. **Carmeliet, P., Schoonjans, L., Kiechens, L., Ream, B., Degen, J., Bronson, R., Vos, R.D., van den Oord, J.J., Collen, D., and Mulligan, R.C.,** Physiological consequences of loss of plasminogen activator gene function in mice, *Nature*, 368, 419, 1994.
4. **Koleske, A.J. and Young, R.A.,** An RNA polymerase II holoenzyme responsive to activators, *Nature*, 368, 466, 1994.

5. **Wang, Z., Svejstrup, J.Q., Feaver, W.J., Wu, X., Kornberg, R.D., and Friedberg, E.C.,** Transcription factor b (TFIIH) is required during nucleotide-excision repair in yeast, *Nature*, 368, 74, 1994.
6. **Zelenin, A.V., Titomirov, A.V., and Kolesnikov, V.A.,** Genetic transformation of mouse cultured cells with the help of high-velocity mechanical DNA injection, *FEBS Lett.*, 244, 65, 1989.
7. **Shillito, R.D., Saul, M.W., Paszkowski, J., Muller, M., and Potrykus, L.,** High efficiency direct gene transfer to plants, *BioTechnology*, 3, 1099, 1985.
8. **Johnston, S.A.,** Biolistic transformation: microbes to mice, *Nature*, 346, 776, 1990.
9. **Moore, K.A. and Belmont, J.W.,** Analysis of gene transfer in bone marrow stem cells, in *Gene Targeting: A Practical Approach*, Joyner, A.L., Ed., Oxford University Press, New York, 1993.
10. **Walmsley, M.E. and Patient, R.K.,** Quantitative and qualitative analysis of exogenous gene expression by the S1 nuclease protection assay, in *Gene Transfer and Expression Protocols*, Murray, E.J., Ed., Humana Press, Clifton, NJ, 1991.
11. **Hasty, P. and Bradley, A.,** Gene targeting vectors for mammalian cells, in *Gene Targeting: A Practical Approach*, Joyner, A.L., Ed., Oxford University Press, New York, 1993.
12. **Wurst, W. and Joyner, A.L.,** Production of targeted embryonic stem cell clones, in *Gene Targeting: A Practical Approach*, Joyner, A.L., Ed., Oxford University Press, New York, 1993.
13. **Papaioannou, V. and Johnson, R.,** Production of chimeras and genetically defined offspring from targeted ES cells, in *Gene Targeting: A Practical Approach*, Joyner, A.L., Ed., Oxford University Press, New York, 1993.
14. **Finer, M.H.,** The *RNase* protection assay, in *Gene Transfer and Expression Protocols*, Murray, E.J., Ed., Humana Press, Clifton, NJ, 1991.
15. **Gossler, A. and Zachgo, J.,** Gene and enhancer trap screens in ES cell chimeras, in *Gene Targeting: A Practical Approach*, Joyner, A.L., Ed., Oxford University Press, New York, 1993.

Chapter 19

Gene Transfer and Expression in Plants

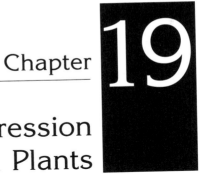

William Wu, Leland J. Cseke, and Peter B. Kaufman

Contents

I. Background 410
II. Protocols 411
    A. Cloning, Isolation, Characterization, and Subcloning of the Gene of Interest 411
    B. Purification of the Gene of Interest from the Recombinant Plasmids 411
        1. Elution of DNA from Agarose Gel Slices 412
        2. Purification of the Target DNA Using NA45 DEAE Membranes 413
    C. Selection of Appropriate Vector, Promoter, Poly(A) Signal, Reporter Gene, and Selectable Marker Gene 413
    D. Preparation of Blunt-End DNA 416
    E. Preparation of DNA Insert Lacking the 5′ Phosphate Groups 416
    F. Construction of Chimeric Genes by Ligating Plasmid Vector and Insert-DNA 417
    G. Transformation of an Appropriate Strain of Bacteria to Amplify the Recombinant Plasmids 418
        1. Introduction of Recombinant Plasmids into *Agrobacterium tumefaciens* (Biotype 1: Ach5, A6, B6, C58, T37, Bo542, and 15955) by Transformation 418
        2. Introduction of Plasmids into *Agrobacteria* by Conjugation 418
    H. Transformation of Plants with Recombinant Gene Constructs 419
        1. Gene Transfer by *Agrobacterium tumefaciens* 419
        2. Direct Gene Transfer to Protoplasts by Electroporation 420
        3. Direct Gene Transfer by Microprojectile Bombardment 423
    I. Proof of Stable Transformation 425
        1. Phenotypic and/or Functional Examinations 426
        2. Analysis of the Expression of the Reporter Gene 426
        3. Analyses of Southern Blotting and Sequencing, and Determination of the Copy Number of the Gene 427

4.	Northern Blot Analysis	428
5.	Western Blot Hybridization or Immunocytochemical Localization	428
Reagents Needed		429
References		431

I. Background

Gene manipulation technology has been widely used in plant systems.[1–10] The process has made it possible to select "superior" plants carrying genes for desired traits. In recent years, the application of gene-transfer techniques has produced significant achievements in generating transgenic plants with desired traits.[2,4–15] The general reasons for producing transgenic plants can be summarized as follows: (1) to improve the resistance of plants to insect pests and to fungal and viral diseases; (2) to modify the amino acid composition of storage protein in cereals, potatoes, and legumes to increase their nutritional value; (3) to improve the composition and storage life of fruits and vegetables; (4) to generate transgenic plants that are herbicide-resistant; (5) to increase the capacity of crop plants to fix atmospheric nitrogen; (6) to increase the rate of photosynthesis in green plants; (7) to modify the phenotypic characteristics (color, height, flowering time, and size) of food plants; (8) to reduce photorespiration in C_3 plants, with resultant increase in carbon fixation; (9) to introduce new pigment colors in petals of flowers such as blue flower color in roses; and (10) for research purposes and the study of gene function.

One of the unique characteristics of plant cells, as compared with animal cells, is that plants have cell walls, which act as barriers to the introduction of foreign deoxyribonucleic acid (DNA). As a result, the gene-transfer methods used in plants are not identical to those applied to animal systems. The initially used gene-transfer technique primarily relies on a natural plant-pathogenic bacterium, *Agrobacterium tumefaciens*, to introduce recombinant plasmid DNA into the plant genome. However, this *Agrobacterium*-mediated gene transfer is largely limited to dicotyledonous flowering plants (dicots). Many of the major crops, however, are monocotyledons (monocots),

TABLE 19.1
Advantages and Disadvantages of Gene-Transfer Techniques

Method	Advantages	Disadvantages
Agrobacterium (Ti and Ri)-mediated gene transfer (to leaf discs, to protoplasts, to roots)	Well established and highly efficient for dicots	Not suitable for monocots in most cases
Polyethylene (PEG) Promotes protoplast fusion and the uptake of DNA	Relatively simple	PEG is toxic to humans
Electroporation The use of short electrical impulses of high field strength that increases the permeability of protoplast membranes	Highly efficient and used for both dicots and monocots	Up to 50% of cells die due to high-voltage damage to plasma membranes of protoplasts
Microinjection Precisely injects recombinant DNA into specific compartments of protoplasts	Specific, highly efficient, used for all plants	Requires special skills, special equipment, and high cost
Particle bombardment Particle gun shoots recombinant DNA into protoplasts or cells	Highly efficient, and DNA is delivered simultaneously into many cells	Requires special skills, special equipment, and high cost

which are not readily susceptible to infection by *Agrobacterium*. For this reason, some other techniques have been developed for gene transfer in monocots. Whichever method is used, however, the general principles are the same. Recombinant DNA carrying the gene(s) of interest is transferred by an appropriate method into topitotent cells or tissues that are capable of regenerating into entire plants with leaves, stems, roots, flowers, and fruits. The foreign-DNA construct will integrate into the plant genome, generating transgenic plants in which the foreign genes are expressed. This chapter describes in detail several well-established gene-transfer techniques that are used in transformation of plants, with emphasis on dicots. Each method has advantages and disadvantages that are summarized in Table 19.1.

The entire process of genetic engineering in plants can be subdivided into several procedures:

1. Cloning, isolation, characterization, and subcloning of the gene of interest to be transferred
2. Selection and/or purification of promoter, enhancer, poly(A) signal, reporter gene, selectable marker gene, and the gene of interest from recombinant plasmids
3. Construction of chimeric or fusion genes
4. Transformation of plant cells or tissue(s) with the chimeric genes
5. Selection and regeneration of transgenic plants
6. Analysis of the expression of the introduced gene in transgenic plants

II. Protocols

A. Cloning, Isolation, Characterization, and Subcloning of the Gene of Interest

The general procedures described in **Chapter 9, Chapter 10,** and **Chapter 18** are not included within the scope of this chapter.

B. Purification of the Gene of Interest from the Recombinant Plasmids

The gene to be transferred is usually precharacterized, subcloned in an appropriate plasmid, and propagated in an appropriate bacterial strain. In that case, the gene should be removed from the recombinant plasmid by using an appropriate restriction enzyme(s), depending on what kind of endonuclease(s) are used for subcloning. The general process is as follows:

1. Set up, on ice, a standard restriction enzyme digestion as follows:
   Single-restriction enzyme digestion:
   - Plasmids (20 µg)
   - 10X Appropriate restriction enzyme buffer (10 µl)
   - 1 mg/ml Acetylated bovine serum albumin (BSA) (optional) (10 µl)
   - Appropriate restriction enzyme (3.4 units/µg DNA)
   - Add distilled, deionized water (ddH$_2$O) to a final volume of 100 µl.

   Double-restriction enzyme digestion:
   - Plasmid DNA or DNA to be inserted (20 µg)
   - 10X Appropriate restriction enzyme buffer (10 µl)
   - 1 mg/ml Acetylated BSA (optional) (10 µl)
   - Appropriate restriction enzyme A (3.4 units/µg DNA)
   - Appropriate restriction enzyme B (3.4 units/µg DNA)
   - Add ddH$_2$O to a final volume of 100 µl.

Note: *For double-restriction enzyme digestions, the appropriate 10X buffer containing higher NaCl concentration than the other buffer may be chosen for the double-enzyme digestion buffer.*

2. Incubate at appropriate temperature, depending on the restriction enzyme, (e.g., 37°C) for 2 to 3 h.
3. Purify the DNA fragment of interest by one of the two following methods.

1. Elution of DNA from Agarose Gel Slices

This is a simple, fast, and effective protocol that is successfully used in our laboratory. The recovery is usually 85 to 95% of the amount of DNA that is loaded in the gel. The DNA eluted by this method is very pure and can be directly used for ligation, cloning, and/or labeling.

1. Carry out electrophoresis by using 1% low-melting-point agarose instead of normal agarose.
2. Stain the gel with EtBr solution (10 μl of 10 mg/ml of EtBr in 100 ml ddH$_2$O) for 10 min.
3. Transfer the gel to a long-wavelength ultraviolet (UV) transilluminator (305 to 327 nm) to visualize DNA bands. Excise the band(s) of interest with a clean razor blade and place the slices in a microcentrifuge tube.

Note: *(1) The gel should not be placed on a short-wavelength (e.g., <270 nm) UV transilluminator because the UV light may cause breakage within the DNA fragment, and thus significantly inhibit the subcloning of the insert of interest. (2) UV light is damaging to the human body. Protective glasses, gloves, and a lab coat should be worn when using UV light. (3) The gel slices containing the DNA bands of interest should be trimmed of excess unstained gel areas as much as possible.*

4. Add two to four volumes of TE buffer to the gel slices and completely melt the gel in a 60 to 65°C water bath.

Note: *The gel slices can be directly melted without adding any TE buffer. The DNA concentration is usually high, but total yield of eluted DNA fragment is much lower than when one adds TE buffer to the gel slice.*

5. Immediately chill the tube on dry ice or place it at –70°C for at least 20 min. Alternatively, add one volume of Tris-buffer-saturated (pH 8.0) phenol (pH 7.0), not phenol/chloroform, to the melted gel solution. Vigorously mix by inversion for 2 to 4 min and chill the tube in dry ice or place it at –70°C for at least 20 min.
6. Centrifuge at 12,000 g at room temperature for 15 min. An alternative way is to thaw the gel mixture by tapping the tube vigorously. It takes about 5 to 10 min to thaw the gel to a resuspension status, followed by centrifuging at 12,000 g for 5 min at room temperature.
7. Carefully transfer the liquid phase (supernatant) containing the elution DNA fragment to a fresh tube.

Note: *(1) Although the DNA solution can be directly used, except that the concentration is low, we strongly recommend that the DNA be precipitated with ethanol and resuspended in a small volume of TE buffer or ddH$_2$O. (2) Phenol serves to extract agarose and EtBr from the stained DNA. If one does not add Tris-buffered phenol to the melted gel solution, EtBr should be extracted from the DNA in the supernatant three times with three volumes of water-saturated n-butanol.*

8. Add 0.5 volume of TE buffer or ddH$_2$O to the phenol phase, mix well, and centrifuge as in step 5. Transfer the supernatant to the tube in step 6.

9. Precipitate the DNA by adding 0.05 volume of 3 M sodium acetate buffer (pH 5.2) and 2.5 volumes of chilled 100% ethanol to the pooled supernatant. Allow precipitation to occur at −70°C for 1 h and centrifuge at 12,000 g for 10 min at room temperature.
10. Decant the supernatant. Briefly wash the DNA pellet with 1 ml of 70% ethanol and centrifuge at 12,000 g for 5 min. Decant the supernatant and dry the DNA under vacuum for 15 min. Dissolve the DNA in an appropriate volume of ddH$_2$O or TE buffer. Store at −20°C until use.

2. Purification of the Target DNA Using NA45 DEAE Membranes

1. Carry out electrophoresis by using 1% low-melting-point agarose containing EtBr (10 μl of 10 mg/ml EtBr in 100 ml of agarose gel solution). Use a long-wavelength UV lamp to monitor the separation of the target fragment and the vector as compared with known sizes of DNA standards.
2. Soak a membrane (Scheicher and Schuell, Keene, NH) in 10 mM EDTA buffer (pH 7.6) for 15 min at room temperature and store at 4°C until use.
3. During the electrophoresis, monitor the migration of DNA bands stained by EtBr in the gel using a long-wavelength UV lamp. Make an incision below the DNA band of interest and insert a piece of the prepared membrane into the incision in the gel.
4. Continue electrophoresis, while monitoring the migration of the stained band, until the DNA fragment migrates onto the membrane.
5. Remove the membrane strip from the gel and place it in a solution containing 20 mM Tris-HCl (pH 8.0), 0.15 M NaCl, and 0.1 mM EDTA. Shake it for 1 min to remove any agarose.
6. Transfer the strip to 0.2 of elution buffer containing 20 mM Tris-HCl (pH 8.0), 1 M NaCl, and 0.1 mM EDTA.
7. Incubate the strip at 55 to 68°C for 30 to 60 min with shaking at 60 rpm.
8. Rinse the strip with 0.1 ml of elution buffer and pool the elution fluids together.
9. Extract EtBr four times with three volumes of water-saturated n-butanol.
10. Precipitate and dissolve the eluted DNA as described in **Section II.B.1**.

C. Selection of Appropriate Vector, Promoter, Poly(A) Signal, Reporter Gene, and Selectable Marker Gene

Transformation of plants was initiated by *Agrobacterium tumefaciens*-mediated gene-transfer methodology. *Agrobacterium tumefaciens* is a gram-negative soil bacterium that can cause crown gall disease in the infected plants. The bacterium contains a unique plasmid called the Ti (tumor-inducing) plasmid that contains oncogenes (tumor-inducing genes). The Ti plasmids are approximately 200 to 250 kb (kilobase-pairs) in size. After infection, a portion of the DNA of the Ti plasmid can transfer into the host cell and integrate into the genome of the host cell. That DNA fragment is called transferred DNA or T-DNA, and it contains oncogenes. The T-DNA codes for enzymes that synthesize auxin and cytokinin, which cause abnormal plant growth by causing the proliferation of undifferentiated wound tissue that forms the crown gall or tumor. The infected plant cells also produce nitrogenous compounds, named opines, by opine synthase genes that are present in the T-DNA. There are more than six different classes of opines, but the most well-known Ti plasmids are the nopaline- and octopine-type plasmids. The nopaline-type T-DNA is 23 kb in length and flanked by 25 bp (base-pair) direct repeats called the left and the right border (LB and RB) sequences. The octopine-type T-DNA is approximately 21 kb in length and contains 25 bp direct-repeat (LB and RB) sequences that are quite similar to the LB and RB sequences of the nopaline-type T-DNA. It has been demonstrated that the border sequences play an essential role in the transfer of T-DNA. In order to make the Ti plasmids nononcogenic and widely useful as vectors for gene transfer into plants, the oncogenes in the T-DNA are deleted except for the LB and RB regions. Such plasmids are said to contain "disarmed" T-DNA. The removed oncogene fragment is replaced with a region of pBR322 or other bacterial plasmid DNA fragment, resulting in

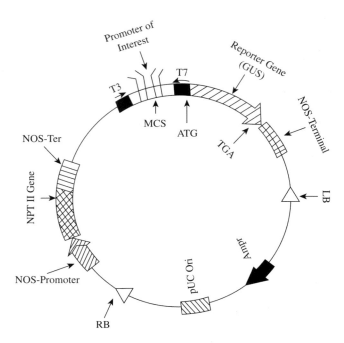

FIGURE 19.1
Structural map of an expression vector designed for insertion and analysis of the activity of a specific promoter in plants. (From Kaufman, P.B., Wu, W., Kim, D., and Cseke, L.J., *Handbook of Molecular and Cellular Methods in Biology and Medicine,* 1st ed., CRC Press, Boca Raton, FL, 1995.)

modification of the Ti plasmids. Since the Ti plasmids were large plasmids that are usually difficult to transfer, binary vectors have been developed. Later, Ti plasmids were extensively modified and fused as a part of chimeric genes that include a promoter, a selectable marker, a reporter gene, and polylinker sites for the insertion of the gene of interest.

In addition to the modified vectors, other regulatory sequences or genes should be prepared for the gene transfer and analysis of the expression of the introduced gene(s). The cauliflower mosaic virus (CaMV) 35S promoter, the 3′ poly(A) signal from the nopaline synthase gene of the Ti plasmid, the chloramphenicol acetyl transferase (CAT) gene or the β-glucuronidase (GUS) gene as a reporter gene, and the neomycin phosphotransferase (NPT II) gene or the hygromycin phosphotransferase (HPT) gene as a selectable marker gene are recommended to be appropriately fused together in the recombinant constructs to be transferred (Figure 19.1 and Figure 19.2). These fragments and/or genes are well characterized, have been subcloned in plasmids, and are commercially available. The fragments can be purified from the recombinant plasmid using appropriate restriction enzyme(s).

Example: Using *Sma* I restriction enzyme to prepare the vector:

1. Set up, on ice, a standard enzyme digestion as follows:
   - Plasmid vector (10 µg)
   - 10X *Sma* I buffer (10 µl)
   - 1 mg/ml Acetylated BSA (optional) (10 µl)
   - *Sma* I (30 units)
   - Add ddH$_2$O to a final volume of 100 µl.
2. Incubate at 37°C for 2 to 3 h.

Gene Transfer and Expression in Plants

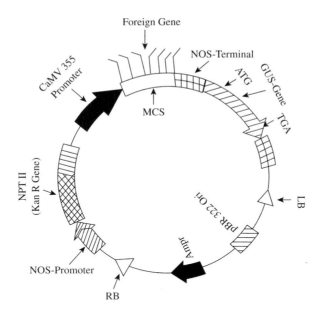

FIGURE 19.2
Structural map of an expression vector with a strong promoter, CaMV 35S, that is used for cloning and expression of a foreign gene of interest. (From Kaufman, P.B., Wu, W., Kim, D., and Cseke, L.J., *Handbook of Molecular and Cellular Methods in Biology and Medicine*, 1st ed., CRC Press, Boca Raton, FL, 1995.)

Note: The digestion efficiency can be checked by loading 1 µg of the digested DNA (10 µl) with loading buffer to a 1% agarose minigel. After electrophoresis, the undigested plasmid DNA may reveal multiple bands because of different levels of supercoiled plasmid DNA. However, one band will be visible after Sma I digestion.

3. Carry out CIAP treatment to remove the 5´ phosphate groups in order to prevent the recircularization of linearized plasmid:
   - Linearized plasmids (90 µl)
   - 10X CIAP buffer (15 µl)
   - CIAP diluted in 10X CIAP buffer (0.01 unit/pmol ends)
   - Add ddH$_2$O to a final volume of 150 µl.

Note: (1) CIAP and 10X CIAP should be kept at 4°C, and CIAP treatment should be set up at 4°C. (2) Calculation of the amount of ends is as follows. There are 9 µg of digested DNA left after taking 1 µg of 10 µg digested DNA for checking on the agarose gel. If the vector is 3.2 kb, the amount of ends can be calculated by the use of the formula below:

$$\text{pmol ends} = \frac{\text{amount of DNA}}{\text{base pairs} \times 600 \text{ / base pair}} \times 2$$

$$= \frac{9}{3.2 \times 1000 \times 660}$$

$$= 4.2 \times 10^{-6} \times 2$$

$$= 8.4 \times 10^{-6} \mu M$$

$$= 8.4 \times 10^{-6} \times 10^{-6} = 8.4 \text{ pmol ends}$$

4. Incubate at 37°C for 1 h and add 2 µl of 0.5 M EDTA buffer (pH 8.0) to stop the reaction.
5. Extract with one volume of TE-saturated phenol/chloroform. Mix well by vortexing for 1 min and centrifuge at 11,000 g for 5 min at room temperature.
6. Carefully transfer the top, aqueous phase to a fresh tube and add one volume of chloroform:isoamyl alcohol (24:1) to the supernatant. Mix well and centrifuge as in step 5.
7. Carefully transfer the upper, aqueous phase to a fresh tube and add 0.1 volume of 3 M sodium acetate buffer (pH 5.2) or 0.5 volume of 7.5 M ammonium acetate to the supernatant. Briefly mix and add 2 to 2.5 volumes of chilled 100% ethanol to the supernatant. Allow to precipitate at –70°C for 1 h or at –20°C for 2 h.
8. Centrifuge at 12,000 g for 10 min and carefully decant the supernatant. Briefly rinse the DNA pellet with 1 ml of 70% ethanol and dry the pellet under vacuum for 20 min. Dissolve the DNA pellet in 20 to 40 µl ddH$_2$O. Take 4 µl of the sample to measure the concentration of the DNA at 260 nm. Store the sample at –20°C until use.

Note: *Adding 0.5 volume of 7.5 M ammonium acetate to the supernatant at step 7 yields a higher amount of DNA precipitation than adding 0.1 volume of 3 M sodium acetate buffer (pH 5.2).*

D. Preparation of Blunt-End DNA

If both the vector and DNA insert have sticky or overhanging ends at the right sites, they can readily ligate to each other. If not, it is usually necessary to generate blunt ends on the DNA fragments prior to their being ligated to the blunt-ended vectors. Blunt ends can be made by filling in 5′ overhangs with DNA polymerase I (Klenow fragment).

1. Set up the 5′ overhang reaction in a microcentrifuge tube on ice as follows:
   - DNA fragment or vector DNA (2 to 4 µg in 4 µl)
   - 10X 5′ overhang buffer (2 µl)
   - 2 mM dNTP mixture (2 µl)
   - DNA polymerase I (Klenow fragment) (2 to 3 units)
   - Add ddH$_2$O to a final volume of 20 µl.
2. Incubate at room temperature for 20 min and heat at 70°C for 5 min to stop the reaction.
3. Add one volume of TE buffer (pH 7.5) and carry out phenol:chloroform extraction and precipitation as described in **Section I.C**.

Note: *While the use of Klenow fragment is the most common method used to produce blunt ends, other methods were discussed in* **Chapter 12, Chapter 14,** *and* **Chapter 18**.

E. Preparation of DNA Insert Lacking the 5′ Phosphate Groups

If the DNA insert lacks the 5′ phosphate groups, the following reaction should be carried out prior to its being ligated to blunt, dephosphorylated vectors.

1. Add phosphate groups to the 5′ end of the oligonucleotides by kinase reaction.
   - Blunt-end DNA insert (15 µl)
   - 10X Polynucleotides kinase buffer (3 µl)
   - 10 mM ATP (1.5 µl)
   - 100 mM DTT (1.5 µl)
   - T4 Polynucleotide kinase (15 units)
   - Add ddH$_2$O to a final volume of 30 µl.
2. Incubate at 37°C for 40 min and heat at 70°C for 10 min to inactivate the enzyme.

Gene Transfer and Expression in Plants

3. Add one volume of TE buffer (pH 7.5) and carry out extraction and precipitation as described in **Section I.C.**

F. Construction of Chimeric Genes by Ligating Plasmid Vector and Insert-DNA

In order to achieve optimal ligation, the ratio of vector to insert-DNA (1:1, 1:2, 1:3, and 3:1 molar ratios) is strongly recommended to be optimized using a small-scale reaction. The following reaction protocol is standard for the ligation of a 3.2-kb plasmid vector and a 2.533-kb DNA insert.

1. Calculate the molar weight of vector and DNA insert:

$$1\ M\ \text{plasmid vector} = 3.2 \times 1000 \times 660 = 2.112 \times 10^6$$

$$1\ M\ \text{insert DNA} = 2.533 \times 1000 \times 660 = 1.672 \times 10^6$$

where 660 refers to the average molecular weight of a nucleotide.

2. Calculate the molar ratio of vector to insert DNA:

Vector DNA:Insert DNA	Amount of DNA (µg)	
	Vector	Insert
1:1	1	0.792
1:2	1	1.584
1:3	1	2.376
3:1	1	0.264

3. Set up the ligation reactions on ice:

Components	Ligation Reactions			
	1 (1:1)	2 (1:2)	3 (1:3)	4 (3:1)
Plasmid DNA as vector (µg)	1	1	1	1
DNA insert (µg)	0.792	1.584	2.376	0.244
10X Ligase buffer (µl)	1	1	1	1
T4 DNA ligase (Weiss units)	4	4	4	4
Add ddH$_2$O to achieve final volume (µl)	10	10	10	10

Note: Restriction enzyme digested plasmid (vector) and DNA insert should be dissolved in ddH$_2$O (nuclease-free) to 0.5 to 1.0 µg/µl. If the DNA is less than 0.4 µg/µl, the DNA should be precipitated to dissolve in about 1 µg/µl.

4. Incubate the reactions at 4°C for 12 to 24 h, at 16°C for 4 to 6 h, or at room temperature (22 to 25°C) for 1 to 2 h.

Note: After the ligations are finished at the above temperatures, the mixture can be stored at 4°C until use.

5. Check the efficiency of ligation by 1% agarose electrophoresis. When the electrophoresis is complete, photograph the gel stained with EtBr under UV light. Compared with unligated vector or insert-DNA wells, a highly efficient ligation should allow one to visualize less than approximately 10% unligated vector and insert-DNA by estimating the intensity of fluorescence. Approximately 90% of the vector and insert-DNA are ligated to each other and show strong band(s) with molecular-weight shifts compared with vector and insert-DNA sizes. By comparing the efficiency of ligation using different molar ratios, the optimal conditions can be determined with ease. This can then be used as a guide for large-scale ligation.

Note: *The above small-scale ligation is optional, but it is strongly recommended.*

6. Large-scale ligation of vector and insert-DNA is based on the optimal conditions determined by the small-scale ligation. For example, if one uses a 1:2 molar ratio of plasmid DNA:insert DNA as the optimal ligation, a large-scale ligation reaction is carried out as follows:
   - Plasmid DNA as vector (3 µg)
   - DNA Insert (4.75 µg)
   - 10X Ligase buffer (3 µl)
   - T4 DNA ligase (15 Weiss units)
   - Add ddH$_2$O to a final volume of 30 µl.
7. Incubate the ligation reaction at 4°C for 12 to 24 h, or at 16°C for 4 to 6 h, or at room temperature (22 to 25°C) for 1 to 2 h. Store at 4°C until use.

G. Transformation of an Appropriate Strain of Bacteria to Amplify the Recombinant Plasmids

The general procedures for transformation of bacteria and selection of transformants are as described under the section on subcloning of plasmid DNA in **Chapter 10**, except for the transformation of *Agrobacteria*.

1. Introduction of Recombinant Plasmids into *Agrobacterium tumefaciens* (Biotype 1: Ach5, A6, B6, C58, T37, Bo542, and 15955) by Transformation

1. Inoculate bacteria on a fresh LB (Luria-Bertaini) plate and incubate at 29°C overnight.
2. Inoculate a single colony of the bacteria in 50 ml of LB medium and culture at 29°C with shaking at 150 rpm.
3. Dilute the culture to an OD$_{660}$ of 0.02 by adding fresh medium and incubating at 29°C for 4 h with shaking at 150 rpm.
4. Centrifuge at 5000 g for 5 min and wash the cell pellet twice with five volumes of 10 mM Tris-HCl buffer (pH 7.5). Repeat centrifugation.
5. Decant the supernatant and resuspend the cells in 0.1 volume of rich medium.
6. Mix 10 µl (3 to 5 µg) of recombinant plasmids with 20 µl of cell suspension and incubate the mixture at –70°C or in a dry ice–ethanol bath for 5 to 8 min.
7. Transfer the mixture to 37°C for 2 min and incubate at room temperature for 30 min.
8. Dilute the suspension with 70 µl of fresh medium and incubate at 29°C for 1 h followed by plating onto a selection medium containing appropriate antibiotics.
9. Select transformants and culture them for isolation of plasmids.

2. Introduction of Plasmids into *Agrobacteria* by Conjugation

The following is a modification of triparental mating procedures that make use of an *E. coli* strain (pRK 2013) to facilitate conjugation. This modified protocol works fine in our lab.
1. Inoculate the bacteria onto a fresh LB plate and incubate overnight at 29°C.

2. Individually inoculate donor and recipient cells into 10 ml of rich liquid medium in a 125-ml sterile flask and culture overnight at 29°C with shaking at 150 rpm.
3. Dilute the culture to an OD_{660} of 0.05 to 0.1 by adding fresh medium and incubating at 29°C for 4 h with shaking at 150 rpm.
4. Mix equal parts of the donor and recipient cultures.
5. Spread 50 to 100 µl of the mixture onto a sterile membrane filter on top of an agar layer with rich growth medium in a petri dish.
6. Incubate for 10 to 12 h at 29°C and remove the filter with sterile forceps. Suspend the bacteria in a 0.9% NaCl solution by vigorous shaking.
7. Plate the dilutions of the suspension onto selective medium plate containing appropriate antibiotics.
8. Purify the transconjugants by single-colony isolation and characterize the transconjugants by the plasmid content.

H. Transformation of Plants with Recombinant Gene Constructs

Due to the complexity of the variables behind the transformation of different plant species, it is impossible for any one chapter to detail them all. The following sections, however, are good examples for the primary methods used to generate transgenic plants. For media preparation and tissue culture techniques, please refer to **Chapter 17** for more details.

1. Gene Transfer by *Agrobacterium tumefaciens*

Leaf disc transformation, mediated by *A. tumefaciens* containing recombinant Ti plasmids is a classic, well-established method used for gene transfer in plants. It has the added benefit of introducing intact transgenes into the plant genome without the problem of fragmentation sometimes associated with other methods. Procedures vary greatly from species to species. A good general protocol is as follows:

a. Preparation of *A. tumefaciens*

1. Three days prior to transformation, plate the *A. tumefaciens* strain harboring the desired construct on LB media containing the appropriate antibiotics for both the *A. tumefaciens* strain used and the contained construct (e.g., 25 µg/ml gentamycin for Agro C58 and 50 µg/ml kanamycin for the construct). Allow the colonies to grow for 2 days at 29°C.
2. Inoculate 30 ml of LB media containing the appropriate antibiotics with a single colony of *A. tumefaciens* from the plate. Allow this culture to grow overnight with shaking at 29°C.
3. Just prior to transformation, dilute the cells 1:20 in medium A to obtain a culture with approximately 5×10^8 cells/ml.

b. Transformation of Leaf Disks

1. In a Laminar flow hood, remove six to eight of the youngest, fully expanded leaves from plants grown in a greenhouse or under sterile conditions. If plants were originally grown from sterilized seeds under sterile conditions, proceed to step 3.
2. For greenhouse-grown plants, sterilize the leaves by immersing them in 5 to 10% Clorox solution containing 2 to 3 drops of Tween-20 surfactant for 15 to 20 min followed by thoroughly rinsing with 40 ml sterile distilled water four to five times to remove the Clorox.
3. Punch leaf discs with a sterile paper punch or sterile cork borer. Metal hole punches and cork borers can be sterilized in an autoclave prior to use in the Laminar flow hood.
4. Immerse the leaf discs in the overnight-cultured *A. tumefaciens* (1:20 dilution in medium A, 5×10^8 cells/ml) bearing recombinant Ti plasmids. Allow to inoculate overnight at 28°C followed by gentle blotting to dryness with sterile filter paper.
5. Prepare nurse culture plates by adding 1.5 ml of a tobacco cell suspension culture as a "feeder layer" to 25 ml of solid medium B in a Petri plate. Swirl the plate so that the medium is evenly covered with the cell suspension and cover with an apropriate size of sterile Whatman No. 1 filter paper.

6. Transfer the inoculated leaf discs upside-down on the nurse culture plates and allow to incubate for 2 to 4 days at 28°C. If *Agrobacterium* growth is highly noticeable, then this co-cultivation should be terminated. In this case, proceed to the next steps.

c. **Selection and Regeneration of Plants**

1. Transfer the explants to medium C and incubate at 28°C and illumination at 500 lux for several weeks depending on the plants' species. Subculture the explants every 2 weeks until shoots become visible.

Note: The appropriate selection antibiotic and antibiotics to kill the Agrobacteria need to be added to media C and D. For example, one can add 100 mg/L kanamycin sulfate for selection as well as 500 mg/L carbenicillin and 100 mg/L cefotaxime to remove Agrobacteria.

2. Transfer the entire explants to medium D when shoots appear and incubate at 28°C and 500 lux until young shoots have elongated.
3. Individually cut shoots from the callus and place upright in medium D to induce root formation at 28°C.
4. Remove the plantlets from the plates, wash the agar, and plant in sterile soild in 2.5-in. (6.35 cm) diameter pots. Place the pots in Magents™ boxes and close tightly to keep humidity for 8 to 10 days at 3000 to 5000 lux.
5. Transfer the boxes to a greenhouse, slowly open the lid to gradually reduce the humidity, and allow the plants to acclimatize to the ambient light and humidity.
6. Fertilize plants and allow to grow in a standard greenhouse.
7. Select stable transgenic lines as described in **Section II.H.3.b**.

TABLE 19.2
Preparation of Media (per liter)

Components	Medium			
	A	B	C	D
Inorganic salts (Gibco BRL)	4.3 g	4.3 g	4.3 g	4.3 g
MS vitamins	—	—	2.5 ml	2.5 ml
B5 vitamins	5 ml	5 ml	—	—
myo-Inositol	—	—	100 mg	100 mg
Sucrose	30 g	30 g	40 g	40 g
Gel-rite	—	—	1.5 g	2 g
Phytagar	8 g	8 g	—	—
Hormones				
1-Naphthyleneacetic acid (NAA) (1 mg/ml)	—	0.1 ml	0.1 ml	—
6-Benzylaminopurine (BAP) (1 mg/ml)	—	1.0 ml	1.0 ml	—

Note: Dissolve well in 700 ml distilled water, adjust the pH to 5.6 with 1 N KOH for each medium, and add distilled water to 1 L. Autoclave. When the medium has cooled to approximately 55°C, add the appropriate selection antibiotic and antibiotics to kill the *Agrobacteria* to media C and D. For example, add 100 mg/L kanamycin sulfate, 500 mg/L carbencillin, and 100 mg/L cefotaxime to media C and D.

2. Direct Gene Transfer to Protoplasts by Electroporation

The plant cell wall is a barrier to the introduction of foreign DNA into plant cells. To increase the efficiency of transformation, electroporation is widely employed to transfer genes of interest into plant protoplasts, which are without cell walls. Media preparation is outlined in Table 19.3.

TABLE 19.3
Preparation of Media

Components	Medium			
	A	B	K3	H
1. Macroelements (mg/ml)				
KNO_3	1010	900	2500	1900
KH_2NO_3	136	170	—	170
NH_4NO_3	800	1650	250	600
$NaH_2PO_4 \cdot H_2O$	—	—	150	—
$CaCl_2 \cdot 2H_2O$	440	440	900	600
$MgSO_4 \cdot 7H_2O$	740	370	250	300
$(NH_4)_2 \cdot SO_4$	—	—	134	—
$NH_4 \cdot$Succinate	50	—	—	—
2. Microelements (µg/ml)				
Na_2EDTA	74.6	74.6	74.6	74.6
$FeCL_3 \cdot 6H_2O$	27	27	27	27
H_3BO_3	3	6.2	3	3
KI	0.75	0.83	0.75	0.75
$MnSO_4 \cdot H_2O$	10	16.9	10	10
$MnSO_4 \cdot 7H_2O$	2	8.6	2	2
$CuSO_4 \cdot 5H_2O$	0.025	0.025	0.025	0.025
$Na_2MoO_4 \cdot 2H_2O$	0.25	0.25	0.25	0.25
$CoCl_2 \cdot 6H_2O$	0.025	—	0.025	0.025
$CoSO_4 \cdot 7H_2O$	—	0.03	—	—
3. Vitamins and organics (µg/ml)				
myo-Inositol	100	100	100	100
Biotin	—	—	—	0.01
Pyridoxine HCl	1	—	1	1
Thiamine HCl	10	0.04	10	10
Nicotinamide	—	—	1	—
Nicotinic acid	1	—	1	—
Folic acid	—	—	0.4	—
D-Ca-Pantothenate	—	—	1	—
p-Aminobenzoic acid	—	—	0.02	—
Choline chloride	—	—	1	—
Riboflavin	—	—	0.2	—
Ascorbic acid	—	—	2	—
Vitamin A	—	—	0.01	—
Vitamin D3	—	—	0.01	—
Vitamin B12	—	—	0.02	—
Glycine	—	—	0.1	—

(continued)

TABLE 19.3 (CONTINUED)
Preparation of Media

Components	Medium			
	A	B	K3	H
4. Sugars (mg/ml)				
Sucrose	30	30	103	0.25
Glucose	—	—	68.4	—
Mannitol	50	—	0.25	—
Sorbitol	—	—	0.25	—
Cellobiose	—	—	0.25	—
Fructose	—	—	0.25	—
Mannose	—	—	0.25	—
Rhamnose	—	—	0.25	—
Ribose	—	—	0.25	—
Xylose	—	—	0.25	—
5. Hormones ($\mu g/ml$)				
2,4-Dichlorophenoxy-acetic acid (2,4-D)	—	—	0.1	0.1
1-Naphthyleneacetic acid (NAA)	0.1	—	1	1
6-Benzylaminopurine (BAP)	1	—	0.2	0.2
6. Adjust pH				
Final pH	5.7	5.8	5.8	5.8

Note: Prepare media in the order shown.

a. Isolation of Protoplasts

1. Remove six to eight of the youngest, fully expanded leaves from plants grown in a greenhouse or under sterile conditions. Peel off the lower epidermis using a pair of forceps and place the leaf tissue in a petri dish.
2. Sterilize leaves from greenhouse-grown plants by immersing them in 5 to 10% Clorox solution for 5 to 10 min followed by thoroughly rinsing with 40 ml sterile distilled water four to five times to remove the Clorox.
3. Add 5 to 10 ml of sterile enzyme medium, mix, and incubate in the dark at room temperature (24 to 25°C) for 18 to 20 h without shaking.

Note: At this stage, the cell walls are hydrolyzed by hydrolytic enzymes such as cellulase, hemicellulase, and pectinase. This can be monitored by looking at them under a microscope. Hydrolyzed cells have only protoplasts.

4. Add 15 ml of washing medium and gently shake to loosen the protoplasts from undigested leaf material.
5. Filter through a nylon mesh (50-μm pore diameter) to remove undigested materials. The protoplasts are in the filtrate solution.
6. Centrifuge the protoplasts at 100 g for 5 min at room temperature and carefully decant the supernatant.
7. Resuspend the protoplasts in 4 ml of washing medium and centrifuge as in step 6.
8. Resuspend the protoplasts in 1 ml washing medium, add 1 ml of 18% sucrose as an underlayer in the protoplast suspension, and centrifuge at 120 g for 5 min at room temperature.
9. Carefully transfer the protoplasts from the interface, using a wide-bore Pasteur pipette, to a fresh centrifuge tube and add 1 ml of cell suspension buffer.

10. Count the protoplasts under a microscope using a haemacytometer.
11. Centrifuge at 100 g for 5 min at room temperature and resuspend the protoplasts at approximately 5×10^7 protoplasts/ml in the cell suspension buffer. Proceed to **Section II.H.2.b**, "Transfer of Gene Constructs into Protoplasts by Electroporation."

b. Transfer of Gene Constructs into Protoplasts by Electroporation

1. Measure the resistance of the protoplast suspension by adding 0.35 ml of the suspension to the chamber of the electroporator. Add an appropriate amount of $MgCl_2$ solution to adjust the resistance from an initial 1 to 4 kV to 1 to 1.1 kV.
2. Heat shock the protoplasts at 45°C for 5 min and cool to room temperature, after which they should be placed on ice.
3. Aliquot 250 µl of the protoplast suspension into three to five sterile disposable tubes. Add 20 µl of DNA solution to be transferred and 125 µl of PEG solution. Mix well and let stand for 10 to 15 min.
4. Transfer the samples to the chamber of the electroporator and pulse three times at 10-sec intervals with an initial pulse field strength of 1.4 kV/cm.
5. Transfer each sample to a 6-cm-diameter petri dish in a laminar-flow hood and let stand for 10 min.
6. Add 3 ml of a 1:1 mixture of K3 and H media containing 0.6% (w/v) SeaPlaque™ agarose to the dish. Gently mix the protoplasts in the agarose medium and allow to harden without any disturbance. Proceed to **Section II.H.2.c**, "Selection and Regeneration of Transgenic Plants."

c. Selection and Regeneration of Transgenic Plants

1. Seal the dishes containing the electroporated protoplasts with Parafilm™ to prevent potential contamination. Incubate at 24°C for 24 h in the dark followed by 4 to 6 h of continuous dim light (500 lux), depending on the plant species.
2. Cut the agarose-containing protoplasts into small quadrants using a clean razor blade and place the agarose blocks from one dish into a culture vessel (10 cm in diameter and 5 cm in depth) containing 50 ml of medium A with 50 µg/ml of kanamycin sulfate for selection of transformants. Culture at 24 to 28°C in continuous dim light (500 lux) with shaking at 80 rpm. Resistant clones will be visible 3 to 4 weeks later, depending on the plant species.
3. When the clones are 2 to 3 mm in diameter (5 to 6 weeks after culture), transfer the clones to medium A containing 0.8% (w/v) agar with 30 g/ml mannitol and 50 µg/ml of kanamycin sulfate. Allow the colonies to grow for 2 to 5 weeks, depending on the plant species.
4. Transfer the colonies to medium A without mannitol and allow to grow for 2 to 3 weeks.
5. Induce the root formation by culturing the colonies in medium A without mannitol but with 20 µg/ml sucrose and 0.25 µg/ml BAP (6-benzylaminopurine) hormone present. Incubate the dishes in the dark for 1 week, followed by illumination at 3000 to 5000 lx until shoots are generated from the callus.
6. Cut off the shoots (1 to 2 cm long) from the callus and place on medium B lacking hormones in order to produce roots. When the shoots are 3 to 5 cm long, proceed to step 7.
7. Gently wash away the agar once the root system is established and transfer the plantlets to pots of soil. They can then be gradually grown in a regular greenhouse.
8. Select stable transgenic lines using methods similar to those found in **Section II.H.3.b**.

3. Direct Gene Transfer by Microprojectile Bombardment

Microparticle bombardment, using a biolistic or particle gun, is still a powerful technique used in direct gene transfer into plant cells. This method involves the acceleration of heavy microparticles coated with recombinant genes. It has several major advantages over other methods: (1) it is relatively simple and easy to handle; (2) one shot can simultaneously transfer genes into many cells; (3) target cells can be cultured cells, pollen, and those in differentiated tissues or meristems of different plant species; and (4) random hits of microparticles can reach competent cells and increase the frequency of stable transformation. One disadvantage, however, is that this method requires expensive bombardment equipment. Also, when bombardment is carried out with tungsten particles, rather than with gold particles, tungsten is not as stable as gold after washing with ethanol. Consequently, while gold particles can be stored for long periods of time, tungsten particles must be made fresh each time. Media preparation is outlined in Table 19.4.

TABLE 19.4
Preparation of Media (for 1 L)

Components	Medium			
	A	B	C	D
Murashige and Skoog				
(MS) inorganic salts (Gibco BRL)	4.3 g	4.3 g	4.3 g	4.3g
MS vitamins	—	2.5 ml	2.5 ml	2.5 ml
B-5 vitamins	5 ml	—	—	—
myo-Inositol	—	100 mg	100 mg	100 mg
Sucrose	30 g	40 g	40 g	40 g
Gel-rite	—	1.5 g	2 g	2 g
Phytagar	8 g	—	—	—
Hormones				
1-Naphthyleneacetic acid (NAA) (1 mg/ml)	0.1 ml	0.1 ml	—	—
6-Benzylaminopurine (BAP) (1 mg/ml)	1.0 ml	1.0 ml	—	—

Note: Dissolve well in 700 ml distilled water, adjust the pH to 5.6 with 1 *N* KOH for each medium, and add distilled water to 1 L. Autoclave. When the medium has cooled to approximately 55°C, add kanamycin sulfate to each selection medium: 100 mg to media B, 50 mg to medium C for root induction and 200 mg to medium C for selection of transgenic seeds.

a. **Transfer of Genes into Leaf Tissue**

1. Remove two to six fully expanded, young leaves from plants of interest that have grown under sterile conditions or in a greenhouse. In the latter case, the leaves must be surface-sterilized in five volumes of 10% Clorox solution for 10 to 15 min in a laminar-flow hood, followed by thorough rinsing four to five times in five volumes of sterile distilled water. Peel off the lower epidermis of the leaf using a pair of jeweler's forceps (optional step).
2. Slice the leaves into approximately 1×0.5-cm pieces using a clean razor blade. Transfer the excised leaf pieces onto Grade 617 Whatman filter paper in disposable petri dishes (60×20 mm) containing 15 ml callus medium with 100 µg/ml kanacycin (medium A). Orient the leaf pieces in the center of each dish for maximal exposure to bombardment.
3. Carry out macroprojectile bombardment as follows: (Steps a through j should be completed prior to step 3.)
    a. Sterilize 100 mg tungsten (1.2-µm microprojectiles) in 1.5 ml of 95% ethanol in a sterile 15-ml centrifuge tube for 5 min.
    b. Sonicate on ice for 10 min with a continuous pulse using a 20% duty cycle at level 2 output.
    c. Transfer the sonicated microprojectiles into a microcentrifuge tube and centrifuge at 12,000 g for 2 min. Decant the ethanol supernatant and gently resuspend the pelleted microprojectiles in 1.5 ml ddH$_2$O.
    d. Centrifuge as in step c. Decant the supernatant, add 1.5 ml ddH$_2$O, and recentrifuge.
    e. Remove the supernatant, resuspend in 1.5 ml ddH$_2$O, and sonicate the vial. Aliquot 25 µl of the samples into microcentrifuge tubes. Resonicate after every two aliquots to maintain uniform bead concentration for each aliquot.
    f. Add 10 µl of gene constructs (1 µg/µl) to each aliquot of microprojectiles and mix well.
    g. Add 25 µl of 2.5 *M* CaCl$_2$ to each DNA/microprojectile mixture and mix well.
    h. Add 10 µl of 100 m*M* spermine to the mixture, mix by vortexing, and let it set for 20 to 30 min.
    i. Centrifuge at 12,000 g for 2 min and carefully remove the supernatant to a final volume of 30 µl.
    j. Sonicate the DNA/microprojectile mixture and pipette 1.5 to 2.0 µl onto a sterile macroprojectile. Resonicate after every two aliquots.
    k. Place the macroprojectiles in the gun barrel and the power level 1 blank in the chamber. After inserting the stopping plate and tissue sample in place, attach the detonator and draw vacuum.

Gene Transfer and Expression in Plants

Fire the gun when the vacuum reaches 68 to 71 cm Hg. (See the manufacturer's instructions for details; DuPont Co., Wilmington, DE.)

4. After bombardment, transfer the dishes to a growth room at 28°C with a 12- to 16-h day length at 100 $\mu E \cdot m^{-2} sec^{-1}$, depending on the plant species, and maintain the bombarded leaf strips on the callus medium (medium A) for 2 to 3 weeks. Callus colonies should be visible.
5. Transfer the leaf strips to regeneration medium (medium B) and allow to grow for 2 to 4 weeks with transfer to fresh medium every 2 weeks. Plantlets will develop from the callus.
6. In a laminar flow hood, separate the plantlets from the callus by cutting their bases with a sterile razor blade. Transfer the plantlets in sterile Flow boxes containing 50 ml of root-inducing medium with 50 μg/ml kanamycin (medium C). Wrap the boxes with Parafilm to prevent contamination. Allow the plantlets to grow in the boxes until root formation occurs and shoot growth is sufficient for the explants to be transferred to a greenhouse, depending on the particular plant species. Changing to fresh medium is not necessary.

b. Selection of Stably Transgenic Plants

1. Transfer the generated plants (Ro) from the tissue culture environment (Step 6 in **Section II.H.3.a**, "Transfer of Genes into Leaf Tissue") under water to 10-cm diameter pots containing a porous soil mix of Terra-lite/Redi-Earth/Peat-lite mix:Perlite™ (1:2). Allow the plants to grow and gradually acclimatize by placing them in a mist chamber set at 25 to 28°C day/20°C night with a 12- to 16-h day length at 200 $\mu E \cdot m^{-2} sec^{-1}$ light intensity, depending on the particular plant species.
2. Transfer the established plants to pots and grow in a greenhouse to maturity. Depending on the plant species, fertilize the plants two to three times a week with fertilizer (e.g., Peters' 20-20-20 N/P/K fertilizer) and feed blood meal once a week.
3. Allow flower buds from each plant to self-pollinate and others to cross as males or females to control plants. Initiate test cross-pollinations by removing the anthers from the designated female parent and pollen from the designated male parent. Transfer the dehisced anther with forceps and touch it to the receptive stigma.
4. Harvest seeds from self- and cross-pollinated plants and carry out surface sterilization with Clorox as previously described.
5. Allow the seeds (16) to germinate in 100 × 25 mm petri dishes containing 15 to 25 ml of medium C with 200 μg/ml kanamycin. Plant seeds on medium D lacking kanamycin to check germination frequency.
6. Transfer the dishes to a growth room and allow to germinate over a period of 2 to 3 weeks at 28°C under approximately 200 $\mu E \cdot m^{-2} sec^{-1}$ of light and a 12- to 16-h day length.
7. Score the phenotypes of the germinated seedlings as kanamycin-resistant (e.g., green) and control without kanamycin (e.g., white). Calculate the genetic ratios and analyze the data using rhe χ^2 (Chi square) statistical method. For example, the phenotypic ratios for differing numbers of independently segregating genes can be expected as follows:
   - For self-pollinated plants: one gene, 3 green:1 white; two genes, 15 green:1 white; three genes, 63 green:1 white.
   - For cross-pollinated plants: one gene, 1 green:1 white; two genes, 3 green:1 white; three genes, 15 green:1 white.

I. Proof of Stable Transformation

In order to avoid possible artifacts of transformation, stable transgenic plants should be established for several generations. Some genes transferred into the nucleus of a plant cell can be expressed to produce messenger ribonucleic acid (mRNA) and proteins in the first generation of transgenic plants. However, the introduced genes may not be integrated into the genome of plant cell, resulting in the absence of their expression in the second or further generations. Therefore, it is necessary to establish stable transformation using appropriate antibiotics in the medium and then find strong evidence to verify the existence of transgenic plants. Generally, the following analyses should be carried out for clear proof of transformation.

1. Phenotypic and/or Functional Examinations

This is a straightforward observation of transgenic and nontransgenic plants. For example, if an insect pest-resistant gene was transferred into crop plants, stable transgenic plants (usually two to four generations) should reveal resistance to attack by the insect as compared with control plants. If a red pigment leaf color gene is introduced into a plant, the stable transgenic plants should have red leaves. In another words, if the phenotype and functions of stable transgenic plants are expected, the gene of interest has been successfully transferred into and expressed in the transgenic plants. However, this is not the only proof, because the introduced gene may somehow not be expressed in transgenic plants. In that case, further evidence should be provided.

2. Analysis of the Expression of the Reporter Gene

A gene that encodes for ß-glucuronidase (GUS) is usually used as a reporter gene in the chimeric gene constructs for transformation especially those used to test promoter function and activity. The enzyme hydrolyzes 4-methyl-umbelliferyl-ß-D-glucuronide (MUG) and produces 4-methyl-umbelliferone (4-MeU), generating a blue fluorescence. The enzyme can be assayed by quantitative measurement or by tissue staining.

a. Quantitative Assay of GUS Activity

1. Homogenize or grind 5 to 10 mg of tissue from transgenic and nontransgenic plants in 0.1 to 0.2 ml of lysis buffer until a fine homogenate is obtained. The lysis buffer is made up of 50 mM sodium phosphate (pH 7.0), 10 mM EDTA, 0.1% (v/v) Triton X-100, 10 mM 2-mercaptoethanol, and 0.1% (w/v) sarcosyl.
2. Transfer the homogenate to a microcentrifuge tube and centrifuge at 12,000 g for 15 min at 4°C. Transfer the supernatant to a fresh tube and store at 4°C until use.
3. Carry out protein measurement (see **Chapter 3**).
4. Determine the background fluorescence from nontransgenic tissue, which may reveal nonspecific hydrolysis of the substrate. Determine the maximum amount of protein from the protein extracts, which can be used for the GUS assay using pure GUS as a positive control.
5. Carry out the fluorescence assay for the sample as follows:
    a. Add protein extract to two microcentrifuge wells containing 2 and 4 µg total proteins, respectively.
    b. Add lysis buffer to final volume of 45 µl in each well.
    c. Add 5 µl of MUG to start the reaction.
    d. Cover the plate and incubate at 37°C for 30 min.
    e. Add 150 µl of 0.2 M Na$_2$CO$_3$ to terminate the reaction.
    f. Read the fluorescence and substract the blank value.
    g. Calculate the specific activity (units/ng protein/min) of GUS in the tissue based on the total proteins and incubation time.

b. Staining Assay Followed by Light Microscopy

1. Fix whole plant seedlings or slice tissues in 25 mM sodium phosphate buffer (pH 7.0) containing 0.1 to 1% (v/v) glutaraldehyde for 30 min at 4°C or at room temperature.
2. Wash the tissue for 3 min five times in 25 mM sodium phosphate buffer (pH 7.0).
3. Quickly and completely cover the tissue with the substrate mixture and vacuum filter. Incubate in the dark at 37°C overnight or for 1 day. The substrate mixture can be made from stock solution to a working concentration of 10 mM sodium phosphate buffer (pH 7.0), 0.5 mM potassium ferricyanide, 0.5 mM potassium ferrocyanide, and 1 mM X-glucuronide (5-bromo-4-chloro-3-indolyl glucuronide).
4. Rinse the tissue for 5 min twice in sodium phosphate buffer until the tissue shows an intense blue color.
5. Directly observe and photograph the stained tissue under a microscope without further processing. To improve the image, the pigments (chlorophyll) can be removed by passing the tissue through 25, 50, 70, 95, 100, 100, and again 100% ethanol with 15 min per step. The tissue can be rehydrated through ethanol series and progressively infiltrated with glycerol. Remove any air bubbles by final vacuum filtration.

3. Analyses of Southern Blotting and Sequencing, and Determination of the Copy Number of the Gene

Genomic DNA should be isolated from transgenic plants and digested with appropriate restriction enzymes. The fragments are separated by agarose gel electrophoresis followed by transfer to a nitrocellulose or a nylon membrane. The membrane is then hybridized with a labeled DNA probe that is the introduced gene or gene fragment. Stable transgenic plants should show a positive signal as compared with nontransgenic plants as control. Further, the positive fragment(s) of DNA may be sequenced. If the sequence shows both host and foreign DNA sequences, the introduced gene has been integrated into the chromosomes of the host cell. This is strong evidence of stable transformation. The detailed procedures for Southern blot hybridization and DNA sequencing are described in **Chapter 5** and **Chapter 11**, respectively.

Method a: Determination of the Copy Number of the Gene Integrated in the Genome of Transgenic Plants

In most cases, a standard Southern blot using genomic DNA from the transgenic plant and a probe specific for the transgene will reveal the copy number of the transgene through the number of individual bands seen on the Southern blot. Each integration event normally results in one copy of the introduced gene at a specific location in the plant genome. However, in cases where the transgene copies are very close to each other, as with tandem repeats, a standard Southern blot may not be able to distinguish the copies. In these cases, this method and the methods that follow are of use.

1. Isolate genomic DNA from transgenic plants and measure the concentration of DNA in the sample. The general procedures are described in **Chapter 1**.
2. Calculate the quantity of DNA that corresponds to a specific copy number. One copy of the gene per haploid genome is equivalent to the size of the transferred gene (g) in base pairs divided by the size of the genome (G) in base pairs, or g/G. Therefore, if 1 μg of the genomic DNA is used for the experiment, the quantity of the transferred gene will be equal to g/G × 1 μg.
3. Set up copy-number standards of known gene equivalents of the query sequence. If the query sequence, the gene, or cDNA insert (I) is purified from a recombinant plasmid, the amount of micrograms (N) that is equivalent to one copy per genome of the target sequence can be calculated as: I × N = g/G × 1 μg. If I is in the cloning vector such as plasmid (V), the N can be calculated as: I/V × N = g/G × 1 μg. Prepare a set of copy-number standards that is equivalent to 1, 2, 3, 4, 5, 10, and 15 copies of the target or transferred gene.
4. Carry out restriction enzyme digestion of 1 μg of genomic DNA using an enzyme that cannot cut within the target sequence.
5. Load the digested genomic DNA and a set of copy-number standards into separate wells of an agarose gel. Carry out electrophoresis and Southern blot hybridization using labeled cDNA, the monomeric sequence of a repetitive DNA family, or a linearized plasmid DNA containing the insert as a probe.
6. Analyze the detected band(s) by scanning with an integrating densitometer. The copy number of the transferred gene in the genome can be estimated by measuring the extent of hybridization signals of genomic bands and comparing the intensity of the signal with those of known gene standards. Thus, the total copy number can be determined with ease. With a family of tandem repeats, the sum of the extent of all hybridized bands is the total number of gene copies, as determined from the copy-number standards.

Method b: Determination of the Copy Number of the Dispersed, Repetitive DNA Sequence

1. Carry out dot blot or slot blot hybridization as described in **Chapter 5**.
2. Scan the autoradiograph with a densitometer.
3. Calculate the percent (T%) of DNA that is complementary to the probe or gene transferred as follows:

$$T\% = [Sg/Svi] \times [Avi/Ag] \times [Lv/Li] \times 100$$

where Sg represents the signal intensity of genomic DNA; Svi refers to the signal intensity of vector including the DNA insert; Avi is the amount (μg) of vector plus DNA insert used for the dot blotting; Ag is the amount (μg) of genomic DNA used for the dot blotting; Lv represents the size (bp) of vector DNA minus the insert; and Li is the size (bp) of insert DNA used as the probe in the dot blot hybridization.

4. Determine the copy number of dispersed, repetitive DNA sequences as follows:

$$\text{Copy number} = T\% \times \text{genome size (bp)/monomer size (bp)}$$

Method c: Determination of Multigene Families

1. Digest the genomic DNA with an appropriate restriction enzyme that does not cut within the target gene or DNA insert used as a probe.
2. Carry out electrophoresis, Southern blot hybridization, and autoradiography as described in **Chapter 5**.
3. Analyze the data as follows:
   a. If one band is detected, the target gene or DNA exists once in the genome, or it may be a highly homogeneous multigene family.
   b. If multiple, discrete bands appear, and if the distance between the bands is larger than the size of the known gene, a multiple gene family may exist. But the multiple bands may not derive from a multiple gene family if the sizes of the bands are very close to each other, or if the restriction enzyme cuts within introns.
   c. If there are no discrete bands, but rather, a smear appears, the target gene or DNA is a highly repetitive, dispersed DNA family.
   d. If the sizes of a ladder of bands are equal to multiples of the smallest fragment, the target gene or DNA is composed of a tandemly organized set of repeated units, in which the monomeric length is probably equal to the smallest DNA fragment size.
   e. If a multiple set of bands is superimposed on a smear, the probe may detect a tandemly repeated gene family that has dispersed units.

4. Northern Blot Analysis

The purpose of gene transfer is to obtain expected gene expression in transgenic plants. This can be carried out by Northern blot hybridization. RNA can be isolated from transgenic plants and separated by formaldehyde-denaturing agarose gel electrophoresis followed by transfer to nitrocellulose or nylon membrane. The membrane is then hybridized with a labeled DNA probe that is the introduced gene or gene fragment. Stable transgenic plants should show a positive signal as compared with nontransgenic control plants. The detailed procedures for Northern blot hybridization analysis are described in **Chapter 6**. When assessing the expression of a reporter gene caused by the introduction of a promoter of interest, Northern blot analysis is usually not required since the expression can be assessed by the reporter gene product (see **Section II.I.2**).

5. Western Blot Hybridization or Immunocytochemical Localization

This method provides direct evidence for the expression of a transferred gene in transgenic plants. Protein extracts can be prepared from transgenic and nontransgenic plants and separated by SDS-PAGE (sodium dodecyl sulfate- polyacrylamide gel electrophoresis). The separated proteins are then transferred onto a nitrocellulose or PVDF membrane followed by probing with an antibody against the protein encoded by the introduced gene. A positive signal should be evident in stable transgenic plants, but not in nontransgenic plants. In addition, the expression of the introduced gene can be checked by immunocytochemical localization *in situ*. That will give detailed information as to where the protein is localized in the cell or the tissue. The detailed procedures for Western blot and immunolocalization analyses are described in **Chapter 8** and **Chapter 22**, respectively.

Reagents Needed

7.5 M Ammonium Acetate
Dissolve 57.8 g ammonium acetate in 50 ml ddH$_2$O and adjust volume to 100 ml.

Calf Intestinal Alkaline Phosphatase (CIAP)
Mix and store at room temperature.

10X CIAP Buffer
0.5 M Tris-HCl, pH 9.0
10 mM MgCl$_2$
1 mM ZnCl$_2$
10 mM Spermidine

Cell Suspension Buffer
10 mM Tris-HCl, pH 7.6
100 mM EDTA, pH 8.0
20 mM NaCl

Chloroform:Isoamyl Alcohol (24:1)
48 ml Chloroform
2 ml Isoamyl alcohol

Clorox Solution
5 to 10% (v/v) Clorox in ddH$_2$O

CPW-Salts Solution (1 L)
KH$_2$PO$_4$ (27.2 g)
KI (0.16 mg)
CuSO$_4$·5H$_2$O (0.025 mg)
KNO$_3$ (0.101 g)
MgSO$_4$·7H$_2$O (0.246 g)

10 mM dATP

T4 DNA Ligase
2 Weiss units/μl

DNA Polymerase I (Klenow Fragment)
1 unit/μl

2 mM dNTP Mmixture
2 mM Each of dATP, dCTP, dGTP, and dTTP in 20 mM Tris-HCl, pH 7.5

0.5 M EDTA, pH 8.0
Dissolve 93.06 g Na$_2$ EDTA·2H$_2$O in 300 ml ddH$_2$O. Adjust pH to 8.0 with 2 N NaOH solution. Adjust final volume to 500 ml with ddH$_2$O and store at room temperature.

Enzyme Medium

    9% (w/v) Mannitol
    3 mM 2-(N-Morpholino)-ethane-sulfonic acid (MES)-KOH, pH 5.8
    1% (w/v) Cellulase
    0.2% (w/v) Macerozyme
    Make up in CPW-salts solution.

LB Medium (Rich Medium)

    1% (w/v) Tryptone
    0.5% (w/v) Yeast extract
    0.8% (w/v) NaCl
    Adjust the pH to 7.0. Autoclave.

5X Ligase Buffer

    0.25 M Tris-HCl, pH 7.5
    50 mM MgCl$_2$
    5 mM ATP
    5 mM DTT
    25% (v/v) PEG, 8000

10X 5′ Overhang Buffer

    500 mM Tris-HCl, pH 7.2
    100 mM MgSO$_4$
    1 mM DTT
    500 µg/ml BSA

PEG Solution

    24% (w/v) Polyethylene glycol (molecular weight 6000) in 0.4 M mannitol buffer (pH 5.6) containing 20 mM MgCl$_2$

Phosphate-Buffered Saline (PBS)

    NaCl (8 g)
    KCl (0.2 g)
    Na$_2$HPO$_4$ (1.44 g)
    KH$_2$PO$_4$ (0.24 g)
    Dissolve well after each addition in 800 ml ddH$_2$O.
    Adjust pH to 7.4 with 2 N HCl and add ddH$_2$O to 1 L.
    Autoclave and store at room temperature.

T4 Polynucleotide Kinase

    1 unit/µl

10X Polynucleotide Kinase Buffer

    500 mM Tris-HCl, pH 7.5
    100 mM MgCl$_2$
    50 mM DTT
    1 mM Spermidine
    1 mM EDTA

10X Sma I Buffer

    0.1 M Tris-HCl, pH 7.8
    0.5 M KCl
    70 mM MgCl$_2$
    10 mM dithiothreitol (DTT)

3 M Sodium Acetate Buffer, pH 5.2

    Dissolve sodium acetate in ddH$_2$O and adjust pH to 5.2 with 3 M glacial acetic acid. Autoclave and store at room temperature.

Sucrose Solution

    18% (w/v) Sucrose
    3 mM MES-KOH, pH 5.8
    Make up in CPW-salts solution. Autoclave.

TE Buffer

    10 mM Tris-HCl, pH7.6
    1 mM EDTA, pH 7.6

TE-Saturated Phenol/Chloroform

B-5 Vitamins (for 100 ml)

    Inositol (2000 mg)
    Thiamine-HCl (200 mg)
    Pyridoxine-HCl (20 mg)
    Nicotinic acid (20 mg)

MS Vitamins (for 100 ml)

    Thiamine-HCl (2 mg)
    Pyridoxine-HCl (10 mg)
    Nicotinic acid (10 mg)
    Glycine (40 mg)
    Aliquot into appropriate containers and store at –20°C until use.

Washing Medium

    3 mM MES-KOH, pH 5.8
    2% (w/v) KCl
    Make up in CPW-salts solution. Autoclave.

References

1. **Gruber, M.Y. and Crosby, W.L.,** Vectors for plant transformation, in *Methods in Plant Molecular Biology and Biotechnology,* Glick, B.R. and Thompson, J.E., Eds., CRC Press, Boca Raton, FL, 1993.
2. **Miki, B.L., Fobert, P.F., Charest, P.J., and Iyer, V.N.,** Procedures for introducing foreign DNA into plants, in *Methods in Plant Molecular Biology and Biotechnology,* Glick, B.R. and Thompson, J.E., Eds., CRC Press, Boca Raton, FL, 1993.
3. **Greenberg, B.M. and Glick, B.R.,** The use of recombinant DNA technology to produce genetically modified plants, in *Methods in Plant Molecular Biology and Biotechnology,* Glick, B.R. and Thompson, J.E., Eds., CRC Press, Boca Raton, FL, 1993.

4. **Potrykus, L.,** Gene transfer to plants: assessment of published approaches and results, *Annu. Rev. Plant Physiol. Plant Mol. Biol.*, 42, 205, 1991.
5. **Gasser, C.S. and Fraley, R.T.,** Genetically engineering plants for crop improvement, *Science,* 244, 1293, 1989.
6. **Heikkila, J.J.,** Use of *Xenopus* oocytes to monitor plant gene expression, in *Methods in Plant Molecular Biology and Biotechnology,* Glick, B.R. and Thompson, J.E., Eds., CRC Press, Boca Raton, FL, 1993.
7. **Brunke, K.J. and Meeusen, R.L.,** Insect control with genetically engineered crops, *Trends in Biotechnol.,* 9, 197, 1991.
8. **Saul, M.W., Shillito, R.D., and Negrutiu, L.,** Direct DNA transfer to protoplasts with and without electroporation, in *Plant Molecular Biology Manual,* Vol. A1, Kluwer Academic Publishers, Dordrecht, 1988.
9. **Christou, P., Ford, T.L., and Kofron, M.,** Production of transgenic rice (*Oryza sahva*, L.) plants from agronomically important Indica and Japonica varieties via electric discharge particle acceleration of exogenous DNA into immature zygotic embryos, *BioTechnology,* 9, 957, 1991.
10. **Vaeck, M., Regnaerts, A., Hofte, H., Jansens, S., De Beuckeleer, M., Datn, C., Zabeau, M., Van Montagu, M., and Leemans, J.,** Transgenic plants protected from insect attack, *Nature,* 328, 33, 1987.
11. **Cullis, C.A., Rivin, C.J., and Walbot, V.,** A rapid procedure for the determination of the copy number of repetitive sequences in eukaryotic genomes, *Plant Mol. Biol. Rep.,* 2, 24, 1984.
12. **Koncz, C., Martini, N., Mayerhofer, R., Koncz-Kalman, Z., Korber, H., Redei, G.P., and Schell, J.,** High-frequency T-DNA-mediated gene tagging in plants, *Proc. Natl. Acad. Sci., U.S.A.,* 86, 8467, 1989.
13. **Jefferson, R.A.,** Assaying chimeric genes in plants: the GUS gene fusion system, *Plant Mol. Biol. Rep.,* 5, 387, 1987.
14. **Fobert, P.R., Miki, B.L., and Iyer, V.N.,** Detection of gene regulatory signals in plants revealed by T-DNA-mediated fusions, *Plant Mol. Biol.,* 17, 837, 1991.
15. **McCabe, D.E., Swain, W.F., Martinell, B.J., and Christou, P.,** Stable transformation of soybean *(Glycine max)* by particle acceleration, *BioTechnology,* 6, 923, 1988.

Chapter 20

Inhibition of Gene Expression

Jaegal Shim, Junho Lee, Leland J. Cseke, and William Wu

Contents

I.	Comparison of Gene Suppression Technologies	434
II.	Antisense Oligonucleotides	435
	A. Synthesis of Antisense Oligonucleotides	435
	B. Treatment of Cultured Cells Using the Antisense Oligomers and Determination of the Optimium Dose of the Oligomers	437
	C. Analysis of the Inhibition of Gene Expression Using the Optimium Dose of the Oligomers	437
	Reagents Needed	438
III.	Antisense Orientation of the Gene of Interest	438
	A. Preparation of Target DNA, Strong Promoters, Enhancers, Poly(A) Signals, and Vectors	441
	1. Preparation of Target DNA	441
	2. Selection of Promoter, Enhancer, Poly(A) Signal, and Appropriate Vectors	442
	3. Preparation of Blunt-End DNA	443
	4. Preparation of DNA Insert Lacking the 5′ Phosphate Group	444
	B. Blunt-End Ligation of Plasmid Vectors and Inserts of Antisense and Sense Orientation	444
	C. Transformation of Appropriate Strain of Bacteria To Amplify the Recombinant Plasmids	445
	D. Selection of Plasmids with Antisense and Sense Orientations	446
	E. Gene Transfer and Expression of Antisense RNA	446
	Reagents Needed	447
IV.	The New Potential of Double-Strand RNA for the Suppression of Gene Expression	449
	A. Insights into the Mechanisms of Double-Strand RNA	449
	B. RNA Interference Using *in vitro* Transcribed Double-Strand RNA	450
	1. Preparation of DNA Template	450
	2. RNA Preparation	451

	3.	Preparation of Host Animals	451
	4.	Microinjection of dsRNA	451
	5.	Analysis of Progeny	452
C.	Alternative Methods of RNA Interference		453
	1.	Feeding of Bacterial Cells That Produce dsRNA	453
	2.	Soaking in dsRNA Solution	454
D.	Inducible RNA Interference		454
	1.	Heat Shock-Induced RNA Interference	454
	2.	Hypoxia-Induced RNA Interference	455
E.	Troubleshooting Guide		455
Equipment and Reagents Needed			456
References			458

I. Comparison of Gene Suppression Technologies

There are currently a variety of approaches used to silence the expression of endogenous genes. For years it has been demonstrated that gene expression can be suppressed by either synthetic antisense oligonucleotides or by expression of the antisense orientation of the gene of interest.[1-4] Although the detailed mechanisms of antisense deoxyribonucleic acid/ribonucleic acid (DNA/RNA) inhibition are not well understood, they have been widely applied in eukaryotes in which the antisense oligonucleotides or antisense RNA (asRNA) transcripts inhibit the expression of specific sense RNAs. It is believed that this inhibition is due to the interaction of the antisense sequence with the endogenous complementary or sense RNA transcripts through hydrogen bonding. This interaction is believed to block the processing of translation of the sense RNA, and the duplex of antisense DNA and sense RNA may also be rapidly degraded within the cells. Therefore, the basic principle of using antisense DNA and RNA can be directed toward partially or completely inhibiting the expression of a target gene.[2-7]

Unfortunately, there are no useful rules or guidelines for engineering effective antisense constructs. Consequently, antisense gene silencing tends to be an unpredictable technology,[8] and there is little current research directed at understanding and improving the use of antisense gene silencing. Three limitations are commonly encountered with antisense gene silencing: (1) antisense constructs often silence the target gene in only a small minority (10 to 30%) of transformants; (2) antisense constructs typically cause only partial gene silencing, i.e., they produce "leaky" phenotypes; and (3) antisense gene silencing often produces epigenetically unstable phenotypes, necessitating extensive screening and vigilant monitoring to identify and maintain stable lines of transformants. Leaky phenotypes, for example, are thought to be produced when antisense RNA does not accumulate to levels sufficient to interfere with all target transcripts in the appropriate cellular compartments and cell types. Another possible cause of partial silencing is that many antisense RNAs do not always pair efficiently with the target RNA due to interference by secondary structures in the asRNA and/or the target RNA.[9,10] Despite these potential difficulties, antisense technology has been profoundly useful in elucidating gene functions when the proper attention to detail is maintained.

There are, however, other alternatives to antisense gene suppression. A large amount of attention has been given to the phenomenon of cosuppression by sense transgenes. Here, a construct typically designed to cause constitutive overexpression of a specific gene has the surprising affect of suppressing the endogenous homologs of the gene. The phenomenon may be particularly useful in plant species. However, although sense transgenes can be much more efficient at gene silencing than antisense transgenes, the cosuppression phenomenon still only occurs in a small percentage of transformants, even in the best conditions (~30%). In addition, efficient silencing requires that sense transgenes be engineered for efficient transcription and translation.[9-11] This leads to a potential complication when applied toward gene silencing, i.e., ectopic expression of the transgene product. Although ectopic expression will not occur if silencing occurs in all tissues expressing the transgene, ectopic expression will potentially occur in any tissue in which transgene expression does not reach the threshold necessary to trigger silencing. Thus, with sense suppression, there is a trade-off between efficiency of silencing in a population of transformants and the potential for undesirable side effects.

More recently, it has been reported that the expression of inverted repeat sequences results in highly efficient silencing of homologous, endogenous transcripts in as much as 90% of the transformants.[11-13] The published experiments in plants on inverted-repeat transcripts suggest that such transcripts are highly efficient inducers of post-transcriptional gene silencing (PTGS). This effectiveness appears to be due to the production of double-strand RNA via interaction between sense and antisense transcripts in a similar fashion as RNA interference (RNAi) in *Caenorrhabditis elegans* and quelling in *Neurospora crassa*.[14,15] Two distinct inverted repeat-transcript approaches have been seen to cause silencing: (1) an inverted repeat in the 5′ or 3′ UTR, not homologous to the target RNA,[11] and (2) an inverted duplication of coding sequences of the transcript.[12,15] However, such constructs are known to inhibit the expression of homologous genes with as little as 70% nucleic acid identity with the target gene. Thus, like with antisense technology, the resulting transgenics must undergo extensive screening and vigilant monitoring of the true effects of the transgenes.

The mechanisms of each of these gene-suppression systems seem to be based on the fact that most organisms have defense systems that limit the effects of abnormal or multicopy gene expression, such as expression caused by viruses.[15,16] Indeed, there appears to be a common mechanism behind the phenomena of PTGS, RNAi, quelling, and cosuppression (see **Section IV** for more details). Consequently, while the antisense approach continues to be useful in addressing fundamental questions in animals, human diseases, and plants, the new technologies based on PTGS and RNAi also promise to provide excellent results for silencing specific genes. This chapter focuses on the techniques using exogenous antisense oligonucleotides in animal cell cultures (**Section II**) and antisense gene constructs (**Section III**) to control gene expression. We then turn to the newer technology of RNAi in animal systems as a means to reduce the expression of specific genes (**Section IV**).

II. Antisense Oligonucleotides

Antisense oligomers can be chemically synthesized based on any target sequence of DNA or RNA of interest. Consequently, gene expression might be regulated at different steps, such as DNA replication and transcription; retroviral RNA reverse transcription; pre-mRNA (messenger RNA) splicing and polyadenylation; tRNA (transfer RNA) translation; and mRNA transport, degradation, and translation. This section, however, primarily focuses on the blocking or inhibition of translation of specific mRNAs.

A. Synthesis of Antisense Oligonucleotides

1. Appropriate antisense oligomers are designed based on the target sequence of mRNA (sense RNA) or its first-strand cDNA. For example, if the target sequence is 5′-CAUGCCCCUCAACGUUAGC-3′, the antisense oligonucleotides should be designed as 3′-GTACGGGGAGTTGCAATCG-5′. They are complementary to each other and can form a hybrid as follows:

5'-CAUGCCCCUCAACGUUAGC-3' mRNA

3'-GTACGGGGAGTTGCAATCG-5' antisense oligomer

    a. **Target region(s):** Oligonucleotides can be designed to correspond to the following regions of the target sequence:
        i. The upstream region of the initiation codon AUG, including the first codon AUG, has proved to be an effective target site when designing an antisense oligomer. The hybridization of the antisense oligomers may prevent the ribosome from interacting with the upstream region of the initiating codon AUG, thus blocking the initiation of translation.
        ii. Antisense oligonucleotides can also be designed to be complementary to the coding regions, such as conserved motif(s) to block the chain elongation of the polypeptide of interest. However, it is less effective as compared with (i).
        iii. The terminal region of translation can be chosen as a target to block the termination of translation. The partial-length polypeptide may not be folded properly and may then be easily degraded.

    The most effective results appear to come from designing antisense oligomers that are complementary to the upstream region, including the first codon AUG, and to the terminal site of the coding region. The efficiency of inhibition of specific gene expression can be as high as 95 to 100%.

    b. **Oligomer modification:** Conventional or unmodified oligonucleotides are usually substrates for nucleases (*DNases* and *RNases* such as *RNase* H), and therefore they cannot effectively block the expression of the gene of interest. To prevent oligonucleotides from being degraded by nucleases, the antisense oligomers should be chemically modified prior to use. Active compounds such as cross-linking and cleaving reagents can be coupled to the oligomers. This causes the oligomers to be much more stable by causing irreversible damage to the target sequence, and subsequently resulting in very effective blocking of translation. For example, oligomers can be conjugated to an alkylating group or to photosensitizers. An antisense oligomer linked to either a nuclease or chemical reagents can generate sequence-dependent cleavage of the target RNA following the binding of the oligomer to its complementary sequence. The cleavage site in the target RNA is usually in the area surrounding the modified antisense oligomer. This prevents the translation of the target RNA. Chemical groups linked to the end of the oligomers can introduce strand breaks either directly (e.g., metal chelates and ellipticine) or indirectly (e.g., photosensitizers) after alkaline treatment. Metal complexes such as Fe^{2+}-EDTA, Cu^+-phenanthroline, and Fe^{2+}-porphyrin can produce hydroxyl radicals (OH) that can oxidize the sugar and result in phosphodiester bond cleavage. Phosphodiester internucleoside linkages can be substituted with phosphotriester, methylphosphonate, phosphorothioate, or with phosphoroselenoate. Oligomers that are modified to phosphorothioates and phosphoroselenoates are resistant to nuclease (*RNase* H) action and can induce cleavage of the target RNA. For this reason, phosphorothioate analogs have been widely applied in intact cells.

Note: *There are also reports of modified oligonucleotides designed to produce site-specific mutagenic disruption of gene function through the formation of triple helical structures in the endogenous DNA. This technology is not discussed in this chapter, since it works through a different mechanism, but it may be of interest to the reader.*

    c. **Oligomer size:** Oligomers between 15 to 26 base-pairs (bp) have been found to be most effective at gene silencing, depending on the particular sequences.
    d. **Oligomer amount:** A range from 10 to 100 μg total oligomer is effective in gene silencing, depending on the particular experiments.

2. Synthesize and purify appropriate antisense oligonucleotides according to the instructions supplied with the synthesizer. Alternatively, oligonucleotides can be commercially produced by ordering them from a multitude of DNA technology companies. Such companies may also be able to provide

Inhibition of Gene Expression

the desired oligonucleotide modifications discussed above, or there are a number of kits available to introduce these modifications (e.g., the OligofectAMINE™ reagent from Invitrogen, Carlsbad, CA).

B. Treatment of Cultured Cells Using the Antisense Oligomers and Determination of the Optimium Dose of the Oligomers

1. Under sterile conditions, cultured cells of interest are grown in a prewarmed medium such as modified Dulbecco's modified Eagle medium (DMEM) containing 10 to 15% (v/v) heat-inactivated fetal bovine serum (FBS) or 5 to 10% (v/v) heat-inactivated newborn calf serum with appropriate antibiotics such as 10^5 units/l penicillin and 0.01 and 0.05% (w/v) streptomycin. Incubate the cells at 37°C in 5% CO_2 and 95% air and saturated with water vapor. Maintain the cells in logarithmic phase.
2. Spin down the cells at 100 to 1000 g for 5 min under sterile conditions. In a sterile cabinet, carefully remove the supernatant and resuspend the cells in a minimum volume of fresh medium. Determine the viability and density of the cells by counting Trypan blue-excluding cells in a hemocytometer.
3. Carry out dose-response assays of synthetic antisense oligomers as follows in order to optimize the amount of the oligomers to be used to block the expression of the gene of interest:
   a. Resuspend each sample of 2×10^6 cells in 2 to 5 ml fresh medium contained in 24-well culture plates or T-flasks.
   b. Add the oligomers to the medium for the following samples:
      - Sample no. 1: no antisense oligomer as a negative control
      - Sample no. 2: 6 μM antisense oligomers
      - Sample no. 3: 4 μM antisense oligomers
      - Sample no. 4: 2 μM antisense oligomers
      - Sample no. 5: 1 μM antisense oligomers
      - Sample no. 6: 0.5 μM antisense oligomers
      - Sample no. 7: 0.25 μM antisense oligomers

Note: *The initial concentration of appropriate oligomers varies with the particular experiment.*

   c. Incubate the cells for 2 to 6 days. Then, add the same amount of the oligomers to appropriate samples every day, and gently mix in the medium.
   d. Centrifuge at 1000 g for 5 to 10 min. If the protein of interest is a secretory protein, the supernatant should be used for extraction of proteins. If not, the pelleted cells of each sample should be washed with appropriate buffer, lysed, and extracted for protein analysis. The general procedures for protein extraction from cultured cells are described in **Chapter 3**. The protein of interest may be purified from the total proteins with a specific antibody against this protein by the immunoprecipitation method. We recommend using total proteins for inhibition analyses and employing an antibody against an appropriate housekeeping gene product as a control.
   e. Analyze the inhibition of the protein of interest by Western blotting. Detailed procedures for doing this are given in **Chapter 8**. An antibody against actin or tubulin should be used as the control that should show equal amounts of protein in each sample.

C. Analysis of the Inhibition of Gene Expression Using the Optimium Dose of the Oligomers

The example that follows assumes that the optimum dose of oligomers was determined to be 2 μM in the optimization procedure detailed in **Section II.B**.

1. Resuspend each sample of 2×10^6 healthy cells in 2 to 5 ml fresh medium contained in 24-well culture plates or T-flasks.
2. Add the optimium amount of antisense oligomers (e.g., 2 μM) to the medium of each sample and then culture for the appropriate periods of time as follows:
   - Sample nos. 1 to 8: no oligomer as controls (0, 0.5, 1, 2, 3, 4, 5, and 6 day)
   - Sample no. 9: 2 μM antisense oligomers for 0 day
   - Sample no. 10: 2 μM antisense oligomers for 0.5 day
   - Sample no. 11: 2 μM antisense oligomers for 1 day
   - Sample no. 12: 2 μM antisense oligomers for 2 days
   - Sample no. 13: 2 μM antisense oligomers for 3 days
   - Sample no. 14: 2 μM antisense oligomers for 4 days
   - Sample no. 15: 2 μM antisense oligomers for 5 days
   - Sample no. 16: 2 μM antisense oligomers for 6 days
3. Harvest the cells at the appropriate time-points by centrifuging at 1000 g for 10 min, and analyze for the presence of the protein of interest using the methods detailed in **Section II.B** steps 3.d and 3.e in the optimization procedure above.

Reagents Needed

DMEM, Heat-inactivated Fetal Bovine Serum (FBS), and Heat-inactivated Newborn Calf Serum (NCS)

Each of these is best ordered from Sigma Chemical Co. (St. Louis, MO) and used within two months of receiving.

III. Antisense Orientation of the Gene of Interest

Synthetic antisense oligonucleotides can effectively inhibit the expression of a specific gene. However, the effect of inhibition is only temporary and is usually applied to cultured cells. To continuously down-regulate the gene expression, it is necessary to keep adding synthetic antisense oligonucleotides, which is time-consuming and expensive. In order to produce stable transformants that can express antisense RNA, constructs containing the antisense orientation of a specific gene have been quite useful. Generally, the cloned target DNA is first altered *in vitro* either by complete inversion of the target sequence within the existing vector or by transferring an appropriate promoter from upstream to downstream of the coding region.

The cloning method is typically as follows: (1) The DNA of interest has usually already been cloned into an appropriate vector (usually a plasmid-based vector) and propagated in an appropriate *E. coli* strain for analysis. (2) Recombinant plasmids will then be isolated, and the DNA fragments containing the target sequence can be removed from the plasmids using restriction endonucleases that produce either overhanging or blunt ends. (3) These fragments are religated to the vector, which is then used to transform *E. coli* again (Figure 20.1). (4) The sense and antisense orientations of new plasmid preparations can be selected on the basis of appropriate restriction enzyme(s) digestion patterns (Figure 20.2). (5) The altered constructs or appropriate sections of these constructs will then be transferred into appropriate cell lines, totipotency cells, or binary vectors that can then be used to generate transgenic animals or plants in which expression of the target gene should be significantly inhibited by the overexpressed antisense RNA. The transfer and analysis of the expression of transferred genes are described in **Chapter 18** and **Chapter 19**.

Inhibition of Gene Expression

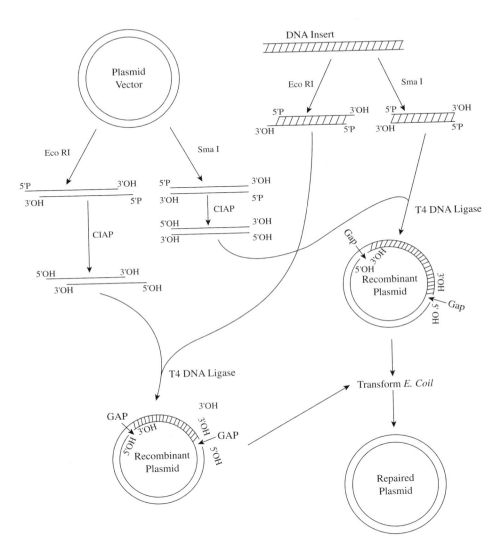

FIGURE 20.1
Schematic representation of strategies for DNA recombination and transformation of *E. coli* with recombinant constructs. (From Kaufman, P.B., Wu, W., Kim, D., and Cseke, L.J., *Handbook of Molecular and Cellular Methods in Biology and Medicine*, 1st ed., CRC Press, Boca Raton, FL, 1995.)

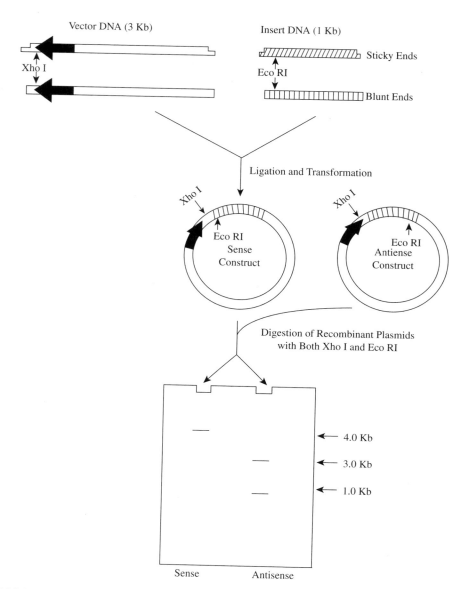

FIGURE 20.2
Strategies for making both sense and antisense DNA constructs and for identifying both constructs by patterns of appropriate restriction enzyme digestions. (From Kaufman, P.B., Wu, W., Kim, D., and Cseke, L.J., *Handbook of Molecular and Cellular Methods in Biology and Medicine*, 1st ed., CRC Press, Boca Raton, FL, 1995.)

Inhibition of Gene Expression

A. Preparation of Target DNA, Strong Promoters, Enhancers, Poly(A) Signals, and Vectors

1. Preparation of Target DNA

The target DNA should already be subcloned and characterized, and the internal restriction enzyme sites should be well established. For eukaryotic genes, the target DNA selected is usually the coding region of the cDNA of interest in order to prevent potential interference of intron sequences. The target DNA should first be removed from the subcloned plasmids using appropriate restriction enzyme(s) that cannot cut within the target sequence.

1. Set up, on ice, a standard restriction enzyme digestion reaction as follows:
   Single-restriction enzyme digestion:
   - Plasmids (20 µg)
   - 10X Appropriate restriction enzyme buffer (10 µl)
   - 1 mg/ml Acetylated bovine serum albumin (BSA) (optional) (10 µl)
   - Appropriate restriction enzyme (3.4 units/µg DNA)
   - Add distilled, deionized water (ddH$_2$O) to a final volume of 100 µl.

   Double-restriction enzyme digestion:
   - Plasmid DNA or DNA to be inserted (20 µg)
   - 10X Appropriate restriction enzyme buffer (10 µl)
   - 1 mg/ml Acetylated BSA (optional) (10 µl)
   - Appropriate restriction enzyme A (3.4 units/µg DNA)
   - Appropriate restriction enzyme B (3.4 units/µg DNA)
   - Add ddH$_2$O to a final volume of 100 µl.

Note: For double-restriction enzyme digestions, the appropriate 10X buffer containing higher NaCl concentration than the other buffer may be chosen.

2. Incubate at an appropriate temperature, depending on the restriction enzyme used, (e.g., 37°C) for 2 to 3 h.
3. Purify the target DNA from the digested vector by gel elution following protocols in **Chapter 1** or the following protocol:
   a. Carry out electrophoresis as described in Southern blotting (**Chapter 5**), except use 1% low-melting-point agarose containing ethidium bromide (EtBr) instead of normal agarose. Monitor the separation of the target fragment and vector as compared with known DNA sizes.
   b. Transfer the gel onto a long-wavelength ultraviolet (UV) transilluminator (305 to 327 nm) to visualize the DNA bands. Excise the band(s) of interest with a clean razor blade and place the slices in a microcentrifuge tube.

Note: (1) The gel should not be placed on a short-wavelength (e.g., <270 nm) UV transilluminator, as the UV light may cause internal breaking of the DNA fragment, and thus significantly inhibit the subcloning of the DNA insert of interest. (2) UV light is harmful. Protective glasses, gloves, and a lab coat should be worn when working with UV light. (3) Unstained regions of the gel slices containing the DNA bands of interest should be trimmed as much as possible.

   c. Add two volumes of TE buffer to the gel slices and completely melt the gel in a 60 to 70°C water bath.

Note: *The gel slices can be directly melted without adding any TE buffer. The DNA concentration is usually high, but the total yield of the eluted DNA fragment is much lower than when TE buffer is added.*

d. Immediately chill the melted gel solution on dry ice and place the tube at −70°C for at least 20 min.
e. Thaw the gel mixture by tapping the tube vigorously. It takes about 5 to 10 min to thaw the gel to obtain a suspension.
f. Centrifuge at 11,000 g for 5 min at room temperature.
g. Carefully transfer the liquid phase containing the elution DNA fragment to a fresh tube. The DNA solution can be directly used for labeling except at low concentration.
h. Extract EtBr three to four times with three volumes of water-saturated *n*-butanol.
i. Precipitate the DNA by adding 0.1 volume of 3 *M* sodium acetate buffer (pH 5.2) and 2.5 volumes of chilled 100% ethanol to the DNA solution. Allow the DNA to precipitate at −70°C for 1 h and centrifuge at 12,000 g for 5 min at room temperature.
j. Discard the supernatant and briefly rinse the pellet with 1 ml of 70% ethanol. Dry the DNA under vacuum for 15 min and dissolve the DNA in an appropriate volume of ddH$_2$O or TE buffer. Store at −20°C until use.

2. Selection of Promoter, Enhancer, Poly(A) Signal, and Appropriate Vectors

In order to obtain highly efficient inhibition of the target RNA, the antisense RNA transcripts should be overexpressed using a strong promoter. An enhancer, inserted upstream of the promoter, can also be used, and 3′ polyadenylation signal should be used to provide polyadenylation of the transcript. For animal systems, an SV40 promoter, enhancer, and poly (A) signal, as well as retrovirus long-terminal repeats (LTRs) containing both a promoter and an enhancer, are recommended. For plant systems, the cauliflower mosaic virus (CaMV) 35S promoter and a poly(A) signal for the nopaline synthase gene of *Agrobacterium tumefaciens* Ti plasmid are recommended. These regulatory fragments are well-characterized, subcloned in plasmids, and are available from Promega Corp. (Madison, WI), Clontech Laboratories, Inc. (Palo Alto, CA), or other companies and laboratories. Such regulatory fragments can also be purified from plasmids using appropriate restriction enzyme(s) and subcloned into new vectors, but this is not typically necessary, since the appropriate promoter, enhancer, or even a reporter gene are built into the commercial plasmid vectors such as pBI121 and pBI221 for plant systems (Clontech Laboratories, Inc.) and retrovirus-based plasmid vectors for animal systems. Generally, there is a unique restriction enzyme site such as *Sma* I (pBI121 and pBI221) or *Hind* III (pGL2-promoter vector) downstream of the promoter, which allows the insertion of antisense- or sense-oriented DNA. In this case, the plasmids can be linearized with *Sma* I or *Hind* III for the ligation with the DNA fragments isolated using the protocol in **Section III.A.1**. In addition, some commercial vectors contain the gene neo^r, or NPT II for plants, as a selectable marker of stable transformants by using appropriate antibiotics (e.g., G418 or Kanamycin) in the culture medium.

1. Set up, on ice, a standard enzyme digestion mixture as follows:
   Sma I
   - Plasmid vectors or DNA to be inserted (10 μg)
   - 10X *Sma* I buffer (10 μl)
   - 1 mg/ml Acetylated BSA (optional) (10 μl)
   - *Sma* I (30 units)
   - Add ddH$_2$O to a final volume of 100 μl.

   Hind III
   - Plasmid vectors or DNA to be inserted (10 μg)
   - 10X *Hind* III buffer (10 μl)
   - 1 mg/ml Acetylated BSA (optional) (10 μl)
   - *Hind* III (30 units)
   - Add ddH$_2$O to a final volume of 100 μl.

2. Incubate at 37°C for 2 to 3 h.

Note: The digestion efficiency can be checked by loading 1 µg of the digested DNA (10 µl) and undigested plasmids with loading buffer to a 1% agarose minigel. After electrophoresis, the undigested plasmid DNA may reveal multiple bands because of different supercoiled DNA. However, one band will be visible after Sma I or Hind III digestion.

3. Carry out CIAP (calf intestinal alkaline phosphatase) treatment, as described below, to remove the 5′ phosphate group in order to prevent the recircularization of the linearized plasmid.
   - Linearized plasmids (90 µl)
   - 10X CIAP buffer (15 µl)
   - CIAP diluted in 10X CIAP buffer (0.01 unit/pmol ends)
   - Add ddH$_2$O to a final volume of 150 µl.

   The calculation of the amount of ends is as follows:
   If there are 9 µg digested DNA left after taking 1 µg of 10 µg digested DNA for checking on an agarose gel, and if the vector is 3.2 kb (kilobase-pairs), then the amount of ends can be calculated by using the following formula:

$$\text{pmol ends} = \frac{\text{amount of DNA}}{\text{base pairs} \times (600 / \text{base pair})} \times 2$$

$$= \frac{9}{3.2 \times 1000 \times 660} \times 2$$

$$= 4.2 \times 10^{-6} \times 2$$

$$= 8.4 \times 10^{-6} \mu M$$

$$= 8.4 \times 10^{-6} \times 10^{-6} = 8.4 \text{ pmol ends}$$

Note: CIAP and 10X CIAP should be kept at 4°C. CIAP treatment should be set up at 4°C.

4. Incubate at 37°C for 1 h and add 2 µl of 0.5 M EDTA buffer (pH 8.0) to stop the reaction.
5. Extract with one volume of phenol/chloroform/isoamyl alcohol (25:24:1). Mix well by vortexing for 1 min and centrifuge at 11,000 g for 5 min at room temperature.
6. Carefully transfer the top, aqueous phase to a fresh tube and add one volume of chloroform:isoamyl alcohol (24:1) to the supernatant. Mix well and centrifuge as in step 5.
7. Carefully transfer the upper, aqueous phase to a fresh tube and add 0.1 volume of 3 M sodium acetate buffer (pH 5.2) or 0.5 volume of 7.5 M ammonium acetate to the supernatant. Briefly mix and add 2 to 2.5 volumes of chilled 100% ethanol to the supernatant. Allow precipitation to occur at −70°C for 1 h or at −20°C for 2 h.
8. Centrifuge at 12,000 g for 10 min and carefully decant the supernatant. Briefly rinse the DNA pellet with 1 ml of 70% ethanol and dry the pellet under vacuum for 15 min. Dissolve the DNA pellet in 20 to 40 µl ddH$_2$O. Take 4 µl of the sample to measure the concentration of the DNA at 260 nm. Store the sample at −20°C until use.

Note: Addition of 0.5 volume of 7.5 M ammonium acetate to the supernatant in step 7 yields a greater amount of DNA precipitation than by adding 0.1 volume of 3 M sodium acetate buffer (pH 5.2).

3. Preparation of Blunt-End DNA

If both vectors and digested DNA fragments have compatible sticky or overhanging ends at the right sites, they can be readily ligated to each other. If not, it is usually necessary to generate blunt ends on the vector and digested DNA fragments prior to their being ligated. For example, if the vector has sticky ends generated by *Hind* III and the DNA fragment has sticky ends generated by *Bam* HI, they will not be compatible for ligation. Both may be converted to blunt ends prior to any CIAP treatment (of the vector) by filling in 5′ overhangs with DNA polymerase I (Klenow fragment).

1. Set up the 5′ overhang reaction mixture in a microcentrifuge tube on ice as follows:
   - DNA fragment or vector DNA (1 to 4 µg in 4 µl)
   - 10X 5′ Overhang buffer (2 µl)
   - 2 mM dNTP mixture (0.5 to 2 mM each) (2 µl)
   - DNA polymerase I (Klenow fragment) (2 to 3 units)
   - Add ddH$_2$O to a final volume of 20 µl.
2. Incubate at room temperature for 20 min and heat at 70°C for 5 min to stop the reaction.
3. Add one volume of TE buffer (pH 7.5) and carry out phenol:chloroform:isomyl alcohol extraction and precipitation as described in steps 5 through 8 of **Section III.A.2**. The vector may now undergo CIAP treatment as detailed in the previous section.

4. Preparation of DNA Insert Lacking the 5′ Phosphate Group

If during preparation the DNA insert lacks the 5′ phosphate group, the following reaction should be carried out prior to carrying out ligation to blunt-ended and dephosphorylated vectors.

1. Add the phosphate group to the 5′ end of the DNA insert by the following kinase reaction:
   - Blunt-end, DNA insert (15 µl)
   - 10X Polynucleotides kinase buffer (3 µl)
   - 10 mM ATP (1.5 µl)
   - 100 mM DTT (1.5 µl)
   - T4 Polynucleotide kinase (15 units)
   - Add ddH$_2$O to a final volume of 30 µl.
2. Incubate at 37°C for 40 min and heat at 70°C for 10 min to inactivate the enzyme.
3. Add one volume of TE buffer (pH 7.5) and carry out extraction and precipitation as described in steps 5 through 8 of **Section III.A.2**.

B. Blunt-End Ligation of Plasmid Vectors and Inserts of Antisense and Sense Orientation

To achieve optimal ligation, the ratio of vector to DNA insert (1:1, 1:2, 1:3, and 3:1 molar ratios) is strongly recommended to be optimized by using a small-scale reaction. The following reaction protocol is standard for the ligation of 3.2 kb plasmid vector and 2.533 kb DNA insert.

1. Calculate molar weight of vector and DNA insert:
   1 M plasmid vector = 3.2 × 1000 × 660 = 2.112 × 10^6
   1 M DNA insert = 2.533 × 1000 × 660 = 1.672 × 10^6
2. Calculate the molar ratio of vector to insert DNA as follows:

Ratio of Vector DNA:Insert DNA	Amount of DNA (µg)	
	Vector	Insert
1:1	1	0.792
1:2	1	1.584
1:3	1	2.376
3:1	1	0.264

Inhibition of Gene Expression

3. Set up the ligations on ice:

Components	Ligation reactions			
	1 (1:1)	2 (1:2)	3 (1:3)	4 (3:1)
Plasmid DNA as vector (μg)	1	1	1	1
DNA insert (μg)	0.792	1.584	2.376	0.244
10X Ligase buffer (μl)	1	1	1	1
T4 DNA ligase (Weiss units)	4	4	4	4
Add ddH$_2$O to achieve final volume (μl)	10	10	10	10

Note: Restriction enzyme-digested plasmid (vector) and insert DNA should be dissolved in ddH$_2$O (nuclease-free) to 0.5 to 1.0 μg/μl. If the DNA is less than 0.4 μg/μl, it is recommended that the DNA be precipitated in order to dissolve in about 1 μg/μl.

4. Incubate the reactions at 4°C for 12 to 24 h, or at 16°C for 4 to 6 h, or at room temperature (22 to 25°C) for 1 to 2 h.

Note: After ligations are completed at one of the above temperatures, the mixture can be stored at 4°C until later use.

5. Check the efficiency of the ligations by performing 1% agarose gel electrophoresis. When the electrophoresis is complete, photograph the EtBr-stained gel under UV light. As compared with unligated vector or insert DNA wells, a highly efficient ligation should show less than approximately 10% unligated vector and insert-DNA by estimation of the intensity of fluorescence. Approximately 90% of the vector and insert DNA are ligated to each other and show strong band(s) with molecular-weight increases compared with the vector and insert DNA sizes. By comparing the efficiency of ligations using different molar ratios, the optimal conditions for carrying out the ligation can be determined with ease. These can then be used as a guide for large-scale ligation.

Note: The above small-scale ligation is often found to be optional.

6. Large-scale ligation of vector and insert DNA is based on determinations of the optimal conditions by small-scale ligations. For example, if one uses a 1:2 molar ratio of plasmid DNA:insert DNA as the optimal condition for ligation, a large-scale ligation can be carried out as follows:
   - Plasmid DNA as the vector (3 μg)
   - Insert DNA (4.75 μg)
   - 10X Ligase buffer (3 μl)
   - T4 DNA ligase (15 Weiss units)
   - Add ddH$_2$O to a final volume of 30 μl.
   - Incubate the ligation mixture at 4°C for 12 to 24 h, or at 16°C for 4 to 6 h, or at room temperature (22 to 25°C) for 1 to 2 h. Store at 4°C until use. Proceed now to transformation of bacteria strains (**Section III.C**).

C. Transformation of Appropriate Strain of Bacteria To Amplify the Recombinant Plasmids

The general procedures for transformation of bacteria and selection of transformants are described under the section on subcloning of plasmids in **Chapter 10**, and revisited in **Chapter 18** and **Chapter 19**.

D. Selection of Plasmids with Antisense and Sense Orientations

During the blunt-end ligations, each of the antisense or sense orientations of DNA typically has a 50% probability to be ligated to the vectors. However, they can be separated from each other on the basis of appropriate restriction enzyme(s) digestion patterns. For example, if *Xho* I is the closest enzyme to the blunt-end, and if the first restriction enzyme at the 5' end of the coding region of the sense orientation gene is *EcoR* I, the recombinant plasmids can be digested with both *Xho* I and *EcoR* I. After electrophoresis and comparison with vector and insert DNA as controls, the plasmids containing the sense-oriented DNA will show one large band that is almost the size of the vector plus the inserted DNA. One very small fragment that spans the short distance between *Xho* I and *EcoR* I may also be visible. However, the plasmids containing the antisense-oriented DNA should show two bands. One is almost the same size as the vector DNA, while the other is approximately the same size as the DNA insert (see Figure 20.2).

1. Isolate the recombinant plasmid DNA for transfection by the DNA miniprep method. The detailed procedures are described under the section on isolation and purification of plasmid DNA in **Chapter 1**.
2. Set up a double-restriction enzyme digestion using the miniprep DNA:
   - Plasmid DNA or DNA to be inserted (20 µg)
   - 10X Appropriate restriction enzyme buffer (10 µl)
   - 1 mg/ml Acetylated BSA (optional) (10 µl)
   - Appropriate restriction enzyme A (3.4 units/µg DNA)
   - Appropriate restriction enzyme B (3.4 units/µg DNA)
   - Add ddH$_2$O to a final volume of 100 µl.
3. Incubate at the appropriate temperature, depending on the restriction enzyme, (e.g., at 37°C) for 2 to 3 h.
4. Carry out electrophoresis on a 1% agarose gel as described in **Chapter 5, Section I.B** and identify antisense-oriented, recombinant plasmids. Culture the bacteria containing the antisense-oriented plasmids under large-scale conditions. Isolate the plasmids and store at –20°C until use for gene transfer.

E. Gene Transfer and Expression of Antisense RNA

After the recombinant constructs containing the antisense orientation of the gene of interest and selectable marker gene (e.g., neo^r or NPT II) have been obtained, the next procedure is to transfer the constructs into specific cell lines, animals, or plants of interest. Transformant cell lines, transgenic animals, or transgenic plants can be selected by using appropriate antibiotics. If all works well, the antisense RNA transcripts expressed in transformants will hybridize with the endogenous, sense RNA in a proportion of the final transformants. The translation of mRNA in these lines will then be blocked. The detail protocols of gene transfer and expression are described in **Chapter 18** and **Chapter 19**. When it comes to antisense technology, stable transformants must be identified and analyzed very carefully using the following criteria:

1. Successful growth of potential transgenics on antibiotics is not enough to conclude that integration of the entire expression construct is present. Evidence of successful integration of the antisense gene in the host genome must also be obtained using genomic polymerase chain reaction (PCR) and/or Southern blot analysis using the transferred gene as a probe. PCR is a quick method for identifying the inserts, but primers must be carefully selected in order to distinguish the transgene from the endogenous copy of the gene. This can be done by choosing primer pairs with one primer in the DNA of interest and the other in the surrounding vector regions. The results of such PCR should always be checked using genomic Southern blots, which provide additional information, such as an estimate of the copy number of the inserted DNA.

2. Evidence of the expression of the antisense RNA is also necessary. This can be done by extracting RNA from the potential transformants and performing Northern blot analysis using the transferred gene as a probe. If double-strand probes are produced, there is a possibility that endogenous gene transcripts may also be identified if the antisense procedure is not working. If it is desired to avoid this, then the production of single-strand probes specific to the antisense mRNA is highly recommended.
3. Evidence of the inhibition of the translation of the target sense RNA at the protein level is also necessary. While Northern blot analysis may suggest that the endogenous transcripts are reduced in quantity, there are many cases where very small levels of transcripts are still able to generate functional levels of the protein. Western blot analysis using an antibody raised against the gene product is the most critical proof of the inhibition of gene expression by antisense RNA.
4. Functional evidence resulting from changes in the phenotype of the transgenic system as a result of antisense RNA expression (as compared to controls) will strongly support that the suppression is working.

Finally, subsequent generations of the transformants will have to be carefully monitored for the above evidence before one concludes that each transgenic line is truly stable. The procedure and supporting evidence may seem tedious, but the method has been highly effective in a wide array of systems.

Reagents Needed

10 M Ammonium Acetate
Dissolve ammonium acetate in 50 ml ddH$_2$O and adjust volume to 100 ml.

Buffered Phenol
We strongly recommend purchase of saturated phenol (pH 8.0) from a commercial vendor. Most buffered phenol comes with a separate bottle of equilibration buffer to adjust the phenol phase to pH 8.0. To use, add the entire bottle of equilibration buffer to the phenol bottle and 0.1% (w/v) hydroxyquinoline (as an antioxidant). Stir the mixture for 15 min and allow the phases to separate (requires several hours) before use. Store at 4°C.

Caution: *Phenol is highly corrosive and can be absorbed through skin. Wear gloves and work in a fume hood when handling phenol.*

Calf Intestinal Alkaline Phosphatase (CIAP)

10X CIAP Buffer
0.5 M Tris-HCl, pH 9.0
10 mM MgCl$_2$
1 mM ZnCl$_2$
10 mM Spermidine

Chloroform:Isoamyl Alcohol (24:1)
48 ml Chloroform
2 ml Isoamyl alcohol
Mix and store at room temperature.

Dephosphorylated Linearized Plasmid Vectors

T4 DNA Ligase
2 Weiss units/μl

DNA Polymerase I (Klenow Fragment)
1 unit/μl

2 mM dNTP Mixture
2 mM Each of dATP, dCTP, dGTP, and dTTP in 20 mM Tris-HCl, pH 7.5

0.5 M EDTA
Dissolve 93.06 g Na$_2$EDTA·2H$_2$O in 300 ml ddH$_2$O. Adjust pH to 8.0 with 2 N NaOH solution. Adjust final volume to 500 ml with ddH$_2$O and store at room temperature.

10X Hind III Buffer
60 mM Tris-HCl, pH 7.5
0.5 M NaCl
100 mM MgCl$_2$
10 mM DTT

5X Ligase Buffer
0.25 M Tris-HCl, pH 7.5
50 mM MgCl$_2$
5 mM ATP
5 mM DTT
25% (v/v) PEG, 8000

10X 5' Overhang Buffer
500 mM Tris-HCl, pH 7.2
100 mM MgSO$_4$
1 mM DTT
500 μg/ml BSA

Phenol:Chloroform:Isoamyl Alcohol
Mix one part of buffered phenol with one part of chloroform:isoamyl alcohol (24:1). Top the organic phase with TE buffer (about 1 cm height) and allow the phases to separate. Store at 4°C in a light-tight bottle.

T4 Polynucleotide Kinase Buffer
1 unit/μl

10X T4 Polynucleotide Kinase buffer
500 mM Tris-HCl, pH 7.5
100 mM MgCl$_2$
50 mM DTT
1 mM Spermidine
1 mM EDTA

10X Sma I Buffer

    0.1 M Tris-HCl, pH 7.8
    0.5 M KCl
    70 mM MgCl$_2$
    10 mM DTT

3 M Sodium Acetate Solution (pH 5.2)

    Dissolve sodium acetate in ddH$_2$O and adjust pH to 5.2 with 3 M glacial acetic acid. Autoclave and store at room temperature.

TE Buffer

    10 mM Tris, pH 8.0
    1 mM EDTA, pH 8.0

Water-Saturated n-Butanol

IV. The New Potential of Double-Strand RNA for the Suppression of Gene Expression

A. Insights into the Mechanisms of Double-Strand RNA

Introduction of double-strand RNA (dsRNA) into cells has been shown to induce specific gene silencing in many diverse species. This phenomenon is termed RNA interference (RNAi). Ranging from trypanosomes to mice, the phenomenon of RNAi is widespread in eukaryotes and has sparked great interest from an applied standpoint as a method for facilitating temporary gene-targeted silencing. RNAi is effective in many invertebrate species, such as insects, trypanosomes, planaria and hydra, as well as in vertebrates, such as the mouse.[16–19] The mechanism of RNAi has been well documented in *Drosophila* and mammalian cell cultures, and it is now generally accepted that RNAi in *C. elegans*, quelling in *Neurospora crassa*, and posttranscriptional gene silencing (PTGS) or cosuppression in plants act posttranscriptionally, targeting sequence-specific RNA transcripts for degradation.[14–16] Indeed, their mechanisms are based on the fact that most organisms seem to have defense systems to limit the effects of abnormal or multicopy gene expression, such as viruses and transposons from dangerous pathogens.[15,16]

    RNAi, quelling, PTGS, and possibly even cosuppression have similarities in their molecular mechanisms in that they form dsRNA in one way or another. Recent experimental evidence shows that both strands of the dsRNA are processed to RNA segments 21 to 23 nucleotides in length.[19–21] These small interferring RNAs (siRNA) are generated by means of sequence-specific ribonuclease III cleavage from longer dsRNAs, and it is these siRNAs that trigger the silencing response.

    Besides the biological significance of the existence of the RNAi phenomenon, the significance of RNAi in terms of application in biological studies is immense. In systems where specific gene targeting is not available, performing RNAi can copy loss-of-function mutations in genes of interest, enabling researchers to postulate the biological functions of the genes. The methods of RNAi are largely divided in two ways: first, by introducing RNAs from the exterior, and second, by inducing dsRNA production *in vivo*. Introducing dsRNA from the exterior can be achieved by: (1) direct microinjection, (2) bacterial cell-mediated RNAi, and (3) soaking the animal in RNA solution (which may function as simply as allowing the animal to ingest the dsRNA). All three methods have been shown to be effective in *C. elegans*. The methods of inducing RNA production *in vivo* includes transformation of the organism with sense and antisense constructs simultaneously, or transformation of the organism with inverted repeat constructs containing the sense sequence connected with antisense sequence mediated by a loop. Such inverted repeat constructs can be expressed (1) under the control of a strong constitutive promoter, or (2) under promoters inducible by various conditions.

Most RNAi techniques were developed in the nematode, *C. elegans*, and these methods can be modified to the needs of investigators studying other biological fields, such as plant biology. Consequently, this section focuses on the methods of RNAi-mediated gene silencing in the *C. elegans* system.

B. RNA Interference Using *in vitro* Transcribed Double-Strand RNA

RNA interference using *in vitro* transcribed double-strand RNA is a rapid and effective method. The general procedure includes: (1) preparation of template DNA; (2) *in vitro* transcription and RNA preparation for microinjection; (3) preparation of the host animal for microinjection of double-strand RNA (dsRNA); (4) microinjection of dsRNA into the host animal; and (5) analysis of progeny.

1. Preparation of DNA Template

The following procedure must be done for both the sense and antisense orientations of the cloned cDNA or genomic DNA of interest.

1. Subclone an adequate length of cDNA or genomic DNA into a proper vector that has a T3 or T7 promoter near the multicloning site (usually the pBluescript vector will do, but there are many others that can be used).

Note: *(1) The RNAi effect depends on exonic sequences, homology, and length; so if one uses genomic DNA, it is good to include exonic sequence regions as much as possible. The length of the cloned region should be at least 200 bp (base-pairs) in C. elegans, and the proper range is between 500 to 1000 bp. (2) In the case of mammals, it is known that long RNA fragments can cause nonspecific mRNA degradation. Thus, using short dsRNA oligos with a length of 21 to 25 nucleotides is recommended for gene-specific RNAi.*

2. Digest at least 2 μg of the cloned plasmid DNA with a proper restriction enzyme to linearize it.
3. Add an equal volume of phenol/chloroform (1:1). After vortexing, centrifuge at 14,000 rpm for 5 min.
4. Transfer supernatant to a new eppendorf tube.
5. Add 1/10 volume of 3*M* sodium acetate and 2.5 volumes of 100% ethanol. Put the tube on ice or at −20°C for 10 min.
6. Centrifuge at 14,000 rpm for 10 min, and pour off the ethanol mixture.
7. Add 150 μl of ice-cold 70% ethanol and centrifuge at 14,000 rpm for 5 min.
8. Remove the 70% ethanol with a micropipet and dry the pellet briefly under vacuum for 10 min.
9. Resuspend the DNA pellet in 10 μl of *RNase*-free TE.
10. Check the concentration of the DNA by running 1 μl on a 1% agarose gel.

Note: *(1) PCR using T3 and T7 primers is more convenient in preparing the linearized template DNA than by digestion with restriction enzymes. In this case, in vitro transcription of RNA using a small amount of template DNA is more efficient. See* **Chapter 9** *and* **Chapter 10** *for more details on PCR.*

2. RNA Preparation

1. For each of the sense and antisense orientations of the clone, incubate 1 μg of linearized template DNA with a proper *in vitro* transcription mixture at 37°C for 1 to 4 h (usually the total reaction volume is 20 μl). Use of a commercial *in vitro* transcription kit is convenient and highly recommended. For example:

5X T7 or T3 RNA polymerase reaction buffer	4 μl
dNTPs (25 mM dATP, dGTP, dCTP, dUTP)	6 μl
T3 or T7 RNA polymerase enzyme mixture	2 μl
Linearized template DNA	1 μg
Total	up to 20 μl

2. After the incubation period, add 1.0 μl of *DNase* I (*RNase*-free), and incubate the sample(s) at 37°C for 15 min.

Note: *This step is not critical because the template DNA does not affect RNA interference.*

3. Perform phenol/chloroform extraction and ethanol precipitation (steps 3 to 8 in **Section III.B.1**).

Note: *Because the presence of free dNTPs and enzyme buffer does not affect RNAi experimentally, omission of phenol/extraction is also acceptable and convenient. In this case, one can centrifuge the in vitro transcribed RNA at 4°C for 10 to 20 min and remove the precipitates for microinjection.*

4. Resuspend the RNA pellet in a proper volume of *RNase*-free TE (to a final volume of 10 to 20 μl).
5. Check the concentration of the RNA.
6. Mix equal amounts of the sense and antisense RNA and incubate at 37°C for 10 min to anneal the dsRNA. (Alternatively, one can raise the incubation temperature up to 65°C and cool to facilitate the formation of double-strand RNA, but in our experience, mixing without incubation is as effective.)
7. Aliquot the dsRNA into small volumes (about 2 μl) and keep them at –70°C until use to protect the RNA from degradation.

3. Preparation of Host Animals

Unlike microinjection of DNA for the transformation of *C. elegans,* in which case DNA should be accurately delivered into the gonad of the host animal, injection of dsRNA for RNAi can be done either into the gonad or other tissues. It is known that injection of dsRNA in any tissue works as effectively as injection into the gonad. Although any adult worm can be used for injection, well-fed, clean, and healthy young hermaphrodites are best because they produce more progeny. To keep healthy worms, pick 20 to 50 well-fed L4 worms and place them onto a freshly-seeded NGM plate the evening before injection. dsRNA injection into larvae is also possible, and in this case, one may be able to see the RNAi effect in the injected Po animals. While the wild-type N2 strain is the strain of choice for the analysis of normal phenotypes, the *rde-1* mutant animals are the strain of choice when triggering zygotic RNAi. The *rde-1* homozygous mutants are resistant to RNAi; thus, if RNAi-affected *rde-1* mutant animals are mated with N2 males, then the progeny will show zygotic gene disruption without disrupting maternal gene expression.[22]

4. Microinjection of dsRNA

Microinjection of dsRNA follows the same procedure as DNA microinjection, except that it is not necessary to inject the RNA into the gonad.[23,24] Preparation for injection includes: (1) making agarose pads for injection, (2) preparation of an injection needle, (3) loading of dsRNA into the injection needle, (4) mounting worms on agarose pads, (5) injection, and (6) recovery of injected worms. It is critical to protect the RNA from degradation by *RNase* during microinjection. Because microinjection of dsRNA into any tissue including the intestine is as effective as injection into the gonad, the procedure is really quite simple.

1. **Making agarose pads:** Prepare 2% regular gelling-temperature agarose in water and melt the agarose completely. Using a Pasteur pipette or micropipette, add a drop of melted agarose on a coverslip, then immediately put another coverslip on top of the agarose, and press gently on the upper coverslip to flatten the agarose. Separate the two coverslips either by lifting the upper one or sliding the two apart. Leave the agarose pad in an 80°C drying oven for an hour or longer. Alternatively, air-dry the pads overnight at room temperature or at 37°C for several hours.

Note: The remaining 2% agarose can be stored for several days. However, one must cover the opening and add an appropriate amount of water to obtain the starting agarose concentration.

2. **Preparing an injection needle:** Ready-to-use injection needles are commercially available. Alternatively, needles for microinjection can be manually pulled using a needle puller (for example, PC10, Narishige, Japan). The end of a pulled injection needle is blocked; so, it must be broken to the proper size for injection. There are three methods of breaking the ends of injection needles. The first way is to break them with dried worms. When an animal is left on a plate for about 5 min, it dies and the body becomes stiff. One can break needles by contacting the needle to the dead body. The second way is to break them using the edge of the coverslip; and the third way is to break the needle using hydrofluoric acid. When using hydrofluoric acid, a drop of hydrofluoric acid:water (1:1) and a drop of water are placed on the cover of a plate. The needle is put on a drop of hydrofluoric acid under high air pressure, and as soon as a bubble is seen, the needle is moved to a drop of water.
3. **Loading dsRNA:** Transfer dsRNA solution into the injection needle with a mouth pipette or weak vacuum. Due to the viscosity of the dsRNA solution, it may take several minutes or more to load the needle with 0.2 to 0.5 μl of dsRNA solution. This allows the injection of more than 100 animals.
4. **Mounting worms:** Moisturize the dried agarose pad by briefly breathing over the surface. This process will also make it obvious on which side of the coverslip the agarose pad is fixed. Add a drop of halocarbon oil (see **Equipment and Reagents Needed** in **Section IV.F**) to the pad. Place from one to four worms into the oil under the dissecting microscope. Let the worm(s) thrash about in the oil to remove attached bacteria. Using a worm-pick, gently force the worm onto the pad while maintaining the dorsal and ventral sides of the worm in the same focal plane. Once the worm is close to the surface of the pad, press gently in order to stick the worm onto the pad. Once one part of the body is stuck to the pad, then press along the rest of the body so that the entire worm sticks to the pad.
5. **Injection:** With a well-made needle, touching the cuticle should suffice to insert the needle into the worm. However, usually a gentle tap on the micromanipulator, while the tip of the needle is firmly against the cuticle, is needed to make penetration. Once the tip of the needle is in the desired tissue, proceed with injection of the dsRNA. The amount of dsRNA used is somewhat arbitrary, but enough dsRNA is usually injected to swell the gonad of the worm. Be careful not to inject so much of the dsRNA that it causes the worm to burst.
6. **Recovery of the worm:** It is important to remove the worm from the dry agarose pad as soon as possible. Add a single drop of M9 buffer onto the surface of the oil and bring the M9 buffer into contact with worms using a mouth-pick so that the worms float away from the agarose pad. Immediate wriggling of the worm indicates a good recovery. The worms can then be transferred onto a freshly-seeded NGM plate by using a worm-pick, mouth pipette, or by simply dropping the worms onto a freshly-seeded NGM plate by inverting the injection coverslip.

5. Analysis of Progeny

The injected worms are transferred to fresh plates at 4-, 6-, 8-, or 12-h intervals to facilitate analysis. Optimal time intervals will have to be determined on a trial-and-error basis. Generally, one should first check expression patterns of the genes before analyzing phenotypes. If one knows the expression pattern of the gene, one should focus on the tissues that express the gene. The method of analysis is variable according to the purpose of the experiments and may include Northern blot analysis (**Chapter 6**), Western blot analysis (**Chapter 8**), or *in situ* RNA localization (**Chapter 22**).

C. Alternative Methods of RNA Interference

The method of direct injection into worms is most effective when using a small volume of *in vitro* transcribed dsRNA; however, the technique requires the use of special equipment and is limited by the number of worms that can be injected. For experiments requiring a lot of worms, such as Western and Northern analysis or mutant screening,[25] bacterial cell-mediated RNAi and the soaking method may prove to be better.

1. Feeding of Bacterial Cells That Produce dsRNA

This method makes use of bacteria deficient for *RNase* III allowing them to produce high quantities of specific dsRNA segments. When fed to *C. elegans*, such bacteria produce populations of RNAi-affected animals with phenotypes corresponding to the loss-of-function mutants. This method was shown to be most effective in inducing RNAi for nonneuronal tissue of late larval and adult hermaphrodites, with decreased effectiveness in the nervous system, in early larval stages, and in males. Bacteria-induced RNAi phenotypes can be maintained (if not lethal) for several generations by continuous feeding. This allows for convenient analysis of the biological consequences of specific genetic interference.[26]

1. Subclone the gene fragment of interest between the T7 promoters in the "double T7" plasmid (see *Note* below). (A slightly more difficult alternative is to place the gene of interest in a hairpin/inverted repeat configuration behind T7 promoter in pBluescript. This method, however, has many factors that may inhibit its success, but if it works, it tends to work extremely well.)

Note: *The double T7 promoter-containing plasmid as well as control plasmids for use in feeding experiments can be obtained from A. Fire (Carnegie Institute). Information and kit request forms can be accessed through the Carnegie web site (http://www.ciwemb.edu/).*

2. Transform the plasmid ligation into competent BL21 (DE3) or HT115 (DE3) bacterial cells and plate onto standard LB (Luria-Bertaini) medium + tetracycline + antibiotic plates.

Note: *The HT115 (DE3) strain is tetracycline (TET)-resistant; nonetheless, care must be taken not to contaminate your bacterial stock. First, plate the cells onto TET plates, immediately freeze an aliquot upon receipt, and also freeze any transformed strains in order to have a reliable backup. In addition, the only reliable way to verify the presence of the DE3 lysogen is by PCR, since T7 phage will not grow in the RNase III background of this cell. The HT115 (DE3) strain is now available from the Caenorhabditis Genetics Center (CGC) (www.cbs.umn.edu/CGC/-2K).*

3. Grow up a liquid culture from a single colony on the selection plates. Grow bacteria with shaking up to OD_{595} = 0.4, and add isopropyl β-D-thiogalactopyranoside (IPTG) up to a final concentration of 0.8 mM. Incubate at 37°C with shaking for 4 h.
4. Induce dsRNA in HT115 (DE3) cells with a T7-promoter-containing plasmid:
   a. Inoculate an overnight culture of HT115 (DE3) + plasmids in LB medium + antibiotics. Incubate at 37°C with shaking overnight (add 75 to 100 μg·ml^{-1} ampicillin for amp-resistant plasmids and 12.5 μg·ml^{-1} tetracycline).

b. Dilute the culture 1:100 in 2xYT + antibiotics and grow it to OD_{595} = 0.4 (a 25-ml culture is usually enough for a small experiment of ~20 small plates).
c. Induce the culture by adding sterile IPTG to 0.4 mM. Incubate at 37°C with shaking for 4 h.
d. Spike the culture with additional antibiotics (another 100 μg·ml^{-1} AMP and 12.5 μg·ml^{-1} TET) and IPTG (to a final total concentration of 0.8 mM). (Seed the plates using the culture as is, or concentrate the cells by centrifugation.)

5. Seed NGM plates with the induced culture. The culture can be used as is (for small plates containing small numbers of handpicked worms), or the cells can be concentrated by centrifugation and spotted onto plates (for large plates containing chunked worms). The proportion of bacteria to worms is important — if the plates starve out, RNAi will not be effective. In addition, the bacterial lawn should NOT be allowed to continue to grow. Cells that do grow on plates after induction are generally cells that have lost the plasmid, cells that have lost the ability to produce T7 polymerase, or cells that are contaminants. The inclusion of tetracycline in the plates significantly improves the results; the addition of ampicillin also helps in the case of amp-resistant plasmids (50 μg·ml^{-1} AMP, 12.5 μg·ml^{-1} TET). Add the worms to the plate and incubate at appropriate temperature. Worms can be added by handpicking or by adding chunks onto wet, freshly-seeded NGM plates, or onto plates that have been allowed to dry after seeding. Although older seeded plates containing IPTG can also induce RNAi phenotypes with good success, researchers prefer to use freshly-seeded NGM plates. One can observe phenotypes at temperatures from 16 to 25°C, although the expressivity and penetrance of the phenotype can vary, depending upon the incubation temperature and gene. It can take 3 days before an RNAi phenotype is observed. Results vary, depending on the dsRNA and the worm strains used.

Note: *The same parameters that give variable results in protein expression using the T7 promoter system are also variable in this system. The variables to consider are: (1) induction temperature (37°C vs. 30°C), (2) induction time (2 h, 4 h, or overnight), (3) concentration of IPTG, (4) induction volume, (5) medium used (LB vs. 2xYT or other media), and (6) additives to induction medium (uracil, lactose, etc.). In addition, fresh cells tend to work best. Bacteria that have been stored on plates at 4°C for a long period of time often lose effectiveness.*

2. Soaking in dsRNA Solution

The soaking method of RNAi follows the same *in vitro* transcription procedure that is used in the direct microinjection method. This method consists of simply soaking the worms in dsRNA solution in an eppendorf tube.[27]

1. Wash L4-stage hermaphrodites in 0.2 M sucrose and phosphate-buffered saline (PBS), and transfer into 10 μl of the same buffer in a siliconized tube.
2. Mix 4 μl of dsRNA (3 to 5 mg·ml^{-1}) and 1 μl liposome (Lipofectin, Invitrogen Life Technologies, Carlsbad, CA) vigorously in another siliconized tube.
3. Add 15 worms to the RNA-liposome mixture, resulting in a total volume of 15 μl and a final RNA concentration of 1 mg·ml^{-1}.
4. Transfer the worms to an agar plate with *E. coli* after 24 h of soaking and culture until mid-adulthood.

D. Inducible RNA Interference

1. Heat Shock-Induced RNA Interference

The method of *in vivo* RNAi using a heat shock promoter can make heritable and inducible genetic interference possible. The most important factor in this method is making an efficient DNA construct. An efficient DNA construct needs an inverted repeat (IR) gene, which is composed of sense DNA, a loop, antisense DNA, and the 3' UTR following a heat shock promoter.[17]

1. Subclone an IR gene construct. The loop must be at least 150 bp for the inverted-repeat section to be viable in bacteria. Without this loop, the bacteria will die.[12,15] In addition, a loop containing an intron tends to allow more-efficient suppression.
2. Make a minipreparation of subcloned plasmid flowing protocols, as in **Chapter 1**, or buy a high-quality miniprep kit (Qiagen, Valencia, CA).
3. Microinject plasmids with an appropriate marker DNA (for example, *rol-6*) into the gonad or the intestine of well-fed hermaphrodite worms, and obtain a stable transgenic line. To obtain stable lines, F1 roller animals are selected and transferred onto new plates, and plates that contain roller F2 animals are kept for further experiments.
4. Place the transgenic worms at an appropriate heat-shock temperature for an appropriate length of time. We recommend that the heat shock take place from 1 to 3 h at 30°C; however, the optimal time and temperatures will have to be determined for each experiment. After heat shock, transfer the animal into a 20°C incubator. The F1 progeny or Po animals are observed under the stereomicroscope or Nomarski optics for their phenotypes.

2. Hypoxia-Induced RNA Interference

Unlike yeast and *Drosophila*, inducible systems are limited in *C. elegans*. Using heat shock has been the most effective inducible system in *C. elegans*, despite the possible damage by heat shock alone. Therefore, we developed a hypoxia-inducible promoter as a new type of inducible system in *C. elegans*. Hsp-41, a conventionally used heat shock protein, is responsive to heat shock only, but hsp16-1 is responsive to hypoxia as well as heat shock.[28] The use of hypoxia as an inducible system is most attractive in that hypoxia can cause only a small amount of damage to worms. The fact that hsp16-1 is induced by hypoxia was found during the study of gene expression profiling by ethanol treatment. This method follows the same procedure as in the heat shock protocol (**Section IV.D.1**), with the exception of using hypoxia shock in place of heat shock. The hypoxia condition is achieved by soaking a large pool of worms in M9 solution so that they consume an adequate amount of oxygen, thus causing the hypoxia condition. The animals are soaked for 6 to 12 h without shaking. After soaking, the animals are dumped to new NGM plates and held there for several hours so that M9 buffer is absorbed into the agar of the plates. Then every one to three animals are transferred onto new plates. The progeny from these animals are examined for their phenotypes.

E. Troubleshooting Guide

Symptom	Solutions
Difficulty in injecting the dsRNA	High concentration or low purity of the RNA causes poor loading into the needle and needle clogging. Therefore, using a proper concentration of RNA and adequate cleansing of the needle is required.
Needles clog during or between injections	Needles are easily clogged by cytoplasm, cuticle, or bacteria around the worm. To protect the needle from clogging, remove the needle from the worm while under injection pressure. This may seem like a waste of injection mixture, but in fact, it helps to keep the worm alive. Between each injection, use the high pressure to clear the tip of the needle.
In vivo RNAi processes did not work	There are not many cases of successful experiments of *in vivo* RNAi in *C. elegans*. The most successful results are from Driscoll and her colleagues' experiments,[17] which showed that the *in vivo* RNAi method is effective not only in general tissues, but also in neuronal cells.[10] Their heat shock conditions were 4 h at 35°C. Generally, the heat shock promoter operates well for 1 to 3 h at 30°C, judging from the experiments in our laboratory. Because the worm can be damaged from heat shock itself, it is necessary to find the appropriate heat shock conditions and perform control experiments.

Equipment and Reagents Needed

Agar Pad (2% agarose and coverslip)
48 × 60 #1 (PGC Scientific, Gaithersburg, MD)

HT115 (DE3) Bacterial Strain
Obtained from CGC (CGC web site: http://biosci.cbs.umn.edu/CGC/CGChomepage.htm)

Halocarbon Oil 700
(Lot #17H0189, Sigma, St. Louis, MO)

IPTG (800 mM)
Dissolve IPTG in distilled water and sterilize by passing it through a 0.22 μm disposable filter. Dispense the solution into 1 ml aliquots and store at −20C.

10X Injection Buffer (if used)
200 mM KPO_4
30 mM K citrate
20% Polyethylene glycol (PEG), molecular weight 6000, pH 7.5

Liposomes
Lipofectin (Invitrogen Life Technologies, Carlsbad, CA)

LB (Luria-Bertaini) Medium
Tryptone (10 g)
Yeast extract (5 g)
NaCl (5 g)
Dissolve in ddH_2O and adjust the volume to 1 L. Adjust pH to 7.5 with 2 N NaOH. Autoclave.

LB Plates
Add 15 g of agar to 1 L of LB medium and autoclave. When the medium cools to 50 to 60°C, add appropriate antibiotics, mix well, and pour 20 to 25 ml of medium into 100 mm petri dishes. Allow the medium to harden in a laminar-flow hood. Store at room temperature for 10 days or at 4°C for up to 2 months.

M9 Buffer
KH_2PO_4, 3 g
Na_2HPO_4, 6 g
NaCl, 5 g
1 M $MgSO_4$, 1 ml
H_2O, up to 1 L

Microinjector with the Compressor
IM-30 (Narishige, Japan) is a cost-effective apparatus.

Micromanipulator
MM151 (Narishige, Japan) is a cost-effective apparatus.

Inverted Compound Light Microscope (Differential Interference Contrast)

Zeiss Axiovert 135 (Germany) is recommended.

Standard Dissecting Microscope

Several brands are availbale. Zeiss SV12 (Germany) is one of the top-line models.

NGM Plates

3.0 g NaCl
2.5 g Bactopeptone
5 mg Cholesterol
17.5 g Agar

Bring up to 1 L with ddH$_2$O and autoclave. Cool to 50°C and add the following from 1 M stocks:

1 mM CaCl$_2$
1 mM MgSO$_4$
25 mM Potassium phosphate (pH 6)
appropriate antibiotics if applicable

Pour the plates and let them solidify. Inoculate ("seed") the NGM plates with a drop (about 25 μl) per plate of an overnight culture of *E. coli* bacteria (usually OP50 obtained from the Caenorhabditis Genetics Center, but this may vary depending on the experiment), spread the drops using a flame sterilized spreader, and leave the plates on the bench overnight. The plates will be ready for feeding the worms the next day.

Needle Puller and Microcapillary (thin glass tube)

PC10 (Narishige, Japan) is an economical version.

Platinum Worm Pick and Mouth Pipette

Aspirator tube assembly (Sigma, St. Louis, MO) is used for mouth pipette.

Double T7 Promoter Vector

pPD129.36 (A. Fire, Carnegie Institute of Embryology)

Phosphate Buffered Saline (PBS)

137 mM NaCl
2.7 mM KCl
10 mM Na$_2$HPO$_4$
2 mM KH$_2$PO$_4$
Sterilize the solution by autoclaving or filtering.

Siliconized Tubes

Can be purchased from Sigma (St. Louis, MO) or Falcon (Lexington, TN).

2 M Sucrose Solution

Add 68.46 g of sucrose to a total volume of 100 ml ddH$_2$O. Autoclave and store at 4°C.

References

1. **Robert, L.S., Donaldson, P.A., Ladaigue, C., Altosaar, I., Arnison, P.G., and Fabijanski, S.F.,** Antisense RNA inhibition of β-glucuronidase gene expression in transgenic tobacco can be transiently overcome using a heat-inducible b-glucuronidase gene construct, *BioTechnology*, 8, 459, 1990.
2. **Rezaian, M.A., Skene, K.G.M., and Ellis, J.G.,** Antisense RNAs of cucumber mosaic virus in transgenic plants assessed for control of the virus, *Plant Mol. Biol.*, 11, 463, 1988.
3. **Hemenway, C., Fang, R.-X., Kaniewski, W.K., Chua, N.-H., and Tumer, N.E.,** Analysis of the mechanism of protection in transgenic plants expressing the potato virus X coat protein or its antisense RNA, *EMBO J.*, 7, 1273, 1988.
4. **Dag, A.G., Bejarano, E.R., Buck, K.W., Burrell, M., and Lichtenstein, C.P.,** Expression of an antisense viral gene in transgenic tobacco confers resistance to the DNA virus tomato golden mosaic virus, *Proc. Natl. Acad. Sci. U.S.A.*, 88, 6721, 1991.
5. **Zelenin, A.V., Titomirov, A.V., and Kolesnikov, V.A.,** Genetic transformation of mouse cultured cells with the help of high-velocity mechanical DNA injection, *FEBS Lett.*, 244, 65, 1989.
6. **Perlak, F.J., Fuchs, R.L., Dean, D.A., McPherson, S.L., and Fischhoff, D.A.,** Modification of the coding sequence enhances plant expression of insect control protein genes, *Proc. Natl. Acad. Sci. U.S.A.*, 88, 3324, 1991.
7. **Morrison, H.G. and Desrosiers, R.C.,** A PCR-based strategy for extensive mutagenesis of a target DNA sequencing, *BioTechniques*, 15, 454, 1993.
8. **Sczakiel, G.,** The design of antisense RNA, *Antisense Nucleic Acid Drug Dev.*, 7, 439, 1997.
9. **Jorgensen, R.A., Cluster, P.D., English, J., Que, Q., and Napoli, C.A.,** Chalcone synthase cosuppression phenotypes in petunia flowers: comparison of sense vs. antisense constructs and single-copy vs. complex T-DNA sequences, *Plant Mol. Biol.*, 31:957, 1996.
10. **Que, Q., Wang, H.-Y., English, J., and Jorgensen, R.A.,** The frequency and degree of cosuppression by sense chalcone synthase transgenes are dependent on transgene promoter strength and are reduced by premature nonsense codons in the transgene coding sequence, *Plant Cell*, 9, 1357, 1997.
11. **Hamilton, A.J., Brown, S., Yuanhai, H., Ishizuka, M., Lowe, A., Alpuche Solis, A.-G., and Grierson, D.,** A transgene with repeated DNA causes high frequency, posttranscriptional suppression of ACC-oxidase gene expression in tomato, *Plant J.*, 15, 737, 1998.
12. **Waterhouse, P.M., Graham, M.W., and Wang, M.-B.,** Virus resistance and gene silencing in plants can be induced by simultaneous expression of sense and antisense RNA, *PNAS*, 95, 13959, 1998.
13. **Chuang, C.-F.C. and Meyerowitz, E.M.,** Specific and heritable genetic interference by double-stranded RNA in *Arabidopsis thaliana*, *PNAS*, 97, 4985, 2000.
14. **Fagard, M., Boutet, S., Morel, J.-B., Bellini, C., and Vaucheret, H.,** AGO1, QDE-2, and RDE-1 are related proteins required for posttranscriptional gene silencing in plants, quelling in fungi, and RNA interference in animals, *PNAS*, 97, 11650, 2000.
15. **Waterhouse, P.M., Wang, M.-B., and Lough, T.,** Gene silencing as an adaptive defense against viruses, *Nature*, 411, 834, 2001.
16. **Bosher, J.M. and Labouesse, M.,** RNA interference: genetic wand and genetic watchdog, *Nat. Cell. Biol.*, 2, E31, 2000.
17. **Tavernarakis, N., Wang, S.L., Dorovkov, M., Ryazanov, A., and Driscoll, M.,** Heritable and inducible genetic interference by double-stranded RNA encoded by transgenes, *Nat. Gen.*, 24, 180, 2000.
18. **Caplen, N.J., Parrish, S., Imani, F., Fire, A., and Morgan, R.A.,** Specific inhibition of gene expression by small double-stranded RNAs in invertebrate and vertebrate systems, *PNAS*, 98, 9742, 2001.
19. **Matzke, M., Matzke, A.J.M., and Kooter, J.M.,** RNA: Guiding gene silencing, *Science*, 293, 1080, 2001.
20. **Zamore, P.D., Tuschl, T., Sharp, P.A., and Bartel, D.P.,** RNAi: double-stranded RNA directs the ATP-dependent cleavage of mRNA at 21 to 23 nucleotide intervals, *Cell*, 101, 25, 2000.
21. **Elbashir, S.M., Harborth, J., Lendeckel, W., Yalcin, A., Weberm, K., and Tuschl, T.,** Duplexes of 21-nucleotide RNAs mediate RNA interference in cultured mammalian cells, *Nature*, 411, 494, 2001.

22. **Kwon, J.-Y. et al.,** in West Coast *C. elegans* Meeting Abstract, UCLA, Los Angeles, CA, 2000.
23. **Henry, F., Epstein, D., and Shakes, C.,** *Methods in Cell Biology — Caenorrhabditis elegans — Modern Biological Analysis of an Organism*, Academy Press, San Diego, CA, 1995.
24. **Hope, I.A.,** *C. elegans: A Practical Approach*, Oxford University Press, New York, 1999.
25. **Tabara, H. et al.,** The *rde*-1 gene, RNA interference, and transposon silencing in *C. elegans*, *Cell*, 99, 123, 1999.
26. **Timmons, L., Court, D.L., and Fire, A.,** Ingestion of bacterially expressed dsRNAs can produce specific and potent genetic interference in *Caenorhabditis elegans*, *Gene*, 263, 103, 2001.
27. **Tabara, H., Grishok, A., and Mello, C.C.,** RNAi in *C. elegans*: soaking in the genome sequence, *Science*, 282, 430, 1998.
28. **Kwon, J., Hong, M., Shim, J., and Lee, J.,** Orientation-dependent hypoxia response of the hsp-16 promoters in the nematodes. In preparation.

Chapter 21

Microscopy: Light, Scanning Electron, Environmental Scanning Electron, Transmission Electron, and Confocal

Casey R. Lu, Peter B. Kaufman, and Leland J. Cseke

Contents

I.	Light or Optical Microscopy (LM)	462
II.	Scanning Electron Microscope (SEM)	464
	A. Principles of Operation	464
	B. Preparation of Biological Tissues for SEM	471
	1. Fresh Biological Material	471
	2. Fixed Biological Material	471
	C. SEM Applications	471
III.	Environmental Scanning Electron Microscopy (ESEM)	474
	A. Principles of Operation	474
	B. ESEM Applications	476
IV.	Transmission Electron Microscopy (TEM)	477
	A. The Conventional Transmission Electron Microscope	477
	B. Preparation of Tissues for TEM	479
	C. TEM Applications	480
V.	Confocal and Multiphoton Microscopy (CM)	482
References		486

Major conceptual breakthroughs have occurred in cellular and molecular biology with the development of light microscopy (LM), scanning electron microscopy (SEM), environmental scanning electron microscopy (ESEM), transmission electron microscopy (TEM), and confocal microscopy

(CM). For investigators conducting studies on *in situ* localization of specific mRNAs or reporter-gene expression in animal or plant tissues and organs, light microscopy and confocal microscopy protocols are required. Scanning electron microscopy allows one to visualize three-dimensional images of the surfaces of whole organisms, organs, tissues, and cells or parts of cells, made possible by the large depth-of-field capabilities of the SEM. X-ray analysis of the distribution of elements present in the cells or tissues can be performed with the SEM as well. The ESEM allows one to observe uncoated, hydrated, living organisms and tissues in the SEM. Subcellular immunolocalization of specific gene products using antibody labels requires knowledge of transmission electron microscopy and specialized techniques of tissue fixation, embedding, and ultrathin sectioning, as well as light microscopy techniques applied via confocal microscopy. In addition to *in situ* localization, confocal microscopy allows researchers to generate three-dimensional optical sections of living tissues and cells using computer programs for data and section analyses, fluorescent markers, and filter systems that allow investigators to observe fluorescence images of specific proteins, nucleic acids, and ions within whole tissues or organs. Multiphoton microscopy is a new technique that allows the same kind of three-dimensional optical sections that one can produce with a confocal microscope, but without the problems of photobleaching, and it allows for deeper penetration due to the longer wavelengths used in this microscope.

In this chapter, we explore the principles underlying how the LM, SEM, ESEM, TEM, and CM work, how to operate them, and a number of applications in biology and medicine where these techniques have been used successfully.[1-16]

I. Light or Optical Microscopy (LM)

Although optical microscopy was developed and refined hundreds of years ago, beginning with the invention of the compound microscope by the Janssens in 1590, it has undergone a remarkable resurgence of interest in its applications in biology and medicine. Witness the advent of Nomarski, phase contrast, and bright-field/dark-field optics to obtain better contrast images by the use of the polarizing optics with a red plate to determine orientation of cellulose microfibrils in cell walls; digital image processing for contrast enhancement (video-enhanced contrast microscopy or VECM); ultraviolet (UV) fluorescence microscopy; and optical sectioning via confocal microscopy (see later in this chapter). In this section, we cover the basic structural features of light or optical microscopy, illustrate a typical research light microscope in use today, and then show several applications in the use of these techniques with biological systems.

Figure 21.1 illustrates the basic external and internal components of a commercial research light microscope. Figure 21.2 shows an Olympus Vanox research-quality light microscope in use in the authors' laboratory. It is fitted with photomicrography capabilities with electronic controls for making photomicrographs; Nomarski, dark-field, and bright-field optics; polarizing filters and red-plate filter; and a UV power supply for UV fluorescence microscopy. We have successfully used polarized light and the red plate to analyze the orientation of cellulose microfibrils in the guard cells of plants. We also have used UV fluorescence microscopy to study callose (β-1, 3-linked glucan) in pollen tubes, using aniline blue stain, as well as cellulose deposition in developing cell walls that form in isolated plant protoplasts using Calcofluor, a stain specific for cellulose (ß-1, 4-linked glucan).

Figure 21.3 and Figure 21.4 illustrate one application of optical light microscopy, namely the analysis of starch-containing chloroplasts in graviresponsive organs (pulvini) in oat shoots. The stain used is iodine-potassium-iodide ($I_2 \cdot KI$), which is specific for staining starch. From this study, it was discovered by the authors that these starch-containing chloroplasts are the gravisensors that initiate an upward-bending response in oat shoots as well as shoots of all members of the grass family (Poaceae).

Microscopy

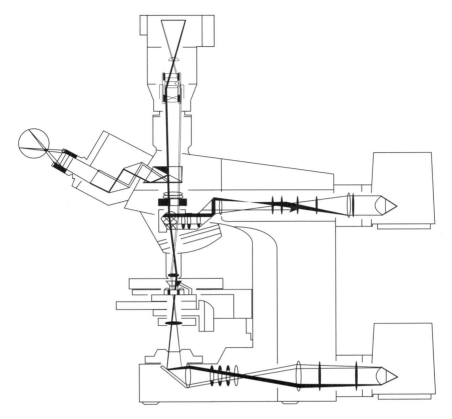

FIGURE 21.1
Diagram of an Axioplan Universal Microscope (courtesy of Carl Zeiss Inc., Thornwood, NY) illustrating the light sources and internal optics of their research light microscope. (Poster diagram contributed by Shelly Almburg, Department of Biology, University of Michigan; from Figure 19.1 in Kaufman, P.B., Wu, W.W., Kim, D., and Cseke, L.J., *Handbook of Molecular and Cellular Methods in Biology and Medicine*, CRC Press, Boca Raton, FL, 1995. With permission.)

The theoretical resolution (how closely spaced two objects can be and still be imaged as two separate objects) of both the LM and the TEM can be calculated from the simple formula shown below.

$$rp = \frac{0.61\lambda}{n(\sin\alpha)}$$

where
- rp = the minimum distance that two objects can be spaced and still imaged as two separate objects
- 0.61 = a constant
- λ = wavelength
- n = refractive index
- α = aperture angle

Substituting 400 nm for wavelength, 1.5 for the refractive index of immersion oil, and 64° for α (typical aperture angle for a glass lens), one obtains a resolving minimum of approximately 200 nm for light microscopes. This limit can be improved slightly by using ultraviolet radiation (λ shorter than 400 nm) and special detectors that are sensitive to UV. However, this resolving

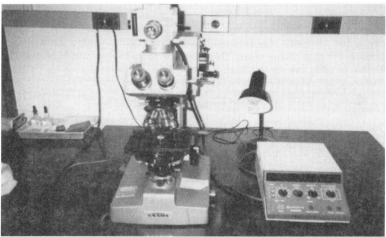

FIGURE 21.2
Side view (top) and front view (bottom) of an Olympus Optical Company, Ltd., Vanox research light microscope with ancillary electronic controls. (Photos by Peter Kaufman; from Figure 19.2 in Kaufman, P.B., Wu, W.W., Kim, D., and Cseke, L.J., *Handbook of Molecular and Cellular Methods in Biology and Medicine*, CRC Press, Boca Raton, FL, 1995. With permission.)

power limit cannot match that of the TEM, which utilizes an electron beam possessing a wavelength 100,000 times shorter than visible light (at 60 keV). Today's modern TEMs have resolving-power limits around 1.5 Å (1×10^{-10} m), or 1000 times better than the light microscope. Some instruments have been constructed that resolve below 1.0 Å.

II. Scanning Electron Microscope (SEM)

A. Principles of Operation

Knoll and Van Ardenne in Germany constructed the first SEM in the late 1930s. Figure 21.5 illustrates a modern scanning electron microscope (SEM). This instrument differs fundamentally from the

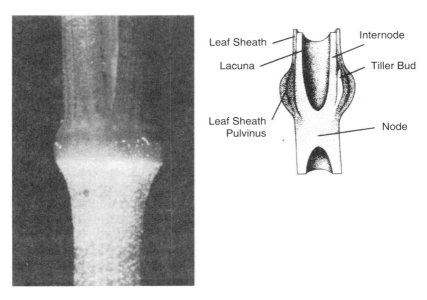

FIGURE 21.3
The leaf-sheath pulvinus of oat (*Avena sativa*) from an upright plant. The papery margin of the leaf sheath can be seen above the pulvinus. (From Brock, T.G. and Kaufman, P.B., *The Pulvinus: Motor Organ for Leaf Movement*, © American Soc. of Plant Physiologists, 1990. With permission; Kaufman, P.B., Wu, W.W., Kim, D., and Cseke, L.J., *Handbook of Molecular and Cellular Methods in Biology and Medicine*, CRC Press, Boca Raton, FL, 1995. With permission.)

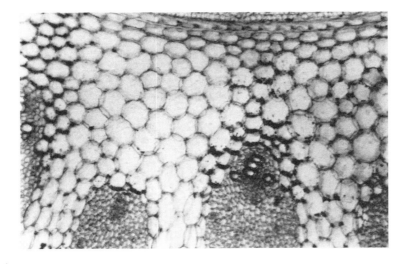

FIGURE 21.4
Light micrograph of cross section through oat shoot leaf-sheath pulvinus illustrating starch gravisensors in bases of cells next to each vascular bundle (×330). (From Brock, T.G. and Kaufman, P.B., *The Pulvinus: Motor Organ for Leaf Movement*, © American Soc. of Plant Physiologists, 1990. With permission; Kaufman, P.B., Wu, W.W., Kim, D., and Cseke, L.J., *Handbook of Molecular and Cellular Methods in Biology and Medicine*, CRC Press, Boca Raton, FL, 1995. With permission.)

conventional transmission electron microscope, and all other common optical instruments, in that no lens acts on the image-forming radiation after it strikes the specimen. Instead, the electron beam is focused to a very small diameter by two or three electromagnetic lenses, which are functionally analogous to the condenser lenses of a TEM, and then caused to scan the specimen in a raster pattern by a system of beam-deflecting coils. The image is then formed in a point-by-point manner using

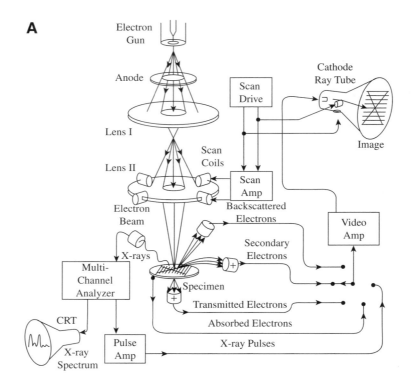

FIGURE 21.5
A. The basic components of the electron optical system of a scanning electron microscope. (From Kaufman, P.B., Labvitch, J., Anderson-Prouty, A., and Ghosheh, N.S., *Laboratory Experiments in Plant Physiology*, Macmillan College Publishing Co, New York, Copyright © 1975. With permission; Kaufman, P.B., Wu, W., Kim, D., and Cseke, L.J., *Handbook of Molecular and Cellular Methods in Biology and Medicine,* 1st ed., CRC Press, Boca Raton, FL, 1995). B. Field-emission SEM and operator, Dwight Bailey. (Photograph courtesy of Delilah Wood.)

various types of signals generated by interaction of the electron beam with the specimen. The image is displayed on a cathode ray tube (CRT) whose electron beam is driven by the same scanning drive generator that scans the beam over the specimen. There is thus a one-to-one correspondence between points on the viewing screen of the cathode ray tube and points on the specimen. Magnification is changed simply by an amplifier, which changes the size of the area scanned on the specimen relative to that of the CRT.

$$\text{magnification} = \frac{\text{area on CRT}}{\text{area on specimen}}$$

Images are produced on the CRT by feeding the signals from the specimen to a video amplifier that modulates the brightness of the beam of the CRT in a manner related to the strength of the signal from the specimen. Several different signals are available for use in image formation; however, the one most frequently used is secondary electrons. These are valence electrons that are dislodged from atoms in the outer 5 to 10 nm of the specimen's surface by the incident electron beam. The secondary electrons emerge from the surface with very low velocities and are generally collected by applying a potential of about +10 keV to the surface of a scintillator. This potential accelerates the secondary electrons enough, so that when they strike the scintillator, they produce minute flashes of light, which are detected by a photomultiplier tube. The video amplifier for image formation then uses the output of the photomultiplier. The number of secondary electrons generated at a given point depends basically on the angle at which the beam strikes the specimen surface at that point. More electrons are emitted when the surface is inclined sharply to the beam than when the surface is perpendicular to it. Consequently, image contrast changes as the beam moves over the surface in a manner related to the topography of the specimen surface.

Some of the electrons of the incident beam interact strongly with the atoms in the specimen and are scattered back out of its surface with essentially the same velocity they had upon striking it. Various types of detectors can also collect the backscattered electrons, and the resulting signal is amplified for image formation. Likewise, the incident electrons that are not backscattered produce a minute current (10^{-6} to 10^{-12} A) of electricity through the sample, which can be amplified for image formation. The fraction of electrons backscattered depends basically on the average atomic number of the atoms in the specimen surface, and it ranges from about 10% for an average atomic number of about 10 to nearly 50% for an average atomic number of about 75. Consequently, contrast in backscattered electron and sample-current images depend on local variations in composition, although it is very difficult to determine, from image characteristics alone, the elements involved in causing these variations. If the specimen is thin enough, electrons transmitted through it can also be used for image formation. For crystalline specimens, transmitted electrons can produce channeling patterns, diffraction patterns, and Kikuchi patterns. Some materials produce fluorescent light when struck by the electron beam. This light can be detected by a photomultiplier tube and used for image formation.

The versatility of the SEM is further increased because the electron beam causes x-ray photons to be emitted from atoms in the specimen. These photons have energies and wavelengths that are determined by the electron energy levels of the atoms. Measurement of either their energies or wavelengths makes it possible to identify the kinds of atoms present in the specimen (qualitative chemical analysis). Measurement of the rate at which they are emitted under carefully controlled conditions, and comparison with standards of known compositions, permits determination of the concentrations of the different kinds of atoms (quantitative analysis).

These measurements can be made with a solid-state x-ray detector in combination with a multichannel analyzer. The solid-state detector is a special silicon diode that produces an electrical pulse each time it receives an x-ray photon. The amplitudes of these pulses are linearly proportional to the energies of the photons. The multichannel analyzer receives the pulses from the detector, and after proper amplification, it measures their amplitudes and records their receipt in memory positions (channels) that are assigned numbers proportional to the photon energies. The counts recorded in each channel can be displayed on a CRT, read out numerically, or printed out, giving an x-ray spectrum of the specimen. In this way, it is possible to determine the major elements present in a specimen and to gain a rough idea of their concentrations (semiquantitative). Detection of elements present in minor amounts, and accurate concentration determinations, both require more complicated and time-consuming procedures. For this purpose, it is often preferable to use a crystal spectrometer, which gives considerably better sensitivity and selectivity than the solid-state detector but requires more time and effort because only one element can be detected at a time. The electron microprobe analyzer is an instrument that has an electron optical system similar to that of the SEM, but it is otherwise designed especially for carrying out accurate quantitative analyses with crystal spectrometers. If the multichannel analyzer of the crystal spectrometer is set to feed pulses from only a single element to the video amplifier, each pulse will produce a single dot of light on the CRT. As the electron beam scans over portions of the specimen where the concentration of the selected element is high, the pulses will be received more rapidly, and the number of dots per unit area on the CRT will increase. In this way, an image can be produced showing the overall distribution of the element in the specimen. When compared with secondary electron or sample current images of the same area of a specimen, such characteristic x-ray images are very useful for showing the overall compositional variations in the specimen and their relationships to various structural features (Figure 21.6). X-ray images usually cannot reveal small changes in the concentrations of major elements, and they are otherwise difficult to interpret quantitatively.

The theoretical resolution of the SEM is determined by the diameter of the electron beam as it strikes the specimen. The beam diameter is limited by the spherical aberration of the second lens, the electron beam accelerating voltage and current, and the brightness and size of the beam at the electron gun. In practice, a number of other factors, such as the mechanical stability of the specimen stage, the quality of the detectors and amplifiers, and the cleanliness of the electron optical system, also affect the resolution. Many commercial instruments are available that can give resolutions below 5 nm, whereas resolution below 1 nm has been achieved with specially designed experimental instruments. The resolution is also dependent on the nature of the interaction between the electron beam and the specimen that generated the image-forming signal. In general, resolution with backscattered electrons and sample current is considerably greater than the diameter of the incident electron beam because the high-energy electrons may be backscattered from a depth of several hundred nanometers beneath the surface of the specimen, depending on the electron accelerating voltage used and the density of the specimen. In traveling this far into the specimen, the electrons may be scattered laterally by collisions with the atoms of the specimen before finally being backscattered. Best resolution is usually obtained with secondary electrons, because these are produced only in the outer 1 to 5 nm of the specimen, before there has been appreciable lateral scattering of the electron beam.

The lateral scattering of the electron beam within the specimen, and its penetration in depth, permits x-ray photons to be emitted from a teardrop-shaped volume of the specimen whose size depends on the density (ρ, $[g] \cdot [cm^3]^{-1}$) of the specimen and the accelerating voltage (keV). The depth of this volume (μm) is roughly as shown in Table 21.1 for different types of materials.

It is evident that the resolution of x-ray images is severely limited by this phenomenon. For thick specimens, resolution is essentially independent of electron beam diameter, provided the beam diameter is less than about 0.1 μm. Resolution can be better for thin specimens, where the beam penetrates through the specimen before it has a chance to spread laterally to any great extent; however, the x-ray intensity falls off sharply because of the smaller volume of specimen being irradiated. This penetration and spreading of the electron beam within the specimen must also be considered when interpreting quantitative data obtained with the electron beam in a fixed spot on the specimen.

Microscopy

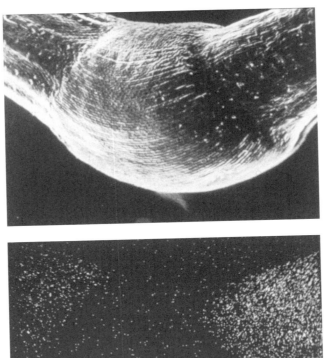

FIGURE 21.6
SEM image of pulvinus (above), and x-ray map of silicon (below) from same tissue. Pulvinus tissue does not contain silicon, and so its cells can enlarge if a plant becomes lodged. The mature internodal tissue contains silicon and lignin, and its cells can no longer enlarge. (Image courtesy of P. Dayanandan.)

TABLE 21.1
Depth of Volume in Various Materials, μm

Material Type	Accelerating Voltage (keV)				
	5	10	15	20	25
Biological ($\rho = 1$)	1	3	6	10	15
Ceramic ($\rho = 3$)	0.3	1	2	3	5
Metallic ($\rho = 7$)	0.1	0.5	0.8	1	2

Perhaps the most remarkable feature of the SEM is the unusually great depth of field it provides. This is a consequence of the characteristics of the electron lenses used to focus the beam. These lenses perform acceptably only when the electron beam has a diameter of less than 500 μm when it passes through the lens. When such a small beam is focused to the finest possible diameter, there is a considerable distance above and below the focus point where it changes diameter very little.

Over this range of distance, the beam will give acceptably sharp images. This depth of field varies with the minimum beam diameter, which in turn limits the maximum beam current and magnification, as follows:

Minimum beam diameter	10 μm	200 nm	10 nm
Depth of field	0.5 mm	100 μm	10 μm
Magnification	20×	1000×	20,000×
Maximum beam current	10^{-6} A	10^{-9} A	10^{-12} A

The significance of these values for the depth of field can be appreciated by comparing them with the following typical values for optical microscopes at comparable magnifications and resolutions:

Magnification	20×	100×	1000×
Depth of field	100 μm	10 μm	0.5 μm
Resolution	10 μm	1 μm	200 nm

Note that at 1000×, which is the maximum useful magnification for most optical microscopes, the SEM has a depth of field 200 times greater than the optical microscope, whereas at 20,000×, its depth of field is about the same as that of an optical microscope at 100×. This means that the SEM is ideally suited for studies of rough and irregular specimens that are too difficult to examine by optical microscopy or conventional transmission electron microscopy. Typical examples are microfossils, textile fibers and fabrics, catalyst particles, pollen grains, fracture surfaces, wood fibers, and the external surfaces of plants and insects. Procedures for preparation of specimens for examination in a SEM vary considerably, depending on the type of specimen material and the results desired. Metallic specimens are good conductors of electricity and can usually be mounted for examination without any special preparation, although it may be desirable to treat them in an ultrasonic cleaner to remove traces of oil and adhering dust particles.

Specimens of ceramics and minerals, fossils, teeth and bone, fibers, and self-supporting or "hard" biological tissues, which are nonconductors, must be coated with a thin layer of conducting material to prevent accumulation of static electric charges that arise from the electron beam and cause image distortion and spurious variations in image intensity. In some cases, satisfactory coating can be obtained by evaporating 10 to 50 nm of gold or some other metal on the surface of the specimen in a conventional vacuum evaporator. This technique is generally unsatisfactory for specimens that are very rough, porous, or fibrous, because the metal will not cover the undersides of fibers and the walls of pores and asperities. The use of devices that rotate and rock the specimen during evaporation will give better results with such specimens. However, a glow-discharge coater is perhaps simpler and faster and gives best overall results.

The preparation of soft biological tissues presents many difficult problems because they become severely distorted during handling and upon exposure to the vacuum and electron beam inside the SEM. Various procedures, including chemical fixation, freeze-drying, and critical-point drying, have been successfully used to stabilize such tissues for examination. The details of these procedures, which vary with the type of tissue and the objectives of the investigation, can be found in the technical literature. When x-ray methods are to be used, the conductive coating should be as thin as possible, and a low-atomic number material, such as aluminum or carbon, is preferred to minimize attenuation of the x-rays by absorption in the coating layer. If accurate quantitative analyses are to be performed, the specimen surface must be smooth, flat, and undistorted to allow for accurate control of experimental variables and to obtain accurate knowledge of the various physical parameters involved in computation of results.

B. Preparation of Biological Tissues for SEM

In contrast to the techniques for the TEM, tissue preparation procedures for work with the SEM are much easier. Very often, no fixation of the tissue is necessary. Peels or sections, made by hand or with a freezing microtome, may be examined in the SEM directly or lyophilized (freeze-dried) and then examined. The basic procedures are as follows:

1. Fresh Biological Material

1. Prepare small pieces 1 to 2 mm in diameter. Pollen grains and spores can be mounted directly.
2. Mount specimens on double-sticky Scotch™ tape or directly with electrically conductive silver paint (in butyl acetate solvent) or carbon paint (TV tube-coat) onto 1-cm aluminum or carbon stubs. Carbon stubs are used if one intends to conduct elemental analysis with the multichannel analyzer attached to the SEM; otherwise, use aluminum stubs (can be prepared from ring stand rods). Air dry with a hair dryer at room temperature.
3. If available, use a critical-point drying apparatus; samples will be much better preserved for scanning with the SEM.
4. Coat specimens with gold (Au), chromium (Cr), or other metal. If this is not done, bright areas will appear in your image as a result of charge buildup on the specimen.
5. Place specimen in the SEM to scan.

2. Fixed Biological Material

1. Fix the sample in 2 to 4% glutaraldehyde in buffer (for example, 50 mM sodium cacodylate buffer at pH = 7.0) for 30 min to 1 h. Wear gloves when using fixative solutions. This treatment chemically fixes (holds cellular structures in place) cells and tissues.
2. Rinse the fixative solution from the tissue with buffer for 10 min.
3. Repeat the buffer rinse.
4. Begin dehydrating the tissue in 50% EtOH (ethyl alcohol) for 10 min.
5. Move tissue to 70% EtOH for 15 min.
6. Move tissue to 95% EtOH for 15 min.
7. Repeat 95% EtOH step with fresh EtOH for another 15 min.
8. Finally, dehydrate the tissue in 100% EtOH for 10 min and repeat twice (total time 30 min).
9. Place in HMDS (hexamethyldisilazane) for 10 min and repeat one time. This replaces all water in tissue, fixes tissue, and stops shrinking.
10. Place the sample in an empty vial in a desiccator and store overnight under full vacuum. Place under vacuum for 20 min before mounting.
11. Mount samples on a stub with carbon-paint adhesive. Clean the aluminum stub with emery paper before mounting. Carbon adhesive prevents the buildup of charge. Dry with a hair dryer.
12. After drying, place the stub with sample in a container with Drierite™ (absorbs moisture) or anhydrous $CaCl_2$.

*Note: Tissues for SEM can also be prepared in a microwave tissue processor as described for the TEM (described below in **Section IV**).*

C. SEM Applications

Figure 21.7 through Figure 21.11 illustrate applications of the SEM to studies of tissues involved in the gravitropic response in cereal grasses and starch-filled chloroplasts (gravisensors) in oat (*Avena sativa*) shoots (Figure 21.7), starch in rice kernels (Figure 21.8), fungi (Figure 21.9 and Figure 21.10), and pollen (Figure 21.11).

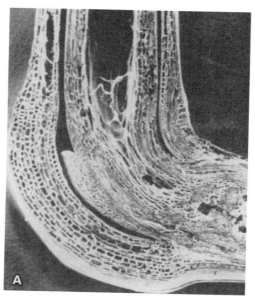
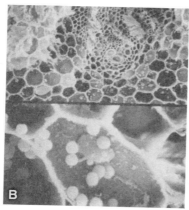
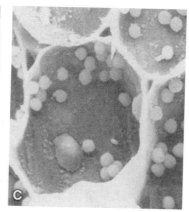

FIGURE 21.7
SEM images of graviresponse in pulvinus oat shoot (A) and starch-filled chloroplast gravisensors in leaf-sheath pulvinus cells of oat shoot (B and C). A, ×40; B, ×220; C, ×3000. SEM micrographs provided by Peter Kaufman from Kaufman, P.B., Wu, W., Kim, D., and Cseke, L.J., *Handbook of Molecular and Cellular Methods in Biology and Medicine*, 1st ed., CRC Press, Boca Raton, FL, 1995.)

Microscopy

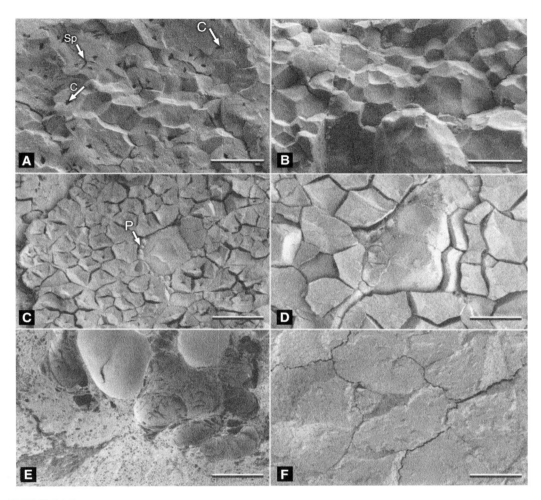

FIGURE 21.8
Scanning electron micrographs of starch in rice kernels. A and B, raw; C and D, raw soaked; E and F, cooked 10 min. Scale bars: A, B, and C, 10 μm; D, 5 μm; E, 100 μm; F, 50 μm. (From Wood, D.F. et al., *30th Annual UJNR Symposium Proceedings*, 2001. With permission.)

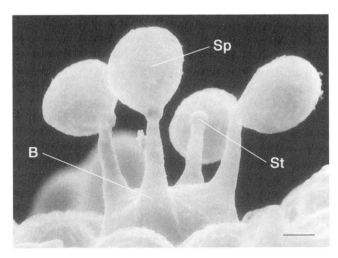

FIGURE 21.9
Spores on the edge of a mushroom gill. Each basidium (B) bears a basidiospore (Sp) attached by a sterigma (St). Scale bar equals 1 μm. (Micrograph courtesy of E. Brooke Danielson-Haskell.)

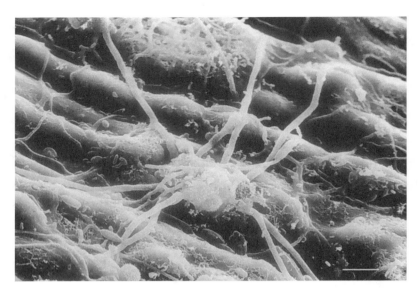

FIGURE 21.10
Fungal colonization on the underside of a Douglas-fir needle. Scale bar equals 10 μm. (Micrograph courtesy of E. Brooke Danielson-Haskell.)

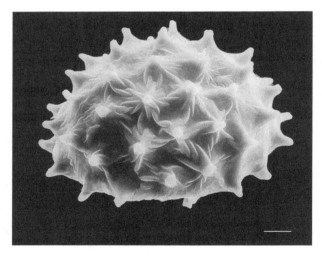

FIGURE 21.11
Hibiscus pollen grain. Scale bar equals 20 μm. (Micrograph courtesy of E. Brooke Danielson-Haskell.)

III. Environmental Scanning Electron Microscopy (ESEM)

A. Principles of Operation

The first ESEM was developed by ElectroScan in the late 1980s. These instruments allow investigators to view the specimen, not as a dead, dried sample, but rather, to image it while hydrated, and in some cases alive (Figure 21.12). ESEMs function by differentially pumping the column, which contains the electron gun and electromagnetic lenses, from the specimen chamber. The ESEM employs pressure limiting apertures that allow the gun to be maintained at 10^{-4} Pa, while the

FIGURE 21.12
FEI's ESEM XL40 environmental scanning electron microscope. (Photograph courtesy of the FEI Company, Hillsboro, OR.)

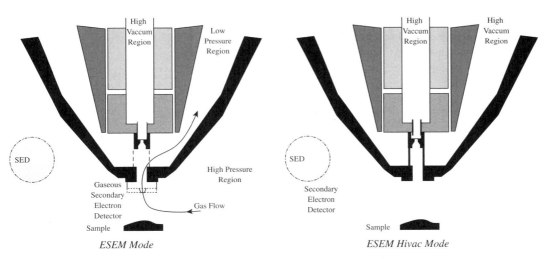

FIGURE 21.13
Philips' XL ESEM twin pressure limiting aperture (PLA) system used to minimize gas flow from sample chamber to high vacuum regions of the electron column. (Figure courtesy of the FEI Company, Hillsboro, OR.)

specimen chamber can be maintained at a much higher pressure, such as 665 Pa (5.0 torr) (Figure 21.13). This differential pumping system allows the specimen to remain in a hydrated condition, while the electron gun and lenses operate at the usual high vacuum conditions.

A gaseous secondary electron detector (GSED) is also employed in the ESEM (Figure 21.14). This detector carries a bias of several hundred volts positive, which accelerates secondary electrons from the specimen toward it for image formation. At the same time, gas molecules between the specimen and GSED are ionized, producing electrons and positive ions. These positive ions serve to eliminate charge buildup on uncoated specimens. The FEI Company (Hillsboro, OR) quotes resolution limits ranging from 2.0 to 3.5 nm for its various XL ESEMs.

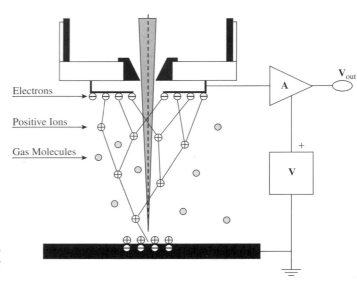

FIGURE 21.14
Principle of the gaseous secondary electron detector (GSED). (Figure courtesy of the FEI Company, Hillsboro, OR.)

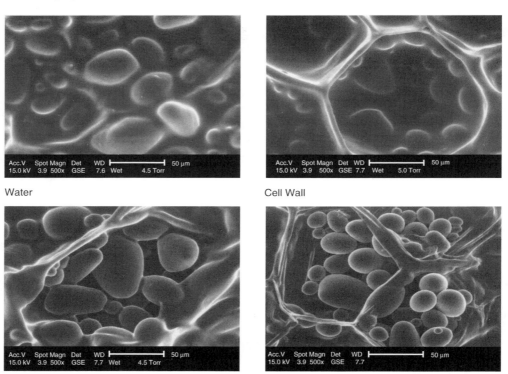

Water

Cell Wall

Meniscus

No cracking

FIGURE 21.15
Demonstration of ESEM. Image of starch grains in plant tissue with water slowly removed by adjusting temperature and pressure of specimen chamber/stage. (Micrograph courtesy of the FEI Company, Hillsboro, OR.)

B. ESEM Applications

Figure 21.15 (above) through Figure 21.17 illustrate application of the ESEM to studies of plant tissues shown fully hydrated through drying (Figure 21.15), flower tissue (Figure 21.16), and infected tissue (Figure 21.17).

FIGURE 21.16
Flower pistil imaged via ESEM. (Micrograph courtesy of the FEI Company, Hillsboro, OR.)

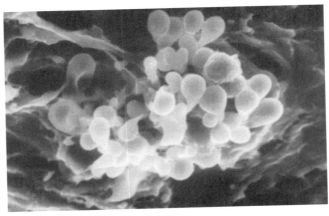

FIGURE 21.17
Infected tissue image via ESEM. (Micrograph courtesy of the FEI Company, Hillsboro, OR.)

IV. Transmission Electron Microscopy (TEM)

A. The Conventional Transmission Electron Microscope

The first electron microscope was built in 1932 by Ruska and Knoll in Germany, and the first commercial TEM was introduced in 1938 to 1939 by Siemens Corp. (Berlin, Germany). The basic components of the electron optical system of the conventional transmission electron microscope (TEM) are shown schematically in Figure 21.18. The electron beam is typically generated by drawing electrons away from a heated tungsten filament by means of a high accelerating voltage, which is usually in the range from 50,000 to 100,000 volts, but may be as great as 3 million volts in high-voltage systems. Other electron beam sources include lanthanum hexaboride filaments and field-emission systems. While these electron sources produce a much brighter beam, they entail a considerably greater cost.

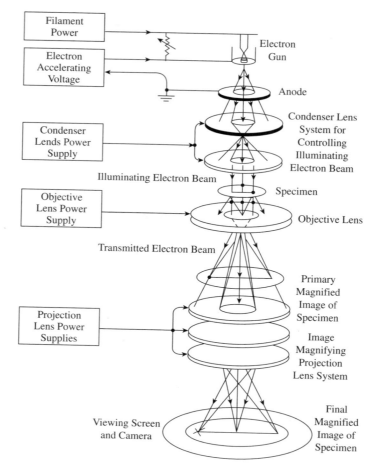

FIGURE 21.18A

Diagram illustrating inner working components of a transmission electron microscope. (From Kaufman, P.B., Labvitch, J., Anderson-Prouty, A., and Ghosheh, N.S., *Laboratory Experiments in Plant Physiology*, Macmillan College Publishing Co, New York, Copyright © 1975. With permission.)

    The electron beam is initially focused to a small diameter by the electron gun. It then passes through the positively charged anode into and out of the electromagnetic condenser lens system and interacts with the specimen.

    After passing through the specimen, the electrons enter the most important component of the electron optical system: the objective lens. This lens serves to produce the primary magnified image of the specimen; hence, its optical quality determines the basic quality of the image produced by the instrument. The highly refined objective lenses of most current electron microscopes can provide resolutions of 0.2 nm or better. These lenses are designed with extremely short focal lengths so as to minimize aberrations. For routine biological specimens, which do not require such extreme resolution, lenses with a somewhat longer focal length may be advantageous because they provide somewhat higher contrast.

    This primary image is further enlarged by a series of two or three projector lenses to give the final image, which is displayed on a fluorescent viewing screen and recorded photographically or electronically via a digital camera. These lenses, as well as the objective and condenser lenses, of most current TEMs are electromagnetic lenses. They consist of coils of wire wound on highly refined bobbins of soft iron, which contain magnetic gaps that shape the magnetic fields along the axes of the lenses so that they produce high-quality electron images. By varying the electric currents

FIGURE 21.18B
Modern TEM, FEI's Morgagni, designed for biological and medical applications. (Image courtesy of the FEI Company, Hillsboro, OR.)

in the coil windings, the strengths of the magnetic fields can be changed; this, in turn, changes the focal lengths of the lenses, making it possible to vary the magnification of the final image continuously from less than 100× to as much as 500,000×.

The fact that the focal length of the lenses can be changed continuously over such wide ranges and with such great ease gives the TEM greater flexibility than any other optical instrument. Thus, it is possible to use the TEM to perform a variety of functions in conjunction with basic image formation. These include several special diffraction techniques, operation in scanning mode, and x-ray spectral analysis. A variety of accessories are also available for special types of specimens and for treating specimens in special ways during examination.

B. Preparation of Tissues for TEM

Fixation, dehydration, and sectioning of tissues for TEM require about a week when carried out on the bench top. Recently developed and refined microwave tissue processing techniques have reduced tissue processing time to as little as 4 h, while affording ultrastructural preservation equal to or better than that produced via bench-top processing (Table 21.2). Microwaves cause dipolar molecules, such as water, to rapidly flip back and forth, and they cause ions to move rapidly back and forth. This increased motion greatly speeds penetration of fixatives and resins, and it also apparently speeds up the rate of chemical reactions, such as reactions between fixatives and biological molecules.

The basic processing steps include the following:
1. Tissue is chopped into small pieces (usually less than or equal to 1 mm in any dimension) in buffered (for example, 50 mM sodium cacodylate buffer at pH = 7.0) fixative. The most common fixative employs 2 to 4% glutaraldehyde. Aldehyde fixatives chemically bind proteins, thus cross-linking cellular components in a stable three-dimensional matrix.
2. Tissue is left in fixative for 2 h to overnight and is followed by washing in two to three changes of buffer. Following washing, tissues are often postfixed in 1% to 2% osmium tetroxide (OsO_4) for 1

TABLE 21.2
Comparison of Bench-Top Tissue Processing and Microwave-Oven Tissue Processing

	Bench Top	Microwave Oven
Initial tissue harvesting and processing in fixative	10 to 15 min	10 to 15 min
Aldehyde fixation	2 to 4 h	5 to 10 min
Osmium tetroxide	1 to 2 h	5 to 10 min
Ethanol dehydration	1 to 2 h	5 min
Liquid resin infiltration	overnight	45 min
Polymerization	overnight at 60°C	45 min, or overnight at 60°C

to 2 h and then washed with distilled water. Osmium tetroxide binds many cellular macromolecules; however, it is especially noted for its ability at binding unsaturated fatty acids, thus fixing and staining lipids. This helps reduce lipid extraction during alcohol dehydration.

3. Following OsO_4 fixation, the tissue is dehydrated in an ethanol series, such as 25, 50, 75, 100, 100, and again 100%. Each ethanol solution is employed for 10 to 20 min. This prepares the tissue for infiltration with hydrophobic epoxy resins.
4. An intermediate solvent may be applied. This is often propylene oxide, in which epoxy resins are highly soluble, thus aiding complete infiltration.
5. Once dehydrated, tissues are infiltrated with the liquid resin mixture for several hours to overnight. Resin is first mixed 1:1 with solvent (for example, 100% ethanol or propylene oxide), applied to tissue samples, and then changed to 100% resin with at least one change to a second aliquot of fresh resin.
6. Resin is polymerized overnight in a 60°C oven, often under vacuum to prevent incomplete polymerization due to interference by atmospheric water or other contaminants.
7. The hardened blocks must be carved with faces (surface of block) shaped into a trapezoid. This is usually performed with a sharp single-edge razor blade, using a dissecting microscope to observe one's work.
8. Carved blocks are next thick sectioned (1 to 2 µm sections), most often with glass knives on an ultramicrotome. The thick sections are examined in the LM, and the block face is further carved if necessary. The trapezoid shape allows the investigator to orient his/her tissue in the block face using the light microscope to target more precisely the desired area of study within the tissue.
9. Thin sections (50 to 90 nm) are cut with glass or diamond knives on the ultramicrotome and collected on 3-mm grids, often made of copper.
10. Thin sections are usually stained with uranyl acetate followed by lead citrate. These two heavy metal-containing compounds, along with osmium tetroxide that was applied during fixation, stain cellular macromolecules with electron dense heavy metal deposits.
11. After the sections are stained, they are ready for examination in the TEM.

C. TEM Applications

Figure 21.19 through Figure 21.21 illustrate examples of TEM applications. The first is an example of starch-filled chloroplasts (gravisensors in cereal grass shoots) in barley shoot leaf-sheath pulvini (Figure 21.19). The second shows control Indian mustard tissue (*Brassica juncea*) as compared with tissue that has been treated with the synthetic chelator, EDTA (Figure 21.20). The third is an example of TEM applied to diagnosing kidney disease (Figure 21.21).

Microscopy

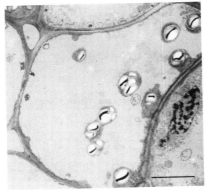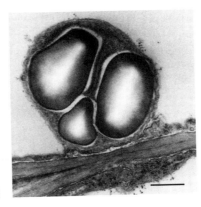

FIGURE 21.19
TEM view of starch-filled chloroplasts of a barley (*Hordeum vulgare*) plant. These plastids, filled with starch, act as the gravisensors in graviresponding cereal grass shoots. Scale bars: left, 5 μm; right, 0.5 μm.(Micrographs courtesy of Casey Lu.)

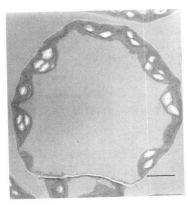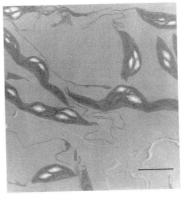

FIGURE 21.20
Cells and chloroplasts from Indian mustard (*Brassica juncea*) control (left) and treated with EDTA (right). Older leaves from plants treated with EDTA show ultrastructural damage, including plasmolysis and chloroplast membrane degredation. Scale bars equal 5 μm. (Micrographs courtesy of Todd Bouchier.)

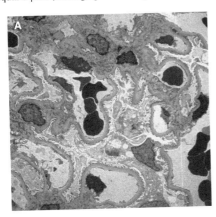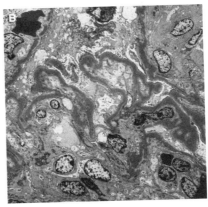

FIGURE 21.21
Kidney disease diagnosed via TEM. (A) Normal kidney glomerulus tissue; note the overall uniform gray appearance of the basement membrane. (B) Glomerulus showing dense deposit disease; note the extensive transformation of the glomerular basement membrane by very electron-dense material characteristic of this disease. (Images courtesy of Trish Bestmann and the Path Logic Company, Fair Oaks, CA.)

V. Confocal and Multiphoton Microscopy (CM)

Confocal microscopy is a relatively new technique that has proved to be a highly valuable tool for the cell biologist and for medical researchers. Confocal microscopy was conceptualized by Minsky in the 1950s, and the first laser confocal microscope became available in 1988, which was soon followed by many commercial versions.

Confocal microscopy allows one to image thick biological tissues in three dimensions by means of "optical sectioning." This contrasts with conventional microscopy, where specimens (smears, peels, sections, or imprints) are observed in a two-dimensional projection mode. It allows one to conduct quantitative image analysis by obtaining three-dimensional images from a succession of thin, independent "optical sections" that are obtained at various levels of unsectioned cell or tissue specimens. These images are obtained by laser beam light sources with scanning capabilities in instruments called confocal scanning laser microscopes (CSLM). Coupled with CSLM is a computer system that allows one to store successive optical images obtained from a given specimen.

Confocal microscopy operates by focusing a visible light beam or a laser beam on a small spot within a specimen rather than illuminating the whole objective field of view as in bright-field and epifluorescence microscopy (Figure 21.22). One problem with standard epifluorescence microscopy is the out-of-focus blur that results from light emitted throughout the specimen from a fluorochrome entering the imaging lenses. This out-of-focus blur (noise) is screened out by the apertures of the confocal microscope, leaving only light that contains information (signal) (Figure 21.23).

Some common applications of the CSLM involve the use of fluorochromes (stains) specific for investigating intracellular pH, lipoprotein (FAD), Ca^{2+} distribution and concentration, DNA, and green fluorescent protein (GFP) tags. Table 21.3 indicates some of the common fluorochromes used for specific applications.

A particularly valuable fluorescent system involves the use of green fluorescent protein, a fluorescent protein isolated from coelenterates, such as the Pacific jellyfish, *Aequoria victoria*. In nature, it transduces blue chemiluminescence, produced by the protein, aequorin, into green fluorescent light by energy transfer. The GFP gene has been isolated and is now used as a common tool used for making chimeric proteins of GFP linked to other proteins. This chimera allows researchers to follow the movement and interactions of the tagged protein via fluorescence emitted by GFP. It can be used as a noninvasive fluorescent marker in living cells, and it allows for a wide range of applications where it can function as a cell lineage tracer, reporter of gene expression, or as a measure of protein-protein interactions.

Another valuable fluorescent technique is known as fluorescent resonance energy transfer (FRET). In this technique, one fluorescently tagged biomolecule transfers its energy to a second tagged receiver molecule at a distance greater than intra-atomic distances (tens of angstroms) and does not require that molecules collide with each other. This technique allows researchers to investigate intermolecular interactions and distributions.

Multiphoton microscopy is the latest advancement in microscopy systems. It allows researchers to produce three-dimensional images of live and fixed tissues as is produced by confocal microscopy. However, multiphoton microscopy does not come with the limitations of confocal microscopy and will likely replace and/or be combined with confocal microscopes in the near future.

Multiphoton microscopy uses short, femtosecond pulses of lower-energy infrared laser light (produced by a titanium:sapphire laser) to excite fluorochromes instead of the shorter, higher-energy wavelengths used in confocal microscopes. When two lower-energy photons strike a fluorochrome within femtosecond time periods, the fluorochrome will be excited. The multiphoton technique will excite fluorochromes that may also be excited by shorter wavelength UV light, but excitation is limited to a small spot in the specimen at the focal plane. Therefore, photobleaching and phototoxicity only occur at a point in the focal plane rather than throughout the specimen, as occurs in confocal and traditional epifluorescence microscopy. This is a major drawback of confocal microscopy, which requires the use of a more intense energy beam to provide enough

Microscopy

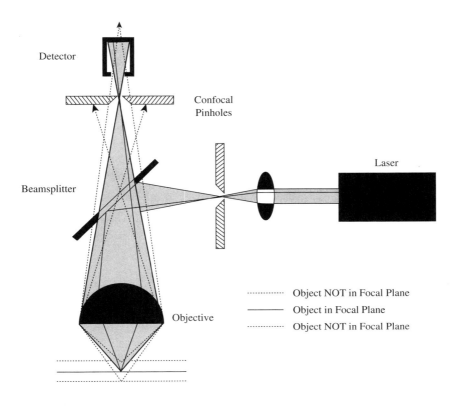

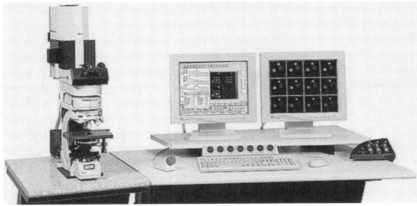

FIGURE 21.22
Confocal basics and Leica Microsystem's TCS SP2 RS Confocal/Multiphoton microscope. (Diagram and image courtesy of Leica Microsystems, Wetzler, Germany.)

returning fluorescent light for image formation via the small pinhole, which is required to remove out-of-focus light. This leads to rapid (sometimes within minutes) photobleaching and phototoxicity. Another advantage of the multiphoton system is that it uses a longer wavelength, which penetrates more deeply into specimens than the laser light employed by standard confocal microscopy. This allows investigators to probe more deeply into samples. These advantages make it likely that standard confocal microscopes will be converted to multiphoton systems in the near future, and that multiphoton systems will become standard for researchers needing to do three-dimensional imaging and living-state imaging.

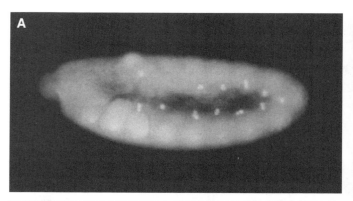
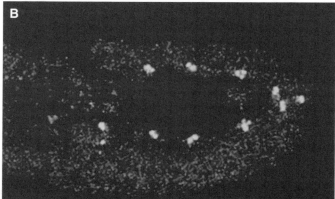

FIGURE 21.23
Standard fluorescent microscopy image (A) of a *Drosophila* embryo compared with the image of same specimen produced by confocal microscopy (B). The images are of whole *Drosophila* embryos stained for Even-skipped (eve) using a Cy3 conjugated secondary antibody. The cells that are staining are the Even-skipped pericardial cells. (Images courtesy of Susan Klinedinst and Rolf Bodmer.)

TABLE 21.3
Fluorochromes and Their Application

Use	Fluorochrome	Excitation peak (nm)	Emission peak (nm)
Covalent labeling	Fluorescein-isothiocyanate (FITC)	490–494	520–521
	Eosin-ITC	524–525	548–550
	Erythrosin-ITC	535–540	558–560
	Tetramethylrhodamine-ITC (TRITC)	541–554	572–573
	Rhodamine X-ITC (XRITC)	578–582	601–604
	Texas Red sulfonyl chloride	596	615–620
	Lissamine rhodamine B sulfonyl chloride	567–570	584–590
	Cascade blue	375–378	423–425
	Cascade blue	398–399	423–425
	Phycoerythrin-R	480–565	578
	Allophycocyanin	650	660
	Coumarin-phalloidin	387	470
	Bodipy phallicidin	505	512

(continued)

TABLE 21.3 (continued)
Fluorochromes and Their Application

Use	Fluorochrome	Excitation peak (nm)	Emission peak (nm)
Covalent labeling *(continued)*			
	Nitrobenzoxadiazole (NBD)	478	520–550
	Dansyl hydrazine	336	531
	Indotrimethinecyanines (CY3)	530–550	575
	Indopentamethinecyanines (CY5)	630–650	670
Nucleic acid labeling	Hoechst 33342	340–343	450–483 (DNA)
	DAPI	358	461 (DNA)
	Propidium iodide	493–536	623–630 (DNA)
	Ethidium bromide	482–510	595–616 (DNA)
	ACMA	430	474 (DNA)
	Chromomycin A3	420–445	474–580 (DNA)
	Mithramycin	420–445	575–580 (DNA)
	7-Aminoactinomycin-D	523	647 (DNA)
	Acridine orange	500	525 (DNA)
	Acridine orange	460	650 (RNA)
	Pyronin Y	497	563 (RNA)
Thiol conjugates	Lucifer yellow	426–490	525–540
	Cascade blue	375–378	423–425
	Cascade blue	398–399	423–425
Membrane potential	Thiazole orange	453	480
	$DiOC_n(3)$	482–485	500–510
	Tetramethyl-rhodamine ethylester	549	574
Mitochondria	Rhodamine 123	505–511	534
	2-di$_1$ASP(DASPMI)	461	518
Lipid content	Nile red	485	525
	NBD phosphocholine	460	534
	Diphenylhexatriene (DPH)	351	430
pH	BCECF	500	530 (basic pH)
	SNAFL-2	485, 514	543 (acidic pH)
	SNARF-1	518–548	587 (basic pH)
	FD	490	515 (acidic pH)
Ca^{2+}	Fura-2	340, 380	510
	Fura-3	506	526
	Indo-1	350	482–485 (low Ca)
	Indo-1	330	390–410 (high Ca)
	Quin-2	339	492 (high Ca)
	Rhod-2	552	581 (low Ca)
	Fluo-3	488	526 (low Ca)

Source: Modified from Table 2 in Häder, D.-P., Ed., *Image Analysis: Methods and Applications,* CRC Press, Boca Raton, FL, 2001.

References

1. **Bozzola, J.J. and Russell, L.D.**, *Electron Microscopy*, 2nd ed., Jones and Bartlett, Boston, MA, 1999.
2. **Cherry, R.J.**, Ed., *New Techniques of Optical Microscopy and Microspectroscopy: Topics in Molecular and Structural Biology*, CRC Press, Boca Raton, FL, 1991.
3. **Häder, D.-P.**, Ed., *Image Analysis: Methods and Applications*, CRC Press, Boca Raton, FL, 2001.
4. **Hayat, M.A.**, *Correlative Microscopy in Biology, Instrumentation and Methods*, Academic Press, New York, 1987.
5. **Hayat, M.A.**, *Principles and Techniques of Electron Microscopy: Biological Applications*, 4th ed., Cambridge University Press, Cambridge, U.K., 2000.
6. **Login, G.R. and Dvorak, A.M.**, *The Microwave Tool Book*, Beth Israel Hospital, Boston, MA, 1994.
7. **Matsumoto, B.**, Ed., *Cell Biological Applications of Confocal Microscopy*, Vol. 38, Methods in Cell Biology, Academic Press, New York, NY, 1993.
8. **Monl, G.**, *Hybridization Techniques for Electron Microscopy*, CRC Press, Boca Raton, FL, 1993.
9. **Ogawa, K. and Barka, T.**, *Electron Microscopic Cytochemistry and Immunocytochemistry in Biomedicine*, CRC Press, Boca Raton, FL, 1993.
10. **Rasmussen, N.**, *Picture Control, The Electron Microscope and the Transformation of Biology in America, 1940–1960*, Stanford University Press, Stanford, CA, 1997.
11. **Rizzuto, R. and Fasolato, C.**, Eds., *Imaging Living Cells*, Springer-Verlag, New York, NY, 1999.
12. **Rost, F. and Oldfield, R.**, *Photography with a Microscope*, Cambridge University Press, Cambridge, U.K., 2000.
13. **Sheppard, C.J.R. and Shotton, D.M.**, *Confocal Laser Scanning Microscopy*, Royal Microscopical Society, Microscopy Handbook No. 38, Springer-Verlag, New York, 1997.
14. **Shotton D.**, *Electronic Light Microscopy: Techniques in Modern Biomedical Microscopy*, John Wiley & Sons, New York, 1993.
15. **Smith, R.F.**, *Photomicroscopy and Photomicrography: A Working Manual*, 2nd ed., CRC Press, Boca Raton, FL, 1993.
16. **Watt, I.M.**, *The Principles and Practice of Electron Microscopy*, 2nd ed., Cambridge University Press, Cambridge, U.K., 1997.

Chapter 22

Localization of Gene Expression

*Scott Harding, Chung-Jui Tsai, Leland J. Cseke,
Peter B. Kaufman, Soo Chul Chang, and Feng Chen*

Contents

I.	Subcellular Localization	488
	A. Preparation of Organelles	489
	B. Sucrose Gradients	489
II.	Tissue Localization by Tissue Printing	490
	A. General Procedures of Tissue Printing	491
	1. Materials	491
	2. Procedures	491
	B. Example with Germinating Tomato (*Lycopersicon esculentum*) Seeds	491
III.	Localization by *In Situ* Hybridization	493
	A. Tissue Preparation	493
	1. Tissue Collection and Fixation	493
	2. Dehydration Using an Ethanol/tert-Butanol/H_2O Series	493
	B. Infiltration, Embedding, and Section Preparation	494
	1. Paraffin Infiltration and Embedding	494
	2. Plastic Infiltration and Embedding	495
	3. Sectioning	495
	C. Hybridization	496
	1. Paraffin or Plastic Removal and Section Rehydration	496
	2. Heat Denaturation and *Protease* Treatment	496
	3. Acetylation for Nonspecific Blocking	496
	D. Nonradioactive Signal Detection	499
	1. Post-Hybridization Washing and Blocking	499
	2. Immunodetection of DIG Probes	500
	3. Post-Antibody Washing and Color Development	500

E. Radioactive Signal Detection 501
   1. Post-Hybridization Washing 501
   2. Autoradiography 502
   3. Staining and Viewing 503
F. Troubleshooting Guide 504
Reagents Needed 504
References 507

I. Subcellular Localization

In order to study cells, their structures and functions should be observed. The structure can be observed by using light and electron microscopy (see **Chapter 21**). The functions can be studied by obtaining fractionated organelles that are relatively pure. Each organelle has its own characteristics, such as size, shape, and density. These characteristics make one organelle different from other organelles within a cell. The procedures of fractionation consist of two consecutive stages, namely, (1) homogenization, which breaks cells open and releases organelles, and (2) centrifugation, which separates the individual organelles based on the organelles' characteristics.[1] Cells should be broken open gently, after which each of its organelles can be subsequently isolated. This homogenization is accomplished by physical breaking, ultrasonification, osmotic stress, and compression.[2] Separation of organelles from one another is achieved by differential centrifugation. Centrifugal force is used to separate organelles as a function of their sedimentation coefficients, which are based on their size, shape, and density. Cells are homogenized and subjected to several steps of centrifugation. At each step, the pellet obtained contains an organelle fraction, and the supernatant is collected for the next step. For instance, the first step precipitates the nuclei, and the resulting supernatant contains the other cellular organelle fractions. At the second step, mitochondria, chloroplasts, and lysosomes are collected in the pellet, and the supernatant is used for the next step. Further fractionation of cell organelles, such as microsomes, is possible by the method of rate-zonal centrifugation.[3] Sucrose or Percoll density gradients can be developed using the differential centrifugation methodology. In this case, the centrifugation medium is characterized by an increase in density. The reason why sucrose and Percoll are used for density gradients is due to their chemical natures. They are inert toward the centrifuge tubes, do not interfere with monitoring of the sample material, are easy to separate from each fraction after centrifugation, are easily used to observe the concentration of the gradient medium, are stable in solution, are available in a pure and analytical form, and exert minimal osmotic pressure. Organelle particles separate according to their own centrifugal field, size, shape, and density differences between the particles and the suspending medium. In order to analyze the relative purity of each fraction, performance of activity assays of "marker enzymes" is recommended (see Table 22.1). A marker enzyme is an enzyme that is characteristic of each particular organelle in a given fraction.

TABLE 22.1
Marker Enzymes for Respective Cellular Organelles

Organelles	Marker Enzymes
Plasma membrane	5-Nucleotidase
Golgi apparatus	Thiamine pyrophosphatase, Glycosyl transferase
Mitochondria (inner membrane)	Succinate-INT-reductase, Cytochrome oxidase
Mitochondria (outer membrane)	Monoamine oxidase
Lysosomes	Acid phosphatase, Amyl sulfatase, Glucuronidase
Endoplasmic reticulum	Glucose-6-phosphatase
Peroxisomes	Catalase, Uric acid oxidase

Localization of Gene Expression

A. Preparation of Organelles

1. Put a 3.0-g sample (e.g., rat liver) in a petri dish. The tissue should be on ice.

Note: It is strongly recommended that all procedures described below be performed on ice, unless otherwise indicated. All media and apparatus should be precooled and maintained at 4°C.

2. Mince the tissue with a razor blade, carefully.
3. Put the tissue into a homogenizer and add 9.0 ml of homogenization buffer (HB) and homogenize the tissue.
4. Pour the homogenate through a layer of cheesecloth into a 50-ml centrifuge tube.

Notes: (1) It is possible to obtain more homogenate by adding another 9.0 ml of HB to the homogenizer and grinding the tissue some more. Pour this through cheesecloth into the centrifuge tube. (2) In the case of plant cells, shearing forces required to break the cell walls can cause unwanted organelle damage.

5. Cap the tube and centrifuge it at 1000 g for 5 min at 4°C.

Note: One more tube should be prepared if you have just one tube. Make sure that the two tubes are balanced.

6. Pour the supernatant into a new 50-ml centrifuge tube.

Note: The pellet is loosely attached to the tube. Be careful pouring off the supernatant; otherwise, the pellet will be detached from the tube.

7. Add 5.0 ml HB into the tube containing the pellet and resuspend the pellet. This is the nuclei-enriched fraction. In plants, this fraction may also contain peroxisomes and chloroplasts containing starch.
8. Cap the tube and centrifuge the supernatant at 20,000 g for 15 min at 4°C.
9. Pour this supernatant into another 50-ml centrifuge tube. Resuspend the pellet with 5.0 ml HB. This is the mitochondria-enriched fraction.
10. Cap the tube and centrifuge the supernatant at 25,000 g for 30 min at 4°C.
11. Again, pour the supernatant into a new 50-ml centrifuge tube and resuspend the pellet with 5.0 ml HB. This is the lysosome-enriched fraction.
12. Cap the tube and centrifuge the supernatant at 40,000 g for 60 min at 4°C.
13. Pour the supernatant and resuspend the pellet with 5.0 ml HB. This is the microsome-enriched fraction.
14. Cap the tube and centrifuge the supernatant at 150,000 g for 3 h at 4°C.
15. Collect the supernatant (cytosol fraction) into a new 50-ml centrifuge tube and resuspend the pellet with 5.0 ml HB. This is the ribosome fraction.
16. Cap all of the tubes containing resuspended pellets and store the tubes at −20°C until use.

B. Sucrose Gradients

1. Perform steps 1 through 7 as listed in **Section I.A**.
2. Mark five points in a centrifuge tube with appropriate gaps. Put sucrose solutions into the tube layer by layer, starting with 55%, and then add 50, 44, 33, and 20% sucrose sequentially up to the marks. In order to minimize mixing of adjacent solutions, place the tip of the pipette on the inside of the tube just above the surface of the previous solution and pour the new solution gently. Store the completed gradient upright on ice.

Notes: *(1) Each sucrose solution can also be frozen between additions. Then the whole step gradient can be thawed to yield very clear layers. (2) Use a different Pasteur pipette for each solution; otherwise, the sucrose density cannot be accurate. The pellet from step 6 is fairly loose, so decant the supernatant with care.*

3. Layer the supernatant onto the top of the sucrose gradient.
4. Carefully dry the outside of your tube and place it in the appropriate centrifuge bucket.

Note: *Tubes should be dried before being put into centrifuges in order to make them easy to remove from the rotor or bucket. Check the weight of the tube, being sure to include the cap on the balance as well.*

5. Cap the tubes, put the tubes into a bucket, and centrifuge at 25,000 g for 4 h at 4°C.
6. Remove the tubes from the bucket carefully and examine the interfaces between the different concentrations of sucrose.
7. Carefully remove the bands, one by one, using syringes. From the bottom of each tube each fraction may contain peroxisomes, lysosomes, mitochondria, smooth endoplasmic reticulum with plasma membranes, and Golgi apparatus. You can use a refractometer to measure the actual density of the samples.
8. Perform marker-enzyme assays on appropriate enzymes (such as those in Table 22.1) on each of the collected fractions. When the data is plotted per fraction, clear peaks should be seen for each marker enzyme with specificity to each type of organelle.

II. Tissue Localization by Tissue Printing

Section I described techniques for obtaining subcellular localization of organelles from cells. However, in order to understand the physiological function of a gene, determination of gene expression and its protein product at the tissue level is often necessary. A simple and rapid technique called tissue printing is available for this purpose.[4] Tissue printing is a well-suited technique for use in experiments involving many samples and when relatively large samples or whole seedlings are being examined.[5] In comparison with techniques used specifically for messenger RNA (mRNA) (e.g., Northern blot) or protein (e.g., Western immunoblotting), tissue printing requires no extraordinary expertise or equipment.[5] Therefore, it is one of the best choices for determination of tissue localization of gene expression at both mRNA and protein levels. By cutting fresh tissue in order to obtain a smooth surface and placing the cut surface of a tissue on a nitrocellulose or nylon membrane, the contents of the cut cells will transfer to the membrane. Because of the presence of rigid cell walls, a physical image of the cut surface is usually produced on the membrane. This makes the anatomy of tissues visible under the microscope without any further treatment. With appropriate probes, it is possible to visualize the cell contents on the membrane. Since cellular mRNAs or proteins show little lateral diffusion during or after printing on the membrane, the precise localization of a specific mRNA or protein can be determined by comparing the signal with a physical print on the same membrane. The tissue print technique has been used to successfully determine tissue-level localization of proteins, enzymes, mRNA, viral DNA, and metabolites.[6]

Localization of Gene Expression

A. General Procedures of Tissue Printing

1. Materials

Fresh plant tissues
Double-edged razor blade
Paper towels
Membrane (nitrocellulose or nylon)
Parafilm™
Forceps
Plastic gloves
Light microscope
Reagents for Northern hybridization and Western immunoblotting

2. Procedures

1. Place several layers of paper towels on a level, hard surface. Then, place a blotting membrane of appropriate size on top of the stack of paper towels.
2. Use a clean double-edged razor blade to hand-cut the tissue in order to obtain a smooth cut surface. Depending on the particular tissue sample, the tissue can be cut by a single-cross section, or by 0.2- to 2-mm tissue sections. If the cut cells contain excess tissue exudates, gently touch the cut surface on a filter paper.
3. Using forceps, gently grasp the edge of the tissue section, and transfer it to the membrane. To avoid smearing, do not move the tissue after it is in contact with the membrane.
4. Place a small piece of Parafilm (or nonabsorbent paper) over the tissue. This can protect the membrane from contamination from one's fingers. Press the tissue with the thumb for a given amount of time using sustained pressure. This varies and depends on the particular tissue sample. For example, to obtain an image of the cut surface of a germinating tomato seed, the section is usually pressed above the membrane for 10 to 15 sec. In some cases, a second image from the same tissue section is desirable. For instance, in order to check mRNA expression of a gene, antisense and sense RNA probes will be hybridized with the same tissue. Therefore, a second image from the same tissue is necessary for this purpose. To do this, transfer the tissue section to a second membrane and press it for longer time (e.g., 20 to 30 sec for tomato seeds).
5. Remove the protective Parafilm and the tissue using forceps. Let the membrane dry in the air. The print image can be observed by using a dissecting light microscope.
6. Repeat steps 2 to 6 for the next printing.
7a. Tissue-print hybridization: If checking for mRNA expression, the membrane should be ultraviolet (UV)-cross-linked before proceeding with prehybridization. All of the following protocols for hybridization, washing, and signal detection are essentially the same as those for Northern blot, as described in **Chapter 6**. Either radioactive probes or nonradio probes could be used for hybridization.
7b. Tissue-print immunolocalization: The membrane can also be used to detect protein following the protocols for Western immunoblotting, as described in **Chapter 8**.

B. Example with Germinating Tomato (*Lycopersicon esculentum*) Seeds

The tissue-printing technique was used recently for a study of tomato seed germination.[7] As seen from Figure 22.1, a germinating tomato seed was bisected into two parts. The anatomical structures are labeled. By pressing the cut surface of the seed onto a nylon membrane, the rigid cell walls leave a clear physical image, as seen in Figure 22.2. In this figure, panel A shows the localization of an expansin gene, *LeEXP4*, during tomato seed germination. It is specifically localized to the endosperm cap region. In panel B, the print hybridized with the sense probe shows no signal.

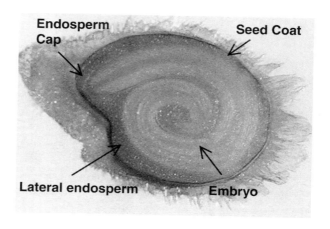

FIGURE 22.1
Anatomy of the cut surface of a germinating tomato seed. A tomato seed was imbibed in water for 24 h and then cross-sectioned. The tomato seed is composed of embryo completely surrounded by a relatively hard and brittle endosperm and a thin testa (seed coat). The radicle tip of the embryo is enclosed by a specialized endosperm tissue, called endosperm cap. The rest of the endosperm is called the lateral endosperm.

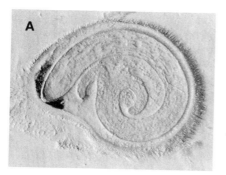 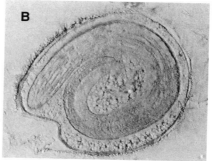

FIGURE 22.2
Tissue localization of expansin gene, *LeEXP4*, in germinating tomato seeds. Water-imbibed seeds were bisected, and the cut surfaces were printed onto two membranes. In panel A, this membrane was hybridized with an antisense RNA probe. In panel B, this membrane was probed with a sense RNA probe and used as a control. The results show that the *LeEXP4* mRNA is specifically localized to the endosperm cap region.

Notes: (1) Gloves are recommended to be worn during the entire procedure so as to avoid contamination. It is especially important for Northern tissue printing. Contamination with RNase should be avoided. (2) The tissue print should be observed to make sure there is no double image. (3) Before probing with an RNA probe or immunublotting, it may be necessary to gently preblot the membrane with 5X SSC in order to wash away excess tissues that stick on the membrane during the printing process. (4) Since a control is needed to confirm the hybridization signal, both a sense probe and an antisense probe are needed in tissue printing to check for mRNA localization. This is done using RNA probes. The segment of the gene is inserted into a vector (e.g., pBlueScript) that contains promoters at both ends of multiple cloning sites. Then, the plasmid is linearized at both ends. T7 RNA polymerase or SP6 RNA polymerase can be used to transcribe either the sense or the antisense RNA probe. There is no difference in the methods used for labeling probes for tissue printing from those used for Northern blot hybridization. Either radioactive or nonradioactive methods can be used for labeling, as described in **Chapter 4**.

III. Localization by *In Situ* Hybridization

Hybridization of paraffin- or plastic-embedded tissue sections with labeled probes to visualize the cellular distribution of mRNA transcripts is known as *in situ* hybridization (ISH). The following ISH protocol is based on that of Jackson,[8] but it includes a number of modifications to improve results from woody plant tissues. Special features of this protocol include the use of full-length RNA probes at high hybridization temperatures, followed, when colorimetric detection is used, by color development in buffer containing polyvinyl alcohol (PVA) to reduce chromagen diffusion.[9] These modifications reduce background from nonspecific hybridization and reduce color development time while improving signal localization. The protocol works well with soft tissues and eliminates background due to nonspecific probe sticking in secondary vascular tissues of woody plant specimens. The method is applicable to animal tissues as well. *RNase*-free technique (see **Chapter 2**) should be employed throughout. Procedural modifications for performing ISH with [^{35}S]-labeled probes[8,10,11] are included where appropriate. However, most steps of the ISH protocol are applicable to both detection systems.

A. Tissue Preparation

Tissue preparation can be flexibly scheduled prior to hybridization experiments, since paraffin or plastic tissue blocks can be stored indefinitely. The success of any ISH protocol depends on the care given to tissue collection, fixation, and dehydration. In general, blocks of solid tissue should not exceed linear dimensions of ~4 mm. Woody stem sections should be excised with a sharp double-edged razor (which is sharper than single-edged razor blades often used in laboratories), and they should not be more than 2 mm thick. Best fixation, dehydration, and sectioning are achieved with 1-mm-thick woody stem sections, although cutting sections this thin requires a good eye and patience. An ethanol/tert-butanol dehydration series is recommended instead of ethanol *per se* in order to reduce brittleness of woody tissues.

1. Tissue Collection and Fixation

1. Immediately submerge specimen into cold fixative of freshly prepared, filtered, 4% buffered formaldehyde in a straight-walled glass vial. The fixative should include 1 mM dithiothreitol (DTT) if [^{35}S]-labeled probes are to be used.
2. Gently, using mild aspiration or a hand pump, vacuum infiltrate the submerged specimen so that it sinks following gentle de-aspiration to release the vacuum.

Notes: *(1) Intact shoot apices and tissues that are hairy or resinous can be harvested into FA solution to facilitate entry of fixative into the tissue. To minimize RNA loss and to preserve resolution, these tissues should be transferred to 4% buffered formaldehyde immediately following brief vacuum infiltration. (2) Specimens that float can be submerged by fashioning a hollow piston from a plastic transfer pipette, and then using that to wedge a piece of nylon screen into the sample vial, pushing the sample below the surface of the fixative. Apply vacuum with the sample submerged.*

3. Allow infiltrated tissues to become fixed (under a restored vacuum if sections do not sink) at 4°C for 6 to 18 h, depending on size and penetrability of fixative into the tissue.

2. Dehydration Using an Ethanol/tert-Butanol/H$_2$O Series

1. Remove fixative and wash specimen twice (5 min each) at 4°C with 5 ml 20% ethanol.

2. Replace with 5 ml of solution 1 and swirl every 15 min for 2 h. At intervals of 2 to 3 h, replace solutions according to the following graded ethanol/tert-butanol/H$_2$O solution series. The first three steps in this series should be conducted at 4°C.
   Solution 1: 40/0/60
   Solution 2: 45/15/40
   Solution 3: 50/28/22
   Solution 4: 45/55/0
   Solution 5: 25/75/0
   Solution 6: 0/100/0
3. Repeat Solution 6 and store tightly capped sample at 28 to 37°C (tert-butanol will crystalize at 20°C) until paraffin infiltration is initiated. If convenient, dehydration steps from Solution 3 on can be allowed to proceed overnight.

Note: *Solutions should be prepared using DEPC-treated distilled, deionized water (ddH$_2$O) containing 0.85% (w/v) NaCl. Tissues that become colorless can be stained with 0.1% Eosin Y (prepared as a 10% stock in ethanol and protected from light) during dehydration from solutions 4 to 6 to help in orientation later. If you choose to dehydrate in an ethanol series, paraffin infiltration (**Section III.B.1**) is initiated after gradually replacing ethanol with xylene.*

B. Infiltration, Embedding, and Section Preparation

In addition to adequate fixation and dehydration, high-quality section preparation also depends on thorough tissue infiltration with paraffin or a plastic resin. Paraffin is the most commonly used and least expensive material for preparation of sections 6 to 15 μm thick. Plastic embedding resins, on the other hand, are harder and thus allow ultrathin sections (1 to 3 μm) to be made for better resolution of structural details. However, it should be understood that thinner sections will yield a weaker ISH signal. Prepare *RNase*-free work areas for slide and section handling. Wear gloves. Dust or particulates in solutions can be difficult to remove from slides, may contain *RNases*, and certainly interfere with image documentation. Paraffin or plastic tissue blocks and sections can be stored indefinitely, but deparaffinized or plastic-depolymerized sections should be used immediately.

1. Paraffin Infiltration and Embedding
1. Add five chips of Paraplast Plus™ (Fisher Scientific, Chicago, IL) to dehydrated sample in 5 ml tert-butanol.
2. Cap and place into a 60°C oven to melt chips (4 h to overnight).
3. Add ten more chips of Paraplast Plus, loosely cap and incubate at 60°C overnight to evaporate tert-butanol.
4. Replace with 2 to 3 ml molten Paraplast Plus, and return loosely capped vials to 60°C oven for 4 h to overnight. Repeat this step three to five times over 2 to 3 days.
5. Orient the specimen in a prewarmed mold, add molten wax, and quickly chill on ice water.
6. Block trimming. After hardening, use a single- or double-edged razor blade to trim the specimen block. The object is to trim the block so that the sections obtained from the microtome are trapezoidal. The trimmed block is mounted onto the microtome specimen chuck so that the side of the block contacting the knife first represents the base of the trapezoid. The opposite (top) edge of the trimmed block should parallel the base to facilitate sectioning.

Notes: *(1) Conduct the initial infiltration steps (1 to 3) at 42°C if dehydrated tissues are stored in xylene. (2) Step 5 can be carried out using bunsen burners or similar heat sources, but care must be taken to avoid overheating the wax. An alternative homemade approach is*

Localization of Gene Expression

to set up on a 65°C slide warmer and place an incandescent light so ca. 30 cm) above the slide warmer to provide a uniform warm zone. 12 in. "table tops" (with temperature control for the work surface) are also available.

2. Plastic Infiltration and Embedding

Although ISH signals are not as strong as in paraffin sections, cellular and morphological features are better preserved in plastic sections. A commonly used, reversible plastic embedding for ISH is butyl-methyl methacrylate. Although the basic process of tissue infiltration and embedding is the same, each plastic resin requires its own catalyst and specific polymerization conditions. Therefore, the manufacturer's instructions should be adhered to for tissue embedding in plastic resin. Briefly, paraformaldehyde-fixed and dehydrated tissues are infiltrated and then embedded in liquid plastic solution in appropriate embedding molds. Polymerization of the plastic is initiated by addition of a catalyst and/or is allowed to proceed under specified conditions for several hours to overnight to form a hard block. The plastic block is trimmed as described above for paraffin blocks (**Section III.B.1**) and can be stored indefinitely.

3. Sectioning

One of the most difficult aspects of sectioning paraffin-embedded samples is static attraction between newly cut sections and microtome or tissue-block surfaces. This problem is exacerbated with sections from woody tissues or from complex and sometimes sclerified tissues, such as intact apical shoots that may also have retained microscopic pockets of air throughout fixation. Some of these problems can be eliminated by purchasing a retracting microtome that moves the tissue block away from the knife edge between cuts. Increasing the local humidity with dampened towels, or manually grounding the face of the tissue block by touching it with a damp, rubber-gloved fingertip before each cut may also help. Woody stem sections with zones of variable hardness sometimes exhibit internal tearing along textural discontinuities. This may be due to differential fixation. We have found that decreasing fixation time or increasing section thickness can sometimes overcome the problem. Extreme overfixation can affect infiltration, leading to separation between paraffin and tissue. When this leads to sectioning failure, one remedy is to immerse the tissue block in ice water for 30 min before resuming sectioning.

1. Tissue sections 6 to 10 μm in thickness are cut from the paraffin block and gently transferred, shiny face down, to a drop of ddH_2O on Superfrost Plus™ slides (Fisher Scientific, Chicago, IL) using a moistened wood stick.
2. Place the slide on a 42°C slide warmer to allow specimen sections to flatten out. Excess water is carefully removed as soon as section flattening is observed, and the slides are left on the warmer to dry for 12 h.

Notes: (1) Superfrost Plus slides can be used for ISH without additional preparation. (2) Avoid high slide warmer temperatures. Sample bubbling or shifting of section components due to paraffin softening can occur at temperatures as low as 50°C, and these problems will be exacerbated if excess water is not removed after section flattening. Protect from airborne particulates. (3) Plastic blocks should be sectioned on an ultramicrotome with a glass knife to obtain thin (0.5 to 3 μm) sections.

3. Paraffin/plastic sections can be stored indefinitely in standard slide storage boxes at room temperature. Slides should be used immediately after the matrix is removed.

Hybridization

1. Paraffin or Plastic Removal and Section Rehydration

Immerse slides in CitriSolv™ cleaning reagent (Fisher Scientific, Chicago, IL) or equivalent for 30 min. Repeat once. Drain excess on Kimwipes™.

If sections are embedded in butyl-methyl methacrylate, use acetone (30 sec) in place of CitriSolv. Rehydrate slides through the following series:

100% ethanol, 2 min
100% ethanol, 1 min
70% ethanol, 30 sec
50% ethanol, 30 sec

CitriSolv cleaning agent is a less toxic and biodegradable substitute for xylene.

Heat Denaturation and *Protease* Treatment

1. Treat slides at 70°C for 40 min to denature proteins prior to proteolysis. Place slides in a prewarmed, covered Pyrex™ casserole dish or plastic storage container lined with H_2O-soaked Kimwipes and incubate in an oven.
2. Cool to room temperature. Shake excess liquid from slides.
3. Perform proteolysis by laying slides flat on clean surface and covering sections dropwise with a solution of 10 µg·ml^{-1} freshly diluted *Protease K* in phosphate-buffered saline (PBS). Incubate for 30 min to 1 h at 30°C. This is a very critical step and duration should be optimized empirically for each tissue.
4. After draining protease solution from slides, cover sections with 2% PBS-buffered formaldehyde and postfix 10 min.
5. Remove excess formaldehyde and cover sections with 3X PBS. Incubate for 5 min at room temperature.
6. Remove excess buffer and replace with 1X PBS. Incubate for 5 min at room temperature.

3. Acetylation for Nonspecific Blocking

1. Use a glass staining dish for this step. Place slides in the dish and incubate in 250 ml 100 mM triethanolamine (TEA), pH 8.1 for 5 min at room temperature.
2. Remove slides, add a stir bar to the TEA-containing dish, and place on a magnetic stir plate.
3. Stir vigorously for 5 to 10 sec while adding 1 ml acetic anhydride.
4. Turn off the stirrer and quickly place slides into the solution. Incubate with occasional stirring for 5 min.
5. Remove slides, replace solution with fresh TEA, and repeat steps 3 to 4.
6. Rinse slides in 2X SSPE for 2 min.
7. Rinse in 50% ethanol for 2 min.
8. Rinse in 70% ethanol for 2 min.
9. Rinse in 100% ethanol for 2 min.
10. Air dry, then transfer to a desiccation chamber until hybridization.

a. RNA Probe Preparation

Single-strand (ss) RNA rather than DNA probes are recommended for *in situ* detection of RNA transcripts. RNA-RNA probe-target duplexes are more stable and produce better signals than RNA-DNA probe-target duplexes under 50% formamide hybridization conditions[12] that are normally used during ISH. ss-Probes are generally preferred over double-strand (ds) probes because the complementary strands of ds-probes can hybridize to form intraprobe hybrids, which reduce probe interaction with target sequences. ISH was originally developed with the use of radioactive labeled probes (^{35}S or ^{3}H), but nonradioactive probes (such as digoxigenin [DIG]-labeled probes) are now commonly used. Signals from radioactive probes usually take several weeks to detect, whereas DIG-labeled probes can produce results within hours of hybridization without having to handle hazardous materials. Moreover, DIG-labeled RNA probes can remain stable for 6 to 12 months when stored in formamide aliquots at –80°C. Normally, a control hybridization to evaluate the specificity of the interaction between a labeled antisense RNA probe and its target sequence is set up using a second probe that will not hybridize with the RNA of interest. For this purpose, many researchers synthesize a sense RNA probe from the same template that was used for the antisense probe. The sense probe should not hybridize with the target sense mRNA transcripts, and any signal obtained by a sense probe can be a sign of inadequate hybridization or wash stringency. Although rare, naturally occurring antisense transcripts exist and could interfere with use of this control.[13,14] Alternatively, an antisense probe hybridizing to an unrelated RNA with its own distinct expression pattern (based on prior Northern hybridization, RT-PCR [reverse transcription-polymerase chain reaction], or ISH data) can be used as a control.

1. Follow standard molecular protocols to obtain *RNase*-free DNA templates for *in vitro* transcription reactions. Use 1 μg of appropriately digested plasmid DNA or 200 to 500 ng of PCR fragment as the template for a 20-μl *in vitro* transcription reaction.

Notes: *(1) Vector sequences can contribute to high background, especially when using short probes less than 400 base-pairs (bp) in length, and should be cleaved off at the non-RNA polymerase priming end before being used as a template for in vitro transcription. (2) Alternatively, RNA polymerase binding sequences (such as SP6, T3, or T7 primers) can be used in conjunction with gene-specific primers to PCR amplify probe template sequences without terminal vector sequence. If gene-specific primers are not available, two RNA polymerase primers flanking the target sequence may be used for PCR, but the terminal vector sequence should be cleaved off by restriction enzyme digestion prior to use.*

2. In an *RNase*-free 1.5-ml microcentrifuge tube, set up *in vitro* transcription reactions by adding components in the following order.
   - DNA template in ddH$_2$O, up to 9 μl
   - 10X transcription buffer, 2 μl
   - *RNase* inhibitor (40 U·μl^{-1}), 1 μl
   - 10 mM ribonucleotide mix (ATP, CTP, GTP), 2 μl
   - 6 mM UTP, 2 μl
   - 4 mM DIG-11-UTP, 2 μl (or [α-^{35}S]UTP if using radioactive detection)
   - RNA polymerase (5 to 20 U·μl^{-1}), 2 μl
3. Tap bottom of tube to mix reaction components and spin briefly to collect the contents.
4. Incubate in a 37°C incubator for at least 1 h. Prolonged incubations do not increase probe yield substantially.

Notes: *(1) DIG-11-UTP is available from Roche Applied Science (Indianapolis, IN). (2) For radioactive probe preparation, evaporate 25 μl of [α-^{35}S]UTP (650 Ci·mmol^{-1}, 10 mCi·ml^{-1}) down to 2 or 3 μl for one reaction. Set up a 20-μl in vitro transcription reaction as above.*

5. Incubate with 10 units of *RNase*-free *DNase* at 37°C for 15 min to remove template.
6. Examine 10% of recovered probe on a 1% nondenaturing TAE agarose gel (see **Chapter 2**). Skip this step for radioactive probe synthesis.
7. Prepare a chromatography spin column of Sephadex G-50, equilibrated with 50% formamide/2X SSPE, as described in **Chapter 4**. Prespin the column at 400 g for 2 min just before use.
8. Remove unincorporated label. Add 30 μl of 50% formamide/2X SSPE to the probe reaction and load the entire reaction onto the prespun column. Position the column on a 1.5 ml microcentrifuge tube and centrifuge at 400 g for 2 min to collect the probe. A less efficient alternative is to clean labeled probes by phenol:chloroform:isoamyl alcohol (25:24:1) and chloroform:isoamyl alcohol (24:1) extraction, followed by ethanol precipitation in the presence of carriers (e.g., glycogen).

Note: *Do not use transfer RNA (tRNA) as the carrier if you intend to quantify your RNA probe yield (step 9).*

9. (Optional) The probe can be partially hydrolyzed to reduce fragment length. Initiate ethanol precipitation of the probe (see **Chapter 2**) and resuspend in 50 µl DEPC-ddH$_2$O. Add 50 µl of 0.2 M carbonate buffer and incubate at 60°C. Neutralize by adding 5 µl of 10% acetic acid, precipitate with 3 µl sodium acetate and 300 µl ethanol, and resuspend in 10 µl ddH$_2$O for quantitation. Hydrolysis time (min) is determined based on the following equation:[8]

    time = (full-length probe length − desired fragment length)/[0.11 (full length probe × desired fragment length)]

10. For DIG-labeled probes, quantitate 1 µl of probe in a fluorometer using RiboGreen™ fluorescent dye (see **Chapter 2**). Probe yield from a 20-µl reaction is typically 2 to 4 µg. Therefore, 1 µl contains 100 to 200 ng RNA, which can readily be fluorimetrically quantitated but may not be detected spectrophotometrically. Dilute remaining probe with formamide to a concentration of 10 to 20 ng·µl^{-1}. DIG-labeled probes can be aliquoted and stored in formamide at −80°C for at least 6 months.
11. For radioactive labeled probes, check 1 µl of probe in a scintillation counter to estimate the specific activity.

Note: *Limited hydrolysis of the probe to obtain 200 to 250 base fragments is widely considered necessary to facilitate entry of the probe into the fixed plant tissue matrix.[8,9,15] However, we have found that intact probes up to 2 kb (kilobase-pairs) in length can be used to obtain strong hybridization signals with less nonspecific hybridization than with hydrolyzed probes.[16] We therefore regard this as an optional step, but one that should be considered if there are difficulties obtaining signal with certain tissues.*

b. Hybridization

Calculate in advance the amount of hybridization solution with denatured probe that is required for your slides. Approximately 0.75 µl solution is sufficient for each 10 mm² slide area. Thus, 60 µl hybridization solution is sufficient for an area of slide covered by a 22 × 40 mm coverslip. Optimal DIG-labeled probe concentration in hybridization solution is usually 0.3 to 0.75 ng·µl^{-1} probe. Therefore, if making a probe solution for 5 slides, and using 22 × 40 mm coverslips, you will need no more than (5 slides × 60 µl × 0.75 ng·µl^{-1} probe = 225 ng probe) or 11 to 12 µl of 20 ng·µl^{-1} probe stock. If using [^{35}S]-labeled probes, approximately 2 to 4 × 10^6 incorporated cpm is adequate for one slide. However, the optimum probe concentration may need to be determined empirically.

1. The following should be ready prior to placing the probe on slides:
   - Hybridization oven stabilized at 40 to 60°C
   - Water bath set at 70°C.
   - Humidity-equilibrated incubation minichambers set up for hybridization
   - HybriSlip™ (Midwest Scientific, Valley Park, MO) or trimmed Parafilm pieces ~2 mm narrower than the width of the slide.
2. Prepare 2X hybridization mix as follows. Prior to mixing with formamide and probe (step 3 below), suspend settled 2X components with warming and gentle pipetting. To avoid frothing, *do not* vortex.
   - 20X SSPE, 1.2 ml (or 1.0 ml of 20X SSC)
   - 10% blocking reagent, 0.4 ml
   - 20% sodium dodecyl sulfate (SDS), 40 µl
   - 5 mg·ml^{-1} tRNA, 100 µl
   - 2 mg·ml^{-1} heparin, 200 µl
   - DEPC-ddH$_2$O, to 2 ml

Localization of Gene Expression

Notes: *(1) SSPE provides better buffering than SSC. Buffer acidification occurs due to formamide degradation during prolonged incubation at temperatures above 37°C.[12] (2) Include 140 mM DTT if using ^{35}S-labeled probe. (3) Other blocking solutions[8–10] containing Denhardt's solution and dextran sulfate may be used if desired. In this case, prewarm the buffer before pipetting, as dextran sulfate is very viscous. See* **Reagents Needed** *for solution preparation.*

3. Prepare an appropriate volume of denatured probe mix by diluting stored RNA probe with additional formamide, and incubate for 5 min at 70°C to denature secondary structure. Store on ice prior to mixing with 2X hybridization solution. Prior to setting up hybridizations, mix one volume 2X hybridization solution with one volume formamide-probe solution.
4. Apply hybridization solution to sections and overlay with plastic HybriSlip coverslips or with a piece of Parafilm. Avoid trapping gas bubbles.

Notes: *(1) We have found that hybridization signals are stronger when Parafilm is used. (2) Set up one slide at a time to minimize drying. Avoid trapping bubbles by using a 1-ml insulin syringe needle to lower the coverslip, occasionally rocking to exclude bubbles. (3) Some investigators suggest including a prehybridization step, but it must be remembered that slight differences in removal of prehybridization fluid can significantly affect final probe concentration and subsequent hybridization. If prehybridization is desired, apply 1X hybridization solution without probe onto a slide, cover with a coverslip, and incubate 1 to 2 h at 45°C in a humid chamber.*

5. There are many ways to set up a hybridization minichamber. We transfer covered slides to a clean, RNase-free, dry plastic petri-dish bottom (150 × 15 mm), pour 25 ml chamber equilibration solution containing 50% formamide and 5X SSPE (prewarmed in a capped Falcon™ tube or bottle) into a Rubbermaid™ round food storage container (~17 cm diameter × 7 cm height, 1.32-l capacity), and quickly place the petri dish with slides into the "chamber." The chamber is immediately covered and placed into a prewarmed incubator set 5 to 10°C above the planned hybridization temperature. A 7-cm-deep chamber can accommodate a stack of three or four petri-dish bottoms separated by bamboo splints (15 cm). Up to five slides can be placed into each petri dish. Once all slides are in the incubator, lower temperature appropriately.
6. Incubate 12 to 16 h at 40 to 60°C, depending on the probe.

D. Nonradioactive Signal Detection

1. Post-Hybridization Washing and Blocking

1. Add 40 to 50 ml of 50% formamide/2X SSPE to each petri dish. This will lift or "float" off all of the coverslips, which can then be removed with clean forceps.
2. Wash for 15 to 30 min with gentle rocking at 45 to 52°C. Repeat once.
3. Replace with fresh 50% formamide/2X SSPE and repeat step 2.

Notes: *(1) An optional step to reduce background is to treat slides after this step with 5 to 20 µg·ml^{-1} RNase A in RNase A buffer for 30 min at 37°C. The optimum RNase A concentration needs to be determined empirically, as excess RNase A treatment may result in loss of signal. (2) The container used for RNase treatment should be clearly labeled and separated from other RNase-sensitive applications.*

4. Wash slides with 2X maleic acid (MA) buffer on an orbital shaker (75 rpm) for 15 min at room temperature.
5. Replace with 1/2X MA and wash for 1 h at 65°C with slow rocking.

6. Replace with 1X MA containing 1% blocking reagent and 0.3% Tween 20™. Cover and incubate on orbital shaker (75 to 100 rpm) for 1 h at room temperature.
7. Place slides into a rinse of 1X MA buffer containing 0.1% blocking reagent and prepare for addition of antibody.

2. Immunodetection of DIG Probes

1. Assemble preequilibrated minichambers, as before, but this time, use 1X MA to equilibrate the chamber.
2. Dilute anti-DIG antibody conjugated with alkaline phosphatase (Roche Applied Science, Indianapolis, IN) 1:750 in 1 ml of 1X MA containing 0.1% blocking reagent.
3. Remove excess buffer from slides (washed in previous steps) by standing slide on edge on a Kimwipe until most of buffer is removed. Work quickly and use Kimwipes to blot the back and the sides of the slide as well.
4. Immediately apply 50 µl of diluted antibody mix to the slide and cover with paraffin or a standard glass coverslip.
5. Incubate for 2 h at room temperature or 12 h at 4°C.

3. Post-Antibody Washing and Color Development

1. Add 40 to 50 ml 1X MA with 1% blocking reagent to float off the coverslips. Use forceps to remove coverslips.
2. Wash slides on orbital shaker (75 to 100 rpm) for 15 min at room temperature.
3. Replace buffer with 1X MA containing 1% blocking reagent and wash for 1 h. Repeat two more times with fresh 1X MA and reduced concentration (0.1%) of blocking reagent.
4. Replace buffer with AP (alkaline phosphatase) buffer. Allow slides to equilibrate for 15 min at room temperature.
5. Prepare color-development buffer (AP buffer containing PVA, NBT [nitroblue tetrazolium], and BCIP [5-bromo-4-chloro-3-indolylphosphate]). Mix thoroughly, but avoid vortexing, because that will introduce bubbles that will take time to remove from this viscous developer.
6. Remove excess AP buffer from slides, apply 50 µl of color-development buffer, and cover with a coverslip.
7. Place slides into AP-buffer-preequilibrated minichamber and incubate in the dark, checking slides hourly under a microscope without removing coverslip until localized color development begins to show.

Notes: *(1) Work in low light during color-development steps. (2) Hybridized sections often develop a purple hue that can be mistaken as color development when viewing by eye. For specific signal development, always view slides under magnification before terminating color development.*

8. Stop color development by submerging slides with coverslips in a solution of 10 mM EDTA in a petri dish. Remove coverslips after they begin to float free. This may take a few minutes if the color-development step requires more than 6 to 7 h and PVA solutions are used.
9. Rinse slides in ddH$_2$O and allow them to dry thoroughly. Keep dust from collecting on slides.
10. Add glycerol or an aqueous mounting solution, apply coverslip (if appropriate), and observe slides under a microscope. Aqueous mounting solutions, such as Crystal/Mount™ (Biomeda, Foster City, CA), are used without coverslips.
11. Images can be captured using bright-field microscopy with either a dissecting or compound microscope equipped with an appropriate camera. An example is shown in Figure 22.3.

Localization of Gene Expression

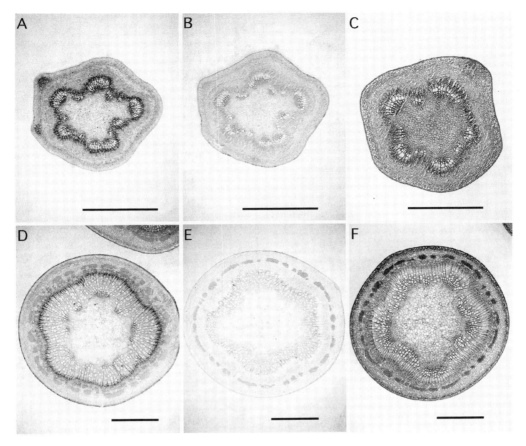

FIGURE 22.3

In situ mRNA localization of a vascular cambium specific transcript in the stem sections of aspen trees (*Populus tremuloides*). Young stem cross-sections were hybridized with a DIG-labeled antisense probe in panels A and D, while sense probe was used in panels B and E as controls. Panels A, B, and C are stage-2 stems, while Panels D, E, and F are stage-6 stems. Bright-field microscopy shows the hybridization coloration as dark regions that form a ring along the vascular cambium when incubated with antisense probe. (The color of this signal would be a bluish-purple if the image were in color.) Panels C and F represent safranin O/fast green differential staining to allow easy identification of the tissue types. The size bars are 1 mm.

E. Radioactive Signal Detection

1. Post-Hybridization Washing

1. Remove coverslips, as described in **Section IV.A**, using 4X SSPE and 5 mM DTT. Dispose of wash solution as radioactive waste according to local radioactive safety procedures.
2. Rinse slides in fresh 4X SSPE and 5 mM DTT solution with gentle rocking. Repeat two more times.
3. (Optional) *RNase* A treatment at 37°C for 30 min can be used to reduce background, as described in **Section IV.A**. Label and separate the *RNase*-treatment labware from other *RNase*-sensitive applications. Rinse slides several times with 2X SSPE and 5 mM DTT.
4. Wash slides in 2X SSPE and 5 mM DTT for 30 min at room temperature with gentle shaking.
5. Replace solution with prewarmed 0.1X SSPE and freshly added 5 mM DTT and wash for 1 h at 55 to 66°C with gentle rocking. Repeat once.

6. Process the slides through the following ethanol series: 15, 30, 50, 70, 85, 95, and 100%.
7. Dry the slides in a desiccator under vacuum for 1 h.
8. (Optional) Quick autoradiography. This step is optional, but it is recommended that unacceptably high/uneven background be tested for by exposing slides overnight on a Biomax™ MR x-ray film. If the background is not excessive, prepare solutions, and set up autoradiography

Caution: *Proper radioisotope safety handling procedures should be followed throughout.*

2. Autoradiography

Note: *The following steps should be carried out in complete darkness. Safelight can be used, but avoid aiming safelight directly onto the work area.*

1. Prewarm Kodak™ NTB-2 emulsion (VWR Scientific, Chicago, IL) in a 45°C water bath (30 to 60 min) and add an equal volume of prewarmed ddH$_2$O. Swirl gently to mix.
2. Aliquot the diluted emulsion into small plastic containers (slide mailers #HS15986 from Fisher Scientific, Chicago, IL, are convenient for this purpose), wrap with aluminum foil, and store at 4°C.

Note: *The emulsion aliquots can be stored for several days at 4°C.*

3. Expose slides to emulsion. Prior to use, melt emulsion in a 45°C water bath and invert the container gently to mix. Dip two blank slides into the emulsion to remove any bubbles and to ensure that the emulsion level is sufficient to cover tissue section areas. Inspect slides outside of the darkroom.
4. Dip each slide into liquid emulsion and withdraw slowly. Allow the slide to drain for a few seconds and place in a rack.

Note: *Make sure the emulsion temperature remains warm. If multiple slides are to be processed, rewarm the emulsion solution in the 45°C water bath frequently.*

5. Place the slide rack in a lightproof box to dry the slides in the darkroom for 60 min.
6. Transfer slides to a slide box with desiccant wrapped in Kimwipes (to avoid exposing slides to dust). Wrap the slide box with heavy-duty aluminum foil and store at 4°C for an appropriate exposure time. Use separate boxes for tester slides.
7. For unknown probes, develop testers after 2 and 4 weeks, and develop remaining slides once the signal becomes visible.
8. Develop slides. Set up in light as long as slides are protected. Fill three glass staining dishes with developer (freshly diluted 1:1 in ddH$_2$O), ddH$_2$O, and fixer, respectively. Chill the solutions in ice water to 15°C.

Note: *Warm the dark box to room temperature prior to developing the slides to avoid background caused by condensation.*

9. In the dark, remove slides from the dark box and place into a slide rack. Place the slide rack into developer, dip gently up and down five times, and leave for 4 min.
10. Remove the rack and drain briefly. Rinse in ddH$_2$O for 30 sec with gentle agitation.
11. Drain and transfer the rack to the fixer. Dip up and down five times and leave for 5 min. Remove slides from fixer and drain briefly.
12. Rinse several times with ddH$_2$O and leave in ddH$_2$O until ready to stain.

Note: *Slides can now be exposed to light.*

3. Staining and Viewing

1. Stain racked slides in 0.05% toluidine blue O solution for 30 to 60 sec. The exact time may vary, depending on tissue type.
2. Rinse slides and soak in water to destain the emulsion. The tissue sections will not destain in water. This step may take up to 1 h.
3. Dehydrate the tissues through the following ethanol series: 25, 50, 75, 100%, and again 100%. Dip the slide rack up and down for 30 sec in each solution. Since the tissue becomes destained in ethanol solutions, the degree of destaining can be controlled by the speed of dehydration.

Note: *Do not allow tissue to be overstained, or the hybridization grains will be obscured.*

4. Rinse slides in CitriSolv for 30 sec. Repeat one more time with fresh CitriSolv.
5. Drain briefly and lay the slides on a paper towel to dry. Add two drops of Permount™ over sections and apply coverslips.
6. Dry overnight to 1 day. Scrape off emulsion residues from the back of the slides using a razor blade.
7. Phase-contrast photography is recommended for documentation of the radioactive *in situ* results (see **Chapter 21**). Generally, three independent pictures are advised for the final results: one bright field only, and two with both bright field and dark field (double exposures), with varying exposure times for dark field. Since there are many means for capturing images from various microscopes, the discussion of this topic is left to the appropriate user manuals. An example of dark-field microscopy *in situ* localization and appropriate controls is shown in Figure 22.4.

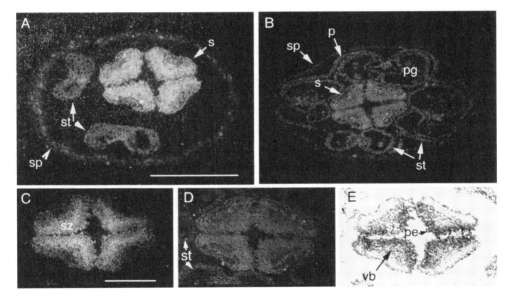

FIGURE 22.4

In situ mRNA localization of a stigma-specific transcript in the flower buds of *Clarkia breweri* one day before anthesis (opening of the flower). Flower bud cross-sections were hybridized with a radioactive antisense probe in panels A and C, while sense probe was used in panels B and D as controls. Dark-field microscopy shows the hybridization signal as the white spots that result in the overall white appearance of the stigma when incubated with antisense probe. Panel E represents a toluidine blue stained section to allow easy identification of the tissue types. The size bar in panel A is 1 mm; the size bar in panel C is 50 μm. Legend: p, petal; pe, papillate epidermis; pg, pollen grain; s, stigma; sp, sepal; st, stamen; sz, secretory zone; vb, vascular bundle. (From Dudareva, N., Cseke, L.J., Blanc, V.M., and Pichersky, E., Evolution of floral scent in *Clarkia*: novel patterns of S-linalool synthase gene expression in the *C. breweri* flower, *Plant Cell*, 8, 1137, 1996. ©American Society of Plant Biologists. With permission.).

F. Troubleshooting Guide

Symptom	Cause and Solutions
Weak or no hybridization signal	May be due to loss of RNA during fixation or dehydration, to low target abundance, or to the introduction of *RNase* activity at almost any step prior to formation of stable RNA-RNA hybrids. *RNase*-free technique must be strictly adhered to. Make certain that all labware and solutions are *RNase*-free. Color development can be prolonged for several days in order to detect rare RNA transcripts. Maintain chamber humidity, especially if PVA is used in the developing solution. Signal-amplification kits are also available to improve weak signals, although we have found that extending color development accomplishes the same result as signal amplification. The probe size may need to be optimized (by varying hydrolysis conditions). The proteolysis step may have been insufficient. Increase *Protease K* concentration to as high as 500 µg·ml^{-1}. *Protease* incubation can also be extended for several hours without causing loss of signal. However, excessive proteolysis can damage section morphology.
Excessive background	May be a sign of poor washing after probe hybridization or antibody incubation, excessive probe concentration, ubiquitous transcript distribution, or nonoptimized hybridization temperature. Probe size may need to be optimized. Make sure vector sequences are removed from probe template prior to *in vitro* transcription.
Uneven signal	Identical sections on different areas of the slide produce unacceptably different signal strengths. Sections that become partially loosened from the slide will produce overly strong signals. Sections near the edge of the slide will produce falsely weak or strong signals if the equilibration buffer used to maintain chamber humidity is significantly less or more osmotic, respectively, than the hybridization buffer. A poly(U) probe, or a constitutively expressed gene probe, can be synthesized to assess whether the RNA within a section is uniformly accessible to probe.[8,11]
Spots or debris on slide	Filter solutions. Minimize dust. Do not use PVA solutions that are more than 1 month old.

Reagents Needed

All solutions should be made in DEPC-ddH$_2$O using *RNase*-free labware and filter sterilized or autoclaved. Solutions used after the hybridization step do not need to be *RNase*-free.

Caution: *DEPC (diethylpyrocarbonate) is a carcinogen and should be handled with care. Gloves should be worn when working with this reagent.*

Alkaline Phosphatase (AP) Buffer

    100 mM Tris, pH 9.5
    150 mM NaCl
    50 mM MgCl$_2$

Prepare 1 M Tris stock (pH 9.5) in autoclaved, DEPC-ddH$_2$O using *RNase*-free labware and autoclave again. Prepare stock solutions of 5 M NaCl and 1 M MgCl$_2$ in ddH$_2$O. Treat the solutions with 0.1% DEPC overnight with stirring in a fume hood. Autoclave and store at room temperature.

Prepare AP buffer by adding appropriate amounts of stock solutions and make up the volume with DEPC-ddH$_2$O. Store at room temperature.

Alternative Hybridization Buffer (with Denhardt's Solution and Dextran Sulfate)

    50% Formamide
    10 mM Tris, pH 7.5
    1 mM EDTA, pH 8.0
    300 mM NaCl
    1X Denhardt's solution
    10% Dextran sulfate (omit for prehybridization)
    70 mM DTT
    150 µg·ml^{-1} tRNA
    500 µg·ml^{-1} Poly(A)

10% Blocking Reagent

Dissolve 10% (w/v) blocking reagent (Roche Applied Science, Indianapolis, IN) in 1X MA buffer by stirring and heating. Autoclave and store at room temperature.

2% Buffered Formaldehyde

Dilute 4% solution described below with 1X PBS.

4% Buffered Formaldehyde

Add 4 g paraformaldehyde pearls to 10 to 15 ml ddH$_2$O with one to two drops 5 N NaOH. Stir at 65°C in a fume hood to dissolve. Bring to 65 ml with ddH$_2$O.
Add 35 ml of 3X PBS.
Check to be sure pH is ~7.2, adjusting with very small amounts of NaOH or H$_2$SO$_4$ if necessary (do not use HCl for this step).
Filter before use and store at room temperature.

0.2 M Carbonate Buffer, pH 9.3

0.2 M sodium carbonate (Na$_2$CO$_3$)
0.2 M sodium bicarbonate (NaHCO$_3$)
Titrate the pH of sodium carbonate to 9.3 with sodium bicarbonate. Store at room temperature for 1 to 2 weeks.

Color Development Buffer

1 ml AP buffer containing 10% polyvinyl alcohol (PVA)
4.5 µl NBT (nitroblue tetrazolium, 50 mg·ml^{-1})
4.5 µl BCIP (5-bromo-4-chloro-3-indoyl-phosphate, 50 mg·ml^{-1})
Dissolve 5 g PVA (MW 70,000 to 100,000) in 50 ml AP buffer by stirring and heating at 90 to 95°C. Store at room temperature up to one month.
Prepare 50 mg·ml^{-1} NBT in 70% dimethylformamide (DMF) and 50 mg·ml^{-1} BCIP in DMF. Store at 4°C and protect from light. BCIP may precipitate during storage and should be warmed at room temperature to dissolve.

DEPC-ddH$_2$O

Add 0.1% DEPC to ddH$_2$O in an *RNase*-free glass bottle. Stir vigorously in a fume hood overnight. Autoclave and store at room temperature.

50X Denhardt's Solution

Dissolve 5 g Ficoll 400, 5 g polyvinylpyrrolidone, and 5 g bovine serum albumin (BSA) (Fraction V) in 500 ml. Filter and store in 25 to 50 ml aliquots at –20°C.

Ethanol/tert-Butanol/H$_2$O Dehydration Solution

Prepare solutions using DEPC-ddH$_2$O, when necessary, containing 0.85% NaCl (from 5 M stock).

FA Solution

5 ml Formaldehyde (37%)
20 ml Ethanol
75 ml 150 mM NaCl

Homogenization Buffer (HB)

    10 ml of 0.5 M Tris-HCl (pH 7.5)
    25 ml of 1.0 M sucrose
    10 ml of 50 mM MgCl$_2$
    10 ml of 50 mM EDTA (dissolved in alkaline solution)
    10 ml of 50 mM EGTA (dissolved in alkaline solution)
    10 ml of 30 mM DTT
    1 ml of 0.1 M phenylmethanesulfonyl fluoride (PMSF) (dissolved in propanol)
    1 ml of 0.1 g/ml leupeptin, pepstatin, and aprotinin
    Add ddH$_2$O to a final volume of 100 ml.

Note: *If the buffer is to be used for homogenization of plant cells, it is advisable to add 1.5% (w/v) polyvinylpyrrolidone (PVP) to remove phenols or quinines, which are commonly found in plant cells.*

Kodak D-19 Developer

Slowly add contents of one packet of developer to 1 gal of water (3.8 L), while stirring. Incubate at 65°C for several hours to completely dissolve.
Use freshly diluted developer (1:1 in ddH$_2$O) for each experiment.

Kodak Fixer

Slowly add contents of one packet of fixer to 1 gal of water (3.8 L) and stir at room temperature to dissolve.

2X MA Buffer

200 mM Maleic acid, pH 7.5

300 mM NaCl

Dissolve maleic acid in DEPC-ddH$_2$O containing NaCl (from 5 M stock), and adjust pH to 7.5 with NaOH ($\sim$16 g·l^{-1}). Autoclave and store at room temperature.

3X PBS Buffer

30 mM Sodium phosphate, pH 7.2
390 mM NaCl
Prepare 1 M stock solutions of Na$_2$HPO$_4$ and NaH$_2$PO$_4$ using DEPC-ddH$_2$O
Prepare 1 M sodium phosphate buffer, pH 7.2, by mixing 68.4 ml Na$_2$HPO$_4$ and 31.6 ml NaH$_2$PO$_4$.
Dilute the stocks in DEPC-ddH$_2$O to make 3X PBS. Autoclave and store at room temperature.

RNase Buffer

10 mM Tris, pH 7.5
500 mM NaCl
1 mM EDTA, pH 8.0

3 M Sodium Acetate Solution (pH 5.2)

Dissolve sodium acetate in ddH$_2$O and adjust pH to 5.2 with 3 M acetic acid. Treat the solutions with 0.1% DEPC overnight with stirring in a fume hood. Autoclave and store at room temperature.

20X SSC

> 3 M NaCl
> 0.3 M Sodium citrate
> Dissolve 175.3 g of NaCl and 88.2 g of sodium citrate in ~700 ml DEPC-ddH$_2$O. Adjust the pH to 7.0 with HCl and the volume to 1 L. Autoclave and store at room temperature.

20X SSPE

> 3.0 M NaCl
> 0.2 M NaH$_2$PO$_4$
> 2 mM EDTA, pH 8.0
> Dissolve 175.3 g NaCl in ~500 ml DEPC-ddH$_2$O. Add 200 ml of 1M NaH$_2$PO$_4$ stock and 4 ml of 0.5 M EDTA stock. Adjust the pH to 7.4 with NaOH and adjust the volume to 1 L. Autoclave and store at room temperature.

Sucrose Gradient

> 60% sucrose
> 50 mM MgCl$_2$
> 0.1 M Tris-HCl (pH 7.5)
> Add 60% sucrose and distilled water to make 55, 50, 44, 33, and 20% sucrose solutions and add 50 mM MgCl$_2$ and 0.1 M Tris-HCl (pH 7.5) to make 5 mM MgCl$_2$ and 10 mM Tris-HCL (pH 7.5).

TE Buffer

> 10 mM Tris-HCl, pH 8.0
> 1 mM EDTA, pH 8.0

Toluidine Blue O Staining Solution

> Dissolve 0.05 g toluidine blue O in 100 ml of 0.1% sodium borate. Filter before use.

10 mM Triethanolamine (TEA)

> Dissolve triethanolamine in ddH$_2$O and adjust pH to 8.1 with NaOH. Prepare fresh.

References

1. **Graham, J.,** Isolation of subcellular organelles and membranes, in *Centrifugation, a Practical Approach*, 2nd ed., Rickwood, O., Ed., IRL Press, Oxford, 1989.
2. **Morrê, D.J.,** Isolation of Golgi apparatus, *Meth. Enzymol.*, 22, 130, 1971.
3. **Yunghan, S. and Morrê, D.J.,** A rapid and reproducible homogenization procedure for the isolation of plasma membranes from rat liver, *Prep. Biochem.*, 3, 301, 1988.
4. **Varner, J. E.,** Introduction in *Tissue Printing*, Reid, P.D., Pont-Lezica, R.F., Campillo, E., and Taylor R., Eds., Academic Press, San Diego, CA, 1992.
5. **McClure, B.A. and Guilfoyle T.J.,** Tissue print hybridization: a simple technique for detecting organ- and tissue-specific gene expression, *Plant Mol. Biol.*, 12, 517, 1989.
6. **Song, Y.R., Ye, Z.H., and Varner, J.E.,** Tissue-print hybridization on membranes for localization of mRNA in plant tissue, *Methods Enzymol.*, 218, 671, 1993.
7. **Chen, F. and Bradford, K.J.,** Expression of an expansin is associated with endosperm weakening during tomato seed germination, *Plant Physiol.*, 124, 1265, 2000.
8. **Jackson, D.,** *In situ* hybridization in plants, in *Molecular Plant Pathology, a Practical Approach*, Vol. 1, Gurr, S.J., McPherson, M.J., and Bowles, D.J., Eds., IRL Press, New York, 1991.

9. **DeBlock, M. and Debrouwer, D.,** RNA-RNA *in situ* hybridization using digoxigenin-labeled probes: the use of high molecular weight polyvinyl alcohol in the alkaline phosphatase indoyl-nitroblue tetrazolium reaction, *Anal. Biochem.*, 215, 86, 1993.
10. **Meyerowitz, E.M.,** *In situ* hybridization to RNA in plant tissue, *Plant Mol. Biol. Rep.*, 5, 242, 1987.
11. **Drews, G.N., Bowman, J.L., and Meyerowitz, E.M.,** Negative regulation of the *Arabidopsis* homeotic gene AGAMOUS by the APETALA2 product, *Cell*, 65, 991, 1991.
12. **Sambrook, J. and Russell, D.W.,** *Molecular Cloning, a Laboratory Manual*, 3rd ed., Cold Spring Harbor Press, Cold Spring Harbor, NY, 2001.
13. **Vanhe-Brossollet, C. and Vaquero, C.,** Do natural antisense transcripts make sense in eukaryotes? *Gene*, 211, 1, 1998.
14. **Terryn, N. and Rouz, P.,** The sense of naturally transcribed antisense RNAs in plants, *Trends Plant Sci.*, 5, 394, 2000.
15. **Moench, T.R., Gendelman, H.E., Clements, J.E., Narayan, O., and Griffin, D.E.,** Efficiency of *in situ* hybridization as a function of probe size and fixation technique, *J. Virol. Methods*, 11, 119, 1985.
16. **Harding, S.A., Leshkevich, J., Chiang, V.L., and Tsai, C.-J.,** Differential substrate inhibition couples kinetically distinct 4-coumarate:coenzyme A ligases with spatially distinct metabolic roles in quaking aspen, *Plant Physiol.*, 128, 428, 2002.
17. **Dudareva, N., Cseke, L., Blanc, V.M., and Pichersky, E.,** Evolution of floral scent in *Clarkia*: novel patterns of S-linalool synthase gene expression in the *C. breweri* flower, *Plant Cell*, 8, 1137, 1996.

Chapter 23

Bioseparation Techniques and Their Applications

Kefei Zheng, Daotian Fu, Leland J. Cseke, Ara Kirakosyan, Sara Warber, and Peter B. Kaufman

Contents

I. Methods of Bioseparation of Organic Molecules 510
   A. Traditional Methods of Extraction 510
   B. Contemporary Methods of Extraction 511
II. Collection, Vouchering, and Storage of Biological Specimens 511
   A. Collection of Plant and Fungal Samples in the Field 511
   B. Vouchering of Samples Collected in the Field 511
      1. Preparation of Dry Specimens 511
      2. Keeping a Good Log Book with Plant Collection Inventory 512
      3. Representative Living Plant Specimens from Field Collections 512
   C. Germplasm Storage Banks for Biological Samples or Organisms 512
III. Grinding and Extraction Protocols 513
   A. General Extraction Protocols for Biologically Important Organic Compounds 513
   B. Hot-Water and Organic-Solvent Extraction of Water-Soluble and Organic-Solvent-Soluble Medicinal Compounds from Plants 513
      1. Hot-Water Extraction of Water-Soluble Medicinal Compounds from Plants 513
      2. Organic-Solvent Extraction of Organic-Solvent-Soluble Medicinal Compounds from Plants 513
      3. Case Studies on the Purification of Crude Extracts prior to Chromatographic Separation: Taxol® and Cuticular Wax Extractions from *Taxus* (Yew) Plants 514
IV. Chromatographic Separation of Organic Molecules: Analytical Protocols 517
   A. Chromatographic Separation Techniques 517
      1. Overview 517
      2. Adsorption Chromatography 518

		3.	Partition Chromatography	523
		4.	Gel-Filtration or -Permeation Chromatography	524
	B.	Capillary-Zone Electrophoresis		525
V.	Use of Mass Spectrometry to Identify Biologically Important Molecules			526
	A.	Ionization Techniques		526
		1.	Electron Ionization	527
		2.	Electrospray Ionization	527
		3.	Matrix-Assisted Laser Desorption/Ionization	527
	B.	Analyzers		528
		1.	Quadrupole Mass Spectrometer	528
		2.	Ion-Trap Mass Spectrometer	528
		3.	Time-of-Flight Mass Spectrometer	529
	C.	Tandem Mass Spectrometry		529
VI.	Nuclear Magnetic Resonance Spectroscopy			530
	A.	Continuous-Wave NMR Spectroscopy		532
	B.	Pulsed NMR		532
	C.	Pulsed Fourier-Transform NMR		532
References				532

I. Methods of Extraction of Organic Molecules

This chapter focuses on the primary methods used to extract, separate, and identify biologically important molecules. We have already covered methods used to separate proteins by one-dimensional and two-dimensional gel electrophoresis (sodium dodecyl sulfate- polyacrylamide gel electrophoresis [SDS-PAGE]) and to visualize them using silver and Coomassie blue stains in **Chapter 3**. In **Chapter 5** and **Chapter 6**, we discussed how ribonucleic acid (RNA) and deoxyribonucleic acid (DNA) are separated by agarose and polyacrylamide gel electrophoresis. In this chapter, we examine: (1) extraction methods for primary and secondary metabolites derived from microbial, plant, and animal cells; (2) purification methods for crude extracts prior to chromatographic separation; (3) chromatographic separation of organic molecules by adsorption chromatography, partition chromatography, and gel filtration or permeation chromatography; and (4) use of mass spectrometry (MS) and nuclear magnetic resonance spectrometry (NMRS) to identify biologically important molecules. We also include the very recent bioseparation techniques in use today. These include capillary zone electrophoresis and high-performance immobilized metal-ion affinity chromatography.

A. Traditional Methods of Extraction

Traditional methods used for bioseparation of metabolites from plants include the use of hot-water extracts to make teas or natural plant dyes. Salves and decoctions are often made from a single plant source (e.g., the monoterpenes from mints to make mint tea or the oleoresin terpenes from pitch of balsam fir used directly to treat burns) or more than one plant (as with many commercial herbal teas that utilize chamomile, *Echinacea*, mints, beebalm, lavendar, and other plants or upregulators of the immune system such as *Echinacea* and goldseal). The important point here is that such extracts or preparations rely on the synergistic action of several plant metabolites that are more effective than any one alone (see **Chapter 6**). An excellent discussion of these traditional methods is found in Penelope Ody's book, *The Complete Medicinal Herbal*.[2]

B. Contemporary Methods of Extraction

Modern methods of extraction utilize principles of separation that are based on the polarity (relative solubility in organic solvents), solubility in water, and various alterational solubilities based on salts and pH (relative acidity or alkalinity). A good discussion of these methods is found in Kaufman et al., *Natural Products from Plants*.[2]

II. Collection, Vouchering, and Storage of Biological Specimens

A. Collection of Plant and Fungal Samples in the Field

When collecting plants in the field for natural product extractions, it is important to be properly prepared. Based on our experience, we suggest that you do the following:

- Wear field clothes and cover yourself from head to toe if collecting is to be in the cold of winter or when mosquitos or deer/black flies are in abundance.
- Take along a note pad and pencil to record information about the collecting site location, soil conditions, ecological habitat, date of collection, plant identity, and who collected the plant(s). A digital camera is also very useful.
- Take along a pocket-size field guide (with photos, drawings, and good, usable identification keys) to the local flora and a hand lens to help you identify the plant.
- If you are collecting live plants, take along some Zip-lock™ plastic bags of various sizes in which to put the samples after collecting plants on site. Slips of recycled paper are good to have for notes on plant identity with your collected specimens that match up with your field notes about the respective collections.
- Take soil samples from each site so as to later get information on soil nutrient, soil pH, and soil type where each plant grows.
- When you collect plants for extracts, it is important to get representative samples of all parts available: roots, vegetative shoots, bark from stems (if woody plant), flowers, fruits, and seeds (if mature).
- When collecting plants in the field, do not take every last plant in the population, especially if the plant is rare, threatened, or endangered.
- In the process of collecting herbaceous perennial plants (plants that come from the same mother plant year after year), leave some of the original plant intact where it is growing so that it can reproduce during the current and following years. Many of these plants take years to produce even a small amount of new biomass every year.
- If you are collecting mushrooms or puffballs in the field, wrap the fruiting bodies in wax paper and place them in a collecting basket or other suitable container where they will not become squashed. This will help for later identification and/or making spore prints from the fruiting bodies. This is impossible with giant puffballs (*Calvatia gigantea*); these can be collected intact and placed in large paper shopping bags. Some of these mushrooms attain a diameter of 0.5 m.

B. Vouchering of Samples Collected in the Field

1. Preparation of Dry Specimens

Dried plant specimens are prepared in order to have them available at any time as voucher specimens representing typical plants that were collected in the field and used for plant extracts. They are also called herbarium specimens. Dried plant specimens are prepared by placing a fresh specimen between single newspaper sheets and then placing that in a sandwich consisting of a dry blotter above and below the newspaper sheets. A piece of corrugated cardboard (with air spaces present) is then placed above and below each blotter. Successive sandwiches are placed atop one another and then compressed between two wood-slatted frames and tied together tightly with straps. The entire assembly is then placed upright on its side over a heat source such as a radiator or a plant drier with the heat on a moderate temperature (e.g., 35 to 40°C). The specimens are allowed to dry this way for 48 h or longer. Rapid drying assures that plant pigments are well preserved; if the drying process is slow, chlorophylls will degrade, and the leaves will appear yellow; flower pigments also fade badly with slow drying. If plant specimens are very high in water content, it is a good idea to replace the blotters with dry ones several times during the drying process.

Once plant specimens are dry, they can be mounted flat (with glue or cement) on heavy paper of sufficient size to accommodate the specimen and a label with information about the plant, location of collecting site, collector, date of collection, genus and species of the plant, and the family to which the plant belongs. The label can be placed in the lower right-hand corner of the sheet of heavy paper. To avoid damage to the dry, mounted specimen, the entire sheet can be covered with SaranWrap™. Dried sheets with plants mounted on them are usually stored in an airtight cabinet in which one can use natural insect repellents such as dried lavender (*Lavandula officinalis*) or neem (*Azadirachta indica*).

If a scanner is available it is also useful to scan leaves, flowers, roots, etc. to obtain a digital record of the collection. Images in the field can be obtained with a digital camera.

2. Keeping a Good Log Book with Plant Collection Inventory

One should always keep a log book of the plants that have been collected and then frozen. A backup inventory on the computer is also a good idea. Why is this important? Sometimes labels with collected plant material become lost. Sometimes, one needs to quickly examine lists of collected plants without going through all of the frozen material. And sometimes, one needs to send such lists to others involved in the project.

3. Representative Living Plant Specimens from Field Collections

In our experience, we have found it to be a good idea to collect seeds and/or living specimens of the plants that are to be used for extraction of natural products. We do this in order to have the living plants on hand, e.g., medicinal plants, dye plants, or culinary herbs, in a garden or in a greenhouse. These can then be used for later extractions or experimental treatments to enhance metabolite biosynthesis. This is especially important when access to the original collecting site is not possible or convenient.

C. Germplasm Storage Banks for Biological Samples or Organisms

Once one comes in from collecting plants in the field, it is a good idea to freeze them immediately in a freezer at −20°C or in a commercial freezer held at −80°C. This prevents any degradation of the plant material or any enzymatic changes that alter or degrade naturally occurring metabolites. Natural drying of the plant material can also be done if yield of metabolites is not critical. This is usually the case for plants used for dyeing fibers, but for extraction of medicinal compounds from plants, the use of dried plant material is not desirable due to degradation of naturally occurring metabolites during the drying process. Rather, it is best to rely on the use of frozen plant material. The only exception is with seeds. They are usually dried to a low moisture content to prolong seed viability. If the drying process is slow and the temperature is at ambient level, very little degradation of stored metabolites in the seeds occurs.

III. Grinding and Extraction Protocols

A. General Extraction Protocols for Biologically Important Organic Compounds

The primary way of extracting organic molecules of interest to biologists and medical investigators involves breaking open the cells of the organism under investigation. Cell rupture is accomplished in a variety of ways. The method used depends on the type of organism being considered and the type of tissue used. For bacterial cells, one usually uses a French press to break open the cell walls. This involves the use of a heavy cylinder with high pressure applied to a piston that compresses the cells into a successively smaller volume within the free cylinder. As the cells leave the cylinder, the rapid drop in pressure causes the cells to lyse. Such a procedure can also be used for plant cells grown in suspension culture or for plant callus tissue. A sonicator can also be used for this purpose. In this case, repeated high-frequency pulses of ultrasonic vibrations rupture the cell membranes. Animal cells and plant cells grown in culture can be ruptured with a glass tissue homogenizer. Highly lignified or silicified plant tissues within organs, such as leaves, stems, roots, seeds, or fruits, are usually frozen and pulverized using liquid nitrogen in a mortar and a pestle. Softer plant tissues can be ground in a small volume of buffer in a mortar, using washed white sand and a pestle to rupture the cells.

B. Hot-Water and Organic-Solvent Extraction of Water-Soluble and Organic-Solvent-Soluble Medicinal Compounds from Plants

1. Hot-Water Extraction of Water-Soluble Medicinal Compounds from Plants

First, weigh out a 0.5-g sample of a given plant. Then, using liquid nitrogen, grind the sample to a fine powder. Next, place the powder into a Corex™ centrifuge tube with hot (80°C) water and place the tube into a hot-water bath for 10 min. The tube is centrifuged at 3000 g for 10 min. By centrifuging it, all of the particulate plant materials from the grinding gets pelleted to the bottom of the tube, leaving a relatively clear liquid (the supernatant) containing the water-soluble compounds of interest. This liquid is filtered to make sure that no plant particulates remain in the filtrate. Next, the filtrate is placed in a petri dish and frozen by placing it on dry ice. After the sample is completely frozen, it is placed in a freeze-drying apparatus and lyophilized. The sample could take anywhere from 2 to 12 h to lyophilize. Freeze-drying is done in order to remove all moisture. This yields a powdered residue.

2. Organic-Solvent Extraction of Organic-Solvent-Soluble Medicinal Compounds from Plants

As mentioned above, this procedure uses an organic solvent to extract the compounds of interest. At least one very toxic solvent (methylene chloride) is used, which requires careful monitoring and caution. The temperature of this extraction must also be carefully watched due to the special equipment that is used. We use a Soxhlet extractor, which is basically a specialized glass refluxing unit that is used for organic-solvent extractions. The temperature of the vapor within the unit must be maintained at 80°C for 18 h in order to obtain complete extraction of a given sample. If the temperature falls below this, extraction will be slow. If the temperature goes above this, the risk of degrading the compounds of interest becomes great.

Note: *Methylene chloride is a suspected carcinogen and should be handled with care in an exhaust hood.*

3. Case Studies on the Purification of Crude Extracts prior to Chromatographic Separation: Taxol® and Cuticular Wax Extractions from *Taxus* (Yew) Plants

For bioseparation, cleanup procedures are usually necessary before samples are analyzed by high-performance liquid chromatography (HPLC) or by any of the other above-mentioned chromatographic techniques. One application that we have employed is for the analysis of taxol, a unique taxane diterpene amide that possesses antitumor and antileukemic properties. Kilograms of this cancer chemotherapeutic agent are needed for clinical treatment of patients having breast cancer. However, taxol exists in only minute quantities — 0.01% of the inner bark and needles of yew (*Taxus*) species. Until recently, taxol could not be synthesized, and even now, the most economical source of taxol is still from the Pacific yew (*Taxus brevifolia*). For this reason, large areas of Pacific yew forests in the Pacific northwest of the U.S. were destroyed in order to obtain this anticancer drug. The taxol-extraction methods used by researchers commonly involve complicated partitioning techniques in which the plant is first extracted with methanol and H_2O and then partitioned using methylene chloride to remove chlorophyll and other unwanted compounds. In the process, taxol molecules move from the aqueous methanol to the more hydrophobic methylene chloride. Unfortunately, methylene chloride is a suspected carcinogen, and it seems counterproductive, in our opinion, to extract a cancer chemotherapeutic with a substance that could cause cancer! Hence, it has been our motive in studying taxol to find a cheap, efficient, and easy way to separate taxol from the thousands of other organic compounds in yew tissues. One of the primary difficulties, in this case, is the removal of chlorophylls from methanolic extracts. Chlorophylls absorb at the same wavelength as taxol and often occur in such large quantities that their resulting peaks interfere with taxane peaks. For this purpose, we use a C-18 reverse-phase Sep-Pak™ column. The protocol for doing this follows. The main point here, however, is that preparatory columns are also useful in removing many unwanted compounds before performing the actual chromatographic analysis.

A new, simple, and rapid method that successfully works for extraction and HPLC separation of taxol from crude extracts of *Taxus cuspidata* (Japanese yew) needles and stems has recently been developed in our laboratory and tested by ten groups independently with repeated success. It requires 2 h to perform steps 1 through 21 and 70 min per sample to run through an automated Shimadzu HPLC apparatus (Shimadzu, Columbia, MD). This long run time is used to separate the multitude of peaks that result at 228-nm spectrophotometric monitoring when not using the C-18 Sep-Pak cleanup method. The advantage of not using cleanup procedures is that the researcher saves time in preparing extracts. In this case, the researcher can go home and sleep while the Shimadzu Sil-6a autoinjector does the work.

In some analytical situations, or for cases where publication is desired, it may be necessary to clean the crude extract with a C-18 Sep-Pak column. The cleanup procedure is given in **Section IV.B.3.6**. (Note: This is by no means the only cleanup procedure for extracts. Any form of chromatography can be used to clean up the extracts [see discussion of chromatographic methods in **Section IV**].) The basic trick is to find the fraction eluted from the cleanup column that contains the compound of interest. In the case of taxol, this was accomplished experimentally using purified taxol standards dissolved in 10% methanol. At this concentration of methanol, it was found that taxol binds to the C-18 packing material in the column. Then, the concentration of methanol was arbitrarily increased in several steps up to 100% methanol. A fraction of eluted mobile phase (2 to 3 ml) was collected at each concentration increase, and HPLC was performed on each fraction to determine where taxol had eluted off the column. After repeated experiments, the specific concentration of methanol that elutes taxol was identified.[3-5]

In some cases, chromatography is not necessary for cleanup of the sample. Addition of adsorptive particles such as activated charcoal to the crude extracts followed by filtration may be all that is necessary to remove unwanted compounds.

a. **Taxol Extraction from Fresh *Taxus* spp. (Yew) Tissue Resulting in Crude Extract Samples That Can Be Used for HPLC**
   1. Obtain *Taxus* spp. plant specimens.
   2. For each sample, weigh out 0.5 g of needles + 0.5 g of stems.
   3. Grind to a fine powder using liquid N_2 in a mortar and pestle.
   4. Let powder warm to room temperature.
   5. Add 1 ml of 100% MeOH and grind vigorously in an exhaust hood.
   6. Transfer this slurry to a 15-ml Corex centrifuge tube.
   7. Add 2 ml of 100% MeOH and grind the remaining plant material in the mortar.
   8. Add this to the tube and label.
   9. Add another 1 ml of 100% MeOH to the mortar and rinse again.
   10. Add this to the tube and label.
   11. Place Parafilm™ over the tube and vortex or shake for 10 min.
   12. Cool at –80°C for 10 min.
   13. Balance the tubes by adding 100% MeOH to the lighter tube.
   14. Centrifuge at 12,000 rpm for 15 min.
   15. Take off the supernatant and place in a clean, acid-washed tube using a Pasteur pipette; then label.
   16. Put this tube into a waterbath at 55°C.
   17. Blow N_2 gas gently into tubes in an exhaust hood until the samples are completely dry.
   18. Add 0.5 ml of ice-cold HPLC-grade 100% MeOH and vortex for several minutes with periodic placement in an ice bath. This allows one to later quantify the amount of taxol extracted because a known volume of sample has been created.
   19. When the MeOH is very cool, place the 0.5 ml into a syringe and filter through a 0.2-μm filter into an HPLC bottle.
   20. Add an aluminum septum to the top of the bottle and cap.
   21. Parafilm well to prevent evaporation and label.
   22. Inject sample into HPLC and run through the Curosil G column (from Phenomenex Corp., Torrence, CA) and collect data at 228 nm. Run each sample for 70 min at 1 ml/min in 36.5% acetonitrile + 63.5% 10 mM ammonium acetate at pH 4.0.
   23. Use purified taxol to make several known concentrations of taxol. Run these standards on HPLC, as in step 22, to make a standard curve.
   24. Compare your data with the standard curve for taxol and calculate the percent of taxol per unit (gram) of fresh weight of tissue.

Note: Methanol is toxic and should be handled with care in an exhaust hood.

b. **Taxol Extraction from Fresh *Taxus* Spp. (Yew) Tissue Resulting in Cleaned Extracts That Can Be Used for HPLC: Use of the C-18 Sep-Pak Column**
   1. Obtain *Taxus* spp. plant specimens.
   2. Weigh out 0.5 g of needles + 0.5 g of stems.
   3. Grind to a fine powder using liquid nitrogen in a mortar and pestle.
   4. Let powder warm to room temperature.
   5. Add 1 ml of 100% MeOH and grind vigorously in an exhaust hood.
   6. Transfer this slurry to a 15-ml Corex centrifuge tube.
   7. Add 2 ml of 100% MeOH and grind the remaining plant material in the mortar.
   8. Add this to the tube and label.
   9. Add another 1 ml of 100% MeOH to the mortar and rinse again.
   10. Add this to the tube and label.
   11. Place Parafilm over the tube and vortex or shake for 10 min.
   12. Cool at –80°C for 10 min.
   13. Balance the tubes by adding more 100% MeOH to the lighter tube.
   14. Centrifuge at 12,000 rpm for 15 min.
   15. Take off the supernatant and place in a clean, acid-washed tube using a Pasteur pipette; then label.
   16. Dilute the volume of MeOH supernatant ten-fold with ddH_2O to produce a 10% MeOH sample.

17. Run all of this diluted sample through a C-18 Sep-Pak column (activated as per instructions from supplier). Taxol-type molecules will bind to the column at this concentration of MeOH.
18. In the case of taxol itself, wash the column first with 3 ml of 35% MeOH, then with 3 ml of 55% MeOH. This washes compounds that have less affinity for the C-18 absorbent than taxol out of the column, while the taxol remains bound.
19. Now wash the C-18 column with 2 ml of 65% MeOH and collect the sample. All taxane species of compounds (including taxol, 7-epi-taxol, 7-epi-10 deacetyl taxol, and cephalomanine) are released from the column during this elevation in MeOH concentration, leaving almost all of the problematic chlorophyll still bound to the column adsorbent phase.
20. The 2 ml of taxol-containing sample is completely dried under a stream of N_2 gas in a 55°C water bath in an exhaust hood while the C-18 column can be cleaned with 100% MeOH and reused. (Note: No detectable amount of taxol is lost during this procedure.)
21. Add 0.5 ml of ice-cold HPLC-grade 100% MeOH and vortex for several minutes with periodic placement in an ice bath.
22. When the MeOH is very cool, place the 0.5 ml into a syringe and filter through a 0.2-μm filter into an HPLC bottle.
23. Add an aluminum septum to the top of the bottle and cap.
24. Parafilm well to prevent evaporation; then label.
25. Inject sample into HPLC and run through a Curosil G column (from Phenomenex Corp.) and collect data at 228 nm. Run each sample for 30 min at 1 ml/min in 40% acetonitrile + 60% 10 mM ammonium acetate at pH 4.0.
26. Use purified taxol to make several known concentrations of taxol. Run these standards on HPLC as in step 22 to make a standard curve.
27. Compare your data with the standard curve for taxol and calculate the percent of taxol per unit (gram) of fresh weight of tissue.

Note: *The run time can be reduced to as little as 10 min by increasing the percentage of acetonitrile. The taxol peak is still clearly resolved in this case.*

c. Extraction of Cuticular Wax from Needles of Yew (*Taxus*) Plants

Waxes are lipids synthesized by plants and animals. Their function is to keep out infectious organisms (e.g., ear wax in animals) and to prevent desiccation and serve as a barrier against fungal and bacterial pathogens in plants. In yew plants (*Taxus* spp.), waxes are synthesized in increased amounts in response to physical stresses. The protocol we have successfully used to quantitatively analyze cuticular waxes from *Taxus* needles is listed. This procedure should be adaptable for any species of plant or animal.

1. Weigh out 1.0 g of *Taxus* needles.
2. Dice needles with a razor blade and place in a mortar and pestle.
3. Add 2:1 mixture of methanol:chloroform to the mortar and grind thoroughly.
4. Transfer the resulting slurry to a centrifuge tube and centrifuge at 12,000 rpm for 15 min.
5. Remove the supernatant and place in a separate collection flask. This supernatant contains the lipids from the *Taxus* needles.
6. Place the pellet back into a mortar and pestle.
7. Add a 2:1:0.8 mixture of methanol:chloroform:water, and grind thoroughly.
8. Centrifuge this slurry at 12,000 rpm for 15 min.
9. Transfer the supernatant to the above collection flask.
10. Regrind the pellet from the centrifuge tube using the original 2:1 mixture of methanol:chloroform mixture, and repeat steps 8 and 9.
11. While collecting the rinse liquid, vacuum filter the pellet, rinsing with 100% chloroform.
12. Transfer this rinse liquid to the collection flask.
13. Transfer all of the collected liquid to a separatory funnel and add a 1:1 mixture of chloroform:water. This will produce two phases. The lower phase will contain the dissolved lipid.

14. Empty the bottom layer of chloroform from the separatory funnel and dry it over solid sodium acetate. This absorbs water from the sample.
15. Weigh a clean rotary evaporator flask (round-bottom flask) and record the weight.
16. Pipette the chloroform solution into the rotary evaporator flask.
17. Vacuum filter the solid sodium acetate, rinsing with chloroform. Add the collected rinse liquid to the rotary evaporator flask.
18. Evaporate the chloroform using a rotary evaporator until the sample is completely dry. Weigh the rotary evaporator flask with the sample and subtract the recorded initial weight for the empty flask. This will give the amount of cuticular wax from the original sample of *Taxus* needles.

Note: Methanol is toxic and chloroform is a carcinogen. Both should be handled with care in an exhaust hood.

IV. Chromatographic Separation of Organic Molecules: Analytical Protocols

A. Chromatographic Separation Techniques

1. Overview

Chromatography is a method of separation with the use of two phases: stationary phase and mobile phase. A mixture of components called analytes is introduced to the system by either directly applying it to the stationary phase or indirectly through the use of mobile phase. As the mobile phase passes through the stationary phase, the components of the mixture partition or equilibrate between the two phases, resulting in differential migration rates through the system. The differential rate of migration of the analytes is determined by their relative tendencies to interact with the stationary phase and mobile phase. Separation can be accomplished by selective adsorption on the solid phase, such as particle clay, silica gel, alumina, cellulose, or other solid support. The stationary phase is usually packed into a column. Consequently, each substance appears in the adsorbent medium eluting from the column at a different time. This time difference is called the retention time (Rt). The technique is very powerful because it allows separation of compounds of very similar physical and chemical properties, including isomers, within a given extract. It is also fast, easy, and economical.

The mobile phase can be either a gas or a liquid, while the stationary phase generally is a solid. When the stationary phase is contained in a column, the term *column chromatography* applies. Column chromatography can also be further divided into gas chromatography (GC)[6,7] and liquid chromatography (LC)[8,9] to reflect the physical state of the mobile phase. Based on the above concept, there are several primary types of chromatographic methods in common use today. These include paper chromatography (PC), thin-layer chromatography (TLC),[10,11] liquid-column chromatography, gas chromatography, and supercritical fluid chromatography (SFC). PC and TLC were named based on the stationary phase, while GC, LC, and SFC were named based on the physical state of the mobile phase. With PC and TLC, the stationary phases are primarily paper (made of cellulose) and silica, respectively, while the mobile phases are composed of a mixture of organic solvents. For LC and GC, the stationary phases are often in a form of silica or other synthetic polymer support, which often are derivatized with other modifying groups to suit the property of analytes. The mobile phases, on the other hand, are often mixtures of different aqueous and organic solvents. Coupled with each mode of chromatographic separation is a unique group of detection techniques. For example, thin-layer and paper chromatographic separations are generally carried out in an open medium instead of a column format. Therefore, the commonly used techniques for detection are colorimetric methods, where the analytes react with a specific reagent to form a colored product.

These colored products (bands) can then be detected or quantified by densitometry. The detection techniques for liquid and gas chromatographic separations, however, are often quite different. Due to the unique nature of these separations, where analytes elute off the column in sequence, an on-line type of detection would be most suited. For LC, the most commonly used detection techniques are ultraviolet (UV) absorbance and refractive index. For GC, the analytes are primarily in the gas state, so they often are detected by using a flame-ionization detector (FID). GCs also have other types of detectors for selective monitoring, such as a thermal conductivity detector or even a mass spectrometer.

The size of the particle used for the stationary phase affects the separation: the smaller the particle, the better is the resolution. This is largely because smaller particle size gives rise to a larger surface area per unit weight of the solid phase and, consequently, a more efficient partition of the analytes. However, because traditional columns use gravity to pull the mobile phase around the particles, smaller particle sizes often result in higher back pressure. Thus there is a limit to how small the particle size can be before the column essentially stops flowing. High-performance liquid chromatography (HPLC) was developed to overcome this limitation, to shorten the analysis time, and to enhance the separation. In an HPLC system, the sample is introduced through the injector into the system and is then pushed through the column by the constant pumping of mobile phase from the reservoir through the system. HPLC has two major advantages. First, it uses a pump to force the mobile phase through a given type of column at high pressure. The column can be made (or usually purchased) using any of the solid adsorbent phases. Hence, HPLC is commonly used to shorten the running times of all types of chromatography, which are usually time-restricted by gravity or capillary action. Second, because high pressure is used, much smaller particle sizes of adsorbent solid support can be used in the columns in conjunction with much smaller column internal diameters. Together, these two factors allow much better resolution of the compounds passing through the column due to the increased contact of the compound with the solid support adsorbent. Today, HPLC has become the premier analytical technique in the chemical and pharmaceutical industries, and it is widely used in all phases of drug discovery, development, and quality control. By using shorter columns (30 to 100 mm) and smaller particles (3 μm) — in contrast to conventional 100- to 150-mm columns and 5-μm particles — HPLC allows highly efficient separation and faster analysis, with separation times as short as a few minutes.

In addition to the different types of chromatographic methods, for each type there are also various modes of separation mechanism. For examples, analytes can be separated based on their molecular shape or size. This mode of separation is often referred to as gel-permeation or size-exclusion chromatography. Analytes can also be separated based on their charge (ion-exchange chromatography), hydrophobicity (reverse-phase or hydrophobic-interaction chromatography), biological function (affinity chromatography), and metal binding (immobilized-metal-ion affinity chromatography). In general, all of these methods fall into two categories of separation mechanism: (1) adsorption (including partition), in which the analyte molecules are adsorbed onto the stationary phase through binding, and (2) nonadsorption (such as gel filtration). Each mode of chromatography has its own type of bonded phases, as discussed below in greater detail (**Sections IV.A.2, IV.A.3, and IV.A.4**).

2. Adsorption Chromatography

In adsorption chromatography, the compounds of interest are separated by allowing them to adsorb (bind to) to the surface of a solid phase. The compounds are desorbed (removed) from the solid phase by an eluting solvent. The solid phase is poured as a slurry into a column. The column typically has a length-to-width ratio of 10:1 (of the actual poured solid phase). The components of the sample are resolved into distinct bands in the column and eluted as separate bands sequentially. These bands of interest are then detected by UV absorbance at a wavelength that is unique to its property. Other detection techniques, such as refractive index and electrochemical detection, can also be used. If sample preparation is the intended purpose of separation, a fraction collector can be used to collect fractions of desired volume for later analysis. These fractions can be analyzed for enzyme activity (using enzyme assays), for salt composition and concentrations (using a conductivity meter or a spectrophotometer), and for protein composition and concentrations (using Bradford or Lowry or other protein assays in conjunction with a spectrophotometer).

a. Normal-Phase Chromatography

Normal-phase chromatography is a technique that uses a polar stationary phase and a nonpolar mobile phase. The analytes interact with the stationary phase, typically through hydrogen bonding or polar interactions. The stationary phase is either an inorganic adsorbent (silica or alumina) or a polar bonded phase (cyano, diol, or amino) on a silica support. They are compatible with nonaqueous solvent systems and are suitable for the separation of many organic compounds. The actual separation occurs by differential polar interactions of the analytes with the polar groups of the stationary phase. Analytes that are very polar strongly interact with the column and tend to elute at later retention times, whereas molecules that are more nonpolar tend to weakly interact with the normal-phase column and elute quickly from the column at earlier retention times. Normal-phase separations are usually run under gradient conditions using two different mobile phases; however, isocratic conditions are not uncommon. The starting solvent is a very nonpolar organic solvent such as hexane. The eluting solvent is a polar organic solvent such as dichloromethane or a short-chain alcohol.

Although normal-phase chromatography was discovered early, it is less commonly used in today's research and industrial laboratories. Its applications are hampered by the limited versatility with which the polarity of the stationary phase can be modified and controlled.

b. Reversed-Phase and Hydrophobic-Interaction Chromatography

In contrast to normal-phase chromatography, reversed-phase and hydrophobic-interaction chromatography use a nonpolar stationary phase and a polar mobile phase. Both methods contain hydrophobic ligands. The matrix in reversed-phase chromatography is more hydrophobic than that used in hydrophobic-interaction chromatography. When applying proteins on reversed-phase columns, the interaction between the protein and the stationary phase is too strong for the protein to be eluted with aqueous buffers. Organic solvents are typically used, and proteins are often denatured. In hydrophobic-interaction chromatography, the interaction is less strong, and aqueous buffers can be used for elution.

Reversed-phase chromatography is widely used in today's research and industrial laboratories (Figure 23.1 through Figure 23.4). Due to the enormous flexibility in which the hydrophobicity of the bonded phase can be modified and controlled, it has proved to be one of the most versatile and popular chromatographic method in use today. Its high selectivity makes it a highly suitable method for separating the most similarly structured small molecules. It is also highly effective in separating small proteins or peptides.

Hydrophobic-interaction chromatography, on the other hand, is based on weak hydrophobic interaction and is most effectively used in separating large and hydrophobic proteins. It is most often run in high concentrations of salt, which can help stabilize the proteins. Hydrophobic-interaction columns are usually based on a polymeric support matrix with a hydrocarbon moiety attached (e.g., phenyl, butyl, or methyl).

c. Affinity Chromatography

Affinity chromatography involves the use of a receptor, such as an antibody, that is linked by covalent bonds to an inert solid support phase. The receptor has a high binding affinity for one of the compounds (its ligand) in the mixture of compounds from the prepared extract. Such binding is both selective and reversible. Inert solid support phases include cross-linked dextrans, cross-linked polyacrylamide, cellulose, and agarose. Due to its great selectivity, affinity chromatography offers a powerful means of achieving excellent separation and purification of biological molecules

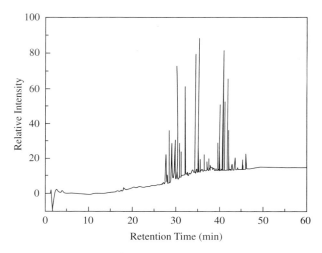

FIGURE 23.1
Chromatogram of a cell culture mixture on a packed C-18 capillary column.

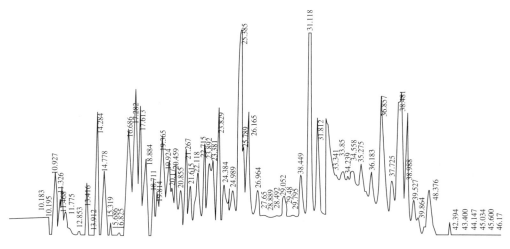

FIGURE 23.2
HPLC separation of extracts from medicinal plants.

in as little as one step. Applications include purification of proteins (including antibodies and enzymes), nucleic acids, or any compound that acts as a ligand for a given bindable receptor.

For example, some drugs, such as taxol, can be purified by using their monoclonal antibody as the bound receptor. After eluting unwanted compounds from the column, the bound ligand is then easily eluted from the column by shifting the mobile phase to a low (or in some cases a high) pH. This procedure can be done by immobilizing the ligand or by immobilizing the ligand's receptor to purify the ligand. Another example of protein purification involves the use of immobilized protein A, a protein commonly found in bacterial cell wall that has a high affinity for antibody. With its high selectivity and efficient binding capacity, affinity chromatography is widely used in the biotechnology industry as a one-step purification process for large-scale production of therapeutic monoclonal antibody.

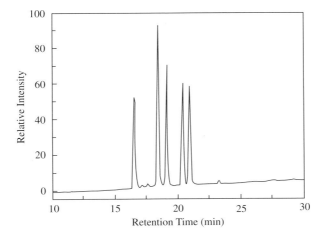

FIGURE 23.3
Chromatogram of a standard peptide mixture on a packed C-18 capillary column.

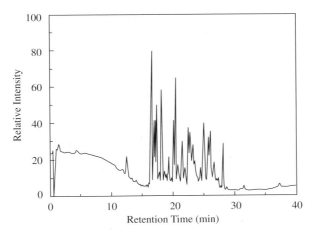

FIGURE 23.4
Chromatogram of B-casein digests on a packed C-18 capillary column.

d. Immobilized-Metal-Ion Affinity Chromatography (IMAC)

Immobilized-metal-ion affinity chromatography is a hybrid modification of affinity chromatography and ion-exchange chromatography. The technique, which is used to separate peptides and other organic molecules, utilizes several types of stationary phases: synthetic polymers, silica, or cross-linked agarose. Basically, chromatographic separation involves a metal-ion chelator bound to the solid support phase. This in turn binds positively charged metal ions such as Zn^{2+}, Mg^{2+}, and Cu^{2+}. Negatively charged side groups on a given protein (or other negatively charged species such as polysaccharides, nucleotides, or nucleic acids) bind to the positively charged metal ions via ionic interactions. Nonbound compounds simply pass through the column and elute off, whereas the bound compounds stick to the solid phase until they are eluted off by use of pH shifts or a selective change in ionic strength. With its high selectivity and ease to scale up, this type of chromatography is widely used in the biotechnology industry for large-scale purification of complex recombinant therapeutic proteins.[12]

e. Dye-Ligand Chromatography

Dye-ligand chromatography was initiated from a blue dye known as Cibacron Blue F3GA. Some enzymes were found to have an affinity for this blue dye. Indeed, most enzymes that bind a purine nucleotide will show an affinity for Cibacron Blue F3GA. Later, many other related dyes were found to bind to various enzymes. The specificity of dye adsorbents is such that they can qualify as affinity adsorbents, even though the dye used bears little resemblance to the true ligand. The selectivity and specificity of dye-ligand chromatography is not as high as antibody-antigen-based affinity chromatography. However, the binding capacity of dye ligands is relatively high, and they are cheap to produce.

f. Ion-Exchange Chromatography

In ion-exchange chromatography, the solid adsorbent phase has charged groups that are linked chemically to an inert solid matrix. What happens during the chromatography is that ions become electrostatically bound to the charged groups of the solid adsorbent. These ions can then be exchanged for ions in the mobile aqueous phase. This is accomplished by changing the ionic strength (pH) of the eluting solvent. Two types of ion exchangers are used in ion-exchange chromatography: (1) cation exchangers, which are exchangers with chemically bound negative charges, and (2) anion exchangers, which are exchangers with chemically bound positive charges. On the exchangers, the charges are balanced by counterions. For this purpose, chloride ions (Cl^-) are used for anion exchangers, and positively charged metal ions are used for cation exchangers. To elute the molecule of interest from such ion-exchange columns, one can use (1) changes in pH of the eluting buffer, (2) increasing ionic strength of salt (e.g., [NaCl] or [KCl]) in solution, and (3) affinity selection, which depends on both charge (opposite to that of the bound macromolecule) and specific affinity for the bound macromolecule.

Ion-exchange chromatography has proved to be an effective separation tool both for small molecules and for biological molecules such as proteins and nucleic acids. It not only has been commonly used as an analytical tool, due to its ease of scaling up, it is often used for preparative purposes. For example, in the pharmaceutical industry, ion-exchange chromatography has been widely used for large-scale purification of active pharmaceutical ingredient. In the meanwhile, due to the mild condition of its mobile phase (in most cases composed of a buffer and a high concentration of salts), it is uniquely used in the biotechnology industry for large-scale purification of biologically active therapeutic proteins.[13]

g. Chromatofocusing

Chromatofocusing is a hybrid technique of both ion-exchange chromatography and isoelectric focusing in a single chromatographic focusing procedure. Proteins are separated as a result of the isocratic formation of internal pH gradients on ion-exchange columns. As proteins of similar net surface charge may still vary in surface-charge distribution, it is possible to resolve proteins during chromatofocusing that are not well separated by isoelectric focusing. The technique is useful for the analysis of protein surface-charge heterogeneity. It is mostly used as an analytical tool for separation of proteins with different surface-charge distributions.

h. Chiral Chromatography

Chiral chromatography is used to separate enantiomers, which are isomeric compounds that have one or more chiral centers in the molecule caused by a carbon atom substituted with four different groups. These compounds are important in the pharmaceutical field, as the biological activity of two enantiomers may be different. For example, the D-isomer of penicillamine is therapeutic, while the L-isomer is toxic. However, enantiomers have the same physical and chemical properties and can not be resolved in most LC columns. Separation of enantiomers usually involves the following three pathways:

1. Using a chiral stationary phase by coating the stationary phase with a chiral reagent. This is the most straightforward method and is widely used today.

2. Transforming into two diastereomeric derivatives that can be separated by reversed-phase or normal-phase chromatography. This approach was popular in the 1970s and 1980s, but its use has declined since the emergence of the new chiral stationary phase. Also, the derivatization is time consuming, and racemization or other side reactions are possible.
3. Adding chiral mobile phase to reversed-phase or normal-phase separation. This approach is used infrequently because of the high cost of chiral reagents and poor resolution. However, due to its unique selectivity, this form of chromatography has been used to separate an active form of a pharmaceutical ingredient from an inactive one. For example, the (D)-isomer of albuterol has been successfully separated from the inactive (L)-isomer to improve the drug's efficacy.

3. Partition Chromatography

a. Paper Chromatography

Paper chromatography is usually carried out in a large glass tank or cabinet and involves either ascending or descending flow of the mobile-phase solvents. Descending paper chromatography is faster because gravity facilitates the flow of solvents. Large sheets of Whatman No. 1 or No. 2 filter paper (the latter is thicker) are cut into long strips (e.g., 22 to 56 cm) for use in descending paper chromatography, while a wide strip of paper (e.g., 25 cm wide) of variable height is used for ascending paper chromatography.[14,15]

For descending liquid-paper chromatography, substances to be separated are applied as spots (e.g., 25 mm apart) along a horizontal pencil line placed down from the V-trough-folded top of the paper. The V-trough-folded paper is placed in a glass trough, held down by a glass rod, and when the tank has been equilibrated (vapor-saturated) with "running solvents" (mobile phase), the same solvent is added to the trough via a hole in the lid covering the chromatography tank. The lid is sealed onto the chamber with stopcock grease to make the chamber airtight. After the mobile phase trails to the base of and off the paper sheet, the paper is hung up to dry in a fume hood, where it can then be sprayed with reagents (e.g., Ninhydrin reagent for amino acids) that give color to the separated compounds of interest in white or UV light. Some compounds of interest have their own distinctive colors, such as chlorophylls, and such compounds can be purified using this technique. In other cases, the dyes used to stain the location of the compound or protein of interest cause irreversible covalent changes to the compound. In these cases, purification is not possible.

In ascending paper chromatography, the same basic setup and principles apply, with the exception that the mobile phase is placed at the bottom of the tank. Separation is achieved when the mobile phase travels up the paper via capillary action.

b. Thin-Layer Chromatography

Thin-layer chromatography (TLC) has several advantages over paper chromatography: (1) greater resolving power, (2) faster speed of separation, and (3) availability of a diverse array of adsorbents. The first two advantages are attributed to the fine particle size of the solid support adsorbent (less than 0.1 mm diameter particles). This allows greater contact of the solid support with the compounds of interest as they travel up the plate. The adsorbents (e.g., silica gel, alumina, cellulose, and derivatives of cellulose) are available commercially on glass plates of various sizes. TLC plates are used in glass tanks for use in ascending or, in some cases, descending chromatography. For the former, one to ten samples of interest are spotted at 2- to 3-cm intervals across a line 15 to 20 mm from the base of the plate. The spots are allowed to dry; then the plate is placed in a glass chromatographic tank containing solvent placed in the bottom of the tank to a depth of 10 mm. Often it is necessary to equilibrate the vapor in the tank by placing filter paper around the sides of the tank. Next, a lid is sealed to the top of the tank with stopcock grease, and the solvent is allowed to rise by capillary flow to the top of the plate. Once the mobile phase reaches the top of the plate, the plate is removed and allowed to dry. Then, the spots are developed with appropriate reagents for the types of compounds being separated and assessed. However, this procedure, as in paper chromatography, may result in sample destruction.[10,11]

TLC can be run in two dimensions, using different solvent systems, as can paper chromatography, to provide better separation of compounds. This procedure is very similar to two-dimensional electrophoresis, which is discussed in **Chapter 8**. TLC is widely used to separate lipids, fructans, sugars, and hormones.

c. Gas–Liquid Chromatography

Gas chromatography (GC) is another type of partition chromatography where a high-boiling-point liquid is the stationary phase and an inert gas is the mobile phase. There is also an inert solid packing used in columns, where these two phases are separated. Separation of compounds of interest is achieved through the differentiated solubility of the compounds in the mobile and stationary phases. Thus, as the carrier gas passes through the column, the compounds in the sample come off the column at different times (Rt). A GC apparatus basically consists of a tank of carrier gas (e.g., helium or nitrogen), an oven containing a coiled metal or glass chromatography column, a sample-injection port, and a detector (e.g., a flame-ionization detector [FID] or other detectors such as thermal conductivity and electron capture).

With FIDs, hydrogen gas is used to provide fuel for the flame. The hydrogen is coupled to a flow of air to the detector to provide oxygen, which allows the hydrogen to burn. A wire loop is positioned above the flame to detect compounds that pass from the column to the flame; the wire, in turn, is connected to the recorder. FIDs are very sensitive to most organic compounds, but not to water, carbon monoxide, carbon dioxide, or the inert gases. Obviously, samples are destroyed when using this type of detector. Thus, GC is usually not used for purification of compounds. Thermal-conductivity detectors, on the other hand, are less sensitive than FIDs. However, they are nondestructive to the samples, which allows complete recovery of a sample. Other detectors commonly used with GC include the electron capture detector, which is often used for detection of halogen-containing compounds.

GC is a commonly used analytical technique both in research laboratories and in the chemical, pharmaceutical, and environmental monitoring industries. It can be used both as a qualitative identification technique for compounds and a quantitative measurement tool. For example, GC is often used in the pharmaceutical industry to monitor the level of process intermediates or to control the level of residual solvents in a drug product. Environmental monitoring laboratories use GC to monitor the level of undesired chemical contaminants.

4. Gel-Filtration or -Permeation Chromatography

Gel-filtration chromatography is often referred to as size-exclusion chromatography. This technique involves the use of porous gel molecules of agarose, cross-linked dextran, or polymers of acrylamide that facilitate the separation of compounds based on their molecular sizes or weights. The pore size or pore diameter of the beads determines the molecular size range or molecular weight range that can be separated with a particular column. The actual separation occurs by repeated diffusion of the analytes into and out of the pores of the beads. Sample molecules with sizes greater than the pore diameter of the support matrix can not enter the pores, and these molecules are excluded and rapidly eluted from the column in the void volume. Molecules with sizes smaller than the pore diameter enter the pores and differentially elute in volumes greater than the void volume at various retention times. Molecules normally elute in order of decreasing size.

All size-exclusion separations are run under isocratic conditions. The three major factors affecting size-exclusion separation are ionic strength, organic modifier, and detergent concentration. The technique is widely used for the separation and characterization of biological macromolecules, especially proteins. Unlike other modes of liquid chromatography, the interaction between sample molecules and the stationary phase is minimized, which is likely to keep the sample intact and result in high recoveries. It is usually used as the first of step of liquid-chromatography purification. One commercial series called Sephadex™ (Amersham Biosciences Corp., Piscataway, NJ) is used for this purpose. These types of column packings must be hydrated before they are functional as

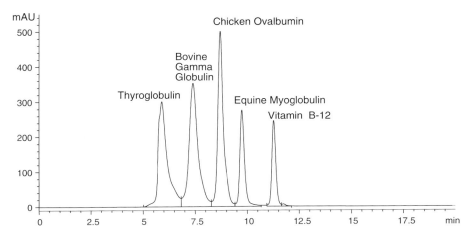

FIGURE 23.5
Separation of different proteins (thyroglobulin, bovine gamma globulin, chicken ovalbumin, equine myoglobulin and vitamin B-12) by gel permeation chromatography and UV detection at 215 nM.

a separation medium. The hydration process causes the pores in the Sephadex to swell to the appropriate size for the given Sephadex type. For example, G-10 Sephadex, during hydration, gains 1 ml of water per gram of dry gel; G-200 Sephadex gains 20 ml of water per gram of dry gel. Bio-Gel beads from Bio-Rad Laboratories (Hercules, CA) consist of long polymers of acrylamide that are cross-linked to *N,N'*-methylene-*bis*-acrylamide. These gels have a larger range of pore sizes than the Sephadex G series. Still another porous gel with an even wider pore size is agarose, which consists of the neutral polysaccharide fraction from agar. Agarose and polyacrylamide are used to separate viruses, ribosomes, nucleic acids, and proteins. Sephadex is widely used in purification of proteins and in determining their molecular weights.

The general rationale for separation is as follows: (1) a gel having an appropriate pore size is chosen in relation to the size of the molecule of interest; (2) samples are added to the top of the gel column and are washed through using an appropriate mobile phase that is based on the solubility of the molecule of interest; (3) molecules that are too big to fit in the pores of the solid support will travel around the gel particles and, hence, elute first from the column; (4) molecules that fit in the pores will elute at different times according to their differential mobilities through the gel pores.

Gel permeation can also be effectively used as an analytical tool under high pressure for the separation of proteins and other molecules based on size. In these cases, the columns are usually packed with porous silica or other synthetic polymer beads that can withstand high pressure. With its ability to separate proteins or other large biological molecules by size, gel permeation has been widely used in the biotechnology industry to monitor the level of aggregates in a protein-based therapeutic. Figure 23.5 shows an example of size-exclusion HPLC used to monitor the percent aggregate formation in a recombinant protein therapeutic.

B. Capillary-Zone Electrophoresis

Capillary-zone electrophoresis is not a form of chromatography, but it deserves recognition as an extremely powerful technique for separating compounds of interest via electrophoresis. The term electrophoresis refers to the movement (migration) of a charged molecule under the influence of an electric field. Basically, capillary electrophoresis utilizes small-bore open capillary tubes (e.g., 200 μm internal diameter) in a system equipped with a grounded high-voltage power supply, solvent reservoir in a Plexiglas™ box connected to the capillary tube, a detector, a solvent reservoir after the detector, and a power supply for current flow to ground.[16]

A very small sample (e.g., a nanogram quantity) is loaded into the capillary tube at one end, and negatively charged species, such as the negatively charged side groups of proteins, are separated by the same mechanism as ordinary electrophoresis. The major advantage here is that the capillary tube has a large surface-to-volume ratio, allowing rapid dissipation of the heat produced by the electric current. Consequently, much higher voltages can be used in capillary-zone electrophoresis than can be used in normal electrophoresis. High voltages in normal electrophoresis tend to cause heat convection within the gel, resulting in distortions and blurring of the separation bands. The high voltages in capillary-zone electrophoresis allow much better resolution of related species of compounds as well as much faster running times. In addition, the capillary tube's inner surface is negatively charged, and thus it attracts positively charged species of molecules. As the buffer travels through the capillary tube via the electrical current, there is an electroosmotic flow produced that carries these positively charged species of molecules in the same direction as the negatively charged species. Hence, both negatively and positively charged species can be separated and analyzed at the same time. Capillary-zone electrophoresis utilizes several types of detectors, including spectrophotometers, mass spectrometers, electrochemical detectors, and radiometric detectors, instead of the cumbersome stains used in ordinary electrophoresis. Its high sensitivity, speed of analysis, high resolution, flexibility of pH, and the ability to utilize a variety of separation matrices make it an attractive separation method for a variety of applications.

V. Use of Mass Spectrometry to Identify Biologically Important Molecules

It is very difficult to say exactly what compound is contained in a given peak from a purified extract that shows up from HPLC. Mass spectrometry (MS) is very useful in clearing up such ambiguities.

Mass spectrometry is an analytical technique that is used to identify unknown compounds, to quantify known compounds, and to elucidate the structure and chemical properties of molecules. It has become very powerful in the analysis of complex biological samples, including proteins, DNAs, or extracts from plants, where information regarding both the identities and structures of the analytes is required. Mass spectrometry can give unambiguous results in such applications. For example, the combination of chromatography with mass spectrometry as a hyphenated technique gives researchers a tool that has the dual advantage of chromatography as a separation method and mass spectrometry as a detection technique. A mass spectrometer is an analytical device that measures the masses of individual molecules that have been converted into ions. Furthermore, it is not the molecular mass that is measured, but rather the mass-to-charge ratio of the ions formed from the molecules. The ions are generated by inducing either the loss or the gain of a charge (e.g., protonation or deprotonation, electron ejection or capture).

A. Ionization Techniques

Ionization techniques have undergone dramatic changes in recent years. Early ion sources required a sample to be a gas — usually accomplished by thermal desorption, such as electron ionization (EI). Two significant developments in the late 1980s and early 1990s — electrospray ionization (ESI) and matrix-assisted laser desorption/ionization (MALDI) — have made it possible to examine samples in liquid solutions or embedded in a solid matrix. These ionization techniques have become the methods of choice because of their greater sensitivity and the availability of a wider mass range for analysis.

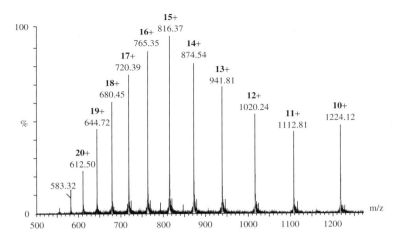

FIGURE 23.6
Mass spectrum of cytochrome C using electron ionization.

1. Electron Ionization

In the commonly used EI source, ions are generated by bombarding the gaseous sample molecules with a beam of energetic electrons. The energy of the bombarding electrons is generally much greater than that of the bonds that hold the molecule together. Thus when high-energy electrons interact with the molecules, not only does ionization occur, but bonds are broken and fragments are formed. It is a "hard" process and may lead to extensive fragmentation that leaves very little or no trace of a molecular ion.

2. Electrospray Ionization

Electrospray ionization (ESI) is used to generate gaseous ionized molecules from liquid solution. The ESI source requires a high electric field at the capillary tip and leads to the formation of highly charged liquid droplets. The electric field causes charge accumulation on the liquid surface at the capillary terminus and disrupts the liquid surface. As the droplet decreases in size, the electric field density on its surface increases. The mutual repulsion between like charges on this surface becomes so great that it exceeds the forces of surface tension; the ions begin to leave the droplet and are directed into an orifice to the mass analyzer. ESI produces multiply charged ions. This is a stunning facet of ESI as mass spectrometers measure the mass-to-charge ratio, it provides high mass range which is accommodated by mass analyzers having a relatively small mass range. Figure 23.6 shows the mass spectrum of cytochrome C, which has a molecular weight over 11 kDa. In addition, tandem mass spectrometry (discussed in **Section V.C**) has been shown to be useful for dissociation studies of the multiply-charged biomolecules produced by ESI, allowing much greater dissociation efficiencies than for singly-charged molecular ions. It allows one to couple liquid chromatography or other separation techniques to ESI mass spectrometry.

3. Matrix-Assisted Laser Desorption/Ionization

Matrix-assisted laser desorption/ionization (MALDI) is accomplished by directing a pulsed laser beam onto a sample mixed in a matrix. The matrix absorbs the laser light energy, resulting in vaporization of the matrix and ionization of the analyte molecules. In contrast to ESI, MALDI usually produces singly charged molecular ions. The most striking feature of MALDI is its very wide mass range and high sensitivity. Figure 23.7 shows an example of a popular MALDI instrument.

FIGURE 23.7
MALDI instrument.

B. Analyzers

The following three mass analyzers are widely used and relatively inexpensive. They all measure the ions according to their mass-to-charge ratio (m/z).

1. Quadrupole Mass Spectrometer

The quadrupole mass spectrometer consists of four parallel poles or rods. Mass sorting depends on ion motion resulting from simultaneously applied constant direct current (DC) and radio frequency (RF) electric fields. Scanning is accomplished by systematically changing the field strengths, thereby changing the m/z value that is transmitted through the analyzer. The quadrupoles function as a mass filter. When field is applied, ions moving into this field region will oscillate according to their mass-to-charge ratios. At a certain RF field, only ions of a particular m/z can pass through the quadrupoles and reach the detector. The m/z of an ion is determined by correlating the field applied to the quadrupoles. A mass spectrum can be obtained by scanning the RF field.

2. Ion-Trap Mass Spectrometer

The ion-trap mass spectrometer uses fields generated by RF (and sometimes DC) voltages applied to electrodes arranged in a sandwich geometry: a ring electrode in the middle with cap electrodes on each end. Within a selected range of mass-to-charge ratios determined by the applied voltages, ions are trapped in the space bounded by the electrodes. A mass spectrum is produced by scanning the applied RF voltages to eject ions of increasing mass-to-charge ratio through an end-cap opening for detection.

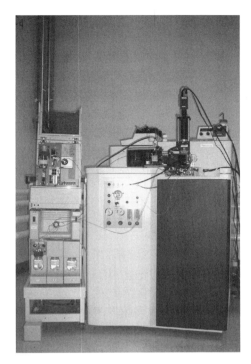

FIGURE 23.8
Q-TOF instrument.

3. Time-of-Flight Mass Spectrometer

The time-of-flight (TOF) mass spectrometer (Figure 23.8) separates ions by their different flight times over a certain distance. Ions are accelerated so that these ions have equal kinetic energy and then are directed into a flight tube. Since kinetic energy is equal to $1/2\ mv^2$, where m is the mass of the ion and v is the ion velocity, the lower the ion's mass, the greater is its velocity and the shorter is its flight time. The travel time from the ion source through the flight tube to the detector, measured in microseconds, can be transformed to the m/z value. Because all ion masses are measured for each ion burst, TOF mass spectrometers offer high sensitivity as well as rapid scanning. They can provide mass data for very-high-mass biomolecules.

C. Tandem Mass Spectrometry

Both ESI and MALDI are used for mass spectrometric analysis of large involatile molecules such as proteins, DNA, or other biomolecules, which are rather soft. Therefore, the fragment ion formation is limited and, as a consequence, the structural information is limited. In contrast, EI produces a lot of fragment ions. In order to obtain more-detailed structural information, it is important that additional fragmentation of selected ions can be induced and analyzed by MS/MS techniques, termed *tandem mass spectrometry* (abbreviated MS^n, where n refers to the number of generations of fragment ions being analyzed). This is accomplished by collisionally generating fragments from a particular ion (parent ion or precursor ion) and then analyzing the fragment ions in a second stage of mass analysis. It could be done with tandem-in-space devices such as a triple-quadrupole mass spectrometer, or with tandem-in-time devices such as an ion-trap mass spectrometer.

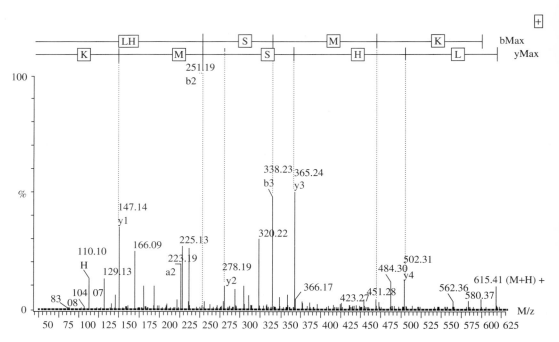

FIGURE 23.9
Peptide mapping by MS/MS.

In a triple-quadrupole mass spectrometer, the second quadrupole is used as a collision cell to generate fragment ions. The target ion with a particular m/z is selected by the first quadrupole, Q1. Only the selected ion is allowed to enter the second quadrupole, Q2, where fragmentation is achieved by ion–argon-atom collisions. These fragments are then analyzed by the third quadrupole, Q3, to obtain structural information. In recent years, hybrid mass spectrometers — a combination of two analyzer types — have emerged as a better alternative to stand-alone devices. A hybrid combines the strengths of each type of analyzer while minimizing compromises that might arise from interfacing the two technologies. If the third quadrupole in a triple-quadrupole mass spectrometer is replaced by a time-of-flight analyzer, the quadrupole–TOF combination will provide higher resolution and wider mass range than conventional a triple-quadrupole mass spectrometer.

In an ion-trap mass spectrometer, ions can be selected, fragmented, and analyzed in the ion trap. This setup is said to be "tandem in time," in contrast to the triple quadrupole, which is "tandem in space." The important feature of ion trap is its ability to perform MS^n experiments, whereas tandem-in-space mass spectrometers can only perform MS^2. Figure 23.9 shows the MS/MS spectrum of a peptide fragment.

VI. Nuclear Magnetic Resonance Spectroscopy

Nuclear magnetic resonance (NMR) spectroscopy is a very powerful method for elucidating chemical structures that is widely used in many of today's research and industrial laboratories. It has also proved to be an indispensable tool for many scientists seeking to delineate the complete sequences of groups or arrangements of atoms within molecules. A typical NMR spectrum contains information regarding what groups there are, how many there are, where they are located in the molecule, what their neighbors are, and how they are related to those neighboring groups. Moreover, the method is nondestructive, and the sample after analysis could be used for other purposes.

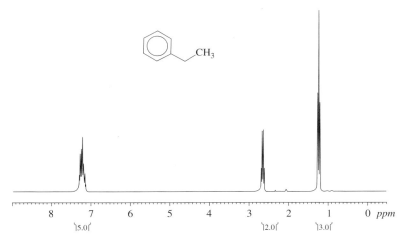

FIGURE 23.10
NMR spectroscopy application with ethylbenzene.

Although NMR is a powerful analytical method, its principle of operation is fairly straightforward, based on the fact that the nuclei of many atoms are constantly spinning. The spinning of these charged particles — the circulation of charge — generates a magnetic moment along the axis of spin, so that these nuclei act like tiny bar magnets. One fine example of such nuclei is the proton, the nucleus of ordinary hydrogen, ^1H. When such a nucleus is placed in an external magnetic field, its magnetic moment, according to quantum mechanics, can be aligned in one of two ways: with or against the external magnetic field. One could easily imagine that when the nucleus is aligned with the external magnetic field, it would represent a more stable state, while aligning against the external magnetic field would represent a less stable or high-energy state. In order to "flip" the tiny magnetic particle (nucleus) from a more stable state to a high-energy and less stable state, energy is absorbed. Of course, the amount of energy it takes to flip the nucleus over depends, as we might expect, on the strength of the external magnetic field and the environment this particular nucleus is in. The stronger the magnetic field, the greater is the tendency for the nucleus to remain aligned with the field, and the higher is the energy or frequency of the radiation needed to do the "flip" job.

In principle, one could place a sample of interest in a magnetic field of constant strength and pass radiation of steadily changing frequency through the sample and observe the frequency at which radiation is observed, which would represent the "flip" frequency. In practice, however, it is found more convenient to keep the radiation frequency the same and to vary the strength of the magnetic field. At some value of the field strength, the energy required to flip the nucleus matches the energy of radiation, flip occurs, a signal is observed, and an NMR spectrum is generated.

If the situation were as simple as what we described, all of the protons in an organic molecule would absorb at the same field strength, and the NMR spectrum would simply be a tool to count the number of protons in the molecules. Fortunately, the situation is far more complex than that. This is because all nuclei are also affected by their neighboring nuclei, and the magnetic fields they all absorb are slightly different, depending on the exact environment they are in. Therefore, at a given radio frequency, all of the protons absorb at different applied field strengths. It is this applied field strength that is measured and plotted in a NMR spectrum. This type of NMR spectroscopy is called continuous-wave NMR. The result is a spectrum showing many absorption peaks, whose relative positions can give a wealth of information about molecular structure. Figure 23.10 shows an example of NMR application with ethylbenzene.

A. Continuous-Wave NMR Spectroscopy

Continuous-wave NMR spectroscopy relies on application of a continuous and steady change of applied magnetic field to cover the entire spectrum of different strengths of magnetic field required by a molecule. Although this technique worked well in the early years after its introduction, it quickly reached its performance limit, most notably sensitivity. In order to obtain a readable spectrum of a small amount of sample, many scans would need to be collected. In addition to the longer time it takes to perform multiple scans, the noise also multiplies as more scans are collected. Therefore, there is a physical limit to the maximum number of scans one could effectively collect. It is precisely this limitation of continuous-wave NMR that has led to the birth of pulsed (Fourier-transform) NMR spectroscopy.

B. Pulsed NMR

Pulsed NMR relies on a simple fact that all excited nuclei eventually need to give up the energy they absorbed and return to the more stable state. This process is called relaxation or free-induction decay. In pulsed NMR, all the nuclei are excited simultaneously. A strong pulse of energy is applied to the sample for a very short time window (1 to 100 μsec). In the interval between the pulses, the excited nuclei gradually relax and decay by giving up the energy. This decay signal following each repetitive pulse is digitized by a fast analog-to-digital converter, and the successive digitized transient signals are coherently added in the computer until a sufficient signal-to-noise ratio is reached. The computer then performs a Fourier transform to convert the time-domain decay data to a frequency-domain spectrum, namely the normal plot of NMR spectrum.

C. Pulsed Fourier-Transform NMR

Pulsed Fourier-transform NMR makes possible the study of less-sensitive nuclei, such as ^{13}C, ^{14}N, ^{15}N, ^{17}O, and ^{31}P. It also serves as a foundation for the more complex and extremely powerful multidimensional NMR spectroscopy.

References

1. **Ody, P.**, *The Complete Medicinal Herbal*, Dorling Kindersley Publishing, London, 1993.
2. **Kaufman, P.B., Cseke, L.J., Warber, S., Duke, J.A., and Brielmann, H.L.**, *Natural Products from Plants*, CRC Press, Boca Raton, FL, 1999.
3. **Falzone, C.J., Benesi, A.J., and Lacomte, J.T.J.**, Characterization of taxol in methylene chloride by NMR spectroscopy, *Tetrahedron Lett.*, 33, 1169, 1992.
4. **McClure, T.D., Schram, K.H., and Reiner, M.L.J.**, The mass spectrometry of taxol, *J. Am. Soc. Mass Spectrom.*, 3, 672, 1992.
5. **Ketchum, R.E.B. and Gibson, D.M.**, Rapid isocratic reversed phase HPLC of taxanes on new columns developed specifically for taxol analysis, *J. Liquid Chromatogr.*, 16, 2519, 1993.
6. **Berezkin, V.G.**, *Chemical Methods in Gas Chromatography*, Elsevier, New York, 1983.
7. **Jennings, W.**, *Analytical Gas Chromatography*, Academic, New York, 1987.
8. **Horvath, C.**, Ed., *High Performance Liquid Chromatography*, Academic, Orlando, FL, Vol. 1, 1980; Vol. 2, 1980; Vol. 3, 1983.
9. **Hancock, W.S.**, Ed., *High Performance Liquid Chromatography in Biotechnology*, J. Wiley & Sons, New York, 1990.

10. **Fried, B. and Sherma, J.,** *Thin Layer Chromatography: Techniques and Applications*, 3rd ed., Dekker, New York, 1994.
11. **Fennimore, D.C. and Davis, C.M.,** High performance thin layer chromatography, *Anal. Chem.*, 53, 252A, 1981.
12. **Porath, J.,** High-performance immobilized-metal-ion affinity chromatography of peptides and proteins, *J. Chromatogr.*, 443, 3, 1988.
13. **Fritz, J.S.,** Ion chromatography, *Anal. Chem.*, 59, 335A, 1987.
14. **Miller, J.M.,** *Chromatography: Concepts and Contrasts*, J. Wiley & Sons, New York, 1988.
15. **Robyt, J.F. and White, B.J.,** *Biochemical Techniques: Theory and Practice*, Brooks/Cole Publishing, Monterey, CA, 1987.
16. **Ewing, A.G., Wallingford, R.A., and Olefirowicz, T.M.,** Capillary electrophoresis, *Anal. Chem.*, 61, 271, 1989.

Chapter 24

Combinatorial Techniques

Harry Brielmann

Contents

I.	History	536
II.	Parallel Synthesis	537
III.	Solution-Phase Combinatorial Chemistry	539
	A. Reaction	539
	B. Workup	540
	C. Purification	540
IV.	Solid-Phase Combinatorial Chemistry	541
	A. Pin Technology	541
	B. Split-Pool Method	541
	C. Iterative Deconvolution Using the "Tea-Bag" Method[9]	542
	D. Encoded Libraries	542
	E. Irori Microreactor Methodology[41–43]	542
	F. Issues	542
VI.	Appendix	543
	A. List of Combinatorial and Contract Synthesis Companies	543
	B. Sources for Detailed Solid- and Solution-Phase Combinatorial Procedures	546
	1. Scientific Journals	546
	2. Books	546
References		546

In many scientific fields, including biology, medicine, and especially in the pharmaceutical chemistry, it is highly desirable to have available a large number of individual organic compounds that are structurally related. With such a library, a scientist can compare the properties of these materials and, by doing so, reveal how the chemical structure affects these properties. Such a structure-activity relationship (SAR) is at the heart of combinatorial chemistry, which provides efficient methods for the preparation of these libraries. Not only can such a library provide the compounds that are needed for a given application, but it can also act as a predictive tool, suggesting how to improve the properties of a given organic compound, regardless of the final application. Several fundamentally different methods are available for making these libraries, and the method of choice will often depend on both the size and quality of the library being generated.

In this chapter, the various combinatorial chemistry techniques will be introduced, surveyed and analyzed. References for detailed experimental procedures are provided at the end of the chapter. Since many of the readers of this handbook may have primarily a biological or medical background, an attempt has been made to describe the chemistry involved here with them in mind. Note that for those who may require a chemical library for their research, there are many alternatives to in-house synthesis. For commercially available chemicals, these may be simply purchased from several available suppliers. Structure-based search programs for commercially available compounds are particularly helpful in this regard. There is also a small but growing list of suppliers that specialize in small quantities of otherwise unavailable compounds for the generation of chemical libraries. The next step up is to consider contract synthesis in order to have a library prepared by a company that specializes in combinatorial library preparation. This can be an expensive proposition, but it may be a viable alternative when no in-house capability exists. Lists of commercial chemical suppliers, as well as combinatorial chemical suppliers, are provided as an appendix to this chapter. The remainder of this chapter is for those who wish to perform combinatorial chemistry at their own facility.

I. History

The synthesis of large numbers of structurally similar organic compounds (libraries), and the evaluation of the structural relationships between them, has been a scientific endeavor long before the term combinatorial chemistry was ever coined. As it is used today, combinatorial chemistry is understood to involve the efficient preparation of a large number of structurally similar and relevant chemical compounds. The term relevant is important here, and is in stark contrast to the basic combinatorial paradigm first written in 1927 by Eddington: "If an army of monkeys were strumming on typewriters, they might write all the books of the British Museum."[1] The field has evolved from considering what *can* be done, to what *should* be done. For example, publications indicate that the percentage of discovery libraries has decreased by two-thirds from 1992 to 1999, while the percentage of targeted libraries during the same time period has doubled.[2]

The initial chemical libraries that were developed in the early to mid 20th century often involved a large number of plant extracts and compounds from natural-product isolation.[3] A bewildering array of fascinating chemical structures emerged from these investigations, often with potent biological activities. In fact some of the most complex organic structures known to man, including palytoxin[4] and maitotoxin,[5] were discovered from natural sources. As the field matured, it became possible to identify rapidly the structures of highly complex molecules, and to assay them quickly against several biological templates. Both of these developments were due in large part to advances in instrumentation, a pattern that shows no sign of abating. In fact, it was not long before the assaying capability rapidly outpaced the ability to provide compounds for testing. Individual chemicals that were not isolated from natural sources were synthesized individually from simple chemical building blocks, or semisynthetically from natural sources. The process was slow, and the need for a more rapid method for the construction of libraries of relevant and synthetically similar organic compounds was born.

Merrifield is largely credited with inventing the field of automated chemical synthesis. In 1963 he developed the approach of using a few repeating reactions to couple amino acids together to form amide bonds that produce peptides.[6] Typically, an amino acid that has been protected at the amino group is covalently attached to a modified polystyrene resin at the carboxyl group. A repetitive deprotection-condensation sequence is then employed, which adds one amino acid at a time to the growing peptide, which is ultimately cleaved from the resin. Although protection–deprotection schemes such as this were nothing new, attaching the molecule to a polymeric support was a radical new idea, and like many new ideas, it was not appreciated for a long time. A simplified view of the overall process (leaving out the amine protection–deprotection process) is shown in Figure 24.1.

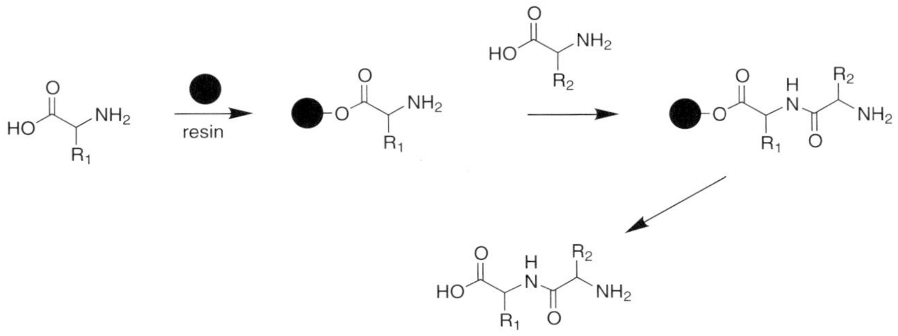

FIGURE 24.1
Merrifield's solid-phase peptide synthesis technique.

In 1984 Geysen developed a technique for synthesizing peptides (it was still peptides) on pin-shaped solid supports.[7,8] In 1985, Houghten developed a technique in which tiny mesh packets or "tea bags" act as a reaction chamber and filtration device for solid-phase parallel peptide synthesis.[9] Seminal work in the area of split synthesis, which involves creative methods for working with mixtures of compounds by pooling them together and splitting them apart, was performed by Furka in the late 1980s.[10] In the early 1990s, these techniques captured the interest of the pharmaceutical industry, and from there, the field became diverse, as numerous methods competed for primacy in this arena. These involved resin-based (including encapsulated resins)[11] methods described above, as well as parallel solution-phase methods as described below in **Section III**. Even today there is no clear consensus on which combinatorial methods will dominate the field, although a growing number of reports on solution-phase libraries have appeared in the literature.

II. Parallel Synthesis

As mentioned earlier, the traditional synthetic method for the preparation of each organic compound that would comprise one component of a chemical library can sometimes be a painfully slow and tedious process. These individual components are synthesized one at a time through a series of individual chemical reactions. Each reaction takes place in an individual flask, and typically involves three separate procedures for each step of a chemical reaction. In the first step, the reaction takes place: chemicals and reagents to be combined are mixed in an inert solvent under an inert gas such as nitrogen. They are allowed to react at a given time and temperature, and the reaction is monitored to determine the time of completion. In the second step, the workup takes place, which typically involves removing any aqueous-soluble impurities or reagents by extraction, and the removal of the solvent in which the reaction takes place. In the final step, which is often the most laborious, the reaction product is purified, often by chromatography. Note that each of these three processes is used for each step of a chemical reaction, and some procedures require a few to as many as ten steps (or more!) to prepare the final compound. A professional chemist, working hard, may only be able to perform one or two complete chemical transformations, including purification, per day. Thus, it is clear to anyone who may need, say, a thousand compounds for a given library that this may take a while, and be quite expensive, using traditional methods. *The great promise of combinatorial chemistry is that it can effectively address these issues.*

As alluded to earlier, two major competing techniques are involved in combinatorial chemistry: solution-phase and solid-phase synthesis. Both of these methods attempt to simplify and, whenever possible, automate the synthesis process just outlined. They also are normally performed in parallel (Figure 24.2). This means that whereas, in a classical organic synthesis, reagent A is mixed with reagent B in a single flask to make one compound, a parallel synthesis involves several reactions

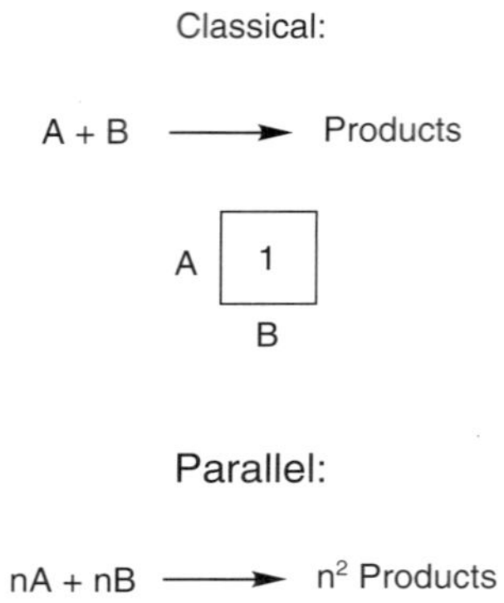

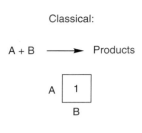

FIGURE 24.2
Classical vs. parallel synthesis.

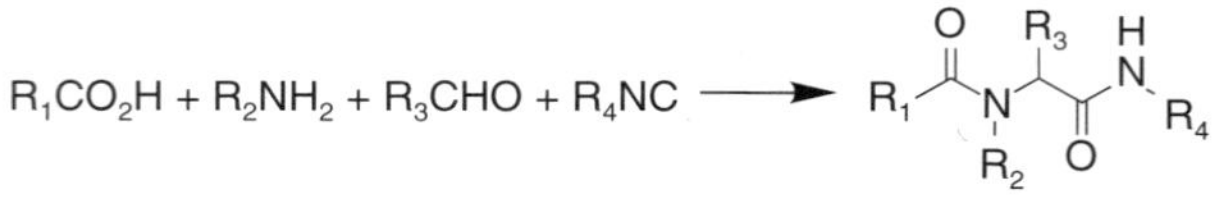

FIGURE 24.3
The Ugi reaction.

$$R_1CO_2H + R_2NH_2 + R_3CHO + R_4NC \longrightarrow$$

TABLE 24.1
A 10 × 10 Parallel Synthesis Matrix

	1A	2A	3A	4A	5A	6A	7A	8A	9A	10A
1B	1	2	3	4	5	6	7	8	9	10
2B	11	12	13	14	15	16	17	18	19	20
3B	21	22	23	24	25	26	27	28	29	30
4B	31	32	33	34	35	36	37	38	39	40
5B	41	42	43	44	45	46	47	48	49	50
6B	51	52	53	54	55	56	57	58	59	60
7B	61	62	63	64	65	66	67	68	69	70
8B	71	72	73	74	75	76	77	78	79	80
9B	81	82	83	84	85	86	87	88	89	90
10B	91	92	93	94	95	96	97	98	99	100

taking place simultaneously and independently to make many products in the same amount of time. This is best envisioned by considering a classical synthesis to be a 1 × 1 matrix of A and B to make one compound, and a parallel synthesis to be, for example, a 10A × 10B matrix to make 100 individual compounds.

Note that this method makes, or should make, 100 individual compounds in 100 separate reaction containers, be they flasks, vials, or wells on a plate. This is, therefore, an example of a "one compound per well" technique that is favored by many skilled in the art. The identity of each of the 100 compounds made can be determined by its position: it is *spatially addressable*. For example, as shown in Table 24.1, well number 33 is the product formed by mixing 3A and 4B. The alternative technique would be to mix all of the 20 reagents of the 10 × 10 matrix in one flask, and then if necessary, separate (or "deconvolute") each of the products. This alternative, or "many compounds per well" technique, is a tempting method, and there are those who still employ it. The large mixtures it provides can be rapidly screened *en masse*, and active compounds can then be deconvoluted from the mixture. Because of the difficulties associated with deconvolution and cross-contamination, discrete methods have become dominant in the industry. Preparing libraries of discrete individual compounds also enables quality assurance and informatics for each singleton. This is crucial for data mining. It is discussed in **Chapter 25**.

Combinatorial Techniques

In any event, a parallel synthesis greatly increases the speed by which a library can be prepared. It is useful here to briefly mention a useful concept, namely, that of a multiple component coupling.[12] Note that in the aforementioned examples, it was assumed that two reagents (A and B) are required to make compound C. This is often, but not always, the case. Certain chemical transformations (for example the Ugi reaction,[13] shown in Figure 24.3) use three or more reagents, each of which becomes a part of a final individual component. In other words, occasionally A + B + C are used to form D.

This can exponentially increase the utility of the individual reagents (or "*fragments*"). Whereas formerly 20 fragments were required to make 100 compounds for our 10 × 10 matrix, for our multicomponent coupling, we can generate 1000 compounds using only 30 fragments in our 10 × 10 × 10 matrix. This will still involve 1000 individual reactions when using the one-compound per well method. However, often the preparation of the fragments is done on a large scale and by hand, so that in this regard, multicomponent couplings can still provide an enormous advantage.

So it is clear, and perhaps a no-brainer, that performing chemical reactions in parallel can speed things up. But in order to really get things moving, one must address the three laborious steps of reaction/workup/purification. Imagine, for example, performing a 10 × 10 matrix and then having to perform 100 individual chromatography columns, with monitoring by thin-layer chromatography, for each. It would be absurd. Numerous creative and practical methods have been designed to deal with these issues. Let us deal with solution-phase techniques first.

III. Solution-Phase Combinatorial Chemistry

Solution-phase combinatorial chemistry is in many ways simply a streamlined version of classical organic chemistry, with a great deal of simplification, and is performed in parallel. As the term implies, the reaction is performed in solution, that is to say, dissolved in an inert solvent. In this respect, solution-phase combinatorial chemistry is the same as classical organic synthesis. Generally, there are no polymer-bound components. An exception to this is the use of polymer-bound scavenging reagents,[14] where a polymer-bound material is added to a reaction to conveniently remove unwanted reaction materials. The key timesaving aspect to solution-phase combinatorial chemistry is that it is performed in parallel, and each of the reaction/workup/purification steps can be automated and monitored, or perhaps avoided altogether. Let us look at each part of the process.

A. Reaction

Reactions are ideally chosen that tend to give a clean conversion to the product. For those that require extensive purification, automated high-performance chromatographic techniques have recently become available.[15] Sterically hindered or unstable reagents are often avoided whenever this is possible. Reactions that are not stoichiometric (that may require an excess of a reagent) can be used, but only if the excess can be removed easily at the latter two stages. Using a reagent that is either polymer bound or that is highly polar and can be removed as baseline during chromatography may simplify the purification process. Scavenging reagents[14] may also easily remove residual starting materials or reagents. Reaction scales vary dramatically, but they are typically performed in individual septa-topped vials, usually on a 0.01 to 0.1 millimolar scale or smaller. Thus, for a product with a molecular weight of 300, you may be looking at 3 to 30 mg of product. This is often far larger than solid-phase techniques. Obviously, these amounts may vary widely, depending on the application. Reactions that require high temperatures or pressures may require specialized equipment. An important criterion for automation is that each component should be transportable as a solution, so that all transformations can be handled as liquids. Quantitative robotic handling of solids may be possible, but it is rarely convenient. Numerous robotic liquid-handling devices are commercially available (see **Section V**, "Appendix"). Occasionally, insoluble materials can be handled as fixed-concentration slurries using large-bore liquid-handling apparatus.

The reaction parameters for parallel reactions are determined by experiment. Typically, this involves varying the reaction conditions for a representative cross-section of the reagents in order to find a single set of conditions that will be used in parallel. If necessary, these must be subdivided to cover widely varying reactivities. The extent of validation is normally proportional to the number of compounds being generated.

B. Workup

An ideal workup involves simple solvent removal, for which positive-pressure evaporative devices, such as a turbovap,[16] can be used. Inorganic by-products, if there are any, can be removed during purification. If high-boiling water-soluble solvents (note that evaporators are available for these as well)[17] such as the polar aprotic types need to be removed, or if it is advantageous to remove a component of the mixture by partitioning it into the aqueous phase, then an extractive workup may be necessary. This normally involves simply adding the aqueous material, effectively agitating the mixture, for example by shaking, and removing the aqueous layer. This can be done robotically. Final drying, if necessary, can often be accomplished in-line during purification. Other extractive techniques, including solid-supported liquid-liquid extraction (SLE) have been developed.[18]

C. Purification

Purification requirements vary with the application required. In addition, continuous advances in instrumentation, notably LC-MS (liquid chromatography-mass spectrometry) equipment, have made for rapid changes in this arena. For assays, crude samples can be used for initial screening, but these can lead to false positive results and poor resulting informatics. As previously stated, the trend in the combinatorial arena is toward discrete individual compounds with quantified levels of purity. Two popular methods toward this end are high-performance liquid chromatography (HPLC)[19] and disposable liquid chromatography, including solid-phase extraction (SPE).[20]

HPLC applied to combinatorial chemistry usually involves an automated system, whereby each individual reaction mixture is robotically purified using either normal (silica gel) or more commonly reverse-phase (organosilane) columns, with peak selection determined based on physical or chemical properties. A particularly appealing system relies on peak selection based on the highly selective parameter of molecular weight. Such LC-MS (liquid chromatography-mass spectrometry) systems may give high levels of quantifiable purity even for complex mixtures. Their chief disadvantages are their expense and speed, which can be prohibitive for parallel efforts. This method has been recently reviewed.[12]

A more affordable method with decent results involves running individual samples through inexpensive disposable SPE chromatography columns. The technique is relatively rapid (on the order of minutes on a 0.1 mmolar scale), and the eluant can be spectroscopically monitored if desired. A wide array of stationary phases is now available.[20] Although separation may be only moderate (perhaps R_f differences of 0.3 or higher may be available for normal phase separations), this can be offset by careful choice of reagents. There is a range of devices that fall in between LC-MS and SPE. Note that several low-, medium-, and high-throughput parallel chromatography devices are now commercially available. By choosing products that are likely to have a similar polarity profile, eluant conditions may be uniform, and no monitoring of the eluant stream may be necessary. Rather, compounds that are not of sufficient purity may be rejected during quality control performed after purification. This involves simple spectroscopic analysis of the eluant by either gas chromatography, mass spectroscopy, liquid chromatography, or combinations thereof, prior to final solvent removal.

IV. Solid-Phase Combinatorial Chemistry

That there is a practical alternative to performing a reaction in solution at all may seem surprising. It is, after all, an effective method for combining reagents. The price to be paid is that the desired product must be somehow isolated from the reaction mixture when the reaction is complete. This may be a straightforward process if the reaction is clean and stoichiometric, or the compound may have to be "fished out" if the situation is otherwise. Polymer-bound, or solid-phase reactions provide a viable alternative to the classical synthetic techniques. The idea, as alluded to earlier in Merrifield's peptide synthesis,[3] is to covalently bind the reagent to a polymer (the solid phase), and by so doing, provide a handle so that it may be far more easily worked with. Visually, one can imagine attaching the molecule to a polymeric substrate, for example a plastic bead. Note that this polymeric support may have identifying features associated with it ("tagged;" see **Section IV.D**, "Encoded Libraries") so that chemical identification and spatial addressability of the final product is avoided. Also, note that the covalent attachment ("link") of the polymer can be done in numerous ways, some of which leave no obvious chemical sign of their attachment ("traceless linkers"). When placed in a reaction mixture, the polymer swells, and the reaction takes place, often on a similar time frame to solution-phase chemistry. In fact, one may use an excess of solution-phase reagent and help push the reaction to completion, since the excess reagent can be washed off at the end of the reaction A big selling point of this technique is that the product can be removed simply by filtration. Though certainly oversimplified, here again, one may visualize placing the molecule on a bead of plastic into the reaction mixture, and then simply removing it when the reaction is complete. For multistep transformations, the polymeric solid phase goes to the next reaction medium, and so on. At the end of the final transformation, the molecule is removed from the polymer. Before analyzing the pros and cons of this method, let us look at a few of the historical variations to this method that have been developed.

A. Pin Technology

Geysen[7] demonstrated that peptides can be synthesized in numbers several orders of magnitude greater than are obtained by conventional one-at-a-time methods. The peptides are synthesized on polyethylene rods arranged in a microtiter plate format. This allows 96 or more separate peptides to be simultaneously synthesized at the tips of the rods. The pin technology is representative of techniques that generate libraries of single compounds in a spatially differentiated manner. An alternative approach, to prepare large mixtures of compounds rapidly, was made possible by the introduction of the split-pool approach.

B. Split-Pool Method

This method, developed by Furka,[10] can provide combinatorial libraries in just a few reaction vessels. Activated resin is distributed into reactors (pools). Each pool undergoes reaction with a unique substrate. Pools are recombined and then split again, so that three pools can give 27 products. Because of the lack of addressability to this method, it has been largely supplanted in favor of high-speed, spatially addressable synthesis of discretes.[21]

C. Iterative Deconvolution Using the "Tea-Bag" Method[9]

Small amounts of resins representing individual peptides are enclosed in porous polypropylene containers. The bags are immersed in individual solutions of the appropriate activated amino acids while deprotections and washings are carried out by mixing all of the bags together. The bags are then reseparated for subsequent coupling steps (the split-pool method). Removal of the peptides from the resins affords peptides in soluble form. This method has the advantage that it affords fully characterizable, nonmodified, solution-phase peptides. This may give more realistic interaction results than solid-support bound peptides. The method also provides variable quantities of soluble peptides for testing against virtually any target, and can be used with unnatural amino acids.

D. Encoded Libraries

No encoding is necessary for spatially addressable discrete systems, such as the one-compound-per-well approach. The split-and-mix techniques can rapidly generate very large libraries, where only one compound is synthesized per bead. However, since these reactions are not spatially addressable, and since classical chemical analysis is slow, numerous tagging encoding techniques have been developed. Often, an "identifier" tag is attached to the solid support material coincident with each monomer, again using a split-pool synthesis procedure. The structure of the molecule on any bead identified through screening is obtained by decoding the identifier tags. Numerous chemical and nonchemical methods of encoding have now been reported, including radiofrequency methods,[22–25] oligonucleotides,[26–30] peptides,[31–33] haloaromatics,[34–38] isotopic labels,[39] encapsulation,[2] and MS/MS techniques,[40] among others.

E. Irori Microreactor Methodology[41–43]

Irori microreactors are miniaturized devices that contain both a functionalized solid-phase support and a unique radio-frequency tag identifier. By splitting and pooling microreactors using a process known as "directed sorting," one discrete compound can be synthesized in each microreactor and can be nonchemically identified using the radio-frequency tag.

F. Issues

Regardless of which particular type of solid-phase combinatorial method is used, there are several issues that must be addressed. These include the following:

1. A solid-phase technique is analogous to a protection–deprotection method, which will add two steps to any given synthetic protocol. Each of these steps needs to take place without otherwise disturbing the molecule to which it is attached. This is normally referred to as orthogonality. In the first step, the molecule is covalently attached to a polymeric support. In the final step, it is removed from that support. This often means that a common functionality may exist on each final molecule at the same position, which can be a disadvantage. Several types of traceless linkers have been developed that leave no functionality (other than carbon) after removal from the resin.
2. Since the compound exists in the solid phase during its formation, monitoring of the reaction may be difficult. Classical chromatographic methods, including thin-layer and liquid chromatography, require removal of the polymeric support. Many on-bead methods, including solid-phase spectroscopic techniques, have been developed. These include magic-angle NMR (nuclear magnetic resonance) spinning[44–46] and IR (infrared) on bead.[47] As shown above, various creative tagging methods

may be employed. These can reveal what the structure of the compound is assumed to be, which may not always be what it actually is. In practice, it is often simplest during method development to cleave an aliquot of the compound from the spectroscopic support, and then handle it as a simple solution.
3. Several of the solid-phase techniques described above are designed for mixture libraries, which require deconvolution.
4. Purification of a given intermediate or final product normally requires removal from the polymeric support.
5. The conditions required for a given solid-phase conversion may vary from the corresponding solution-phase technique. Therefore solution-phase methods can use the vast body of existing synthetic chemistry methodology directly, while the solid-phase methods must be individually tested. In some cases, there are literature techniques for solid-phase chemistries, and this body of information is growing. For those interested, far more exhaustive comparisons are available.[48]

As mentioned previously, a creative and useful variant to solution-phase synthesis involves the use of solid-phase quench or scavenger reagents.[14] Here, the idea is not to attach the compound of interest to the polymeric support but, rather, to attach the coupling reagents — the parts of the reaction that normally stay in solution for a solid-phase synthesis — to a resin. By doing so, one has all the advantages of solution-phase chemistry, and some from solid-phase chemistry as well. No protection–deprotection, linking, or tracing needs to be worked out. Reactions can be monitored using normal chromatographic techniques. As is the case with solid-phase reactions, reagents can be used in excess to help drive a reaction to completion rapidly, since they are scavenged. And since they are filtered off, workup procedures are greatly simplified. These techniques have been rapidly adopted, not only for combinatorial chemistry, but also for traditional synthetic organic chemistry as well.

VI. Appendix

A. List of Combinatorial and Contract Synthesis Companies

Advanced ChemTech, Inc.
5609 Fern Valley Road
Louisville, KY 40228-1075
Telephone: (800) 456-1403 (U.S. only) or (502) 969-0000
Fax: (502) 962-5368

AnaSpec, Inc.
2149 O'Toole Avenue, Suite F
San Jose, CA 95131
Telephone: 800-452-5530 (toll-free)
Telephone: (408) 452-5055
Fax: (408) 452-5059
E-mail: service@anaspec.com

Affymax Research Institute
4001 Miranda Avenue
Palo Alto, CA 94304
Telephone : (415) 496-2300
Fax: (415) 424-0832

ChemBridge Corporation
16981 Via Tazon, Suite G
San Diego, CA 92127
Telephone: (858) 451-7400
Fax: (858) 451-7401
E-mail: support@chembridge.com

ChemDiv, Inc.
11575 Sorrento Valley Road, Suite 210
San Diego, CA 92121
Telephone: (858) 794-4860
Fax: (858) 794-4931
E-mail: chemdiv@chemdiv.com

ChemRx Advanced Technologies
385 Oyster Point Blvd., Suite 1
South San Francisco, CA 94080
Telephone : (650) 829-1400
E-mail: biz_dev@chemrx.com

Coelacanth Corporation
279 Princeton-Hightstown Road
East Windsor, NJ 08520
Telephone: (609) 448-8200
Fax: (609) 448-8299
E-mail: Michael_Eagen@coelacorp.com

DuPont Pharmaceuticals Research Laboratories
4570 Executive Drive, Suite 400
San Diego, CA 92121
Telephone: (302) 992-5000

Maybridge PLC
Trevillett, Tintagel, Cornwall PL34 OHW, U.K.
Telephone: 44 (0)1840 770453
Fax: 44 (0)1840 770111
E-mail: enquiries@maybridge.com

MediChem Research Inc.
12305 S. New Avenue
Lemont, IL 60439
Telephone: (630) 257-1500
Fax: (630) 257-1505

MicroSource Discovery Systems, Inc.
21 George Washington Plaza
Gaylordsville, CT 06755-1500
Telephone: (203) 350-8078
Fax: (203) 354-5300

Nanosyn
> 625 Clyde Ave.
> Mountain View, CA 94043-2213
> Telephone: (650) 404-8050
> Fax: (650) 428-1770

PharmaCore, Inc.
> 4170 Mendenhall Oaks Parkway, Suite 140
> High Point, NC 20265
> Telephone: (866) 841-5250 (toll-free)
> Telephone: (336) 841-5250
> Fax: (336) 841-5350
> E-mail: info@pharmacore.com

Pharm-Eco Laboratories Inc.
> 128 Spring Street
> Lexington, MA 02421
> Telephone: (781) 861-9303
> Fax: (781) 861-9386

SIDDCO Inc.
> 9040 S. Rita Road, No. 2338
> Tucson, AZ 85747
> Telephone: (520) 663-4001
> Fax: (520) 663-0795

SPECS and BioSPECS B.V.
> Fleminglaan 16, 2289 CP Rijswijk,
> The Netherlands
> Telephone: 31 70 319 0038
> Fax: 31 70 319 0011
> E-mail: jan.schultz@specs.net

TimTec Inc.
> 100 Interchange Boulevard
> Newark, DE 19711
> Telephone: (302) 292-8500
> Fax: (302) 292-8520
> E-mail: info@timtec.net

Vertex Pharmaceuticals, Inc.
> 40 Alston Street
> Cambridge, MA 02139-4211
> Telephone: (617) 576-3111
> Fax: (617) 499-2480 or (617) 576-2109

Zelinsky Institute
> 1300 First State Blvd., Suite E-1
> Wilmington, DE 19804
> Telephone: (302) 292-8515
> Fax: (302) 292-8520
> E-mail: info@zelinsky.com

B. Sources for Detailed Solid- and Solution-Phase Combinatorial Procedures

1. Scientific Journals

Scientific journals that include detailed experimental procedures in combinatorial chemistry include the *Journal of Combinatorial Chemistry*, *Journal of the American Chemical Society*, *Tetrahedron Letters*, *Journal of Organic Chemistry*, *Tetrahedron*, *Molecular Diversity*, *Organic Letters*, and *Bioorganic & Medicinal Chemistry*.

2. Books

Useful books for detailed combinatorial experimental procedures include:

> Abelson, J.N., *Combinatorial Chemistry*, Academic Press, San Diego, 1996.
> Bannwarth, W. and Felder, E., *Combinatorial Chemistry: A Practical Approach*, Wiley-VCH, New York, 2000.
> Bunin, B.A., *The Combinatorial Index*, Academic Press, San Diego, 1998.
> Burgess, K., *Solid-Phase Organic Synthesis*, John Wiley & Sons, New York, 2000.
> Cabilly, S., *Combinatorial Peptide Library Protocols*, Humana Press, Totowa, NJ, 1997.
> Chaiken, I.M. and Janda, K.D., *Molecular Diversity and Combinatorial Chemistry, Libraries and Drug Discovery*, Americal Chemical Society, Washington, DC, 1996.
> Cortese, R., *Combinatorial Libraries: Synthesis, Screening and Application Potential*, Walter de Gruyter, New York, 1996.
> Czarnik, A.W. and DeWitt, S.H., *A Practical Guide to Combinatorial Chemistry*, American Chemical Society, Washington, DC, 1997.
> Dorwald, F.Z., *Organic Synthesis on Solid Phase: Supports, Linkers, Reactions*, Wiley-VCH, New York, 2000.
> Epton, R., *Innovation and Perspectives in Solid Phase Synthesis & Combinatorial Libraries*, Mayflower Scientific Limited, Birmingham, AL, 1997.
> Fields, G.B. and Colowick, S.P., *Solid-Phase Peptide Synthesis*, Academic Press, San Diego, CA, 1997.
> Gordon, E.M. and Kerwin, J.F.J., *Combinatorial Chemistry and Molecular Diversity in Drug Discovery*, John Wiley & Sons, New York, 1998.
> Harvey, A.L., *Advances in Drug Discovery Techniques*, John Wiley & Sons, New York, 1998.
> Jung, G., *Combinatorial Chemistry: Synthesis, Analysis, Screening*, Wiley-VCH, Weinheim, 1999.
> Jung, G., *Combinatorial Peptide and Nonpeptide Libraries: A Handbook*, Wiley-VCH, Weinheim, 1996.
> Lloyd-Williams, P., Albericio, F., and Giralt, E., *Chemical Approaches to the Synthesis of Peptides and Proteins*, CRC Press, Boca Raton, FL, 1997.
> Miertus, S. and Fassina, G., *Combinatorial Chemistry and Technology: Principles, Methods, and Applications*, Marcel Dekker, New York, 1999.
> Schneider, C.H., *Peptides in Immunolgy*, John Wiley & Sons, New York, 1997.
> Seneci, P., *Solid-Phase Synthesis and Combinatorial Technologies*, John Wiley & Sons, New York, 2001.
> Terrett, N.K., *Combinatorial Chemistry*, Oxford University Press, New York, 1998.
> Wilson, S.R. and Czarnik, A.W., *Combinatorial Chemistry: Synthesis and Applications*, John Wiley & Sons, New York, 1997.

References

1. **Eddington, A.S.,** *The Nature of the Physical World: The Gifford Lectures, 1927*, Macmillan, New York, 1929, p. 72.
2. **Appell, K., Baldwin, J., and Willian, J.,** Combinatorial chemistry and high-throughput screening in drug discovery and development, in *Handbook of Modern Pharmaceutical Analysis*, Ahuja, S. and Scypinski, S., Eds., Academic Press, New York, 2001, chap. 2.
3. http://www.lybradyn.com

4. **Moore, R.E. and Scheuer, P.J.,** Palytoxin: new marine toxin from a coelenterate, *Science*, 172, 495, 1971.
5. **Nonomura, T. et al.,** Complete structure of maitotoxin, part II; configuration of the C135-C142 side chain and absolute configuration of the entire molecule, *Angew. Chem.*, 108, 178, 1996; *Angew. Chem. Int. Ed. Engl.*, 35, 1675, 1996.
6. **Merrifield, B.,** Solid phase synthesis, *Science*, 232, 341, 1986.
7. **Geysen, H.M., Meloen, R.H., and Barteling, S.J.,** Use of peptide synthesis to probe viral antigens for epitopes to a resolution of a single amino acid. *Proc. Natl. Acad. Sci. U.S.A.*, 81, 3998 1984.
8. **Geysen, H.M., Barteling, S.J., and Meleon, R.H.,** Small peptides induce antibodies with a sequence and structural requirement for binding antigen comparable to antibodies raised against the native protein, *Proc. Natl. Acad. Sci. U.S.A.*, 82, 178, 1985.
9. **Houghten, R.A.,** General method for the rapid solid phase synthesis of large numbers of peptides: specificity of antigen-antibody interaction at the level of individual amino acids, *Proc. Natl. Acad. Sci. U.S.A.*, 82, 5131, 1985.
10. **Sebestyen, F. et al.,** Chemical synthesis of peptide libraries, *Bioorg. Med. Chem. Lett.*, 3, 413, 1993.
11. http://www.discoverypartners.com.
12. **Dolle, R.E. and Nelson, K.H.,** Comprehensive survey of combinatorial library synthesis, *J. Comb. Chem.*, 1, 235, 1999.
13. **Ugi, I., Domling, A., and Horl, W.,** Multicomponent reactions in organic chemistry, *Endeavour*, 18, 115, 1994.
14. **Booth, R.J.,** Polymer-supported quenching reagents for parallel purification, *J. Am. Chem. Soc.*, 119, 4882, 1997.
15. **Kassel, D.B.,** Combinatorial chemistry and mass spectrometry in the 21st century drug discovery laboratory, *Chem Rev.*, 101, 255, 2001.
16. http://www.zymark.com.
17. http://www.genevac.co.uk.
18. **Breitenbucher, J.G., Arienti, K.L., and McClure, K.J.,** Scope and limitations of solid-supported liquid-liquid extraction for the high-throughput purification of compound libraries, *J. Comb. Chem.*, 3, 528, 2001.
19. http://www.merck.de/english/services/chromatographie/lachrom/htp.htm.
20. http://info.sial.com/Graphics/Supelco/objects/11300/11264.pdf.
21. **Parlow, J.J. and Flynn, D.L.,** Solution-phase parallel synthesis of a benzoxazinone library using complementary molecular reactivity and molecular recognition purification technology, *Tetrahedron*, 54, 4013, 1998.
22. **Nicolaou, K.C. et al.,** Radiofrequency encoded combinatorial chemistry, *Angew. Chem. Int. Ed. Engl.*, 34, 2289, 1995.
23. **Moran, E.J. et al.,** Radio frequency tag encoded combinatorial library method for the discovery of tripeptide-substituted cinnamic acid inhibitors of the protein kinase tyrosine phosphatase, *J. Am. Chem. Soc.*, 117, 10787, 1995.
24. **Urbas, D.J. and Ellwood, D.,** U.S. Patent 5,252,962: System Monitoring Programmable Implantable Transponder, 1993.
25. **D'Hont, L., Tip, A., and Meier, H.,** U.S. Patent 5,351,052: Transponder Systems for Automatic Identification Purposes, 1994.
26. **Gallop, M.A. et al.,** Applications of combinatorial technologies to drug discovery: 1. Background and peptide combinatorial libraries, *J. Med. Chem.*, 37, 1233, 1994.
27. **Janda, K.D.,** Tagged versus untagged libraries: methods for the generation and screening of, *Proc. Natl. Acad. Sci. U.S.A.*, 91, 10779, 1994.
28. **Nielsen, J., Brenner, S., and Janda, K.D.,** Synthetic methods for the implementation of encoded combinatorial chemistry, *J. Am. Chem. Soc.*, 115, 9812, 1993.
29. **Brenner, S. and Lerner, R.A.,** Encoded combinatorial chemistry, *Proc. Natl. Acad. Sci. U.S.A.*, 89, 5381, 1992.

30. **Needels, M.C. et al.**, Generation and screening of an oligonucleotide-encoded synthetic peptide library, *Proc. Natl. Acad. Sci. U.S.A.*, 90, 10700, 1993.
31. **Felder, E.R. et al.**, A new combination of protecting groups and links for encoded synthetic libraries suited for consecutive tests on the solid phase and in solution, *Mol. Diversity*, 1, 109, 1995.
32. **Kerr, J.M., Banville, S.C., and Zuckerman, R.N.**, Encoded combinatorial peptide libraries containing non-natural amino acids, *J. Am. Chem. Soc.*, 115, 2529, 1993.
33. **Lebl, M. et al.**, WO Patent 94/28028: Topologically Segregated, Encoded Solid Phase Libraries, 1994.
34. **Nestler, H.P., Bartlett, P.A., and Still, W.C.**, A general method for molecular tagging of encoded combinatorial chemistry libraries, *J. Org. Chem.*, 59, 4723, 1994.
35. **Ohlmeyer, M.H. et al.**, Complex synthetic chemical libraries indexed with molecular tags, *Proc. Natl. Acad. Sci. U.S.A.*, 90, 10922, 1993.
36. **Eckes, P.,** Binary coding of compound libraries, *Angew. Chem. Int. Ed. Engl.*, 33, 1573, 1994.
37. **Baldwin, J.J. et al.**, Synthesis of a small molecule combinatorial library encoded with molecular tags, *J. Am. Chem. Soc.*, 117, 5588, 1995.
38. **Burbaum, J.J. et al.**, A paradigm for drug discovery employing encoded combinatorial libraries, *Proc. Natl. Acad. Sci. U.S.A.*, 92, 6027, 1995.
39. **Wagner, D.S. and Geysen, H.M.**, Isotope or mass encoding of combinatorial libraries, *Chem. Biol.*, 3, 679, 1996.
40. **Brummel, C.L. et al.**, Evaluation of mass spectrometric methods applicable to the direct analysis of non-peptide bead-bound combinatorial libraries, *Anal. Chem.*, 68, 237, 1996.
41. **Czarnik, A.W.,** Solid-phase synthesis supports are like solvents, *Biotechnol. Bioeng. (Comb. Chem.)*, 61, 77, 1998.
42. **Czarnik, A.W.,** Encoding strategies in combinatorial chemistry, *Proc. Natl. Acad. Sci. U.S.A.*, 94, 12738, 1997.
43. **Czarnik, A.W. and Nova, M.P.,** No static at all: electronic encoding in combinatorial organic synthesis, *Chem. Brit.*, 33, 39, 1997.
44. **Keifer, P.A.,** NMR tools for biotechnology, *Curr. Opin. Biotech.*, 10, 34, 1999.
45. **Shapiro, M.J. and Wareing, J.R.,** NMR methods in combinatorial chemistry, *Curr. Opin. Chem. Biol.*, 2, 372, 1998.
46. **Keifer, P.A.,** New methods for obtaining high resolution NMR spectra of solid phase synthesis resins, natural products, and solution-state combinatorial libraries, *Drugs Future*, 2, 468, 1998.
47. **Haap, W.J., Walk, T., and Jung, G.,** FTIR mapping — a new tool for spatially resolved characterization of polymer-bound combinatorial libraries with IR microscopy, *Angew. Chem. Int. Ed. Engl.*, 23, 3311, 1998.
48. http://www.5z.com/divinfo/

Chapter 25

Computational Data-Mining Methods Applied to Combinatorial Chemistry Libraries

Shirley Louise-May

Contents

I.	Chemoinformatics, Data Mining, and the Combinatorial Age	549
II.	Computational Paradigms for Building and Storing Combinatorial Libraries	551
III.	Property Design of Combinatorial Libraries	552
IV.	High-Throughput Screening Data	554
V.	Generation of Molecular Property Descriptors	555
VI.	*In Silico* Screening of Virtual Combinatorial Libraries	556
VII.	Data-Mining Methods	556
	A. Similarity/Diversity/K-Nearest Neighbors	556
	B. Decision Trees/Recursive Partitioning	558
	C. Artificial Neural Networks and Genetic Algorithms	559
	D. Structure-Based Pharmacophore Methods	561
VIII.	Summary	562
References		564

I. Chemoinformatics, Data Mining, and the Combinatorial Age

Chemoinformatics is the representation and manipulation of chemical information that describes molecular fragments, molecules, and compound libraries. The correlation of chemoinformatic data to biological response data — such as activity toward a target receptor, molecular or cellular toxicity, and the pharmaceutically relevant processes of absorption, distribution, metabolism, and excretion (collectively called ADME properties) — can be achieved via data-mining techniques. Data-mining techniques assume that a pattern exists between a population of molecules, their molecular properties, and their biological behavior. Data-mining algorithms use different strategies to uncover or "mine" these patterns. High-throughput methods geared toward pharmaceutical application and therapeutic-target research, such as combinatorial chemistry (CC) and high-throughput screening (HTS), produce large populations of molecules. The associated data overwhelm traditional quantitative structure activity relationship (QSAR) and computational modeling techniques. Thus, high-throughput computational modeling techniques, such as data mining, have become a necessary and natural complement to the present high-throughput combinatorial age. Prioritization of tractable chemical libraries from a large virtual chemical space of possible synthesizable compounds is a necessary and key step in using high-throughput methods to search for biological activity of a given receptor target. Prioritization of biological screening of existing-compound libraries can be based on computational model scores of similarity to known active compounds or predictions of activity, selectivity, and other relevant properties. Strategic application of data-mining techniques aims to focus high-throughput synthesis and screening resources efficiently on biologically relevant or enriched libraries at the "virtual" stage, thus enabling a higher return on efforts.

Current data-mining strategies are moving beyond design and prioritization of libraries for potency toward virtual optimization of libraries in terms of pharmaceutical properties. Computational filters and models can be applied to address the druglike properties along with the issues of selectivity, solubility, metabolism, and toxicity liabilities of potential libraries prior to synthesis, in addition to maximizing biological activity at a particular receptor. Increasing the screening hit rate of a target maximizes the data content of each HTP run and can help aid in optimizing assay protocols. Data-mining models, in combination with two-dimensional descriptors, are particularly adept at picking up a structurally diverse, biologically enriched population, thus broadening the chemical series space hits towards a given receptor. Rapid exploration of the structure-activity relationships (SARs) around multiple series can thus proceed simultaneously with data-mining models as assay information on actives is amplified in subsequent models built on the growing population of compounds screened through HTS.

Combinatorial libraries can consist of hundreds to millions of compounds and are generated to systematically assess the effects of structural variation on interactions with target receptors. Analysis of the patterns of interaction of a combinatorial library in a high-throughput *in vitro* pharmacology screening assay can provide a wealth of information about the nature of the interactions being studied. Computational methods that are employed to encode, organize, extract, and interpret these patterns, and to allow predictions and/or design of compounds or compound libraries with improved properties, fall under the category of data-mining tools.

The utility of these computational methods can even begin before synthesis when they are used to design combinatorial libraries with particular properties. Encoding and organization of the process of reagents → products can streamline the synthesis automation by robotic computers and enable not only efficient data tracking, but also deconvolution of product-specific activity to fragment-specific activity. Correlation of chemical properties to biological activity occurs through the generation of molecular-property descriptors and the use of data-mining modeling methods. Generally, hundreds to thousands of descriptors are generated for each compound, and hundreds to hundreds of thousands of compounds are used to train a data-mining model. The key to the success of these models lies in population analysis and other common statistical measures that allow the subtle patterns and rules in the data to emerge, and that are also helpful in evaluating the balance of gains to risks associated with future strategies. Following a "numbers in, numbers out" policy, i.e., using the results of a data-mining model to design a new population of compounds, results in the highest gains. When this process is applied as an iterative cycle, each turn of the cycle has the potential to yield more information than the last. This allows for amplification of signal, reduction of noise, better predictions, better design, and ultimately higher gains. Applied to combinatorial chemistry in the design of compounds with biological activity and druglike properties, the process is called *library enrichment*. This process has been applied successfully in the pharmaceutical industry over a wide range of therapeutic targets.

The basic principles and methods of combinatorial-chemistry-related data-mining tools are presented in the remainder of this chapter. The prerequisite to generating data-mining models for library enrichment is to have an initial library with biological screening data. Thus, the sections addressing "library encoding," "library design," and "high-throughput screening data" — **Sections II, III**, and **IV**, respectively — are discussed first. Generation of chemical descriptors and *in silico* virtual screening concepts — covered in **Sections V** and **VI** — follow in order to help outline the process. Four distinct data-mining methods that have received considerable development and application are discussed in **Section VII** in the order of increasing complexity with their particular requirements, advantages, and disadvantages. In the summary (**Section VIII**), the process as a whole is examined, and the power of applying data-mining methods to combinatorial chemistry, particularly in an iterative cycle, is discussed. The references section contains seminal methods papers as well as some excellent reviews in the field.

II. Computational Paradigms for Building and Storing Combinatorial Libraries

There are several common computational paradigms for building and storing combinatorial libraries. Each has its own advantages. The aim is to build a chemically *relevant* library of products from available reagents using *accessible* synthetic methods. The key emphasis should be on relevance, specifically biological relevance, rather than on accessible protocols. The importance of focusing primarily on the concept of biological relevance in combinatorial endeavors, rather than synthetic accessibility, in biological discovery projects can be likened to searching for and recognizing functional gene patterns in a genome analysis rather than sampling from all possible sequence combinations. Conserved binding motifs are recurring chemical arrangements with potent biological activity found among, or engineered between, distinct and disparate receptor types. Smaller combinatorial libraries that are constructed from conserved-binding-motif concepts have shown much higher rates of biological activity, even against new receptor targets, than vast combinatorial libraries constructed to sample from the diversity of accessible chemical space. Once a relevant chemical space has been conceptualized, chemoinformatic encoding of the virtual space will allow enumeration of molecules within that space, which can then be screened *in silico* by computational models.

Minimally, the available reagents and some form of a generic product need to be encoded. Two main methods have emerged in the field, namely, linked products and core products. Linked products constitute an example of convergent synthesis strategies, wherein the main body of the resulting product is formed by the reaction. Greater diversity can be achieved with linked products, in which reagents are linked together spatially as shown in Figure 25.1A. Alternatively, focused libraries result from core methods in which a generic product is represented by a "core" of conserved atoms. Here, reaction with reagents occurs at substituent positions along the extremeties of the core, as shown in Figure 25.1B. A third emerging technique is to encode the synthetic transformation of a reaction and allow mixing and matching of different reagent types in order to create different product classes, as shown in Figure 25.1C. This technique mirrors most closely the traditional synthetic practices and allows a "toolbox" of reactions to be used with an inventory of reagents so as to form a limitless variety of product libraries. The choice of the building method depends more on the project aims and synthetic capabilities than any intrinsic advantages or disadvantages. There is a variety of software available for each approach, spanning stand-alone menu-driven systems to

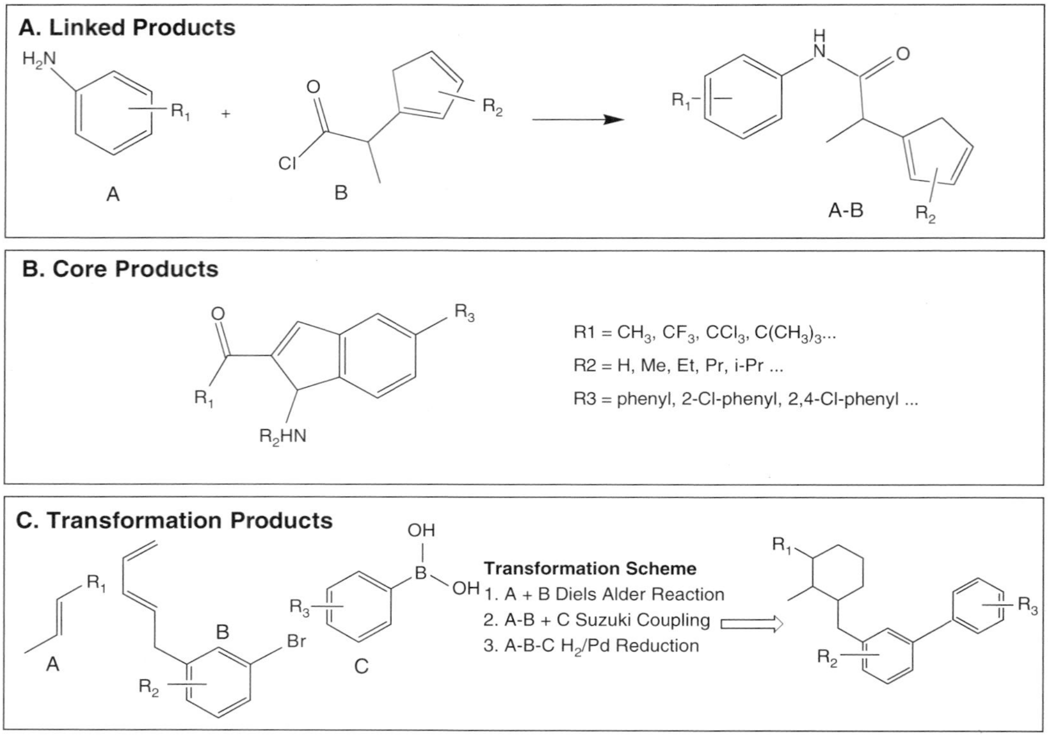

FIGURE 25.1
Combinatorial library informatics schemes.

developable/configurable suites of modules. Some practical issues to consider when evaluating which kind of system to implement are the following:

- **Reagent integration and organization:** Can commercially available sources of reagents be imported in batch? Can relevant reagent properties be encoded (inventory, molecular weight, QC characteristics, commercial ID number, clipping rules)? Can reagents be grouped/organized by a number of different properties (molecular weight range, reactivity, cost, substructural elements)?
- **Product library deconvolution and optimization:** Is the reagent information retained in product specification? Can potential product libraries be filtered based on target constitutive property ranges and structural alerts? Can potential product libraries be optimized based on target statistical properties such as diversity, similarity, size, and distributions?
- **Enumerated virtual library export:** Quite often it is useful to score a potential product library by more advanced computational modeling techniques. In this case, the system must have the ability to export — in an accessible format — the virtual library of all possible product molecules within a combinatorial synthesis paradigm.

III. Property Design of Combinatorial Libraries

Although the endgame of utilizing combinatorial libraries is to aid in the design of molecules that have improved and specific interactions with their targets in biological systems, the initial considerations of what should be synthesized are more concerned with getting "a foot in the door." Molecules encounter numerous obstacles, namely, the biological system's defense mechanisms, which rely on bulk properties and nonspecific interactions as filters against unwanted foreign agents. Successfully passing by these defense filters should be the first design consideration. Pharmacokinetics is the study of what the body does to a drug over time. It can be broken down into several discrete processes: absorption, distribution, metabolism, and excretion (commonly termed ADME properties). Each process imposes a set of restrictions on a compound that controls how much drug is passed on to the next process and, ultimately, how much of a compound can reach the target site of interaction. Although not all of the requirements of these processes are fully understood, some of the following properties have been found to be particularly relevant: molecular weight, number of hydrogen bond donors and acceptors, lipophilicity, solubility, pKa properties, number of rotatable bonds, and polar surface area. Given an enumerated virtual library, values or estimates of the values for these properties can be used to filter out products with property profiles that diminish the ability of a compound to reach the target site.

Some commonly used filter strategies are described below:

- **Lipinski's rule of five for improved permeability:** Molecular weight maximum of 500 amu, number of hydrogen bond donors less than five, number of hydrogen bond acceptors less than ten, calculated CLogP value maximum of 5.0.[1]
- **Polar surface area and permeability:** The polar surface area (PSA) of a molecule is the surface area of polar atoms such as oxygen, nitrogen, and hydrogens attached to them. Many recent studies have confirmed the presence of a strong link between the PSA of a molecule and its permeability via oral absorption, passive transport across the endothelium, and transport across the blood–brain barrier. In general, a molecule having a polar surface area greater than 140 $Å^2$ will exhibit poor permeability for oral absorption, and compounds targeted for central nervous system (CNS) indications that must cross the blood–brain barrier are further restricted to having PSA values less than 80 $Å^2$.
- **Central nervous system (CNS) indications limitations:** Lipinski et al.[1] found an even stricter set of guidelines for molecules that must pass the blood-brain barrier to reach the site of interaction. Such molecules should possess a molecular weight less than 400 amu, six or fewer hydrogen bond acceptors, three or fewer hydrogen bond donors, and as stated above, a PSA less than 80 $Å^2$.
- **Solubility/permeability guidelines and theraputic indexing:** Lipinski et al.[1] have calculated the necessary solubility requirements for three theraputic-dose classes and three permeability classes. Potent compounds that can be administered at 0.1 mg·kg^{-1} possessing high permeability need only 1 μg·ml^{-1} solubility in a phosphate buffer (pH 7.4) medium whereas, at the same dose, compounds with poor permeability need to have a solubility of 21 μg·ml^{-1}. At a 1-mg·kg^{-1} dose, the solubility range for high- to low-permeability compounds is 10 to 207 μg·ml^{-1}. At a 10-mg·kg^{-1} dose, the solubility for even a high-permeability compound must be at least 100 μg·ml^{-1}. Most small-molecule organic compounds have sparing solubilities in the 0 to 100 μg·ml^{-1} range, which in turn imposes higher requirements for potency and permeability. The balance point between these three properties, in terms of effort needed to effect a desired outcome, is a medium-potency compound (1 mg·kg^{-1} dose) with medium permeability having a solubility of 65 μg·ml^{-1} or less.
- **Conformational flexibility limitations:** The number of available conformations of a molecule increases dramatically with the number of rotatable bonds. The probability of a molecule encountering the receptor in an "active" conformation goes down as the number of accessible conformations goes up. Optimization of lead compounds to development candidates often involves strategies of "fixing" the molecule to its active conformation through bioisosteric replacement or replacement of substructures with equivalent groups possessing desired properties. For druglike compounds, retrospective analyses show a mean value of 7 and maximum value of 15 rotatable bonds.[2] A recent study on over 1100 proprietary compounds found that a maximum value of ten rotatable bonds correlated highly with bioavailability values of greater than 20%.[3]
- **Synthetic complexity limitations:** Limiting the number of chiral centers and stereochemically controlled reactions is a desirable practice in terms of reagent cost, synthetic success, and process scale-up. Four or less chiral centers are usually synthetically tolerable.

IV. High-Throughput Screening Data

When a combinatorial library is screened biologically in a high-throughput format, the information content of each compound differs considerably from biological information generally obtained on individual compounds. For individual compounds, the extent of binding to the receptor, or inhibition or enhancement of some signal, is determined at several concentrations of the molecule so as to determine an intrinsic interaction constant. In the high-throughput format, the compounds are usually assumed to be at a fixed concentration, and a single-point determination of interaction is measured. The concentration of the compound is estimated from the amount of reagents used and the stoichiometry of the reaction. The actual concentration of the product depends on the extent of the reaction and workup conditions. For combinatorial compounds that have become part of a stored-compound archive, the product concentration may also depend on the storage medium, the length of time stored, and the inherent stability of the product. The extent of interaction in a biological assay, of course, depends directly on the concentration of the product. Discrepancies limit the resolution of the QSAR that can be determined. Mixtures of compounds are quite often used in high-throughput format screening assays to reduce the costs of assay materials and speed up the time that it takes to screen a large number of compounds. In this case, a deconvolution step must be undertaken to determine which compounds in a mixture are giving rise to the signals of interaction. If there is interaction among the molecules, the signal might not be additive.

A false positive result can occur when a molecule gives rise to a significant interaction signal in a single measurement but does not replicate, or does not show a sufficient interaction signal over a more biologically relevant concentration range. False positives must be managed to avoid draining resources, but they can be tolerated. A false negative result — a molecule that gives rise to insignificant signal in a single test but that would retest at a significant and therefore relevant level — cannot be tolerated, since the actual rate of positive results is so low (generally much less than 1%). Some high-throughput practices that minimize false negatives and reduce the overall noise of a data set are a more rigorous quality control (QC) analysis prior to testing. This facilitates accurate determination of the molecule concentration and the presence of reaction by-products or compound degradation and enables the use of replicates and controls in the testing process.

The following are general guidelines for estimating the information content of a high-throughput screening data set:

- **Use of controls:** Generally a control is used to verify that the assay is behaving as designed and to quantify activity responses. If the control generally corresponds to a high-activity signal (100% at a single concentration), the assay experiments in which the signal of the control drops to below 80%, or is in excess of 120%, of its normal value are usually rejected, and the experiments are redone.
- **Use of replicates:** If possible, replicate data on each compound within the same assay experiment can give valuable information on the compounds and assay conditions. When the covariance between replicates is greater than 30%, there is the danger of there being more noise than signal, and the data point is usually rejected. Another guideline for replicates is that at 30 to 40% signal, the replication across all compounds should fall between 5 to 15% covariance. In excess of this, the assay conditions need be tuned to a greater signal-to-noise ratio.
- **Hit dosing:** Generally, compounds that produce interesting levels of signal in the single-dose experiment are run at multiple doses to determine the intrinsic binding or activity coefficient. A well-designed assay will have a decent correlation between percent activity at a single dose and the intrinsic-activity coefficient for the signal range of 40 to 85%. This is the most linear part of the interaction curves. Finally, compounds with interesting levels of signal at a single concentration should generally dose down to interesting intrinsic-activity coefficients.

V. Generation of Molecular Property Descriptors

There is an almost infinite number of ways that molecules can be described that is relevant to their ability to interact with their target receptors involving size, shape, and spatial and chemical properties. These properties can also be represented in a variety of ways: presence/absence encoded as bit strings, counts encoded as integers, quantitative properties encoded as real values, qualities encoded as text strings, physical fields encoded as numerical vectors, and any mix of these. Some descriptors can be calculated from one-dimensional molecular data, such as molecular weight from a molecular formula, and from constitutive properties, such as the counts of atoms. Many more properties can be calculated from two-dimensional molecular data, which refer to the atoms of a molecule and a map of how they are connected to one another specifying bond types. In fact, a two-dimensional representation of a molecule is sufficiently descriptive of its likely chemical properties that three-dimensional data can often be approximated from it, such as shape, lipophilicity, dipole fields, and surface properties. Three-dimensional data are often used to relate spatial arrangements of molecular recognition elements, such as hydrogen bond donors and acceptors, molecular fields, or shape characteristics. They are often employed when complementary structural information is known about the target receptor. Three-dimensional data are the most computationally expensive data to obtain and are often prohibitive for large regions of chemical space, say 10,000 to 1,000,000 molecule spaces. Also, quite often, as the dimensionality of the molecular information is increased (one-dimensional → two-dimensional → three-dimensional), the dimensionality of the property vector increases dramatically, making some data-mining approaches intractable.

As a rule of thumb, the simpler data-mining techniques can accommodate a larger number of molecular property descriptors (on the order of 10^4 to 10^6), while the more sophisticated techniques require the property vector dimensionality to be much smaller (on the order of 10^3). Descriptors can be generated from the products themselves, but they are much more accessible and relevant for combinatorial chemistry spaces when computed from the reagent fragments that make up the product. This enables deconvolution of specific chemical and biological properties to specific reagents. It also enables approximation of molecular properties from unenumerated libraries, resulting in a huge computational savings.

There are at least two guiding philosophies concerning molecular property descriptors and their application to specific biological receptor targets. The philosophy of traditional QSAR modeling is to find a small number of descriptors that are generally applicable across a wide range of biologically relevant target properties and that yield results that are relatively invariant to the type of modeling method applied.[4] The philosophy of more recent, high-throughput virtual-screening-based applications is one of enrichment; a set of descriptors is evaluated positively if it can selectively enrich a population in a given property relative to its natural rate of occurrence. The main difference between traditional QSAR descriptors and data-mining descriptors is specificity. Traditional QSAR methods are quite successful within an analog series, but some descriptors lose their sense when applied to a structurally distinct analog series. Conversely, high-throughput SAR and data-mining models benefit from a degree of "fuzziness" in the descriptors that allows for recognition of more than one structural arrangement leading to similar biological activity. Both philosophies aim for the principle of strong causality[5] or the definition of a property space in which small changes in properties result in small changes in a target property, such as biological activity, and large changes in properties reflect a large change in the target property.

VI. *In Silico* Screening of Virtual Combinatorial Libraries

In **Chapter 24**, it was stressed that the central issue in combinatorial chemistry library synthesis is not one of "What can be made?" but, rather, one of "What should be made?" One way to answer this question is to construct as large and relevant a virtual combinatorial chemistry library as possible and then to use some amount of biological data to guide the answer of what should be made. The biological data, along with appropriate descriptors, are used to either model desired features and properties of compounds of known activity or to model the response surface of the receptor–ligand interaction. The model must have the following properties when applied to virtual screening of a large combinatorial library:

- **Generalizability:** This speaks to the model's ability to return a diverse set of possible solutions.
- **Specificity:** This speaks to the model's ability to suggest a reasonable number of possible solutions.
- **Predictability:** Given a test set, the model should demonstrate a predictive advantage over other methods of selecting which compounds to synthesize. Specifically, the "hits" returned should have a high probability of being "active."

VII. Data-Mining Methods

Data-mining methods have long been applied in the fields of engineering, market research, and process control. Recent technological advances in chemistry and biology have increased the number of discrete entities that can be experimented upon by many orders of magnitude. Thus, the successful development and application of data-mining techniques has entered these fields. The data-mining techniques listed below are some of the more commonly applied methods in the analysis of high-throughput biological screening data on combinatorial libraries. They are listed in order of increasing complexity and computational resource requirements. Some model types are untrained, that is, an inherent property of the set of molecules is being used as a probe in mining molecules that are similar or dissimilar in the property space. Then, *a priori* knowledge of some compounds possessing the desired target property can be used to "mine" the library space. Trained models use some form of biological response information to try and learn what areas of property space maximize (or minimize) the biological response. The choice of data-mining method to use depends on the type of question being asked and the statistical success desired for the final outcome. Application of data-mining methods is often directional; queries returning breadth are distinct from those returning depth. As a rule, they are most successfully applied at statistically relevant levels. That is to say, the accuracy of correctly predicting one compound of ten is substantially reduced over correctly predicting 100 compounds out of 1000. (In mathematical terms, the probability outcome statistics are more stable with increasing N.)

A. Similarity/Diversity/K-Nearest Neighbors

Similarity, diversity, and K-nearest neighbors are untrained data-mining methods that compare the property vector of a probe with the property vectors of each molecule in a library and assign a score based on the relatedness of the probe to the library molecule. The relatedness is computed via a distance metric. Distance metrics are usually normalized on a scale of 0.0 to 1.0, with 1.0 being identical to the probe and 0.0 indicating that there is nothing in common between the probe and molecule to be scored. The probe can be a molecule, a subset of molecules, or a composite of the molecular set or any subset.

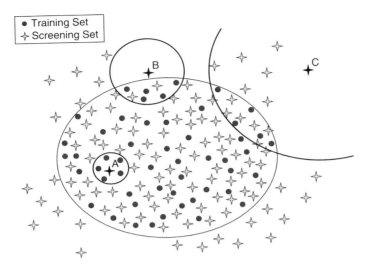

FIGURE 25.2
K-nearest neighbors.

In a similarity data-mining exercise, top scoring molecules that have maximally similar property vectors are generally sought. These molecules will be clustered in property space. The assumption is that if the property space of a set of molecules is highly homogeneous, the biological response of each molecule in the set will be similar. The result is, of course, highly dependent on the information content of the property space with respect to the biological receptor, the distance metric used, the specificity of the receptor, and the pharmacokinetic properties of the molecules. As a general rule of thumb, greater than 80% similarity is needed for there to be a 50% chance of finding a molecule of similar activity.[6] For extremely potent molecules that have fine-tuned molecular–receptor interactions, a much higher degree of similarity is required. Similarity approaches are often used on the small set of lead molecules in initial stages of a data-mining project so as to generate a sufficient "signal" in a subsequent population of tested molecules to which more-sophisticated data-mining algorithms can then be applied.

K-nearest neighbors (KNN) algorithms find the set of molecules of size K closest to the molecule to be scored, and then they relate the "local" biological response of the set to the score. The score in this method is a biological activity prediction that is based on the biological activities of molecules in the local neighborhood. It can be seen from Figure 25.2 that the KNN algorithm yields prediction results that vary according to the distance and distribution of a screened compound to its neighborhood. In this figure, the circles represent the training-set compounds, and the stars represent the library that is being screened. The property space is represented in two spatial dimensions, the x- and y-axes. Three compounds, A, B and C, are shown in the center of a circle that encloses their five nearest neighbors in the training data set. In this example, the prediction for compound A should be well approximated by its five closest neighbors in the training set because they form a small cluster in which compound A is centrally located. The prediction may be mediocre for compound B because of the uneven distribution of its five closest neighbors in x-axis of the property space. Finally, the prediction for compound C would be expected to be poor because of the large distances to its five nearest neighbors as well as their wide distribution over the x and y property space.

In attempts to address the variability in prediction quality inherent in the simplest KNN algorithms, which specify the number of nearest neighbors irrespective of distance or distribution, KNN variations have been developed that define the "neighborhood" in different ways: by specifying a distance cutoff (in essence a similarity threshold); by weighting the values of neighbors according to their distance; or by taking into account the extent of variations in the target property, relative to the variations in the local property space, in the scoring process. For efficiency gains in computations, approximated KNN algorithms are often used for property spaces larger than dimensions of $N = 8$.

The goal of diversity data mining is to find the subset of molecules within a population that are most dissimilar from one another. This is often employed as a means of efficiently sampling some portion of chemical space by selecting or synthesizing only a subset of a chemical compound library. Distance metrics are used in the scoring of molecules analogous to similarity searches that return maximally dissimilar property vectors. However, additional steps are needed to effect "mutual" dissimilarity. As distinct molecules are "found," they are added to a pool of molecules that become the reference set for the next molecule to be scored. If the molecule being scored is above a certain similarity value threshold to any member of the reference set, it is not distinct and is discarded. For a very large chemical space, this approach is intractable, and many algorithms have been developed that attempt to approximate the end result in a computationally efficient manner. Diversity is often employed when no lead matter is known for a receptor target or when the available lead matter has undesirable properties that cannot be easily engineered out (e.g., patent issues, toxicity). The chemical space that is being sampled should be a biologically and pharmaceutically relevant space in order for there to be much chance of finding new activity.

B. Decision Trees/Recursive Partitioning

Decision trees and recursive partitioning are generally applied as classifiers, i.e., they are applied to predict the "class" of a compound from a discrete set of classes rather than a value from a continuum value range. They have been used extensively and successfully in medical diagnostics to determine risk levels of diseases such as heart attack and cancer. Using training data consisting of a property vector and target property value for each input, sets of rules are determined, based on individual properties of the property vector, that result in a statistically meaningful prediction of a class. The rules are determined recursively by partitioning subpopulations of the training data into classes for each input property. The initial split is based on the property that best partitions the data into the desired classes. Next, each subpopulation is examined independently to determine the next most relevant property for the split. This continues for each successive subpopulation until the least relevant property remains or until the splitting fails to meet established statistical criteria. The successive "splits" into layers of subpopulations result in a tree structure, within which each branch represents a rule for a particular property in a particular path. A "path," the route from the initial population (root) to the final one (leaf or node), is a rule set that terminates in a class prediction. Significantly, the same final class prediction can be arrived at by many different paths. These concepts are illustrated in Figure 25.3.

The advantages of these methods are that they are simple to implement, intuitive in interpretation, allow for many unique simultaneous solutions and proceed from a global property influence (most relevant property) to a local property influence (subtle yet important variations). The disadvantages for application to pharmaceutically relevant problems are poor model performance for classes representing less than 1% of the initial population (0.05 to 0.3% "active" molecules of those tested is a common rate for therapeutic targets) and intractable computations for property spaces greater than dimension $N = 500$ and training data sets of more than 25,000 compounds. However, the disadvantages can be easily managed by careful construction of the input training data set to achieve "acceptable" if not "optimal" results.

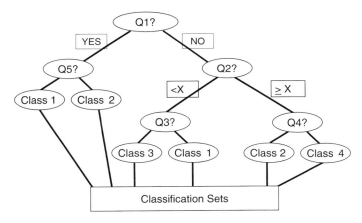

FIGURE 25.3
Recursive partitioning.

C. Artificial Neural Networks and Genetic Algorithms

Adaptive learning data-mining methods, such as artificial neural networks (ANNs) and genetic algorithms (GAs), use initially random input distributions to "learn" the pattern presented in the training data over many cycles or generations. Supervised adaptive learning uses a reference data set with known target values to train a model. The process can be generally described by the following: Inputs(descriptors) from the training set are sent through a processing function that returns an output that is a prediction of the target value. The predicted value is compared with the target value for a training compound, and the "residual" or difference between the processing function output and the target value is communicated to the system, causing a dynamic change in the processing function on the next cycle or generation, which tends to minimize the residual.

Artificial neural networks derive from Hebb (1949),[7] based on models of cognition, who proposed that a mass of computing units or "neurons" could "learn" if the strength of the connections between them could change in a purposeful way. Mathematically, ANNs are universal function approximators that imitate the biological nervous system in that they comprise a large number of densely interconnected processing elements, analogous to neurons, that are tied together with weighted connections, analogous to synapses, and they learn adaptively.

ANN architecture specifies the number of inputs in the input layer, the number of outputs in the output layer, the number of hidden layers, and the number of processing elements or neurons per hidden layer. The number of inputs is the number of descriptors being used. The number of outputs is equal to one for continuous prediction and equal to the number of classes specified for classification. Within the hidden layers is where the "adaptive learning" takes place, and the number of hidden layers and number of neurons per hidden layer is tailored to the system being studied. ANNs are then constructed by forming a connection matrix of neurons between each layer that relates the inputs to outputs. Each neuron is a nonlinear vector function that takes in inputs and produces outputs. Most ANNs are fully connected, i.e., all of the inputs from one layer are connected to all of the outputs in the subsequent layer. Most ANNs are also reductionist (Figure 25.4), in that a large number of initial inputs (descriptors) is converted to a small number of outputs (a continuous or class prediction). Both the vector function of the neuron and the connections between neurons have adjustable parameters that allow the network almost infinite ability to "adapt" to a desired output.

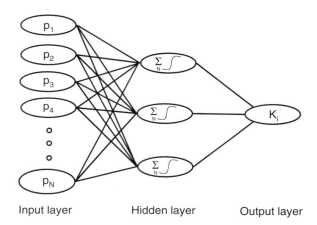

FIGURE 25.4
Artificial neural network.

Supervised learning networks utilize a delta rule or a learning function that adjusts the connection weights between neurons at each cycle based on the training-set residuals, such that large residuals diminish connections and small residuals strengthen connections. Over many cycles, a subset of neurons that best reproduces the training-data-set pattern emerge dominant. A stable ANN, if allowed an infinite number of cycles, will eventually memorize the training-data-set pattern. While this is an amazing result, it is not useful as a predictive model for virtual screening of pharmaceutical libraries. For this reason, a validation set is often employed to determine when to stop training an ANN, and this is defined as the point at which the root mean square error (RMSE) between the predicted and target values for the validation set is at a minimum.

There are two properties of ANNs that make them particularly well-suited to data-mining problems which typically suffer in most traditional analysis methods from low signal to noise; they are distributed and reductionist. Because they are distributed, ANNs tolerate a significant amount of "noise" in the data. The noise may even help ANNs in pattern completion tasks, that is, to generalize to input descriptor values that they have not been trained on, analogous to "fuzzy logic" concepts. Additionally, because the networks are reductionist, errors present in one layer are reduced on proceeding to the next layer with fewer neurons.

There are no general rules of thumb regarding optimal ANN architectures. However, some useful discussions and some anecdotal information can be found on the ratio of inputs to connections,[8] the best number of hidden layers and neurons in a hidden layer,[9] and the general behavior and performance evaluation of ANNs.[10]

Genetic algorithms are evolutionary algorithms that can be used to optimize the size and other properties of a target combinatorial library from a large virtual combinatorial library by utilization of a fitness function encoded for biological activity and druglike properties. Figure 25.5 illustrates the conceptual elements and process of a GA applied as a data-mining tool on a combinatorial virtual library. The genome, which represents all possible solutions, is the virtual library. A chromosome, or one possible solution, represents a particular subset of molecules from the virtual library. Genes represent individual molecules from the population that can be identified by their specific patterns of descriptor properties or alleles. In the GA optimization, initially random populations of molecules are subjected to crossover (swapping of subpopulations between two libraries) and mutation (randomly changing the bits of the descriptor vector for a particular molecule to that of another molecule). Then the new populations are scored and selected according to a fitness function that can use the residuals of the predicted properties of the population to select "improved" populations with total lower residuals. The selection of a new population completes one generation of the GA. GAs can evolve over many generations to a set of potential optimized

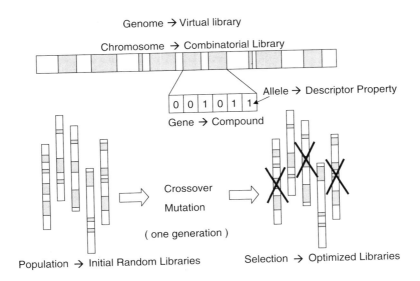

FIGURE 25.5
Pharmacophore methods.

libraries. Note that the GA itself does not predict molecular properties and that other models, potentially data-mining models, must be employed in the fitness function to perform this task. Several dated publications explore the application of genetic algorithms to optimize combinatorial libraries.[11–15] For a more recent application, see the Senomyx Senobase modeling prescription described in the Daylight MUG'02 presentation of Binun.[16]

D. Structure-Based Pharmacophore Methods

A pharmacophore is a specification of the spatial arrangement of atoms and functional groups that dominate the binding interaction between a molecule and its target receptor. A pharmacophore hypothesis can be constructed from known key elements of the receptor structure, determined via mutation studies, by crystal structures of receptors with bound ligands, or by inference — when the structure of the receptor is unknown — in a three-dimensional structural comparison of a number of distinct molecules that show similar binding behavior. Once a pharmacophore hypothesis has been constructed, the spatial elements can be used to screen a virtual library of compounds for three-dimensional structures that show similar spatial features and thus represent a good "fit" to the pharmacophore (see Figure 25.6). The number of features, types of features, and allowed spatial ranges are all parameters that can be optimized for the aims and goals of each particular study.

In general, two features are too few to be useful as a three-dimensional query, returning too many "hits" to be useful. Moreover, the hits likely will not have enough specificity to provide a high probability of being active. Four features is computationally expensive when screening a large virtual library and may be too specific to return many hits. The diversity of the hits will also be diminished by overspecification. In general, three pharmacophore features best balances computational expense and optimizes both diversity and the probability of finding activity.

In order to allow for diversity in the resulting hits, pharmacophore features are not specific functional groups, like C=O, but are types or classes of functional groups. Some commonly used pharmacophore types are hydrogen bond donors, hydrogen bond acceptors, positively charged groups, aromatic rings, and hydrophobic groups.

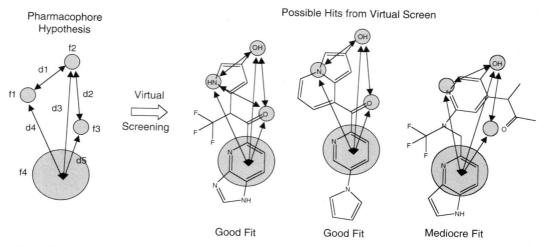

FIGURE 25.6
Genetic algorithms.

There are several mitigating factors to the successful application of structure-based pharmacophores to data-mining problems. The root of all the factors comes back to computational expense. Three-dimensional data provide more than just one additional dimension over two-dimensional data. One cannot invoke a spatial arrangement of atoms in a molecule without reference to its energy state and the potential range of its dynamic structure. This holds true for both the ligand and the receptor. The concept of "structural tolerance" is an important factor in modeling of the ligand–receptor interaction that has two facets. Regarding the ligand, the amount of energy a molecule can spend to adopt a favorable binding conformation needs be considered. Additionally, the intrinsic dynamic flexibility, or lack thereof, of the receptor also has a large influence on the binding interaction. These factors can be accounted for in the following ways:

- **Pharmacophore feature ranges:** The spatial positions of the pharmacophore can be represented by allowed ranges that represent the dynamic flexibility of the receptor.
- **Multiple ligand conformations:** For each structure in the large virtual library, not one but many energetically accessible conformations can be screened for fit to the pharmacophore.
- **Pharmacophore feature ranges and multiple ligand conformations:** The combination of these two methods is very powerful but computationally prohibitive.

Pharmacophore descriptor sets have evolved out of the need to include as much of the dynamic nature of the ligand–receptor interaction as possible while maintaining a computationally tractable solution. The descriptor sets are predefined bits that represent combinations of pharmacophore features usually binned according to distance. They can be precalculated for one or many conformations of a molecule and stored as one vector. This usually results in a very large binary string that can be efficiently stored and scored for fit to a probe pharmacophore hypothesis vector.

VIII. Summary

The use of computational data-mining methods to design and enrich combinatorial chemistry library synthesis is a strategy that is increasingly being applied successfully to projects designed to discover

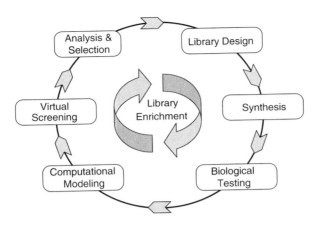

FIGURE 25.7
Discovery data-mining cycle.

biological activity. The chemoinformatic concepts and computational data-mining methods described in this chapter can be used in combination to identify rapidly a diverse set of molecules of biological interest, to design combinatorial libraries around these molecules that have druglike properties built in, and to explore the SAR of synthesized series efficiently. Additionally, these methods can be used to address potential liabilities such as toxicity, unfavorable ADME properties, and selectivity issues.

The process usually requires, and in most cases benefits from, many iterations of a cycle of library design → synthesis → biological testing → computational modeling → virtual screening → analysis and selection → library design (Figure 25.7). Each tour of the cycle has the potential to teach the computational model more about the biological activity landscape. As the model learns more, its predictions can become more accurate. With more accurate predictions come better analyses and selection, which can amplify the biological signal in the resulting library design. In a series of carefully executed cycles, an initial population with weak biological activity can evolve into a directed combinatorial library expansion yielding potent, druglike, selective, bioavailable series with a well-explored SAR.

Construction of a relevant virtual chemical space on which to perform *in silico* screening will rely heavily on inclusion of conserved binding motifs and application of druglike filter strategies, irrespective of which informatics scheme is used to encode the library. The quality of the biological data associated with the training set will strongly affect the quality of models that will be used to mine the virtual library, and data quality should be assessed and optimized in the earliest discovery cycles. Generation of a variety of property descriptor sets to be used in multiple combinations with a portfolio of data-mining methods can be a very effective enhancement strategy.[17]

There is an entire field of data-mining methods devoted to bioinformatics and the extraction and decoding of relevant, functional biological receptors from gene sequences. The inevitable marriage of bioinformatics data mining to the methods described in this chapter concerning chemoinformatics data mining cannot be long in coming. The success of that marriage depends on the current courtship between researchers in the fields of biology, chemistry, computational informatics, robotics, and pharmaceuticals. The inclusion of this particular chapter, "Computational Data-Mining Methods Applied to Combinatorial Chemistry Libraries," in the *Handbook of Molecular and Cellular Methods in Biology and Medicine* is certainly a positive step toward that inevitable future union.

References

1. **Lipinski, C.A. et al.**, Experimental and computational approaches to estimate solubility and permeability in drug discovery and development settings, *Advanced Drug Delivery Rev.*, 23, 3, 1997.
2. **Ghose, A.K., Viswandhan, V.N., and Wendoloski, J.J.**, A knowledge based approach in designing combinatorial or medicinal chemistry libraries for drug discovery, 1: A qualitative and quantitative characterization of known drug databases, *J. Combinatorial Chem.*, 1, 55, 1999.
3. **Veber, D.F. et al.**, Molecular properties that influence the oral bioavailability of drug candidates, *J. Medicinal Chem.*, 45, 2615, 2002.
4. **Labute, P.**, A widely applicable set of descriptors, *J. Molecular Graphics Modeling*, 18, 464, 2000.
5. **Rechenberg, I.**, *Evolutionstrategie — Optimierung technischer Systeme nach Prinzipien der biologischen Evolution*, Frommann-Holzboog, Stuttgart, 1973.
6. **Martin, Y.M.**, A similarity investigation, http://www.daylight.com/meetings/mug02/Martin/index.html, 2002.
7. **Hebb, D.O.**, *The Organization of Behavior: A Neuropsychological Theory*, John Wiley & Sons, New York, 1949.
8. **Boger, Z. and Guterman, H.**, Knowledge extraction from artificial neural networks models, in *IEEE Systems, Man and Cybernetics Conference Proceedings* (Orlando, October 1997), 3030, 1997. http://216.239.53.100/search?q=cache:4B6oc0z9M-8J:209.68.240.11:8080/2ndMoment/970504110/970897708/KNOWL6.doc+knowledge+Extraction+from+artificial+neural+network+models&hl=en&ie=UTF-8, 1998.
9. **Neural Net FAQ**, ftp://ftp.sas.com/pub/neural/FAQ3.html#A_hl and ftp://ftp.sas.com/pub/neural/FAQ3.html#A_hu, May 21, 2001.
10. **Funahashi, K.I.**, On the approximation realization of continuous mappings by neural networks, *Neural Networks*, 2, 183, 1989.
11. **Goodacre, R., Neal, M.J., and Kell, D.B.**, Quantitative analysis of multivariate data using artificial neural networks: a tutorial review and applications to the deconvolution of pyrolysis mass spectra, http://gepasi.dbs.aber.ac.uk/roy/tutorial/pymstut.htm, 1996.
12. **Sheridan, R.P. and Kearsley, S.K.**, Using a genetic algorithm to suggest combinatorial libraries, *J. Chemical Inf. Computer Sci.*, 35, 310, 1995.
13. **Weber, L. et al.**, Optimization of the biological activity of combinatorial compound libraries by a genetic algorithm, *Angewandte Chemie Int. Ed.*, 34, 2280, 1995.
14. **Singh, J. et al.**, Application of genetic algorithms to combinatorial synthesis: a computational approach for lead identification and lead optimization, *J. Am. Chemical Soc.*, 118, 1669, 1996.
15. **Brown R.D. and Martin, Y.C.**, Designing combinatorial library mixtures using genetic algorithms, *J. Medicinal Chem.*, 40, 2304, 1997.
16. **Binun, M.**, The senomyx discovery process, http://www.daylight.com/meetings/mug02/Senomyx/, 2002.
17. **Manly, C.J., Louise-May, S., and Hammer, J.D.**, The impact of informatics and computational chemistry on synthesis and screening, *Drug Discovery Today*, 6, 1101, 2001.

Index

A

Abscisic acid, 360, 363
ABTS, see 2,2-Azino-di(3-ethylbenzthiazoline sulfonic acid)
ACD, see Acid citrate dextrose
Acid citrate dextrose (ACD), 10
Acid guanidinium thiocyanate, 26
Acid-phenol extraction, purification of supercoiled plasmid DNA by, 273
Acrylamide gel, recipe for regular, 250, 251, 258
Adenovirus vector-mediated transfer, 391
ADME properties, 549–550, 553
Adsorption chromatography, 518
Advantage Genomic Polymerase Mix, 233
Aequoria victoria, 482
Affinity chromatography, 52, 56, 519, 520
Agarase, hydrolysis of agarose by, 226
Agarose
    gel
        electrophoresis, 15, 36, 87, 326, 327
        elution of DNA fragments in wells of, 16
        formaldehyde, 106
        purification of DNA fragments from, 15
        slices, 386, 412
    hydrolysis of, 226
    partial restriction enzyme digestion of DNA, 224
Agrobacterium
    rhizogenes, 373
    tumefaciens, 373, 410, 413, 418, 419, 442
Aliquot components, 215
Alkaline
    agarose gel electrophoresis of, cDNA, 156
    blotting, 89
    lysis, isolation of plasmid DNS by, 2
    phosphatase (AP), 80, 98, 138
    transfer, 91
Amino acid incorporation, TCA assay for, 350
Aminoglycoside 3′-phosphotransferase gene, 389
Ampicillin, 215, 284, 371
Amplification, conditions for, 188
AMV, see Avian myeloblastosis virus
α-Amylase, 57
ANNs, see Artificial neural networks
Antibody(ies)
    activity, 206
    –antigen complex, 128
    definition of, 113
    monoclonal, 114
    polyclonal, 114
    verification of by Western blot analysis, 121
Antisense gene silencing, 434
Antisense RNA (asRNA), 210, 434
    gene transfer and expression of, 446
    hybridization, 105
AP, see Alkaline phosphatase
Aqueous hybridization buffer, 96
Arabidopsis, 320, 341
ARS, see Autonomous replicating sequence
Artificial neural networks (ANNs), 559, 560
Ascending paper chromatography, 523
asRNA, see Antisense RNA
Automated sequencing, 259, 262, 268
Autonomous replicating sequence (ARS), 222
Autoradiography, 255, 308, 502
    gel, 299
    visualization of DNA by, 292
Auxin, 358, 360, 363
Avena sativa, 465
Avian myeloblastosis virus (AMV), 148, 157, 243
Azadirachta indica, 512
2,2-Azino-di(3-ethylbenzthiazoline sulfonic acid) (ABTS), 121

B

Bacillus thuringiensis, 373
Backscattered electrons, 467
Bacteria
    applications of DNA microarrays in, 341
    -to-bacteriophage ratio, 8
    recombinant plasmid amplification and, 418, 445
Bacterial contamination, 52
Bacterial EST clones, growing of, 322
Bacteriophage
    double-strand, 240
    lambda library, 195
    lambda linear duplex DNA, sequencing of, 246

lambda vectors, 152
T4 DNA polymerase, 272
T4 polynucleotide kinase, 74
Bal 31 deletions, 277, 279
BCA, *see* Bicinchoninic acid
B-cell lymphoma, 342
BCIP detection method, 207
Bench-top tissue processing, 480
Bicinchoninic acid (BCA), 51
Biological activity assay, 347
Biological tissues
    hard, 470
    preparation of for SEM, 471
Bioseparation techniques and applications, 509–533
    chromatographic separation of organic molecules, 517–526
        capillary-zone electrophoresis, 525–526
        chromatographic separation techniques, 517–525
    collection, vouchering, and storage of biological specimens, 511–512
        collection of plant and fungal samples, 511
        germplasm storage banks for biological samples or organisms, 512
        vouchering of samples collected, 511–512
    grinding and extraction protocols, 513–517
        general extraction protocols for biologically important organic compounds, 513
        hot-water and organic-solvent extraction of water-soluble and organic-solvent-soluble medicinal compounds from plants, 513–517
    methods of extraction of organic molecules, 510–511
        contemporary, 511
        traditional, 510
    nuclear magnetic resonance spectroscopy, 530–532
        continuous-wave NMR spectroscopy, 532
        pulsed Fourier-transform NMR, 532
        pulsed NMR, 532
    use of mass spectrometry to identify biologically important molecules, 526–530
        analyzers, 528–529
        ionization techniques, 526–527
        tandem mass spectrometry, 529–530
Blotting
    alkaline, 89
    RNA, 105, 108
    semidry, 137
    wet, 137
Blunt end(s)
    cloning, 314
    DNA, preparation of, 416, 443
    unique enzymes for generating, 275
BPB, *see* Bromophenol blue
Bradford assay, 52
Brassica juncea, 480
Bromophenol blue (BPB), 197, 254
BSA, 196

C

Caenorhabditis elegans, 435
Calcium phosphate transfection, 391
Calf intestinal alkaline phosphatase (CIAP), 74, 211, 212, 280, 385, 443
Callus cultures
    greening of, 366
    induction of shoots and roots in, 367
    initiation of cell-suspension cultures from, 368
Callus substructure
    carrot, 365
    rice, 364
    tobacco, 365
Callus tissue
    growth, measurement of, 365
    regeneration of plants from, 358
CaMV, *see* Cauliflower mosaic virus
Cancer
    chemotherapeutic, 514
    formation, genes involved in, 378
Capillary transfer, nucleic acid, 89–90, 91
Capillary zone electrophoresis, 510, 525
Carbenicillin, 371
Carcinogen, suspected, 26
Carrot
    callus substructure, 365
    somatic embryogenesis of, 369
CAT, *see* Chloramphenicol acetyl transferase
Cathode ray tube (CRT), 467, 468
Cauliflower mosaic virus (CaMV), 414, 442
CB, *see* Coomassie blue
CC, *see* Combinatorial chemistry
cDNA, *see* Complementary DNA
cDNA libraries, 147–191
    cDNA cloning and analysis by polymerase chain reaction, 180–190
        analysis of gene expression by semiquantitative PCR, 186–189
        cDNA cloning by RT-PCR, 184–186
        general amplification of double-strand DNA by PCR, 181–184
        reagents needed, 189–190
        selection of oligonucleotides, 181
    construction of cDNA library by subtractive hybridization techniques, 171–175
        hybridization of cDNA to mRNA, 172
        making cDNA library from subtracted first-strand cDNA, 173–174
        reagents needed, 174–175
        removal of mRNA template, 172
        separation of cDNA/mRNA hybrids from single-strand cDNA by hydroxyapatite chromatography, 173
        synthesis of first-strand cDNA, 171
    construction of fractional cDNA libraries using *Xenopus* oocytes as expression system, 175–180
        protocol, 176–179
        reagents needed, 180

Index

construction and screening of cDNA library using lambda DNA as vector, 152–171
    immunoscreening of lambda expression cDNA library, 163–165
    large-scale ligation and *in vitro* packaging, 162–163
    protocols, 153–161
    reagents needed, 167–171
    screening of cDNA library using labeled DNA probes, 166–167
    titering and amplification of packaged phage, 161–162
    troubleshooting guide for immunoscreening, 166
    vectors used, 152–153
principles and strategies for construction, 148–152

Cell
  dry weight, determination of, 366
  lines
    freezing of, 122
    thawing of, 122
  numbers, counting of, 365
  -suspension cultures, 359

Centrifugation speed, swinging-bucket rotors, 28
Cerenkov radiation, 155
Cesium chloride gradients, 238
Chaotropic agents, 29
CHEF gel, size fractionation of DNA by, 225
Chemically competent cells, transformation using, 215
Chemiluminescence detection, 139, 140
Chemoinformatics, 549
Chimeric genes
  construction of, 389, 411, 417
  preparation of for microinjection, 402
Chinese cabbage, cotyledonary-petioles of, 368
Chiral chromatography, 522
Chloramphenicol acetyl transferase (CAT), 379
  enzyme assay, 394
  gene, 414
Chloroplasts, 12, 14
Chromogen diffusion, 493
Chromatofocusing, 522
Church hybridization buffer, 96
CIAP, *see* Calf intestinal alkaline phosphatase
Clarkia breweri, 503
Clones, bacterial EST, 322
Cloning
  blunt-end, 314
  cDNA, 184
  directional, 150, 159
  efficiency, 313
  preparation of double-strand oligonucleotides for, 388
  preparation of YAC vectors for, 225
  random, 148
  T/A, 313
  YAC, 224
CM, *see* Confocal microscopy
CM Sephadex chromatography, 55
Column chromatography, 517
Combinatorial chemistry (CC), 550
Combinatorial chemistry libraries, 549–564
  chemoinformatics, 549–551

computational paradigms for building and storing combinatorial libraries, 551–552
data-mining methods, 556–562
  artificial neural networks, 559–561
  decision trees/recursive partitioning, 558
  similarity/diversity/K-nearest neighbors, 556–558
  structure-based pharmacophore methods, 561–562
generation of molecular property descriptors, 555
high-throughput screening data, 554
in silico screening of virtual combinatorial libraries, 556
property design of combinatorial libraries, 552–553
Combinatorial techniques, 535–548
  appendix, 543–546
    list of combinatorial and contract synthesis companies, 543–545
    sources for detailed solid- and solution-phase combinatorial procedures, 546
  history, 536–537
  parallel synthesis, 537–539
  solid-phase combinatorial chemistry, 541–543
    encoded libraries, 542
    Irori microreactor methodology, 542
    issues, 542–543
    iterative deconvolution using tea-bag method, 542
    pin technology, 541
    split-pool method, 541
  solution-phase combinatorial chemistry, 539–541
    purification, 540–541
    reaction, 539–540
    workup, 540
Complementary DNA (cDNA), 32, 194, 306, 347, 383
  alkaline agarose gel electrophoresis of, 156
  amplification, 188, 321
  cloning
    method, classical, 148
    by RT-PCR, 184
  double-strand, 154, 158
  hybridization of to mRNA, 172
  insert(s)
    PCR amplification of, 324
    subcloned, 350
  labeling, materials for, 335
  library(ies)
    construction of, 149, 151, 152
    lambda expression, 163
    screening of, 46, 166
  making cDNA library from subtracted first-strand, 173
  microarrays, 321
  /mRNA hybrids, separation of, 173
  packaging, 156
  subcloning of, 75
  synthesis, 35, 148, 153
ConA-Sepharose affinity chromatography, 56
Confocal microscopy (CM), 461–462, 482
Continuous-wave NMR spectroscopy, 532
Coomassie blue (CB), 59, 510, 134, 135
Copper–phenanthroline, gel retardation assay using, 302
Cosmids, 195
Crataegus monogyna, 372
Critical-point drying, 470

CRT, *see* Cathode ray tube
Cryopreservation, preservation of plant tissue cultures by, 374
CsCl centrifugation
    isolation of plasmid DNA by, 5
    large-scale purification of λDNA by, 9
CsCl-ethidium bromide gradients, equilibrium sedimentation in, 273
Cycle sequencing, 259
Cytokinin, 358, 360, 363
Cytosolic proteins, extraction of from eukaryotic cells, 50

D

Data
    -mining methods, *see* Combinatorial chemistry libraries
    normalization and analysis, 339
Daucus carota, 371
DD, *see* Differential display
ddNTP, *see* 2′,3′-Dideoxynucleotide 5′-triphosphate
DEAE
    cellulose, *see* Diethylaminoethyl cellulose
    dextran-mediated transfection, 391, 392
Decision trees, 558
Degenerate primers, 185, 187
Dehydrogenase alkaline phosphatase, 57
Deletion(s)
    Bal 31, 277, 279
    exonuclease III, 273
    mutagenesis by, 272
    subclones, estimation of sizes of, 277, 281
    unidirectional, 275
DEPC, *see* Diethylpyrocarbonate
Descending liquid-paper chromatography, 523
Detergents, 47
Dideoxynucleotide chain termination, 238, 243
2′,3′-Dideoxynucleotide 5′-triphosphate (ddNTP), 243
Diethylaminoethyl (DEAE) cellulose, 54
Diethylpyrocarbonate (DEPC), 26, 106, 152
Differential display (DD), 305–317
    confirmation of differential expression, 315
        Northern hybridization, 315
        reverse Northern hybridization, 315
    denaturing polyacrylamide gel electrophoresis, 310–311
    electrophoresis, 310–311
    gel preparation, 310
    isolation of differentially displayed bands, 311–315
        analysis of transformants, 314–315
        cloning, 313–314
        excision and elution of DNA bands, 311–313
        PCR reamplification, 313
    reagents needed, 316–317
    RNA preparation, 306–308
        DNase treatment, 307–308
        total RNA extraction, 307
    RT-PCR, 308–310
        radioactive PCR amplification, 309–310
        reverse transcription, 308–309

    troubleshooting guide, 316
DIG, *see* Digoxigenin
Digestion
    double-enzyme, 208
    efficiency, 415
    rate, dependence of on temperature, 276
    reactions
        large-scale, 200
        small-scale, 198
    single-enzyme, 208
Digoxigenin (DIG), 94, 496
    -dUTP, 80
    -labeled DNA probe, 98
    -NHS-ester, 82
    probes, immunodetection of, 500
Dimethylsulfoxide (DMSO), 163, 374
Directed sorting, 542
Directional cloning, 150, 159
Dithiothreitol (DTT), 47, 76, 152, 350, 493
DMSO, *see* Dimethylsulfoxide
DNA (deoxyribonucleic acid), 2, 46, 67, 85
    agarose
        concentration for separation of, 87
        gel electrophoresis of, 87
    bacteriophage lambda linear duplex, sequencing of, 246
    bands, excision and elution of, 311
    blotting of onto nylon membranes, 89
    blunt-end, 416, 443
    chain elongation, 243
    chloroplast, isolation of by sucrose gradient, 14
    cleavage of recombinant phage, 179
    clones, positive recombinant bacteriophage, 208
    constructs, microinjection of into oocytes, 403
    degradation, 19
    dot/slot blots, 92
    elution of from agarose gel slices, 386, 412
    fragment(s)
        blunt-end, 387
        double-strand, 240
        elution of, 15, 16, 387
        large-scale ligation of partially filled-in, 203
        purification of, 15, 384
        subcloning of, 209
    genomic
        extraction of with formamide, 11
        isolation of from animals, 10
        isolation of from plants, 12
        restriction enzyme digestion of, 86
    insert
        ligation of, 212
        preparation of, 212
        subcloning of, 273
        unidirectional deletion of, 275
    introduction of foreign, 410
    in vitro packaging of ligated, 161, 201
    isolation of high-molecular-weight, 223
    isolation of intact yeast, 223
    isolation of from mouse tails, 404
    isolation and purification, 2
    lambda, 152
    manipulation, 25, 379

Index

mapping, 86
markers, 207
microinjection, 391
partial restriction-enzyme digestion of in agarose, 224
PCR-based mutation of, 282
plasmid, 2, 4
preparation of unidirectional deletion of target, 240
probe(s)
    DIG-labeled, 98
    fluorescent-labeled, 321
    screening of cDNA library using labeled, 166
    stripping of, 99
–protein complexes
    formation of, 299
    stability of, 301
quality and quantity, determination of, 17
replication, 379, 435
samples, contaminated, 254
sequence(s)
    determination of copy number of dispersed, 427
    devices, chip-based, 259
    probability of fishing out, 194
    techniques, high-throughput, 320
single-strand phagemid, generation of, 284
target, preparation of, 441
template(s)
    preparation of, 450
    use of during sequencing, 238
tumor virus, 379
DNA, isolation and purification of, 1–24
    determination of DNA quality and quantity, 17–19
        fluorometric measurement, 18–19
        spectrophotometric measurement, 17–18
    isolation of λDNA, 7–10
        large-scale purification of λDNA by CsCl gradient centrifugation, 9–10
        mini-preparation of λDNA, 9
        phage lysate preparation by liquid method, 8
        phage lysate preparation by plate method, 7–8
    isolation of genomic DNA from animals, 10–12
        formamide, 11–12
        organic solvents, 10–11
    isolation of genomic or organelle DNA from plants, 12–14
        isolation of chloroplast DNA by sucrose gradient, 14
        isolation of plant genomic DNA, 12–13
    isolation of plasmid DNA, 2–7
        alkaline lysis, 2–5
        CsCl gradient centrifugation, 5–7
    purification of DNA fragments from agarose gels, 15–16
        elution of DNA fragments by freeze-and-thaw method, 15–16
        elution of DNA fragments in wells of agarose gel, 16
    reagents needed, 19–23
    troubleshooting guide, 19
DNA footprinting and gel retardation assay, 291–304
    footprinting following gel retardation assay using copper–phenanthroline, 302–303
    footprinting protocols, 293–296
        DNase I protection assay, 293–294
        electrophoresis, 294–295
        preparation of denaturing polyacrylamide gels, 293
        preparation of single-end labeled DNA probe, 293
        reagents needed, 296–297
        troubleshooting guide, 296
    gel retardation assay protocol, 298–301
        autoradiography of gel, 299–300
        formation of DNA–protein complexes and electrophoresis, 299
        preparation of nondenaturing polyacrylamide gels, 298–299
        preparation of protein sample for assay, 298
        preparation of single-end labeled DNA probe, 298
        reagents needed, 301–302
        troubleshooting guide, 301
DNA microarray technology, functional genomics and, 319–346
    applications of DNA microarrays in various groups of organisms, 341–344
        humans, 342
        microorganisms, 341
        plants, 341–342
    DNA microarray protocols, 321–340
        array hybridization, 335–336
        array postprocessing, 337
        growing bacterial EST clones, 322
        image processing and data analysis, 339–340
        isolation of plasmid templates for PCR, 323–324
        overview of process, 321–322
        PCR amplification of cDNA inserts, 324–327
        preparation of poly-L-lysine coated slides, 330–331
        printing of microarrays, 331–335
        purification and quantification of PCR products, 327–328
        RNA preparation from samples, 329–330
        scanning DNA microarrays, 337–338
        slide scanning, 338–339
    pros and cons of DNA macroarrays, 321
    troubleshooting guide, 340–341
    what technology can tell us, 320–321
DNase, 10, 293, 307
DNA sequencing and analysis, 237–269
    cycle sequencing and automated sequencing, 259–265
        DNA sequencing using automated sequencer, 262–265
        DNA synthesis and termination reactions by PCR, 261
        selection of oligonucleotides, 260
    DNA sequencing by dideoxynucleotides chain termination, 243–259
        denaturing polyacrylamide gel electrophoresis, 250–257
        extending sequence far from primer, 259
        general considerations and strategies, 243–244
        sequencing reactions, 244–247
    preparation of DNA for sequencing, 238–242
        preparation of double-strand DNA fragments, 240
        preparation of double-strand plasmid DNA, 238

preparation of single-strand template DNA, 239–240
preparation of unidirectional deletions of target DNA, 240–241
purification of double-strand λDNA, 240
reagents needed, 241–242
troubleshooting guide, 266–269
automated sequencing, 268–269
manual sequencing, 266–267
dNTPs, 79
Dot blotting, RNA, 108, 396
Dot/slot blot analysis, 86
Double restriction enzyme digestion, 275, 385, 441
Double-strand DNA (dsDNA), 68, 72, 328
amplification, 180, 181
denaturing of labeled, 205
mechanisms of, 449
microinjection of, 451
nick-transition labeling of, 81
random-primer labeling of, 71, 80
RNA interference using *in vitro* transcribed, 450
Drosophila, 449
Drug-selection marker gene, 389, 397
dsDNA, *see* Double-strand DNA
DTT, *see* Dithiothreitol
Dye-ligand chromatography, 522
Dynabeads, 34

E

EB, *see* Equilibrating/binding buffer
ECGS, *see* Endothelial cell growth supplement
Echinacea, 510
EDTA, *see* Ethylenediaminetetraacetic acid
EI, *see* Electron ionization
Electroelution, protein/enzyme, 60
Electron ionization (EI), 526, 527
Electron microscope, first, 477
Electrophoresis, 58, 133, 253, 308
agarose gel, 36
buffer, 87
DNA footprinting, 294
efficient, 130
gel
alkaline agarose, 156
two-dimensional, 131
prolonged, 87
protein analysis and, 354
RNA, 106
Electroporation
direct gene transfer to protoplasts by, 420
preparation of competent cells for, 215
transfection by, 391, 393
transformation by, 216
Electrospray ionization (ESI), 526, 527
ELISA, *see* Enzyme-linked immunosorbent assay
Enantiomers, separation of, 522
Endonuclease, linearizing recombinant plasmids using appropriate, 279

Endothelial cell growth supplement (ECGS), 117
Environmental scanning electron microscopy (ESEM), 461, 474, 476
Enzyme(s)
assay
chloramphenicol acetyl transferase, 394
preparation of cytoplasmic extract for, 394
in situ immunocytochemical localization of, 114
Klenow, 80, 243
-linked immunosorbent assay (ELISA), 47, 116
determination of mouse Ig concentration by, 116
screening of hybridoma supernatant by, 120
marker, 488
stability, 46
Equilibrating/binding buffer (EB), 34
Escherichia coli, 320
DNA polymerase I, 68, 71, 243, 272
expression of protein fragments in, 123
extract, phage-infected, 161
recombinant plasmids propagated in, 383
short-life-cycle, 209
transformation, 214, 390
ESEM, *see* Environmental scanning electron microscopy
ESI, *see* Electrospray ionization
Esterase, 57
Ethanol precipitation, alternative to, 265
Ethylenediaminetetraacetic acid (EDTA), 46, 156, 174, 199, 334
Eukaryotic cells
extraction of cytosolic proteins from, 50
gene expression in, 291
Exonuclease, 246, 273
Exonuclease III
deletions, 273
-resistant 3′ protruding termini, enzymes producing, 275
series deletions of linearized DNA with, 276

F

FCS, *see* Fetal calf serum
Fertilizer, 425
Fetal calf serum (FCS), 119, 120
FID, *see* Flame-ionization detector
Field-inversion gel electrophoresis (FIGE), 222
FIGE, *see* Field-inversion gel electrophoresis
Flame-ionization detector (FID), 518, 524
Flowers, new pigment colors in petals of, 410
Fluorescein diacetate, 375
Fluorescence
assay, PCR DNA concentration, 328
detection, 263, 264
maximum intensity of, 199
Fluorescent resonance energy transfer (FRET), 482
Fluorochromes, 484–485
Fluorometric measurements, 18, 38
Food plants, phenotypic characteristics of, 410
Formaldehyde agarose gels, electrophoresis of RNA using, 106

Index

Formamide
  -based hybridization buffer, 95
  gels, use of for sequencing of G-C rich templates, 255
Founder mice, generation of, 403
Free-induction decay, 532
Freeze-and-thaw method, elution of DNA fragments by, 15
FRET, see Fluorescent resonance energy transfer
β-Fructofuranosidase, 57
Fungal samples, collection of, 511

G

β-Galactosidase, 216, 217, 260, 380, 395
GAPDH, see Glyceraldehyde-3-phosphate dehydrogenase
GAs, see Genetic algorithms
Gas chromatography (GC), 517, 524
Gaseous secondary electron detector (GSED), 475
GC, see Gas chromatography
Gel(s)
  acrylamide, recipe for regular, 250, 251, 258
  autoradiography, 299
  destaining, 134
  DNA-sequencing, 292
  electrophoresis, 86
    alkaline agarose, 156
    denaturing, 302
    two-dimensional, 131, 347
  filtration, 52, 524
  formamide, use of for sequencing of G-C rich templates, 255
  identifying bands in unstained, 59
  isoelectric focusing, 131
  loading of protein standard markers onto, 353
  loading of samples into, 58
  manual sequencing, 259
  mixture, Lone Ranger, 250, 251, 255, 257
  mobility shift assay, 292
  permeation, 525
  polyacrylamide
    denaturing, 293
    nondenaturing, 298
  retardation
    assay, see DNA footprinting and gel-retardation assay
    schematic representation of, 300
    types of, 301
  separating, 57, 352
  stacking, 57, 353
  staining, 134
Gene(s)
  antisense orientation, 438
  cloning, 86
  defects, 378
  determination of copy number of, 427
  expression
    analysis of inhibition of, 437
    linear regression analysis for, 340
    patterns, plant–fungal interaction, 341–342
  functions unknown, 320
  luciferase, 380
  manipulation technology, 410
  -prediction algorithms, 320
  recombination, 378
  subcloning, 411
  suppression, antisense, 434
  transfer
    advantages and disadvantages of methods of, 391
    purpose of, 428
    techniques, advantages and disadvantages of, 410
  tumor-inducing, 413
Gene expression, inhibition of, 433–459
  antisense oligonucleotides, 435–438
    analysis of inhibition of gene expression using optimum dose of oligomers, 437–438
    reagents needed, 438
    synthesis of antisense oligonucleotides, 435–437
    treatment of cultured cells using antisense oligomers and determination of optimum dose of oligomers, 437
  antisense orientation of gene of interest, 438–449
    blunt-end ligation of plasmid vectors and inserts of antisense and sense orientation, 444–445
    gene transfer and expression of antisense RNA, 446–447
    preparation of target DNA, strong promoters, enhancers, poly(A) signals, and vectors, 441–444
    reagents needed, 447–449
    selection of plasmids with antisense and sense orientations, 446
    transformation of appropriate strain of bacteria to amplify recombinant plasmids, 445
  comparison of gene suppression technologies, 434–435
  new potential of double-strand RNA for suppression of gene expression, 449–457
    alternative methods of RNA interference, 453–454
    equipment and reagents needed, 456–457
    inducible RNA interference, 454–455
    insights into mechanisms of double-strand RNA, 449–450
    RNA interference using in vitro transcribed double-strand RNA, 450–453
    troubleshooting guide, 455
Gene expression, localization of, 487–508
  localization by in situ hybridization, 493–507
    hybridization, 496–499
    infiltration, embedding, and section preparation, 494–495
    nonradioactive signal detection, 499–501
    radioactive signal detection, 501–503
    reagents needed, 504–507
    tissue preparation, 493–494
    troubleshooting guide, 504
  subcellular localization, 488–490
    preparation of organelles, 489
    sucrose gradients, 489–490
  tissue localization by tissue printing, 490–492
    example with germinating tomato seeds, 491–492
    general procedures of tissue printing, 491
Genetic algorithms (GAs), 559

Gene transfer and expression in animals, 377–408
  mammalian cells, 379–401
    analysis of expression of reporter gene or inserted gene of interest in transfected cells, 394–397
    construction of chimeric genes for transfection, 389–391
    preparation of double-strand oligonucleotides for cloning, 388
    purification of plasmid vectors and DNA fragments or genes to be used for transfer, 384–388
    reagents needed, 398–401
    selection of stable transformants, 397–398
    selection of vectors, promoters, and enhancers for gene transfer and expression, 379–383
    transfection of reporter DNA constructs into cultured cells, 391–394
  mice, 402–407
    identification of transgenic mice, 404–405
    microinjection of DNA constructs into oocytes, 403
    preparation of chimeric genes for microinjection, 402
    preparation of oocytes, 402
    reagents needed, 406–407
    reimplantation of injected eggs into recipient female mice and generation of founder mice, 403–404
    selection of transgenic lines by breeding of transgenic mice, 405
Gene transfer and expression in plants, 409–432
  cloning, isolation, characterization, and subcloning of gene of interest, 411
  construction of chimeric genes by ligating plasmid vector and insert-DNA, 417–418
  preparation of blunt-end DNA, 416
  preparation of DNA insert lacking 5′ phosphate groups, 416–417
  proof of stable transformation, 425–428
    analysis of expression of reporter gene, 426
    analysis of Southern blotting and sequencing, and determination of copy number of gene, 427–428
    Northern blot analysis, 428
    phenotypic and/or functional examinations, 426
    Western blot hybridization or immunocytochemical localization, 428
  purification of gene of interest from recombinant plasmids, 411–413
    elution of DNA from agarose gel slices, 412–413
    purification of target DNA using NA45 DEAE membranes, 413
  reagents needed, 429–431
  selection of appropriate vector, promoter, poly(A) signal, reporter gene, and selectable marker gene, 413–416
  transformation of appropriate strain of bacteria to amplify recombinant plasmids, 418–419
    introduction of plasmids into *Agrobacteria* by conjugation, 418–419
    introduction of recombinant plasmids into *Agrobacterium tumefaciens*, 418
  transformation of plants with recombinant gene constructs, 419–425
    direct gene transfer by microprojectile bombardment, 423–425
    direct gene transfer to protoplasts by electroporation, 420–423
    gene transfer by *Agrobacterium tumefaciens*, 419–420
Genome sequencing, rapid developments in, 320
Genomic DNA libraries, 193–236
  bacteriophage lambda library, 195–221
    construction of genomic DNA library, 195–203
    construction of partial genomic libraries, 207–208
    reagents needed, 217–221
    restriction mapping of positive recombinant bacteriophage DNA clones, 208–209
    screening of genomic DNA library, 203–207
    subcloning of DNA fragment of interest, 209–217
  general strategies and applications, 194–195
  genomic cloning using PCR, 232–236
    isolation of flanking sequences by inverse PCR, 235–236
    PCR amplification of genomic DNA and genomic libraries, 233–235
    selection of oligonucleotides, 232–233
  troubleshooting guide, 232
  YAC libraries, 221–231
    amplification and storage of YAC library, 227
    hydrolysis of agarose by agarase, 226
    isolation of high-molecular-weight DNA, 223
    isolation of intact yeast DNA, 223–224
    isolation of yeast DNA for PCR screening, 224
    ligation of partially digested genomic DNA insert to pYAC4 vector, 225
    partial restriction enzyme digestion of DNA in agarose, 224
    preparation of cells or tissues for isolation and purification of high-molecular-weight DNA, 222–223
    preparation of spheroblasts for transformation, 226
    preparation of YAC vectors for cloning, 225
    reagents needed, 228–231
    screening of YAC library, 227–228
    size fractionation of DNA by CHEF gel or other PFGE, 225
    transformation of spheroblasts with recombinant YAC/DNA insert, 226
    verification of YAC transformants, 227
GFP, *see* Green fluorescent protein
Gibberellin, 360, 363
Glow-discharge coater, 470
β-Glucuronidase (GUS), 382, 414, 426
Glyceraldehyde-3-phosphate dehydrogenase (GAPDH), 108
Glycine max, 372
Glycoproteins, 114
Green fluorescent protein (GFP), 482
GSED, *see* Gaseous secondary electron detector
GUS, *see* β-Glucuronidase

H

HAP, *see* Hydroxyapatite
HAT, *see* Hypoxanthine, aminopterin, and thymidine
HB, *see* Homogenization buffer
HCG, *see* Human chorionic gonadotrophin
HCMV, *see* Human cytomegalovirus
Helper phage R408, 239
Hemerocallis sp., 371
Herbicide-resistant plants, 358, 410
Hexanucleotides, random, 71
HGRPT, *see* Hypoxanthine guanine phosphoribosyl transferase
High-performance liquid chromatography (HPLC), 52, 514, 518, 540
High-throughput screening (HTS), 550, 554
Homogenization buffer (HB), 489
Housekeeping genes, 187, 340
HPLC, *see* High-performance liquid chromatography
HPT gene, *see* Hygromycin phosphotransferase gene
HTS, *see* High-throughput screening
Human cancers, microarray-based analysis of, 342
Human chorionic gonadotrophin (HCG), 402
Human cytomegalovirus (HCMV), 342
Hybridization
　buffer
　　aqueous, 96
　　Church, 96
　　formamide-based, 95
　materials and reagents for, 337
　procedures, 94
　solution, volume of, 336
Hybridoma supernatant, screening of by ELISA, 120
Hydrolases, 57
Hydrophobic-interaction chromatography, 519
Hydroxyapatite (HAP), 171
Hydroxyapatite chromatography, 173
Hygromycin, 371
　B phosphotransferase gene, 389
　phosphotransferase (HPT) gene, 397, 414
Hypericum perforatum, 372
Hypoxanthine, aminopterin, and thymidine (HAT), 119, 120, 397
Hypoxanthine guanine phosphoribosyl transferase (HGRPT), 119

I

IEF, *see* Isoelectric focusing
Ig, *see* Immunoglobulin
IMAC, *see* Immobilized-metal-ion affinity chromatography
Image processing, stages of, 339
Immobilized-metal-ion affinity chromatography (IMAC), 521
Immunoglobulin (Ig), 116, 119
Immunoprecipitation, 56, 347
Immunoscreening, troubleshooting guide for, 166
In situ hybridization (ISH), 38, 493, 494

Inverse PCR, isolation of flanking sequences by, 235
Inverted repeat (IR) gene, 454
In vitro transcription, 4, 75
Ion-exchange chromatography, 52, 54, 522
Ion-trap mass spectrometer, 528
IPTG, *see* Isopropyl β-D-thiogalactopyranoside
IR gene, *see* Inverted repeat gene
Irori microreactors, 542
ISH, see *In situ* hybridization
Isoelectric focusing (IEF), 131, 132
Isomerases, 57
Isopropyl β-D-thiogalactopyranoside (IPTG), 453

J

Janus Green B, 375

K

Kanamycin, 371, 425
Kidney disease, 480
Kikuchi patterns, 467
Kinase reaction, 75
Klenow enzyme, 80, 243
K-nearest neighbors (KNN), 556, 557, 558
KNN, *see* K-nearest neighbors

L

Lambda DNA, 152
Lambda vectors, 195
Lavandula officinalis, 512
LC, *see* Liquid chromatography
LC-MS, *see* Liquid chromatography-mass spectrometry
Leaf tissue, transfer of genes into, 424
Library
　enrichment, 551
　screening, 86
　space, mining of, 556
Ligand–receptor interaction, 562
Ligation(s)
　efficiency of, 213
　reactions, components of, 213
　small-scale, 213, 214
Light microscopy (LM), 426, 461, 462
Linkers, ligation of, 207
Liquid chromatography (LC), 517
Liquid chromatography-mass spectrometry (LC-MS), 540
Liquid method, phage lysate preparation by, 8
Liriodendron tulipifera, 371
Lithospermum erythrorhizon, 372, 373
LM, *see* Light microscopy
Lone Ranger gel mixture, 250, 251, 255, 257
Long terminal repeat (LTR), 380, 382, 442

LTR, *see* Long terminal repeat
Luciferase gene, 380
Luria-Bertaini medium, 2
Luria-Bertaini plates, titering of packaged phage on, 202
Lyases, 57
Lycopersicon esculentum, 491

M

Macroarrays, 321
Magic-angle NMR spinning, 542
Magnetic oligo(dT) beads, purification of poly(A)$^+$ RNA using, 33
Maitotoxin, 536
MALDI, *see* Matrix-assisted laser desorption/ionization
Manual sequencing, 259, 266
Marker enzymes, organelle, 488
Mass spectrometry (MS), 510, 526
    tandem, 529
    use of to identify biologically important molecules, 526
Matrix-assisted laser desorption/ionization (MALDI), 526, 527
MBM, *see* Modified Barth's medium
MCS, *see* Multiple cloning sites
Medicinal compounds, plant, 513
Messenger RNA (mRNA), 148, 306, 320, 425, 435, 489
    hybridization of cDNA to, 172
    in vitro translation of, 349
    template, removal of, 172
4-Methyl-umbelliferyl-β-D-glucuronide (MUG), 426
Microarrays, printing of, 331
Microbial contamination, sterile techniques to prevent, 370
Microinjection, preparation of chimeric genes for, 402
Microorganisms, applications of DNA microarrays in, 341
Microprojectile bombardment, direct gene transfer by, 423
Microscopy, 461–486
    confocal and multiphoton microscopy, 482–485
    environmental scanning electron microscopy, 474–477
        applications, 476–477
        principles of operation, 474–475
    light or optical microscopy, 462–464
    scanning electron microscope, 464–473
        applications, 471–473
        preparation of biological tissues, 471
        principles of operation, 464–470
    transmission electron microscopy, 477–481
        applications, 480–481
        conventional transmission electron microscope, 477–479
        preparation of tissues, 479–480
Microwave-oven tissue processing, 480
Modified Barth's medium (MBM), 177
Molar ratios, calculation of, 213
Molecular property descriptors, generation of, 555
Moloney murine leukemia virus (MoMLV), 382
MoMLV, *see* Moloney murine leukemia virus
Monoclonal and polyclonal antibodies, preparation of against specific proteins, 113–126
    monoclonal antibodies, 114–122

        determination of mouse immunoglobulin concentrations by ELISA, 116–117
        freezing and thawing cell lines under sterile conditions, 122
        fusion of spleen cells and myeloma cells under sterile conditions, 119
        in vivo immunization of mice with purified antigen, 114–116
        isotyping of monoclonal antibodies, 120–121
        preparation of peritoneal exudate cells under sterile conditions, 117–118
        preparation of spleen cells and myeloma cells for fusion, 118–119
        purification of antigen for immunization, 114
        purification of monoclonal antibodies using affinity chromatography, 121
        screening of hybridoma supernatant by ELISA and harvesting of monoclonal antibodies, 120
        selection and propagation of hybridoma cells, 119–120
        verification of antibodies by Western blot analysis, 121
    polyclonal antibodies, 122–126
        in vivo immunization of female rabbits, 123–124
        production of antigens of interest, 123
        purification of antigen for immunization, 123
        purification of final polyclonal antibodies, 124
        reagents needed, 125–126
mRNA, *see* Messenger RNA
MS, *see* Mass spectrometry
MUG, *see* 4-Methyl-umbelliferyl-β-D-glucuronide
Multichannel analyzer, 467
Multigene families, determination of, 428
Multiphoton microscopy, 462, 482
Multiple cloning sites (MCS), 379
Mung bean nuclease, 272, 388
Murashige-Skoog medium, 361, 362
Mutagenesis, DNA site directed and deletion, 271–290
    mutagenesis by deletion, 272–281
        Bal 31 deletions, 277–281
        exonuclease III deletions, 273–277
    oligonucleotide site-directed mutagenesis, 281–289
        generation of single-strand phagemid DNA, 284
        PCR-based mutation of DNA, 282
        reagents needed, 285–289
        site-directed mutagenesis on single-strand phagemid DNA, 284–285
Mutations, loss-of-function, 449
Mycobacterium tuberculosis, 341
Myeloma cells
    fusion of under sterile conditions, 119
    preparation of for fusion, 118

N

Neomycin phosphotransferase (NPT II) gene, 414
Neurospora crassa, 435, 449
Nick translation, 68, 73

NMR spectroscopy, *see* Nuclear magnetic resonance spectroscopy
Nonradioactive labeling, 79, 82
Nonradioactive probe(s)
  hybridization to, 98
  screening of genomic library using, 206
Nonradioactive signal detection, 499
Normal-phase chromatography, 519
Northern blot analysis, 33, 428
Northern blot hybridization, 85, 105–111
  blotting RNO onto nylon or nitrocellulose membranes, 108
    after electrophoresis, 108
    dot/slot blots, 108
  controls, 108
  electrophoresis of RNA using formaldehyde agarose gels, 106–108
  hybridization, 109
  quantitative analysis of RNA, 109–110
  reagents needed, 110–111
  troubleshooting guide, 110
Northern hybridization, reverse, 315
NPT II gene, *see* Neomycin phosphotransferase gene
Nuclear magnetic resonance (NMR) spectroscopy, 510, 530
Nuclease, 272, 388
Nucleic acid
  hybridization, 67
  labeling, 485
Nucleic acid probes, preparation of, 67–84
  nonradioactive labeling methods, 79–84
    3′-end labeling with DIG-11-ddUTP, 81
    5′-end labeling with DIG-NHS-ester, 82
    nick-translation labeling of dsDNA with digoxigenin-11-dUTP, 81
    random-primer labeling of dsDNA, 80–81
    reagents needed, 82–84
    RNA labeling, 82
    3′ tailing of DNA with DIG-11-dUTP/dATP, 81
  radioactive labeling methods, 68–79
    3′-end labeling to fill recessed 3′ ends of ds DNA, 72–73
    3′-end labeling of ssDNA with terminal transferase, 74
    5′-end labeling using bacteriophage T4 polynucleotide kinase, 74–75
    labeling of dsDNA, 72
    labeling of RNA by *in vivo* transcription, 75–77
    nick-translation labeling of dsDNS, 68–71
    random-primer labeling of dsDNA, 71–72
    reagents needed, 77–79

O

Oligomer(s)
  modification, 436
  treatment of cultured cells using antisense, 437
Oligonucleotide(s), 181, 232, 244, 260
  antisense, 435
  preparation of for cloning, 388

site-directed mutagenesis, 272, 282
Optical sectioning, 482
Organelles
  marker enzymes for, 488
  separation of, 488
Organic carbon, plant tissue growth and, 360
Ouabain-resistance gene, 119

P

PAGE, *see* Polyacrylamide gel electrophoresis
Palytoxin, 536
Paper chromatography (PC), 517, 523
Parallel synthesis matrix, 538
PBS, *see* Phosphate-buffered saline
PC, *see* Paper chromatography
pCAT vectors, Bluescript-based series of, 284
PCR, *see* Polymerase chain reaction
PECs, *see* Peritoneal exudate cells
PEG, *see* Polyethylene glycol
Pepstatin A, 47
Peptide synthesis, 541
Peritoneal exudate cells (PECs), 117
PFGE, *see* Pulsed-field gel electrophoresis
pGEM Teasy vector, 183
Phage
  lysate preparation, 7
    liquid method, 8
    plate method, 7
  titering of packaged, 202
Phagemid vectors, 239
Pharmacophore
  definition of, 561
  feature ranges, 562
  hypothesis, 561
Phenol, 27
Phenol/chloroform purification, 238
Phenylmethanesulfonyl fluoride (PMSF), 46
Phosphate-buffered saline (PBS), 10, 47, 114, 136, 222, 454
Phosphodiester internucleoside linkages, 436
Phospho-imager, 140
Photomultiplier tube (PMT), 338, 467
Photosynthesis, rate of in green plants, 410
Photosynthetic carbon assimilation, 360
Photosynthetic cell cultures, 366
Plant(s)
  applications of DNA microarrays in, 341
  growth regulators, uses and stock preparation for, 363
  organ extracts, inoculation of into sterile media, 364
  samples, collection of, 511
  specimens
    dry, 511
    living, 512
  tissue culture
    applications with, 371
    media, nutrients in, 361
  tissue growth, organic carbon and, 360
Plant tissue and cell culture, 357–375
  applications with plant tissue cultures, 371–375

induction of somatic embryogenesis from callus cultures, 371
preservation of plant tissue cultures by means of cryopreservation, 374–375
somatic cell hybridization, 371–372
synthesis of useful secondary metabolites, 372–373
basic cell and tissue culture techniques, 360–271
induction of greening of callus cultures, 366–367
induction of shoots and roots in callus cultures, 367–368
initiation of cell-suspension cultures from callus cultures, 368–370
inoculating seeds or plant organ explants into sterile media, 364–365
measurement of growth of callus tissues in culture, 365–366
media preparation and media kits, 360–364
troubleshooting guide, 370–371
basic concept of totipotency of plant cell and tissue culture, 358–360
Plaque-forming units, 202
Plasmid(s)
advanced features of, 153
circular, double-restriction enzyme digestions of, 275
deleted, 280
DNA
double-strand, 244
isolation of, 2, 5
large-scale preparation of, 4
preparation of double-strand, 238
purification of supercoiled, 273
restriction digestion of, 315
recircularized, transformation using, 277
recombinant, selection of transformants containing, 216
selection of with antisense and sense orientations, 446
subcloning, 209
transformation of recircularized, 281
vector
construction of chimeric genes by ligating, 417
ligation of, 212
Plastic infiltration, 495
Plate method, phage lysate preparation by, 7
PMS, see Pregnant mare's serum
PMSF, see Phenylmethanesulfonyl fluoride
PMT, see Photomultiplier tube
Polar surface area (PSA), 553
Polyacrylamide gel electrophoresis (PAGE), 47, 56, 250, 293, 310
Polyamines, 363
Poly(A)+ RNA, purification of, 32, 33
Poly(A) signals, 379, 411, 413, 441
Polyclonal antibodies, see Monoclonal and polyclonal antibodies, preparation of against specific proteins
Polyethylene glycol (PEG), 9, 228, 238
Poly-L-lysine coated slides, preparation of, 330, 331
Polymerase chain reaction (PCR), 4, 154, 80, 272, 446
amplification, 181, 185, 308
genomic DNA, 233
materials for, 325
radioactive, 309
touchdown, 234
analysis of gene expression by semiquantitative, 186
cDNA
amplification by, 321
cloning and analysis by, 180
colony, 314
DNA concentration, fluorometric determination of, 328
genomic cloning using, 232
inverse, isolation of flanking sequences by, 235
isolation of plasmid templates for, 323
method for preparing terminating-sequence reactions using, 261
products
purification and quantification of, 327
Taq polymerase-amplified, 313
reamplification, 313
screening, isolation of yeast DNA for, 224
use of to clone genomic fragments, 195
Polyvinyl alcohol (PVA), 493
Polyvinylidene difluoride (PVDF), 128, 136
Populus tremuloides, 306, 501
Post-transcriptional gene silencing (PTGS), 435, 449
Pregnant mare's serum (PMS), 402
Primers, design of sequencing, 260
Protease inhibitors, 46
Protein(s), see also Proteins, extraction and purification of
analysis, 210, 352
binding of to DNA sequences, 291
diffusion feature of, 164
DNA-binding, 292
–DNA complex, measurement of protein size in, 301
fluorescence images of, 462
fragments, expression of in *E. coli*, 123
green fluorescent, 482
immunodetection of specific, 138
inactivation, 47
in situ immunocytochemical localization of, 114
quantitative analysis of after Western blot hybridization, 140
samples, preparation of for gel retardation assay, 298
separation of by SDS-PAGE, 128
stability, 46
standard markers, 129, 353
synthesis of from purified mRNAs, 348
transfer of from gel to membranes, 136
Protein analysis by gel electrophoresis, *in vitro* translation of mRNA and, 347–356
analysis of labeled proteins by SDS-PAGE, 352–355
electrophoresis, 354–355
loading of samples and protein standard markers onto gel, 353–354
preparation of separating gel, 352–353
preparation of stacking gel, 353
reagents needed, 355–356
synthesis of proteins from mRNA transcribed *in vitro*, 350
synthesis of proteins from purified mRNAs using rabbit reticulocyte lysates, 348
synthesis of proteins from purified mRNAs using wheat germ extract, 348–349

in vitro translation of mRNAs using wheat germ extract, 349
  preparation of wheat germ extract, 348–349
  TCA assay for amino acid incorporation, 350–352
Proteinase K, 10, 11, 14
Proteins, extraction and purification of, 45–65
  general considerations, 46–47
  protein concentration determination, 51–52
    bicinchoninic acid assay, 51–52
    Bradford assay, 52
  protein extraction protocols, 47–51
    extraction of cytosolic proteins from eukaryotic cells, 50
    extraction of proteins from cultures of *E. coli*, 49–50
    extraction of proteins from fresh animal or plant tissues, 47–49
    extraction of proteins from frozen animal or plant tissues, 49
    isolation of proteins from cellular membranes in cell suspensions and animal and plant tissues, 50–51
  purification of protein(s) of interest from protein mixtures, 52–60
    affinity chromatography, 56
    gel filtration, 52–54
    immunoprecipitation of proteins, 56
    ion-exchange chromatography, 54–56
    purification of proteins/enzymes by nondenaturing gel electrophoresis, 56–60
  reagents needed, 60–64
Protoplast(s), 223
  direct gene transfer to, 420
  fusion, 391
  isolation of, 372, 422
Provirus, 383
PSA, *see* Polar surface area
PTGS, *see* Post-transcriptional gene silencing
Pueraria montana, 372
Pulsed-field gel electrophoresis (PFGE), 222, 224, 225
Pulsed Fourier-transform NMR, 532
PVA, *see* Polyvinyl alcohol
PVDF, *see* Polyvinylidene difluoride

Q

QC, *see* Quality control
QSAR, *see* Quantitative structure activity relationship
Quadrupole mass spectrometer, 528
Quality control (QC), 554
Quantitative structure activity relationship (QSAR), 550, 554, 555

R

Radioactive labeling, reagents needed for, 77
Radioactive signal detection, 501

Radioimmunoassay (RIA), 47
Random cloning, 148
Recombinant plasmids, selection of transformants containing, 216
Recursive partitioning, 558
Reporter gene expression, analysis of, 405
Restriction enzyme(s)
  digestion, 15, 209, 210, 313
    plasmid vector, 384
    reaction, 441
  selection of, 384
  sites, 195
  star activity of, 86
Restriction mapping, positive recombinant bacteriophage DNA clones, 208
Retroviruses, 382, 391
Reversed-phase chromatography, 519
Reverse genetics, 46
Reverse Northern hybridization, 315
Reverse transcriptase
  avian myeloblastosis virus, 148, 157
  fluorescent labeling by, 329
Reverse-transcription PCR (RT-PCR), 180, 184, 497, 308
RIA, *see* Radioimmunoassay
Ribosomal RNA (rRNA), 347
Rice callus subculture, 364
RMSE, *see* Root mean square error
RNA (ribonucleic acid), 25, 67, 306, 320
  analysis of by dot-blot hybridization, 396
  antisense, 210
  blotting, 105
  degradation, 39
  dot blotting, 108
  electrophoresis, 106
  extraction
    buffer, 26
    guanidinium thiocyanate–CsCl gradient method, 28
    kits, commercial, 31
  fluorescent labeling of, 333
  integrity, assessment of, 37
  interference (RNAi), 435, 449
    alternative methods of, 453
    heat shock-induced, 454
    hypoxia-induced, 455
  isolation, phenol-chloroform/LiCl method of, 29
  preparation, differential display, 306
  probe(s)
    DIG-labeled, 487
    preparation, 496
  purification, 321, 330
RNA, isolation and purification of, 25–44
  determination of RNA quality and quantity, 36–38
    electrophoresis, 36–37
    fluorometric measurements, 38
    spectrophotometric measurements, 37–38
  extraction of total RNA from animals, 26–29
    acid guanidinium thiocyanate method, 26–27
    guanidinium thiocyanate–CsCl gradient method, 28–29
    LiCl/urea method, 27–28
    phenol-chloroform/LiCl method, 29

extraction of total RNA from plants, 29–31
    alkaline tris/sarkosyl method, 29–30
    commercial RNA extraction kits, 31
    CTAB method, 31
fractionation of RNA using denaturing sucrose
    gradients, 35–36
purification of poly(A)⁺ RNA from total RNA, 32–35
    magnetic oligo(dT)-beads, 33–35
    mini-purification using oligo(dT)-cellulose
        columns, 33
    oligo(dT)-cellulose column, 32–33
reagents needed, 39–43
troubleshooting guide, 39
RNAi, see RNA interference
RNase
    activity, inactivation of, 105
    contamination, 105
    degradation, 452
    stability of, 25
RNasin ribonuclease inhibitor, 349
Robotic arrayer, 321
Robotic liquid-handling devices, 540
Root mean square error (RMSE), 560
Rous sarcoma virus (RSV), 382
rRNA, see(rRNA), 347
RSV, see Rous sarcoma virus
RT-PCR, see Reverse-transcription PCR

S

Saccharomyces cerevisiae, 341
SAR, see Structure-activity relationship
*Sau*3A I dilutions, preparation of, 198
Scanning densitomety, 108
Scanning DNA microarrays, 337
Scanning electron microscopy (SEM), 461, 464
Scavenging reagents, 539
SDS, see Sodium dodecyl sulfate
SDS-PAGE, see Sodium dodecyl sulfate-polyacrylamide
    gel electrophoresis
Search programs, structure-based, 536
SEM, see Scanning electron microscopy
Semidry blotting, 137
Sense RNA, 105, 210
Separating gel, 352
Sequence
    alignments, 265
    chromatograms, analysis of, 265
    DNA polymerase, 243
    motifs, identification of, 265
Sequencing
    automated, 259, 262, 268
    manual, 259, 266
    reactions, 263
    reagent, 264
SFC, see Supercritical fluid chromatography
Shoot regeneration, 368
Signal detection, 94, 99
    nonradioactive, 499

radioactive, 501
Silver staining, 135
Single-strand DNA (ssDNA), 71, 239
    3′-end labeling of, 74
    labeling of, 72
    sequencing of, 246
Single-strand RNA (ssRNA), 496
siRNA, see Small interferring RNA
Size-exclusion chromatography, 524
SLE, see Solid-supported liquid-liquid extraction
Slide
    printing, materials and reagents for, 333
    scanning, 338
Small interferring RNA (siRNA), 449
Sodium dodecyl sulfate (SDS), 47, 96, 323
Sodium dodecyl sulfate-polyacrylamide gel electrophoresis
    (SDS-PAGE), 47, 123, 301, 347, 510
    analysis of labeled proteins by, 352
    separation of proteins by, 128
Solid-phase combinatorial chemistry, 541
Solid-phase extraction (SPE), 540
Solid-supported liquid-liquid extraction (SLE), 540
Solution-phase combinatorial chemistry, 539
Somatic cell hybridization, 371
Southern blot hybridization, 85–104
    blotting DNA onto nylon membranes, 89–92
        alkaline transfer method, 91
        capillary transfer method, 89–91
        vacuum transfer method, 91–92
    hybridization procedures, 94–99
        hybridization to nonradioactive probes, 98–99
        hybridization to ³²P-labeled probes, 94–97
        stripping DNA probes, 99
    preparation of DNA dot/slot blots, 92–94
    preparation of DNA samples, 86–88
        agarose gel electrophoreses of DNA, 87–88
        restriction-enzyme digestion of genomic DNA,
            86–87
    reagents needed, 101–103
    troubleshooting guide, 100
Southern blotting, genomic DNA, 207
Soxhlet extractor, 513
SPE, see Solid-phase extraction
Spectrophotometric measurements, 17, 37
Spheroblasts, preparation of for transformation, 226
Spleen cells
    fusion of under sterile conditions, 119
    preparation of for fusion, 118
Split-pool synthesis procedure, 542
ssRNA, see Single-strand RNA
Stacking gel, 353
Standing culture procedure, 3
Structure-activity relationship (SAR), 550, 535, 555
Subcloning, 411
    deletion, 281
    DNA insert, 210, 273
    plasmids, 209
Subtractive hybridization techniques, 171
Subtractive library screening, 306
Succinic anhydride, 333
Sucrose gradients, 489

fractionation of RNA using denaturing, 35
isolation of chloroplast DNA by, 14
Supercritical fluid chromatography (SFC), 517
Suspension cultured plant cells, 358, 359
Swinging-bucket rotors, centrifugation speed and time for, 28
Synthetic antisense oligonucleotides, 438

T

T/A cloning, 313
TAE, see Tris-acetate-EDTA
Tandem mass spectrometry, 529
Taq polymerases, 182, 282
Taxol, 514, 515
Taxus brevifolia, 514
TCA, see Trichloroacetic acid
TEA, see Triethanolamine
Tea-bag method, iterative deconvolution using, 542
TEL, see Telomeres
Telomeres (TEL), 222
TEM, see Transmission electron microscopy
TEMED, see Tetramethylethylenediamine
Template-primer annealing reaction, 245
Tetracycline, 284, 453
Tetramethylethylenediamine (TEMED), 129, 132, 251, 310, 352
Thermal cycling profiles, 261
Thermus aquaticus, 180
Thin-layer chromatography (TLC), 380, 517, 523
Thymidine kinase gene, 389, 397
Ticarcillin, 215
Time-of-flight (TOF) mass spectrometer, 529
Tissue
    printing
        general procedures of, 491
        tissue localization by, 489
    processing
        bench-top, 480
        microwave-oven, 480
TLC, see Thin-layer chromatography
TOF mass spectrometer, see Time-of-flight mass spectrometer
Transcription vectors, 150
Transferases, 57
Transfer RNA (tRNA), 382, 435, 498
Transgenic mice
    identification of, 404
    selection of transgenic lines by breeding, 405
Transmission electron microscopy (TEM), 461, 477
    applications, 480
    conventional, 477
    preparation of tissues for, 479
Trichloroacetic acid (TCA), 51, 349
    assay, amino acid incorporation, 350
    precipitation, 69, 154, 174
Triethanolamine (TEA), 496
Triphenyltetrazolium chloride (TTC), 375
Tris-acetate-EDTA (TAE), 36

Triticum aestivum, 372
tRNA, see Transfer RNA
Tryptophan synthetase gene, 389, 398
TTC, see Triphenyltetrazolium chloride
Tumor
    formation, genes involved in, 378
    -inducing genes, 413
    virus, 379

U

Unidirectional deletion, 240, 241

V

Vacuum transfer, 91
VECM, see Video-enhanced contrast microscopy
Vector(s)
    phagemid, 239
    preparation of, 211
    restriction enzyme digestion for, 210
    structural maps of, 151, 197
    SV40-based, 379
    transcription, 150
    YAC, 225
Video-enhanced contrast microscopy (VECM), 462
Vinblastine, 373
Vincristine, 373
Virtual library export, 552

W

Western blot analysis, verification of antibodies by, 121
Western blot hybridization, 85, 127–145, 428
    immunodetection of specific proteins, 138–140
        alkaline phosphatase, 138–139
        chemiluminescence detection, 139–140
    quantitative analysis of proteins, 140
    reagents needed, 141–145
    separation of proteins, 128–133
        SDS-PAGE, 128–131
        two-dimensional gel electrophoresis, 131–133
    staining and destaining of gel, 134–135
        Coomassie blue staining and destaining method, 134–135
        silver staining method, 135
    transfer of proteins from gel to membranes, 136–138
        semidry blotting, 137–138
        wet blotting, 136–137
    troubleshooting guide, 140–141
Wet blotting, 136
Wheat, cell-suspension cultures of, 372

X

Xenopus oocytes, construction of fractional cDNA libraries using, 175

Y

YAC, *see* Yeast artificial chromosome
Yeast
    applications of DNA microarrays in, 341
    spheroplasts, 226

Yeast artificial chromosome (YAC), 195, 222
    cloning, 224
    /DNA insert, transformation of spheroblasts with recombinant, 226
    library(ies), 221
        amplification and storage of, 227
        screening of, 227
    transformants, verification of, 227

Z

Zymolyase, 226